Annual Review of Ecology,
Evolution, and Systematics

Annual Review of Ecology, Evolution, and Systematics

Volume 43, 2012

Douglas J. Futuyma, *Editor*
State University of New York, Stony Brook

H. Bradley Shaffer, *Associate Editor*
University of California, Los Angeles

Daniel Simberloff, *Associate Editor*
University of Tennessee

www.annualreviews.org • science@annualreviews.org • 650-493-4400

Annual Reviews
4139 El Camino Way • P.O. Box 10139 • Palo Alto, California 94303-0139

Annual Reviews
Palo Alto, California, USA

International Standard Serial Number: 1543-592X
International Standard Book Number: 978-0-8243-1443-9
Library of Congress Catalog Card Number: 71-135616

TYPESET BY APTARA
PRINTED AND BOUND BY SHERIDAN BOOKS, INC., CHELSEA, MICHIGAN

Annual Review of
Ecology, Evolution,
and Systematics

Volume 43, 2012

Contents

Indexes

Errata

An online log of corrections to *Annual Review of Ecology, Evolution, and Systematics*
articles may be found at http://ecolsys.annualreviews.org/errata.shtml

Related Articles

Scaling Up in Ecology: Mechanistic Approaches

Mark Denny[1] and Lisandro Benedetti-Cecchi[2]

[1]Hopkins Marine Station of Stanford University, Pacific Grove, California 93950; email: mwdenny@stanford.edu

[2]Department of Biology, University of Pisa, 56126, Pisa, Italy; email: lbenedetti@biologia.unipi.it

Annu. Rev. Ecol. Evol. Syst. 2012. 43:1–22

First published online as a Review in Advance on August 28, 2012

The *Annual Review of Ecology, Evolution, and Systematics* is online at ecolsys.annualreviews.org

This article's doi: 10.1146/annurev-ecolsys-102710-145103

Keywords

criticality, dispersal distance, environmental bootstrap, extreme events, pattern formation, random walks, response function, scale transition, self-organization

Abstract

Ecologists have long grappled with the problem of scaling up from tractable, small-scale observations and experiments to the prediction of large-scale patterns. Although there are multiple approaches to this formidable task, there is a common underpinning in the formulation, testing, and use of mechanistic response functions to describe how phenomena interact across scales. Here, we review the principles of response functions to illustrate how they provide a means to guide research, extrapolate beyond measured data, and simplify our conceptual grasp of reality. We illustrate these principles with examples of mechanistic approaches ranging from explorations of the ecological niche, random walks, and macrophysiology to theories dealing with scale transition, self-organization, and the prediction of extremes.

INTRODUCTION: RELATING PHENOMENA ACROSS SCALES

In a classic perspective on the role of modeling in ecology, Levin (1992) noted that "the problem of relating phenomena across scales is the central problem in biology and in all of science." Currently, there are three predominant ecological perspectives on this fundamental issue. Proponents of macroecology (e.g., Brown 1995) begin with a search for scaling "laws" through an examination of empirical data measured over wide ranges of scale. If substantial correlations exist—between body mass and metabolic rate, for instance—macroecologists then proceed to test hypotheses concerning the mechanisms that underlie these correlations (McGill & Nekola 2010). In the converse of this "top-down" approach, community ecologists traditionally take a "bottom-up" perspective, using small-scale field observations and experimental manipulation to infer pattern at larger scales (Underwood & Paine 2007). Both top-down and bottom-up perspectives begin with empirical data. In contrast, ecological theorists begin with a conceptual map of how scales might be related. Only after these initial heuristic models have been validated qualitatively are subsequent quantitative models tested empirically.

As distinct as they are, these perspectives are not independent. Both macroecologists and experimental ecologists are guided by theoretical concepts, and theorists rely on these researchers to provide observations from which concepts are born and data against which models can be tested. Indeed, it is through the interaction of these perspectives that ecology can best relate phenomena across scales.

Interaction is easiest on common ground. Although ecologists of all stripes seem to find pleasure in promoting the differences among their approaches, there is a philosophical unity underlying these disparate perspectives. Each assumes that the connections of biology at one scale of time or space to that at other scales can be described by response functions, mathematical expressions of how systems respond to the conditions in which they find themselves. Although widely used, the concept of a response function appears under a confusing variety of names. For example, in ecology, functional response describes how the rate of prey capture varies with changes in prey density. In physiology, thermal reaction norms quantify the fitness of an organism as a function of its body temperature. In physics, Boyle's law specifies how the volume of a gas varies as a function of pressure, and Newton's second law of motion predicts how the acceleration of a mass depends on the force applied to it. Despite the differences in nomenclature, each of these mathematical descriptors is a response function.

Biological response functions take three general forms. Conceptual response functions—the grist of theoretical ecology—describe, usually in a simple, abstract, and mathematically tractable form, how a system might respond in an idealized world. They are often used as the first step toward defining connections across scales. By organizing key concepts and processes and providing qualitative answers, they can be used to identify which parameters at one scale are likely to have greatest impact at others. The second category of functions is that of measured response functions, i.e., the measured response to a physical variable, as defined by Huey & Stevenson (1979). Measured response functions provide a practical means by which to connect a process at one level of organization (e.g., escape speed) to that at a higher level (e.g., survivorship). However, without a mechanistic explanation, the connection of measured response functions to processes at lower levels of organization (e.g., muscle physiology) is uncertain. To obtain full understanding of a system across multiple scales, it is therefore necessary to define and test the potential mechanisms that account for measured responses. These are response functions in the sense of the third category defined by Holling (1965): a mechanistic, quantitative explanation of the response of an environmental factor, physiological process, whole organism, population, or community as a function of some physical variable. At their most fundamental, mechanistic response functions are explicit applications of the principles of physics, engineering, and chemistry; at higher levels, they may apply well-established principles of physiology, behavior, and statistical mechanics. Although their connections can be complex—and unexpected properties may emerge from their interactions—mechanistic response functions form the basis for our understanding of biology.

RESPONSE FUNCTIONS: PRINCIPLES AND CHARACTERISTICS

Two examples illustrate the formulation and utility of mechanistic response functions. The first describes response to a single variable and the second to multiple variables.

Predator-Prey Response Functions

In an elegant experiment, Holling (1959) blindfolded a student and had her tap her finger on a wall to locate randomly placed sandpaper discs. When a disc was located, the student removed it, set it aside, and searched again. The rate at which discs were "captured" was analyzed as a function of their spatial density. Noting that the student's time was divided between searching for and handling discs, Holling derived a simple equation that accurately described how the student (a "predator") responded to "prey" density (spatial density of disks). The resulting Type II response function (**Figure 1**) has since been found to accurately describe the response of a wide variety of invertebrate predators (Maynard-Smith 1974).

The utility of the Type II function extends beyond its ability to accurately reproduce empirical results. For example, handling and searching times can be quantified directly and used to parameterize a Type II model independent of field measurements of predation. If this independently parameterized model fits field-measured data, we can be reasonably assured that we have discovered the pertinent mechanisms governing prey capture, whereas deviations from measured data provide evidence that the simple assumptions of the Type II model are inadequate. If, for example, refuges are available in which low-density prey can hide, the response function may be sigmoidal (Type III, **Figure 1**) (Holling 1965). As Nisbet et al. (2000) note, "In the search for mechanisms, deviations from model predictions are at least as instructive as data that support" a model.

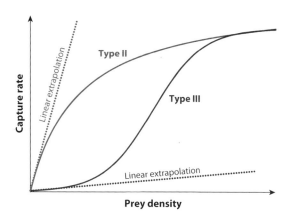

Figure 1

The rate at which predators capture prey varies with prey density in patterns that depend on the underlying mechanics. For both the Holling Type II and Type III response functions, linear extrapolation from the response measured at low prey densities substantially deviates from the actual rate of capture at high densities.

Once it has been confirmed that a predator conforms to the assumptions of a Type II or III response, the function can be used to extrapolate prey densities outside the measured range. These mechanistic extrapolations can differ substantially from phenomenological extrapolations made on the basis of field data alone. For example, at low prey density, both Types II and III response functions closely approximate straight lines. Linear extrapolation to high prey density would either overestimate (Type II) or underestimate (Type III) actual rates of prey capture (**Figure 1**).

Multiple Inputs, Single Output: Wave-Induced Hydrodynamic Force

The example above describes the response of an organism to variation in a single factor. It is often extremely useful to define the mechanistic response of an organism (or other physical system) to the simultaneous variation of multiple factors. Fluid dynamic theory of ocean waves provides an example. Wave-swept rocky shores are subjected to severe hydrodynamic forces, which can break or dislodge plants and animals, inhibit foraging, and deter settlement of larvae (e.g., Denny 1988). These forces vary with the height of the waves as they reach the shore. In turn, inshore wave height is a function of wave height offshore, water depth (affected by the tides), and wave period. Measuring hydrodynamic force (the system's "response") as a function of each of these variables separately is a futile task because it is their interaction that matters, and there is an infinite number of combinations that potentially impose biologically meaningful forces (Denny et al. 2009). However, fluid dynamic theory allows one to mechanistically combine the effects of water depth, wave height, and wave period, translating these variables into a single imposed force (Denny et al. 2009, Helmuth & Denny 2003, Madin & Connolly 2006, Madin et al. 2006, Massel 1996, Massel & Done 1993), which can in turn be used to predict the intensity of physical disturbance in a wave-washed community.

Useful Characteristics

The following characteristics summarize the nature of mechanistic response functions:

1. Coherence between reality and the predictions of a mechanistic response function is evidence that the function contains all pertinent details of the process under consideration.
2. If predictions differ from measured data, the form of the deviations can guide the search for additional information.
3. A validated mechanistic response function allows one to extrapolate accurately beyond measured conditions.
4. Mechanistic functions can translate multiple input variables into a single, biologically relevant response, thereby simplifying our understanding of complex systems.

We list these characteristics to highlight the utility of mechanistic response functions, but we do not mean to imply that this utility is either universal or absolute. Only with infinite data and perfect coherence between function and reality could one be absolutely sure that all pertinent details have been included. Similarly, the only way to be certain that extrapolation from existing data is valid is to test that extrapolation directly, preferably through manipulative experiments. But to dwell here on these considerations would miss our point: Though the results may be neither perfect nor absolute, the formulation of mechanistic response functions provides ecologists with valuable tools that phenomenological measurements cannot.

RESPONSE FUNCTIONS: A SECOND LOOK

Simple response functions suffice for many applications. However, the complexity of some biological systems requires that additional factors be taken into account.

Uncertainty

The response functions of **Figure 1** are deterministic, that is, exact. In reality, there is always some uncertainty (noise) in how a system responds, and this uncertainty can be incorporated in at least two ways.

First, in addition to specifying a deterministic response, one can specify the probability distribution of the system's noise. In this case, calculation of response proceeds through two steps. For a given value of the input variable, one first calculates the expected (average) response. A value is then chosen at random from the noise distribution and added to the expectation to give one random realization of the actual response. This approach forms the basis for wide-ranging theoretical studies in population dynamics (e.g., Lande et al. 2003, Melbourne & Hastings 2008), and is a standard approach to the study of turbulent fluid dynamics (Tennekes & Lumley 1972). Gaylord et al. (1994) and Denny (1995) used two-part response functions of this sort to quantify the force required to break intertidal organisms as a function of their size.

Second, in some extreme examples, the system under consideration is driven by chance alone. In turbulent flow, for example, small particles (e.g., gametes, seeds, spores, or phytoplankton) move in a random walk, and the distance they travel as a function of time can only be described statistically (Berg 1984, Denny & Gaines 2000, McNair et al. 1997). Alternatively, random walks can be driven by an organism's behavior (e.g., de Jaeger et al. 2011, Viswanathan et al. 2000). In both cases, the response function relating distance to time is an explicit probability function, specifying (exactly) what the probability is that, after a given period, an object will have traveled less than a certain distance.

Change Through Time

Response functions often change through time. Thermal tolerance in animals can shift, for instance, as repeated exposure to sublethal high temperatures result in an increase in lethal temperature (Angilletta 2009). Similarly, response can change through time due to evolution (Schoener 2011, Whitehead 2012). For example, as the climate warms, arctic red squirrels give birth earlier in the year, an effect due in part to a shift in gene frequencies (Berteaux et al. 2004). In such cases, it is necessary to specify the time at which a response function is to be applied or to express response as an explicit function of both time and the input variable.

Hysteresis

The simple functions of **Figure 1** imply that response is independent of the direction of change in input variable, but this is not always true. For example, the response of plants to sunlight can exhibit hysteresis (Cullen et al. 1992, Long et al. 1994). Prolonged exposure to irradiance above a threshold can damage cells; as a result, photosynthesis follows a lower trajectory as irradiance declines (**Figure 2**). Other examples abound: Metabolic rate increases as an organism's body temperature rises, but if temperature is sufficient to cause damage to ATP-producing machinery, metabolic rate may take a different, lower trajectory as temperature subsequently falls. Likewise, the response of a desert community to gradually increasing rainfall can take a different path than that followed as rainfall subsequently decreases (Gutschick & BassiriRad 2003), and recovery of mussel beds from disturbance can take a different path than that followed as beds are disturbed (Guichard et al. 2003).

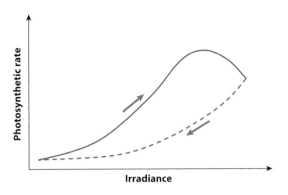

Figure 2

Photosynthetic rate increases with increasing solar irradiance, but above some limit, increased irradiance damages cells. Because of this damage, if irradiance subsequently declines, the rate of photosynthesis is reduced below that of undamaged tissue.

Models

When complexities such as uncertainty, temporal evolution, and hysteresis are incorporated, the distinction between a response function and a mechanistic "model" becomes blurred. Indeed, there is really no need to make a formal distinction between the two: Response functions are a simple form of model, and models are often, in essence, compound response functions.

A central issue for mechanistic models is the trade-off between mechanistic detail and heuristic generality (Baskett 2012, Rastetter et al. 2003). A model based on mechanistic response functions can be so parameter rich that it applies only to a specific organism; generality is gained only by making the model more abstract. This trade-off is exemplified by bioenergetic models, in which individual response functions (both measured and mechanistic) are combined to predict the rate and allocation of energy flow through an organism. The proximal goal of such models is to construct a whole-organism response function that takes as its inputs environmental variables such as temperature and food concentration and gives as its outputs predictions for growth rate, ultimate body size, age of reproductive maturity, and lifetime reproductive output (Nisbet et al. 2000). To date, this effort has been carried farthest by a variety of dynamic energy budget (DEB) models (reviewed by Kooijman 2000, Nisbet et al. 2000, van der Meer 2006), but this progress has come at the expense of mechanistic detail—DEB models depend on a set of simplifying assumptions and incorporate variables that cannot be measured directly. Efforts are under way to retain the generality of DEB models while incorporating mechanistic details (Nisbet et al. 2012).

SCALING UP

Having outlined the nature of mechanistic response functions, we now review six examples in which they can assist ecologists' efforts to scale up from knowledge of small-scale interactions to the prediction of large-scale patterns. These examples are drawn from a broad array of systems (**Table 1**). However, we draw added insight by highlighting the diverse response functions operating within a common system—the intertidal zone of wave-washed shores.

Mechanism versus Correlation: The Ecological Niche

The concept of the niche has been central in ecology since the 1950s (Hutchinson 1957, MacArthur & Levins 1967), and it has recently been used to predict species range limits (Buckley et al. 2008,

Table 1 Representative applications of mechanistic response functions

Response function(s)	Representative application	Complexity	Key references
Functional response of predators to varying density of prey	Predator-prey interactions	Individual response functions	Holling 1959
Mechanistic scaling relationship of metabolism with body mass	Explaining macroecological patterns in body and range size		Chown & Gaston 1999, Chown et al. 2007, Gaston et al. 2009, Kozlowski et al. 2003, West et al. 1997
Probabilistic response functions	Modeling of chance events such as return times, distance traveled, disturbance		Gaylord et al. 2006; Katul et al. 2005; Levin et al. 2003; McNair et al. 1997; Nathan 2006; Nathan et al. 2002, 2005; Zimmer et al. 2009
Fluid dynamic theory translating multiple physical variables into a single hydrodynamic force	Physical disturbance on wave-swept rocky shores		Denny 1988, Denny et al. 2009, Helmuth & Denny 2003, Madin & Connolly 2006, Madin et al. 2006, Massel 1996, Massel & Done 1993
Response functions describing local population processes (e.g., competition, predation)	Scale-transition theory: predicting large-scale dynamics through nonlinear averaging		Benedetti-Cecchi et al. 2012, Chesson et al. 2005, Melbourne & Chesson 2006, Melbourne et al. 2005
Response functions describing local population processes (e.g., disturbance, recruitment)	Self-organization: predicting large-scale patterns and system stability		de Jaeger et al. 2011, Guichard et al. 2003, Pascual & Guichard 2005, Solé & Bascompte 2006
Bioenergetic models (e.g., dynamic energy budget or heat-budget models)	Prediction of organism's vital rates as a function of its environment (e.g., temperature, food concentration)		Bell 1995; Campbell & Norman 1998; Denny & Harley 2006; Gates 1980; Helmuth 1998; Kooijman 2000; Nisbet et al. 2000, 2012; van der Meer 2006
Fluid dynamic theory translating multiple physical variables into a single hydrodynamic force, heat-budget models	Predicting ecological and evolutionary extreme events in the presence of stochastic environmental fluctuations		Denny & Dowd 2012, Denny et al. 2009
Heat-budget models, dynamic energy budget models, measured response functions for locomotion	Delineation of the ecological niche	Compound response functions	Kearney et al. 2008, 2010, 2012

2010; Kearney et al. 2010, 2012). In this approach, a suite of measured environmental variables is correlated with presence or absence of a given species, and the resulting "climate envelope" is used to delineate the species' potential range. For example, the toxic cane toad *Bufo marinus* was introduced into northeastern Australia in 1935 and has spread worrisomely south and west. Climate envelope models suggest that toads might eventually reach the urbanized southeastern coast of Australia (Urban et al. 2007). However, by relying on correlation rather than mechanism, climate envelope models can miss the true causes of range limitation (Davis et al. 1998, Elith & Leathwick 2009, Gaylord & Gaines 2000, Kearney & Porter 2009). In the case of cane toads, Kearney et al. (2008) combined response functions for the animal's interactions with its environment to calculate the rate at which adult toads could invade new territory. Due primarily to the animal's temperature-dependent ability to hop, the rate of invasion is predicted to approach zero as toads move into the cooler south, a prediction that matches the measured rate of southerly range extension. Adult toads could live on the southern coast if they were accidentally (or intentionally) introduced there—and hence, climate envelope models include the coast in the toad's potential range—but mechanistic analysis reveals that toads do not have the locomotory capacity to reach those locations. Furthermore, by identifying the most relevant axes of the niche, the mechanistic approach sets the stage for measurement of an organism's ability to evolve in response to the limiting factor (e.g., Kuo & Sanford 2009).

Macrophysiology

At the interface between physiology and macroecology, macrophysiology explores variation in physiological traits at large spatial and temporal scales. When based on mechanistic response functions, macrophysiological results can have ecological implications regarding patterns in body size, abundance, range size, and species richness (Gaston et al. 2009). For example, the mechanistic basis for the increase in range size with latitude—Rapoport's rule—involves physiological adaptations to increased climate variability toward the poles (Gaston et al. 2009). Reductions in metabolism at the elevated temperature of low latitudes allow species to reduce range size while preferentially allocating resources to growth and development (Chown & Gaston 1999). In contrast, elevation of metabolic rates as an adaptation to low temperatures has been postulated to enable species, particularly insects, to maintain high growth rates (thereby requiring larger ranges) in regions where the growing season is short, such as at high latitudes (Chown & Gaston 1999).

Macrophysiological investigations have also used mechanistic scaling relationships of metabolism to explain macroecological patterns in body size (Chown & Gaston 1999, Chown et al. 2007). Competing physiological models make distinct predictions for the scaling exponent of metabolism as a function of mass, such that, in principle, it should be possible to discriminate among theories using metabolic scaling data. For example, the nutrient supply network model of the metabolic theory of ecology predicts that metabolic rate scales as mass$^{3/4}$ (West et al. 1997). This prediction—based on the assumption that nutrients are supplied through space-filling fractal networks—in theory applies generally from molecules to organisms. An alternative model posits that body size changes as a consequence of variation in cell size and/or number under natural selection pressure and predicts an exponent of 3/4 for interspecific scaling relationships and values of 2/3 to 1 for intraspecific scaling relationships (Kozlowski et al. 2003). The precise value of the intraspecific exponent depends on the relative contribution of variation in cell size and cell number to body mass. Isometric scaling (an exponent of 1) is expected when changes in body size are determined entirely by variation in cell number, whereas an exponent of 2/3 is expected from the exclusive contribution of variation in cell size.

Chown et al. (2007) compared these models using metabolic rate data from 391 species of insects. They found a scaling exponent of 3/4 at the interspecific (phylogenetic) level and 1 at the intraspecific level, results more consistent with the cell-size model than the nutrient supply network model, although large confidence intervals associated with estimated scaling exponents complicate definitive distinction between models.

Macroecological consequences may accrue from these macrophysiological results if developmental processes underlying variation in cell number and size have prevalence over other proximate determinants of body mass. For example, to maintain a constant energy budget, cell size should decrease with increasing temperature. Therefore, if variation in cell size is not compensated by changes in cell number, as the results of Chown et al. (2007) suggest, one may predict a decrease in body size with increasing temperature (Chown & Gaston 2010).

Scale-Transition Theory

A potentially important role for mechanistic response functions has been formalized by Chesson and coworkers in scale-transition theory (Chesson et al. 2005, Melbourne et al. 2005, Melbourne &

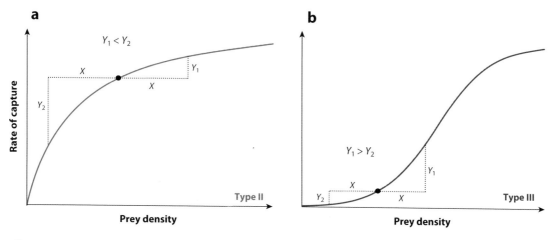

Figure 3

Examples of Jensen's inequality. (*a*) The Holling Type II response function decelerates such that an increase of X in prey density results in a relatively small increase in capture rate (Y_1). In contrast, a decrease of X in density results in a relatively large decrease in capture rate (Y_2). Thus, if prey density varies at $\pm X$ around an average value (*solid dot*), then the overall rate of prey capture is less than if density remained at that average. (*b*) At low prey density, the Holling Type III response function accelerates. In this case, variation of $\pm X$ results in a net benefit to the predator because $Y_1 > Y_2$.

Chesson 2006), a scaling "recipe" stemming from recognition that response functions are typically nonlinear and that interaction of local nonlinearities with spatial or temporal variation distorts the prediction of large-scale patterns. Consider a heuristic example. As we have seen, capture rate can vary nonlinearly with prey density (**Figure 1**). For each predator, there is a particular prey density at which energy input from prey just offsets energy expended in respiration. At this compensation density, the predator's growth rate is zero. Consider, now, a scenario in which the spatial average of prey density equals compensation density, but density varies from place to place. If the response function is Type II, capture rates decelerate with increasing prey density (i.e., the second derivative of the function is <0). In this case, if predators wander randomly across the landscape, the increased energy influx from encounters with above-average prey density is small compared to the reduced energy influx from periods spent sampling below-average densities (**Figure 3*a***). Thus, even in an area where average density equals compensation density, a predator will eventually starve when confronted with sufficient variance in prey availability. In contrast, at low prey densities, a Type III response function accelerates (the local second derivative is >0; **Figure 3*b***). In this case, if it experiences sufficient variation in density, a predator can actually grow even though average prey density equals only compensation density.

These phenomena are examples of Jensen's inequality (Jensen 1906): For any monotonic (but nonlinear) function f, the function of the mean value $f(\bar{x})$ (e.g., capture rate at mean population density) differs from the mean value of the function $\bar{f}(x)$ (e.g., the "true" capture rate averaged across individuals as they encounter variation in prey density). In these examples, because the Type II response decelerates, $\bar{f}(x) < f(\bar{x})$. At low prey density, the Type III response accelerates, so within this range, $\bar{f}(x) > f(\bar{x})$. Ruel & Ayres (1999) and Benedetti-Cecchi (2005) provide other ecological examples of Jensen's inequality.

Often, the variation that invokes application of Jensen's inequality depends on the scale at which a process occurs. For example, the variation in prey density described above depends on temporal patterns of foraging and spatial scale(s) of prey distribution. Scale transition theory provides a means to combine these scale-dependent factors with mechanistic descriptions of local dynamics to accurately predict an average response at large scales.

The procedure starts with a definition of the response function for the process in question. Parameters of the function are estimated at small scale from empirical (mensurative or experimental) data. Large-scale predictions are then expressed as the sum of the local-scale average (the so-called mean field model) and correction terms that take into account distortion introduced by nonlinear processes (the scale-transition terms). In its simplest form (Chesson et al. 2005),

$$\bar{f}(x) \approx f(\bar{x}) + \frac{1}{2} f''(\bar{x}) Var(x). \qquad 1.$$

Here, $f(\bar{x})$ is the mean field estimate, $f''(\bar{x})$ is the second derivative of f (measured at the mean, $\bar{x}$), and $Var(x)$ is the variance in x (spatial or temporal) at the largest scale of interest.

Continuing our predator-prey example, if the nonlinearity of capture rate is due to density dependence, one needs to estimate how much variance in prey density occurs at the largest scale over which a prediction is required. This quantity is then used to correct the inaccurate prediction obtained by simply averaging local measurements. Consider a comparison between Type II and III responses to prey density (**Figure 4a**). In this hypothetical example, the Type II response is more effective at uniform low densities, but less effective at uniform high densities. Increasing the variance of prey density decreases the advantage of a Type II response at low densities, but reduces the advantage of the Type III response at high densities (**Figure 4b**).

This simple example ignores the effects of predation on prey density. If consumer and resource interact, one needs to estimate the covariance between consumer location and resource density at the large scale and use this covariance to correct the average obtained from local measurements. Melbourne et al. (2005) and Melbourne & Chesson (2006) illustrate this process with examples based on empirical data, focusing largely on nonlinear population and community processes. But there is a formidable opportunity to apply the theory to other levels of biological organization (e.g., the physiological level) and to physical processes (e.g., disturbance and thermal stress).

Case studies illustrating the application of scale-transition principles typically focus on discrete spatial scales. For example, Melbourne et al. (2005) show how the dynamics of periphyton algae on individual cobbles can be scaled up to an entire stream, and how metacommunity dynamics of intertidal crabs and desert annual plants depend on variation at discrete small scales. Benedetti-Cecchi et al. (2012) illustrate how the dynamics of algal turfs and their interactions with canopy-forming algae can be scaled up from individual plots (20 cm × 20 cm) to an entire island (tens of kilometers). Scale-transition theory can, however, be extended to incorporate continuous measures of environmental variance, enabling great flexibility for scaling up local dynamics over multiple spatial or temporal scales. This may prove particularly useful when, for example, deciding upon the size of a network of protected areas (e.g., McLeod et al. 2009). In this case, the ability to scale up local dynamics as a function of network size may constitute a valuable decision-making tool.

Self-Organization

Ecological literature is replete with examples of pattern at spatial scales much larger than individual organisms: waves, spirals, and labyrinthine clumps in arid vegetation (e.g., Couteron & Lejeune 2001, Klausmeier 1999, Rietkerk et al. 2002); power-law distributions of gap sizes in forests (e.g., Malamud et al. 1998); and traveling waves of lynx, voles, and bud moths (Bjørnstad et al. 2002, Kaitala & Ranta 1998, Ranta & Kaitala 1997, Ranta et al. 1997). These patterns can have functional advantages for the organisms involved, increasing their productivity and stability in variable environments. Efforts to explain these emergent patterns and their ecological role(s) have

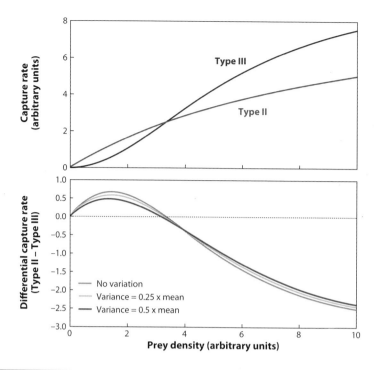

Figure 4

A hypothetical example of the effect of spatial variation in prey density on the rate of predation. (*a*) In the absence of spatial variability, the Type III response results in higher rates of prey capture at all but the lowest prey densities. (*b*) The difference between Type II and Type III capture rates differs depending on spatial variability in prey density. The higher the variance (set equal to a given fraction of the mean density), the smaller the difference between the types of response.

been led by theoretical ecologists drawing on insights from developmental biology and physics (Solé & Bascompte 2006). We review two examples.

Facilitation-inhibition models. In 1952, Alan Turing outlined a theory in which facilitation and inhibition interacted at different spatial scales to produce wave- or spot-like patterns. Turing (1952) developed his theory in the context of organismal development, but the principles apply equally well in ecological contexts. Patterns observed in bed-forming mussels provide an instructive example. When growing on soft substrata, young *Mytilus edulis* form distinctive rows or clusters of near-uniform size and spacing (**Figure 5a**). van de Koppel et al. (2005, 2008) propose that this large-scale pattern is due to the small-scale, local response of individual mussels to both hydrodynamic forces and suspended food concentrations. They posit that it is advantageous for mussels to adhere to each other in clusters to avoid dislodgment by drag (facilitation at the individual scale), but it is disadvantageous to be in the middle of too large a clump because the available food has already been eaten by mussels on the periphery (inhibition at the group scale). van de Koppel et al. (2008) demonstrated the efficacy of their explanation through field observations and experiments: The predicted pattern of clusters matched that observed, and cluster size increased when food was artificially provided to mussels in the center. Furthermore, per capita growth rate was higher for observed patterns of clustering than it was for either solitary mussels (which are susceptible to disturbance) or a continuous bed (in which many mussels are underfed). Increased growth rate

Drag: hydrodynamic force acting in the direction of relative flow between an object and the surrounding fluid

a

b

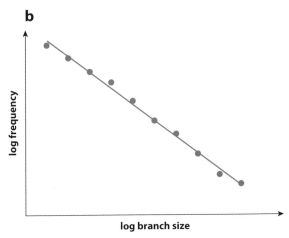

Figure 5

(*a*) Mussels on the soft substratum of Menai Straight (Wales) form characteristic clumps and rows indicative of self-organization through the interaction of facilitation and inhibition (photo by J. Widdows). The scale and pattern of clumps and rows depend on flow speed, food availability, and the properties of the substratum (van de Koppel et al. 2005, 2008; Widdows et al. 2009). (*b*) A notional example of scale independence: the frequency of snowflake branches decreases with increasing branch size. When plotted on log-log axes (as we have done here), the power function indicative of self-organized criticality forms a straight line.

of clumped mussels thus provides a potential selective factor in the evolution of cluster-forming behavior. De Jaeger et al. (2011) showed that rapid formation of clumps is abetted by the tendency of mussels to crawl in Lévy walks (a particular form of random motion) and that this behavior forms an evolutionarily stable strategy. Similar large-scale patterns are found in the vegetation of arid ecosystems, although the mechanism of pattern formation is different (Rietkerk et al. 2002).

Criticality. At the critical temperature of 0°C, water can abruptly change from liquid to solid. The mechanics of this and similar threshold transitions is such that, near the critical point (that is, at criticality), small-scale (in this case, molecular) interactions result in large-scale patterns (for reviews, see Pascual & Guichard 2005, Solé & Bascompte 2006). For instance, snowflakes, which form near water's critical temperature, have delicate branching patterns visible to the naked eye, i.e., at scales far removed from those of individual molecules. In many systems, the patterns that emerge at criticality are characterized by scale independence. For example, the branching pattern in one small "twig" of one arm of a snowflake is similar to the branching pattern of the whole arm, which itself may be repeated between arms. If one were to take an entire snowflake and record the number of branches of different sizes, scale independence of pattern would emerge as a power-law relationship between frequency and size (Guichard et al. 2003, Solé & Bascompte 2006): The ratio of the number of branches at one size to that *n* times that size is constant regardless of the size chosen, a relationship that produces a straight line on a log-log plot (**Figure 5*b***). In this example, the system's behavior is related to temperature, but the same ideas apply for critical values of any factor.

Because the capacity for small-scale interactions to produce scale-independent patterns is manifest only at criticality, presence of a scale-independent pattern might signal that a biological system (a population or community) is poised near a threshold and, therefore, susceptible to abrupt switches between alternative states or large temporal swings, either of which presents challenges for conservation and management. However, recent work suggests that under certain conditions scale-independent patterns can be formed in systems that are resistant to abrupt shifts in state (Pascual & Guichard 2005). Indeed, in these cases of robust criticality the mechanisms that lead to pattern formation can contribute to the system's overall stability.

Mussel beds serve as a model system for investigating robust criticality. On wave-swept rocky shores, gaps are produced in otherwise continuous mussel beds as hydrodynamic lift dislodges mussels (Denny 1987, Paine & Levin 1981). The size distribution of resulting gaps follows the power-law relationship indicative of a process near its critical point, and Guichard et al. (2003) have devised a model that accounts for this observation. In short, when one mussel is dislodged, adjacent mussels are initially weakened and regain their adhesive strength only after a period of susceptibility. As a result, once a gap is initiated, it can grow from its edges if waves apply sufficient force within the period of susceptibility. Conversely, gaps can shrink as mussels recruit at gap edges. Empirical estimates of the period of susceptibility and rates of recruitment and gap formation showed that for mussels in Oregon disturbance and recovery occurred at similar temporal and spatial scales. When Guichard et al. (2003) parameterized their model with these measured values they were able to accurately recreate the observed distribution of gap sizes, suggesting that their model indeed captured the dynamics of the system.

Furthermore, they note that a scale-independent pattern of gaps has the tendency to maintain the system near its critical point. In their model, if the shore is initially uniformly covered with contiguous mussels, the first gap formed can rapidly propagate through the entire bed as one susceptible mussel after another peels off at the gap's edges. Once all mussels are removed, the bed cannot recover because there is no edge to which mussels can recruit. In contrast, a bed with a distribution of gap sizes is relatively stable because the emergent pattern of gaps inhibits propagation of catastrophic disturbance, allowing time for sufficient recruitment. Models suggest that this tendency toward stability is inherent in systems in which disturbance and recovery are local in space and intermittent in time and have the same spatial and temporal scales (Pascual & Guichard 2005).

The role of mechanism. In the past two decades, predictions of large-scale, self-organized patterns (as outlined above) have moved swiftly from formulation of abstract theories to their demonstration in the field. The next step—prediction of when and where a given type of pattern formation will manifest—requires the sort of mechanistic approach we espouse here. Again, consider mussel beds. In one environment, mussels form uniform-sized clusters and rows indicative of facilitation-inhibition dynamics; in another, cluster size is distributed in the scale-independent pattern indicative of criticality. In their examinations of mussel dynamics, Guichard et al. (2003), van de Koppel et al. (2008), and de Jaeger et al. (2011) were content to demonstrate that their theories worked for a given set of environmental conditions, refraining from speculation as to how differences in the environment might favor one model over the other or when and where those different conditions might occur. It is here that the mechanistic approach can play an important role.

We know a great deal about the mechanics of mussels' attachment to the substratum (e.g., Carrington 2002a,b; Carrington et al. 2008, 2009; Moeser & Carrington 2006; Moeser et al. 2006) and the hydrodynamics of mussels in beds (Denny 1987). Similarly, fluid dynamic theory can account for the ability of mussels to reduce the local density of suspended food (e.g., Frechette et al. 1989). Application of this information allows one to predict when and where the aforementioned models are operative.

The mechanism proposed by van de Koppel et al. (2008) requires flow slow enough that mussels resting on mud or sand can resist hydrodynamic forces. Denny (1987) has shown that in mussel

Lift: hydrodynamic force acting perpendicular to the direction of relative flow between an object and the surrounding fluid

beds, lift F_L (in newtons) is the dominant hydrodynamic force:

$$F_L \cong 451 U^2 A, \qquad \qquad 2.$$

where A is the area of bed occupied by a mussel (in square meters) and U is velocity (in meters per second). For mytilid mussels, $A \cong 0.08 L^{1.77}$, where L is the maximum length (in meters) of the mussel shell (M. Denny, personal observation). Thus,

$$F_L \cong 36 U^2 L^{1.77}. \qquad \qquad 3.$$

For a mussel on unconsolidated substratum, the primary force resisting lift is the mussel's weight in water, W, which (from unpublished measurements) is

$$W\,(\text{in } N) \cong 354 L^{2.93}. \qquad \qquad 4.$$

Combining Equations 3 and 4 and solving for U, we find that, for a typical mussel 5 cm long, velocities greater than 0.55 m s^{-1} are sufficient to dislodge the organism from the substratum. Mussels' byssal attachment to mud or sand might provide some slight added resistance to lift, but the critical value at which U disrupts a bed should nonetheless be less than approximately 1 m s^{-1}. Indeed, the organized pattern of mussels is disrupted during winter storms (van de Koppel et al. 2005). To resist higher velocities, mussels must attach to solid substrata—that is, to rock—with a concomitant stepwise increase in resistance. Minimum force per area required to dislodge a bed mussel from rock is approximately 4×10^4 N m^{-2} (Denny et al. 2009), which, in conjunction with Equation 2, implies that velocities greater than approximately 10 m s^{-1} are required to dislodge mussels from rock.

These calculations thus suggest that the mechanism of van de Koppel et al. is viable only where $U < 1$ m s^{-1} and that of Guichard et al. (2003) only where $U > 10$ m s^{-1}. A mechanistic approach to the process of self-organization thus confirms the results of existing models, but in addition predicts that there is a range of conditions—mussels attached to rock with 1 m s$^{-1} < U < 10$ m s^{-1}—under which neither model of self-organization should apply.

Random Walks and Dispersal

The ability to disperse is often a key aspect of the interaction among species. For example, if the competitive dominant in a community cannot disperse sufficiently to occupy patches of new space created by physical disturbance, competitive inferiors with superior dispersal strategies can persist (Rees et al. 2001, Tilman 1994; but see Clark et al. 2004). Furthermore, the rate and scope of dispersal govern the magnitude of gene flow between populations, thereby affecting population genetics and the potential for local evolution.

In recent years, researchers have combined engineering theories of turbulent mixing and flow with statistical theories of random walks to construct detailed mechanistic models that use small-scale motion to predict the large-scale distance traveled by propagules such as seeds, pollen, ballooning spiders, and aquatic larvae and spores (e.g., Gaylord et al. 2006; Katul et al. 2005; Levin et al. 2003; Nathan 2006; Nathan et al. 2002, 2005; Zimmer et al. 2009). These studies provide insight into a wide variety of ecological and evolutionary processes, from the rate of species' advance to community assembly (Levin et al. 2003). In each case, the output of the model—a form of mechanistic response function—is a probability distribution that quantifies the likelihood that a propagule released at a certain point in space will first impact the substratum at a given distance.

These random-walk models are noteworthy in two respects. First, because they are derived from basic physical principles, they are generalizable. The same approach that leads to accurate predictions for seeds in air (Nathan 2006; Nathan et al. 2002, 2005) leads to accurate predictions

---

In the left margin:

Turbulent mixing: eddies and swirls of turbulent flow that move water in random patterns, thereby mixing the fluid

for algal spores in seawater (Gaylord et al. 2006). Furthermore, models for both air and water lead to similar surprising results. In both media, and across a wide variety of propagules, the probability distribution of impact distances has a "fat tail," meaning that it is considerably more probable than previously assumed that propagules travel great distances from their point of release. The ability to predict the likelihood of this long-distance dispersal has practical importance in many aspects of ecology, e.g., the rate of range extension for an invading species or pathogen (Kinlan et al. 2005), the persistence of "fugitive" species in disturbed patches, and recolonization following extreme events (Morritt et al. 2010, Phillips et al. 2010, Platt & Connell 2003). For example, severe winter storms can locally extirpate kelp populations. If kelps are to reestablish, they must be "seeded" by spores from other less-impacted locations. Knowledge of the dispersal characteristics of kelp spores (Gaylord et al. 2006) allows designers of marine protected areas to bolster the stability of their ecosystems by spacing kelp-bed refugia close enough together to ensure reestablishment after storms (Gaines et al. 2010). Note that dispersal distance predicted by these models depends in mechanistic fashion on relevant variables of the physical environment. As these variables—such as wind and current speeds—change (Solomon et al. 2007), new predictions can be made even for environmental conditions that do not currently exist.

Mechanistic approaches have been used to quantify the dispersal of propagules at even larger scales. For example, high-resolution models of wind- and current-driven water motions have been used to predict the pattern of dispersal of reef-fish larvae in the Caribbean Sea, and these predictions highlight the effect of larval behavior on dispersal distance (Cowen & Sponaugle 2009; Cowen et al. 2000, 2006). Absent behavior, fish larvae are likely to be swept hundreds of kilometers from their natal locations. However, when larvae actively adjust their depth (a mechanistic response), their dispersal can be severely limited, and this can have drastic effects on the "connectedness" of populations among islands. Depth-adjustment behavior can similarly affect local retention of invertebrate larvae (e.g., Morgan & Fisher 2010).

Extreme Events and the Environmental Bootstrap

Extreme events play critical roles in ecology and evolution (e.g., Gaines & Denny 1993, Katz et al. 2005), potentially shifting communities between alternate stable states (e.g., Barkai & McQuaid 1988, Paine & Trimble 2004, Petraitis et al. 2009). Some extreme events are due to simple, unitary causes—factors that either happen or not, such as earthquakes and volcanic eruptions—and they are consequently difficult to predict. Many extreme ecological events, however, are due not to the imposition of a single environmental stressor, but rather to the simultaneous imposition of several, individually benign stressors (Paine et al. 1998), and these compound events are open to prediction. Building on recently developed resampling theory (Efron & Tibshirani 1993), Denny et al. (2009) devised a statistical technique (the environmental bootstrap) that takes as its input a relatively short time series of environmental data—5–10 years—and produces an ensemble of hypothetical year-long realizations of how the environment might by chance have played out differently. Using this technique, it is possible to estimate the probability that even extremely rare events might occur by chance alone, and the environmental bootstrap thus provides a tool to move from short-term measurements to long-term predictions.

Implementation of this tool requires mechanistic response functions. For example, resampling a short record of tidal height, wave height, and wave period can provide an extensive ensemble of hypothetical time series for ocean "waviness." However, as described above, this information must be interpreted by a mechanistic response function to provide biologically meaningful data about the distribution of maximum hydrodynamic forces imposed during hypothetical years. When coupled with the organism's structural response function (the probability of being dislodged by a given

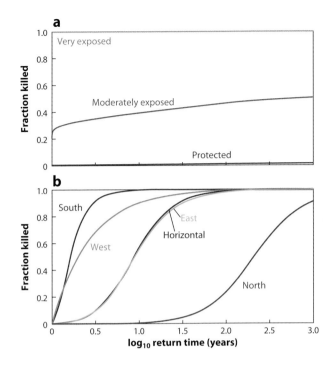

Figure 6

(*a*) The fraction of mussels killed by hydrodynamic forces differs dramatically among sites with different wave exposures, but varies little through time. (*b*) In contrast, the fraction of limpets killed by thermal stress varies substantially with both habitat (the orientation of the rock surface: tilted to the north, east, west, or south) and time. Redrawn from Denny et al. (2009).

hydrodynamic force), this distribution allows one to estimate the likelihood that the organism will be killed by an extreme wave event in a year chosen at random.

Denny et al. (2009) applied this analytic process to predict rates of dislodgment in mussels. In this case, the physics of ocean waves [specifically, the topographically imposed limit to height at breaking (Helmuth & Denny 2003)] renders mussel beds "immune" to extreme events. Risk varies among sites of differing wave exposure (**Figure 6***a*), but at a given site, extreme dislodgment likely to be encountered in a century or millennium is only slightly greater than that likely to be encountered in a decade. In other words, substantial spatial variation is the norm, but catastrophes cannot occur. This immunity to catastrophe may help to explain mussels' widespread competitive dominance, and accords with the study by Guichard et al. (2003) regarding the stability of intertidal mussel beds.

A similar analysis can be carried out using heat-budget models (Bell 1995, Campbell & Norman 1998, Gates 1980, Helmuth 1998) to predict the probability of extreme thermal events. Based on the heat-budget model of Denny & Harley (2006), Denny et al. (2009) used an environmental bootstrap to examine the effect of extreme temperatures on the distribution of a species of intertidal limpet. In contrast to mussels, which are immune to hydrodynamic catastrophes, *Lottia gigantea* is subject to thermal catastrophes. In an average year, the species is unlikely to be thermally stressed by its environment (**Figure 6***b*). However, every decade (on average) greater than 90% of limpets on south-facing rocks will be killed by the chance imposition of a stressful combination of environmental factors. In contrast, it is predicted that limpets on north-facing rocks will not encounter stressful conditions even if they wait a hundred years. Given their rarity, it is not

Heat-budget model: a method for calculating body temperature by taking into account all the ways in which heat can enter or leave an organism

surprising that thermal catastrophes of the type predicted by Denny et al. (2009) have not been directly observed for *L. gigantea* in nature, but their existence is consistent with the measured distribution of limpets high on the shore: Limpets are rare or absent on south-facing rocks but abundant on north-facing rocks (Miller et al. 2009). Building on these studies, Denny & Dowd (2012) used the bootstrap predictions for *L. gigantea*'s body temperature to explore the evolution of thermal tolerance, concluding that it is rare extreme events that set the "safety margin" for these gastropods. The mechanistic approach afforded by the combination of an environmental bootstrap and a heat-budget model thus allows for the prediction of effects that occur on timescales beyond those available for (or at least convenient for) direct measurement. To date, this approach has been applied only in marine habitats, but ample opportunity exists for its extension to terrestrial systems.

The ability to predict the probability of extreme events may prove useful as we explore the effects of global climate change. For example, if field measurements record two extreme thermal events in quick succession when none have been recorded for some time previously, the tendency might be to attribute the increased frequency of events to global warming. The environmental bootstrap, coupled with a heat-budget model, provides a means for testing this hypothesis. By allowing one to estimate the mean probability of extreme events, one can calculate the probability of two such events happening in quick succession by chance alone (Denny et al. 2009). If this probability is unacceptably high, encountering two such events in quick succession cannot be unequivocally attributed to a change in environment.

We emphasize that practical use of the environmental bootstrap requires a mechanistic approach. For example, in their assessment of the probability of thermal death in limpets, Denny et al. (2009) examined 53 million data points for each of six environmental factors. It would be impractical to test empirically the effect of each possible combination of these factors. But a heat-budget model allows for efficient translation of all possible combinations into a single biologically relevant output—body temperature—the effects of which can be readily quantified.

CONCLUSIONS

The mechanistic approach to ecological scaling we advocate here provides a methodological framework to unravel the interplay among processes operating at different scales and how these interactions contribute to the diversity of ecological patterns in nature (Carpenter & Turner 2000, Holling 1992). Cross-scale interactions are becoming more likely in a world that is increasingly connected through the flow of materials and organisms and by climate instabilities (Peters et al. 2004, 2008). Increasing connectivity implies that fine-scale processes, such as point-source pollution, exotic invasions, and epidemics, can propagate to influence large-scale areas. Broad-scale drivers operating at the regional or continental scale (e.g., trends in climate-related variables such as sea-surface and atmospheric temperature and CO_2) can magnify or overwhelm local events (Peters et al. 2008). By linking ecological phenomena across scales, mechanistic response functions have a great potential to explain and ultimately predict how populations and communities respond to the compound effects of scale-dependent, interacting processes.

DISCLOSURE STATEMENT

The authors are not aware of any affiliations, memberships, funding, or financial holdings that might be perceived as affecting the objectivity of this review.

ACKNOWLEDGMENTS

This is contribution 383 of PISCO, the Partnership for Interdisciplinary Studies of Coastal Oceans, a consortium funded by the Gordon and Betty Moore Foundation and the David and Lucile Packard Foundation. We thank the Friday Harbor Working Group (in particular M. Baskett, C.D.G. Harley, P. Jonsson, H. Nepf, and R. Zimmerman) for stimulating discussions and J. Widdows for kind use of his photograph. L.B.C. acknowledges support from the European Community under the FP7 projects VECTORS and CoCoNET.

LITERATURE CITED

Angilletta MJ. 2009. *Thermal Adaptation: A Theoretical and Empirical Synthesis*. New York: Oxford Univ. Press

Barkai A, McQuaid CD. 1988. Predator prey role reversal in a marine benthic ecosystem. *Science* 242:62–64

Baskett ML. 2012. Integrating mechanistic organism-environment interactions into the basic theory of community and evolutionary ecology. *J. Exp. Biol.* 215:948–61

Bell EC. 1995. Environmental and morphological influences on body temperature and desiccation of the intertidal alga *Mastorcarpus papillatus* Kützing. *J. Exp. Mar. Biol. Ecol.* 191:29–55

Benedetti-Cecchi L. 2005. Unanticipated impacts of spatial variance of biodiversity on plant productivity. *Ecol. Lett.* 8:791–99

Benedetti-Cecchi L, Tamburello L, Bulleri F, Maggi E, Gennusa V, Miller M. 2012. Linking patterns and processes across scales: the application of scale-transition theory to algal dynamics on rocky shores. *J. Exp. Biol.* 215:977–85

Berg H. 1984. *Random Walks in Biology*. Princeton: Princeton Univ. Press

Berteaux D, Réale D, McAdam AG, Boutin S. 2004. Keeping pace with fast climate change: Can arctic life count on evolution? *Integr. Comp. Biol.* 44:140–51

Bjørnstad ON, Peltonen M, Liebhold AM, Baltensweiler W. 2002. Waves of larch budmoth outbreaks in the European Alps. *Science* 298:1020–23

Brown JH. 1995. *Macroecology*. Chicago: Univ. Chicago Press

Buckley LB, Rodda GH, Jetz W. 2008. Thermal energetic constraints on ectotherm abundance. *Ecology* 89:48–55

Buckley LB, Urban MC, Angilletta MJ, Crozier LG, Rissler LJ, Sears MW. 2010. Can mechanism inform species distribution models? *Ecol. Lett.* 13:1041–54

Campbell GS, Norman JM. 1998. *An Introduction to Environmental Biophysics*. New York: Springer-Verlag

Carpenter SR, Turner MG. 2000. Hares and tortoises: interactions of fast and slow variables in ecosystems. *Ecosystems* 3:495–97

Carrington E. 2002a. Seasonal variation in the attachment strength of blue mussels: causes and consequences. *Limnol. Oceanogr.* 47:1723–33

Carrington E. 2002b. The ecomechanics of mussel attachment: from molecules to ecosystems. *Integr. Comp. Biol.* 42:846–52

Carrington E, Moeser GM, Dimond J, Mello JJ, Boller ML. 2009. Seasonal disturbance to mussel beds: field test of a mechanistic model predicting wave dislodgment. *Limnol. Oceanogr.* 54:973–86

Carrington E, Moeser GM, Thompson SB, Coutts LC, Craig CA. 2008. Mussel attachment on rocky shores: the effect of flow on byssus production. *Int. Comp. Biol.* 48:801–7

Chesson P, Donahue MJ, Melbourne B, Sears AL. 2005. Scale transition theory for understanding mechanisms in metacommunities. In *Metacommunities: Spatial Dynamics and Ecological Communities*, ed. M Holyoak, MA Leibold, RD Holt, pp. 279–306. Chicago: Univ. Chicago Press

Chown SL, Gaston KJ. 1999. Exploring links between physiology and ecology at macro-scales: the role of respiratory metabolism in insects. *Biol. Rev.* 74:87–120

Chown SL, Gaston KJ. 2010. Body size variation in insects: a macroecological perspective. *Biol. Rev.* 85:139–69

Chown SL, Marais E, Terblanche JS, Klok CJ, Lighton JRB, Blackburn TM. 2007. Scaling of insect metabolic rate is inconsistent with the nutrient supply network model. *Funct. Ecol.* 21:282–90

Clark JS, Shannon L, Ibanez I. 2004. Fecundity of trees and the colonization-competition hypothesis. *Ecol. Monogr.* 74:415–42

Couteron P, Lejeune O. 2001. Periodic spotted patterns in semi-arid vegetation explained by a propagation-inhibition model. *J. Ecol.* 89:616–28

Cowen RK, Lwiza KMM, Sponaugle S, Paris CB, Olson DB. 2000. Connectivity of marine populations: open or closed? *Science* 287:857–59

Cowen RK, Paris CB, Srinivasan A. 2006. Scaling connectivity in marine populations. *Science* 311:522–27

Cowen RK, Sponaugle S. 2009. Larval dispersal and marine population connectivity. *Annu. Rev. Mar. Sci.* 1: 443–66

Cullen J, Neale P, Lesser M. 1992. Biological weighting function for the inhibition of phytoplankton photosynthesis by ultraviolet radiation. *Science* 258:646–50

Davis, AJ, Jenkinson LS, Lawton JH, Shorrocks B, Wood S. 1998. Making mistakes when predicting shifts in species range in response to global warming. *Nature* 391:783–86

de Jaeger M, Weissing FJ, Herman PMJ, Nolet BA, van de Koppel J. 2011. Lévy walks evolve through interaction between movement and environmental complexity. *Science* 332:1551–53

Denny MW. 1987. Lift as a mechanism of patch initiation in mussel beds. *J. Exp. Mar. Biol. Ecol.* 113:231–45

Denny MW. 1988. *Biology and the Mechanics of the Wave-Swept Environment.* Princeton: Princeton Univ. Press

Denny MW. 1995. Predicting physical disturbance: mechanistic approaches to the study of survivorship on wave-swept shores. *Ecol. Monogr.* 65:371–418

Denny MW, Dowd WW. 2012. Biophysics, environmental stochasticity, and the evolution of thermal safety margins in intertidal limpets. *J. Exp. Biol.* 215:934–47

Denny MW, Gaines S. 2000. *Chance in Biology.* Princeton: Princeton Univ. Press

Denny MW, Harley CDG. 2006. Hot limpets: predicting body temperature in a conductance-mediated system. *J. Exp. Biol.* 209:2409–19

Denny MW, Hunt LJH, Miller LP, Harley CDG. 2009. On the prediction of ecological extremes. *Ecol. Monogr.* 79:397–421

Efron B, Tibshirani RJ. 1993. *An Introduction to the Bootstrap.* Boca Raton: Chapman and Hall/CRC

Elith J, Leathwick JR. 2009. Species distribution models: ecological explanation and prediction across space and time. *Annu. Rev. Ecol. Evol. Syst.* 40:677–97

Frechette M, Butman CA, Geyer WR. 1989. The importance of boundary-layer flows in supplying phytoplankton to the benthic suspension feeder, *Mytilus edulis* L. *Limnol. Oceanogr.* 34:19–36

Gaines SD, Denny MW. 1993. The largest, smallest, highest, lowest, longest and shortest: extremes in ecology. *Ecology* 74:1677–92

Gaines SD, White C, Carr MH, Palumbi SR. 2010. Designing marine reserve networks for both conservation and fisheries management. *Proc. Natl. Acad. Sci. USA* 107:18286–93

Gaston KJ, Chown SL, Calosi P, Bernardo J, Bilton DT, et al. 2009. Macrophysiology: a conceptual reunification. *Am. Nat.* 174:595–612

Gates DM. 1980. *Biophysical Ecology.* Mineola, NY: Dover

Gaylord B, Blanchette C, Denny MW. 1994. Mechanical consequences of size in wave-swept algae. *Ecol. Monogr.* 64:287–313

Gaylord B, Gaines SD. 2000. Temperature or transport? Range limits in marine species mediated solely by flow. *Am. Nat.* 155:769–89

Gaylord B, Reed DC, Raimondi PT, Washburn L. 2006. Macroalgal spore dispersal in coastal environments: mechanistic insights revealed by theory and experiment. *Ecol. Monogr.* 76:481–502

Guichard F, Halpin PM, Allison GW, Lubchenco J, Menge BA. 2003. Mussel disturbance dynamics: signatures of oceanographic forcing from local interactions. *Am. Nat.* 161:889–904

Gutschick VP, BassiriRad H. 2003. Extreme events as shaping physiology, ecology, and evolution of plants: toward a unified definition and evaluation of their consequences. *New Phytol.* 160:21–42

Helmuth BST. 1998. Intertidal mussel microclimates: predicting the temperature of a sessile invertebrate. *Ecol. Monogr.* 68:51–74

Helmuth BST, Denny MW. 2003. Predicting wave exposure in the rocky intertidal zone: Do bigger waves always lead to larger forces? *Limnol. Oceangr.* 48:1338–45

Holling CS. 1959. Some characteristic of simple types of predation and parasitism. *Can. Entomol.* 7:385–98

Holling CS 1965. The functional response of predators to prey density and its role in mimicry and population regulation. *Mem. Entomol. Soc. Can.* 45:1–60

Holling CS. 1992. Cross-scale morphology, geometry, and dynamics of ecosystems. *Ecol. Monogr.* 62:447–502

Huey RB, Stevenson RD. 1979. Integrating thermal physiology and ecology of ectotherms: a discussion of approaches. *Am. Zool.* 19:357–66

Hutchinson GE. 1957. Concluding remarks. *Cold Spring Harb. Symp. Quant. Biol.* 22:415–27

Jensen JL. 1906. Sur les functions convexes et les inéqualités entre les valeurs moyennes. *Acta Math.* 30:175–93

Kaitala V, Ranta E. 1998. Travelling wave dynamics and self organization in a spatio-temporally structured population. *Ecol. Lett.* 1:186–92

Katul GG, Porporato R, Nathan R, Siqueira M, Soons MB, et al. 2005. Mechanistic analytical models for long-distance seed dispersal by wind. *Am. Nat.* 166:368–81

Katz RW, Brush GS, Parlange MB. 2005. Statistics of extremes: modeling ecological disturbances. *Ecology* 86:1124–34

Kearney M, Matzelle A, Helmuth B. 2012. Biomechanics meets the ecological niche: the importance of temporal data resolution. *J. Exp. Biol.* 215(8):1422–24

Kearney M, Phillips BL, Tracy CR, Christian KA, Betts G, Porter WP. 2008. Modeling species distributions without using species distributions: the cane toad in Australia under current and future climates. *Ecography* 31:423–34

Kearney M, Porter W. 2009. Mechanistic niche modeling: combining physiological and spatial data to predict species ranges. *Ecol. Lett.* 12:334–50

Kearney M, Simpson SJ, Raubenheimer D, Helmuth B. 2010. Modelling the ecological niche from functional traits. *Philos. Trans. R. Soc. B* 365:3469–83

Kinlan BP, Gaines SD, Lester SE. 2005. Propagule dispersal and the scales of marine community process. *Divers. Distrib.* 11:139–48

Klausmeier CA. 1999. Regular and irregular patterns in semiarid vegetation. *Science* 284:1826–28

Kooijman SALM. 2000. *Dynamic Energy Budget Theory for Metabolic Organisation*. Cambridge, UK: Cambridge Univ. Press

Kozlowski J, Konarzewski M, Gawelczyk AT. 2003. Cell size as a link between noncoding DNA and metabolic rate scaling. *Proc. Natl. Acad. Sci. USA* 100:14080–85

Kuo ESL, Sanford E. 2009. Geographic variation in the upper thermal limits of an intertidal snail: implications for climate envelope models. *Mar. Ecol. Progr. Ser.* 388:137–46

Lande R, Engen S, Saether BE. 2003. *Stochastic Population Dynamics in Ecology and Conservation*. Oxford, UK: Oxford Univ. Press

Levin SA. 1992. The problem of pattern and scale in ecology: the Robert H. MacArthur Award lecture. *Ecology* 73:1943–67

Levin SA, Muller-Landau HC, Nathan R, Chave J. 2003. The ecology and evolution of seed dispersal: a theoretical perspective. *Annu. Rev. Ecol. Evol. Syst.* 34:575–604

Long S, Humphries S, Falkowski P. 1994. Photoinhibition of photosynthesis in nature. *Annu. Rev. Plant Physiol. Plant Mol. Biol.* 45:633–62

MacArthur RH, Levins R. 1967. The limiting similarity, convergence and divergence of coexisting species. *Am. Nat.* 101:377–85

Madin JS, Black KP, Connolly SR. 2006. Scaling water motion on coral reefs: from regional to organismal scales. *Coral Reefs* 25:635–44

Madin JS, Connolly ST. 2006. Ecological consequences of major hydrodynamic disturbances on coral reefs. *Nature* 444:477–80

Malamud BD, Morein G, Turcotte DL. 1998. Forest fires: an example of self-organized critical behavior. *Science* 281:1840–42

Massel SR. 1996. *Advanced Series on Ocean Engineering*, Vol. 11. *Ocean Surface Waves: Their Physics and Prediction*. Singapore: World Sci. 508 pp.

Massel SR, Done TJ. 1993. Effects of cyclone waves in massive coral assemblages on the Great Barrier Reef: meteorology, hydrodynamics and demography. *Coral Reefs* 12:153–66

Maynard-Smith J. 1974. *Models in Ecology*. Cambridge, UK: Cambridge Univ. Press

McGill BJ, Nekola JC. 2010. Mechanisms in macroecology: AWOL or purloined letter? Towards a pragmatic view of mechanism. *Oikos* 119:591–603

McLeod E, Salm R, Green A, Almany J. 2009. Designing marine protected areas networks to address the impact of climate change. *Front. Ecol. Environ.* 7:362–70

McNair JN, Newbold JD, Hart DD. 1997. Turbulent transport of suspended particles and dispersing benthic organisms: how long to hit bottom? *J. Theor. Biol.* 188:29–52

Melbourne BA, Chesson P. 2006. The scale transition: scaling up population dynamics with field data. *Ecology* 87:1476–88

Melbourne BA, Hastings A. 2008. Extinction risk depends strongly on factors contributing to stochasticity. *Nature* 454:100–3

Melbourne BA, Sears ALW, Donahue MJ, Chesson P. 2005. Applying scale transition theory to metacommunities in the field. In *Metacommunities: Spatial Dynamics and Ecological Communities*, ed. M Holyoak, MA Leibold, RD Holt, pp. 307–30. Chicago: Univ. Chicago Press

Miller LP, Harley CDG, Denny MW. 2009. The role of temperature and desiccation stress in limiting the local-scale distribution of the owl limpet, *Lottia gigantea*. *Funct. Ecol.* 23:756–67

Moeser GM, Carrington E. 2006. Seasonal variation in mussel byssal thread mechanics. *J. Exp. Biol.* 209:1996–2003

Moeser GM, Leba H, Carrington E. 2006. Seasonal influence of wave action on thread production in *Mytilus edulis*. *J. Exp. Biol.* 209:881–90

Morgan SG, Fisher JL. 2010. Larval behavior regulates nearshore retention and offshore migration in an upwelling shadow and along the open coast. *Mar. Ecol. Progr. Ser.* 404:109–26

Morritt DM, Nilsson C, Jansson R. 2010. Consequences of propagule dispersal and river fragmentation for riparian plant community diversity and turnover. *Ecol. Monogr.* 80:609–26

Nathan R. 2006. Long-distance dispersal of plants. *Science* 313:786–88

Nathan R, Katul GG, Horn HS, Thomas SM, Oren R, et al. 2002. Mechanism of long-distance dispersal of seeds by wind. *Nature* 418:409–13

Nathan R, Sapir N, Trakenbrot A, Katul GG, Bohrer G, et al. 2005. Long-distance biological transport processes through the air: can nature's complexity be unfolded *in silico*? *Diversity Distrib.* 11:131–37

Nisbet RM, Jusup M, Klanjseek T, Pecquerie L. 2012. Integrating dynamic energy budget (DEB) theory with traditional bioenergetic models. *J. Exp. Biol.* 215(7):1246

Nisbet RM, Muller EB, Lika K, Kooijman SALM. 2000. From molecules to ecosystems through dynamic energy budget models. *J. Anim. Ecol.* 69:913–26

Paine RT, Levin SA. 1981. Intertidal landscapes: disturbance and the dynamics of pattern. *Ecol. Monogr.* 51:145–78

Paine RT, Tegner MJ, Johnson EA. 1998. Compounded perturbations yield ecological surprises. *Ecosystems* 1:535–45

Paine RT, Trimble AC. 2004. Abrupt community change on a rocky shore—biological mechanisms contributing to the potential formation of an alternative state. *Ecol. Lett.* 7:441–45

Pascual M, Guichard F. 2005. Criticality and disturbance in spatial ecological systems. *Trends Ecol. Evol.* 20:88–95

Peters DPC, Groffman PM, Nadelhoffer KJ, Grimm NB, Collins SL, et al. 2008. Living in an increasingly connected world: a framework for continental-scale environmental science. *Front. Ecol. Environ.* 6:229–37

Peters DPC, Pielke RA Sr, Bestelmeyer BT, Allen CD, Munson-McGee S, et al. 2004. Cross-scale interactions, nonlinearities and forecasting catastrophic events. *Proc. Natl. Acad. Sci. USA* 101:15130–35

Petraitis PS, Methratta ET, Rhile EC, Vidargas NA, Dudgeon SR. 2009. Experimental confirmation of multiple community states in a marine ecosystem. *Oecologia* 161:139–48

Phillips BL, Kelehear C, Pizzatto L, Brown GP, Barton D, Shine R. 2010. Parasites and pathogens lag behind their host during periods of host range advance. *Ecology* 91:872–81

Platt WJ, Connell JH. 2003. Natural disturbances and directional replacement of species. *Ecol. Monogr.* 73:507–22

Ranta E, Kaitala V. 1997. Travelling waves in vole population dynamics. *Nature* 390:456

Ranta E, Kaitala V, Lindstrom J. 1997. Dynamics of Canadian lynx populations in space and time. *Ecography* 20:454–60

Rastetter EB, Aber JD, Peters DPC, Ojima DS, Burke IC. 2003. Using mechanistic models to scale ecological processes across space and time. *BioScience* 53(1):68–76

Rees MR, Condit R, Crawley M, Pacala S, Tilman D. 2001. Long-term studies of vegetation dynamics. *Science* 293:650–55

Rietkerk M, Boerlijst MC, van Langevelde F, HilleRisLambers R, van de Koppel J, et al. 2002. Self-organization of vegetation in arid ecosystems. *Am. Nat.* 160:524–30

Ruel JJ, Ayres MP. 1999. Jensen's inequality predicts effects of environmental variation. *Trends Ecol. Evol.* 14:361–66

Schoener TW. 2011. The newest synthesis: understanding the interplay of evolutionary and ecological dynamics. *Science* 331:426–29

Solé RV, Bascompte J. 2006. *Self-Organization in Complex Ecosystems*. Princeton: Princeton Univ. Press

Solomon S, Qin D, Manning M, Chen Z, Marquis M, et al., eds. 2007. *Contribution of Working Group I to the Fourth Assessment Report of the Intergovernmental Panel on Climate Change*. Cambridge, UK: Cambridge Univ. Press. 996 pp.

Tennekes H, Lumley JL. 1972. *A First Course in Turbulence*. Cambridge, MA: MIT Press

Tilman D. 1994. Competition and biodiversity in spatially structured habitats. *Ecology* 75:2–16

Turing A. 1952. On the chemical basis of morphogenesis. *Philos. Trans. R. Soc. B* 237:37–72

Underwood AJ, Paine RT. 2007. Two views on ecological experimentation. *Bull. Br. Ecol. Soc.* 38(3):24–30

Urban MC, Phillips BL, Skelly DK, Shine R. 2007. The cane toad's (*Chaunus* [*Bufo*] *marinus*) increasing ability to invade Australia is revealed by a dynamically updated range model. *Proc. R. Soc. B* 274:1413–19

van de Koppel J, Gascoigne JC, Theraulaz G, Rietkerk M, Mooij WM, Herman PMJ. 2008. Experimental evidence for spatial self-organization and its emergent effects in mussel bed ecosystems. *Science* 322:739–42

van de Koppel J, Rietkerk M, Dankers N, Herman PMJ. 2005. Scale-dependent feedback and regular spatial patterns in young mussel beds. *Am. Nat.* 165:E66–77

van der Meer J. 2006. An introduction to dynamic energy budget (DEB) models with special emphasis on parameter estimation. *J. Sea Res.* 56:85–102

Viswanathan GM, Afanasyev V, Buldryrev SV, Havlin S, da Luz MGE, et al. 2000. Lévy flights in random searches. *Physica A* 282:1–12

West GB, Brown JH, Enquist BJ. 1997. A general model for the origin of allometric scaling laws in biology. *Science* 276:122–26

Whitehead A. 2012. Comparative genomics in ecological physiology: toward a more nuanced understanding of acclimation and adaptation. *J. Exp. Biol.* 215:884–91

Widdows J, Pope ND, Brinsely MD, Gascoigne J, Kaiser MJ. 2009. Influence of self-organised structures on near-bed hydrodynamics and sediment dynamics within a mussel (*Mytilus edulis*) bed in the Menai Strait. *J. Exp. Mar. Biol. Ecol.* 379:92–100

Zimmer RK, Fingerut JT, Zimmer CA. 2009. Dispersal pathways, seed rains, and the dynamics of larval behavior. *Ecology* 90:1933–47

Adaptive Genetic Variation on the Landscape: Methods and Cases

Sean D. Schoville,[1,2] Aurélie Bonin,[2] Olivier François,[1] Stéphane Lobreaux,[2] Christelle Melodelima,[2] and Stéphanie Manel[2,3]

[1] Laboratoire TIMC-IMAG, UMR-CNRS 5525, Université Joseph Fourier, 38041 Grenoble, France; email: sean.schoville@imag.fr, olivier.francois@imag.fr

[2] Laboratoire d'Ecologie Alpine, UMR-CNRS 5553, Université Joseph Fourier, 38041 Grenoble, France; email: aurelie.bonin@ujf-grenoble.fr, stephane.lobreaux@ujf-grenoble.fr, christelle.melo-de-lima@ujf-grenoble.fr, stephanie.manel@univ-amu.fr

[3] Laboratoire Population Environnement et Développement, UMR-IRD 151, Université Aix-Marseille, 13331 Marseille, France

Annu. Rev. Ecol. Evol. Syst. 2012. 43:23–43

First published online as a Review in Advance on August 28, 2012

The *Annual Review of Ecology, Evolution, and Systematics* is online at ecolsys.annualreviews.org

This article's doi: 10.1146/annurev-ecolsys-110411-160248

Keywords

global change, genome scan, ecological selection, landscape genetics, landscape genomics, spatial statistics

Abstract

There is a growing interest in identifying ecological factors that influence adaptive genetic diversity patterns in both model and nonmodel species. The emergence of large genomic and environmental data sets, as well as the increasing sophistication of population genetics methods, provides an opportunity to characterize these patterns in relation to the environment. Landscape genetics has emerged as a flexible analytical framework that connects patterns of adaptive genetic variation to environmental heterogeneity in a spatially explicit context. Recent growth in this field has led to the development of numerous spatial statistical methods, prompting a discussion of the current benefits and limitations of these approaches. Here we provide a review of the design of landscape genetics studies, the different statistical tools, some important case studies, and perspectives on how future advances in this field are likely to shed light on important processes in evolution and ecology.

1. INTRODUCTION

Indeed, every living organism can be viewed as an evolutionary success story.

(Bartholomew 1987)

Adaptive genetic variation (see also the sidebar, Demography and Adaptive Genetic Variation) is defined as the variation found between the genomes of individuals and resulting from natural selection (Holderegger et al. 2006, Lowry 2010). This variability evolves as a result of environmental factors imposing selective pressure on individuals' inhabiting a heterogeneous landscape (see **Figure 1**). Landscape genomics has been proposed as a framework for studying adaptive and neutral genetic variation at the population level in a spatially explicit context (Joost et al. 2007). This framework merges large environmental data sets, spatial statistical methods, and high-resolution sampling of genomic variation to improve our understanding of both species ecology and ecological adaptation (Manel et al. 2003, 2010a). Researchers have developed a suite of genome-scan methods that aim to detect adaptive genetic variation by screening many genetic markers and identifying those that are linked to the loci under selection (Storz 2005). By examining the spatial distribution of alleles at different spatial hierarchies (e.g., individual, population, metapopulation), these new methods can determine the causal relationships between environmental variables and adaptive genetic variability, while controlling for spatial dependence among samples. This provides opportunities to improve our understanding of genetic adaptation (Orr 2005) and to address a number of fundamental questions that remain, such as: Under what circumstances and how frequently does genetic adaptation occur in natural populations? How important are spatial scale and habitat heterogeneity in maintaining adaptive genetic variation? How has recent global change affected patterns of neutral and adaptive genetic variation? Are species likely to adapt to ongoing global change on an ecological timescale?

There is a rich history of studies describing how spatial patterns of environmental variation shape adaptive genetic variation and how spatial scale is an important factor in determining the distribution of this variation (Haldane 1948, Endler 1977). Organisms are distributed over heterogeneous environments, including continuous gradients, discrete habitat patches, and habitats fragmented as a result of human activities. This heterogeneity exposes populations to varying environmental selection pressures, which, depending on the interplay of selection and gene flow, can result in adaptation to local environmental conditions (Lenormand 2002, Savolainen et al. 2007). Although there are some examples of ecological adaptation despite strong gene flow, such

DEMOGRAPHY AND ADAPTIVE GENETIC VARIATION

Demographic change influences the likelihood of adaptation and the maintenance of adaptive genetic variation, albeit in unpredictable ways. Population growth can facilitate rapid local adaptation if selection acts on a small number of founders before an increase in population size (Phillips et al. 2006). Theoretical studies also suggest declining populations can undergo adaptive genetic changes, "genetic revolutions," under certain circumstances (Templeton 2008). Strong selection moves genes more rapidly to fixation in smaller populations, thereby increasing the probability that coadaptive genes will jointly fixate (Crow & Kimura 1965). However, genetic drift is often more likely to overwhelm weak selective forces in a shrinking population and lead to a loss of adaptive genetic variation (Nei et al. 1975). Genetic drift can similarly reduce the likelihood of maintaining adaptive genetic variation in species with fluctuating demographic trends, particularly within a metapopulation structure (Kawecki & Ebert 2004).

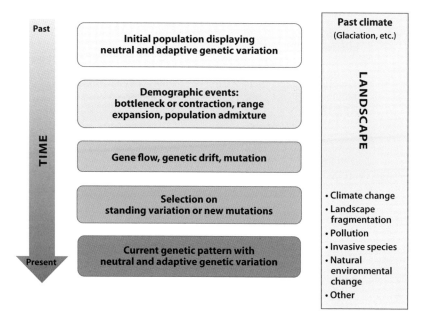

Figure 1

A general template explaining how different factors influence adaptive genetic variation in the landscape over evolutionary time.

as heavy-metal tolerance in plants, host-race divergence in insects, and diversification of fish eco-types in lake ecosystems (Schluter 2000), both the likelihood of adaptation and the maintenance of adaptive variability are dramatically diminished when levels of gene flow exceed "migration-selection equilibrium" (Bridle & Vines 2007, Yeaman & Otto 2011). One of the main strengths of the landscape genetics approach is that it provides a set of tools for incorporating the spatial heterogeneity of the landscape and measures of gene flow with statistical models that elucidate patterns of adaptive variation.

Several reviews have discussed adaptive genetic variation in relation to the landscape (Storz 2005, Holderegger & Wagner 2008, Nielsen et al. 2009, Lowry 2010, Manel et al. 2010a, Schwartz et al. 2010). Here, we attempt to build on those reviews and to distinguish ours by critically examining the existing methods that are used to detect adaptive genetic variation on the landscape. We also review case studies that focus on nonmodel species for which adaptive genetic markers, phenotypes, and selective pressure are often unknown. Finally, we highlight several promising research directions in the field. With the emergence of large genomic data sets, as well as the increasing sophistication of analytical tools in population genetics, it is now possible to identify ecological factors that influence genetic diversity patterns, including both natural and anthropogenic variables driving genetic adaptation. As a consequence, landscape genomics can add another dimension to decision making and prediction in conservation science (Allendorf et al. 2010), particularly in modeling potential adaptive responses of species to global change (Davis et al. 2005, Hoffmann & Willi 2008).

2. STUDY DESIGN

Landscape genetics studies typically try to understand the processes driving genetic variation at the level of individuals by exploring statistical relationships between genetic and environmental

Landscape genetics: a scientific field aiming to study how landscape features interact with microevolutionary processes (e.g., gene flow, genetic drift, selection)

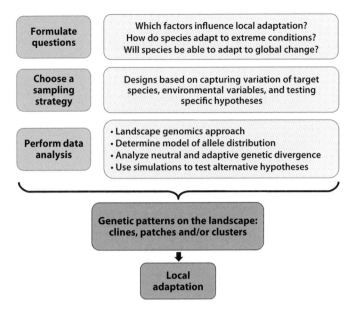

Figure 2

General method and steps for conducting a landscape genomics study.

variation (Manel et al. 2003). Unlike other methods that examine evidence of natural selection in genetic data (Nielsen 2005), landscape genetics addresses the question within a spatially explicit context (Holderegger & Wagner 2008), which is particularly relevant for understanding local adaptation. A typical study might test whether local adaptation to climatic conditions occurs along an altitudinal gradient, but to discriminate effectively between genetic divergence due to selection and that resulting from genetic drift, the study design would require a sample strategy that controls for the effect of drift (e.g., Poncet et al. 2010). At the same time, there are additional challenges in interpreting the relationship between genetic and environmental variation. These include spatial dependence among environmental variables, such as the covariation of latitude and temperature. Therefore, landscape genetics studies are often stratified to account for both spatial and environmental variation (Manel et al. 2010a). To address some of the issues involved in designing landscape genetics studies of adaptive variation (see **Figure 2**), we provide some perspectives by addressing the following: How do spatial sampling, the availability of environmental data, and the choice of molecular markers affect studies of adaptive genetic variation?

2.1. Sampling and Scale

Cluster: a genetic grouping of individuals, often from several nearby sampling locations, that shares a similar multilocus allele-frequency distribution

Until recently, landscape geneticists were mostly interested in the effects of sampling effort (how many individuals to sample per population) (e.g., Muirhead et al. 2008) and did not fully elaborate on sampling designs (where to sample), particularly with respect to environmental variation. Using an appropriate sampling design is fundamental to determining and interpreting patterns of adaptive genetic variation on the landscape. Specifically, the spatial scale of the study should be defined by the ecological factors of interest and should overlap with the spatial distribution and time frame over which assumed phenomena have influenced the spatial genetic structure (Anderson et al. 2010). As selective pressures act mostly on individuals, the sampling unit is the individual. However, sampling may also occur at higher hierarchical levels, such as clusters of individuals or populations, where both physical constraints (such as frogs in ponds or butterflies on meadows) and ecological processes occurring within groups of individuals (e.g., competitive interactions among individuals, local exchanges among clusters) may characterize a particular study system (Anderson et al. 2010). Choosing to sample multiple, potentially closely related individuals from a site will have downstream consequences on the types of statistical models that are appropriate.

Typically we want to test whether the correlation between the distribution of alleles and an environmental factor is different from zero (no correlation), but we must also evaluate whether our experimental design is biased by uneven sampling or hidden population structure (Excoffier et al. 2009). A series of steps can be helpful in developing an appropriate sampling design (Manel et al. 2011). First, define the target population, which should include all individuals that may be affected by a set of environmental factors of interest. Second, define the appropriate space, including the climatic space (e.g., temperature, precipitation gradients) and the biological space (e.g., range limits), taking into account all the aspects related to the environmental gradients of interest. Third, determine a sampling strategy so that independence among sampling units or model residuals can be attained. It will often be important to avoid relatedness by choosing distant individuals and to avoid spatial dependence of ecological factors in environmental space. However, statistical methods that allow for nonindependence of the sampling units are now available (reviewed in Diniz-Filho et al. 2009, Manel et al. 2010a). Fourth, adjust the sampling design as needed to test specific hypotheses, such as whether individuals have more adaptive variation at the core or the edge of their distribution or whether climatic extremes (as opposed to the average values) drive adaptation.

Relatedness: genetic nonindependence due to shared ancestry or kinship

Cline: a gradient in allele frequency in relation to distance and/or to another environmental variable

2.2. Environmental Data

The landscape genetics approach requires environmental data collected in the field or available within existing databases using a geographic information system (GIS). Local or microenvironmental data can be collected using observer measurements or data-recording sensors. Although these often provide the highest resolution (and often most ecologically relevant) environmental data, they typically involve a serious investment of time and resources. Analyzing these data is also challenging, although projects such as the US National Ecological Observatory Network are attempting to provide a framework for collecting and analyzing high-resolution environmental data. Such data can be incorporated into GIS resources for most landscape genomics studies. Existing GIS data include climatic, vegetation, and geological variables and are often freely available [e.g., The Global Map Project (**http://www.globalmap.org/**), WorldClim (**http://www.worldclim.org**)]. Several previous reviews have cataloged some useful GIS resources for landscape genomics (Manel et al. 2010a, Thomassen et al. 2010).

In some cases, it may make sense to measure changes in environmental data to test specific hypotheses about how the environment influences genetic diversity patterns. Such an approach may include testing for effects of recent climate change (Buckley et al. 2011), fragmentation, species invasion, or pollution (Hoffmann & Daborn 2007). Nonclimatic variables may be more ecologically important to some species, for example, pathogen diversity or dietary data for human evolution (Hancock et al. 2010, Fumagalli et al. 2011). Future landscape genomics projects are likely to benefit from rich ecological data sets that measure resource availability and biotic interactions, providing an opportunity to test hypotheses about whether specific ecological factors drive adaptive genetic variation. These projects are also likely to benefit from time-series data and historical samples of ecological and genetic data. For example, Umina et al. (2005) studied the effect of climate change on the classic *Adh* cline in *Drosophila melanogaster*, showing that genetic diversity responded to shifts in climate variation measured over a 20-year period. This type of analysis will be particularly useful in creating predictive maps of how patterns of genetic variation will respond to changing environmental conditions (Thomassen et al. 2010).

2.3. Molecular Markers

Amplified fragment length polymorphisms (AFLPs) were the first molecular markers that could be obtained in large volume and at low cost; they represented a big leap forward in the development of tools aiming to unravel the genetic basis of adaptation. They remain a useful marker system for obtaining hundreds of randomly distributed, polymorphic genetic loci from a large number of individuals and without prior knowledge of the genome. Although AFLPs have been the markers of choice for the implementation of genome scans in nonmodel species (e.g., Poncet et al. 2010), their information content is low because they are dominant and biallelic. More informative markers such as microsatellites and single nucleotide polymorphisms (SNPs) are now being used in genome scans (e.g., Eckert et al. 2010, Meier et al. 2011), although their discovery and genotyping remain labor and cost intensive when traditional sequencers are used. Today, the ongoing next-generation sequencing revolution is transforming genotyping strategies for landscape genomics studies by providing new opportunities for large-scale SNP (and, to a lesser extent, microsatellite) discovery and genotyping (Ekblom & Galindo 2010, Seeb et al. 2011). The purpose of this review is not to discuss the relative merits of these new sequencing platforms (for a recent review, see Glenn 2011), but we do provide a few recommendations in choosing markers for landscape genomics studies.

When resources and genome size are not limiting factors, whole-genome sequencing is the best strategy for identifying loci involved in adaptation. Recently, the throughput of next-generation technologies that generate a large number of short sequences has increased to the point where they have now become a realistic strategy for de novo whole-genome sequencing (Li et al. 2010). Although this approach can be extended to a reasonably large number of individuals during marker development, the quality of genome sequence data is an important parameter to consider. As a general rule of thumb, it is desirable to obtain 10-fold sequencing coverage at a given candidate marker (i.e., at least 10 different sequence reads cover each nucleotide position) to limit the risk of choosing uninformative or unreliable markers. Because costs are often too prohibitive to allow reasonable coverage over the entire genome, especially for species with big genomes, it is sometimes more appropriate to focus the sequencing effort on part of the genome. For example, the RAD (restriction-site-associated DNA) method targets specific restriction fragments for sequencing rather than the entire genome (Hohenlohe et al. 2010a). This approach is becoming an increasingly attractive genotyping alternative to AFLPs for organisms without a reference genome. In the North American pitcher plant mosquito (*Wyeomyia smithii*), only two sequencing lanes of an Illumina GAIIX sequencer were necessary to reveal more than 3,500 SNPs segregating in 21 populations (Emerson et al. 2010). Like AFLPs, however, RAD-tag markers present the major drawback of being random and thus a priori devoid of any genomic information. Sequencing of the RNA transcript pool is another way to obtain high-quality SNPs for an informative subset of the whole genome (Wheat 2011), offering the additional advantage of specifically targeting markers associated with protein-coding sequences. Nevertheless, one has to keep in mind how the choice of tissue, developmental stage, and physiological state of the sampled organism will influence the transcribed genes represented in the final data set. Finally, perhaps the most important factor to consider is the computational resources needed for next-generation sequencing analyses. Even though sophisticated desktop machines are sometimes sufficient to handle next-generation sequencing data sets (typically those with fewer long reads) (Glenn 2011), the ongoing inflation in throughput and sample size will ultimately require massive data storage and curation.

Furthermore, analysis of such large-scale data sets will require a shift to computer clusters and will intensify the need for bioinformatics software (and bioinformaticians) to handle these data.

3. STATISTICAL METHODS

Underlying selective processes can affect patterns of variation in the genome in different ways and are often difficult to distinguish from the background variation inherent in the genome. This has implications for the analytical approaches that can be used to detect adaptive genetic variation. Much of the theory of selection on genetic variation, and the methods to detect it, has been based on the idea that a novel mutation arises in a population experiencing a strong, constant selection pressure, also known as a hard selective sweep (Novembre & Di Rienzo 2009) (see sidebar, Signatures of Selection: Hard Sweeps). However, many selective events in natural populations may be characterized by soft sweeps (also see sidebar, Signatures of Selection: Soft Sweeps), where polymorphism is maintained at linked sites surrounding the favorable allele, as occurs during adaptation from standing genetic variation. Emerging evidence suggests that standing genetic variation plays an important role in adaptation (Innan & Kim 2004, Tennessen & Akey 2011; for an alternate view, see Woodruff & Zhang 2009), which is exemplified in cases where isolated

Selective sweep: the fixation of a selected allele reduces genomic variation in the region surrounding the new mutation

Standing genetic variation: genetic variation that is segregating in a population

Outlier: an observation (or statistical measurement) that has an extreme value compared with the distribution of other observations $F \cong \frac{1}{4Nm+1}$

populations adapt in parallel to the same selective pressures (Kane & Rieseberg 2007, Hohenlohe et al. 2010a). There are several reasons why adaptation from standing genetic variation may be important in natural populations. First, the response to selection does not require a long waiting time; the favorable allele is already in the population, starts at a higher initial frequency, and therefore can rise rapidly to fixation (Innan & Kim 2004). Second, standing genetic variation is more likely to respond to weaker selection pressures, whereas novel mutations have a tendency to be lost to genetic drift under weak selection. Third, natural populations often show patterns of admixture and hybridization (Anderson et al. 2009, De Carvalho et al. 2010), which can increase levels of standing genetic variation and lead to adaptation in novel habitats.

Here we describe how statistical methods can detect adaptive genetic patterns, the methodological assumptions of these methods, and their ability to characterize a selection event when confounding factors exist. Some statistical methods that detect adaptation in the genome involve candidate gene approaches and are not easily connected to patterns of environmental variation, such as tests based on functional changes in proteins (Hohenlohe et al. 2010b). Here we highlight methods that scan large numbers of loci across many individuals and link patterns of genetic variation to environmental variation, noting the many improvements that have recently emerged (see also Storz 2005, Manel et al. 2010a).

3.1. Statistical Methods: Outlier-Detection Methods

Under a neutral island model, populations share genetic variability in proportion to the amount of interpopulation gene flow, $F \cong \frac{1}{4Nm+1}$, where Nm is the product of the effective population size N and the interpopulation migration rate m, and allele frequency differences between populations should arise only as a product of genetic drift (Wright 1931, Haldane 1948). Natural selection is expected to change the pattern of allele frequency differentiation, leading to either strong interpopulation divergence (positive diversifying selection in the case of local adaptation) or extremely low interpopulation divergence (directional or purifying selection, or balancing selection if high heterozygosity is maintained). Using this theory, researchers could measure the population differentiation of allele frequencies across a large number of loci to infer the process of selection acting on a subset of loci (Lewontin & Krakauer 1973). Beaumont & Nichols (1996) implemented the testing procedure FDIST, measuring F_{ST} at multiallelic loci weighted by heterozygosity, to identify outliers that exhibit strong differences from a null distribution of F_{ST}. The null distribution of F_{ST} was determined by assuming an infinite island demographic model, where equilibrium is maintained by symmetrical gene flow and genetic drift. This approach (now FDIST2) has been implemented in the user-friendly software LOSITAN (Antao et al. 2008). Several other methods have been designed to detect statistical outliers while addressing specific problems in the design of an appropriate null distribution and the significance testing procedure. Beaumont & Balding (2004) developed a hierarchical Bayesian model (BayesFST), where correlation among loci, as well as differences in locus-specific and population-specific parameters, could be addressed through a multinomial-Dirichlet likelihood function. More recently, Foll & Gaggiotti (2008) developed BayeScan, which assigns a posterior probability value to each locus by comparing the fit of the observed F_{ST} distribution under a model including selection to a model without selection.

These outlier-detection methods are prone to high false-positive rates if there is hidden population structure or evidence of demographic change, and null distributions may need to be adjusted to reflect these effects (Narum & Hess 2011). Excoffier et al. (2009) were able to reduce the false-positive rate by explicitly modeling hierarchical population structure. To address the influence of demographic history, Bazin et al. (2010) developed an approximate Bayesian computational approach to estimate demographic history and selection simultaneously. However, if populations exhibit extremely complex population structure or demographic history (e.g., serial population expansions, admixture between species, or partial reproductive isolation), developing appropriate null distributions is likely to be challenging and controversial (Teshima et al. 2006, Hermisson 2009, Bierne et al. 2011). As an alternate approach, Vitalis et al. (2001) advocate using a two-population pairwise analysis, where the demographic parameters are treated as nuisance parameters in a population divergence model. This procedure has been implemented in the program DETSEL (Vitalis et al. 2003) and more recently placed in a likelihood framework (Chen et al. 2010), which provides a model-testing procedure to identify candidate loci under selection. Other recent developments have considered measurements of linkage as a way to increase the sensitivity of outlier detection (Tang et al. 2007, Wiehe et al. 2007, Kern & Haussler 2010).

3.2. Statistical Methods: Detecting Correlations Between Allele Distributions and Environmental Variation

The above outlier-detection methods are used to identify genomic regions under selection, but they do not directly integrate tests for environmental factors that cause selection. A number of methods have been proposed to test whether allele distributions are correlated with environmental predictor variables (Manel et al. 2010a). These environmental correlation methods do not always require population samples, but instead can be applied to an individual-based design that draws samples from a large number of different points in space. The basic assumption of allele-distribution models is that natural selection along an environmental gradient, or among patches, generates changes in allele frequencies at loci linked to selected genes. As a result, these methods rely less on overall levels of genetic differentiation and may be more likely to detect adaptation from standing variation (Hancock et al. 2010).

Joost et al. (2007) were the first to formalize an analytical approach to test genetic-environmental correlations across large genomic data sets. They implemented a logistic regression, wherein multiple univariate tests (per locus) are used to detect significant associations with environmental variables. The null hypothesis indicates no correlation between a particular allele and environmental factors (such as temperature or moisture), apart from a background pattern caused by limited dispersal and genetic drift (Manel et al. 2010a). Although Joost et al. (2007) proposed using a logistic regression model, any type of linear or nonlinear regression approach could be used. Recent implementations of this correlative approach have involved more sophisticated models that explicitly consider spatial structure in the data and correct for effects such as isolation by distance, secondary contact, or other demographic effects (see following section). In one example, Manel et al. (2010b) used principal coordinate analysis of neighbor matrices to adjust for spatial autocorrelation and analyzed the resulting eigenvectors in multiple linear regression of environmental variables on AFLP genetic variation. In another approach, Poncet et al. (2010) used generalized estimating equations (GEE) with binomial error to correlate allele frequencies to environmental variables. GEE is a quasi-likelihood extension of a general linear model allowing for possible correlation in the response variable, which Poncet et al. (2010) used to take into account the genetic relatedness of individuals at nearby sampling locations. Similarly, the inclusion of a covariance matrix of allele frequencies between populations can be used as a null model, an approach implemented in the BAYENV program (Coop et al. 2010). A recent study comparing environmental correlation methods to outlier-based methods has shown that the correlation methods often have higher power to detect adaptive genetic variation (S. De Mita, unpublished material). However, the utility of any given method often varies with the sampling strategy and the type of population structure inherent in the study system, and environmental correlation methods sometimes suffer from higher false-positive rates.

3.3. Confounding Effects in Genome-Scan Methods

A fundamental hypothesis in empirical scans for selective sweeps, in which loci with extreme values of a particular statistic are considered to be influenced by positive selection, is that demographic processes have uniform effects across the entire genome, whereas the effects of selection are expected to be locus specific (Lewontin & Krakauer 1973). However, signatures of adaptive processes are not always distinguishable from the genomic background. Population structure causes correlated allele frequencies, which increases the number of false positives generated by genome scans for selective sweeps (Excoffier et al. 2009). It is also widely accepted that demographic history can create patterns resembling selection, as in cases of severe bottlenecks, allele surfing during population expansion, secondary contact, and isolation by distance. Severe population bottlenecks increase genetic drift and population differentiation; they also inflate variation in the genealogical history of different loci. In such cases, a test statistic used in a genome scan may frequently deviate from its empirical average (Hermisson 2009). The allele-surfing phenomenon is another important confounding demographic effect occuring during the colonization of an unoccupied geographic area (Edmonds et al. 2004). The typical genomic signature of a colonization process is a gradual decrease in genetic diversity as a function of the distance from the source, a result that is consistent with the establishment of small founding populations during expansion (Austerlitz et al. 1997). Genetic drift occurs at the wave front of the expanding population and produces clinal variation that resembles the pattern expected under models of positive selection (Excoffier & Ray 2008). Correlative approaches also suffer when population structure and demography are not considered. For example, geographic clines are expected to arise under neutral models if a population is expanding spatially, if two separated populations have recently come into contact, or under various models of isolation by distance (Novembre & Di Rienzo 2009). Ideally, the effect of these demographic processes would be corrected for in statistical tests so that selection could be distinguished from population history.

Another difficulty with interpreting the results of both methodological approaches is that some loci may appear adaptive, even though they actually result from intrinsic genetic incompatibilities or background selection on gene-coding regions. Bierne et al. (2011) argued that intrinsic genetic incompatibilities, which arise during periods of isolation, impede neutral gene flow at linked portions of the genome and lead to spurious statistical outliers that can coincide with environmental variation. However, it is unclear how prevalent such intrinsic genetic incompatibilities are in natural populations or what proportion of the genome they are likely to affect. Background selection, the process in which purifying selection acts to remove deleterious mutations (Charlesworth et al. 1997), can produce patterns of genetic variation similar to those of positive selection, with selected sites and closely linked sites exhibiting reduced levels of diversity and a lower effective population size. Background selection against deleterious mutations is thus expected to increase population differentiation, which could result in an enrichment of high F_{ST} loci in comparison with neutral loci. In humans, patterns of diversity around genic substitutions and of highly differentiated alleles are inconsistent with frequent classic sweeps, but they could result, at least in part, from background selection (Hernandez et al. 2011).

3.4. Testing Hypotheses with Simulation Programs

The inferential power of the outlier-detection methods can be improved through the use of computer simulations designed to generate samples of genetic variation under models of demographic change, local adaptation, and selection on a heterogeneous landscape (Epperson et al. 2010, Balkenhol & Landguth 2011). Genetic samplers can help to distinguish patterns of neutral variation and adaptive genetic variation in a hypothesis-testing framework relevant to landscape genetics. A number of simulation packages have been designed to model selection and spatially explicit population dynamics (Hoban et al. 2012). These simulation programs can also be used in conjunction with approximate Bayesian computation to test the fit of observed data to alternative models of neutral and adaptive processes that may have acted on natural populations (Csilléry et al. 2010). For example, Gerbault et al. (2009) used computer simulations to explore the evolution of the lactase enzyme in Europe. They modeled different demographic hypotheses of gene flow and selection in Europe and tested which model best fit the distribution of lactase alleles over the continent. From these simulations, they found that models that included both demographic history and varying selection intensity fit the observed data better.

Ecological niche modeling: a prediction of a species range based on environmental data; also known as species distribution modeling

3.5. Detecting Environmental Adaptation from Neutral Genetic Variation

Although the most direct way to study adaptive genetic variation on the landscape is to look for outliers or environmental correlations using genome scans, alternative approaches attempt to identify factors influencing genetic differentiation at neutral markers (Pease et al. 2009, Lee & Mitchell-Olds 2011). If the movement of individuals is limited because of strong ecological selection, local adaptation will occur in a form that could impede gene flow via selection against immigrants (Rundle & Nosil 2005). In this case, populations will diverge at both adaptive and neutral loci. For example, Lee & Mitchell-Olds (2011) estimated the relative contributions of environmental adaptation and isolation by distance on genetic variation in *Boechera stricta*, a wild relative of *Arabidopsis*. By comparing patterns of neutral population differentiation to geographical and environmental distance, their study contrasted measures of isolation by distance and "isolation by adaptation" (Nosil et al. 2009). They identified specific environmental variables, including water availability, as the possible cause of differential local adaptation across geographical regions. More generally, identifying environmental adaptation from neutral genetic variation can be accomplished through a variety of statistical tests, including a combination of ecological niche modeling with an analysis of the neutral genetic structure (Freedman et al. 2010). However, nonadaptive factors may also affect gene flow between populations, including impediments to dispersal (habitat matrix or disperser behavior), differences in emigration and immigration rates due to density or local population dynamics, and reproductive barriers.

4. CASE STUDIES OF ADAPTIVE GENETIC VARIATION ON THE LANDSCAPE

Examples of the landscape genomics approach to studying adaptive genetic variation are accumulating rapidly. We begin by illustrating how these methods have been successfully implemented in nonmodel species and end with an example using humans that exemplifies the effectiveness of a high-resolution analysis of adaptive genomic variation.

4.1. Amplified Fragment Length Polymorphisms in Genome Scans in Nonmodel Plants

An outstanding example using AFLP markers in genome scans is the recent paper by Poncet et al. (2010), which utilized a large data set (825 markers) and a novel statistical approach (GEE) and focused on a species inhabiting a complex landscape (see **Figure 3** for the methodology). The perennial plant *Arabis alpina* (Brassicaeae) is common in the European Alps and grows across a large altitudinal range with pronounced climatic gradients. Plant samples were harvested in two different

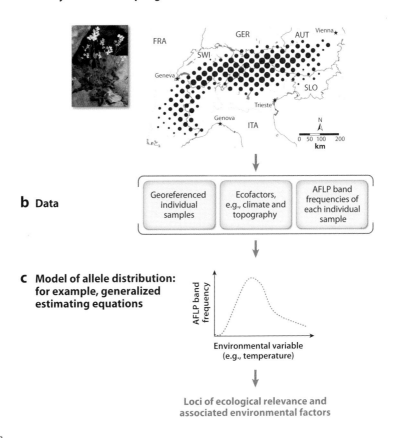

a **Study area and sampling scheme**

b Data

Georeferenced individual samples

Ecofactors, e.g., climate and topography

AFLP band frequencies of each individual sample

c **Model of allele distribution: for example, generalized estimating equations**

AFLP band frequency

Environmental variable (e.g., temperature)

Loci of ecological relevance and associated environmental factors

Figure 3

Approach used to detect loci of ecological relevance in an alpine plant species. (*a*) *Arabis alpina* (photo copyright B. N. Poncet) with a map of the sampling intensity in the European Alps. (*b*) Type of genetic, environmental and spatial data. (*c*) Scheme of analytical approach. Abbreviations: AFLP, amplified fragment length polymorphisms; AUT, Austria; FRA, France; GER, Germany; ITA, Italy; SLO, Slovenia; SWI, Switzerland.

areas, Switzerland and France. Estimates of the population genetics structure suggested that the two areas correspond to two distinct clusters with no admixture and therefore could be considered as independent replicates. Under these conditions, loci considered to have undergone divergent selection between sampling sites could be detected in each area, with higher confidence for loci that were detected in both areas independently. Poncet et al. (2010) performed a variable selection procedure, reducing 20 explanatory environmental variables to four uncorrelated variables by using principal component analysis, an ordination technique that transforms correlated variables into a set of uncorrelated variables called principal components. They identified 78 loci (9%) of ecological relevance in *A. alpina*, which were mainly linked to mean annual minimum temperature. The originality of their approach was their use of GEE models to account for relatedness among neighboring individuals, which can be high in sessile plants collected at nearby sampling locations. Their analysis produced four loci as candidates for adaptive evolution in both the French and Swiss clusters.

4.2. Prospecting for Adaptive Genetic Variation in Large, Nonmodel Genomes

For many species, their large genomes will continue to prevent researchers from developing genomic resources and finding adaptive genetic variation. However, the example of the black spruce, *Picea mariane* (Prunier et al. 2011), with a genome size of approximately 16 Gb, illustrates that successful genome scans can be implemented in these species. To overcome this large genome size, Prunier et al. (2011) focused their work on a set of 656 expressed genes selected from an expressed sequence tag database developed using white spruce (*Picea glauca*). In this targeted approach, genes were selected for their potentially adaptive roles based on the annotation of *Arabidopsis thaliana* homologous genes. A total of 768 SNPs were identified in these targeted genes and genotyped in natural populations across North America. The ultimate objective of the study was to focus on adaptation to specific climatic variables, so populations were partitioned in groups according to temperature and precipitation data. A two-step approach was successfully used to detect SNPs potentially involved in adaptation to temperature and precipitation: (*a*) outlier detection and (*b*) identification of correlations with environmental variables. To detect outliers, researchers used FstSNP [a summary-statistic method similar to that developed by Beaumont & Nichols (1996)] as well as the two-level Bayesian modeling approach BayesFST (Beaumont & Balding 2004). In total, 26 SNPs from 25 genes were detected as outliers; nearly half of the outliers were located in protein-coding genes, and half of those were located at nonsynonymous nucleotide positions. The loci identified using outlier detection were corroborated by applying the regression method of Joost et al. (2007), with the result that 16 of the 26 outlier SNPs (62%) displayed significant regressions between allele frequency and climatic variation.

4.3. Inferring Ecological Selection with Hypothesis Testing

Often, it is necessary to integrate past and present evolutionary processes to understand patterns of neutral and adaptive genetic diversity. The paper by Freedman et al. (2010) provides an elegant example of how genome scans can be used to test the relative roles of drift and selection during evolutionary diversification. The authors compared hypotheses of whether refugial isolation, river barriers to gene flow, or selection along ecological gradients caused population diversification of an African rainforest lizard, *Trachylepis affinis*. This species is characterized by low dispersal rates, and it has several important fitness traits that are associated with temperature gradients. Furthermore, paleoecological data and ecological niche modeling provide evidence for two rainforest refugia that may have affected levels of genetic diversity in *T. affinis*. The authors first performed an AFLP-based genome scan to distinguish between adaptive and neutral loci using the software BayeScan (Foll & Gaggiotti 2008). Genetic analyses were conducted on individuals sampled in three habitat types: lowland rainforest, montane forest, and gallery forests of the rainforest-savanna ecotone. All individuals were genotyped for 191 AFLP markers as well as a fragment of a mitochondrial gene. The genome scan revealed that 7% of the AFLPs were good candidates for divergent selection. Next, the authors contrasted spatial diversity patterns for the neutral and adaptive classes of loci and looked for environmental associations using generalized dissimilarity modeling, a regression-based technique relating a dissimilarity matrix of predictors to a dissimilarity matrix of response variables. Neutral genetic differentiation (from both AFLPs and the mitochondrial gene) was best predicted along a cline between lowland rainforest and the rainforest-savanna ecotone. Ecological adaptation (to temperature and precipitation variables) explained the most variation at the adaptive loci, with a small subset explained by the refugia hypothesis. Taken together, these results suggest that diversification in *T. affinis* is predominately driven by adaptive evolution along the rainforest-ecotone gradient, with a secondary role for historical range expansion.

4.4. Adaptive Genetic Variation in Humans

Studies of humans (Akey et al. 2002, Hancock et al. 2008, Fumagalli et al. 2011) and other model species (Harr et al. 2002, Hancock et al. 2011a) often provide the most rigorous analysis of adaptive genomic variation and show the potential power of genome scans when high-resolution genomic data sets are available. A recent study by Hancock et al. (2011b) examined environmental correlations with climatic variables for a data set of >620,000 SNPs in 61 human populations from around the world. Using the BAYENV model, they tested the association of each SNP against nine separate climatic variables. Given the large number of statistical tests and concern over the false-positive rate, Hancock et al. (2011b) proposed the simple hypothesis that the number of significant loci in protein-coding regions should be enriched relative to sites in noncoding regions. They found that both synonymous and nonsynonymous SNPs were significantly enriched relative to nongenic SNPs and that nonsynonymous SNPs were enriched relative to synonymous SNPs. Some of the strongest associations were with latitude, solar radiation, and temperature and included loci involved in cold tolerance and disease resistance. The authors then repeated the analysis for subsets of the data from two geographical clusters, one representing Africa and western Eurasia and another including Africa, East Asia, and Oceania. Again they found a significant enrichment of nonsynonymous SNPs in both population clusters, but notably the SNPs were different from those identified in the global analysis, suggesting that different loci may underlie adaptation to climate in different populations. Finally, by examining the frequency of significant SNPs in different biological pathways and processes, they found that loci were often associated with disease susceptibility (cancer, cardiovascular and autoimmune diseases) and immunity. To explain the prevalence of SNPs in these functional categories, they suggested that climatic adaptations may often involve fitness trade-offs and provided several examples of loci that increase pathogen resistance but lead to autoimmune disorders later in life.

5. PERSPECTIVES

The emergence of relatively inexpensive genomic-sequencing technologies and environmental databases promises to rapidly increase our knowledge of adaptive genetic variation in nonmodel species. To harness these tools, researchers will have to develop approaches that help us visualize and summarize "genomes on the landscape" (Stapley et al. 2010), thereby providing an opportunity to improve our understanding of landscapes in complex environments for which we often lack information. Such environments include tropical habitats and marine seascapes, where ongoing global change is likely to have a disproportionally negative impact. Here we discuss several major research areas that are likely to emerge as data begin to proliferate on adaptive genetic variation on the landscape.

5.1. New and Improved Analytical Methods

As genomic resources continue to grow, it will be essential to develop statistical methods that are fast and accurate for large numbers of loci. These methods will also be challenged to distinguish true outliers from false positives (Thornton & Jensen 2007). In well-characterized genomes, one way to improve the reliability of outlier methods is to focus on likely candidate genes by incorporating gene function from gene ontology databases or information on gene networks from metabolic pathway databases (e.g., Fumagalli et al. 2011, Hancock et al. 2011a). A second area of improvement requires that additional genetic information be incorporated into landscape genomics methods. Most genome-scan methods currently ignore any effect that nucleotide base composition or recurrent mutation may have in generating bias toward detecting certain loci as outliers. Moving forward, biases that arise from demographic history, hierarchical population structure, or other evolutionary histories also need to be addressed. Finally, significant advances in statistical power are likely to come from methods that include measurements of linkage disequilibrium, even when full linkage maps are not available (Kern & Haussler 2010).

Another major area of improvement will involve the development of methodological approaches that simultaneously look for evidence of adaptation based on selective sweeps and standing variation (Hernandez et al. 2011). Under rapid environmental change, selection based on standing variation, or on polygenic traits, will modify allele frequencies to a lesser extent than will hard selective sweeps. Useful tests include correlations between clines in allele frequencies and environmental variation (Joost et al. 2007), especially those methods that correct for spatial autocorrelation between samples (Coop et al. 2010, Poncet et al. 2010). Other tests for adaptation from standing variation use comparisons between populations that inhabit similar environments to look for parallel adaptation (Hohenlohe et al. 2010a). However, we still need new tests. Providing a better understanding of spatially explicit selection models, which often have specific predictions for the spatial distribution of alleles, is another important focus for theoretical research and may lead to the development of the needed novel tests (Novembre & Di Rienzo 2009).

5.2. Adaptation to a Changing Environment

An important application of the study of adaptive variation on the landscape will be to predict how populations respond to a changing environment. A classical prediction for an adaptive response to climate change states that gene flow from populations in warmer climates will bring preadapted alleles to the leading edge and promote adaptation, whereas populations at the rear edge will likely face local extinction (Davis & Shaw 2001). Garcia-Ramos & Kirkpatrick (1997) found that a species distributed along an environmental gradient will tend to track its optimum via migration, as long as the temporal rate of change does not exceed the spatial change in the environment over the dispersal distance of the species. Predictive models analogous to ecological niche models may be used to model the genetic response of a species to future climate (Aitken et al. 2008, Jay et al. 2012). Future changes in the geographical distribution of a species can be identified by projecting the niche on a map with modified environmental data according to specified climate change scenarios. From a landscape genomics point of view, most of the niche-modeling approaches have the drawback of not accounting for gene migration, population differentiation, and niche variation within the species range (Aitken et al. 2008). Thus, developing methods that include genetic diversity patterns, population structure (Mimura & Aitken 2007), and differences in adaptive variation between leading- and trailing-edge populations (Hampe & Petit 2005) could yield different predictions of species responses to climate change. Landscape genetics will likely play an increasingly important role in the prediction of climate change effects on natural populations (Manel et al. 2010a, Sork & Waits 2010, Thomassen et al. 2010).

5.3. From Genotypes to Phenotypes

Ultimately, statistical correlation with environmental variation does not provide unassailable evidence of natural selection (Barrett & Hoekstra 2011). It remains necessary to link genetic variation to phenotypic and fitness differences, either through direct measurements in the field (e.g., reciprocal transplant studies) or through controlled laboratory experiments. Traditionally, this has been approached through quantitative trait locus (QTL) mapping, which examines correlations between a dense array of markers in a linkage map and phenotypic variation in a controlled breeding experiment. More recently, investigators have developed genome-wide association studies (GWAS), which compare a large number of genomic markers to phenotypic data in a test and control group. Although GWAS and QTL approaches are widely used in humans and other model species, they have only recently been applied in the context of understanding patterns of adaptive genetic variation on the landscape (Rogers & Bernatchez 2005, Fournier-Level et al. 2011). A third strategy is to conduct selection experiments and track the frequencies of alleles across generations; this approach has recently been applied to track adaptation at candidate loci for a climate-related trait (e.g., Rhoné et al. 2010). As genome sequencing of large numbers of individuals becomes more feasible, these approaches are likely to flourish in evolutionary studies of nonmodel species.

SUMMARY POINTS

1. Species are distributed on heterogeneous landscapes, where individuals experience selective pressures from a variety of environmental factors. Landscape genetics provides tools and a framework for identifying patterns of adaptive genetic variation on landscapes.

2. Spatial sampling strategies are an important component in identifying adaptive genetic variation on the landscape. Maximizing environmental heterogeneity and designing experiments that target "candidate" environmental variables at correct spatial and temporal scales can improve tests for adaptive genetic variation.

3. More adaptive loci may be detected using next-generation sequencing technology, especially those techniques in which marker loci can be obtained with high coverage and in a large number of individuals.

4. Two main types of genome-scan methods are available to detect adaptive variation on the landscape: one based on outlier detection and another based on environmental correlations. These statistical methods differ in their ability to detect signatures of natural selection based on hard and soft selective sweeps.

5. A number of confounding factors (demographic history, population structure, etc.) complicate the implementation of these statistical methods and should be taken into account.

FUTURE ISSUES

1. Researchers need to improve their landscape measurements, including the characterization of novel environmental variables involved in potential adaptation (e.g., species interactions).

2. Also in need of improvement are the genome-scan methods, especially those used to detect both soft and hard selective sweeps. New methods should have the flexibility to account for spatial autocorrelation among samples.

3. Niche-modeling tools need to be expanded, and alternative approaches that can be used to incorporate patterns of adaptive genetic variability and to predict whether adaptation is a possible response to global change also need to be developed.

4. Investigators need to expand the purview of landscape genetics moving from the detection of genes with ecological relevance to the determination of the genetic basis of phenotypic and/or fitness differences on the landscape.

5. Landscape genetics approaches need to be integrated with data sources and methods from other disciplines, particularly environmental science.

DISCLOSURE STATEMENT

The authors are not aware of any affiliations, memberships, funding, or financial holdings that might be perceived as affecting the objectivity of this review.

ACKNOWLEDGMENTS

S.D.S. was supported by an international research fellowship from the National Science Foundation (OISE-0965038). S.M. was supported by the Institut Universitaire de France.

LITERATURE CITED

Aitken SN, Yeaman S, Holliday JA, Wang T, Curtis-McLane S. 2008. Adaptation, migration or extirpation: climate change outcomes for tree populations. *Evol. Appl.* 1:95–111

Akey JM. 2009. Constructing genomic maps of positive selection in humans: Where do we go from here? *Genome Res.* 19:711–22

Akey JM, Zhang G, Zhang K, Jin L, Shriver MD. 2002. Interrogating a high-density SNP map for signatures of natural selection. *Genome Res.* 12:1805–14

Allendorf FW, Hohenlohe PA, Luikart G. 2010. Genomics and the future of conservation genetics. *Nat. Rev. Genet.* 11:697–709

Anderson CD, Epperson BK, Fortin M-J, Holderegger R, James PMA, et al. 2010. Considering spatial and temporal scale in landscape-genetic studies of gene flow. *Mol. Ecol.* 19:3565–75

Anderson TM, vonHoldt BM, Candille SI, Musiani M, Greco C, et al. 2009. Molecular and evolutionary history of melanism in North American gray wolves. *Science* 323:1339–43

Antao T, Lopes A, Lopes RJ, Beja-Pereira A, Luikart G. 2008. LOSITAN: a workbench to detect molecular adaptation based on a Fst-outlier method. *BMC Bioinforma.* 9:323

Austerlitz F, Jung-Muller B, Godelle B, Gouyon P-H. 1997. Evolution of coalescence times, genetic diversity and structure during colonization. *Theor. Popul. Biol.* 51:148–64

Balkenhol N, Landguth EL. 2011. Simulation modelling in landscape genetics: on the need to go further. *Mol. Ecol.* 20:667–70

Barrett RDH, Hoekstra HE. 2011. Molecular spandrels: tests of adaptation at the genetic level. *Nat. Rev. Genet.* 12:767–80

Bartholomew GA. 1987. Interspecific comparison as a tool for ecological physiologists. In *New Directions in Ecological Physiology*, ed. ME Feder, AF Bennett, WW Burggren, RB Huey, pp. 11–35. Cambridge, UK: Cambridge Univ. Press

Bazin E, Dawson KJ, Beaumont MA. 2010. Likelihood-free inference of population structure and local adaptation in a Bayesian hierarchical model. *Genetics* 185:587–602

Beaumont MA, Balding DJ. 2004. Identifying adaptive genetic divergence among populations from genome scans. *Mol. Ecol.* 13:969–80

Beaumont MA, Nichols RA. 1996. Loci for use in the genetic analysis of population structure. *Proc. R. Soc. Lond. Ser. B* 263:1619–26

Bierne N, Welch J, Loire E, Bonhomme F, David P. 2011. The coupling hypothesis: why genome scans may fail to map local adaptation genes. *Mol. Ecol.* 20:2044–72

Braverman JM, Hudson RR, Kaplan NL, Langley CH, Stephan W. 1995. The hitchhiking effect on the site frequency spectrum of DNA polymorphisms. *Genetics* 140:783–96

Bridle JR, Vines TH. 2007. Limits to evolution at range margins: When and why does adaptation fail? *Trends Ecol. Evol.* 22:140–47

Buckley J, Butlin RK, Bridle JR. 2011. Evidence for evolutionary change associated with the recent range expansion of the British butterfly, *Aricia agestis*, in response to climate change. *Mol. Ecol.* 21:267–80

Charlesworth B, Nordborg M, Charlesworth D. 1997. The effects of local selection, balanced polymorphism and background selection on equilibrium patterns of genetic diversity in subdivided populations. *Genet. Res.* 70:155–74

Chen H, Patterson N, Reich D. 2010. Population differentiation as a test for selective sweeps. *Genome Res.* 20:393–402

Coop G, Witonsky D, Di Rienzo A, Pritchard JK. 2010. Using environmental correlations to identify loci underlying local adaptation. *Genetics* 185:1411–23

Crow JF, Kimura M. 1965. Evolution in sexual and asexual populations. *Am. Nat.* 99:439–50

Csilléry K, Blum MGB, Gaggiotti OE, François O. 2010. Approximate Bayesian computation (ABC) in practice. *Trends Ecol. Evol.* 25:410–18

Davis MB, Shaw RG. 2001. Range shifts and adaptive responses to Quaternary climate change. *Science* 292:673–79

Davis MB, Shaw RG, Etterson JR. 2005. Evolutionary responses to changing climate. *Ecology* 86:1704–14

De Carvalho D, Ingvarsson PK, Joseph J, Suter L, Sedivy C, et al. 2010. Admixture facilitates adaptation from standing variation in the European aspen (*Populus tremula* L.), a widespread forest tree. *Mol. Ecol.* 19:1638–50

Diniz-Filho JAF, Nabout JC, Telles MPD, Soares TN, Rangel T. 2009. A review of techniques for spatial modeling in geographical, conservation and landscape genetics. *Genet. Mol. Biol.* 32:203–11

Eckert AJ, Bower AD, Gonzalez-Martinez SC, Wegrzyn JL, Coop G, Neale DB. 2010. Back to nature: ecological genomics of loblolly pine (*Pinus taeda*, Pinaceae). *Mol. Ecol.* 19:3789–805

Edmonds CA, Lillie AS, Cavalli-Sforza LL. 2004. Mutations arising in the wave front of an expanding population. *Proc. Natl. Acad. Sci. USA* 101:975–79

Ekblom R, Galindo J. 2010. Applications of next-generation sequencing in molecular ecology of non-model organisms. *Heredity* 107:1–15

Emerson KJ, Merz CR, Catchen JM, Hohenlohe PA, Cresko WA, et al. 2010. Resolving postglacial phylogeography using high-throughput sequencing. *Proc. Natl. Acad. Sci. USA* 107:16196–200

Endler JA. 1977. *Geographic Variation, Speciation, and Clines*. Princeton, NJ: Princeton Univ. Press

Epperson BK, McRae BH, Scribner KIM, Cushman SA, Rosenberg MS, et al. 2010. Utility of computer simulations in landscape genetics. *Mol. Ecol.* 19:3549–64

Excoffier L, Hofer T, Foll M. 2009. Detecting loci under selection in a hierarchically structured population. *Heredity* 103:285–98

Excoffier L, Ray N. 2008. Surfing during population expansions promotes genetic revolutions and structuration. *Trends Ecol. Evol.* 23:347–51

Foll M, Gaggiotti O. 2008. A genome-scan method to identify selected loci appropriate for both dominant and codominant markers: a Bayesian perspective. *Genetics* 180:977–93

Fournier-Level A, Korte A, Cooper MD, Nordborg M, Schmitt J, Wilczek AM. 2011. A map of local adaptation in *Arabidopsis thaliana*. *Science* 334:86–89

Freedman AH, Thomassen HA, Buermann W, Smith TB. 2010. Genomic signals of diversification along ecological gradients in a tropical lizard. *Mol. Ecol.* 19:3773–88

Fumagalli M, Sironi M, Pozzoli U, Ferrer-Admettla A, Pattini L, Nielsen R. 2011. Signatures of environmental genetic adaptation pinpoint pathogens as the main selective pressure through human evolution. *PLoS Genet.* 7:e1002355

García-Ramos G, Kirkpatrick M. 1997. Genetic models of adaptation and gene flow in peripheral populations. *Evolution* 51:21–28

Gerbault P, Moret C, Currat M, Sanchez-Mazas A. 2009. Impact of selection and demography on the diffusion of lactase persistence. *PLoS ONE* 4:e6369

Glenn TC. 2011. Field guide to next-generation DNA sequencers. *Mol. Ecol. Res.* 11:759–69

Haldane JBS. 1927. A mathematical theory of natural and artifical selection. Part V: selection and mutation. *Proc. Camb. Philol. Soc.* 23:838–44

Haldane JBS. 1948. A mathematical theory of natural and artificial selection. VI. Isolation. *Proc. Camb. Philol. Soc.* 26:220–30

Hampe A, Petit RJ. 2005. Conserving biodiversity under climate change: the rear edge matters. *Ecol. Lett.* 8:461–67

Hancock AM, Brachi B, Faure N, Horton MW, Jarymowycz LB, et al. 2011a. Adaptation to climate across the *Arabidopsis thaliana* genome. *Science* 334:83–86

Hancock AM, Witonsky DB, Alkorta-Aranburu G, Beall CM, Gebremedhin A, et al. 2011b. Adaptations to climate-mediated selective pressures in humans. *PLoS Genet.* 7:e1001375

Hancock AM, Witonsky DB, Ehler E, Alkorta-Aranburu G, Beall C, et al. 2010. Human adaptations to diet, subsistence, and ecoregion are due to subtle shifts in allele frequency. *Proc. Natl. Acad. Sci. USA* 107:8924–30

Hancock AM, Witonsky DB, Gordon AS, Eshel G, Pritchard JK, et al. 2008. Adaptations to climate in candidate genes for common metabolic disorders. *PLoS Genet.* 4:e32

Harr B, Kauer M, Schlötterer C. 2002. Hitchhiking mapping: a population-based fine-mapping strategy for adaptive mutations in *Drosophila melanogaster*. *Proc. Natl. Acad. Sci. USA* 99:12949–54

Hermisson J. 2009. Who believes in whole-genome scans for selection? *Heredity* 103:283

Hermisson J, Pennings PS. 2005. Soft sweeps: molecular population genetics of adaptation from standing genetic variation. *Genetics* 169:2335–52

Hernandez RD, Kelley JL, Elyashiv E, Melton SC, Auton A, et al. 2011. Classic selective sweeps were rare in recent human evolution. *Science* 331:920–24

Hoban S, Bertorelle G, Gaggiotti OE. 2012. Computer simulations: tools for population and evolutionary genetics. *Nat. Rev. Genet.* 13:110–22

Hoffmann A, Willi Y. 2008. Detecting genetic responses to environmental change. *Nat. Rev. Genet.* 9:421–32

Hoffmann AA, Daborn PJ. 2007. Towards genetic markers in animal populations as biomonitors for human-induced environmental change. *Ecol. Lett.* 10:63–76

Hohenlohe PA, Bassham S, Etter PD, Stiffler N, Johnson EA, Cresko WA. 2010a. Population genomics of parallel adaptation in threespine stickleback using sequenced RAD tags. *PLoS Genet.* 6:e1000862

Hohenlohe PA, Phillips PC, Cresko WA. 2010b. Using population genomics to detect selection in natural populations: key concepts and methodological considerations. *Int. J. Plant Sci.* 171:1059–71

Holderegger R, Kamm U, Gugerli F. 2006. Adaptive versus neutral genetic diversity: implications for landscape genetics. *Landsc. Ecol.* 21:797–807

Holderegger R, Wagner HH. 2008. Landscape genetics. *Bioscience* 58:199–207

Innan H, Kim Y. 2004. Pattern of polymorphism after strong artificial selection in a domestication event. *Proc. Natl. Acad. Sci. USA* 101:10667–72

Jay F, Manel S, Alvarez N, Durand EY, Thuiller W, et al. 2012. Forecasting changes in population genetic structure of alpine plants in response to global warming. *Mol. Ecol.* 21:2354–68

Joost S, Bonin A, Bruford MW, Després L, Conord C, et al. 2007. A spatial analysis method (SAM) to detect candidate loci for selection: towards a landscape genomics approach to adaptation. *Mol. Ecol.* 16:3955–69

Kane NC, Rieseberg LH. 2007. Selective sweeps reveal candidate genes for adaptation to drought and salt tolerance in common sunflower, *Helianthus annuus*. *Genetics* 175:1823–34

Kawecki TJ, Ebert D. 2004. Conceptual issues in local adaptation. *Ecol. Lett.* 7:1225–41

Kern AD, Haussler D. 2010. A population genetic hidden Markov model for detecting genomic regions under selection. *Mol. Biol. Evol.* 27:1673–85

Kimura M. 1962. On the probability of fixation of mutant genes in a population. *Genetics* 47:713–19

Kimura M, Ohta T. 1969. The average number of generations until fixation of a mutant gene in a finite population. *Genetics* 61:763–71

Lee C-R, Mitchell-Olds T. 2011. Quantifying effects of environmental and geographical factors on patterns of genetic differentiation. *Mol. Ecol.* 20:4631–42

Lenormand T. 2002. Gene flow and the limits to natural selection. *Trends Ecol. Evol.* 17:183–89

Lewontin RC, Krakauer J. 1973. Distribution of gene frequency as a test of the theory of the selective neutrality of polymorphisms. *Genetics* 74:175–95

Li R, Zhu H, Ruan J, Qian W, Fang X, et al. 2010. De novo assembly of human genomes with massively parallel short read sequencing. *Genome Res.* 20:265–72

Linnen CR, Kingsley EP, Jensen JD, Hoekstra HE. 2009. On the origin and spread of an adaptive allele in deer mice. *Science* 325:1095–98

Lowry DB. 2010. Landscape evolutionary genomics. *Biol. Lett.* 6:502–4

Manel S, Albert C, Yoccoz NG. 2011. Sampling in landscape genomics. In *Data Production and Analysis in Population Genomics*, ed. A Bonin, F Pompanon, pp. 93–112. New York: Humana

Manel S, Joost S, Epperson B, Storfer A, Holderegger R, et al. 2010a. Perspectives on the use of landscape genetics to detect genetic adaptive variation in the field. *Mol. Ecol.* 19:3760–72

Manel S, Poncet BN, Legendre P, Gugerli F, Holderegger R. 2010b. Common factors drive adaptive genetic variation at different spatial scales in *Arabis alpina*. *Mol. Ecol.* 19:3824–35

Manel S, Schwartz MK, Luikart G, Taberlet P. 2003. Landscape genetics: combining landscape ecology and population genetics. *Trends Ecol. Evol.* 18:189–97

Maynard Smith J, Haigh J. 1974. The hitch-hiking effect of a favourable gene. *Genet. Res.* 23:23–35

Meier K, Hansen MM, Bekkevold D, Skaala O, Mensberg KLD. 2011. An assessment of the spatial scale of local adaptation in brown trout (*Salmo trutta* L.): footprints of selection at microsatellite DNA loci. *Heredity* 106:488–99

Mimura M, Aitken SN. 2007. Adaptive gradients and isolation-by-distance with postglacial migration in *Picea sitchensis*. *Heredity* 99:224–32

Muirhead JR, Gray DK, Kelly DW, Ellis SM, Heath DD, MacIsaac HJ. 2008. Identifying the source of species invasions: sampling intensity versus genetic diversity. *Mol. Ecol.* 17:1020–35

Narum SR, Hess JE. 2011. Comparison of FST outlier tests for SNP loci under selection. *Mol. Ecol. Res.* 11:184–94

Nei M, Maruyama T, Chakraborty R. 1975. Bottleneck effect and genetic variability in populations. *Evolution* 29:1–10

Nielsen EE, Hemmer-Hansen J, Larsen PF, Bekkevold D. 2009. Population genomics of marine fishes: identifying adaptive variation in space and time. *Mol. Ecol.* 18:3128–50

Nielsen R. 2005. Molecular signatures of natural selection. *Annu. Rev. Genet.* 39:197–218

Nosil P, Funk DJ, Ortiz-Barrientos D. 2009. Divergent selection and heterogeneous genomic divergence. *Mol. Ecol.* 18:375–402

Novembre J, Di Rienzo A. 2009. Spatial patterns of variation due to natural selection in humans. *Nat. Rev. Genet.* 10:745–55

Orr HA. 2005. The genetic theory of adaptation: a brief history. *Nat. Rev. Genet.* 6:119–27

Pease KM, Freedman AH, Pollinger JP, McCormack JE, Buermann W, et al. 2009. Landscape genetics of California mule deer (*Odocoileus hemionus*): the roles of ecological and historical factors in generating differentiation. *Mol. Ecol.* 18:1848–62

Pennings PS, Hermisson J. 2006. Soft sweeps II: molecular population genetics of adaptation from recurrent mutation or migration. *Mol. Biol. Evol.* 23:1076–84

Phillips BL, Brown GP, Webb JK, Shine R. 2006. Invasion and the evolution of speed in toads. *Nature* 439:803

Poncet BN, Herrmann D, Gugerli F, Taberlet P, Holderegger R, et al. 2010. Tracking genes of ecological relevance using a genome scan in two independent regional population samples of *Arabis alpina*. *Mol. Ecol.* 19:2896–907

Prunier J, Laroche J, Beaulieu J, Bousquet J. 2011. Scanning the genome for gene SNPs related to climate adaptation and estimating selection at the molecular level in boreal black spruce. *Mol. Ecol.* 20:1702–16

Rhoné B, Vitalis R, Goldringer I, Bonnin I. 2010. Evolution of flowering time in experimental wheat populations: a comprehensive approach to detect genetic signatures of natural selection. *Evolution* 64:2110–25

Rogers SM, Bernatchez L. 2005. FAST-TRACK: integrating QTL mapping and genome scans towards the characterization of candidate loci under parallel selection in the lake whitefish (*Coregonus clupeaformis*). *Mol. Ecol.* 14:351–61

Rundle HD, Nosil P. 2005. Ecological speciation. *Ecol. Lett.* 8:336–52

Savolainen O, Pyhäjärvi T, Knürr T. 2007. Gene flow and local adaptation in trees. *Annu. Rev. Ecol. Evol. Syst.* 38:595–619

Schluter D. 2000. *The Ecology of Adaptive Radiation*. Oxford, UK: Oxford Univ. Press. 288 pp.

Schwartz MK, Luikart G, McKelvey KS, Cushman SA. 2010. Landscape genomics: a brief perspective. In *Spatial Complexity, Informatics, and Wildlife Conservation*, ed. F Huettman, SA Cushman, pp. 165–74. New York: Springer

Seeb JE, Carvalho G, Hauser L, Naish K, Roberts S, Seeb LW. 2011. Single-nucleotide polymorphism (SNP) discovery and applications of SNP genotyping in nonmodel organisms. *Mol. Ecol. Res.* 11:1–8

Sella G, Petrov DA, Przeworski M, Andolfatto P. 2009. Pervasive natural selection in the *Drosophila* genome? *PLoS Genet.* 5:e1000495

Sork VL, Waits L. 2010. Contributions of landscape genetics: approaches, insights, and future potential. *Mol. Ecol.* 19:3489–95

Stapley J, Reger J, Feulner PGD, Smadja C, Galindo J, et al. 2010. Adaptation genomics: the next generation. *Trends Ecol. Evol.* 25:705

Storz JF. 2005. Using genome scans of DNA polymorphism to infer adaptive population divergence. *Mol. Ecol.* 14:671–88

Tang K, Thornton KR, Stoneking M. 2007. A new approach for using genome scans to detect recent positive selection in the human genome. *PLoS Biol.* 5:e171

Templeton AR. 2008. The reality and importance of founder speciation in evolution. *BioEssays* 30:470–79

Tennessen JA, Akey JM. 2011. Parallel adaptive divergence among geographically diverse human populations. *PLoS Genet.* 7:e1002127

Teshima KM, Coop G, Przeworski M. 2006. How reliable are empirical genomic scans for selective sweeps? *Genome Res.* 16:702–12

Thomassen HA, Cheviron ZA, Freedman AH, Harrigan RJ, Wayne RK, Smith TB. 2010. Spatial modelling and landscape-level approaches for visualizing intra-specific variation. *Mol. Ecol.* 19:3532–48

Thornton KR, Jensen JD. 2007. Controlling the false-positive rate in multilocus genome scans for selection. *Genetics* 175:737–50

Umina PA, Weeks AR, Kearney MR, McKechnie SW, Hoffmann AA. 2005. A rapid shift in a classic clinal pattern in *Drosophila* reflecting climate change. *Science* 308:691–93

Vitalis R, Dawson K, Boursot P. 2001. Interpretation of variation across marker loci as evidence of selection. *Genetics* 158:1811–23

Vitalis R, Dawson K, Boursot P, Belkhir K. 2003. DetSel 1.0: a computer program to detect markers responding to selection. *J. Hered.* 94:429–31

Wheat CW. 2011. SNP discovery using 454 next-generation sequencing. In *Population Genomics: Methods and Protocols*, ed. A Bonin, F Pompanon, pp. 33–35. New York: Humana

Wiehe T, Nolte V, Zivkovic D, Schlötterer C. 2007. Identification of selective sweeps using a dynamically adjusted number of linked microsatellites. *Genetics* 175:207–18

Woodruff RC, Zhang M. 2009. Adaptation from leaps in the dark. *J. Hered.* 100:7–10

Wright S. 1931. Evolution in Mendelian populations. *Genetics* 16:97–159

Yeaman S, Otto SP. 2011. Establishment and maintenance of adaptive genetic divergence under migration, selection, and drift. *Evolution* 65:2123–29

Endogenous Plant Cell Wall Digestion: A Key Mechanism in Insect Evolution

Nancy Calderón-Cortés,[1] Mauricio Quesada,[1] Hirofumi Watanabe,[2] Horacio Cano-Camacho,[3] and Ken Oyama[1]

[1] Centro de Investigaciones en Ecosistemas, Universidad Nacional Autónoma de México (UNAM), 58190, Michoacán, México; email: ncalderon@oikos.unam.mx, mquesada@oikos.unam.mx, akoyama@oikos.unam.mx

[2] Insect-Microbe Research Unit, National Institute of Agrobiological Sciences, Tsukuba, Ibaraki 305-8634, Japan; email: hinabe@affrc.go.jp

[3] Centro Multidisciplinario de Estudios en Biotecnología, Universidad Michoacana de San Nicolás de Hidalgo, 58262, Michoacán, México; email: hcano1gz1@mac.com

Annu. Rev. Ecol. Evol. Syst. 2012. 43:45–71

First published online as a Review in Advance on August 28, 2012

The *Annual Review of Ecology, Evolution, and Systematics* is online at ecolsys.annualreviews.org

This article's doi: 10.1146/annurev-ecolsys-110411-160312

Keywords

lignocellulolytic gut symbionts, insect lignocellulolytic enzymes, insect diversification, plant-insect interactions

Abstract

The prevailing view that insects lack endogenous enzymes for plant cell wall (PCW) digestion had led to the hypothesis that PCW digestion evolved independently in different insect taxa through the establishment of symbiotic relationships with microorganisms. However, recent studies reporting endogenous PCW-degrading genes and enzymes for several insects, including phylogenetically basal insects and closely related arthropod groups, challenge this hypothesis. Here, we summarize the molecular and biochemical evidence on the mechanisms of PCW digestion in insects to analyze its evolutionary pathways. The evidence reveals that the symbiotic-independent mechanism may be the ancestral mechanism for PCW digestion. We discuss the implications of this alternative hypothesis in the evolution of plant-insect interactions and suggest that changes in the composition of lignocellulolytic complexes were involved in the evolution of feeding habits and diet specializations in insects, playing important roles in the evolution of plant-insect interactions and in the diversification of insects.

1. INTRODUCTION

PCW: plant cell wall

Most studies of plant-herbivore interactions have focused on the chemical "arms race" between plants and insects to explain the evolution of plant-insect interactions (Ehrlich & Raven 1964). This hypothesis proposes that secondary compounds produced by plants are the result of defensive responses to the attack of herbivores, and the herbivores in turn avoid, escape, metabolize, or use these chemicals for their own benefit. Therefore, there are selective pressures of plants on insects and vice versa. Alternative explanations suggest that other selective pressures including natural enemies and the recalcitrance of complex molecules such as the cellulose, present in the plant cell wall (PCW), also play a prominent role in regulating plant-insect interactions (Hochuli 1996, Price et al. 1980). The PCW is the major source of energy in the biosphere and an essential food source for insects that thrive on wood, foliage, and detritus. Martin (1991) proposed the hypothesis that insects evolved the ability to digest the PCW through the establishment of symbiotic relationships with microorganisms residing in their digestive tracts. However, this widely accepted hypothesis has been challenged by the report of an endogenous cellulase gene from a termite (Watanabe et al. 1998) and more recent studies of insect PCW-degrading genes (Watanabe & Tokuda 2010). Today, PCW-degrading genes have been reported for all major insect lineages encompassing diverse diets, suggesting that these genes can be involved in the evolution of plant-insect interactions.

In this review, we analyze the molecular and biochemical evidence reporting symbiotic-dependent and symbiotic-independent mechanisms of PCW digestion in insects to understand the role of insect PCW-degrading enzymes in the evolution of plant-insect interactions since ancient times. We propose that the symbiotic-independent mechanism for PCW digestion may be the ancestral mechanism in insects and that insect PCW-degrading enzymes were involved in the evolution of different feeding habits and diet specializations. This alternative hypothesis draws attention to the possibility that insect PCW-degrading enzymes can have remarkable but, thus far, neglected ecological and evolutionary implications in plant-insect interactions.

2. PLANT CELL WALL DIGESTION IN INSECTS

The PCW consists of an interconnected network of cellulose and hemicelluloses, embedded in a complex pectin matrix that includes lignin in secondary PCWs (Gilbert 2010). Hence, PCW digestion requires a complex breakdown process, involving numerous enzymes with diverse substrates, including cellulases, hemicellulases, pectinases, and ligninases (Gilbert 2010).

Cellulases include three classes of hydrolytic enzymes: endo-β-1,4-glucanases (endoglucanases), exo-β-1,4-glucanases (cellobiohydrolases), and β-glucosidases (cellobiases) (Gilbert 2010). The most representative hemicellulases are xylanase and xylooligosaccharidase. However, they also include mannanase, α-glucuronidase, α-L-arabinofuranosidase, laminarinase, and licheninase as well as esterases such as acetyl- and xylan-acetylesterase and arylesterase (Shallom & Shoham 2003, Terra & Ferreira 1994). Pectinases include pectinesterase, pectin/pectate and rhamnogalacturonan lyase, and polygalacturonase (Gilbert 2010). Ligninases consist mainly of phenoloxidase (laccase) and peroxidase (lignin and manganese peroxidase) (Taprab et al. 2005). The PCW degradation process studied in fungi consists of three coordinated steps: depolymerization of lignin or pectin (for primary or secondary PCW, respectively), hemicellulose degradation, and finally cellulose degradation.

The digestive tract of insect herbivores is divided in three main regions: foregut, midgut, and hindgut (Terra 1990). Biochemical transformation occurs in the midgut where the epithelial cells secrete digestive enzymes and absorb the resultant breakdown products (Terra 1990, Terra &

Ferreira 1994). However, in some insects with specialized hindgut structures (e.g., fermentation chambers) such as termites, PCW digestion can also occur in the hindgut via the action of the digestive enzymes of symbiotic microorganisms (Terra 1990).

Different mechanisms have been proposed to explain PCW digestion in insects: (*a*) symbiosis with cellulolytic hindgut protists, (*b*) symbiosis with cellulolytic hindgut bacteria, (*c*) exploitation of the cellulolytic ability of midgut yeasts and bacteria, (*d*) reliance on ingested fungal enzymes, and (*e*) secretion of insect endogenous enzymes (Martin 1991). One method commonly used to identify endogenous PCW-degrading enzymes is the removal of symbionts through the use of antibiotics to eliminate or reduce significantly any possible symbiotic enzymes. However, the methods for removing the symbionts have prompted questions (Slaytor 1992) because the defaunation process can affect a broad spectrum of noncellulolytic bacteria, which can play other important roles in the survival of insects. Insect endogenous enzymes are secreted by the insect midgut epithelial cells or salivary glands, and in some insects the enzymes secreted in the midgut can move forward to the foregut (Terra 1990, Terra & Ferreira 1994). Therefore, researchers also measure enzymatic activities in these particular organs as another method to identify putative endogenous PCW-degrading enzymes. To demonstrate the endogenous origin of isolated genes in insects, researchers also employ other molecular methods including Southern blotting using DNA isolated from tissues free of symbionts as a template, gene expression analyses via reverse transcription polymerase chain reaction (RT-PCR) or in situ hybridization using an insect's excreting organs, and structural/sequence homology analyses based on insect genes previously characterized. To understand the contribution of each mechanism for PCW digestion in insects, recent research has focused on cloning and characterizing the genes encoding PCW-degrading enzymes from both insects and symbionts. We summarize this evidence and discuss the contribution of each mechanism to the PCW digestion process.

Insect endogenous enzymes: enzymes secreted by the insect midgut epithelial cells or salivary glands

2.1. Hindgut Symbiotic Protists

Termites possess a great diversity of symbiotic microorganisms in their hindguts, including protists, bacteria, and Archaea (Hongoh 2011). Anaerobic symbiotic protists (Parabasalia and Preaxostyla: Oximonadidae) are restricted to "lower" termites (Mastotermitidae, Termopsidae, Hodotermitidae, Kalotermitidae, Serritermitidae, and Rhinotermitidae) and wood-feeding cockroaches (Cryptocercidae) (Hongoh 2011). These symbionts possess PCW-degrading genes encoding cellulases and hemicellulases (**Table 1**). The evidence indicates that symbiotic cellobiohydrolases and xylanases are the major expressed enzymes in some species of lower termites and in the cockroach *Cryptocercus punctulatus* (Tartar et al. 2009, Todaka et al. 2010). However, genes encoding endoglucanases are also expressed by symbiotic protists (**Table 1**). Based on enzymatic activities, the total contribution of symbiotic enzymes in the hindgut of lower termites varies from 12 to 40%, 62 to 84%, and 88 to 98% for endoglucanases (Inoue et al. 2005, Tokuda et al. 2004, Zhou et al. 2007), cellobiohydrolases (Veivers et al. 1982, Zhou et al. 2007), and xylanases (Arakawa et al. 2009, Zhou et al. 2007), respectively. Considering that cellobiohydrolases and xylanases represent key enzymes that act together to effectively degrade lignocellulose (Gilbert 2010), the main contribution of hindgut symbiotic protists is related to the digestion of highly polymerized (hemi)cellulose. Endogenous cellobiohydrolases have never been found in termites (Watanabe & Tokuda 2010), and hemicellulases play relatively minor roles in termites (Arakawa et al. 2009, Zhou et al. 2007). Thus, the symbiotic protist enzymes may account for an effective extraction of energy from the termite partially digested wood particles (Hongoh 2011), explaining the specialization of lower termites on diets consisting mainly of wood.

Table 1 Biochemical and molecular evidence of symbiotic-mediated mechanisms for plant cell wall digestion in insects

INSECT ORDER Family Species	Symbiont species	Symbiont[a]: tissue[b]	Enzymes[c]	Biochemical evidence[d]	Molecular evidence[e]	Reference(s)
ORTHOPTERA						
Gryllotalpidae						
Gryllotalpa orientalis	*Cellulosimicrobium* sp. HY-12	B: G	X	–	GHF10	Oh et al. 2008
BLATTARIA						
Cryptocercidae						
Cryptocercus punctulatus	uncultured symbiotic protists	P: G	C, X	–	GHF: 5, 7, 10, 45	Todaka et al. 2010
ISOPTERA						
Mastotermitidae						
Mastotermes darwiniensis	*Koruga bonita, Deltotrichonympha nana*	P: H	C	–	GHF45	Li et al. 2003
	uncultured symbiotic protists	P: G	C, X		GHF: 5, 7, 10,45	Todaka et al. 2010
Hodotermopsis sjoestedti	uncultured symbiotic protists	P: G	C, X	–	GHF:5,7,10, 11,45	Todaka et al. 2010
Zootermopsis angusticollis	*Cellulomonas* sp., *Microbacterium* sp.	B: H	C	EA	–	Wenzel et al. 2002
	Bacillus spp., *Paenibacillus* sp., *Sphingomonas* sp.	B: H	C	EA	–	Wenzel et al. 2002
Kalotermitidae						
Neotermes koshunensis	uncultured symbiotic protists	P: G	C, X	–	GHF: 5, 7, 10	Todaka et al. 2010
Rhinotermitidae						
Coptotermes formosanus	*Pseudotrichonympha grassii*	P: H	C, X	AP	GHF7	Nakashima et al. 2002a, Watanabe et al. 2002
	Spirotrichonympha leidyi	P: H	C	EA	GHF5	Inoue et al. 2005
	Holomastigotoides mirabile	P: H	C, X	EA	GHF: 7, 11	Arakawa et al. 2009, Watanabe et al. 2002
Coptotermes lacteus	*P. grassii, H. mirabile*	P: H	C	EA	GHF7	Watanabe et al. 2002
Reticulitermes flavipes	uncultured symbiotic protists	P: G	C, X	–	18 GHFs	Tartar et al. 2009, Zhou et al. 2007
Reticulitermes speratus	*Trichonympha agilis, Teranympha mirabilis*	P: H	C	–	GHF45	Ohtoko et al. 2000
	uncultured symbiotic protists	P: G	C, X	–	GHF: 5, 7, 11, 45	Todaka et al. 2010
	Comamonas odontotermitis, Citrobacter spp.	B: G	C	EA	–	Cho et al. 2000

(Continued)

Table 1   (*Continued*)

INSECT ORDER Family *Species*	**Symbiont species**	**Symbiont[a]: tissue[b]**	**Enzymes[c]**	**Biochemical evidence[d]**	**Molecular evidence[e]**	**Reference(s)**
	Klebsiella pneumoniae, Serratia marcescens, Bacillus spp., *Chryseobacterium* spp., *Dyella ginsengisoli*	B: G	C	EA	–	Cho et al. 2000
Termitidae						
Macrotermes natalensis	*Termitomyces* sp.	F: E	C	EP		Martin & Martin 1978
Macrotermes gilvus	*Termitomyces* sp.	F: E	L	EA		Taprab et al. 2005
Nasutitermes sp.	Spirochaetes, *Fibrobacter* spp.	B: H	C, X, P	–	30 GHFs	Warnecke et al. 2007
Nasutitermes walkeri	*Clostridium termitidis* sp. nov	B: M-H	C	–	–	Hethener et al. 1992
Nasutitermes takasagoensis	SR *Clostridium thermocellum, B. subtilis*	B: M-H	C	–	–	Tokuda et al. 2000
HEMIPTERA						
Cicadellidae						
Homalodisca vitripennis	*Erwinia* sp., *Serratia* sp., *Bacillus* sp.	B: G (M)	P, C	–	–	Hail et al. 2011
COLEOPTERA						
Scarabaeidae						
Melolontha melolontha	Clostridiales and Bacteroidetes	B: H	C, X	–	–	Egert et al. 2005
Pachnoda ephippiata	*Clostridium* spp., *Staphylococcus* sp.	B: H, M	C, X	–	–	Egert et al. 2003
	Promicromonospora sp., *Arthrobacter* sp.	B: H, M	C, X	–	–	Egert et al. 2003
Pachnoda marginata	*Cellulomonas pachnodae, P. pachnodae*	B: H	C, X	EA	GHF: 6, 10,11	Cazemier et al. 1999a, 1999b, 2003
Passalidae						
Odontotaenius disjunctus	*Pichia stipitis*-like yeast	F: G	X	–	–	Suh et al. 2003
Verres sternbergianus	*Pichia stipitis*-like yeast	F: G	X	–	–	Suh et al. 2003
Cerambycidae						
Anoplophora glabripennis	*Brevibacterium* sp., *Micrococcus* sp., *Streptomyces* sp, *Bacillus* sp., *Staphylococcus* sp.	B: G	C	AP	–	Geib et al. 2009

(*Continued*)

Table 1   (*Continued*)

INSECT ORDER Family *Species*	Symbiont species	Symbiont[a]: tissue[b]	Enzymes[c]	Biochemical evidence[d]	Molecular evidence[e]	Reference(s)
Batocera horsfieldi	*Sphingobacterium* sp. TN19	B: G (M)	X	EP, EA	GHF10	Zhou et al. 2009
Corymbia rubra	Gammaproteobacteria	B: G (M)	X, P	EA	–	Park et al. 2007
Psacothea hilaris	Actinobacteria, Gammaproteobacteria	B: G (M)	X, P	EA	–	Park et al. 2007
Massicus raddei	Actinobacteria, Gammaproteobacteria	B: G (M)	X, P	EA	–	Park et al. 2007
Mesosa hirsute	Actinobacteria, Gammaproteobacteria	B: G (M)	X, P	EA	–	Park et al. 2007
Moechotypa diphysis	Actinobacteria, Gammaproteobacteria	B: G (M)	X, P	EA	–	Park et al. 2007
	Paenibacillus sp. HY-8	B: G (M)	X	EP, EA	GHF11	Sunyeon et al. 2006
Monochamus alternatus	Gammaproteobacteria	B: G (M)	X	EA	–	Park et al. 2007
Monochamus marmorator	*Trichoderma harzianum*	F: E	C	EP	–	Kukor & Martin 1986
Olenecamptus clarus	Gammaproteobacteria	B: G (M)	X	EA	–	Park et al. 2007
Prionus insularis	Actinobacteria	B: G (M)	X	EA	–	Park et al. 2007
Rhagium inquisitor	Actinobacteria, Gammaproteobacteria	B: M	C	–	–	Grünwald et al. 2010
Saperda vestita	*Sphingobium yanoikuyae*	B: G	C	EA	–	Delalibera et al. 2005
	SR *Nectria haematococca*	F: G	C	AP	–	Delalibera et al. 2005
	SR *Fusarium culmorum*, SR *Penicillium* spp.	F: G	C	AP	–	
Curculionidae						
Dendroctonus frontalis	SR *Penicillium* spp.	F: G	C	AP	–	
Ips pini	SR *Penicillium* spp.	F: G	C	AP	–	Delalibera et al. 2005
HYMENOPTERA						
Formicidae						
Atta colombica	NI	F: E	P	EA	–	Rønhede et al. 2004
Acromyrmex echinatior	*Leucoagaricus gongylophorus*	F: E	P	EA	GHF: 28, 51, 53, PL1, CE8, CE12	Rønhede et al. 2004, Schiøtt et al. 2010
Siricidae						
NI species	*Amylostereum chailletii*	F: E	C, X	EP	–	Kukor & Martin 1983

(Continued)

Table 1   (*Continued*)

INSECT ORDER Family *Species*	Symbiont species	Symbiont[a]: tissue[b]	Enzymes[c]	Biochemical evidence[d]	Molecular evidence[e]	Reference(s)
Xiphydriidae						
NI species	*Amylostereum chailletii*	F: E	C, X	EP	–	Kukor & Martin 1983
LEPIDOPTERA						
Bombycidae						
Bombyx mori	*K. pneumoniae, Proteus vulgaris, Citrobacter freundii*	B: G	C, X	EA	–	Anand et al. 2010
	Erwinia sp., *Serratia liquefaciens, B. circulans, Aeromonas* sp.	B: G	C, X, P	EA	–	Anand et al. 2010
Saturniidae						
Likely, *Rothschildia lebeau*	SR rumen fungi and bacteria	NI: G(M)	X	EA	GHF: 8, 11	Brennan et al. 2004
Lasiocampidae						
Samia cynthia pryeri	*Aeromonas* sp.	B: G	X	EP, EA	–	Roy et al. 2003
DIPTERA						
Stratiomyidae						
Hermetia illucens	*Bacillus* spp.	B: G	C	–	–	Jeon et al. 2011
Tipulidae						
Tipula abdominalis	Proteobacteria. Actinobacteria. Firmicutes	B: H	C, X, P	AP	–	Cook et al. 2007

[a]Type of symbiont: B, bacteria; F, fungi; NI, nonidentified; P, protozoa; SR, strain related.

[b]Tissue: G, whole gut; G(M), likely midgut; E, ectosymbiosis; H, hindgut; M, midgut.

[c]Type of enzyme: C, cellulase; P, pectinase; X, xylanase.

[d]Biochemical evidence: AP, enzymatic activity presence; CE, carbohydrate esterases; EA, enzymatic activity; EP, enzyme purification; PL, polysaccharide lyases.

[e]Molecular evidence: GHF, glycosyl hydrolase family.

2.2. Hindgut Symbiotic Bacteria

Higher termites (Termitidae) comprising 75% of all termite species, lack cellulolytic protists, but they harbor diverse bacterial communities, including Archaea, Proteobacteria, *Bacteroidetes*, and Spirochaetes (Hongoh 2011). However, for a long time the participation of these bacterial communities in cellulose digestion was controversial (Martin 1983). Recently, it was reported that cellulase activity on highly polymerized cellulose in the bacterial insoluble fraction on the hindgut of *Nasutitermes takasagoensis* and *Nasutitermes walkeri* contributed up to 59% of total cellulase activity (Tokuda & Watanabe 2007). Additionally, in a metagenomic analysis of the hindgut bacteria of *Nasutitermes* sp. more than 100 PCW-degrading enzymes were identified (Warnecke et al. 2007). Hence, hindgut symbiotic bacteria can contribute significantly to PCW digestion in higher termites.

Cellulolysis by hindgut bacteria has also been reported for some Orthoptera, Coleoptera, and Diptera (**Table 1**), but the isolation of cellulolytic genes has been reported only for *Cellulomonas pachnodae* and *Cellulosimicrobium* sp. HY-12 isolated from a scarabaeid beetle and a cricket, respectively (**Table 1**). Interestingly, Scarabaeidae (Coleoptera) and Tipulidae (Diptera) have a fermentation chamber harboring numerous microorganisms (Terra 1990). Nonetheless, Gryllidae and Gryllotalpidae (Orthoptera) do not have a fermentation chamber; their hindguts contain three to four rows of projecting papillae with long setae densely populated by microorganisms (Nation 1983). This suggests that insects with these morphological adaptations in their hindguts can maintain permanent microbial populations that can contribute significantly to PCW digestion. However, the contribution by these symbionts to digestion processes in insect hosts is poorly understood.

2.3. Exploitation of the Cellulolytic Ability of Midgut Yeasts and Bacteria

Symbiotic yeasts and bacteria housed in specialized cells called mycetocytes represent widespread insect-microbial interactions (Douglas 2009). Mycetocytes can be located in the haemocoel and fat body (hemipterans, cockroaches), in the midgut caeca (cerambycid beetles), and in specialized organs (mycetomes) of the gut (heteropterans: Hemiptera) (Douglas 2009). Because most insects with mycetocytes feed on plant tissues, such symbionts were believed to be involved in PCW digestion (Martin 1983), but evidence supporting this idea is debated. Yeasts have been isolated from the gut contents of 27 beetle families (Berkov et al. 2007; Grünwald et al. 2010; Suh et al. 2003, 2008) and from species of six additional orders (Orthoptera, Blattaria, Dermaptera, Hymenoptera, Neuroptera, and Megaloptera) (Suh et al. 2008). Some of these yeasts, such as *Pichia stipitis* and closely related yeast-like symbionts, ferment xylose (Berkov et al. 2007; Grünwald et al. 2010; Suh et al. 2003, 2008). Additional evidence also shows cellulolytic bacteria in insect midguts for the orders Hemiptera, Coleoptera, Lepidoptera, and higher termites (Termitidae) (**Table 1**). However, cellulolytic genes have been isolated only from the *Paenibacillus* sp. HY8 and *Sphingobacterium* sp. TN19 midgut bacteria of cerambycid beetles and from midgut bacteria of a lepidopteran species (**Table 1**). This evidence confirms that some midgut microorganisms can degrade polysaccharides, although it is not clear if these microorganisms contribute significantly to insect nutrient gain.

2.4. Ectosymbiosis and Acquired Fungal Enzymes

Organisms that cultivate fungi as food are represented by Macrotermitidae termites, Formicidae ants (Myrmicinae: Attini), and ambrosia beetles (Curculionidae: Scolytinae and Platypodinae) (Mueller & Gerardo 2002). In these fungus-farming symbioses, insects provide their fungus symbiont with plant material and a favorable environment to sustain its growth, whereas fungi provide food for the insects and their brood (Moller et al. 2011). In addition to direct nutrition, in the termite-fungi symbiosis, fungi degrade the lignin (Hyodo et al. 2003, Taprab et al. 2005), and pectin (Johjima et al. 2006) contained in fungus combs, which enables termite endogenous enzymes to use cellulose more efficiently. In ant-fungi symbiosis, fungi degrade polysaccharides such as pectin (De Fine Licht et al. 2010, Moller et al. 2011) and hemicellulose (Richard et al. 2005), whereas ants hydrolyze oligosaccharides (Richard et al. 2005). However, much attention has been devoted to the "acquired enzyme hypothesis," which proposes that fungi supply enzymes that are ingested by insects and that remain active in the insect gut (Martin & Martin 1978). Similar isoelectric properties between enzymes independently isolated from the gut of some insects and from their symbiotic fungi (Kukor & Martin 1983, 1986; Martin & Martin 1978; Rønhede et al.

2004) have been considered to support this hypothesis, which further suggests that acquired fungal enzymes are widespread among xylophagous and detritivorous insects (Kukor & Martin 1983, 1986). However, this generalization has been questioned because, for some insects, acquired fungal enzymes play little, if any, role in cellulose digestion (Scrivener & Slaytor 1994a, Slaytor 1992). Recent molecular studies (cDNA cloning and mass spectrometry protein analysis) found seven active pectinases encoded by cDNAs isolated from the fungal symbiont *Leucoagaricus gongylophorus* in fecal droplets of the ant *Acromyrmex equinatior* (Schiøtt et al. 2010), demonstrating that, in some obligate fungus-farming symbioses, the insects acquire fungal enzymes for pectin degradation. Nevertheless, more studies are needed to confirm that acquired fungal enzymes are widespread among insects, particularly for insects that are not involved in obligate fungus-farming symbioses.

2.5. Symbiotic-Independent Mechanism

After years of debate, cloning and characterization of genes encoding PCW-degrading enzymes have finally demonstrated symbiotic-independent digestion of the PCW in insects. Endogenous insect PCW-degrading enzymes include cellulases, hemicellulases, pectinases, and ligninases (**Table 2**).

2.5.1. Cellulases. The insect cellulase complex consists of endoglucanases and β-glucosidases, but it lacks cellobiohydrolases (Martin 1983, 1991; Scrivener & Slaytor 1994b). The lack of cellobiohydrolases in xylophagous insects may be explained by the presence of (*a*) mandibles and/or proventriculus that physically crush wood into small particles (10–50 μm), allowing enzymes to effectively access substrates (Fujita et al. 2010, Nakashima et al. 2002b); (*b*) a long digestive tract that allows slowness of transit giving endoglucanases enough time to hydrolyze the cellulose (Haack & Slansky 1987, Watanabe & Tokuda 2010); and (*c*) numerous endoglucanases to compensate for their inefficiency against crystalline cellulose (Scrivener & Slaytor 1994a).

Endoglucanases appear to be widespread among insects, as molecular and/or biochemical evidence for the presence of these enzymes exists for 16 insect orders (**Table 2**). All feeding habits are represented by insects possessing endoglucanase enzymes and/or genes (**Table 2**). However, for some insects such as the body louse (Phthiraptera), the fruit fly (Diptera), and the ants *Acromyrmex echinatior* and *Camponotus floridanus* (Hymenoptera), the reported sequences homologous to cellulases have not been proven to encode active enzymes for digesting cellulose. The presence of this sequence in Phthiraptera is not expected according to its feeding habits (blood feeders). However, discordant results have been previously reported for other insects and invertebrates (Boyd et al. 2002, Monk 1976, Yokoe & Yasumasu 1964) in which the activity of PCW-degrading enzymes and gut physiology reflected a phylogenetic signal rather than an adaptation to a specific feeding habit (Terra & Ferreira 1994, Yokoe & Yasumasu 1964). Cellulase genes have been reported for other paraneopterans such as Hemiptera (**Figure 1**). Basal paraneopterans have also been found to be detritivores (Grimaldi & Engel 2005). Therefore, paraneopteran insects may have inherited a cellulase gene from a common ancestor (i.e., phylogenetic inertia). We do not show information about the reported sequences of β-glucosidases because these enzymes are commonly present in animals (but see the next section for a discussion of the role of these enzymes in plant-insect interactions).

2.5.2. Hemicellulases. Nutritional studies reporting hemicellulase activities have shown that hemicellulose may be utilized by insects (Terra & Ferreira 1994). However, endogenous genes encoding xylanases have been reported for only one cerambycid beetle species (**Table 2**). Even though xylanases appear to be rarely present in insects, xylan can be hydrolyzed by

Table 2 Biochemical and molecular evidence of symbiotic-independent (endogenous) mechanism for plant cell wall digestion in insects

INSECT ORDER Family Species	Feeding habits and guilds[a]	Enzymes[b]	Evidence[c]	Reference(s)/GenBank accession number
ZYGENTOMA				
Lepismatidae				
Thermobia domestica	D	C	EA/DG	Treves & Martin 1994
EPHEMEROPTERA				
Heptageniidae				
Ecdyonurus dispar	D	C	EA/MI, PEA	Monk 1976
Caenidae				
Caenis horaria	D	C	EA/MI, PEA	Monk 1976
PLECOPTERA				
Pteronarcyidae				
Pteronarcys proteus	D	C	EA/FM, PEA	Sinsabaugh et al. 1985
ORTHOPTERA				
Gryllidae				
Allonemobius sp.	O	C	EA/HF, PEA	Oppert et al. 2010
Teleogryllus emma	O	C	GHF9	Kim et al. 2008
Acrididae				
Abracris flavolineata	H (lc)	L	EP/M, PEA	Genta et al. 2007
Chortophaga viridifasciata	H (lc)	C	EA/HF, PEA	Oppert et al. 2010
Dissosteira carolina	H (lc)	C	EP/HM	Willis et al. 2010
Hippiscus ocelote	H (lc)	C	EA/HF, PEA	Oppert et al. 2010
Melanoplus differentialis	H (lc)	C	EA/HF, PEA	Oppert et al. 2010
Melanoplus femurrubrum	H (lc)	C	EA/HF, PEA	Oppert et al. 2010
Schistocerca americana	H (lc)	C	EA/HF, PEA	Oppert et al. 2010
Schistocerca gregaria	H (lc)	C, X	EA/FM, PEA	Cazemier et al. 1997
Syrbula admirabilis	H (lc)	C	EA/HF, PEA	Oppert et al. 2010
PHASMATODEA				
Eurycantha calcarata	H (lc)	C, X	EA/FM, PEA	Cazemier et al. 1997
BLATTARIA				
Polyphagidae				
Polyphaga aegyptiaca	O	C	GHF9	Lo et al. 2000
Blattellidae				
Blatella germanica	O	C	GHF9	Lo et al. 2000
Blaberidae				
Blaberus fuscus	O	C, X	EA/FM, PEA	Cazemier et al. 1997
Calolampra elegans	D	C, X	EA/FM, PEA	Zhang et al. 1993
Geoscapheus dilatatus	D	C, X	EA/FM, PEA	Zhang et al. 1993
Panesthia cribrata	D	C, X	EP, GHF9	Lo et al. 2000; Scrivener & Slaytor 1994a, 1994b

(Continued)

Table 2 (*Continued*)

INSECT ORDER Family *Species*	Feeding habits and guilds[a]	Enzymes[b]	Evidence[c]	Reference(s)/GenBank accession number
Panesthia angustipennis	D	C	GHF9	AB438950–AB438952
Pycnoscelus surinamensis	O	C	EA/FM, PEA	Cazemier et al. 1997
Salganea esakii	D	C	GHF9	AB438946–AB438948
Blattidae				
Gromphadorrhina portentosa	D	C, X	EA/FM, PEA	Cazemier et al. 1997
Periplaneta americana	O	C, X, Li, L, G-L	EP, EA/FM, GHF9, GHF16	Cazemier et al. 1997, Genta et al. 2003, Lo et al. 2000, ABR28480
Periplaneta australasia	O	C, X	EA/FM, PEA	Cazemier et al. 1997
Cryptocercidae				
Cryptocercus clevelandi	D (xy)	C	GHF9	Lo et al. 2000
ISOPTERA				
Mastotermitidae				
Mastotermes darwiniensis	D (xy)	C, X	EA, GHF9	Cazemier et al. 1997, Lo et al. 2000
Termopsidae				
Hodotermopsis sjoestedti	D (xy)	C	GHF9	Tokuda et al. 2004
Kalotermitidae				
Neotermes koshunensis	D (xy)	C	GHF9	Tokuda et al. 2004
Rhinotermitidae				
Coptotermes formosanus	D (xy)	C, X	EA, GHF9	Arakawa et al. 2009, Nakashima et al. 2002b
Reticulitermes speratus	D (xy)	C	EP, GHF9	Watanabe et al. 1998
Reticulitermes flavipes	D (xy)	C, La	EA, GHF9	Tartar et al. 2009
Termitidae				
Odontotermes formosanus	D (xy)	C	GHF9	Tokuda et al. 2004
Nasutitermes takasagoensis	D (xy)	C	EP, GHF9	Tokuda et al. 1999
Nasutitermes walkeri	D (xy)	C	GHF9	Tokuda et al. 1999
Sinocapritermes mushae	D	C	GHF9	Tokuda et al. 2004
HEMIPTERA				
Heteroptera				
Miridae				
Adelphocoris lineolatus	H (ss)	Pg	EA, PEA	Fratti et al. 2006
Closterotomus norwegicus	H (ss)	Pg	EA, PEA	Fratti et al. 2006
Deraeocoris nebulosus	C	Pg	EA/S, PEA	Boyd et al. 2002
Lygus hesperus	H (ss)	Pg	EP, GHF28	Celorio-Mancera et al. 2008
Lygus lineolaris	H (ss) & C	Pg	GHF28	Allen & Mertens 2008
Lygus pratensis	H (ss)	Pg	EA, PEA	Fratti et al. 2006
Lygus rugulipennis	H (ss)	Pg, Pho	EA, PEA	Fratti et al. 2006
Orthops kalmi	H (ss)	Pg	EA, PEA	Fratti et al. 2006

(*Continued*)

Table 2   (*Continued*)

INSECT ORDER Family *Species*	Feeding habits and guilds[a]	Enzymes[b]	Evidence[c]	Reference(s)/GenBank accession number
Sternorrhyncha				
Aphididae				
Acyrthosiphon pisum	H (ss)	C, Pg, Pm, Pho, Po	EA/S, GHF9	Cherqui & Tjallingii 2000, XM_001944739
Myzus persicae	H (ss)	Pg, Pm, Pho, Po	EA/S	Cherqui & Tjallingii 2000
Schizaphis graminum	H (ss)	Pg, Pm, Pho	EA/S	Cherqui & Tjallingii 2000
PHTHIRAPTERA				
Pediculidae				
Pediculus humanus humanus	P	C	GHF9	XM_002426420
COLEOPTERA				
Tenebronidae				
Tenebrio molitor	H (sp)	G-L	EP, GHF16	Genta et al. 2009
Tribolium castaneum	H (sp)	C	EP, GHF9	Willis et al. 2011
Chrysomelidae				
Callosobruchus maculatus	H (se)	C, Pg, M	GHF5, GHF28	Pauchet et al. 2010
Chrysomela tremulae	H (lc)	C, Pg, M	GHF5, GHF28, GHF45	Pauchet et al. 2009b, 2010
Diabrotica virgifera	H (ro)	C, Pg	GHF28, GHF45	Pauchet et al. 2010
Gastrophysa viridula	H (lc)	C, Pg	GHF5, GHF28, GHF45	Pauchet et al. 2010
Leptinotarsa decemlineata	H (lc)	C, Pg	GHF28, GHF45	Pauchet et al. 2010
Phaedon cochleariae	H (lc)	C, X, Pg	EA/M, GHF11, GHF28, GHF45	Girard & Jouanin 1999
Cerambycidae				
Psacothea hilaris	H (b-xy, lc)	C, X, L, Li, Pg	EA; GHF5	Scrivener et al. 1997
Hylotrupes bajulus	H (b-xy)	C, X	EA/FM, PEA	Cazemier et al. 1997
Apriona germari	H (b-xy, lc)	C	EP, GHF5, GHF45	Lee et al. 2005, Wei et al. 2006
Oncideres albomarginata chamela	H (b-xy)	C	EA, GHF5, GHF45	Calderón-Cortés et al. 2010
Curculionidae				
Dendroctonus ponderosae	H (b-xy)	C, Pg, Pm	GHF28, GHF45, CE8	Aw et al. 2010, Pauchet et al. 2010
Diaprepes abbreviatus	H (ro)	C, Pg, Pm	EA, GHF28, GHF45	Doostdar et al. 1997, DN200030, DN200477
Sitophilus oryzae	H (sp)	C, Pg, Pm	EP, GHF28, GHF45, GHF48, CE8	Pauchet et al. 2010; Shen et al. 2003, 2005
Ips pini	H (b-xy)	C, Pg	GHF28, GHF45	Pauchet et al. 2010
Hypothenemus hampei	H (b-xy)	C, Pg	GHF28, GHF45	FD662948–FD662950, FD663237

(*Continued*)

Table 2   (Continued)

INSECT ORDER Family Species	Feeding habits and guilds[a]	Enzymes[b]	Evidence[c]	Reference(s)/GenBank accession number
MEGALOPTERA				
Sialidae				
Sialis lutaria	C	C	EA/MI, PEA	Monk 1976
HYMENOPTERA				
Apidae				
Apis mellifera	H (po-n)	C	GHF9	Kunieda et al. 2006
Bombus terrestris	H (po-n)	C	GHF9	XP_003402778
Formicidae				
Acromyrmex echinatior	H (lc)	C	GHF9	EGI63652
Camponotus floridanus	H (lc)	C	GHF9	EFN70196
Pteromalidae				
Nasonia vitripennis	P	C	GHF9	XM_001606404
LEPIDOPTERA				
Plutellidae				
Plutella xylostella	H (lc)	G-L	GHF16	Pauchet et al. 2009a
Sesiidae				
Melittia satyriniformis	H (b)	C	EA/HF, PEA	Oppert et al. 2010
Synanthedon scitula	H (b)	C	EA/HF, PEA	Oppert et al. 2010
Crambidae				
Diatrea saccharalis	H (b)	G-L	GHF16	ABR28479
Ostrinia nubilalis	H (b)	G-L	GHF16	Pauchet et al. 2009a
Pyralidae				
Galleria mellonella	H	G-L	GHF16	CAK22401
Bombycidae				
Bombyx mori	H (lc)	G-L	GHF16	Pauchet et al. 2009a
Noctuidae				
Helicoverpa armigera	H (lc)	G-L	GHF16	Pauchet et al. 2009a
Spodoptera frugiperda	H (lc)	G-L, X, P	GHF16; EA/M	Bragatto et al. 2010
Spodoptera littoralis	H (lc)	G-L	GHF16	Pauchet et al. 2009a
Pieridae				
Anthoracharis cardamines	H (lc)	G-L	GHF16	Pauchet et al. 2009a
Delias nigrina	H (lc)	G-L	GHF16	Pauchet et al. 2009a
Pieris rapae	H (lc)	G-L	GHF16	Pauchet et al. 2009a
Sphingidae				
Manduca sexta	H (lc)	La	Multicopper oxidases	Dittmer et al. 2004
TRICHOPTERA				
Polycentropodidae				
Polycentropus flavomaculatus	D	C	EA/MI, PEA	Monk 1976

(Continued)

Table 2 (*Continued*)

INSECT ORDER Family *Species*	Feeding habits and guilds[a]	Enzymes[b]	Evidence[c]	Reference(s)/GenBank accession number
Limnephilidae				
Halesus sp.	D	C	EA/MI, PEA	Monk 1976
Potamophylax sp.	H	C	EA/MI, PEA	Monk 1976
Plectrocnemia geniculata	D	C	EA/MI, PEA	Monk 1976
Pycnopsyche guttifer	O	X, L	EA/MI, PEA	Martin et al. 1981
Phryganeidae				
Agrypnia vestita	D	C, X, L, M	EA/M, PEA	Martin et al. 1981
Phryganea sp.	O	X, L	EA/M, PEA	Martin et al. 1981
DIPTERA				
Drosophilidae				
Drosophila melanogaster	D	C, Pg	GHF28, GHF45	EC068056, CO334668, CO335003

[a]Feeding habits and guilds: b, borer; b-xy, wood borer; C, carnivorous; d, detritivorous; H, herbivorous; lc, leaf chewing; O, omnivorous; P, parasite; po-n, pollen and nectar feeder; ro, root feeder; se, seed feeder; sp, stored products feeder; ss, sap sucker; xy, xylophagous.

[b]Enzymes: C, cellulase; G-L, β-1,3-glucanase-laminarinase; L, laminarinase; La, laccase; Li, licheninase; M, mannanase; Pg, endopolygalacturonase; Pho, phenol oxidase; Pm, pectin methylesterase; Po, peroxidase; X, xylanase.

[c]Evidence: CE, carbohydrate esterases; EA, enzymatic actvity presence; EA/DG, enzymatic activity in defaunated insects; EA/FM, enzymatic activity in foregut and midgut; EA/HF, enzymatic activity in head fluids; EA/M, enzymatic activity in midgut; EA/MI, enzymatic activity in presence of microbial inhibitors; EA/S, enzymatic activity in salivary glands; EP, enzyme purification; EP/HM, enzyme purification from head and midgut; EP/M, enzyme purification from midgut; GHF, glycosyl hydrolase family of insect endogenous genes; PEA, putative endogenous activity.

endoglucanases (Scrivener et al. 1997). Instead of xylanases, the enzymes, termed xyloglucanases, could be involved in the xylan hydrolysis reported for most insect herbivores, although genes encoding xyloglucanases have not yet been cloned.

Other enzymes involved in hemicellulose digestion are β-1,3-glucanases. These enzymes exclusively hydrolyze β-1,3-linkages of β-1,3-glucans and are recognized as components of the laminarinase complex (Pauchet et al. 2009a, Terra & Ferreira 1994). β-1,3-glucans are found in PCWs (i.e., callose), algae, lichens, and fungi (Genta et al. 2009). β-1,3-glucanases can be important enzymes for hemicellulose digestion, given their presence in various insect groups such as Orthoptera, Blattaria, Coleoptera, Trichoptera, and Lepidoptera (**Table 2**), particularly for detritivorous/omnivorous insects, which feed on materials of diverse origins.

2.5.3. Pectinases. The insect pectinase complex comprises endopolygalacturonases, rhamnogalacturonan lyases, and pectin methylesterases; they occur mainly in Hemiptera and Coleoptera (**Table 2**). In piercing-sucking hemipterans, salivary pectinases are involved in plant penetration (Cherqui & Tjallingii 2000) and in the softening of plant material before oviposition (Boyd et al. 2002). However, for most beetles, pectinases are involved in the digestion of young plant tissues and grains. Pectin degradation plays an important role in PCW degradation, as it facilitates the decomposition of (hemi)cellulose and makes the PCW more susceptible to further breakdown by other enzymes. This may explain why pectinases are present in insect herbivores.

2.5.4. Ligninases. Laccases are found in the epidermis of insects, where they are involved in sclerotization of the cuticle (Dittmer et al. 2004). Other phenoloxidases and peroxidases are found in the salivary secretions of aphids (**Table 2**). By neutralizing toxic phenolics, they help aphids

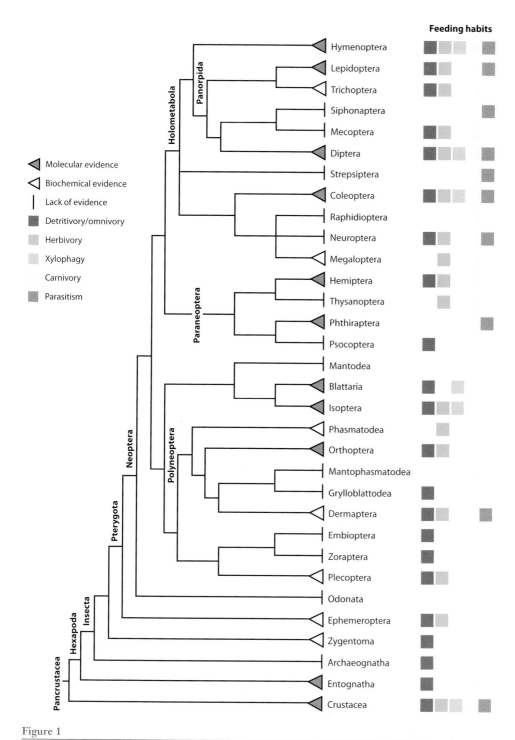

Figure 1

Phylogeny of insects showing molecular and biochemical evidence of the symbiotic-independent mechanism for plant cell wall digestion. Feeding habits are reported for each order. Phylogenetic relationships and feeding habits are based on Grimaldi & Engel (2005).

overcome plant defenses (Cherqui & Tjallingii 2000), although these enzymes do not have digestive functions (but see Tartar et al. 2009).

2.6. Plant Cell Wall Digestion: Symbiotic, Endogenous, or Both?

The balance of current evidence favors a combination of endogenous and symbiotic enzymes for PCW digestion in termites, wood-feeding cockroaches, Scarabaeidae beetles, Tipulidae flies, and likely some Orthoptera (i.e., Gryllidae and Gryllotalpidae) species that possess specialized hindguts to maintain permanent populations of symbionts, as well as in those fungus-growing insects. The participation of enzymes from multiple sources (symbionts and hosts) with different enzymatic properties seems important to enable successful feeding on wood and other cellulose-rich diets, given the high efficiency (65–99%) of (hemi)cellulose digestion found in termites (Hongoh 2011). However, most insects do not have specialized hindguts to harbor permanent microbial populations (Terra 1990). How can these insects digest the PCW present in their diets without the contribution of hindgut symbionts?

One explanation proposes that midgut microorganisms and ingested fungal enzymes digest the PCW in most insects (Martin 1983). The evidence presented here confirms that some of these microorganisms have (hemi)cellulolytic enzymes. Nevertheless, it is not clear if these microorganisms contribute significantly to insect nutrient gain because they are present in low numbers in the midgut or are absent in some insects (Egert et al. 2005, Slaytor 1992). Furthermore, the composition of midgut communities varies with diet and comprises a significant fraction of transient microorganisms ingested with food and voided in feces (Grünwald et al. 2010, Harris 1993). As a result, the question of differentiation between stable and transient gut microorganisms arises: Stable microorganisms are more likely to represent symbiotic associations, whereas transient microorganisms may represent commensalisms (Harris 1993). With the exception of the putative symbiotic association between *P. stipitis* and Passalidae beetles, for which subsociality and proctodeal trophallaxis have essential roles in the vertical transmission of gut symbionts (Suh et al. 2003), the information reviewed shows no evidence supporting the existence of obligate symbiosis between insects and most midgut lignocellulolytic microorganisms (Berkov et al. 2007, Grünwald et al. 2010). Furthermore, none of the sequenced genome of the vertically inherited bacterial symbionts associated to mycetocytes showed lignocellulolytic genes (GenBank accession numbers AP000398, BA000021, CP000238, CP000770, CP001429, and CP001605). Obligate symbiotic associations have proven to be important to their hosts by (*a*) provisioning a supplement of nitrogen and critical micronutrients (e.g., vitamins and essential fatty acids), (*b*) protecting the host against disease agents, (*c*) detoxifying secondary plant compounds, and (*d*) allowing the host to feed on different diets (Douglas 1998, 2009; Oliver et al. 2010). Consequently, the association between midgut microbiota and insects may be mainly related to these benefits, rather than the degradation of refractory PCW polymers. Indeed, a recent study conducted to evaluate the role of lignocellulolytic gut microbiota in PCW digestion in the beetle *Tenebrio molitor* showed that, even when this gut microbiota exhibited some enzymatic activities for PCW degradation, these activities were not essential for the insect because the comparison of enzymatic activities between insects with and without symbionts did not show significant differences (Genta et al. 2006). However, the results of this study showed that the gut microbiota was likely involved in the detoxification of plant allelochemicals and in the production of volatiles (Genta et al. 2006). Similar studies for other insect species harboring transient (i.e., nonsymbiotic) lignocellulolytic midgut microbiota are needed to determine the contribution of the midgut microbiota to PCW digestion.

The alternative explanation lies in the endogenous lignocellulolytic ability of insects. Most insects have endogenous enzymes involved in PCW degradation (**Table 2**). However, insects that depend on their own enzymatic abilities also evolved alternative strategies to assist PCW digestion. For example, detritivorous/omnivorous insects feed on broad diets (Nalepa et al. 2001), and beetles feeding on living wood have long midguts and periods of development (Haack & Slansky 1987).

3. EVOLUTION OF PLANT CELL WALL DIGESTION IN INSECTS

Working with the assumption that endogenous cellulolytic systems are very rare, Martin (1991) argued that "independent-symbiotic cellulolytic ability is a trait in insects rarely advantageous to possess" because insects are nitrogen, rather than carbon, limited. In this context, cellulose digestion in insects was hypothesized to have evolved only through the establishment of symbiotic relationships. Yet, insect-symbiotic interactions have evolved independently in different groups (Douglas 2009, Martin 1991, Mueller & Gerardo 2002). The origin of obligate symbiotic interactions for PCW digestion such as Protist-Dictyoptera symbiosis and ant/termite fungus farming occurred approximately 150–130 Mya (Grimaldi & Engel 2005) (**Figure 2**) and 50–24 Mya (De Fine Licht et al. 2010), respectively. The appearance of the earliest terrestrial insects dates further back to the Devonian, approximately 400 Mya (Grimaldi & Engel 2005) (**Figure 2**). Given these findings, how were leaf litter and other abundant plant materials processed prior to the establishment of symbiotic relationships? More specifically, which is the most ancient mechanism for PCW digestion in insects?

Arthropods marked the beginning of the Paleozoic Era (**Figure 2**), but today most Paleozoic biota are extinct or highly diminished in diversity (Grimaldi & Engel 2005). Basal Paleozoic insect orders that have not significantly diminished include Archaeognatha, Zygentoma, Ephemeroptera, and Orthoptera (Grimaldi & Engel 2005). Thus, understanding how these insects digest the PCW will allow us to determine the ancestral mechanism of PCW digestion. Currently, biochemical evidence suggests the presence of endogenous PCW-degrading enzymes for Zygentoma and Ephemeroptera (**Table 2**, **Figure 1**), and molecular evidence confirms the presence of endogenous PCW-degrading enzymes for Orthoptera (**Table 2**, **Figure 1**). Putative endogenous PCW-degrading enzymes and endogenous genes are also present in six additional Paleozoic orders (Plecoptera, Dermaptera, Phasmatodea, Blattaria, Hemiptera, and Coleoptera) with modern representatives (**Figures 1** and **2**). This suggests that the symbiotic-independent mechanism may be the ancestral mechanism for PCW digestion in insects (**Figure 1**). This hypothesis is further supported by molecular evidence confirming the presence of endogenous PCW-degrading enzymes in Collembola and Crustacea, the arthropod groups most closely related to insects (**Table 3**, **Figure 1**). Furthermore, endogenous genes encoding functional enzymes that degrade polymers present in algae (i.e., cellulase, β-1,3-glucans, xyloglucan, pectin) (for details about algae cell wall composition, see Popper et al. 2011) and/or plant tissues have also been reported for Cnidaria, Nematoda, Mollusca, and Annelida (**Table 3**), suggesting that some cell wall–degrading genes (**Figure 2**) were indeed present in the last common ancestor of bilaterian animals (Calderón-Cortés et al. 2010, Davison & Blaxter 2005, Lo et al. 2003). This is supported by phylogenetic analyses performed with cellulases belonging to glycosyl hydrolase family 9 (Davison & Blaxter 2005) as well as the positional identity of introns across these cellulase genes (Lo et al. 2003). However, it is not currently clear whether other PCW-degrading genes (particularly those belonging to glycosyl hydrolase families 5, 28, and 45) were vertically transferred from a common ancestor of bilaterian animals to insects or whether they were horizontally transferred from bacteria and/or fungi as has been previously reported for nematodes (e.g., Danchin et al. 2010 and references therein). As more sequences become available for invertebrates, reliable phylogenetic analyses should provide more insights into the origins and evolution all PCW-degrading genes in insects.

Age	Period (Mya)	Plant evolution	Insect evolution
Mesozoic	**Cretaceous** (144–65)	Diversification of angiosperms	Diversification of most modern herbivorous insect orders (Coleoptera, Hemiptera, Diptera, Lepidoptera) Origin of eusociality in some insect groups
Mesozoic	**Jurassic** (208–144)	Monocot-dicot divergence Origin of angiosperms and emergence of more digestible and nutritious plant tissues (foliage, fruits, nectar)	Origin of Isoptera Acquisition of hindgut protozoa and evolution of an obligate mutualism by an ancestral wood-feeding cockroach Divergence of Dictyoptera from a roach-like ancestor Occurrence of Lepidoptera and Trichoptera
Mesozoic	**Triassic** (245–208)	Diversification of conifers	Occurrence of Hymenoptera, Coleoptera, and Diptera Divergence of Hymenoptera and Panorpoid insects
Paleozoic	**Permian** (286–245)	Origin of conifers and increase of lignocellulose biomass	Occurrence of Polyneoptera (Phasmathodea, Orthoptera)
Paleozoic	**Carboniferous** (360–286)	Diversification of tall woody plants (horse-tails, club mosses) and emergence of new niches for xylem-feeding insects	Occurrence of Neoptera and divergence of the common ancestors of Polyneoptera, Paraneoptera, and Holometabola lineages Origin of Pterygota (winged insects)
Paleozoic	**Devonian** (408–360)	Origin of seed plants	Occurrence of Apterygota (Archaeognatha) Occurrence of Collembola
Paleozoic	**Silurian** (438–408)	Origin of vascular plants	Divergence of Hexapoda from crustacean ancestors
Paleozoic	**Ordovician** (505–438)	Plant terrestrialization	
Paleozoic	**Cambrian** (540–505)		Acquisition of PCW-degrading genes (GHF9 and GHF16, and likely GHF5 and GHF45 as well) from the common ancestor of the annelid-arthropod lineage

Figure 2

Summary of the major events in the evolution of plants and insects in geological time. Compiled from Grimaldi & Engel (2005) and Taylor et al. (2009). Abbreviations: GHF, glycosyl hydrolase family; PCW, plant cell wall.

3.1. Plant Cell Wall–Degrading Enzymes and Insect-Feeding Habits: Evolutionary Trends

The earliest known hexapod, the collembolan *Rhyniella praecursor* (Paleozoic: Devonian), was probably detritivorous (Chaloner et al. 1991), as are modern springtails (Collembola) and phylogenetically basal insect lineages such as Archaeognatha and Zygentoma (**Figure 1**). This suggests that detritivory was the ancestral feeding habit for insects (Chaloner et al. 1991, Labandeira 1998). Enzymes active to a wide spectrum of substrates (e.g., cellulose, xylan, laminarin, lichenin, callose, curdlan) are needed to degrade detritus: These enzymes make the polysaccharides present in PCWs, algae, lichens, and fungal cells (associated to the detritus) accessible as nutrients (Genta et al. 2003, Sinsabaugh et al. 1985). Thus, the enzymatic complexes of detritivores may include β-1,3-glucanases such as laminarinases and licheninases, xylanases, and particularly cellulases, given that terrestrial detritivorous/omnivorous insects such as silverfish, firebrats, and roaches digest 40–90% of the cellulose they ingest (Martin 1991). Accordingly, our review shows that detritivorous/omnivorous insects (**Table 2**) possess a diverse array of enzymatic activities (cellulases, laminarinases, xylanases, and licheninases). Therefore, if detritivory is the ancestral feeding habit for insects, it is conceivable that a full complement of digestive enzymes was a key trait for early evolving insects feeding on broad diets. Moreover, given that detritivores feed mainly on rotting plant material, detritivory could be related to the success of insects living in the Carboniferous (**Figure 2**). Furthermore, primary production in the Carboniferous was likely overwhelmingly routed through detritivores, and detritivores could have benefited from the large-scale events of vegetative mortality and changes in plant species composition that occurred in the Paleozoic (Nalepa et al. 2001) (**Figure 2**). In this scenario, detritivory would have played an important role in insect evolution as a feeding strategy common among basal hemipteroid and holometabolan insects (Labandeira 1998) that prevailed in several modern insect orders (**Figure 1**). Detritivory also seems to be a prerequisite for the evolution of symbiotic-dependent mechanisms for PCW digestion in insects (Martin 1991, Nalepa et al. 2001). For example, detritivorous insects with diverse communities of noncellulolytic microorganisms in their guts were more likely to evolve symbiont-mediated cellulolytic processes via incidental colonization of hindgut by cellulolytic microorganisms. Consequently, they could specialize on diets rich in lignocellulose (Martin 1991, Nalepa et al. 2001).

The second and most important step in the evolution of PCW-feeding insects was the evolution of herbivory. In this context, the endogenous PCW-digesting enzymes of basal detritivores could have been a preadaptation to herbivory: These enzymes could have favored an evolutionary transition to an herbivorous diet. Modern herbivores (Acrididae, Hemiptera, Coleoptera, Hymenoptera, and Lepidoptera) have endogenous cellulases, hemicellulases, or pectinases (**Table 2**). However, some of these insects have engaged in alternative feeding strategies enabling them to consume different plant organs or tissues including (*a*) consuming more nutritious portions of the plants such as fruits, seeds, pollen, nectar, and young tissues (Abe & Higashi 1991); (*b*) consuming large amounts of plant tissues and developing mouthpart adaptations to maximize the efficiency of processing plant tissues (leaf-chewing insects in particular) (Hochuli 1996); and (*c*) developing gut-physiology adaptations such as high alkalinity to extract PCW polymers and proteins (Panorpoid insects and Scarabaeidae beetles in particular) (Terra 1990). Therefore, if insect herbivores display a diverse array of feeding strategies for PCW digestion, the following questions can be addressed: Is an endogenous lignocellulolytic ability an adaptive trait for insect herbivores? Are insect PCW-degrading enzymes involved in the evolution of plant-insect interactions? Several lines of evidence suggest that an endogenous (hemi)cellulolytic ability may be an adaptive trait for some insect herbivores. The presence and activity of pectinases correlate with the herbivory feeding habit (**Table 2**). Pectinases represent key enzymes enabling insects to feed on tissues rich in primary PCW such as young tissues, fruits, and seeds. Pectinases also play a key role in piercing-sucking feeding (**Table 2**). Similarly, cellulases and hemicellulases appear to be key enzymes for borer beetles feeding on living tissues (**Table 2**). In other insect guilds such as those composed of insects feeding on pollen/nectar as well as in some leaf-chewing insects, an endogenous (hemi)cellulolytic ability may be complementary to other feeding strategies. Acrididae and most lepidopteran leaf-chewing insects extract soluble carbohydrates and proteins from cell contents by crushing and snipping leaf tissues (Abe & Higashi 1991, Hochuli 1996). Gut alkalinity likely also enables such extraction, as in the case of lepidopterans (Terra 1990). Even though these insects have evolved efficient feeding strategies, (hemi)cellulolytic activities can assist the nutrient extraction process by partially digesting the PCW; such is the case at least for Acrididae, whose ability to digest plant fibers reaches 60% (Cazemier et al. 1997). However, in other leaf-chewing insects such as coleopterans, for which a high abundance of PCW-degrading enzymes has been reported, cellulases, pectinases, and hemicellulases can play a more important role in PCW

Preadaptation: possession of the necessary properties to permit a shift to a new niche, habitat, or function

Table 3 Molecular evidence confirming the presence of endogenous genes for plant cell wall digestion in invertebrates

Invertebrate group (number of species)	Glycosyl hydrolase families	GenBank accession number
Cnidaria (1)	10	XM_001626034
Nematoda (30)	5, 16, 28, 45	AAC15707, AAD45868, AAD56392, AAD56393, AAK21881-AAK21895, AAK85303, AAM50039, AAN03645, AAN03647, AAN03688, AAN32884, AAP33282, AAR37374, AAR37375, AAM28240, ABV54446-ABV54449, ABX79356, ABY52965, ACD12136, ACJ60676, ACM44321-ACM44323, ACO55952, ACP20205, ADM72857, ADV58280-ADV58293, ADV58301-ADV58305, ADV58310-ADV58313, AER27723, AER27760-AER27762, AER27764-AER27790, AER27793-AER27798, AER27800-AER27807, BAB68522, BAB68523, BAD34544, BAD34546, BAD34548, BAE02683, BAE02684, BAI44493, BAI44495, CAC12958, CAC12959, CAJ77137
Mollusca		
Gastropoda (14)	5, 9, 10, 16, 45	AAP31839, AAT76428, AAV91523, AAV91524, AAY46801, ABD24274-ABD24281, ABR92637, ABR92638, AB026608, AB026609, AB135995, AB135996, ACN22491, ACJ12612, ACS15341-ACS15344, ACS15346-ACS15350, AY285999, BAC67186, BAD44734, BAE78456, BAH84971, BAI99559, BAJ60954
Bivalvia (9)	5, 9, 10, 16, 45	AAP74223, AAZ04385, AAZ04386, ACN59490, ACM68926, AEE89455, AWW34372, BAF38757, BAG57455, BAH23793, BAH23794, BAH84829, BAH85844, CAC59694, CAC59695, CAC81056
Annelida		
Polychaeta (1)	9	BAK20401
Oligochaeta (2)	9	AAX92641, ACE75510, ACE75511, BAH22180
Crustacea		
Malacostraca (4)	5, 7, 9	AAD38027, ABA87134, ACY70393, ADB85437-ADB85442, ADE58567-ADE58569
Branchiopoda (1)	9	EFX69372, EFX80603, EFX80604
Entognatha		
Collembola (1)	5, 16, 45	ABV68808, ACD93221, ACV50414

digestion (**Table 2**). For bees, endogenous cellulases (**Table 2**) may be needed to digest the cellulose present in the inner wall of pollen grains and to increase the release of nutrients present in the pollen (Kunieda et al. 2006). Thus, it is likely that other pollen-feeding insects could also have endogenous cellulases to assist the digestion of pollen.

Interestingly, in addition to their roles in digestion, some PCW-degrading enzymes (particularly β-glycosidases, pectinases, phenoloxidases, peroxidases, and β-1,3-glucanases) in insect herbivores may also contribute to the detoxification of secondary compounds (Cherqui & Tjallingii 2000, Terra & Ferreira 1994) and be involved in immune-defense responses (Bragatto et al. 2010,

Pauchet et al. 2009a). The evolution of new functions (i.e., neofunctionalization) in a gene family is frequently associated with a particular selective pressure (Zhang 2006). Undoubtedly, secondary compounds produced by plants and natural enemies represent important selective pressures for insect herbivores (Ehrlich & Raven 1964, Price et al. 1980). Therefore, PCW-degrading genes with new functions may serve as mechanisms to overcome plant toxic compounds and natural enemies. Additionally, some ecological studies have demonstrated that β-glucosidases and pectinases are important mediating factors in plant-insect interactions: Some of the products resulting from the activity of these enzymes elicit signaling pathways that activate both direct (e.g., oxidative responses that activate the induction of defense genes) (Orozco-Cardenas & Ryan 1999) and indirect plant defense responses (e.g., the release of terpenoid volatile blends that attract natural enemies) (Mattiacci et al. 1995).

Neofunctionalization: divergence of duplicate genes whereby one acquires a new function

SUMMARY POINTS

1. Endogenous PCW-degrading enzymes are present in basal insect orders and the most closely related arthropod groups to insects, suggesting that the symbiotic-independent mechanism is the ancestral mechanism for PCW digestion in insects.

2. Obligate symbiotic-dependent mechanisms are complementary to the endogenous mechanism of insects, contributing approximately 40–60% of PCW digestion in some insects such as termites.

3. PCW-degrading enzymes represent key traits for early-evolving insects feeding on broad diets of decaying plant and fungal tissues; they likely also represent a preadaptation to evolution of herbivory.

4. The activities of some PCW-degrading enzymes correlate to a particular feeding habit, particularly pectinases, which are related to herbivory.

5. Some insect PCW-degrading enzymes are involved in detoxifying functions and immune-defense responses, playing important roles in the evolution of plant-insect and insect-pathogen interactions.

FUTURE ISSUES

1. Cloning endogenous PCW-degrading genes of the most basal insects, such as Archaeognatha and Zygentoma, remains a missing critical piece of information needed to confirm that endogenous PCW digestion is a basal trait in insects.

2. Molecular evidence is needed for several insect orders that have not been studied to allow rigorous inferences about the patterns of PCW digestion evolution in insects.

3. Metagenomic and metatranscriptomic studies are needed to understand the contribution of gut microbiota enzymes to insect hosts.

4. Neofunctionalization of some PCW-degrading genes, as well as the ecological importance of these genes in plant-insect and insect-pathogen interactions, are important issues for future studies.

5. The analysis of the enzymatic and structural properties of PCW-degrading enzymes remains a central issue in the study of PCW digestion in insects.

DISCLOSURE STATEMENT

The authors are not aware of any affiliations, memberships, funding, or financial holdings that might be perceived as affecting the objectivity of this review.

ACKNOWLEDGMENTS

This work was financed by the National Autonomous University of México (UNAM: DGAPA-PAPIIT projects IN304308 and IN201011 to M.Q.), National Council of Science and Technology (CONACyT projects U50863Q, 2009-C01-131008, 2010-155016 to M.Q.), and the Graduate Program of Biological Sciences-UNAM and CONACyT (scholarship number 164921 to N.C.C.). We are grateful to D. J. Futuyma, A. G. Stephenson, S. Martén-Rodríguez, A. González-Rodríguez, and P. Hanson for critical reading of the manuscript.

LITERATURE CITED

Abe T, Higashi M. 1991. Cellulose centered perspective on terrestrial community structure. *Oikos* 60:127–33

Allen ML, Mertens JA. 2008. Molecular cloning and expression of three polygalacturonase cDNAs from the tarnished plant bug, *Lygus lineolaris*. *J. Insect Sci.* 8:27

Anand AAP, Vennison SJ, Sankar SG, Prabhu DIG, Vasan PT, et al. 2010. Isolation and characterization of bacteria from the gut of *Bombyx mori* that degrade cellulose, xylan, pectin and starch and their impact on digestion. *J. Insect Sci.* 10:107

Arakawa G, Watanabe H, Yamasaki H, Maekawa H, Tokuda G. 2009. Purification and molecular cloning of xylanases from the wood-feeding termite, *Coptotermes formosanus* Shiraki. *Biosci. Biotechnol. Biochem.* 73:710–18

Aw T, Schlauch K, Keeling CI, Young S, Bearfield JC, et al. 2010. Functional genomics of mountain pine beetle (*Dendroctonus ponderosae*) midguts and fat bodies. *BMC Genomics* 11:215

Berkov A, Feinstein J, Small J, Nkamany M. 2007. Yeasts isolated from Neotropical wood-boring beetles in SE Peru. *Biotropica* 39:530–38

Boyd DW, Cohen AC, Alverson DR. 2002. Digestive enzymes and stylet morphology of *Deraeocoris nebulosus* (Hemiptera, Miridae): a predacious plant bug. *Ann. Entomol. Soc. Am.* 95:395–01

Bragatto I, Genta FA, Ribeiro AF, Terra WR, Ferreira C. 2010. Characterization of a β-1,3-glucanase active in the alkaline midgut of *Spodoptera frugiperda* larvae and its relation to β-glucan-binding proteins. *Insect Biochem. Mol. Biol.* 40:861–72

Brennan Y, Callen WN, Christoffersen L, Dupree P, Goubet F, et al. 2004. Unusual microbial xylanases from insect guts. *Appl. Environ. Microbiol.* 70:3609–17

Calderón-Cortés N, Watanabe H, Cano-Camacho H, Závala-Páramo G, Quesada M. 2010. cDNA cloning, homology modeling and evolutionary insights of novel endogenous cellulases of the borer beetle *Oncideres albomarginata chamela* (Cerambycidae). *Insect Mol. Biol.* 19:323–36

Cazemier AE, Op den Camp HJM, Hackstein JHP, Vogels GD. 1997. Fibre digestion in arthropods. *Comp. Biochem. Physiol. A* 118:101–09

Cazemier AE, Verdoes JC, Rubsaet FAG, Hackstein JHP, van der Drift C, Op den Camp HJM. 2003. *Promicromanospora pachnodae* sp. nov., a member of the (hemi)cellulolytic hindgut flora of larvae of the scarab beetle *Pachnoda marginata*. *Antonie van Leeuwenhoeck* 83:135–48

Cazemier AE, Verdoes JC, van Ooyen AJJ, Op den Camp HJM. 1999a. A β-1,4-endoglucanase-encoding gene from *Cellulomonas pachnodae*. *Appl. Microbiol. Biotechnol.* 52:232–39

Cazemier AE, Verdoes JC, van Ooyen AJJ, Op den Camp HJM. 1999b. Molecular and biochemical characterization of two xylanase-encoding genes from *Cellulomonas pachnodae*. *Appl. Environ. Microbiol.* 65:4099–107

Celorio-Mancera MP, Allen ML, Powell AL, Ahmadi H, Salemi MR, et al. 2008. Polygalacturonase causes lygus-like damage on plants: cloning and identification of western tarnished plant bug (*Lygus hesperus*) polygalacturonases secreted during feeding. *Arthropod Plant Interact.* 2:215–25

Chaloner WG, Scott AC, Stephenson J. 1991. Fossil evidence for plant-arthropod interactions in the Paleozoic and Mesozoic. *Philos. Trans. R. Soc. Lond. Ser. B* 333:177–86

Cherqui A, Tjallingii WF. 2000. Salivary proteins of aphids, a pilot study on identification, separation and immunolocalisation. *J. Insect Physiol.* 46:1177–86

Cho MJ, Kim YH, Shin K, Kim YK, Kim YS, Kim TJ. 2000. Symbiotic adaptation of bacteria in the gut of *Reticulitermes speratus*: low endo-β-1,4-glucanase activity. *Biochem. Biophys. Res. Commun.* 395:432–35

Cook DM, Henriksen ED, Upchurch R, Peterson JBD. 2007. Isolation of polymer-degrading bacteria and characterization of the hindgut bacterial community from the detritus-feeding larvae of *Tipula abdominalis* (Diptera: Tipulidae). *Appl. Environ. Microbiol.* 73:5683–86

Danchin EGJ, Rosso MN, Vieira P, de Almeida-Engler J, Coutinho PM, et al. 2010. Multiple lateral gene transfers and duplications have promoted plant parasitism ability in nematodes. *Proc. Natl. Acad. Sci. USA* 107:17651–56

Davison A, Blaxter M. 2005. Ancient origin of glycosyl hydrolase family 9 cellulase genes. *Mol. Biol. Evol.* 22:1273–84

De Fine Licht HH, Schiøtt M, Mueller UG, Boomsma JJ. 2010. Evolutionary transitions in enzyme activity of ant fungus gardens. *Evolution* 64:2055–69

Delalibera I Jr, Handelsman J, Raffa KF. 2005. Contrasts in cellulolytic activities of gut microorganisms between the wood borer, *Saperda vestita* (Coleoptera: Cerambycidae), and the bark beetles, *Ips pini* and *Dendroctonus frontalis* (Coleoptera: Curculionidae). *Physiol. Ecol.* 34:541–47

Dittmer NT, Suderman RJ, Jiang H, Zhu YC, Gorman MJ, et al. 2004. Characterization of cDNAs encoding putative laccase-like multicopper oxidases and developmental expression in the tobacco hornworm, *Manduca sexta*, and the malaria mosquito, *Anopheles gambiae*. *Insect Biochem. Mol. Biol.* 34:29–41

Doostdar H, McCollum TG, Mayer RT. 1997. Purification and characterization of an endo-polygalacturonase from the gut of West Indies sugarcane rootstalk borer weevil (*Diaprepes abbreviatus* L.) larvae. *Comp. Biochem. Physiol. B* 118:861–67

Douglas AE. 1998. Nutritional interactions in insect-microbial symbioses: aphids and their symbiotic bacteria *Buchnera*. *Annu. Rev. Entomol.* 43:17–37

Douglas AE. 2009. The microbial dimension in insect nutritional ecology. *Funct. Ecol.* 23:38–47

Egert M, Stingl U, Bruun LD, Pommerenke B, Brune A, Friedrich MW. 2005. Structure and topology of microbial communities in the major gut compartments of *Melolontha melolontha* larvae (Coleoptera: Scarabaeidae). *Appl. Environ. Microbiol.* 71:4556–66

Egert M, Wagner B, Lemke T, Brune A, Friedrich MW. 2003. Microbial community structure in midgut and hindgut of the humus-feeding larvae of *Pachnoda ephippiata* (Coleoptera: Scarabaeidae). *Appl. Environ. Microbiol.* 69:6659–68

Ehrlich PR, Raven PH. 1964. Butterflies and plants: a study in coevolution. *Evolution* 18:586–608

Fratti F, Galletti R, De Lorenzo G, Salerno G, Conti E. 2006. Activity of endo-polygalacturonases in mirid bugs (Heteroptera: Miridae) and their inhibition by plant cell wall proteins (PGIPs). *Eur. J. Entomol.* 103:515–22

Fujita A, Hojo M, Aoyagi T, Hayashi Y, Arakawa G, et al. 2010. Details of the digestive system in the midgut of *Coptotermes formosanus* Shiraki. *J. Wood Sci.* 56:222–26

Geib SM, Jimenez-Gasco MM, Carlson JE, Tien M, Hoover K. 2009. Effect of host tree species on cellulase activity and bacterial community composition in the gut of larval Asian longhorned beetle. *Environ. Entomol.* 38:686–99

Genta FA, Bragatto I, Terra WR, Ferreira C. 2009. Purification, characterization and sequencing of the major β-1,3-glucanase from the midgut of *Tenebrio molitor* larvae. *Insect Biochem. Mol. Biol.* 39:861–74

Genta FA, Dillon RJ, Terra WR, Ferreira C. 2006. Potential role for gut microbiota in cell wall digestion and glucoside detoxification in *Tenebrio molitor* larvae. *J. Insect Physiol.* 52:593–601

Genta FA, Dumont AF, Marana SR, Terra WR, Ferreira C. 2007. The interplay of processivity, substrate inhibition and a secondary substrate binding site of an insect exo-β-1,3-glucanase. *Biochim. Biophys. Acta* 1774:1079–91

Genta FA, Terra WR, Ferreira C. 2003. Action pattern, specificity, lytic activities, and physiological role of five digestive β-glucanases isolated from *Periplaneta americana*. *Insect Biochem. Mol. Biol.* 33:1085–97

Empirical study indicating that GHF9 cellulases in Metazoa shared a common ancestor.

Gilbert HJ. 2010. The biochemistry and structural biology of plant cell wall deconstruction. *Plant Physiol.* 153:444–55

Girard C, Jouanin L. 1999. Molecular cloning of cDNAs encoding a range of digestive enzymes from a phytophagous beetle, *Phaedon cochleariae*. *Insect Biochem. Mol. Biol.* 2:1129–42

Grimaldi D, Engel MS. 2005. *Evolution of the Insects*. Cambridge, UK: Cambridge Univ. Press. 784 pp.

Grünwald S, Pilhofer M, Höll W. 2010. Microbial associations in gut systems of wood- and bark-inhabiting longhorned beetles (Coleoptera: Cerambycidae). *Syst. Appl. Microbiol.* 33:25–34

Haack RA, Slansky F. 1987. Nutritional ecology of wood-feeding Coleoptera, Lepidoptera and Hymenoptera. In *Nutritional Ecology of Insects, Mites, Spiders, and Related Invertebrates*, ed. F Slansky, JG Rodriguez, pp. 449–86. New York: John Wiley & Sons

Hail D, Lauziere I, Dowd SE, Bextine B. 2011. Culture independent survey of the microbiota of the glassy-winged sharpshooter (*Homalodisca vitripennis*) using 454 pyrosequencing. *Environ. Entomol.* 40:23–39

Harris JM. 1993. The presence, nature, and role of gut microflora in aquatic invertebrates: a synthesis. *Microbiol. Ecol.* 25:195–31

Hethener P, Braumann A, Garcia JL. 1992. *Clostridium termitidis* sp. nov., a cellulolytic bacterium from the gut of the wood feeding termite, *Nasutitermes lujae*. *Syst. Appl. Microbiol.* 15:52–58

Hochuli FF. 1996. The ecology of plant/insect interactions: implication of digestive strategy for feeding by phytophagous insects. *Oikos* 75:133–41

Hongoh Y. 2011. Toward the functional analysis of uncultivable, symbiotic microorganisms in the termite gut. *Cell. Mol. Life Sci.* 68:1311–25

Hyodo F, Tayasu I, Inoue T, Azuma JI, Kudo T. 2003. Differential role of symbiotic fungi in lignin degradation and food provision for fungus-growing termites (Macrotermitinae: Isoptera). *Funct. Ecol.* 17:186–93

Inoue T, Moriya S, Ohkuma M, Kudo T. 2005. Molecular cloning and characterization of a cellulase gene from a symbiotic protist of the lower termite, *Coptotermes formosanus*. *Gene* 349:67–75

Jeon H, Park S, Choi J, Jeong G, Lee SB, et al. 2011. The intestinal bacterial community in the food waste-reducing larvae of *Hermetia illucens*. *Curr. Microbiol.* 62:1390–99

Johjima T, Taprab Y, Noparatnaraporn N, Kudo T, Ohkuma M. 2006. Large-scale identification of transcripts expressed in a symbiotic fungus (*Termitomyces*) during plant biomass degradation. *Appl. Microbiol. Biotechnol.* 73:195–203

Kim N, Choo YM, Lee KS, Hong SJ, Seol KY, et al. 2008. Molecular cloning and characterization of a glycosyl hydrolase family 9 cellulase distributed throughout the digestive tract of the cricket *Teleogryllus emma*. *Comp. Biochem. Physiol. B* 150:368–76

Kukor JJ, Martin MM. 1983. Acquisition of digestive enzymes by siricid woodwasps from their fungal symbiont. *Science* 220:1161–63

Kukor JJ, Martin MM. 1986. Cellulose digestion in *Monochamus marmorator* Kby. (Coleoptera: Cerambycidae): role of acquired fungal enzymes. *J. Chem. Ecol.* 12:1057–70

Kunieda T, Fujiyuki T, Kucharski R, Foret S, Ament SA, et al. 2006. Carbohydrate metabolism genes and pathways in insects: insights from the honey bee genome. *Insect Mol. Biol.* 15:563–76

Labandeira CC. 1998. Early history of arthropod and vascular plant associations. *Annu. Rev. Earth Planet. Sci.* 26:329–77

Lee SJ, Lee KS, Kim SR, Gui ZZ, Kim YS, et al. 2005. A novel cellulase gene from the mulberry longicorn beetle, *Apriona germari*, gene structure, expression and enzymatic activity. *Comp. Biochem. Physiol. B* 140:551–60

Li L, Fröhlich J, Pfeiffer P, König H. 2003. Termite gut symbiotic Archaezoa are becoming living metabolic fossils. *Eukaryot. Cell* 2:1091–98

Lo N, Tokuda G, Watanabe H, Rose H, Slaytor M, et al. 2000. Evidence from multiple gene sequences indicates that termites evolved from wood-feeding cockroaches. *Curr. Biol.* 10:801–4

Lo N, Watanabe H, Sugimura M. 2003. Evidence for the presence of a cellulase gene in the last common ancestor of bilaterian animals. *Proc. R. Soc. Lond. Ser. B* 270:S69–72

Martin M. 1983. Cellulose digestion in insects. *Comp. Biochem. Physiol. A* 75:313–24

Martin M. 1991. The evolution of cellulose digestion in insects. *Philos. Trans. R. Soc. Lond. Ser. B* 333:281–88

Martin MM, Kukor JJ, Martin JS, Lawson DL, Merrit RW. 1981. Digestive enzymes of larvae of three species of caddisflies (Trichoptera). *Insect Biochem.* 11:501–5

First empirical study demonstrating an ancient origin of cellulase genes in invertebrates.

Martin MM, Martin JS. 1978. Cellulose digestion in the midgut of the fungus-growing termite *Macrotermes natalensis*: the role of acquired digestive enzymes. *Science* 199:1 453–55

Mattiacci L, Dicke M, Posthumus MA. 1995. β-glucosidase, an elicitor of herbivore-induced plant odor that attracts host-searching parasitic wasps. *Proc. Natl. Acad. Sci. USA* 92:2036–40

Moller IE, De Fine Licht HH, Harholt J, Willats WGT, Boomsma JJ. 2011. The dynamics of plant cell wall polysaccharide decomposition in leaf-cutting ant fungus gardens. *PLoS ONE* 6:e17506

Monk DC. 1976. The distribution of cellulose in freshwater invertebrate of different feeding habits. *Freshw. Biol.* 6:471–75

Mueller UG, Gerardo N. 2002. Fungus-farming insects: multiple origins and diverse evolutionary histories. *Proc. Natl. Acad. Sci. USA* 99:15247–49

Nakashima K, Watanabe H, Azuma JI. 2002a. Cellulase genes from the Parabasalian symbiont *Pseudotrichonympha grassii* in the hindgut of the wood-feeding termite, *Coptotermes formosanus. Cell Mol. Life Sci.* 59:1554–60

Nakashima K, Watanabe H, Saitoh H, Tokuda G, Azuma JI. 2002b. Dual cellulose-digesting system of the wood-feeding termite, *Coptotermes formosanus* Shiraki. *Insect Biochem. Mol. Biol.* 32:777–84

Nalepa CA, Bignell DE, Bandi C. 2001. Detritivory, coprophagy, and the evolution of digestive mutualisms in Dictyoptera. *Insect. Soc.* 48:194–201

Nation JL. 1983. Specialization in the alimentary canal of some mole crickets (Orthoptera: Gryllotalpidae). *Int. J. Insect Morphol. Embryol.* 12:201–10

Oh HW, Heo SY, Kim DY, Park DS, Bae KS, Park HY. 2008. Biochemical characterization and sequence analysis of a xylanase produced by an exo-symbiotic bacterium of *Gryllotalpa arientalis, Cellulosimicrobium* sp. HY-12. *Antonie Van Leeuwenhoek* 93:437–42

Ohtoko K, Ohkuma M, Moriya S, Inoue T, Usami R, Kudo T. 2000. Diverse genes of cellulase homologues of glycosyl hydrolase family 45 from the symbiotic protists in the hindgut of the termite *Reticulitermes speratus. Extremophiles* 4:343–49

Oliver KM, Degnen PH, Burke GR, Moran NA. 2010. Facultative symbionts in aphids and the horizontal transfer of ecologically important traits. *Annu. Rev. Entomol.* 55:247–66

Oppert C, Klingeman WE, Willis JD, Oppert B, Jurat-Fuentes JL. 2010. Prospecting for cellulolytic activity in insect digestive fluids. *Comp. Biochem. Physiol. B* 155:145–54

Orozco-Cardenas ML, Ryan CA. 1999. Hydrogen peroxide is generated systemically in plant leaves by wounding and systemin via the octadecanoid pathway. *Proc. Natl. Acad. Sci. USA* 96:6553–57

Park DS, Oh HW, Jeong WJ, Kim H, Park HY, Bae KS. 2007. A culture-based study of the bacterial communities within the guts on nine longicorn beetle species and their exo-enzyme producing properties for degradading xylan and pectin. *J. Microbiol.* 45:394–401

Pauchet Y, Freitak D, Heidel-Fischer HM, Heckel DG, Vogel H. 2009a. Glucanase activity in a glucan-binding protein family from Lepidoptera. *J. Biol. Chem.* 284:2214–24

Pauchet Y, Wilkinson P, van Munster M, Augustin S, Pauron D, French-Constant RH. 2009b. Pyrosequencing of the midgut trancriptome of the poplar leaf beetle *Chrysomela tremulae* reveals new gene families in Coleoptera. *Insect Biochem. Mol. Biol.* 39:403–13

Pauchet Y, Wilkinson P, Chauhan R, ffrench-Constant RH. 2010. Diversity of beetle genes encoding novel plant cell wall degrading enzymes. *PLoS ONE* 5:e15635

Popper ZA, Michel G, Hervé C, Domozych DS, Willats WGT, et al. 2011. Evolution and diversity of plant cell walls: from algae to flowering plants. *Annu. Rev. Plant Biol.* 62:567–90

Price PW, Bouton CE, Gross P, McPheron BA, Thompson JN, Weis AE. 1980. Interactions among three trophic levels: influence of plants on interactions between insect herbivores and natural enemies. *Annu. Rev. Ecol. Syst.* 11:41–65

Richard FJ, Mora P, Errard C, Rouland C. 2005. Digestive capacities of leaf-cutting ants and the contribution of their fungal cultivar to the degradation of plant material. *J. Comp. Physiol. B* 175:297–303

Rønhede S, Boomsma JJ, Rosendahl S. 2004. Fungal enzymes transferred by leaf-cutting ants in their fungus gardens. *Mycol. Res.* 108:101–06

Roy N, Masud MR, Uddin ATMS. 2003. Isolation and some properties of new xylanase from the intestine of a herbivorous insect (*Samia cynthia pryeri*). *J. Biol. Sci.* 4:27–33

Shows that some PCW-degrading enzymes elicit plant defense responses.

Demonstrates the participation of symbiotic and endogenous enzymes for lignocellulose digestion in termites.

Reviews the evolution of hindgut symbiosis in insects.

Shows that some lepidopteran β-1,3-glucanases have glucanase activity and also act as glucan-binding proteins.

Schiøtt M, Rogowska-Wrzesinska A, Roepstorff P, Boomsma JJ. 2010. Leaf-cutting ant fungi produce cell wall degrading pectinase complexes reminiscent of phytopathogenic fungi. *BMC Biol.* 8:156

Scrivener AM, Slaytor M. 1994a. Cellulose digestion in *Panesthia cribrata* Saussure: Does fungal cellulase play a role? *Comp. Biochem. Physiol. B* 107:309–15

Scrivener AM, Slaytor M. 1994b. Properties of the endogenous cellulase from *Panesthia cribrata* Saussure and purification of major endo-β-1,4-glucanase components. *Insect Biochem. Mol. Biol.* 24:223–31

Scrivener AM, Watanabe H, Noda H. 1997. Diet and carbohydrate digestion in the yellow-spotted longicorn beetle *Psacothea hilaris. J. Insect Physiol.* 43:1039–52

Shallom D, Shoham Y. 2003. Microbial hemicellulases. *Curr. Opin. Microbiol.* 6:219–28

Shen Z, Denton M, Mutti N, Pappan K, Kanost MR, et al. 2003. Polygalacturonase from *Sitophilus oryzae*: possible horizontal transfer of a pectinase gene from fungi to weevils. *Insect Sci.* 3:24–32

Shen Z, Pappan K, Mutti N, He QJ, Denton M, et al. 2005. Pectinmethylesterase from the rice weevil, *Sitophilus oryzae*, cDNA isolation and sequencing, genetic origin, and expression of the recombinant enzyme. *J. Insect Sci.* 5:21–30

Sinsabaugh RL, Linkins AE, Benfield EF. 1985. Cellulose digestion and assimilation by three leaf-shredding aquatic insects. *Ecology* 66:1464–71

Slaytor M. 1992. Cellulose digestion in termites and cockroaches: What role do symbionts play? *Comp. Biochem. Physiol.* 103:775–84

Sugimura M, Watanabe H, Lo N, Saito H. 2003. Purification, characterization, cDNA cloning and nucleotide sequencing of a cellulase from the yellow-spotted longicorn beetle, *Psacothea hilaris. Eur. J. Biochem.* 270:3455–60

Suh SO, Mashall CJ, McHugh J, Blackwell M. 2003. Wood ingestion by passalid beetles in the presence of xylose-fermenting gut yeasts. *Mol. Ecol.* 12:3137–45

Suh SO, Nguyen NH, Blackwell M. 2008. Yeast isolated from plant associated beetles and other insects: seven novel *Candida* species near *Candida albicans. FEMS Yeast Res.* 8:88–02

Sunyeon H, Kwak J, Oh HW, Park DS, Bae KS, et al. 2006. Characterization of an extracellular xylanase in *Paenibacillus* sp. HY-8 isolated from an herbivorous longicorn beetle. *J. Microbiol. Biotechnol.* 16:1753–59

Taprab Y, Johjima T, Maeda Y, Moriya S, Trakulnaleamsai S, et al. 2005. Symbiotic fungi produce laccases potentially involved in phenol degradation in fungus combs of fungus-growing termites in Thailand. *Appl. Environ. Microbiol.* 71:7696–704

Tartar A, Wheeler MM, Zhou X, Coy MR, Boucias GG, Scharf ME. 2009. Parallel metatranscriptome analyses of host and symbiont gene expression in the gut of the termite *Reticulitermes flavipes. Biotechnol. Biofuels* 2:25

Taylor TN, Taylor EL, Krings M. 2009. *Paleobotany: The Biology and Evolution of Fossil Plants.* New York: Elsevier. 1230 pp. 2nd ed.

Terra WR. 1990. Evolution of digestive system of insects. *Annu. Rev. Entomol.* 35:181–200

Terra WR, Ferreira C. 1994. Insect digestive enzymes: properties, compartmentalization and function. *Comp. Biochem. Physiol. B* 109:1–62

Todaka N, Inoue T, Saita K, Ohkuma M, Nalepa CA, et al. 2010. Phylogenetic analysis of cellulolytic enzyme genes from representative lineages of termites and a related cockroach. *PLoS ONE* 5:e8636

Tokuda G, Lo N, Watanabe H, Arakawa G, Matsumoto T, Noda H. 2004. Major alteration of the expression site of endogenous cellulases in members of an apical termite lineage. *Mol. Ecol.* 13:3219–28

Tokuda G, Lo N, Watanabe H, Slaytor M, Matsumoto T, Noda H. 1999. Metazoan cellulase genes from termites, intron/exon structures and sites of expression. *Biochim. Biophys. Acta* 1447:146–59

Tokuda G, Watanabe H. 2007. Hidden cellulases in termites: revision of an old hypothesis. *Biol. Lett.* 3:336–39

Tokuda G, Yamaoka I, Noda H. 2000. Localization of symbiotic clostridia in the mixed segment of the termite *Nasutitermes takasagoensis* (Shiraki). *Appl. Environ. Microbiol.* 66:2199–207

Treves DS, Martin MM. 1994. Cellulose digestion in primitive hexapods: effect of ingested antibiotics on gut microbial populations and gut cellulose levels in the firebrat, *Thermobia domestica* (Zygentoma, Lepismatidae). *J. Chem. Ecol.* 20:2003–20

Veivers PC, Musca AM, O'Brien RW, Slaytor M. 1982. Digestive enzymes of the salivary glands and gut of *Mastotermes darwiniensis. Insect Biochem.* 12:35–40

Uses a phylogenetic basis to provide a comparative analysis of the digestive systems of insects.

Provides evidence of endogenous cellulase activity in the basal insects Zygentoma.

Warnecke F, Luginbühl P, Ivanova N, Ghassemian M, Richardson TH, et al. 2007. Metagenomic and functional analysis of hindgut microbial of a wood-feeding higher termite. *Nat. Lett.* 450:560–69

Watanabe H, Nakashima K, Slaytor M. 2002. New endo-β-1,4-glucanases from the parabasalian symbionts, *Pseudotrichonympha grassii* and *Holomastigotoides mirabile* of *Coptotermes* termites. *Cell. Mol. Life Sci.* 59:1983–92

Watanabe H, Noda H, Tokuda G, Lo N. 1998. A cellulase gene of termite origin. *Nature* 394:330–31

Watanabe H, Tokuda G. 2010. Cellulolytic systems in insects. *Annu. Rev. Entomol.* 55:609–32

Wei YD, Lee KS, Gui ZZ, Yoon HJ, Kim I, et al. 2006. Molecular cloning, expression, and enzymatic activity of a novel endogenous cellulase from the mulberry longicorn beetle, *Apriona germari. Comp. Biochem. Physiol. B* 145:220–29

Wenzel W, Schöning I, Berchtold M, Kämpfer P, König H. 2002. Aerobic and facultatively anaerobic cellulolytic bacteria from the gut of the termite *Zootermopsis angusticollis. J. Appl. Microbiol.* 92:32–40

Willis JD, Klingeman WE, Oppert C, Oppert B, Jurat-Fuentes JL. 2010. Characterization of cellulolytic activity from digestive fluids of *Dissosteira carolina* (Orthoptera: Acrididae). *Comp. Biochem. Physiol. B* 157:267–72

Willis JD, Oppert B, Oppert C, Klingeman WE, Jurat-Fuentes JL. 2011. Identification, cloning, and expression of a GHF9 cellulase from *Tribolium castaneum* (Coleoptera: Tenebrionidae). *J. Insect Physiol.* 57:300–6

Yokoe Y, Yasumasu I. 1964. The distribution of cellulase in invertebrates. *Comp. Biochem. Physiol.* 13:323–38

Zhang J. 2006. Evolution by gene duplication: an update. *Trends Ecol. Evol.* 18:292–98

Zhang J, Scrivener AM, Slaytor M, Rose HA. 1993. Diet and carbohydrate activities in three cockroaches, *Calolampra elegans* Roth and Princis, *Geoscapheus dilatatus* Saussure and *Panesthia cribrata* Saussure. *Comp. Biochem. Physiol. A* 104:155–61

Zhou X, Huang H, Meng K, Shi P, Wang Y, et al. 2009. Molecular and biochemical characterization of a novel xylanase from the symbiotic *Sphingobaterium* sp. TN19. *Appl. Microbiol. Biotechnol.* 85:323–33

Zhou X, Smith JA, Oi FM, Kohler PG, Bennett GW, Scharf ME. 2007. Correlation of cellulose genes expression and cellulolytic activity throughout the gut of the termite *Reticulitermes flavipens. Gene* 395:29–39

First empirical study demonstrating the presence of an endogenous cellulase gene in insects.

RELATED RESOURCES

Coutinho PM, Henrissat B. 2001. *Carbohydrate active enzymes server.* **http://afmb.cnrs-mrs.fr/~pedro/CAZY/db.html**

Grimaldi D. 2010. 400 million years on six legs: on the origin and early evolution of Hexapoda. *Arthropod Struct. Dev.* 39:191–203

Watanabe H, Tokuda G. 2001. Animal cellulases. *Cell Mol. Life Sci.* 58:1167–78

New Insights into Pelagic Migrations: Implications for Ecology and Conservation

Daniel P. Costa, Greg A. Breed, and Patrick W. Robinson

Department of Ecology and Evolutionary Biology, University of California, Santa Cruz, California 95060; email: costa@ucsc.edu

Annu. Rev. Ecol. Evol. Syst. 2012. 43:73–96

First published online as a Review in Advance on August 28, 2012

The *Annual Review of Ecology, Evolution, and Systematics* is online at ecolsys.annualreviews.org

This article's doi:
10.1146/annurev-ecolsys-102710-145045

Keywords

animal navigation, animal tracking, Argos, global positioning system (GPS), marine protected area, migration, satellite telemetry, transect survey

Abstract

Highly pelagic large marine vertebrates have evolved the capability of moving across large expanses of the marine environment; some species routinely move across entire ocean basins. Our understanding of these movements has been enhanced by new technologies that now allow us to follow their movements over great distances and long time periods in great detail. This technology provides not only detailed information on the movements of a wide variety of marine species, but also detailed characteristics of the habitats they use and clues to their navigation abilities. Advances in electronic tracking technologies have been coupled with rapid development of statistical and analytical techniques. With these developments, conservation of highly migratory species has been aided by providing new information on where uncommon or endangered species go, what behaviors they perform and why, which habitats are critical, and where they range, as well as, in many cases, better estimates of their population size and the interconnectedness of subpopulations. Together these tools are providing critical insights into the ecology of highly pelagic marine vertebrates that are key for their conservation and management.

1. INTRODUCTION

Many pelagic vertebrates have evolved migratory life histories that allow them to adjust to dynamic marine environments by moving long distances to acquire needed resources that vary predictably in space and time. For many marine species, food resources and suitable breeding habitat are separated by hundreds or thousands of kilometers, necessitating seasonal migrations (Bost et al. 2009b, Boustany et al. 2010, Costa 1991, Le Boeuf et al. 2000, Mate et al. 1998, Rasmussen et al. 2007, Shillinger et al. 2008, Weng et al. 2005). These migrations may be repeated several times each year, as in the case of highly mobile species like albatross, but also annually or across multiple years as is the case for some tunas and sea turtles. Optimal habitat temperature, as much as food resources or breeding habitat, may also drive the migrations of both ectothermic and endothermic marine species (Block et al. 2011, Boustany et al. 2010, Durban & Pitman 2012, Rasmussen et al. 2007, Sleeman et al. 2010, Weng et al. 2005). In some species, the migratory cycle is associated with development; juveniles migrate to distant regions where prey resources are more available and later return to the breeding grounds as adults (Bestley et al. 2009, Boustany et al. 2010, Polovina et al. 2006). In colony-breeding species, the limited availability of predator-free regions, particularly islands, has resulted in many seabirds and pinnipeds evolving life histories that necessitate traveling hundreds or thousands of kilometers from breeding colonies to regions where suitable prey are abundant enough to profitably forage (Bost et al. 2009b, Costa 1991, Egevang et al. 2010, Le Boeuf et al. 2000, Shaffer et al. 2006, Weimerskirch et al. 2012).

The earliest studies of animal migration relied upon the seasonal presence or absence of animals on a breeding colony, region, or their seasonal availability in a fishery. Although this provided an understanding of population-level movement patterns, it was not until investigators applied unique identification tags to the legs, flippers, and bodies of animals, or took advantage of natural marks or scars, that it became possible to measure where, when, and how far individuals were actually migrating (Anderson et al. 2011, Rasmussen et al. 2007). More recently, stable isotopes have proven to be an excellent way to identify where animals forage during migrations because they can be used to track the large-scale displacements and also the geographic range of both extant and ancient animals (Newsome et al. 2010, Zbinden et al. 2011). Passive acoustic methods have also been used to track large-scale movements of vocal species such as whales (Sirovic et al. 2004).

Although conventional tagging and natural marks provide information on the general patterns of arrival and departure for many species, they do not provide information on migration corridors or individual movement patterns. Recent advancements in the size and power efficiency of electronic tracking tags ensures that tags are small enough not to affect natural behavior and robust enough to withstand the rigors of the marine environment; these factors have revolutionized our understanding of marine animal migration. A variety of electronic tagging technologies have been developed and deployed on large scales. Archival and satellite-linked data-logging tags have made possible the study of ocean basin–scale movements, oceanographic preferences, and the fine-scale behaviors of many pelagic species (Bailey et al. 2008; Bailleul et al. 2007; Block et al. 2005, 2011; Bost et al. 2009a; Boustany et al. 2010; Cotté et al. 2007; Dias et al. 2011; Guinet et al. 2001; Jorgensen et al. 2010; Le Boeuf et al. 2000; Shaffer et al. 2006; Shillinger et al. 2008). The most ambitious tagging effort to date has been the Tagging of Pacific Predators (TOPP) project, where 4,306 tags were deployed on 23 species in the North Pacific Ocean (Block et al. 2011) (**Figure 1**). Many of the TOPP animals exhibited clear periodicity in their movement patterns, which corresponded with changes in water temperature and primary production. Two regions stood out as important habitats, the highly productive California Current and the North Pacific Transition Zone. Leatherback sea turtles, Laysan and black-footed albatrosses, sooty shearwaters, bluefin

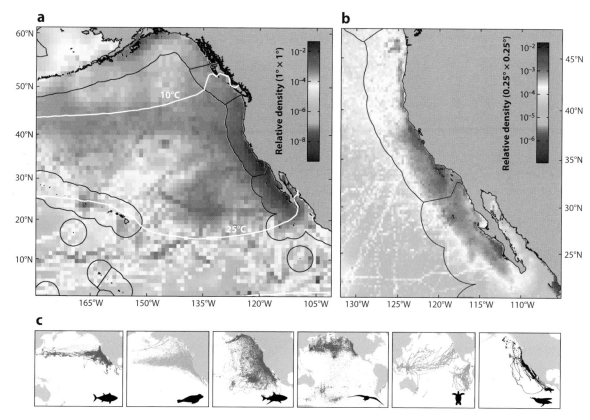

Figure 1

Predator density maps and residency patterns. (*a*) Density of large marine predators within the eastern North Pacific. Densities of the time-weighted and species-normalized position estimates of all tagged individuals were summed within 1° × 1° grid cells. (*b*) Density of large marine predators within the California Current Large Marine Ecosystem at a 0.25° × 0.25° resolution. Sea surface temperature contours in panel *a* are denoted by solid white lines. Exclusive economic zones are delineated by solid black lines. Bottom panels show the daily mean position estimates of the major Tagging of Pacific Predators (TOPP) guilds (from *left*): tunas (yellowfin, bluefin, and albacore), pinnipeds (northern elephant seals, California sea lions, and northern fur seals), sharks (salmon, white, blue, common thresher, and mako), seabirds (Laysan and black-footed albatrosses and sooty shearwaters), sea turtles (leatherback and loggerhead) and cetaceans (blue, fin, sperm, and humpback whales). Figure reproduced from Block et al. (2011).

tuna, and salmon sharks exhibited migrations greater than 5,000 km to regions throughout the Pacific Ocean with a strong affinity to the highly productive waters of the California Current. Those species that remained in the California Current (tunas, salmon, mako and blue sharks, and blue whales) showed an annual north-south migration, which was associated with changes in water temperature and primary productivity. Other species migrated between the California Current and pelagic waters: elephant seals, blue and mako sharks, and leatherback sea turtles traveled to the North Pacific Transition Zone. The subtropical gyre and north equatorial current was an important region for blue and mako sharks and leatherback sea turtles. Similarly, the "Café" region of the eastern Pacific and the Hawaiian Islands was routinely visited by white sharks, albacore tunas, and black-footed albatrosses (see Jorgensen et al. 2010).

Northern elephant seals are a classic example of how such technology has radically changed our understanding of their biology. Using ship and aerial surveys, their range was thought to be

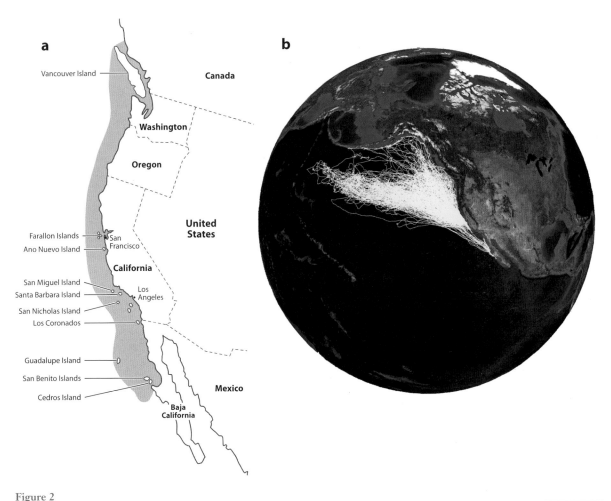

a

Vancouver Island

Canada

Washington

Oregon

United States

Farallon Islands — San Francisco
Ano Nuevo Island

California

San Miguel Island
Santa Barbara Island — Los Angeles
San Nicholas Island
Los Coronados

Guadalupe Island

San Benito Islands

Cedros Island

Mexico

Baja California

b

Figure 2

(*a*) Distribution of northern elephant seals (*orange*) as determined using boat- and plane-based surveys (redrawn from Riedman 1990; islands not shown to scale). (*b*) Distribution of female northern elephant seals determined from their migration tracks (*yellow*) observed using satellite telemetry (from Robinson et al. 2012).

restricted to offshore regions hugging the west coast of North America (**Figure 2a**). As electronic tracking data became available, it was discovered that these animals range throughout the Northeast Pacific Ocean (**Figure 2b**). Ship or plane surveys are limited to where we look, whereas tags carried by the animals provide information wherever the animal goes. Similarly, we knew white sharks periodically appeared along the coast of California, but it was impossible to know where they might be at other times or that they were congregating in a nondescript area between Hawaii and California (the so-called White Shark Café; see Jorgensen et al. 2010) for much of the year (**Figure 3**). Electronic tags have also provided unique insight into the fidelity of migratory paths of individual animals. For example, a female northern elephant seal was found to follow nearly identical migration paths 11 years apart (**Figure 4**). Electronic tracking data have also elucidated in great detail how the greatest migrations on Earth are timed and executed by Arctic terns and sooty shearwaters (Egevang et al. 2010, Shaffer et al. 2006) (**Figure 5**).

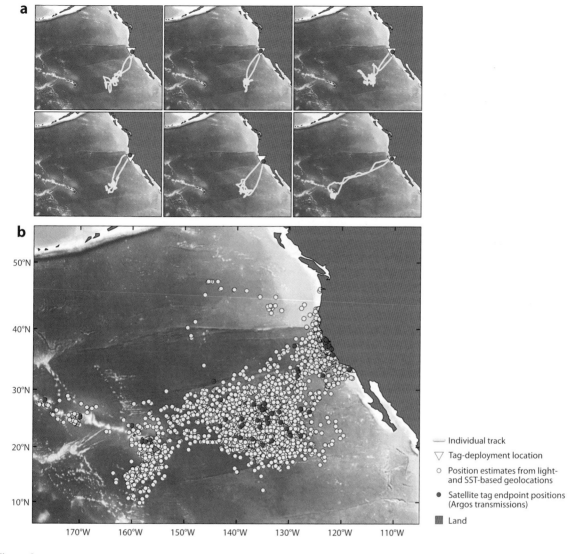

Figure 3

Site fidelity and homing of white sharks tagged along the Central California Coast during 2000–2007 revealed by pop-up archival tag records (PAT). (*a*) Site fidelity demonstrated by six individual tracks (*yellow lines*; based on five-point moving average of geolocations). Triangles indicate tag-deployment locations and red circles indicate the position where the PAT popped up and made an Argos satellite transmission. (*b*) Site fidelity of all satellite-tagged white sharks (*n* = 68) to three core areas in the Northeast Pacific including the North American continental shelf waters, the waters surrounding the Hawaiian Island Archipelago, and the White Shark Café. Yellow circles represent position estimates from light-based and sea surface temperature (SST)–based geolocations. Ocean color indicates depth, from white (shallowest) to dark blue (deepest). Figure reproduced from Jorgensen et al. (2010).

1.1. New Technologies

The primary tools used today for tracking marine animals are global positioning system (GPS) and Argos satellite telemetry. Archival data logging tags that collect light-level data are also extremely popular, as light level can be used to reconstruct positions using day length and clock offset to calculate position. Acoustic tracking on an ocean basin scale is also becoming more widespread.

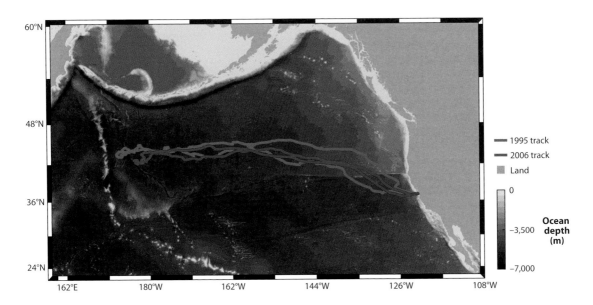

Figure 4

Argos transmission tracks of a female northern elephant seal recorded in 1995 when she was 6 years old (*blue*) and then again in 2006, 11 years later, when she was 17 years old (*red*) (D. P. Costa, P. W. Robinson, J. L. Hassrick, S. E. Simmons, unpublished data).

Finally, accelerometer/magnetometer data enable 3D dead-reckoning calculations to reconstruct true 3D tracks of diving animals through the ocean at incredibly fine resolution (Wilson et al. 2008). These tracks can last from weeks to years, and the tags can also collect ancillary behavioral information that can be used to identify behaviors and associated habitats and collect pressure data to measure the dive pattern (**Figure 6b**). This suite of associated data can be used to describe the environment a tracked animal experiences (**Figure 6c**) including temperature, salinity, and light level. Such behavioral and environmental data are often key in identifying differences in the movement patterns and habitat utilization of different species (Block et al. 2011, Costa et al. 2010a).

1.1.1. Archival tags. Archival tags record data as a time series from sensors that can record depth (pressure), water and/or body temperature, salinity, chlorophyll, three-axis acceleration, orientation, heart rate, stomach temperature, pO_2, GPS positions, and light level. The major limitation of archival tags is that they must be recovered in order to obtain the data they collect. Judicious choice of animals or use on exploited species where a reward is offered for tags collected in commercial fisheries has nonetheless provided a wealth of information on the foraging behavior and habitat use of many marine animals (Block et al. 2005, Johnson et al. 2006, Miller et al. 2004, Shaffer & Costa 2006, Shaffer et al. 2006, Tinker et al. 2007). Movement patterns can be derived with archival tags using light level. Local noon can be used to calculate longitude and day length to calculate latitude. These locations often contain error but can be refined by correcting with sea surface temperature data (Shaffer et al. 2005). Archival tags have the advantage of being relatively inexpensive, and they do not incur fees for accessing the data via satellite networks.

1.1.2. Argos satellite tags. Argos satellite tags provide at-sea locations and have the advantage that the data can be recovered remotely. The Argos receivers fly on National Oceanic and Atmospheric Administration low-orbiting weather satellites; they receive transmissions from tags

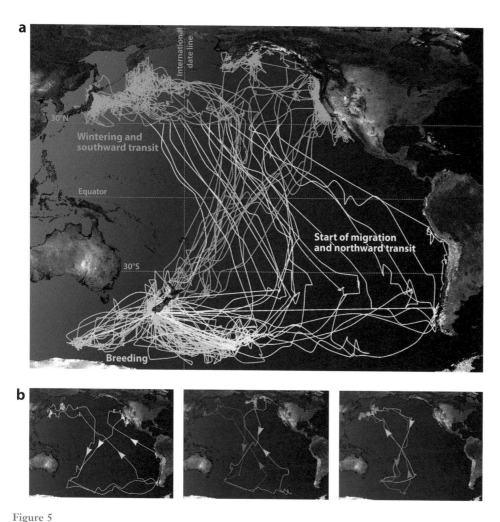

Figure 5

Shearwater migrations originating from breeding colonies in New Zealand. (*a*) Interpolated geolocation tracks of 19 sooty shearwaters during breeding (*light blue*) and subsequent migration pathways (*yellow*, start of migration and northward transit; *orange*, wintering grounds and southward transit). (*b*) Representative figure-eight movement patterns of individual shearwaters traveling to one of three "winter" destinations in the North Pacific. Each panel also represents a breeding pair and their subsequent migration after the breeding season. Figure reproduced from Shaffer et al. (2006).

and are capable of downloading data to Service Argos (Toulouse, France, or Landover, MD). Service Argos uses the Doppler shift of a tag's radio frequency to calculate the geolocation in successive uplinks. Tags with onboard data processing and compression have made it possible to transmit ancillary data through the Argos system, including detailed oceanographic and behavioral information such as dive profiles and ocean salinity and temperature (Boehme et al. 2009).

Argos tracking systems have been available longer than all other satellite tracking technologies and have been used on a wide variety of marine vertebrates, providing insight into the movements of marine birds (Bost et al. 1997, Kappes et al. 2010, Pinaud & Weimerskirch 2007, Weimerskirch et al. 2012), sea turtles (e.g., James et al. 2005, Maxwell et al. 2011a, Polovina et al. 2000), sharks

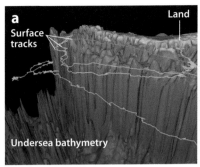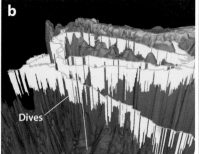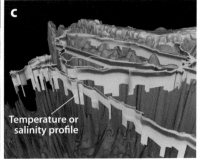

Figure 6

Tracks of three southern elephant seals in the Western Antarctic Peninsula. Panels show (*a*) just the surface track, (*b*) the surface track along with diving behavior, and (*c*) the temperature and salinity profiles that can be obtained to provide data on the physical environment the animals are moving through. Figure reproduced from Costa et al. (2010a).

(Eckert et al. 2002, Hammerschlag et al. 2011, Weng et al. 2005), and marine mammals (Bailey et al. 2009, Costa et al. 2010a, Guinet et al. 2001, Le Boeuf et al. 2000, Mate et al. 1998, Shaffer & Costa 2006). Because the tag's antenna must be out of the water to communicate with the satellites, the technology has mainly been used on air breathers that surface regularly. For animals that remain submerged, pop-up archival tags (PATs) are favored (Block et al. 1998, 2005; Boustany et al. 2010; Campana et al. 2011; Carlson et al. 2010). PATs combine archival tags with satellite transmitters and then send their data to researchers via Argos satellites once the tag is released from the tracked animal and floats to the surface.

1.1.3. Global positioning system tags. Although standard GPS tags have been deployed on seabirds for some time, the long time and high battery demands required to calculate GPS satellite positions delayed their application to marine animals that are only at the surface periodically (Weimerskirch et al. 2005, 2007). However, tags are now available that take a snapshot of the GPS satellite data, which is either stored for later calculation after tag recovery or is used to calculate pseudoranges that can then be transmitted via Argos (Tomkiewicz et al. 2010). Researchers can now track marine animal movements to within 10 meters, a vast improvement over the 1–10-km error currently possible with Argos satellite tags (Costa et al. 2010b, Kuhn et al. 2009). However, communication bandwidth to the Argos satellites is still a barrier and often only a small fraction of collected GPS locations can be remotely recovered. In practice, complete GPS tracks are usually attainable only by recovering tracking tags and downloading the data from them. GPS tags have been developed that can link to cell phone networks, which have enormous bandwidth to upload archived GPS position and behavioral data (McConnell et al. 2004). These tags are, of course, limited to species that regularly enter the range of wireless telecommunication networks.

1.1.4. Acoustic animal tracking. A variety of marine organisms have been tagged with tiny acoustic pingers that can be tracked with fixed or mobile acoustic receiver arrays (Dagorn et al. 2007). Movements of animals ranging from tiny salmon smolt on their migration from rivers into the ocean (Welch et al. 2011) to large sturgeon and sharks have been tracked (Andrews et al. 2010, Lindley et al. 2011). Mobile acoustic transceivers (so-called business card tags) that both send and receive acoustic signals are now being developed and deployed. These tags can listen for signals sent by animals too small to have receivers, chronicling all of the acoustic tags that an animal encounters (Hayes et al. 2012). In addition, both active and passive acoustic methods have been used to track large-scale movements of whales (Sirovic et al. 2004).

1.2. Analyses for Understanding Animal Migration

Animal tracking data are often of immediate qualitative value to identify previously unknown ecological patterns such as migratory pathways or home ranges. Developments in quantitative techniques to process raw tracking data are now helping to extract fine-scale information that was previously obscured by factors such as location error. Analyses of animal movement and migration have used diffusion and random walk processes (Dobzhansky & Wright 1947); Skellam (1951) was the most influential of early studies. Developments and extensions of diffusion models in biology are reviewed and updated by Okubo & Levin (2002). Analyses of movement to understand behavioral processes of individuals were developed in the 1970s and 1980s as the first tracking data became available (Siniff & Jessen 1969). Most analyses of individual tracking data are based upon correlated random walks (CRWs) (Turchin 1998), but fractals, first passage time (FPT) analysis, and Lévy flight methods have also been introduced. As Turchin (1998) provides an excellent guide for analyses of animal movement, we focus on analytical developments after 1998.

Computational power has been the single most important advance and has allowed ecologists to borrow, from the physics and engineering communities, sophisticated Bayesian model fitting methods (Markov chain Monte Carlo, particle filters, expectation maximization, etc.) to fit diffusion, CRW, mixed-effects, and generalized additive models with much more ecological and behavioral complexity. Large sample sizes and high data quality are an ideal combination for advanced methods and are allowing the detection of subtle behavioral signals to reveal deeply complex behavior and ecology in migrating animals.

1.2.1. State-space models and hidden Markov models.
State-space models (SSMs) and hidden Markov models (HMMs) are by far the most powerful and sophisticated new tools for analyzing animal movement and migration from electronic tracking data. These models have been applied to a wide range of engineering, physics, geoscience, and economics problems, where the goal is to infer an unobservable "hidden" or "latent" system state. In animal movement and migration, the hidden condition is an animal's behavioral state (**Figure 7**). That state affects how an animal moves, and both SSMs and HMMs fit CRW models that utilize movement properties such as turn angles, move lengths, and autocorrelation to infer latent behavioral state from telemetry data (Jonsen et al. 2005; Morales et al. 2004; Patterson et al. 2008, 2009). These models can also be structured to detect both the mechanisms and accuracy of animal navigation (Jonsen et al. 2006; Mills-Flemming et al. 2006, 2010).

Both SSMs and HMMs are fit to CRW models; the most powerful are so-called switching models or composite CRWs (Breed et al. 2009; Jonsen et al. 2005, 2006; Mills-Flemming et al. 2006). In these models, two, three, or sometimes more sets of parameters, each parameter set representing a behavioral state, are estimated for a CRW. These parameters control attributes such as move speed, turn angle, cardinal direction, and autocorrelation, as well as the degree of stochasticity. The parameter space can produce many different kinds of movement. Most switching models have a fixed number of discrete states to be inferred, and the model fits a track by inferring which of the two or three sets of parameters is most likely at any point in the track (**Figure 8**). Some are set up as mixing models, where the behavioral state is determined strictly from the movement pattern at a particular time in the track, whereas others have transition equations between states, so that both the current movement pattern and the movement leading up to it affect which state is inferred at any given time.

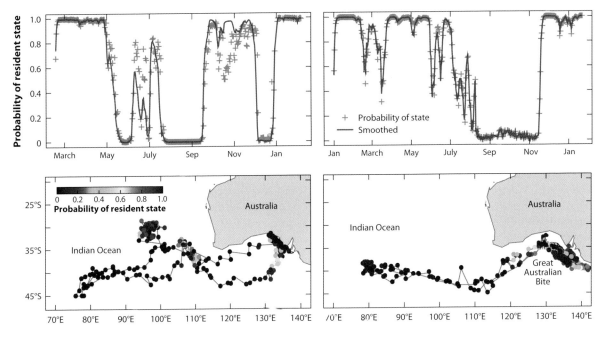

Figure 7

Hidden Markov model fits for two southern bluefin tuna tracks, estimating the probability of being in a "migratory" or "resident" behavioral state. Upper panels show the behavioral state time series; lower panels show those probabilities on a map. Modified from Patterson et al. (2009).

SSMs are a special case of HMMs. In an HMM, the hidden states are a set of discrete, categorical states that must be predefined before a model is fit. In an SSM, the hidden states are distributed on a continuum, or space, and are not discrete, thus the term state space. As implemented, the state-space aspect of SSMs is usually restricted to handling observation error in animal tracking data, particularly in Argos and light-level geolocation data, to estimate the most likely location. This improved location estimate more accurately reflects the movement pattern made by the animal. Thus, switching SSMs actually have an HMM for behavioral state linked to and embedded within an SSM for location. SSMs have two equations: one models behavior and accounts for behavioral stochasticity (the process equation), and the other models observation error (the observation equation). The two equations make SSMs much better for analyzing data with significant degrees of observation error (e.g., Argos, light-level geolocation), but this comes at the cost of being more difficult to implement than HMMs. Many researchers use SSMs solely for track correction (e.g., Tremblay et al. 2009). One practical solution is to implement a simple SSM using a Kalman filter to fit a random walk with no behavior to correct for observation error followed by an HMM to estimate the hidden behavioral state (Patterson et al. 2010). Otherwise HMMs are generally useful only for analyzing GPS quality tracking data; if observation error is not accounted for, it interferes with behavioral state discrimination.

SSMs can be fit in either continuous or discrete time. Discrete-time models estimate locations and behavioral state at regular time steps. These have many favorable properties, but are somewhat more complex to fit (Breed et al. 2009, 2011). Continuous-time SSMs estimate a location for every observation on irregular intervals, and some argue this is a more natural framework (Johnson et al. 2008, Kuhn et al. 2009).

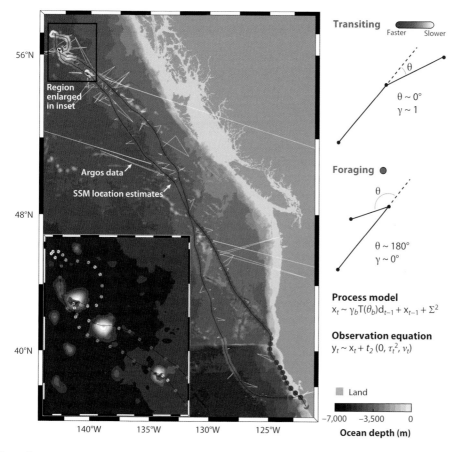

Figure 8

Example of a switching state-space model (SSM) fit to an Argos satellite track collected from a northern elephant seal, in this case revealing intense use of a seamount chain (Maxwell et al. 2011b). The map shows the highly error-prone Argos data overlaid with the SSM best-location estimates; color indicates inferred behavioral state. The switching model fit is shown to the right of the map, with behavioral state inferred from the autocorrelation (γ) to the previous displacement (d_{t-1}) and turn angle θ. The parameters are indexed into two states by the vector b. The Argos error is modeled with the observation equation and assumes t-distributed error, with variance τ and degrees of freedom υ (Jonsen et al. 2005). The nominally "foraging" state had low estimates for γ and estimates of θ near 180°, whereas nominally "transiting" states had high values of γ and estimates of θ near zero.

Finally, there are a few examples of models that fit behavior using a continuum of parameters within state-space or similar time-series frameworks (Breed et al. 2012, Gurarie et al. 2009). This is a promising new direction and should allow more flexible and natural frameworks for inferring a wide range of behavioral states from movement patterns.

1.2.2. Heuristic approaches. Several heuristic approaches have gained traction in recent years. These methods qualify track properties using sets of rules that describe turning angles, time spent in a region, or other movement metrics to infer search intensity or foraging activity. These approaches can be extremely useful and practical in many situations. However, because many are not well grounded in probability theory, objective interpretation can be difficult.

1.2.2.1. Fractals. Fractal analyses have been especially popular and useful in landscape ecology (Halley et al. 2004, Turcotte 1997) and were introduced as a method to quantify animal movement by Dicke & Burrough (1988). A series of influential papers (Johnson et al. 1992, Milne 1991, Wiens et al. 1995) built on the work of Dicke & Burrough (1988) lay out a simple method for calculating the fractal dimension (*d*) of an animal's track. Some researchers also argued that because fractals are scale invariant, *d* could be used to compare directly the behavior of animals of very different sizes moving through landscapes (Wiens et al. 1995).

d represents the 2D space that would be covered by various movement patterns and is calculated by log-log regressing "ruler length" against measured path length. Shorter ruler lengths measure more detail, and the measured path length is thus longer. The slope of this regression is the fractal dimension *d*, and it has been interpreted biologically as the intensity of search or foraging effort. However, Turchin (1996) clearly demonstrated that the log-log regression needed to calculate *d* turns into a curve as the ruler length goes to infinity; to be valid, the relationship must be linear. Since the critique by Turchin (1996), the popularity of fractals for analyzing animal movement has dropped considerably.

A variation of the fractal method has recently been implemented to identify area-restricted search (ARS) behavior in wide-ranging marine animals. Known as the fractal landscape method, tracks are objectively segmented as a moving window of a given size passes over the track, and *d* is calculated for the data subset in the window at any given time (Tremblay et al. 2007). Windows with higher *d* could be considered to have higher search or foraging intensity and represent more important habitat. The caveat from Turchin (1996) notwithstanding, the fractal landscape method is easy to implement and represents a simpler alternative for identifying pelagic ARS and foraging areas than behavior-discriminating SSMs or HMMs.

1.2.2.2. First passage time analysis. Another method similar to fractal analysis, but perhaps more flexible and less prone to bias, is FPT analysis. This method was first suggested for analyzing tracking data by Johnson et al. (1992), but it was not formalized until Fauchald & Tveraa (2003).

The analysis proceeds by linearly interpolating tracking locations so that they are evenly spaced in time, and circles with a radius *r* are placed around each point. The amount of time a tagged animal spends within each circle is calculated: If the animal moves slowly or turns frequently, the amount of time will be greater; if it moves straight or quickly, it will be smaller. The time it takes for the animal to leave the circle of a given *r* is the FPT for the point at the circle's center. It is then possible to see how the FPT index changes along the track. Large FPT values are associated with ARS and small values with directed or migratory movements (Fauchald & Tveraa 2003). The FPT index is very similar to the fractal dimension *d*, with high FPT values corresponding to high *d*.

This process is repeated for a range of *r*'s. For each *r*, an FPT value is generated for each point in the track, and it is possible to calculate a mean and variance of FPT for the entire track. Fauchald & Tveraa (2003) argued that this variance could be log transformed and plotted against *r*, and if a peak in var[log(FPT)] was clear at a particular *r* value, this value of *r* could be interpreted as the "characteristic scale" at which an individual animal's search movements are concentrated.

The FPT method is easy to apply and interpret, and it has been a popular choice since its introduction by Fauchald & Tveraa (2003), especially for marine birds (e.g., Fauchald & Tveraa 2006, Pinaud 2008, Weimerskirch et al. 2007). The method has also been extended to include a vertical dimension for diving animals (Bailleul et al. 2008). As an index of track sinuosity, it is at least as good as fractal analysis, and unlike fractals no mathematical problems associated with the calculation of FPT indices have been discovered. However, the method does require high-resolution data, and many of the published studies to date use GPS data with interpolation to get extremely fine temporal resolution.

1.2.2.3. Lévy flight models. Lévy flights are a particular variant of a random walk model. The random draws of step length come from a long-tailed probability distribution, most often a Pareto distribution with infinite variance, $P(l_j) \sim l_j^{-\mu}$, where $P(l_j)$ describes the probability density of step lengths. To be considered a Lévy flight, the exponent μ must be between 1 and 3 (Viswanathan et al. 1999). When $\mu \leq 1$, motion is ballistic, and when $\mu \geq 3$ it is Brownian.

Lévy flight models have a very simple mathematical form, produce tracks very similar to real animals, and are easily fit using regression and maximum-likelihood methods. Their ease and intuitiveness have made them tractable to many biologists, and they have been fit to movement data collected from a wide range of migratory marine species (e.g., Humphries et al. 2010, Sims et al. 2008, Viswanathan et al. 1999). In addition, it has been suggested that a Lévy flight with $\mu = 2$ is the most efficient search strategy possible (Viswanathan et al. 1999), and thus foragers entering unfamiliar environments should employ this strategy to find food (Reynolds & Rhodes 2009).

These findings, and the use of Lévy flight models for movement data, however, are highly controversial. Numerous papers have pointed out flaws in both the mathematical and logical underpinnings of Lévy flight analyses. Among other factors, these studies have found that Lévy searches are the most efficient only under very special, even peculiar circumstances (James et al. 2008); that Lévy flights have been misidentified in many of the foundational papers (Edwards 2008, 2011; Edwards et al. 2007); and that alternative models, especially composite CRWs, cannot be distinguished from Lévy flights in real data (Auger-Méthé et al. 2011, Benhamou 2007, Plank & Codling 2009). Given the multitude of potential issues, we cannot recommend the use of Lévy flight methods or theory in analysis or understanding of marine animal migration or movement.

1.3. Advances in Animal Navigation

Many species routinely migrate across vast expanses of oceanic habitats, regions seemingly devoid of the cues necessary for accurate navigation (Lohmann et al. 2008). Albatrosses, sea turtles, whales, seals, sharks, and many other taxa rely on well-developed navigation abilities to move between distant foraging and reproduction regions (Mueller & Fagan 2008). Thus, there is a strong selective pressure to maintain and refine navigation ability. Although considerable progress has been made in revealing the underlying mechanisms of animal navigation, particularly in the terrestrial realm (Able 1995), a holistic understanding remains elusive. This is in large part due to the use of multiple redundant or condition-dependent cues by navigating animals (Muheim et al. 2006a).

Pelagic migrants may use one or more of a diverse suite of environmental cues to navigate. Stable visual or bathymetric cues, such as shallow coastal areas or seamounts, may be used during part of a migration, but would not be available in deep pelagic habitats. Oceanographic features, including temperature fields, salinity, and associated fronts, vary reliably on a coarse scale; however, these features are dynamic and would require remarkable detection sensitivity (Bost et al. 2009a, Hays et al. 2001). Olfactory cues, such as aerosolized di-methyl sulfide, may be useful to direct predators toward prey patches (Nevitt & Bonadonna 2005, Nevitt et al. 2008), but again are likely not used for long-distance navigation (Lohmann et al. 1999).

Celestial and geomagnetic cues, however, are appealing candidates for pelagic migrants because they are ubiquitous and potentially very accurate. Diurnal celestial cues, including sunlight polarization and Sun position, are used by migrants (Gould 1998). Similarly, nocturnal celestial cues, including stellar orientation and moon position, are used to navigate by some species (Able & Able 1996, Muheim et al. 2006a). Geomagnetic cues have also been shown to be very important for many species (Lohmann et al. 2007). Geomagnetic inclination and field intensity show significant variation over the surface of Earth and, in many places, are orthogonally aligned (Akesson & Alerstam 1998). If animals can detect these features, the variation may provide enough information to form a cognitive map sufficient for long-distance pelagic navigation (Lohmann & Lohmann 1996).

A priori, it is unlikely that visual, oceanographic, or olfactory cues are the dominant source of navigation information for ocean basin–scale movements. Although identifying the particular cues used during oceanic migration is logistically challenging, recent advances and miniaturization of tracking technologies enable detailed measurement of migratory pathways. These data can be used to test navigation performance (i.e., accuracy in following a particular route) and may give insight into the mechanisms at play. A recent study of migrating humpback whales demonstrated that individuals maintain straight paths by correcting for fine-scale current patterns, implying the use of cues that are both accurate and stable (Horton et al. 2011). In elephant seals, outbound migrations are oriented toward vague dynamic targets, whereas inbound migrations are oriented toward a discrete fixed target (the home colony). Seals swim nearly continuously during these transit phases, as evident by a constant forward trajectory and characteristic "transit" dives (Le Boeuf et al. 2000). The path of the animal during these transit phases is often direct with a stable orientation, although this has yet to be adequately characterized (Le Boeuf et al. 2000).

Despite the logistical difficulty of studying oceanic migrants, they are actually excellent models for the study of navigation because their environment lacks many cues available to terrestrial organisms. Existing studies focus largely on albatrosses and turtles. The remarkable navigation ability of turtles has long been known from conventional tagging studies (Carr 1967), but the precision of island homing was more recently demonstrated with satellite tracking (Papi et al. 1995). Subsequent studies utilizing translocated free-ranging turtles attempt to identify the importance of particular cues but met limited success. For example, experimentally displaced turtles with attached magnets (to disrupt a hypothesized magnetic sense) retained their navigation ability (Papi et al. 2000). Such studies are unable to implicate particular cues, but suggest redundancy. Complementary laboratory-based experiments have shown the importance of several cues including at least two components of the geomagnetic field: intensity and inclination angle (Lohmann et al. 1999).

Investigations into albatross navigation have found similar results. Satellite tracking clearly demonstrates a keen navigation ability (Weimerskirch et al. 2012), but experimental attachment of magnets was inconclusive; like turtles, albatrosses retained their navigation ability (Bonadonna et al. 2003). Controlled studies of other bird species demonstrate detection and potential use of many cues: olfactory, sky polarization or Sun position, and star patterns (Able & Able 1996, Muheim et al. 2006b, Nevitt & Bonadonna 2005). However, detection ability does not necessarily imply use in free-ranging conditions, and demonstration of reliance on particular cues remains elusive for most species. For example, albatross dynamic soaring (Alerstam 1996) or the impact of ocean currents on sea turtles (Gaspar et al. 2006) may mask important trends that would otherwise be apparent in movement patterns. Much additional experimental work is necessary to understand which cues pelagic animals use to navigate and how these are integrated into a cognitive map.

1.4. Animals as Oceanographers

Electronic tags deployed on animals are also providing oceanographic data in areas where conventional methods are limited or absent (Boehme et al. 2009, Charrassin et al. 2008, Costa et al. 2010a) (**Figure 9**). Different water masses have unique temperature and salinity signatures and can be used to describe the hydrographic characteristics of habitat used by tracked animals. Tags are available that measure temperature, salinity, light, fluorescence, and pO_2 while the animal moves through the water column. These hydrographic data are collected at a scale and resolution that is perfectly coincident with the animal's behavior, allowing far better measures of habitat

Figure 9

A Weddell seal is shown wearing a Sea Mammal Research Unit conductivity-temperature-depth tag in McMurdo Sound, Antarctica. Photo by Dan Costa.

than remotely sensed satellite data. Such an approach has been used to define the foraging habitat of elephant and crabeater seals in the Southern Ocean by physical oceanographic characteristics (Biuw et al. 2007, 2010; Costa et al. 2010a). For elephant seals, changes in drift rate measured during periods when the seal was not swimming provide an index of body condition (fatter seals tend to be more buoyant, whereas leaner seals tend to sink). Changes in body condition were then correlated with temperature and salinity data collected by the tag and allowed identification of the specific water masses where elephant seals had the greatest foraging success. These areas were linked with warm, deep water known as circumpolar deep water. This warm and nutrient-rich water is associated with the Antarctic Circumpolar Current and upwells along the continental shelf. The data available from existing oceanographic sampling techniques and/or oceanographic models are too coarse or imprecise to identify the habitat characteristics of individual dives.

An added benefit is that hydrographic tags carried by animals provide highly cost-effective platforms from which detailed oceanographic data can be collected on a scale not possible with conventional methods. Such data are particularly lacking in the polar oceans where ship time is limited (especially in the winter), where cloud cover or sea ice limits the capability of satellite remote sensing, and where current patterns wash oceanographic floats away from the Antarctic continent. An international effort deployed conductivity-temperature-depth (CTD) tags on 85 elephant seals simultaneously at Kerguelen, South Georgia, Macquarie and the South Shetland Islands in the Southern Ocean between January 2004 and April 2006 (Biuw et al. 2007) and increased by ninefold the number of CTD profiles collected by traditional methods (Charrassin et al. 2008). Similar data sets can complement even well-studied regions by adding measurements from mesopelagic depths (Robinson et al. 2012). Animal-derived data are being made available to the general oceanographic community through databases historically reserved for ship-based data [e.g., Autonomous Pinniped Bathythermographs (APBs) in the World Ocean Database].

Oceanographic data collected by elephant seals in the Western Antarctic Peninsula provided insight into the unexpected breakup of the Wilkins Ice Shelf (WIS) in 2008. Two sets of data were collected; the first relied on observations that elephant seals were diving deeper than the known bathymetry and, thus, provided a basis to improve the bathymetry in this region. This refined bathymetry led to the discovery of a series of deep troughs that extend from the outer to the inner continental shelf near the WIS (Padman et al. 2010). These troughs acted as conduits for on-shelf movement of relatively warm (>1°C) upper circumpolar deep water across the continental shelf and under the WIS. The second data set—temperature records recorded from the seals' tags—confirmed the flow of this warm deep water underneath the WIS. The added heat contributed to the breakup of the ice shelf (Padman et al. 2012).

1.5. Including Migration in Marine Conservation

For centuries, fishers and hunters have relied on knowledge of the movement patterns of exploited species such as tuna, whales, seals, sea turtles, and seabirds to predict when and where to harvest them. Today, management and conservation of highly migratory species is a more pressing need and requires detailed information on movements of threatened species. This information is also fundamental to understanding how diseases might be spread and how disease networks might change as animals change their migration patterns (Altizer et al. 2011). Satellite and acoustic tagging of white sharks, combined with a Bayesian model, has been used to provide estimates of white shark populations (Jorgensen et al. 2010). Electronic tags are being used to reveal patterns of habitat utilization and to identify and/or help avoid or mitigate conflicts with oil and gas development, military activities, fisheries interactions, and shipping and research activities (Chilvers 2008, Costa et al. 2003, Goldsworthy & Page 2007, Peckham et al. 2007, Tyack et al. 2011, Žydelis et al. 2011). Tracking data were important in listing black-footed albatrosses as an endangered species by the US Fish and Wildlife Service and by BirdLife International for deliberations within the international Agreement for the Conservation of Albatrosses and Petrels, and tracking data have been crucial in the development of a management plan for the endangered Australian and New Zealand sea lions (Campbell et al. 2006, Chilvers 2008, Goldsworthy & Page 2007). Finally, tracking data are providing insights into the potential impact of climate change on pelagic species (Costa et al. 2010a, Weimerskirch et al. 2012).

Marine animals do not recognize political boundaries, so knowledge of their movement patterns and where they perform vital activities such as foraging and breeding can provide the basis for regional management plans. Such information is critical for identifying key habitat for both implementation of marine protected areas (MPAs) (Maxwell et al. 2011a, Peckham et al. 2007, Schofield et al. 2007, Wallace et al. 2011, Witt et al. 2011, Žydelis et al. 2011) and determination of the spatial and temporal extent of such management measures. For example, Laysan albatrosses tagged at Guadalupe Island, Mexico, are found within the California Current System and within exclusive economic zones of at least three other countries. Pacific bluefin tuna that swam to the Eastern Pacific Ocean from Japan are so overexploited that few tagged fish live long enough to make the return trans-Pacific migration to spawn (Block et al. 2011). Leatherback sea turtles have been observed to use corridors shaped by persistent oceanographic features such as the southern edge of the Costa Rica Dome and the highly energetic currents of the equatorial Pacific (Shillinger et al. 2008). These findings have led to an International Union for Conservation of Nature resolution to conserve leatherback sea turtles in the open seas. Similarly, tracking data were used to develop an MPA off the coast of Baja California to protect loggerhead sea turtles (Peckham et al. 2007) and to assess the efficacy of an implemented MPA to protect olive ridley sea turtles off the coast of Gabon (Maxwell et al. 2011a) (**Figure 10**).

2. FUTURE DIRECTIONS

In recent years, the capability of electronic tags has increased considerably. However, there are a number of technological advances that need further development, including novel ways of powering tags, increased sensor capabilities (including oceanographic sensors and animal behavior and/or physiology), better attachment methods, miniaturization of tags, and alternative methods

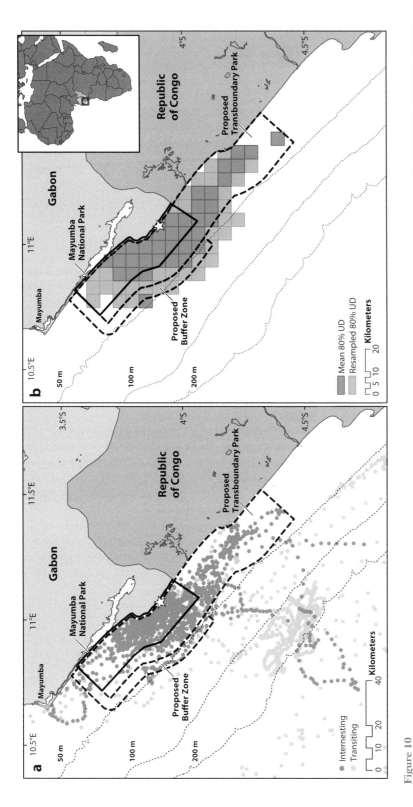

Figure 10

(*a*) State-space modeled (SSM) tracks (*n* = 18) of olive ridley sea turtles tagged from Mayumba National Park. Orange points represent internesting behavioral mode; blue points represent transiting behavioral mode. The star indicates the tagging location. (*b*) Confidence intervals of movements for olive ridley sea turtles derived from the data in panel *a*. Outer error bounds for 80% utilization distribution (UD) for mean SSM estimates (*light green*) and resampled SSM (*dark green*). Reproduced from Maxwell et al. (2011a).

of data recovery. Although new higher capacity batteries may be developed, an alternative is to develop technologies that collect energy from an animal's movement. Conceptually, this seems very straightforward, but the development of reliable power-harvesting systems has so far proven elusive. Other sensors that could be added to the tags include such important oceanographic measures as pH, CO_2, and chlorophyll, as well as measures of animal behavior that include reliable measures of feeding behavior and even active sonar to measure prey fields in front of the animal. Novel methods of data recovery would greatly enhance the range of species that these tags could be deployed on. A major advance would be achieved if the data obtained by electronic tags could be transmitted underwater via an acoustic modem.

Electronic tags have provided a hitherto unprecedented view of the movement patterns and habitat preferences of highly migratory upper trophic–level species. Effective management and conservation requires a better understanding of both how migrants navigate to essential habitat and the processes that make these habitats desirable. Although there has been a significant move toward development of MPAs, the efficacy of such protected areas has yet to be considered. Protection of a species on their foraging or breeding grounds may not enhance population viability if the animals are caught along their migration corridor. Modeling efforts could examine the relative value of developing MPAs around species' movement patterns elucidated using electronic tagging.

Mysteries of long-distance navigation will be solved only by efforts targeting long-distance continuous migrants for controlled field-based manipulations, where navigational cues such as magnetic or visual senses are manipulated. The resulting change in the animal's navigational ability could then be monitored using high-resolution tracking. Such field experiments would be complemented by lab-based experiments (e.g., psychophysical) that allow a greater examination of mechanisms.

Further study of habitat association is also needed to integrate large-scale movement patterns with the lower trophic levels and the biophysical forces that structure ecosystems. Future discoveries will be made when these tools are applied in an integrated manner, providing a seamless understanding of the biophysical processes driving primary production at lower trophic levels to the movements and behavior patterns of apex predators. Such an integrated effort would need to be focused on a number of regions where existing infrastructure is in place or locations that are representative of critical marine habitats. This would provide not just a onetime snapshot of the biodiversity of a marine habitat, but a dynamic view into the processes that maintain biodiversity with a better understanding of how it can be protected. A critically important aspect of this is that we will be able to monitor how life in the ocean is adjusting in response to climate change (Costa et al. 2010a, Weimerskirch et al. 2012). On a warming planet, that insight will be essential if we wish to mitigate the impacts of climate change on pelagic migrants and save those species for future generations.

DISCLOSURE STATEMENT

The authors are not aware of any affiliations, memberships, funding, or financial holdings that might be perceived as affecting the objectivity of this review.

ACKNOWLEDGMENTS

This review was supported by the NSF Office of Polar Programs grant ANT-0838937, the Office of Naval Research grant N00014-08-1-1195, the E&P Sound and Marine Life Joint Industry Project of the International Association of Oil and Gas Producers, and from Steve Blank and the Ida Benson Lynn Chair in Ocean Health. Autumn-Lynn Harrison helped modify **Figure 1**, and Samantha Simmons provided useful comments on an earlier draft.

LITERATURE CITED

Able KP. 1995. Orientation and navigation: a perspective on 50 years of research. *Condor* 97:592–604

Able KP, Able MA. 1996. The flexible migratory orientation system of the Savannah sparrow (*Passerculus sandwichensis*). *J. Exp. Biol.* 199:3–8

Akesson S, Alerstam T. 1998. Oceanic navigation: Are there any feasible geomagnetic bi-coordinate combinations for albatrosses? *J. Avian Biol.* 29:618–25

Alerstam T. 1996. The geographical scale factor in orientation of migrating birds. *J. Exp. Biol.* 199:9–19

Altizer S, Bartel R, Han BA. 2011. Animal migration and infectious disease risk. *Science* 331:296–302

Anderson SD, Chapple TK, Jorgensen SJ, Klimley AP, Block BA. 2011. Long-term individual identification and site fidelity of white sharks, *Carcharodon carcharias*, off California using dorsal fins. *Mar. Biol.* 158:1233–37

Andrews KS, Williams GD, Levin PS. 2010. Seasonal and ontogenetic changes in movement patterns of sixgill sharks. *PLoS ONE* 5:e12549

Auger-Méthé M, St. Clair C, Lewis M, Derocher A. 2011. Sampling rate and misidentification of Lévy and non-Lévy movement paths: comment. *Ecology* 92:8, 1699–1701

Bailey H, Mate BR, Palacios DM, Irvine L, Bograd SJ, Costa DP. 2009. Behavioural estimation of blue whale movements in the Northeast Pacific from state-space model analysis of satellite tracks. *Endanger. Species Res.* 10:93–106

Bailey HR, Shillinger GL, Palacios DM, Bograd SJ, Spotila JR, et al. 2008. Identifying and comparing phases of movement by leatherback turtles using state-space models. *J. Exp. Mar. Biol. Ecol.* 356:128–35

Bailleul F, Charrassin J-B, Ezraty R, Girard-Ardhuin F, McMahon CR, et al. 2007. Southern elephant seals from Kerguelen Islands confronted by Antarctic Sea ice. Changes in movements and in diving behaviour. *Deep-Sea Res. Part II Top. Stud. Oceanogr.* 54:343–55

Bailleul F, Pinaud D, Hindell M, Charrassin JBÎ, Guinet C. 2008. Assessment of scale-dependent foraging behaviour in southern elephant seals incorporating the vertical dimension: a development of the First Passage Time method. *J. Anim. Ecol.* 77:948–57

Benhamou S. 2007. How many animals really do the Lévy walk? *Ecology* 88:1962–69

Bestley S, Gunn JS, Hindell MA. 2009. Plasticity in vertical behaviour of migrating juvenile southern bluefin tuna (*Thunnus maccoyii*) in relation to oceanography of the south Indian Ocean. *Fish. Oceanogr.* 18:237–54

Biuw M, Boehme L, Guinet C, Hindell M, Costa D, et al. 2007. Variations in behavior and condition of a Southern Ocean top predator in relation to in situ oceanographic conditions. *Proc. Natl. Acad. Sci. USA* 104:13705–10

Biuw M, Nost OA, Stien A, Zhou Q, Lydersen C, Kovacs KM. 2010. Effects of hydrographic variability on the spatial, seasonal and diel diving patterns of southern elephant seals in the eastern Weddell Sea. *PLoS ONE* 5(11):e13816

Block BA, Dewar H, Farwell C, Prince ED. 1998. A new satellite technology for tracking the movements of Atlantic bluefin tuna. *Proc. Natl. Acad. Sci. USA* 95:9384–89

Block BA, Jonsen ID, Jorgensen SJ, Winship AJ, Shaffer SA, et al. 2011. Tracking apex marine predator movements in a dynamic ocean. *Nature* 475:86–90

Block BA, Teo SL, Walli A, Boustany A, Stokesbury MJ, et al. 2005. Electronic tagging and population structure of Atlantic bluefin tuna. *Nature* 434:1121–27

Boehme L, Lovell P, Biuw M, Roquet F, Nicholson J, et al. 2009. Technical note: animal-borne CTD-satellite relay data loggers for real-time oceanographic data collection. *Ocean Sci.* 5:685–95

Bonadonna F, Chamaille-Jammes S, Pinaud D, Weimerskirch H. 2003. Magnetic cues: Are they important in Black-browed Albatross *Diomedea melanophris* orientation? *Ibis* 145:152–55

Bost CA, Cotté C, Bailleul F, Cherel Y, Charrassin JB, et al. 2009a. The importance of oceanographic fronts to marine birds and mammals of the southern oceans. *J. Mar. Syst.* 78:363–76

Bost CA, Georges JY, Guinet C, Cherel Y, Puetz K, et al. 1997. Foraging habitat and food intake of satellite-tracked king penguins during the austral summer at Crozet Archipelago. *Mar. Ecol. Progr. Ser.* 150:21–33

Bost CA, Thiebot JB, Pinaud D, Cherel Y, Trathan PN. 2009b. Where do penguins go during the inter-breeding period? Using geolocation to track the winter dispersion of the macaroni penguin. *Biol. Lett.* 5:473–76

Boustany AM, Matteson R, Castleton M, Farwell C, Block BA. 2010. Movements of Pacific bluefin tuna (*Thunnus orientalis*) in the Eastern North Pacific revealed with archival tags. *Progr. Oceanogr.* 86:94–104

Breed GA, Costa DP, Goebel ME, Robinson PW. 2011. Electronic tracking tag programming is critical to data collection for behavioral time-series analysis. *Ecosphere* 2:art. 10

Breed GA, Costa DP, Jonsen ID, Robinson PW, Mills-Flemming J. 2012. State-space methods for more completely capturing behavioral dynamics from animal tracks. *Ecol. Model.* 235–236:49–58

Breed GA, Jonsen ID, Myers RA, Bowen WD, Leonard ML. 2009. Sex-specific, seasonal foraging tactics of adult grey seals (*Halichoerus grypus*) revealed by state-space analysis. *Ecology* 90:3209–21

Campana SE, Dorey A, Fowler M, Joyce W, Wang Z, et al. 2011. Migration pathways, behavioural thermoregulation and overwintering grounds of blue sharks in the Northwest Atlantic. *PLoS ONE* 6(2):e16854

Campbell RA, Chilvers BL, Childerhouse S, Gales NJ. 2006. Conservation management issues and status of the New Zealand (*Phocarctos hookeri*) and Australian (*Neophoca cinerea*) sea lion. In *Sea Lions of the World*, ed. AW Trites, SK Atkinson, DP DeMaster, LW Fritz, TS Gelatt, LD Rea, KM Wynne, pp. 455–71. Fairbanks, AK: Alsk. Sea Grant

Carlson JK, Ribera MM, Conrath CL, Heupel MR, Burgess GH. 2010. Habitat use and movement patterns of bull sharks *Carcharhinus leucas* determined using pop-up satellite archival tags. *J. Fish Biol.* 77:661–75

Carr A. 1967. Adaptive aspects of the scheduled travel of *Chelonia*. In *Animal Orientation and Navigation*, ed. RM Storm, pp. 35. Corvallis, OR: Or. State Univ. Press

Charrassin J-B, Hindell M, Rintoul SR, Roquet F, Sokolov S, et al. 2008. Southern Ocean frontal structure and sea-ice formation rates revealed by elephant seals. *Proc. Natl. Acad. Sci. USA* 105:11634–39

Chilvers BL. 2008. New Zealand sea lions (*Phocarctos hookeri*) and squid trawl fisheries: bycatch problems and management options. *Endanger. Species Res.* 5:193–204

Costa DP. 1991. Reproductive and foraging energetics of high latitude penguins, albatrosses and pinnipeds: implications for life history patterns. *Am. Zool.* 31:111–30

Costa DP, Crocker DE, Gedamke J, Webb PM, Houser DS, et al. 2003. The effect of a low-frequency sound source (acoustic thermometry of the ocean climate) on the diving behavior of juvenile northern elephant seals, *Mirounga angustirostris*. *J. Acoust. Soc. Am.* 113:1155–65

Costa DP, Huckstadt LA, Crocker DE, McDonald BI, Goebel ME, Fedak MA. 2010a. Approaches to studying climatic change and its role on the habitat selection of Antarctic pinnipeds. *Integr. Comp. Biol.* 50:1018–30

Costa DP, Robinson PW, Arnould JP, Harrison AL, Simmons SE, et al. 2010b. Accuracy of ARGOS locations of Pinnipeds at-sea estimated using Fastloc GPS. *PLoS ONE* 5:e8677

Cotté C, Park Y-H, Guinet C, Bost C-A. 2007. Movements of foraging king penguins through marine mesoscale eddies. *Proc. R. Soc. B* 274:2385–91

Dagorn L, Pincock D, Girard C, Holland K, Taquet M, et al. 2007. Satellite-linked acoustic receivers to observe behavior of fish in remote areas. *Aquat. Living Resour.* 20:307–12

Dias MP, Granadeiro JP, Phillips RA, Alonso H, Catry P. 2011. Breaking the routine: individual Cory's shearwaters shift winter destinations between hemispheres and across ocean basins. *Proc. R. Soc. B: Biol. Sci.* 278:1786–93

Dicke M, Burrough P. 1988. Using fractal dimensions for characterizing tortuosity of animal trails. *Physiol. Entomol.* 13:393–98

Dobzhansky T, Wright S. 1947. Genetics of natural populations. XV. Rate of diffusion of a mutant gene through a population of *Drosophila pseudoobscura*. *Genetics* 32:303–24

Durban JW, Pitman RL. 2012. Antarctic killer whales make rapid, round-trip movements to subtropical waters: evidence for physiological maintenance migrations? *Biol. Lett.* 8(2):274–77

Eckert SA, Dolar LL, Kooyman GL, Perrin W, Rahman RA. 2002. Movements of whale sharks (*Rhincodon typus*) in South-east Asian waters as determined by satellite telemetry. *J. Zool. (Lond.)* 257:111–15

Edwards AM. 2008. Using likelihood to test for Lévy flight search patterns and for general power-law distributions in nature. *J. Anim. Ecol.* 77:1212–22

Edwards AM. 2011. Overturning conclusions of Lévy flight movement patterns by fishing boats and foraging animals. *Ecology* 92:1247–57

Edwards AM, Phillips RA, Watkins NW, Freeman MP, Murphy EJ, et al. 2007. Revisiting Lévy flight search patterns of wandering albatrosses, bumblebees and deer. *Nature* 449:1044–48

Egevang C, Stenhouse IJ, Phillips RA, Petersen A, Fox JW, Silk JRD. 2010. Tracking of Arctic terns *Sterna paradisaea* reveals longest animal migration. *Proc. Natl. Acad. Sci. USA* 107:2078–81

Fauchald P, Tveraa T. 2003. Using first-passage time in the analysis of area-restricted search and habitat selection. *Ecology* 84:282–88

Fauchald P, Tveraa T. 2006. Hierarchical patch dynamics and animal movement pattern. *Oecologia* 149:383–95

Gaspar P, Georges JY, Fossette S, Lenoble A, Ferraroli S, Le Maho Y. 2006. Marine animal behaviour: neglecting ocean currents can lead us up the wrong track. *Proc. R. Soc. B: Biol. Sci.* 273:2697–702

Goldsworthy SD, Page B. 2007. A risk-assessment approach to evaluating the significance of seal bycatch in two Australian fisheries. *Biol. Conserv.* 139:269–85

Gould JL. 1998. Sensory bases of navigation. *Curr. Biol.* 8:R731–38

Guinet C, Dubroca L, Lea MA, Goldsworthy S, Cherel Y, et al. 2001. Spatial distribution of foraging in female Antarctic fur seals (*Arctocephalus gazella*) in relation to oceanographic variables: a scale-dependent approach using geographic information systems. *Mar. Ecol. Progr. Ser.* 219:251–64

Gurarie E, Andrews RD, Laidre KL. 2009. A novel method for identifying behavioural changes in animal movement data. *Ecol. Lett.* 12:395–408

Halley JM, Hartley S, Kallimanis AS, Kunin WE, Lennon JJ, Sgardelis SP. 2004. Uses and abuses of fractal methodology in ecology. *Ecol. Lett.* 7:254–71

Hammerschlag N, Gallagher AJ, Lazarre DM. 2011. A review of shark satellite tagging studies. *J. Exp. Mar. Biol. Ecol.* 398:1–8

Hayes SA, Teutschel N, Michel C, Champagne C, Robinson P, et al. 2012. Mobile receivers: releasing the mooring to 'see' where fish go. *Environ. Biol. Fishes.* In press (doi: 10.1007/s10641-011-9940-x)

Hays GC, Dray M, Quaife T, Smyth TJ, Mironnet NC, et al. 2001. Movements of migrating green turtles in relation to AVHRR derived sea surface temperature. *Int. J. Remote Sens.* 22:1403–11

Horton TW, Holdaway RN, Zerbini AN, Hauser N, Garrigue C, et al. 2011. Straight as an arrow: humpback whales swim constant course tracks during long-distance migration. *Biol. Lett.* 7:674–79

Humphries NE, Queiroz N, Dyer JRM, Pade NG, Musyl MK, et al. 2010. Environmental context explains Lévy and Brownian movement patterns of marine predators. *Nature* 465:1066–69

James A, Plank MJ, Brown R. 2008. Optimizing the encounter rate in biological interactions: ballistic versus Lévy versus Brownian strategies. *Phys. Rev. E* 78:051128

James MC, Myers RA, Ottensmeyer CA. 2005. Behaviour of leatherback sea turtles, *Dermochelys coriacea*, during the migratory cycle. *Proc. R. Soc. B* 272:1547–55

Johnson AR, Wiens JA, Milne BT, Crist TO. 1992. Animal movements and population dynamics in heterogeneous landscapes. *Landsc. Ecol.* 7:63–75

Johnson DS, London JM, Lea MA, Durban JW. 2008. Continuous-time correlated random walk model for animal telemetry data. *Ecology* 89:1208–15

Johnson M, Madsen PT, Zimmer WM, de Soto NA, Tyack PL. 2006. Foraging Blainville's beaked whales (*Mesoplodon densirostris*) produce distinct click types matched to different phases of echolocation. *J. Exp. Biol.* 209:5038–50

Jonsen ID, Mills Flemming J, Myers RA. 2005. Robust state-space modeling of animal movement data. *Ecology* 86:2874–80

Jonsen ID, Myers RA, James MC. 2006. Robust hierarchical state-space models reveal diel variation in movement rates of migrating leatherback turtles. *J. Anim. Ecol.* 75:1046–57

Jorgensen SJ, Reeb CA, Chapple TK, Anderson S, Perle C, et al. 2010. Philopatry and migration of Pacific white sharks. *Proc. R. Soc. B* 277:679–88

Kappes MA, Shaffer SA, Tremblay Y, Foley DG, Palacios DM, et al. 2010. Hawaiian albatrosses track interannual variability of marine habitats in the North Pacific. *Progr. Oceanogr.* 86:246–60

Kuhn CE, Johnson DS, Ream RR, Gelatt TS. 2009. Advances in the tracking of marine species: using GPS locations to evaluate satellite track data and a continuous-time movement model. *Mar. Ecol. Progr. Ser.* 393:97–109

Le Boeuf BJ, Crocker DE, Costa DP, Blackwell SB, Webb PM, Houser DS. 2000. Foraging ecology of northern elephant seals. *Ecol. Monogr.* 70:353–82

Lindley ST, Erickson DL, Moser ML, Williams G, Langness OP, et al. 2011. Electronic tagging of green sturgeon reveals population structure and movement among estuaries. *Trans. Am. Fish. Soc.* 140:108–22

Lohmann KJ, Hester JT, Lohmann CMF. 1999. Long-distance navigation in sea turtles. *Ethol. Ecol. Evol.* 11:1–23

Lohmann KJ, Lohmann CMF. 1996. Orientation and open-sea navigation in sea turtles. *J. Exp. Biol.* 199:73–81

Lohmann KJ, Lohmann CMF, Endres CS. 2008. The sensory ecology of ocean navigation. *J. Exp. Biol.* 211:1719–28

Lohmann KJ, Lohmann CMF, Putman NF. 2007. Magnetic maps in animals: nature's GPS. *J. Exp. Biol.* 210:3697–705

Mate BR, Gisiner R, Mobley J. 1998. Local and migratory movements of Hawaiian humpback whales tracked by satellite telemetry. *Can. J. Zool.* 76:863–68

Maxwell SM, Breed GA, Nickel BA, Makanga-Bahouna J, Pemo-Makaya E, et al. 2011a. Using satellite tracking to optimize protection of long-lived marine species: olive ridley sea turtle conservation in Central Africa. *PLoS ONE* 6:e19905

Maxwell SM, Frank JJ, Breed GA, Robinson PW, Simmons SE, et al. 2011b. Benthic foraging on seamounts: a specialized foraging behavior in a deep-diving pinniped. *Mar. Mammal Sci.: Early View* 28:E333–44

McConnell B, Bryant E, Hunter C, Lovell P, Hall A. 2004. Phoning home: a new GSM mobile phone telemetry system to collect mark-recapture data. *Mar. Mammal Sci.* 20:274–83

Miller PJ, Johnson MP, Tyack PL. 2004. Sperm whale behaviour indicates the use of echolocation click buzzes "creaks" in prey capture. *Proc. R. Soc. Lond. Ser. B* 271:2239–47

Mills-Flemming JE, Field CA, James MC, Jonsen ID, Myers RA. 2006. How well can animals navigate? Estimating the circle of confusion from tracking data. *Environmentrics* 17:351–62

Mills-Flemming JE, Jonsen ID, Myers RA, Field CA. 2010. Hierarchical state-space estimation of leatherback turtle navigation ability. *PLoS ONE* 5:e14245

Milne BT. 1991. Lessons from applying fractal models to landscape patterns. In *Quantitative Methods in Landscape Ecology*, ed. MG Turner, RH Gardner, pp. 199–235. New York: Springer-Verlag

Morales JM, Haydon DT, Frair J, Hosinger KE, Fryxell JM. 2004. Extracting more from relocation data: building movement models as mixtures of random walks. *Ecology* 85:2436–45

Mueller T, Fagan WF. 2008. Search and navigation in dynamic environments—from individual behaviors to population distributions. *Oikos* 117:654–64

Muheim R, Moore FR, Phillips JB. 2006a. Calibration of magnetic and celestial compass cues in migratory birds: a review of cue-conflict experiments. *J. Exp. Biol.* 209:2–17

Muheim R, Phillips JB, Akesson S. 2006b. Polarized light cues underlie compass calibration in migratory songbirds. *Science* 313:837–39

Nevitt GA, Bonadonna F. 2005. Sensitivity to dimethyl sulphide suggests a mechanism for olfactory navigation by seabirds. *Biol. Lett.* 1:303–5

Nevitt GA, Losekoot M, Weimerskirch H. 2008. Evidence for olfactory search in wandering albatross, *Diomedea exulans. Proc. Natl. Acad. Sci. USA* 105:4576–81

Newsome SD, Clementz MT, Koch PL. 2010. Using stable isotope biogeochemistry to study marine mammal ecology. *Mar. Mammal. Sci.* 26:509–72

Okubo A, Levin SA. 2002. *Diffusion and Ecological Problems*. New York: Springer-Verlag

Padman L, Costa DP, Bolmer ST, Goebel ME, Huckstadt LA, et al. 2010. Seals map bathymetry of the Antarctic continental shelf. *Geophys. Res. Lett.* 37:L21601

Padman L, Costa DP, Dinniman MS, Fricker HA, Goebel ME, et al. 2012. Oceanic controls on the mass balance of Wilkins Ice Shelf, Antarctica. *J. Geophys. Res. Oceans* 117:C01010

Papi F, Liew HC, Luschi P, Chan EH. 1995. Long-range migratory travel of a green turtle tracked by satellite: evidence for navigational ability in the open sea. *Mar. Biol.* 122:171–75

Papi F, Luschi P, Akesson S, Capogrossi S, Hays GC. 2000. Open-sea migration of magnetically disturbed sea turtles. *J. Exp. Biol.* 203:3435–43

Patterson TA, Basson M, Bravington MV, Gunn JS. 2009. Classifying movement behaviour in relation to environmental conditions using hidden Markov models. *J. Anim. Ecol.* 78:1113–23

Patterson TA, McConnell BJ, Fedak MA, Bravington MV, Hindell MA. 2010. Using GPS data to evaluate the accuracy of state-space methods for correction of Argos satellite telemetry error. *Ecology* 91:273–85

Patterson TA, Thomas L, Wilcox C, Ovaskainen O, Matthiopoulos J. 2008. State-space models of individual animal movement. *Trends Ecol. Evol.* 23:87–94

Peckham SH, Diaz DM, Walli A, Ruiz G, Crowder LB, Nichols WJ. 2007. Small-scale fisheries bycatch jeopardizes endangered Pacific loggerhead turtles. *PLoS ONE* 2:e1041

Pinaud D. 2008. Quantifying search effort of moving animals at several spatial scales using first-passage time analysis: effect of the structure of environment and tracking systems. *J. Appl. Ecol.* 45:91–99

Pinaud D, Weimerskirch H. 2007. At-sea distribution and scale-dependent foraging behaviour of petrels and albatrosses: a comparative study. *J. Anim. Ecol.* 76:9–19

Plank MJ, Codling EA. 2009. Sampling rate and misidentification of Lévy and non-Lévy movement paths. *Ecology* 90:3546–53

Polovina J, Uchida I, Balazs G, Howell EA, Parker D, Dutton P. 2006. The Kuroshio Extension Bifurcation Region: a pelagic hotspot for juvenile loggerhead sea turtles. *Deep Sea Res. Part II* 53:326–39

Polovina JJ, Kobayashi DR, Parker DM, Seki MP, Balazs GH. 2000. Turtles on the edge: movement of loggerhead turtles (*Caretta caretta*) along oceanic fronts, spanning longline fishing grounds in the central North Pacific, 1997–1998. *Fish. Oceanogr.* 9:71–82

Rasmussen K, Palacios DM, Calambokidis J, Saborío MT, Dalla Rosa L, et al. 2007. Southern Hemisphere humpback whales wintering off Central America: insights from water temperature into the longest mammalian migration. *Biol. Lett.* 3:302–5

Riedman M. 1990. *The Pinnipeds: Seals, Sea Lions, and Walruses*. Berkeley: Univ. Calif. Press. 439 pp.

Robinson PW, Costa DP, Crocker DE, Gallo-Reynoso JP, Champagne CD, et al. 2012. Foraging behavior and success of a mesopelagic predator in the northeast Pacific Ocean: insights from a data-rich species, the northern elephant seal. *PLoS ONE* 7(5):e36728

Reynolds AM, Rhodes CJ. 2009. The Lévy flight paradigm: random search patterns and mechanisms. *Ecology* 90:877–87

Schofield G, Bishop CM, MacLean G, Brown P, Baker M, et al. 2007. Novel GPS tracking of sea turtles as a tool for conservation management. *J. Exp. Mar. Biol. Ecol.* 347:58–68

Shaffer SA, Costa DP. 2006. A database for the study of marine mammal behavior: gap analysis, data standardization, and future directions. *IEEE J. Ocean Eng.* 31:82–86

Shaffer SA, Tremblay Y, Awkerman JA, Henry RW, Teo SLH, et al. 2005. Comparison of light- and SST-based geolocation with satellite telemetry in free-ranging albatrosses. *Mar. Biol.* V147:833–43

Shaffer SA, Tremblay Y, Weimerskirch H, Scott D, Thompson DR, et al. 2006. Migratory shearwaters integrate oceanic resources across the Pacific Ocean in an endless summer. *Proc. Natl. Acad. Sci. USA* 103:12799–802

Shillinger GL, Palacios DM, Bailey H, Bograd SJ, Swithenbank AM, et al. 2008. Persistent leatherback turtle migrations present opportunities for conservation. *PLoS Biol.* 6:e171

Sims DW, Southall EJ, Humphries NE, Hays GC, Bradshaw CJA, et al. 2008. Scaling laws of marine predator search behaviour. *Nature* 451:1098–102

Siniff DB, Jessen CR. 1969. A simulation model of animal movement patterns. *Adv. Ecol. Res.* 6:185–217

Sirovic A, Hildebrand JA, Wiggins SM, McDonald MA, Moore SE, Thiele D. 2004. Seasonality of blue and fin whale calls and the influence of sea ice in the Western Antarctic Peninsula. *Deep Sea Res. Part II* 51:2327–44

Skellam JG. 1951. Random dispersal in theoretical populations. *Biometrika* 38:196

Sleeman JC, Meekan MG, Wilson SG, Polovina JJ, Stevens JD, et al. 2010. To go or not to go with the flow: environmental influences on whale shark movement patterns. *J. Exp. Mar. Biol. Ecol.* 390:84–98

Tinker MT, Costa DP, Estes JA, Wieringa N. 2007. Individual dietary specialization and dive behaviour in the California sea otter: using archival time-depth data to detect alternative foraging strategies. *Deep Sea Res. Part II* 54:330–42

Tomkiewicz SM, Fuller MR, Kie JG, Bates KK. 2010. Global positioning system and associated technologies in animal behaviour and ecological research. *Philos. Trans. R. Soc. B* 365:2163–76

Tremblay Y, Roberts AJ, Costa DP. 2007. Fractal landscape method: an alternative approach to measuring area-restricted searching behavior. *J. Exp. Biol.* 210:935–45

Tremblay Y, Robinson PW, Costa DP. 2009. A parsimonious approach to modeling animal movement data. *PLoS ONE* 4:e4711

Turchin P. 1996. Fractal analyses of animal movement: a critique. *Ecology* 77:2086–90

Turchin P. 1998. *Quantitative Analysis of Movement: Measuring and Modeling Population Redistribution in Plants and Animals*. Sunderland, MA: Sinauer Assoc.

Turcotte DL. 1997. *Fractals and Chaos in Geology and Geophysics*. Cambridge, UK: Cambridge Univ. Press

Tyack PL, Zimmer WMX, Moretti D, Southall BL, Claridge DE, et al. 2011. Beaked whales respond to simulated and actual navy sonar. *PLoS ONE* 6:e17009

Viswanathan G, Buldyrev S, Havlin S, da Luz M, Raposo E, Stanley H. 1999. Optimizing the success of random searches. *Nature* 401:911–14

Wallace BP, DiMatteo AD, Bolten AB, Chaloupka MY, Hutchinson BJ, et al. 2011. Global conservation priorities for marine turtles. *PLoS ONE* 6:e24510

Weimerskirch H, Le Corre M, Ropert-Coudert Y, Kato A, Marsac F. 2005. The three-dimensional flight of red-footed boobies: adaptations to foraging in a tropical environment? *Proc. R. Soc. Lond. Ser. B* 272:53–61

Weimerskirch H, Louzao M, de Grissac S, Delord K. 2012. Changes in wind pattern alter albatross distribution and life-history traits. *Science* 335:211–14

Weimerskirch H, Pinaud D, Pawlowski F, Bost CA. 2007. Does prey capture induce area-restricted search? A fine-scale study using GPS in a marine predator, the wandering albatross. *Am. Nat.* 170:734–43

Welch DW, Melnychuk MC, Payne JC, Rechisky EL, Porter AD, et al. 2011. In situ measurement of coastal ocean movements and survival of juvenile Pacific salmon. *Proc. Natl. Acad. Sci. USA* 108:8708–13

Weng KC, Castilho PC, Morrissette JM, Landeira-Fernandez AM, Holts DB, et al. 2005. Satellite tagging and cardiac physiology reveal niche expansion in salmon sharks. *Science* 310:104–6

Wiens JA, Crist TO, With KA, Milne BT. 1995. Fractal patterns of insect movement in microlandscape mosaics. *Ecology* 76:663–66

Wilson RP, Shepard ELC, Liebsch N. 2008. Prying into the intimate details of animal lives: use of a daily diary on animals. *Endanger. Species Res.* 4:123–37

Witt MJ, Bonguno EA, Broderick AC, Coyne MS, Formia A, et al. 2011. Tracking leatherback turtles from the world's largest rookery: assessing threats across the South Atlantic. *Proc. R. Soc. Lond. Ser. B* 278:2338–47

Zbinden JA, Bearhop S, Bradshaw P, Gill B, Margaritoulis D, et al. 2011. Migratory dichotomy and associated phenotypic variation in marine turtles revealed by satellite tracking and stable isotope analysis. *Mar. Ecol. Progr. Ser.* 421:291–302

Žydelis R, Lewison RL, Shaffer SA, Moore JE, Boustany AM, et al. 2011. Dynamic habitat models: using telemetry data to project fisheries bycatch. *Proc. R. Soc. B* 278:3191–200

The Biogeography of Marine Invertebrate Life Histories

Dustin J. Marshall,[1,2] Patrick J. Krug,[3]
Elena K. Kupriyanova,[4] Maria Byrne,[5]
and Richard B. Emlet[6]

[1] School of Biological Sciences, Monash University, Clayton, Victoria 3800, Australia; email: dustin.marshall@monash.edu

[2] School of Biological Sciences, The University of Queensland, Brisbane, Queensland 4072, Australia

[3] Department of Biological Sciences, California State University, Los Angeles, California 90032; email: pkrug@exchange.calstatela.edu

[4] Marine Invertebrates, Australian Museum, Sydney, New South Wales 2010, Australia; email: Elena.Kupriyanova@austmus.gov.au

[5] School of Medical and Biological Sciences, The University of Sydney, New South Wales 2006, Australia; email: mbyrne@anatomy.usyd.edu.au

[6] Oregon Institute of Marine Biology, The University of Oregon, Charleston 97420; email: remlet@uoregon.edu

Annu. Rev. Ecol. Evol. Syst. 2012. 43:97–114

First published online as a Review in Advance on August 28, 2012

The *Annual Review of Ecology, Evolution, and Systematics* is online at ecolsys.annualreviews.org

This article's doi: 10.1146/annurev-ecolsys-102710-145004

Keywords

complex life-cycles, egg size, maternal effects, meta-analysis, offspring size

Abstract

Biologists have long sought to identify and explain patterns in the diverse array of marine life histories. The most famous speculation about such patterns is Gunnar Thorson's suggestion that species producing planktonic larvae are rarer at higher latitudes (Thorson's rule). Although some elements of Thorson's rule have proven incorrect, other elements remain untested. With a wealth of new life-history data, statistical approaches, and remote-sensing technology, new insights into marine reproduction can be generated. We gathered life-history data for more than 1,000 marine invertebrates and examined patterns in the prevalence of different life histories. Systematic patterns in marine life histories exist at a range of scales, some of which support Thorson, whereas others suggest previously unrecognized relationships between the marine environment and the life histories of marine invertebrates. Overall, marine life histories covary strongly with temperature and local ocean productivity, and different regions should be managed accordingly.

1. INTRODUCTION

Geographical patterns in life-history traits such as body size, cell size, and maternal investment have inspired hypotheses for over a century of biological research (Allen 1877, Bergmann 1847, Moles & Westoby 2003, Thorson 1936). In the marine environment, identifying and understanding life-history patterns have a particular significance because of the otherwise bewildering array of modes of reproduction, developmental modes, and life histories in the sea (Strathmann 1985). Terrestrial life histories tend to map strongly to phylogeny, but marine life histories show tremendous variation that can be completely free of phylogenetic constraints. For example, congeners can vary from external fertilization with tiny long-lived, feeding larvae, to internal fertilization and no larval stage and very large offspring (Byrne 2006). Such variation demands exploration and explanation; so from the very beginning of marine larval studies, biologists have sought to identify patterns in life histories (for an excellent review, see Young 1990). Today, both managing and understanding our marine systems rely more than ever on the identification of patterns in life-history variation.

From an ecological perspective, the identification of geographical variation in marine life histories should lead to more effective management (Palumbi 2003). The ecological dynamics of any marine species is affected by its life history: Species with long-lived, far-dispersing larvae can have different population dynamics from species with short-lived larvae (Eckert 2003, Kinlan & Gaines 2003; but see Weersing & Toonen 2009). Similarly, species with highly dispersive larvae are likely to respond to natural and anthropogenic disturbances differently from species with larvae that spend only a few minutes in the water column (Levin 1984). If geographical variation in life histories exists, then management practices developed in one region may be inappropriate for another. For example, it has been suggested that the western coast of the United States is unusual for its preponderance of planktotrophic species (Goddard 1992). Much work on the spatial scales of connectivity in marine systems (e.g., Becker et al. 2007, Kinlan & Gaines 2003) and on marine life-history evolution (Strathmann 1987) comes from this region, but the generality of this research to other regions remains unclear.

From an evolutionary perspective, geographical patterns in life-history strategies may provide clues as to the selection pressures acting upon marine life histories (Thorson 1950). There has been much speculation on the advantages and disadvantages of a larval phase in marine organisms (Pechenik 1999; Strathmann 1974, 1993) and on whether mothers should produce many small larvae or a few large offspring (Smith & Fretwell 1974, Vance 1973). Although many attempts to address these problems have been made and progress has been steady, our understanding of the selection pressures that favor different strategies remains remarkably incomplete (Marshall & Morgan 2011). By identifying the conditions that are more commonly associated with some life-history strategies but not others, we may be able to infer how selection acts on reproduction and dispersal in the sea. For example, recent theory predicts that temperature should have a fundamental influence on marine invertebrate life histories (O'Connor et al. 2007) such that systematic variation among species across temperature gradients should be expected. Caution must be exercised, however. A genuine understanding of the evolutionary processes that generate macroevolutionary patterns must come from intraspecific studies (Bernardo 1996), and it must consider phylogenetic constraints (Collin 2004, Eckelbarger & Watling 1995, McHugh & Rouse 1998). For example, differential extinction or chance colonization events could drive spatial patterns in life history in apparently adaptive ways (Poulin & Feral 1996, Uthicke et al. 2009).

Perhaps the most famous speculation about the geography of marine life histories comes from the great Danish larval biologist Gunnar Thorson, who suggested that species from polar and deep-sea regions rarely, if ever, have planktonic development (Thorson 1936, 1946, 1950). He suggested that "very limited periods of continuous phytoplankton production in connection with very low water temperatures" (Thorson 1950, p. 25) made conditions inhospitable to a larval phase. Because larvae are small and relatively vulnerable, the pelagic environment is a dangerous place (Morgan 1995). Thus, conditions that extend the larval period, low temperatures, or limited food likely increase larval mortality and select against a larval phase (O'Connor et al. 2007, Vance 1973). Thorson also suggested that species with pelagic, nonfeeding larvae were very rare and "constitute a rather small percentage of invertebrate species in temperate and warm seas, but are apparently absent from high-arctic seas" (Thorson 1946, p. 477). Importantly, most discussions of these ideas have centered on considering polar regions versus the rest of the world (Pearse 1994, Poulin & Feral 1996). Although Thorson certainly viewed polar regions as being particularly different, he also suggested that planktotrophy was much more common in the tropics than in temperate regions (Thorson 1946). Some of these ideas gathered widespread appeal, achieving paradigm status, and together they are sometimes referred to as "Thorson's rule" (Mileikovsky 1975). Today, Thorson's rule has less support (Pearse 1994). Indeed, it is now clear that species with pelagic larvae are present at both poles (Thorson believed species with pelagic larvae were absent from Antarctic waters) and in the deep sea (Clarke 1992, Pearse 1994). Perhaps the biggest problem with Thorson's suggestions was that they were so absolute in nature—too often was the term "all" used in describing the patterns that he saw. Such language seems unwarranted given the likelihood of even one exception. Indeed, Thorson himself seemed troubled by the exceptions and noted that despite pelagic larvae being "suppressed" in the high arctic sea, some species with planktotrophic pelagic larvae "are among the most dominant animals of the high arctic coastal zones" (Thorson 1946, p. 434).

Over the ensuing ~70+ years, evidence accumulated that led opinions to shift against (or at least modify) Thorson's rule. Both Young and Pearse give comprehensive accounts of the history of challenges and modifications to Thorson's hypotheses (Pearse 1994, Young 1990). Suffice to say, evidence contradicting Thorson's rule now seems so strong that Pearse suggested, "Thorson's rule should be laid to rest, and Thorson should be remembered for his stimulating hypotheses that generated so many contributions in marine biology" (Pearse 1994, p. 26). It is now generally held that nonfeeding development predominates in both the poles and the deep sea, whereas more species produce feeding larvae in warm or temperate shallow waters (Clarke 1992, Pearse 1994, Pearse & Lockhart 2004). Since the 1990s, the biogeography of marine invertebrate life histories has been largely put aside (for some recent exceptions, see Collin 2003, Fernandez et al. 2009, Laptikhovsky 2006, Poulin & Feral 1996), the matter now seemingly resolved and strong cautions against generalizations advised (Pearse 1994).

We believe, however, that there are now many reasons for revisiting the biogeography of reproduction in the sea. First, in the years since the last comprehensive global review (Emlet et al. 1987), a wealth of new data on the life history of marine invertebrates has accumulated (e.g., Anthes & Michiels 2007, Collin 2003, Fernandez et al. 2009, Kohn & Perron 1994, Marshall & Keough 2008a, McEdward & Miner 2001, Pearse 1994, Strathmann 1987, Wilson 2002), allowing a more comprehensive treatment of groups previously overlooked. Second, there have been few formal statistical treatments of marine invertebrate biogeography. Emlet et al. (1987) and Collin (2003) analyzed correlations between offspring size and latitude with linear regression and chi square tests, but for the most part, assessments of Thorson's rule have lacked a formal statistical framework. Today, more targeted statistical approaches such as logistic regression are available that can directly address the relationship between developmental modes and latitude in a quantitative framework. Instead of viewing polar regions as "all or nothing" with regards to particular developmental modes, we can ask whether there is any statistically significant relationship between latitude and the distribution of alternative developmental modes. Recently,

Fernandez et al. (2009) used a more sophisticated statistical approach to model the molluscan and decapod species richness in different developmental modes along the coast of Chile to great effect. Third, latitude was used as a proxy for specific environmental variables because detailed biophysical data were previously unavailable. Thorson believed that both temperature and food availability decreased at higher latitudes and that one or both of these variables drove the patterns in marine invertebrate life histories (Thorson 1936, 1946); however, the poles are not foodless deserts as Thorson believed, and chlorophyll levels at any one time are not negatively correlated with latitude (see database available at **http://dx.doi.org/10.5061/dryad.m7j72**). Today, we can use remote sensing technology to directly test these hypotheses, using satellite imaging to estimate sea-surface temperatures and phytoplankton concentrations. Indeed, Collin (2003) used sea temperature data to examine latitudinal patterns in a group of gastropods. Fernandez et al. (2009) combined temperature data with chlorophyll a measurements to explore the biogeography of different developmental modes in molluscs and crabs along the Chilean coast and found that both temperature and productivity played a role in molluscs and anomuran crabs but not brachyuran crabs. With access to unprecedented levels of environmental data, more sophisticated statistical analyses, and a wealth of new life-history data, a re-examination of Thorson's rule should provide new insights into patterns of reproduction in the sea. For details of how we compiled life-history and environmental data and defined developmental groups and our analytical approach, see **Supplemental Text 1**. (Follow the **Supplemental Material link** from the Annual Reviews home page at **http://www.annualreviews.org**.)

2. DISTRIBUTION OF DEVELOPMENTAL MODES ACROSS LATITUDES

Before discussing our findings, we should first acknowledge an important limitation of our meta-analysis. **Figure 1** shows that, despite the unprecedented size of our data set, it is extremely restricted geographically; most studies come from just a few, well-studied regions. Our findings are summarized in **Tables 1** and **2**. There is a higher fraction of aplanktonic species in the Southern Hemisphere than the Northern Hemisphere (north, 8%; south, 18%; $\chi^2 = 26.2$, P < 0.001), but of the species with planktonic larvae, identical fractions have feeding larvae (60%) in both hemispheres. The association between latitude and the fraction of species with planktonic larvae differs between hemispheres ($\chi^2 = 27.8$, P < 0.001). The fraction of aplanktonic species increases strongly with latitude in the south ($\chi^2 = 36.6$, P < 0.001), but if exclusively deep-water species are excluded, there is no relationship in the north (**Figure 2**). The latitudinal relationship also varies with phylum, indicated by a strong phylum $\times$ latitude $\times$ hemisphere interaction ($\chi^2 = 20.7$, P < 0.001), driven by a consistent lack of a relationship in the north but variation in the relationship among phyla in the south. Both southern echinoderms and molluscs show a strong relationship with latitude but no relationship is evident in annelids.

Of the species with pelagic larvae, a far lower fraction are feeding at higher latitudes ($\chi^2 = 41.3$, P < 0.0001) (**Figure 3**). This relationship is consistent between hemispheres (latitude $\times$ hemisphere: $\chi^2 = 0.022$, P = 0.882) and among phyla (phylum $\times$ latitude: $\chi^2 = 0.23$, P = 0.627), though the relationship appears strongest in molluscs and echinoderms.

3. VARIATION IN OFFSPRING SIZE AMONG DEVELOPMENTAL MODES ACROSS LATITUDES

As has been demonstrated repeatedly for different taxa, developmental mode is an excellent predictor of offspring size. Aplanktonic species produce the largest offspring, those with planktonic

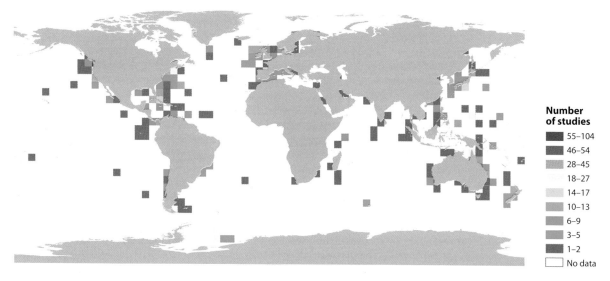

Figure 1

A heat map of the distribution of studies used in this review to examine geographical variation in marine invertebrate life histories. Warmer colors indicate regions from which many species have been studied, cooler colors indicate areas from which only a few species have been studied, and white areas indicate areas from which we have no data. The vast majority of the marine environment remains unstudied, and our view of marine life histories comes from only a small fraction of those studies that exist. It should be noted that some areas are likely to have been studied, but these studies are largely inaccessible to the authors owing to language differences or limited accessibility.

nonfeeding larvae have intermediate offspring sizes, and those with feeding larvae have the smallest offspring sizes (**Figure 4**). Within aplanktonic species and those with nonfeeding larvae, offspring are generally larger in the Southern Hemisphere relative to the Northern Hemisphere, but species with feeding larvae are similar in size (**Figure 5**). The rank differences in offspring size among developmental modes were consistent across phyla (**Figure 4**), but within any developmental mode, there are significant differences in offspring size among phyla ($F_{6,1091} = 34.3$, $P < 0.001$). For example, among the aplanktonic species, echinoderm eggs were the largest and annelids eggs were the smallest; among the planktonic, nonfeeding species, the eggs of echinoderms were again the largest, but mollusc eggs were the smallest.

Table 1 Summary table of the relationship between the fraction of species with each development mode and various biophysical variables in marine invertebrates

	Aplanktonic	Planktonic nonfeeding	Planktonic feeding
Hemisphere	✓	✗	✗
Latitude	✓ (south only)	✓ (+)	✓ (−)
SST	✓ (−)	✓ (−)	✓ (+)
Ch a	✓ (−)	✗	✗
SST × Ch a	✓	✗	✗

Key: Ticks, significant relationship; crosses, no relationship; − and +, negative or positive relationship between the predictor and response variable, respectively. Abbreviations: Ch a, chlorophyll a; SST, sea surface temperature.

Table 2 Summary table of the relationship between offspring size and various biophysical variables in marine invertebrates

	Aplanktonic	Planktonic nonfeeding	Planktonic feeding
Hemisphere	✓	✓	✗
Phylum	✓	✓	✗
Latitude	✓ (+)	✓ (+)	✓ (+)
Sea surface temperature	✓ (−)	✓ (−)	✓ (−)
Chlorophyll a	✗	✓ (−)	✗

Key: Ticks, significant relationship; crosses, no relationship; − and +, negative or positive relationship between the predictor and response variable, respectively.

Offspring size increases with latitude across all developmental modes, but the relationship varies among modes ($F_{2,1035} = 141.3$, $P < 0.001$). The steepest relationship between offspring size and latitude occurred in aplanktonic species, and the shallowest relationship (though still significantly different from zero: $P = 0.005$) occurred in species with feeding larvae (**Figure 6**). Although the relationship between offspring size and latitude was positive in all phyla, it varied in slope among phyla ($F_{4,962} = 23.01$, $P < 0.001$). The steepest relationships occurred in echinoderms and molluscs; among annelids the relationship was much weaker. Taxonomic class further influenced the relationship between latitude and offspring size (class × latitude × development mode: $F_{10,953} = 28.3$, $P < 0.001$); in fact, this interaction was driven by the lack of a relationship between latitude

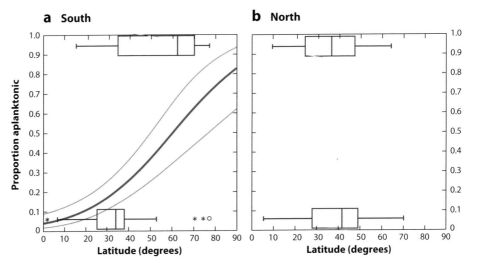

a   South   **b   North**

Latitude (degrees)

Proportion aplanktonic

Figure 2

Proportion of species with aplanktonic development across latitude for (*a*) the Southern Hemisphere and (*b*) the Northern Hemisphere. There is no relationship between latitude and the prevalence of aplanktonic development in the Northern Hemisphere. The dark blue line indicates the line of best fit generated from a logistic regression, and light blue lines indicate upper and lower 95%-confidence intervals. Boxplots indicate the distribution of species across latitude in each developmental mode. Asterisks indicate points outside the interquartile range, and circles indicate points greatly outside of the interquartile range.

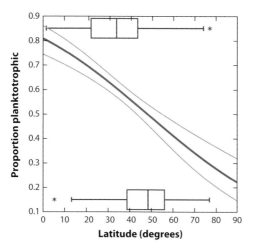

Figure 3

Proportion of species with planktonic development that have feeding larvae across latitude (pooled for both hemispheres). The dark blue indicates the line of best fit generated from a logistic regression, and light blue lines indicate upper and lower confidence intervals. Boxplots indicate the distribution of species across latitude in each developmental mode. Asterisks indicate points outside the interquartile range.

and offspring size in holothuroids and ophiuroids but by more complex associations in polychaetes. In polychaetes, there is a weak, but significantly positive relationship between latitude and offspring size, but this relationship varies in slope between hemispheres (latitude × hemisphere: $F_{1,228} = 8.43$, $P = 0.004$). There is a strong, positive relationship between latitude and offspring size in all developmental modes in the Southern Hemisphere polychaetes ($F_{1,34} = 6.008$, P 0.02), but for Northern Hemisphere polychaetes, the relationship is marginally nonsignificant ($F_{1,194} = 2.87$, $P = 0.09$).

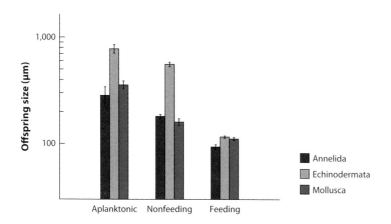

Figure 4

The size of offspring (estimated as egg diameter) from three developmental modes across the three phyla (Annelida, Echinodermata, and Mollusca) for which we have the most complete data. Each bar indicates the mean (± standard error) offspring size for each developmental mode. Note the y-axis shows a $\log_{10}$ scale.

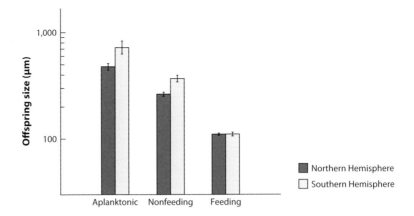

Figure 5

The size of offspring from three developmental modes across two hemispheres. Each bar indicates the mean (± standard error) offspring size for each developmental mode. Note the y-axis shows a $\log_{10}$ scale.

4. REGIONAL EFFECTS

The use of ANCOVA (analysis of covariance) accounts for the effect of latitude, but there were still regional differences in the relative fractions of each developmental mode and in egg sizes (excluding regions that spanned a narrow range of latitudes such as the Caribbean and Antarctica). There were strong region × latitude interactions affecting the fraction of species with pelagic larvae ($\chi^2 = 5.03, P = 0.025$) and the fraction of larvae that were feeding ($\chi^2 = 7.25, P = 0.007$). For both response variables, latitudinal gradients in developmental mode were much steeper in Australia than in North America, and these regions appeared to drive much of the interaction.

There were also regional differences in egg size within developmental modes. Offspring size differed among regions for species with planktotrophic larvae ($F_{5,453} = 7.85, P < 0.001$) and

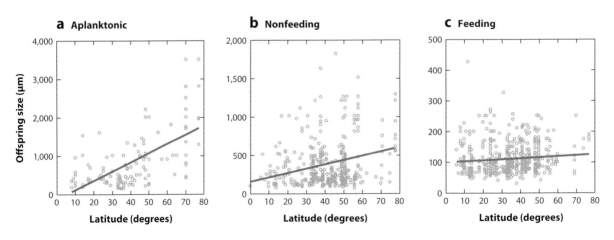

Figure 6

The relationship between offspring size (estimated as egg diameter) and latitude for each developmental mode. Each point represents the mean of individual species and the dark blue line indicates the line of best fit from a linear regression. Panels show (*a*) the relationship for species with aplanktonic development, (*b*) the relationship for species with nonfeeding larvae, and (*c*) the relationship for species with feeding larvae. Note that in all three developmental modes, there is a significantly positive relationship between offspring size and latitude.

also for species with pelagic lecithotrophic development ($F_{5,315} = 3.81, P = 0.002$). The largest nonfeeding larvae are in Australia (mean size = 480 μm), and the smallest nonfeeding larvae are in North America (mean size = 250 μm). The largest feeding larvae are in South America (mean size = 145 μm), and the smallest feeding larvae are in Australia (mean size = 90 μm). There was no influence of region on egg size in aplanktonic species.

5. THE ROLE OF EXTREMES: EXCLUDING THE POLES AND/OR THE TROPICS

The latitudinal patterns of developmental modes are largely unaffected by whether we include the poles and the tropics. Excluding neither the poles (>60°) nor the tropics (<30°) has no effect on the pattern of higher fractions of aplanktonic species at higher latitudes. Similarly, the relationship between the prevalence of species with feeding larvae and latitude remains regardless of whether the tropics and poles are included.

Within each developmental mode, trends in offspring size were also fairly robust to the influence of low and high latitudes. When the tropics were excluded, there was still a strong positive association between offspring size and latitude across all developmental modes. When the poles were excluded (both with and without the tropics), the relationships between offspring size and latitude remain for aplanktonic species and species with nonfeeding larvae, but was not detected in species with feeding larvae.

6. THE ROLE OF SEA SURFACE TEMPERATURE AND OCEAN PRODUCTIVITY

There is a strong association between the prevalence of each developmental mode and local temperature and productivity. The fraction of species with aplanktonic development varied with both temperature and chlorophyll concentration (chlorophyll × temperature: $\chi^2 = 7.22$, $P = 0.007$). Lower temperatures were associated with less planktonic development, and this association was stronger at low productivity levels (**Figure 7**): Planktonic larvae were more common where both temperature and productivity were high, whereas areas with the lowest productivity and temperature had the most aplanktonic species. The fraction of species with feeding larvae increased with temperature ($\chi^2 = 27.0, P < 0.001$) but not with productivity ($\chi^2 = 0.77$, $P = 0.381$).

Within each developmental mode, the relationship between temperature, productivity, and offspring size is variable (developmental mode × temperature: $F_{2,809} = 109.9, P < 0.0001$; developmental mode × chlorophyll: $F_{2,809} = 4.76, P = 0.009$). In aplanktonic species or species with feeding larvae, as temperature increases, offspring size decreases (aplanktonic: $t_{87} = 8.23$, $P < 0.001$; feeding: $t_{408} = 3.37, P = 0.001$). In species with nonfeeding larvae, larger offspring sizes were associated with lower temperatures and also with lower levels of productivity (temperature: $t_{315} = 5.35, P < 0.001$; chlorophyll: $t_{315} = 3.17, P = 0.002$; **Figure 8**).

7. PHYLOGENETICALLY CONTROLLED ANALYSES

For 83 species of sacoglossan sea slugs, evolutionary history had little effect on the patterns we saw. Model fit did not improve when development mode and latitude were allowed to covary (**Supplemental Table 1**), supporting a model of uncorrelated trait evolution and indicating that lecithotrophic species were not more abundant in the tropics. Models of egg-size evolution with no phylogenetic correction ($\lambda = 0$) were preferred over models in which λ was jointly estimated

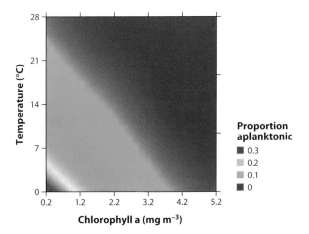

Figure 7

The relationship between the proportion of species showing aplanktonic development and the average sea surface temperature and chlorophyll a levels in the region in which the species was studied. The figure shows a heat plot for the plane of best fit generated from logistic multiple regressions, where warmer colors indicate a higher proportion of species with aplanktonic development and cooler colors indicate a lower proportion of species with aplanktonic development. When temperatures are warmer and food availability is greater, planktonic development is more common.

(log-BF test = 3.7), indicating no phylogenetic effect on egg size exists in this group. Among species with lecithotrophic development, there was a significant and positive relationship between egg size and latitude even when correcting for phylogenetic effects (**Supplemental Table 1**). There was no relationship between egg size and latitude for planktotrophic sacoglossans.

For 40 species of Calyptraeid gastropods, phylogenetic effects were more pronounced. There was no latitudinal pattern in the distribution of development modes (**Supplemental Table 1**).

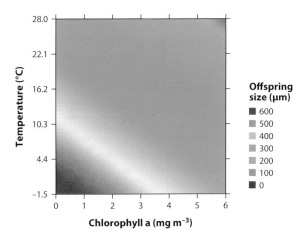

Figure 8

The relationship between offspring size and average sea surface temperature and chlorophyll a levels in the region in which the species was studied for species with nonfeeding larvae. The figure shows a heat plot for the plane of best fit generated from multiple regressions, where warmer colors indicate larger offspring sizes and cooler colors indicate smaller offspring sizes. When temperatures are cooler and food availability is lower, offspring tend to be larger.

Models of egg-size evolution that included a phylogenetic correction were strongly preferred over models in which $\lambda = 0$ (log-BF test $= 12.1$), indicating a strong phylogenetic effect. In contrast to the overall pattern for molluscs, planktotrophic egg size was significantly but negatively correlated with latitude (**Supplemental Table 1**); however, this relationship was apparent with or without the phylogenetic correction. There was no correlation between lecithotrophic egg size and latitude.

8. DISCUSSION

The geographic distributions of the life histories of marine invertebrates differ greatly between the Northern and Southern Hemispheres. The Northern Hemisphere is characterized by low levels of aplanktonic development and much smaller offspring sizes within aplanktonic and nonfeeding pelagic species. The Southern Hemisphere has twice as many aplanktonic species as the Northern Hemisphere and also appears to show much stronger latitudinal trends than the Northern Hemisphere. The increase in the fraction of aplanktonic species at higher southern latitudes could be explained by the general idea that Thorson put forward, that the cold Antarctic waters would be inhospitable to larvae; but it should be remembered that $\sim 20\%$ of the species in the Antarctic have a larval phase, and in the Arctic, more than 70% of species have a larval phase. Our finding of stronger latitudinal patterns in the Southern Hemisphere relative to the Northern Hemisphere echoes earlier work on echinoids and asteroids (Clarke 1992, Emlet et al. 1987, Pearse 1994). It appears that, for a range of phyla, Thorson's idea that species with planktonic larvae are rarer at high latitudes only applies in the Southern Hemisphere. Why there should be a significant relationship in the south but no relationship in the north is unclear—differences in phylogeny, evolutionary history, or oceanography could drive patterns at this scale. We suspect that a combination of oceanography and differences in the representation of higher latitudes among hemispheres drove this pattern. Boundary currents moving along the edges of ocean basins carry water from different latitudes along the coast, reducing the latitudinal effect on temperature (and therefore, presumably offspring size) in some places. In the Northern Hemisphere, boundary currents and the (relatively) poor sampling of very high latitudes resulted in a smaller temperature range being sampled than in the Southern Hemisphere. Thus, latitude is a poorer proxy for temperature in the north than the south in our data set (R^2 for latitude-temperature relationship in the north $= 0.913$, and south $= 0.949$). Importantly, when temperature, rather than latitude, was used as a predictor for the prevalence of planktonic species, we found no differences between hemispheres. Thorson's suggestion that planktotrophy is rarer at the poles does stand the test of time: In both hemispheres, there is a marked decrease in the fraction of species with feeding larvae at higher latitudes.

Thorson's predictions were less accurate for species with nonfeeding larvae. As noted elsewhere (Clarke 1992, Pearse 1994), there is a marked increase in species with pelagic nonfeeding larvae moving from the equator to the poles. This is in stark contrast with Thorson's assertion that pelagic nonfeeding larvae were absent from the poles, a notion that troubled him: "...that lecithotrophic larvae seem to be totally absent...seems at first to be the opposite of what might be expected. At the outset it would seem that precisely this type of larval development...would be well fit for the life in arctic seas" (Thorson 1946, p. 437). As Thorson intuited, the majority of species in the poles indeed do have nonfeeding larvae. More generally, we were surprised by the high fraction of species with nonfeeding larvae: In both hemispheres, 40% of species with planktonic larvae have lecithotrophic development. If we were to include phyla such as sponges and bryozoans (i.e., those with almost exclusively nonfeeding larvae), this fraction would be much higher, suggesting that nonfeeding larval development is the dominant mode of reproduction in coastal marine invertebrates.

Prior consideration of Thorson's predictions and ensuing discussions of latitudinal patterns (Clarke 1992, Pearse 1994, Poulin & Feral 1996) focused largely on patterns driven by the poles (but see Collin 2003, Emlet et al. 1987). However, our analyses support a more nuanced view of how life-history strategies are distributed across the globe. As Thorson first suggested, there are also patterns in temperate and tropical seas: Both pelagic development and feeding larvae are rarer at higher latitudes (Thorson 1946, 1950). This pattern exists even when considering only the temperate latitidudinal band of 30°–60°. Thus, planktonic development and planktotrophy both decrease in prevalence moving poleward, but this pattern is not driven solely by adaptation to the extreme environment of the polar oceans. These latitudinal patterns within the temperate band are perhaps not surprising given similar intraspecific patterns in offspring size (Marshall et al. 2008). Overall, it seems that we must not only account for variation in reproduction and development between the poles and the rest of the world, but also within temperate regions.

It is difficult to account for the differences in offspring sizes between hemispheres. Temperatures for a given latitude are generally cooler in the Southern Hemisphere, but only slightly. Interestingly, the differences in offspring size among hemispheres were restricted to species with nonfeeding larvae, which suggests that simple temperature effects may not explain the differences. Based on the biophysical data included in our analyses, the Southern Hemisphere is also less productive than the Northern Hemisphere (Northern Hemisphere productivity is largely driven by the productive western coast of the US and the limited representation of the productive South American coast in our data set) (**Figure 1**). Southern Hemisphere mothers may, therefore, provide higher levels of provisioning for their offspring to offset the lack of food available to offspring. Alternatively, phylogenetic effects could be driving the differences in offspring size among hemispheres.

The inclusion of biophysical variables in our analyses provided key insights into marine life-history patterns. As we expected, there was a strong negative relationship between offspring size and temperature. Physiologists have long recognized that temperature affects many life-history traits, including offspring size (Von Bertanffy 1960, Woods 1999). It could be that the patterns in offspring size we observed with temperature are simple physiological side effects of cooler temperatures (Bownds et al. 2010, Fischer et al. 2003, Van der Have & de Jong 1996). Van der Have & de Jong (1996) showed that, because growth (cell size increases) and differentiation (cell number increases) rates are differentially affected by temperature, larger sizes at metamorphosis are inevitable. How these physiological processes affect offspring size in marine invertebrates remains unclear, but we suspect that the pattern is driven by more than simple physiology given that the relationship between temperature and offspring size was so variable among developmental modes. Furthermore, the few intraspecific studies that have examined temperature-induced offspring size changes in marine invertebrates and fish suggest that they do indeed have an adaptive basis (Bownds et al. 2010, Burgess & Marshall 2011, Salinas & Munch 2012). Nevertheless, we suggest that future studies examine the complex interplay between temperature, development, and offspring size to delineate the role of physiological processes in generating these patterns (Van der Have & de Jong 1996).

Perhaps the most exciting result of our meta-analysis is the combined associations of temperature and productivity with marine invertebrate life-history patterns. Thorson was incorrect in assuming that the polar waters are completely food limited, but he was correct in thinking that both temperature and food influence the incidence of planktonic development. Our results in this regard are intuitively appealing: Planktonic larvae are more common when food levels and temperature are high, and these are conditions that allow rapid development and, hence, minimize larval mortality due to advection, predation, and starvation (Morgan 1995, Vance 1973). The interaction between temperature and productivity (and by inference, larval food) is particularly interesting: For a given temperature, aplanktonic development is more common under conditions of lower productivity. A surprising result was that food affects the fraction of species with free-swimming pelagic larvae but not the incidence of planktotrophy; this finding suggests that productivity affects selection on the presence of a larval phase but not on larval feeding. This finding is in contrast to that of Fernandez et al. (2009), where productivity affects the species richness of planktotrophic molluscs and anomuran species along the Chilean coast. Our measure of productivity was very coarse; however, based on mean productivity across two summer and two winter months. Future studies of more fine-scale estimates of productivity should provide further insight into how it shapes marine life histories.

Our meta-analysis showed that in species with nonfeeding larvae, lower levels of productivity are associated with much larger offspring sizes. Differences in productivity have previously been implicated in shifts in offspring size and developmental mode over evolutionary time (Uthicke et al. 2009). For example, the rise of the Panamanian Isthmus resulted in high levels of productivity on the Pacific side and low levels in the Caribbean. In geminate pairs of bivalves and echinoderms, Pacific species produced much smaller eggs compared to species in the nutrient-poor Caribbean (Lessios 1990, Moran 2004).

We found no overall effect of phylogeny on egg size in sacoglossan sea slugs, but there was a strong influence of phylogeny on Calyptraeid egg sizes. The lack of phylogenetic effect is perhaps not surprising given the frequent evolutionary transitions between development modes in sacoglossans (Krug 2009). The effect estimated for Calyptraiedae contrasts with results by Collin (2004), who found no such effect; differences in methodology (Markov chain Monte Carlo versus maximum likelihood) or phylogenetic resolution may explain the different outcomes. In each group, however, patterns within a development mode were detectable with or without a phylogenetic correction. Latitude was positively related to egg size for lecithotrophic sacoglossans but was negatively correlated with egg size for planktotrophic Calyptraeidae. These trends were seen in conventional statistical analyses of each group and were unaffected by inclusion of a formal phylogenetic correction in models of trait evolution, suggesting that the trends we report from the full data set are not artifacts of shared history within groups. More comparative studies are clearly needed to identify clades like the Calyptraeidae and *Conus*, within which trends run counter to those described for the rest of a given phylum; such exceptions may prove useful in identifying causal factors that drive trends in egg size. Overall, our efforts to include phylogenetic analyses were hampered by a poor overlap between those species for which we have life-history data and those species for which molecular phylogenies exist, highlighting a need for future efforts to bridge this gap.

9. SELECTION ACROSS THE LIFE HISTORY

Thorson and subsequent investigators sought to explain marine life-history patterns by focusing on selection acting on planktonic larvae (but see Havenhand 1995, Marshall & Morgan 2011). However, selection across the entire life history must be considered to advance our understanding of how larval strategies evolve. High planktonic mortality rates certainly impose strong selection on larvae, but so too must forces acting on pre- and postlarval stages (Marshall & Morgan 2011, Roughgarden 1989). Within a species, offspring size-specific selection can act at both fertilization and postmetamorphic performance (Marshall & Keough 2008a). More generally, organisms with complex life cycles are likely to express traits that are the product of complementary and conflicting selection pressures across the entire life history (Schluter et al. 1991). Multiple lines of evidence in our meta-analysis suggest that environmental influences on multistage selection may be responsible for interspecific variation in offspring size in marine invertebrates that we observed, and below we outline each.

The relationship between offspring size and temperature was not consistent among developmental modes and was steepest in two modes in which no larval feeding takes place. If simple physiological effects drove the relationship between temperature and offspring size, it seems reasonable to expect similar slopes across all developmental modes. Because species with feeding larvae occupy a narrower band of egg sizes, this group may simply be more constrained with regard to the slope of the relationship between temperature and offspring size. One theory predicts metamorphosis should be more costly at cooler temperatures (Rombough 2006), but this idea remains untested in marine invertebrates and should be explored. Alternatively, selection for increased size postmetamorphosis may be stronger in cooler temperatures. In species with no larval feeding stage, there is a positive association between egg size and size at metamorphosis (Collin 2003, Emlet et al. 1987). As noted by several other researchers (Emlet et al. 1987, Pearse et al. 1991), nonfeeding larvae can complete metamorphosis coming from eggs smaller than 200 microns, and so any increase in offspring size beyond this may be used solely to increase performance and resilience postmetamorphosis. We saw a strong association between offspring size and temperature in species with no larval stage and in species with nonfeeding larvae, suggesting that selection favors larger juveniles at lower temperatures. Although this is an intriguing hypothesis, only one study has formally examined selection on offspring size across a latitudinal gradient. Marshall & Keough (2008b) found postmetamorphic selection favored larger offspring at higher latitudes, but the effect of latitude was not due solely to temperature: Selection actually favored smaller offspring in winter than summer. We therefore echo earlier calls (Havenhand 1995, Pearse et al. 1991, Wray 1995) for studies that formally examine variability in postmetamorphic selection on offspring size across environmental gradients.

Productivity affected offspring size in species with nonfeeding larvae; lower levels of productivity resulted in larger offspring sizes in this group. Many studies within species show that mothers that experience limited food often produce larger offspring because selection for increased offspring size intensifies when food is scarce (Allen et al. 2008, Bashey 2006, Fox et al. 1997). In areas with lower productivity, presumably food availability is lower (particularly for filter feeders that rely on phytoplankton), and in these areas, mothers may produce larger eggs in order to compensate for poorer food conditions that their offspring will encounter postmetamorphosis (Allen et al. 2008, Fox et al. 1997). This explanation assumes that lower productivity as measured by chlorophyll a is a good predictor of food availability more generally: This assumption will be more appropriate for some species than others. The pattern of larger offspring in species with nonfeeding larvae in conditions of lower productivity is repeated at the regional scale: Adjusted for latitude, North America has the highest measures of chlorophyll a, whereas Australia has the lowest. These regions were also the site of the smallest and largest nonfeeding larvae, respectively. To test the hypothesis that lower food availability selects for larger size at metamorphosis, formal tests across naturally varying food gradients or experimental manipulations are required (Fox et al. 1997).

10. MANAGEMENT IMPLICATIONS

Our results are relevant for managers: The life histories of marine invertebrates vary predictably among regions and along latitudinal gradients, and different regions should be managed accordingly (Kelly & Eernisse 2007, O'Connor et al. 2007). For most benthic animals, larval development mode determines the spatial scale of migration among demes. Understanding larval connectivity is thus critical for spacing of marine protected areas to ensure genetic and demographic exchange among reserves and replenishment of adjacent areas (Botsford et al. 2001, Palumbi 2003, Shanks et al. 2003, White et al. 2010). A mismatch between the typical larval dispersal kernal and the spacing of marine reserves can threaten population persistence and recolonization dynamics on ecological timescales (Hastings & Botsford 2006). Differences in life-history patterns among regions and latitudes may drive predictable changes in connectivity among reserves, particularly across latitudinal bands; higher latitudes are poor in species with dispersive, feeding larvae, thus high-latitude reserves may experience reduced connectivity compared to tropical regions. Such predictions are overly simplistic, however; for example, nonfeeding larvae that develop in extreme cold can have longer pelagic durations than feeding larvae developing in warm tropical waters (Pearse et al. 1991). On balance, however, we expect cooler regions to have more poorly connected populations than the tropics owing to the increased fraction of species with aplanktonic and nonfeeding larvae. Note that our expectation, at first glance, seems to contradict that of O'Connor et al. (2007), who suggested that, all being equal, connectivity should be greater at the poles relative to the tropics because developmental rates are much faster in the tropics. The conclusions of O'Connor et al. (2007) certainly hold for any one developmental mode: Within each developmental mode, we would expect greater connectivity among distant populations in poles relative to the tropics because of slower developmental rates in cooler clines. Overall, however, because of the differences in representation of different developmental modes across latitudes, we would expect a higher fraction of dispersive, planktotrophic species in the tropics relative to the poles and, therefore, greater connectivity at lower latitudes.

Connectivity is not the only ecological process relevant to management that is likely to be affected by life history; there is some evidence that population variability may also be affected. Eckert (2003) showed that species with aplanktonic development tend to show more variation in abundance over time relative to species with a larval stage. If this pattern applies globally, our results suggest that, because low-temperature regions (as well as places with very low productivity) have more species with aplanktonic development, populations in these regions should exhibit more variability over time and should be managed accordingly.

Our findings have implications for how anthropogenic climate change may affect species distributions in the future. We found that species with planktotrophic development were more common in warmer conditions and that offspring were smaller at higher temperatures across all developmental modes. Assuming that temperature is the causal factor driving these relationships, we might expect that as global temperatures increase with climate change, species with planktotrophic development (and smaller eggs more generally) will become more common. Conversely, we might expect species with nonfeeding larvae to become less common worldwide. Furthermore, we might expect species with planktotrophic development to show range expansions toward the poles and contraction of the ranges of species with nonfeeding toward the poles. Already, the range shifts of some planktotrophic species have been linked to climate change (Ling et al. 2009). Further tests examining the direct effect of temperature on the selective advantage of different developmental modes and offspring sizes are therefore urgently needed.

DISCLOSURE STATEMENT

The authors are not aware of any affiliations, memberships, funding, or financial holdings that might be perceived as affecting the objectivity of this review.

ACKNOWLEDGMENTS

The authors acknowledge the support of the Australian Research Council for financial support while working on this paper. The authors thank Rick Vance, Jeb Byers, and Steve Gaines for excellent comments that improved the manuscript. The authors also acknowledge the tremendous effort in collating the data for the analyses in this manuscript and thank Carly Cook, Hanna Ritchie, Jacquei Burgin, Kurt Davies, and Matthew Thompson for their efforts in this regard.

Richard Fuller assisted in the preparation of **Figure 1**. This paper is dedicated to the memory of Leonard George Drennan (April 1, 1927–November 24, 2011).

LITERATURE CITED

Allen JA. 1877. The influence of physical conditions in the genesis of species. *Radic. Rev.* 1:108–40

Allen RM, Buckley YM, Marshall DJ. 2008. Offspring size plasticity in response to intraspecific competition: an adaptive maternal effect across life-history stages. *Am. Nat.* 171:225–37

Anthes N, Michiels NK. 2007. Reproductive morphology, mating behavior, and spawning ecology of cephalaspid sea slugs (Aglajidae and Gastropteridae). *Invertebr. Biol.* 126:335–65

Bashey F. 2006. Cross-generational environmental effects and the evolution of offspring size in the Trinidadian guppy *Poecilia reticulata*. *Evolution* 60:348–61

Becker BJ, Levin LA, Fodrie FJ, McMillan PA. 2007. Complex larval connectivity patterns among marine invertebrate populations. *Proc. Natl. Acad. Sci. USA* 104:3267–72

Bergmann C. 1847. Ueber die Verhältnisse der Wärmeökonomie der Thiere zu ihrer Grösse. *Göttinger Stud.* 3:595–708

Bernardo J. 1996. The particular maternal effect of propagule size, especially egg size: patterns, models, quality of evidence and interpretations. *Am. Zool.* 36:216–36

Botsford LW, Hastings A, Gaines SD. 2001. Dependence of sustainability on the configuration of marine reserves and larval dispersal distance. *Ecol. Lett.* 4:144–50

Bownds C, Wilson RS, Marshall DJ. 2010. Why do colder mothers produce larger eggs? An optimality approach. *J. Exp. Biol.* 213:3796–801

Burgess SC, Marshall DJ. 2011. Temperature-induced maternal effects and environmental predictability. *J. Exp. Biol.* 214:2329–36

Byrne M. 2006. Life-history diversity and evolution in the Asterinidae. *Integr. Comp. Biol.* 46:243–54

Clarke A. 1992. Reproduction in the cold: Thorson revisited. *Inv. Repro. Dev.* 22:175–84

Collin R. 2003. Worldwide patterns in mode of development in calyptraeid gastropods. *Mar. Ecol. Progr. Ser.* 247:103–22

Collin R. 2004. Phylogenetic effects, the loss of complex characters, and the evolution of development in calyptraeid gastropods. *Evolution* 58:1488–502

Eckelbarger KJ, Watling L. 1995. Role of phylogenetic constraints in determining reproductive patterns in deep-sea invertebrates. *Invertebr. Biol.* 114:256–69

Eckert GL. 2003. Effects of the planktonic period on marine population fluctuations. *Ecology* 84:372–83

Emlet RB, McEdward LR, Strathmann RR. 1987. Echinoderm larval ecology viewed from the egg. In *Echinoderm Studies*, ed. M Langoux, J Lawrence, pp. 55–136. Rotterdam: A.A. Balkema

Fernandez M, Astorga A, Navarrete SA, Valdovinos C, Marquet PA. 2009. Deconstructing latitudinal species richness patterns in the ocean: Does larval development hold the clue? *Ecol. Lett.* 12:601–611

Fischer K, Brakefield PM, Zwaan BJ. 2003. Plasticity in butterfly egg size: Why larger offspring at lower temperatures? *Ecology* 84:3138–47

Fox CW, Thakar MS, Mosseau TA. 1997. Egg size plasticity in a seed beetle: an adaptive maternal effect. *Am. Nat.* 149:149–63

Goddard JHR. 1992. *Patterns of Development in Nudibranch Molluscs from the Northeast Pacific Ocean, with Regional Comparisons*. Eugene, Oregon: Univ. Or. Press

Hastings A, Botsford LW. 2006. Persistence of spatial populations depends on returning home. *Proc. Natl. Acad. Sci. USA* 103:6067–72

Havenhand JN. 1995. Evolutionary ecology of larval types. In *Ecology of Marine Invertebrate Larvae*, ed. LR McEdward, pp. 79–122. Boca Raton: CRC

Kelly RP, Eernisse DJ. 2007. Southern hospitality: a latitudinal gradient in gene flow in the marine environment. *Evolution* 61:700–7

Kinlan BP, Gaines SD. 2003. Propagule dispersal in marine and terrestrial environments: a community perspective. *Ecology* 84:2007–20

Kohn AJ, Perron FE. 1994. *Life-History and Biogeography Patterns in Conus*. Oxford: Clarendon. 106 pp.

Krug PJ. 2009. Not my "type": larval dispersal dimorphisms and bet-hedging in opisthobranch life histories. *Biol. Bull.* 216:355–72

Laptikhovsky V. 2006. Latitudinal and bathymetric trends in egg size variation: a new look at Thorson's and Rass's rules. *Mar. Ecol.* 27:7–14

Lessios HA. 1990. Adaptation and phylogeny as determinants of egg size in echinoderms from the two sides of the Isthmus of Panama. *Am. Nat.* 135:1–13

Levin LA. 1984. Life history and dispersal patterns in a dense infaunal polychaete assemblage: community structure and response to disturbance. *Ecology* 65:1185–200

Ling SD, Johnson CR, Ridgway K, Hobday AJ, Haddon M. 2009. Climate-driven range extension of a sea urchin: inferring future trends by analysis of recent population dynamics. *Glob. Chang. Biol.* 15:719–31

Marshall DJ, Allen RM, Crean AJ. 2008. The ecological and evolutionary importance of maternal effects in the sea. *Oceanogr. Mar. Biol. Annu. Rev.* 46:203–50

Marshall DJ, Keough MJ. 2008a. The evolutionary ecology of offspring size in marine invertebrates. *Adv. Mar. Biol.* 53:1–60

Marshall DJ, Keough MJ. 2008b. The relationship between offspring size and performance in the sea. *Am. Nat.* 171:214–24

Marshall DJ, Morgan SG. 2011. Ecological and evolutionary consequences of linked life-history stages in the sea. *Curr. Biol.* 21:R718–25

McEdward LR, Miner BG. 2001. Larval and life-cycle patterns in echinoderms. *Can. J. Zool.* 79:1125–70

McHugh D, Rouse GW. 1998. Life history evolution of marine invertebrates: new views from phylogenetic systematics. *Trends Ecol. Evol.* 13:182–86

Mileikovsky SA. 1975. Types of larval development in Littorinidae (Gastropoda: Prosobranchia) of the world ocean, and ecological patterns of their distribution. *Mar. Biol.* 30:129–35

Moles AT, Westoby M. 2003. Latitude, seed predation and seed mass. *J. Biogeogr.* 30:105–28

Moran AL. 2004. Egg size evolution in tropical American bivalves: the fossil record and the comparative method. *Evolution* 58:2718–33

Morgan SG. 1995. Life and death in the plankton: larval mortality and adaptation. In *Ecology of Marine Invertebrate Larvae*, ed. L McEdward, pp. 279–322. Boca Raton: CRC

O'Connor M, Bruno JF, Gaines SD, Halpern BS, Lester SE, et al. 2007. Temperature control of larval dispersal and the implications for marine ecology, evolution, and conservation. *Proc. Natl. Acad. Sci. USA* 104:1266–71

Palumbi SR. 2003. Population genetics, demographic connectivity, and the design of marine reserves. *Ecol. Appl.* 13:S146–58

Pearse JS. 1994. Cold-water echinoderms break "Thorson's Rule". In *Reproduction, Larval Biology and Recruitment of the Deep-Sea Benthos*, ed. CM Young, KJ Eckelbarger, pp. 26–44. New York: Columbia Univ. Press

Pearse JS, Lockhart SJ. 2004. Reproduction in cold water: paradigm changes in the 20th century and a role for cidaroid sea urchins. *Deep-Sea Res. II* 51:1533–49

Pearse JS, McClintock JB, Bosch I. 1991. Reproduction of Antarctic benthic marine invertebrates: tempos, modes, and timing. *Am. Zool.* 31:65–80

Pechenik JA. 1999. On the advantages and disadvantages of larval stages in benthic marine invertebrate life cycles. *Mar. Ecol. Progr. Ser.* 177:269–97

Poulin E, Feral J-P. 1996. Why are there so many species of brooding Antarctic echinoids. *Evolution* 50:820–30

Rombough P. 2006. Developmental costs and the partitioning of metabolic energy. In *Comparative Developmental Physiology: Contributions, Tools, Trends*, ed. SJ Warburton, WW Burggren, B Pelster, CL Reiber, J Spicer. Oxford: Oxford Univ. Press

Roughgarden J. 1989. The evolution of marine life cycles. In *Mathematical Evolutionary Theory*, ed. MW Feldman, pp. 270–300. Princeton: Princeton Univ. Press

Salinas S, Munch SB. 2012. Thermal legacies: transgenerational effects of temperature on growth in a vertebrate. *Ecol. Lett.* 15:159–63

Schluter D, Price TD, Rowe L. 1991. Conflicting selection pressures and life history trade-offs. *Proc. R. Soc. Lond. B* 246:11–17

Shanks AL, Grantham BA, Carr MH. 2003. Propagule dispersal distance and the size and spacing of marine reserves. *Ecol. Appl.* 13:S159–69

Smith CC, Fretwell SD. 1974. The optimal balance between size and number of offspring. *Am. Nat.* 108:499–506

Strathmann MF. 1987. *Reproduction and Development of Marine Invertebrates of the Northern Pacific Coast*. Seattle: Univ. Wash. Press

Strathmann RR. 1974. The spread of sibling larvae of sedentary marine invertebrates. *Am. Nat.* 108:29–44

Strathmann RR. 1985. Feeding and non-feeding larval development and life-history evolution in marine invertebrates. *Annu. Rev. Ecol. Syst.* 16:339–61

Strathmann RR. 1993. Hypotheses on the origin of marine larvae. *Annu. Rev. Ecol. Syst.* 24:89–117

Thorson G. 1936. The larval development, growth and metabolism of arctic marine bottom invertebrates compared with those of other seas. *Medd. Grønland* 100:1–155

Thorson G. 1946. Reproduction and larval development of Danish marine invertebrates with special reference to the planktonic larvae in the sound (Oresund). *Medd. Komm. Dan. Fisk. Havunder. Ser. Plankton* 4:1–523

Thorson G. 1950. Reproductive and larval ecology of marine bottom invertebrates. *Biol. Rev.* 25:1–45

Uthicke S, Schaffelke B, Byrne M. 2009. A boom-bust phylum? Ecological and evolutionary consequences of density variations in echinoderms. *Ecol. Monogr.* 79:3–24

Van der Have TM, de Jong G. 1996. Adult size in ectotherms: temperature effects on growth and differentiation. *J. Theor. Biol.* 183:329–40

Vance RR. 1973. On reproductive strategies in marine benthic invertebrates. *Am. Nat.* 107:339–52

Von Bertanffy L. 1960. Principles and theory of growth. In *Fundamental Aspects of Normal and Malignant Growth*, ed. WW Nowinskii, pp. 137–259. Amsterdam: Elsevier

Weersing K, Toonen RJ. 2009. Population genetics, larval dispersal, and connectivity in marine systems. *Mar. Ecol. Progr. Ser.* 393:1–12

White JW, Botsford LW, Hastings A, Largier JL. 2010. Population persistence in marine reserve networks: incorporating spatial heterogeneities in larval dispersal. *Mar. Ecol. Progr. Ser.* 398:49–67

Wilson NG. 2002. Egg masses of chromodorid nudibranchs (Mollusca: Gastropoda: Opisthobranchia). *Malacologia* 44:289–305

Woods HA. 1999. Egg-mass size and cell size: effects of temperature on oxygen distribution. *Am. Zool.* 39:244–52

Wray GA. 1995. Evolution of larvae and developmental modes. In *Ecology of Marine Invertebrate Larvae*, ed. LR McEdward, pp. 413–48. Boca Raton: CRC

Young CM. 1990. Larval ecology of marine invertebrates: a sesquicentennial history. *Ophelia* 32:1–48

Mutation Load: The Fitness of Individuals in Populations Where Deleterious Alleles Are Abundant

Aneil F. Agrawal[1] and Michael C. Whitlock[2]

[1]Department of Ecology & Evolutionary Biology, University of Toronto, Toronto, Ontario, Canada M5S 3B2; email: a.agrawal@utoronto.ca

[2]Department of Zoology, University of British Columbia, Vancouver, British Columbia, Canada V6T 1Z4; email: whitlock@zoology.ubc.ca

Annu. Rev. Ecol. Evol. Syst. 2012. 43:115–35

First published online as a Review in Advance on August 28, 2012

The *Annual Review of Ecology, Evolution, and Systematics* is online at ecolsys.annualreviews.org

This article's doi: 10.1146/annurev-ecolsys-110411-160257

Keywords

genetic load, evolutionary theory, population ecology

Abstract

Many multicellular eukaryotes have reasonably high per-generation mutation rates. Consequently, most populations harbor an abundance of segregating deleterious alleles. These alleles, most of which are of small effect individually, collectively can reduce substantially the fitness of individuals relative to what it would be otherwise; this is mutation load. Mutation load can be lessened by any factor that causes more mutations to be removed per selective death, such as inbreeding, synergistic epistasis, population structure, or harsh environments. The ecological effects of load are not clear-cut because some conditions (such as selection early in life, sexual selection, reproductive compensation, and intraspecific competition) reduce the effects of load on population size and persistence, but other conditions (such as interspecific competition and load on resource use efficiency) can cause small amounts of load to have strong effects on the population, even extinction. We suggest a series of studies to improve our understanding of the effects of mutation load.

INTRODUCTION

New mutations enter a population each generation. Of those that affect fitness, most are deleterious (Keightley & Lynch 2003). Excluding those with severe (lethal) effects, a deleterious allele can persist for multiple generations before it is eradicated by selection. Meanwhile, other mutations arise, so that deleterious alleles are always present.

Humans are no exception to this phenomenon. The average human carries 250 to 300 loss-of-function mutations (1000 Genomes Project Consortium 2010). Inferences from molecular population genetics data indicate that the average person also carries another several hundred less severely deleterious amino acid variants (Eyre-Walker et al. 2006, Charlesworth & Charlesworth 2010). Based on the amount of sequence constraint across the genome (Eöry et al. 2010), it is reasonable to speculate that the number of deleterious alleles carried at noncoding sites at least matches, and is likely several times greater than, the number at nonsynonymous sites. It is thus reasonable to say that the average human carries well over a thousand deleterious mutations.

The genomics era has greatly improved our ability to assess the abundance of deleterious alleles segregating in populations, but the question remains whether these deleterious alleles matter much or at all. This depends on what it means to "matter." Deleterious alleles are not the stuff of adaptation, so they are sometimes regarded as an uninteresting and unimportant part of the evolutionary process. However, some have argued that deleterious mutations provided the selection pressures that led to a variety of major adaptive phenomena, including diploidy (Otto & Goldstein 1992), recombination and sex (Kondrashov 1988, Keightley & Otto 2006), outcrossing (Lande & Schemske 1985, Charlesworth et al. 1990), secondary sexual traits (Rowe & Houle 1996, Houle & Kondrashov 2002), and various aspects of genomic complexity (Lynch & Conery 2003, Lynch 2007).

However, the simplest and most direct way that one might expect deleterious mutations to matter is by causing a reduction in fitness. The loss of fitness resulting from deleterious alleles maintained by mutation-selection balance is known as mutation load.

WHAT IS MUTATION LOAD?

H.J. Muller (1950) wrote a paper famously titled "Our Load of Mutations," in which he argued that deleterious alleles make a substantial contribution to human mortality and disease and, more generally, discussed the reduction in fitness due to recurrent mutation. According to Crow (1993), the term mutation load arose from Muller's title and has been a part of the evolutionary genetics lexicon ever since. Mutation load is sometimes used loosely to describe a variety of consequences of segregating deleterious alleles including the incidence of genetic disorders and the magnitudes of inbreeding depression and standing genetic variance in fitness resulting from mutational input. However, here we use it in its formal sense to refer to the reduction in fitness due to the presence of deleterious mutations segregating at mutation-selection balance.

This concept traces back to Haldane's (1937) classic theoretical paper, in which he quantified the expected "loss of fitness" resulting from mutation-selection balance. For a simple one-locus model with fitnesses of $W_{AA} = 1$, $W_{Aa} = 1 - hs$, and $W_{aa} = 1 - s$, the equilibrium frequency of the deleterious allele is $q \approx \mu/hs$ (assuming $hs \gg \mu$), where μ is the mutation rate for $A \to a$. Mean fitness is $\bar{W} = 1 - 2pqhs - q^2s$, which is closely approximated by $\bar{W} \approx 1 - 2qhs$ when the deleterious allele is rare ($q \ll 1$) and simplifies to $\bar{W} \approx 1 - 2\mu$ at mutation-selection balance. Mutation load is the extent to which individuals are less fit than they would be otherwise because of mutation. Using $\bar{W}_{NoMut} = 1$ to represent the mean fitness in the absence of mutation, mutation load is defined by $L \equiv (\bar{W}_{NoMut} - \bar{W})/\bar{W}_{NoMut}$, which in the one-locus case is $L = 2\mu$.

Haldane extrapolated his single-locus results across the genome by making two simplifying assumptions: no epistasis and no linkage disequilibrium. Under these assumptions, total mean fitness is equal to the product of mean fitnesses with respect to each locus,

$$\bar{W} = \prod \bar{W}_i = \prod (1 - 2\mu_i) \approx \prod e^{-2\mu_i} = e^{-U},$$

where $\bar{W}_i$ is the mean fitness with respect to locus i, and μ_i is the mutation rate for that locus. $U = \Sigma(2\mu_i)$ is the genome-wide rate of deleterious mutation, where the 2 arises because in diploid organisms each individual has two copies of each gene, each with a chance of a mutation. From the equation above, the genome-wide mutation load is $L = 1 - e^{-U}$.

Haldane's prediction for mean fitness $\bar{W} = e^{-U}$ is rather disturbing once we consider realistic values for U. For example, if there is only a single deleterious mutation per genome per generation on average ($U = 1$)—close to the estimate for *Drosophila* (Haag-Liautard et al. 2007)—then mean fitness is less than 40% of what it would be in the absence of mutation. With a recent estimate for humans of $U = 2.2$ (Keightley 2012), Haldane's result predicts an onerous load of 89%. Of course, Haldane's analysis makes a number of assumptions, and below we discuss some of the subsequent development of the theory. Nonetheless, his prediction provides a baseline for thinking about the potential magnitude of mutation load. Was he close to being right? If so, what do such large loads mean?

Before discussing these questions, we return to the one-locus model to clarify some key terms. In this model, the fitness of each genotype is defined relative to the best type. Thus, mean fitness $\bar{W}$, as used in the context of mutation load, refers to the average fitness of individuals relative to the best type; it does not refer to their absolute fitness. Critically, mutation load refers to the reduction in fitness of individuals, not populations, and the relationship between the two can be complex.

In Haldane's model, the extent to which a mutation reduces the mean fitness (i.e., the hs in $\bar{W} \approx 1 - 2qhs$) is exactly the same as the selection pressure that determines the equilibrium allele frequency (i.e., the hs in $q \approx \mu/hs$) so that these two effects of selection cancel each other out, making L independent of s. This has likely contributed to the appeal of mutation load among theoretical population geneticists as it implies that we need to know nothing about the ecology—or even strength—of selection. However, this sentiment is misleading. Haldane's model and most other mutation load models describe population genetics, not demography or ecology. Consequently, we cannot easily link $\bar{W}$ (or L) to any measure of population performance unless we have a specific ecological framework for doing so.

The biggest limitation to empirically measuring mutation load sensu stricto is identifying a mutation-free reference genotype. In fact, such an individual is unlikely to exist if the mutation rate is reasonably high and if most mutations are only weakly deleterious. Without the reference genotype, mutation load cannot be measured. So is the subject worthy of study at all? There are at least three reasons why it is. First, it is interesting to know how much less fit the average individual is than it could be if there were no segregating deleterious alleles. Even if we cannot directly quantify this loss of fitness empirically, we can use theory to inform ourselves about what relevant parameters to measure (e.g., mutation rate, epistasis, inbreeding) so we can estimate the load based on the theoretical models. Second, although we cannot measure absolute load, we should be able to compare the loads under different circumstances (relative load) to test the importance of various factors predicted by theory to influence load. Third, we can study the ecological consequences of load in models and by empirical manipulation of the load.

THE GENOME DELETERIOUS MUTATION RATE, U

The most crucial parameter to predicting the mutation load is the genomic deleterious mutation rate, U. Several recent reviews provide summaries of estimates of U as well as descriptions of the methods by which U is estimated and their associated caveats (Baer et al. 2007, Halligan & Keightley 2009). Consequently, we limit our discussion to a few major points. Microbes (with the exception of viruses) tend to have very small mutation rates ($U < 0.01$, Drake et al. 1998), so mutation load is unlikely to be important in these taxa. However, for multicellular plants and animals, recent mutation rate estimates are often above 0.1 and sometimes greater than 1 (Baer et al. 2007, Halligan & Keightley 2009), i.e., well into the range where Haldane would predict a substantial load. For example, recent estimates in some well-studied taxa include: *Caenorhabditis elegans*, $U \approx 0.25$ to 2.5; *Drosophila melanogaster*, $U \approx 1.2$ to 1.4; humans, $U \approx 2.2$ (Denver et al. 2004, 2009; Haag-Liautard et al. 2007; in each case the higher estimates come from Keightley 2012).

The measures of U reported above were determined by combining direct measures of the per-nucleotide mutation rate μ, the total genome size G, and the fraction of nucleotides constrained by selection c; i.e., $U = \mu G c$ (Kondrashov & Crow 1993). Selective constraint is estimated using between-species comparisons from the proportion of variants of a particular type, for example nonsynonymous amino acid changes that are missing relative to variants presumed to be neutral, such as synonymous changes. Classically, U was inferred from the rates of phenotypic changes in fitness of genetic lines maintained without selection; such estimates of U are typically much smaller. However, phenotypically based methods underestimate U because they are insensitive to mutations that have weak effects on fitness that are difficult to measure in the lab (Halligan & Keightley 2009, Keightley & Halligan 2009). Most mutations have weak effects (Davies et al. 1999, Estes et al. 2004, Bégin & Schoen 2006, Keightley & Eyre-Walker 2007), but they should not be neglected; the per-locus load is independent of s as long as the allele is not nearly neutral. The estimates based on $U = \mu G c$ are not perfect but are likely to be of the correct order of magnitude.

Considerable uncertainty in U persists for several reasons including (*a*) ongoing refinements in measuring constraint, especially for noncoding regions, (*b*) measurement error in μ, and (*c*) variation among genotypes in mutation rates. For example, in a single *Drosophila* study (Haag-Liautard et al. 2007), estimates of U from three genotypes were 0.66, 0.94, and 2.56. If the species-wide average U is closer to the lowest of these estimates, then Haldane's equation gives a load of 48%, whereas with the higher estimate a load of 92% is predicted.

Understanding the intraspecific variation in U (Baer et al. 2005, Haag-Liautard et al. 2007, Conrad et al. 2011) is not only necessary to refine our estimates of the species-wide average but is also interesting in its own right. Recent evidence suggests that an individual's mutation rate may be correlated to its fitness (Goho & Bell 2000, Agrawal & Wang 2008, Sharp & Agrawal 2012). This relationship results in a positive feedback loop whereby individuals of low genetic quality transmit mutations at an elevated rate, creating offspring that are of even worse genetic quality. This feedback loop causes average mutation rate to evolve and differentially affects the equilibrium load as well as the extinction risk of populations with different reproductive modes (Agrawal 2002, Shaw & Baer 2011).

FACTORS AFFECTING RELATIVE FITNESS

Rule of Thumb

The calculation of load, as based on Haldane's model, is relatively simple when there is random mating, no linkage disequilibrium, and no gene interaction. When, as in most realistic scenarios, one or more of these assumptions is violated, the realized load depends on details of selection and reproduction. Fortunately, there are some general principles for understanding how various factors affect load. Extending the earlier work of King (1966), Kondrashov & Crow (1988) provided perhaps the clearest perspective.

At equilibrium, the number of mutations entering the population, U per offspring, must equal the number removed by selection. This selection is manifested through a reduction in relative fitness of loaded genotypes. The "lost" relative fitness, which occurs in the form of death or a partial or complete failure to reproduce, represents the load. From the general model of Kondrashov & Crow, it can be shown that the load is $L = U/(\bar{z} - \bar{y})$, where $\bar{y}$ and $\bar{z}$ are the mean numbers of mutations carried by the winners and losers of selection, respectively. It is easiest to think about this in the special case when selection only acts on survivorship; in that case, the "winners" are the survivors who have the opportunity to reproduce and the "losers" are the individuals who die through selection. More generally, we can calculate $\bar{y}$, the mean number of mutations in winners, by averaging over all individuals weighted by their offspring number, and for losers, $\bar{z}$ can be calculated by weighting each individual by its lost fecundity, i.e., the difference between its fecundity and that possible for an unloaded genotype. If an individual loses all of its fitness due to selection, this is counted as one full "selective death." The difference in the number of mutations between these two groupings of individuals $\bar{z} - \bar{y}$ represents the number of mutations eliminated per selective death. Because the probability of selective death (or, in more general terms, the amount of fitness loss per capita) is the load, L, and the number of mutations removed per selective death is $\bar{z} - \bar{y}$, the average number of mutations eliminated by selection per capita is $(\bar{z} - \bar{y})L$. At equilibrium, the selective loss of deleterious alleles per capita must be balanced by mutation, U. Rearranging, this gives $L = U/(\bar{z} - \bar{y})$. The number of mutations eliminated per selective event (relative to U) can be thought of as the efficiency of selection. When many mutations are eliminated in each selective event, relatively few individuals need to lose fitness in order to balance out the influx of new mutations across the entire population. Thus, any factor that increases the number of mutations eliminated per selective death helps to reduce load. Remember, though, that selective death is just inaccurate shorthand; alleles can be selectively removed through reduced fecundity or lowered mating success, and dying zygotes and offspring are often replaced by healthier siblings. These selective events count as selective deaths without creating visible carnage in the population.

The utility of $L = U/(\bar{z} - \bar{y})$ can be illustrated by considering a simple case. In the single-locus diploid model described above ($U = 2\mu$) with some expression in the heterozygote, a selective loser carries one deleterious allele, whereas selective winners have none, so we obtain $L = 2\mu/(1 - 0) = 2\mu$. In the case of a totally recessive mutation, the deleterious allele remains rare, so selective winners still have, on average, close to zero mutations. However, selective losers carry two deleterious alleles, so $L = 2\mu/(2 - 0) = \mu$, which is the same result obtained by other means (Haldane 1937). In the case of totally recessive mutations, selection is twice as efficient; each selective death removes twice as many mutations, so the load is only half as large as when mutations have heterozygous effects.

Epistasis

Just as interactions between alleles at the same locus (i.e., dominance) can affect the efficiency of selection, so can epistasis (interlocus interaction). When genes interact such that having multiple deleterious alleles together is worse than expected based on their individual effects, i.e., "synergistic" (or negative) epistasis, then selection is more efficient and the load is reduced (Kimura & Maruyama 1966, King 1966, Kondrashov 1982, Charlesworth 1990). Perhaps the simplest case is one in which individuals carrying less than a threshold number of mutations T have high fitness but those with more than T mutations have low fitness. Immediately following selection, the average number of mutations is somewhat less than T. Through recombination, a substantial frequency of offspring is produced carrying more than T mutations. Consequently, the losers of selection can carry, on average, quite a few more deleterious mutations than the winners; multiple mutations are eliminated with each selective death, substantially reducing the load.

This type of threshold selection is one rather extreme form of epistasis. Others have modeled fitness as a quadratic function of mutation number, implying negative pairwise gene interactions (Charlesworth 1990, Howard & Lively 1998). Such models also show a reduction in load but the effect is less dramatic than in the threshold case. Consistent with these results, Hansen & Wagner (2001) studied a somewhat more general model of epistasis and found that higher-order gene interactions potentially have a stronger influence on the load, especially when mutation rates are high, as the effect of kth order epistasis is proportional to U^k.

The empirical evidence for epistasis is mixed. Studies examining the fitness effects from combinations of small numbers of specific mutations have found a mix of both positive and negative interactions with no strong epistasis on average (vesicular stomatitis virus, Sanjuán et al. 2004; *Escherichia coli*, Elena & Lenski 1997; yeast, Jasnos & Korona 2007; *Aspergillus*, de Visser et al. 1997b; *D. melanogaster*, Whitlock & Bourguet 2000, Wang et al. 2009). The observation that epistasis is highly variable among gene-combinations and with a mean close to zero is consistent with a landscape model of mutational effects (Martin et al. 2007). However, there appears to be an interesting, but somewhat tenuous, pattern across taxa in which multicellular eukaryotes (which have higher U) tend to have more evidence of synergistic epistasis than unicellular organisms (Sanjuán & Elena 2006, Agrawal & Whitlock 2010).

There are several caveats that apply to the data sets mentioned above. First, the experimental procedures bias most studies, to varying degrees, against observing interactions with strong negative epistasis (e.g., synthetic lethals). Second, it is possible that synergistic epistasis only becomes prevalent under particular environmental conditions (King 1967, Kondrashov 1988, Peck & Waxman 2000, Peters & Keightley 2000, Kishony & Leibler 2003). However, studies measuring epistasis in different environments have found no evidence that epistasis becomes more synergistic under stress (Jasnos et al. 2008, Wang et al. 2009). Third, the studies above examine interactions among small numbers of mutations, thereby missing out on higher-order interactions. Yet, studies that presumably examine the effects of larger numbers of (unidentified) mutations also tend to show no epistasis or only weak synergism (Mukai 1969, de Visser et al. 1997a, West et al. 1998, Peters & Keightley 2000). Fourth, most studies examine epistasis among random combinations of mutations. There is no reason to expect any directional epistasis among random mutations (Martin et al. 2007, Agrawal & Whitlock 2010), but there are reasons to expect epistasis among genes within functional pathways (Szathmáry 1993, Keightley 1996, Segrè et al. 2005, Sanjuán & Nebot 2008). Rice (1998) argued that pathway epistasis is ignored by existing empirical studies but could contribute to reducing load. He considered a model in which the genome was divided into 100 pathways with synergistic epistasis occurring among genes within pathways but no epistasis among pathways. Even though epistasis would be observed in only 1% of random mutant pairs (and likely to pass undetected in an experiment), Rice found that total load was greatly reduced.

Asexual Reproduction

When selection is multiplicative, there is no difference in mutation load between asexual and sexual populations (assuming random mating for the latter and ignoring drift). However, epistasis changes the load of sexual populations but does not affect asexual populations (Kimura & Maruyama

1966, Kondrashov & Crow 1988). This difference arises because of the important role of sex and recombination, which reduces linkage disequilibria.

Modifying an approach used by Rice (1998), we can express the load as $L = 1 + \beta\sigma^2/U$, where β is the regression of mutation number on fitness relative to the best type (which will be negative) and σ^2 is the variance in mutation number, which depends on allele frequency as well as linkage disequilibria. Considering various forms of epistasis, Rice (1998) found that, in asexual populations, the disequilibria always evolve in such a way as to compensate for changes made to the fitness function (i.e., so that the product $\beta\sigma^2$ remains constant). With recombination, the disequilibria remain relatively closer to zero regardless of how selection changes, resulting in "mismatches" between selection and the genetic variance, which further result in lower or higher loads, depending on the sign of epistasis. The contrast between how the loads of sexual and asexual populations are affected by epistasis illustrates the potential importance of disequilibria in determining load.

Nonrandom Mating

In many if not most sexual species, mating is not completely random. For example, theory suggests that positive assortative mating should be a common outcome whenever male-male competition or female mate choice is costly (Fawcett & Johnstone 2003). This type of mating expands the variance in number of deleterious alleles per offspring, allowing for a greater difference in mutation number between selection's winners and losers. Rice (1998) found that even weak positive assortative mating for fitness could substantially reduce the load. Correlations in body size between mates are reasonably common in nature but explicit tests of nonrandom mating with respect to fitness have been rare (reviewed in Sharp & Agrawal 2009). Moreover, the relevant correlation is the genetic correlation in fitness between mates, which is likely weaker than any observed phenotypic correlation. More information on genetic correlations is required to assess by how much positive assortative mating reduces load in nature.

Inbreeding is another common form of nonrandom mating. In many taxa, matings occur between relatives more often than expected by chance, often simply due to geography. This creates an excess of homozygotes. Returning to the one-locus model, the losers of selection are more often homozygotes than if mating were truly random, increasing the efficiency of selective deaths, because in inbred individuals two alleles are removed per selective death rather than one. This effect is magnified if deleterious alleles are at least partially recessive because then selection falls disproportionately on the homozygotes, further increasing $\bar{z}$, by a fraction proportional to the inbreeding coefficient or more. This homozygosity effect has been noted by several researchers (Crow & Kimura 1970, Whitlock 2002, Glémin et al. 2003, Roze & Rousset 2004) and can cause a substantial reduction in genome-wide load for reasonably low levels of inbreeding (Agrawal & Chasnov 2001), provided deleterious alleles are partially recessive. The available evidence supports partial recessivity (Simmons & Crow 1977, Phadnis & Fry 2005, Agrawal & Whitlock 2011, Manna et al. 2012) though much of the data comes from a small number of taxa (mostly D. *melanogaster* and *Saccharomyces cerevisiae*) and good estimates of h for typical small-effect genes are lacking.

Spatial Effects

In addition to causing inbreeding, spatial structure may also lead to local competition, which counters the homozygosity effect described above (Whitlock 2002, Glémin et al. 2003, Roze & Rousset 2004). Mutant individuals can be sheltered from selection if they tend to compete for resources against their mutant relatives. This lessens the difference in mutational burden between the losers and winners of natural selection ($\bar{z} - \bar{y}$), thereby increasing the mutation load. The importance of the "local competition effect" depends strongly on the ecological details of population regulation and the nature of selection on individual genes (Holsinger & Pacala 1990, Agrawal 2010, Laffafian et al. 2010). The local competition effect can be particularly strong if some juveniles die before reaching reproductive maturity but consume limited local resources, making resources unavailable to others. However, load is only increased if surviving mutants can take advantage of the resources made available by the death of their mutant relatives.

The discussion above assumes that selection is the same across space but this need not be the case. Consider mutations that are deleterious in some environments but neutral in others. Kawecki and colleagues (Kawecki 1995, Kawecki et al. 1997) showed that the buildup of such alleles in habitats where they are neutral can hinder populations from adapting to habitats where these alleles would be selectively eliminated. This result arises from an interaction between ecological and evolutionary effects. Moreover, if the demographic contribution from rarely used habitats is low, then selection within such habitats becomes ineffective, allowing the accumulation of alleles with deleterious effects specific to these habitats. If these types of deleterious alleles prevent establishment in marginal habitats, then this would represent a major unseen effect of load.

Roze (2012) argued that spatial variation in the average strength of selection would reduce the load. Variation in selection intensity among demes creates positive linkage disequilibrium, increasing the efficiency of selection. Moreover, the nonlinear relationship between average mutation number and average fitness means that demes receiving good genotypes benefit more than demes receiving bad ones suffer. These effects occur only when some environments are more selective across the genome than others. Though the change in selection between environments is often variable among genes, several experiments (Kishony & Leibler 2003, Jasnos et al. 2008, Wang et al. 2009) have documented reasonably consistent changes (i.e., selection on the majority of genes is stronger in some environments than in others), although this does not necessarily correlate with the quality of the environment. The reasons why certain environments are more selective remain elusive (Martin & Lenormand 2006, Agrawal & Whitlock 2010).

Temporal Effects

Just as selection can vary over space, it can also vary over time. We are unaware of any formal models of load when selection varies over time. The simplest case can be considered as follows. Imagine that every t generations there is an extreme weather event. There are some loci that, if in a mutated state, are lethal during these events but are neutral otherwise. Immediately following one of these events, the population is free of such alleles. Assuming the mutation rate to such loci is U, then by the next event the frequency of individuals carrying one or more of these alleles is $1 - e^{-Ut}$. Calculating the geometric mean fitness over one complete cycle (i.e., $t - 1$ generations without selection and one generation with selection), we obtain $\bar{W} = \sqrt[t]{e^{-Ut}} = e^{-U}$, which is Haldane's classic result. Of course, this average belies the temporal variation in mean fitness, which is maximal ($\bar{W} = 1$) between extreme weather events but quite low during those events ($\bar{W} = e^{-Ut}$). Moreover, temporal variation may be more important in selective contexts beyond the simple scenario described above. For example, if there is synergistic epistasis on small-effect mutations during extreme weather events, then the period between events allows for more rounds of recombination, increasing the efficiency of selection when it next occurs. If such temporal variation in selection is common, point estimates of load could be very misleading.

Genetic Drift

Genetic drift can also affect the amount of mutation load experienced by a population. First, drift can cause the fixation of deleterious alleles, which in itself causes a reduction in fitness called drift load or finite population load (Crow 1970). This type of load has been considered by some researchers as part of mutation load (Kimura et al. 1963), but not by others (Crow 1970). When drift is strong relative to selection, deleterious alleles can reach fixation provided that the rate of reverse mutation is not too high. The overall genetic load in small populations can be much larger than predicted by deterministic mutation load theory (Kimura et al. 1963, Bataillon & Kirkpatrick 2000, Glémin 2003, Haag & Roze 2007), and such load can potentially even contribute to the extinction of small populations (Lynch et al. 1995). Contrary to intuition, intermediate levels of drift can lead to a net reduction in load from partially recessive alleles because of purging, but this only occurs under very limited combinations of N, h, and s (Kimura et al. 1963, Glémin 2003). Even when drift is strong, the load attributable to segregating mutations (rather than total genetic load) can be reduced simply because the deleterious alleles are fixed rather than segregating.

For alleles that are not nearly neutral, the expected mutation load in a finite population will be similar to the deterministic prediction. However, this is not true at the level of a single locus at any specific time point. In a small population, there may be no deleterious alleles present at any given time, whereas at other loci the frequency of deleterious mutations may be higher than expected by deterministic mutation-selection balance predictions. As a result, it can be very difficult to infer the strength of selection against a particular deleterious mutation in a finite population. However, it is possible to infer distributions of selection coefficients for categories of mutations (e.g., nonsynonymous sites) from population-level sequence data (Eyre-Walker et al. 2006, Keightley & Eyre-Walker 2007, Boyko et al. 2008).

WHEN DOES LOAD HAVE ECOLOGICAL CONSEQUENCES?

It is clear that deleterious mutations can affect the relative fitness of individuals, making them less able to compete against individuals with fewer low-fitness alleles. However, it is not as obvious that such relative declines in fitness affect the absolute mean fitness of the population. Some have implied that a population cannot persist with a heavy mutation load (Kondrashov & Crow 1993, Kondrashov 1995, Nachman & Crowell 2000, Reed & Aquadro 2006). Numerous others (Haldane 1957, Turner & Williamson 1968, Wallace 1968, Mather 1969, Wallace 1970) have argued that mutation load is largely irrelevant to ecology because of density-dependent processes. Wallace (1968, 1975) used the terms soft selection to refer to cases where selective deaths would otherwise be replaced by nonselective (ecological) deaths and hard selection to refer to cases where they would not. Because of the perceived ubiquity of density dependence, many believe selection is typically soft and, therefore, load is ecologically unimportant (i.e., if selection were not removing individuals because of bad genotypes, then more individuals would die because of competition for limited resources). However, the ecological importance of load has remained controversial for several decades, partly because the debate has typically occurred without considering explicit ecological models (but see Clarke 1973a,b).

In fact, ecological processes are unlikely to completely mask the effects of genetic load. The effect of variation in fitness components on ecological success has been studied often (e.g., MacArthur 1972; Schoener 1973, 1976; Abrams 2003; Abrams et al. 2003; Schreiber & Rudolf 2008). These studies have explicitly looked at the effects of resource acquisition rates, fecundity, and survival on equilibrium population size. These models usually find that reductions in fitness components caused a decrease in population size. The key to understanding the effects of load on ecology is intuitive: Load affects a species population size if individuals that die selective deaths remove resources that would otherwise be available to other members of the same species. If an individual dies before consuming any resources and if the resources freed up are consumed by the same species, then load has little effect on population size. However, if a dying individual has consumed resources without reproduction or if the resources freed up are consumed by another species or otherwise lost, load affects population size.

To frame our discussion, we have made explicit calculations about the ecological effects of load (**Supplemental Text 1**). We use a MacArthur (1972) type model of population growth rather than the logistic model (as used by Clarke 1973a,b) because the parameters of the latter (especially "carrying capacity") can be difficult to interpret in terms of individual-level traits (Matessi & Gatto 1984) and can lead to misleading conclusions.

First consider the case where the consumer population is at low density so that its consumption of resources is limited not by a scarcity of resources but by an upper limit of the consumption rate, C_{max}. The conversion efficiency of resources to offspring is b and the death rate is d. In that case, the proportional change in consumer population size is $bC_{max} - d$. This is equivalent to the intrinsic growth rate of the consumer population. Any decline in consumption rate or conversion efficiency, or increase in death rate, leads to a drop in the ecological performance of the consumer species in this situation. Thus, mutation load in any of these parameters would affect a population's growth rate at low density.

At the other extreme, when the consumption of resources is at equilibrium with the resource input, we have to consider the effect of the consumption on the amount of resources. We take MacArthur's (1972) simplest model for the replenishment of the resources; resource R becomes available at rate I and is depleted only by consumption by this consumer, which happens with acquisition rate a. Therefore the dynamics of the resources are given by $\partial R/\partial t = I - aNR$, where N is the population size of the consumer, and the dynamics of the consumer species are given by $\partial N/\partial t = (baR - d)N$. At equilibrium (i.e., the carrying capacity), N will be

$$\hat{N} = \frac{bI}{d}.$$

$\hat{N}$ is higher with greater conversion efficiency and lower with lower death rate, but it is unaffected by the resource capture rate, a. This is because if an individual captures resources but uses them inefficiently or dies, it eliminates some reproductive potential of the species. If it does not capture resources well, then those resources are still available for other individuals of the species to use. In this simple model, load can have either zero effect on N (in the case of acquisition ability) or cause a proportional reduction in N (for load on birth rate, b, or generation time, $1/d$), depending on what traits are affected.

Real organisms pass through different life stages where they feel the selective effects of different mutant alleles and are subjected to different types of ecological regulation. The timing of selection can greatly affect load. For example, Wallace (1991) suggested that load would not reduce a population's productivity if it occurred early in life, e.g., at the zygote stage. Presumably the logic behind this claim is that dying zygotes have used fewer resources than dying adults, and therefore their competitive effect on conspecifics would be reduced.

To investigate this issue further, we expanded our model to include the possibility of early selective deaths. Imagine L of the population dies due to deleterious alleles each generation and does so before reaching full adulthood. If individuals that die selective deaths consume only a fraction β of the resources that a healthy adult would consume, then it can be shown (**Supplemental Text 1**) that the equilibrium population size is

$$\hat{N} = \frac{bI}{d} \frac{(1 - L)}{(1 - L(1 - \beta))}.$$

Selective deaths of adults after consumption but before reproduction reduce the size of the species in proportion to load. However, if the selective deaths occur early in life, such that β is small, the demographic consequences of load can be greatly reduced, at least at the adult stage.

In situations like the example above, load can reduce the equilibrium abundance of juveniles but have little effect on the abundance of adults if density regulation occurs after selective juvenile mortality and prior to the censusing of adults. Thus, the answer to how greatly load affects abundance depends on what life stage is being censused.

Interspecific Competition

As far as we are aware, all previous explicit discussions of the ecological effects of load focus on the effects of intraspecific competition. However, when organisms compete interspecifically for resources, the ecological effects of load can be much more drastic, as shown in **Supplemental Text 1**. Load on most vital rates can greatly reduce equilibrium population size in the presence of a close competitor. If the two species are sufficiently similar (such that their acquisition rates of various resources are similar), a small increase in load can cause extinction of one species in the presence of a competitor (**Figure 1**).

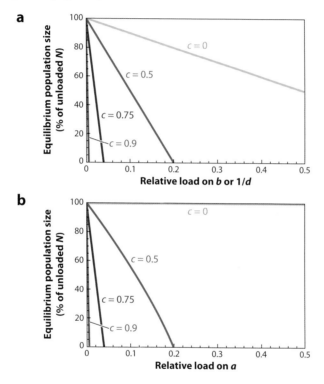

Figure 1

When species compete strongly for resources, a small amount of load in one species can cause a disproportionately large effect in its equilibrium population size. In some cases, relatively small levels of load can cause species extinction. This graph plots the results from the calculations in **Supplemental Text 1** for load in either (*a*) fecundity or longevity or (*b*) acquisition rate. It assumes that two consumer species are identical to each other except that one has a relatively larger load (as plotted on the x-axis) and that the two species are competing for two resources, each of which is better used by one of the species. The effects of load on population size and persistence depend on how ecologically similar the two species are, as measured by *c*, the relative resource acquisition rate of a species on its suboptimal resource, compared to its preferred resource. (These calculations assume that each species acquires one of the resources better than the other by the same ratio as for the unloaded genotypes.) When this *c* ratio is high, very small levels of load can cause extinction. Other parameters used for these calculations for the default unloaded genotypes are: $b = 0.02$, $d = 0.1$, $a = 1$, $I_1 = I_2 = 500$.

In the one-species model, the effects of load on the equilibrium abundance of adults could be masked by ecological compensation if load were expressed as juvenile mortality. This type of masking of load is less likely to occur in species limited by interspecific competition. In a two-species, two-stage model with two resources available (see **Supplemental Text 1**), early deaths free up resources, but those resources are available to both species. As a result, the dynamics of the two-species model continue to show similar effects of load regardless of when in the life cycle that load occurs.

Abrams (2003), with a more complicated model of consumer competition including population dynamics of the resource species and type 2 functional responses, similarly showed that the ecological effects of changing vital rates can be large, whether through mutation or other mechanisms. Counterintuitively, a decrease in consumption rates can, in some cases, cause an increase of the population density of that species, if the consumer species previously were overexploiting the resource. The ecological effects of changing life-history parameters can be complicated to predict and are somewhat model dependent. One of our main conclusions is that we need to be ecologically explicit before we can claim to know the effects of load on population size or persistence.

However, four points are clear. First, density-dependent effects do not necessarily compensate for load, i.e., the mere existence of density dependence does not preclude severe demographic consequences of mutation load (see **Supplemental Figures 1** and **2**). Second, load affecting different life-history traits (e.g., resource capture rate versus conversion efficiency) may or may not differ in the ecological consequences, depending on the circumstances. Third, the demographic consequences of load can differ between life stages (e.g., juveniles versus adults). Fourth, in populations limited by strong interspecific competition, the demographic consequences of load can be particularly strong and could result in competitive exclusion.

To evaluate the importance of load in mediating interspecific competition, we first need to ask how likely it is that relative load varies between species. Point estimates for the rates of evolution at synonymous sites between two closely related species of *Drosophila* differ by ~10% (Cutter 2008). A similar difference occurs between related species of *Caenorhabditis* (Cutter 2008). Because neutral evolution is a function of mutation rate per unit time, such differences could reflect differences in either mutation rate per generation or in generation time. With these caveats, though, estimates of rates of mutational decline indicate that related *Caenorhabditis* species (or even strains within species) may vary substantially in mutation load (Baer et al. 2005, Phillips et al. 2009).

The Effects of Correlations among Fitness Components

Of the models we have described, load on fecundity or longevity always has some effect on equilibrium population size, provided that this size is determined by intra- or interspecific competition. With intraspecific competition only, load on resource acquisition ability does not affect population size (assuming no resource decay). However, there is still selection on acquisition rate, even with pure intraspecific competition, meaning that selection should still be effective on this trait even though it does not affect population size. This raises the possibility that selection on acquisition ability could, through pleiotropy, reduce the load on other vital rates. If new mutations that deleteriously affect fecundity or longevity also reduce acquisition ability, then selection on the latter can also make selection indirectly more effective on the former. As a result, the equilibrium frequency of alleles decreasing fecundity or longevity becomes decreased, and load on those vital rates also becomes reduced. Therefore, with intraspecific competition and positive correlations between the effects of alleles on acquisition rate and fecundity or longevity, the effect of total load on equilibrium population size can be reduced by selection on acquisition ability. A similar effect is seen with sexual selection on males when the effects on male mating success are positively correlated with the effects on female productivity (Whitlock & Agrawal 2009). Sexual selection is a form of intraspecific competition (by males for access to females), so it is not strange that it should behave similarly to the intraspecific competition case presented here.

HOW TO STUDY LOAD EMPIRICALLY

Given that deleterious mutation rates in many multicellular organisms are on the order of $U = 1$ or more (Baer et al. 2007, Halligan & Keightley 2009, Keightley 2012), then there should be substantial mutation loads in these taxa. But are individuals really as much less fit as predicted by Haldane? If not, then why not? Despite the topic being of continued theoretical investigation for more than 70 years, there is very little evidence that Haldane's theory is even approximately correct, even though the predicted effects should be large.

The problem of identifying a mutant-free reference genotype makes measuring the absolute mutation load a near impossible task. However, major tenets of mutation load theory should be testable by comparing individuals from populations that have evolved under different mutation rates. In a number of systems, mutation rate can be manipulated in replicate populations. According to Haldane, if two populations evolved with deleterious mutation rates of U and $U + \Delta U$, then the average fitness of individuals of the second population relative to those of the first should be $e^{-\Delta U}$ (i.e., increasing the mutation rate by $\Delta U = 1$ should result in a ~60% reduction in fitness). By combining manipulations of mutation rate with other types of treatments, experimental evolution could be used to test the roles of sexual selection or population structure in load or whether load affects the outcome of interspecific competition.

However, there are several difficulties in employing this approach. First, the populations should be well adapted initially so that there is little scope for beneficial mutations. (Alternatively, experimental mutagenesis may be stopped after some time followed by a test of whether selection is able to improve fitness by purging the load.) Second, to test the theory quantitatively, it is necessary to estimate the mutation rate in each treatment. Modern advances in sequencing allow direct estimates of the total mutation rate to be obtained more easily. However, discerning the fraction of mutations that are deleterious in the lab is difficult (Davies et al. 1999, Bégin & Schoen 2006). Classic mutation accumulation studies with fitness measures may need to be combined with direct estimates of μ to determine the deleterious mutation rate under each treatment. Third, it can take a long time to approach mutation-selection balance. For example, if the mutation rate is increased by $\Delta U = 1$ deleterious mutation per genome per generation, then it would take ~120 generations for fitness to decline 80% of its expected amount if $s = 0.01$ (and longer if s or ΔU is smaller). However, by obtaining a time series of data, one could infer the equilibrium load after tens or hundreds of generations even if equilibrium is not reached. Finally, adaptation to the experimental mutagenesis procedure may introduce unwanted differences between populations (Nothel 1987). In principle, this can be tested by comparing the sensitivity to mutagenesis in control populations with that in populations that have evolved with it.

Despite these difficulties, experimental evolution offers the most direct way to test mutation load theory. However, few studies have employed this approach (Sankaranarayanan 1964, 1965; Tobari & Murata 1970a,b; Springman et al. 2010), and all suffer from poor replication or lack of proper controls (see also Anderson et al. 2004). Perhaps the best study was conducted by Bruce Wallace. He maintained populations of *D. melanogaster* with and without chronic exposure to gamma radiation for ~120–150 generations (Wallace 1956, 1959, 1991). Though he collected an impressive amount of data over a four-year period, he had several treatments and none of them were replicated, so it is difficult to draw strong conclusions. Taken at face value, the observed fitness reductions were 5–30% (depending on which fitness measure and which reference are used). Unfortunately, it is impossible to assess whether these numbers accord with Haldane's expectation because there is no quantitative estimate of how much radiation elevated the mutation rate (ΔU). Though such studies are challenging, there are clearly a number of fundamental issues that remain to be tested using experimental evolution.

Although direct testing of load remains the primary challenge, there are other avenues requiring further exploration. Mutation rate is the most important parameter affecting load, and sequencing technologies are making this easier to measure. We can now begin more detailed studies of the variation in mutation rate both within species as well as between closely related (and potentially competing) species. Predicting ecological consequences of load will also require an understanding of how different types of ecologically relevant traits, including male reproductive success (Mack et al. 2000, Whitlock & Agrawal 2009, Mallet et al. 2011), are affected by mutation. The extent of pleiotropy between different traits is a key issue but will be challenging to resolve for typical mutations of small effect. Finally, the nature of selection with respect to both gene interaction (epistasis, dominance) and environmental effects (whether selection changes predictably across the genome) will continue to be an issue of importance.

RELEVANCE OF LOAD IN BASIC AND APPLIED PROBLEMS

By its formal definition, mutation load refers to the reduction in the average fitness of individuals due to mutation. As such, it is relevant to those situations where we are interested in the absolute fitness of individuals relative to what it might be otherwise. One obvious area of concern is human health, as highlighted by Muller's classic 1950 paper, as well as by several prominent papers since (Crow 1997, Lynch 2010). To some extent, mutation load is not a treatable problem because it is so diffuse (i.e., many thousands of rare deleterious alleles of small effect affecting fitness in a variety of ways). Nonetheless, Muller (1950) and Crow (1997) warned against escalating the problem by increasing the mutation rate, either through increased exposure to mutagens in our environment or by shifting reproduction to later ages when mutation rates are higher. They also pointed out that the relaxation of selection by medical intervention allows for the accumulation of more mutations. In the long term, the realized mutation load is only reduced so long as medical technology continues to outpace the higher frequency of deleterious alleles.

Of course, mutation load is also relevant to the health of other organisms with high genomic mutation rates. Mutation load likely affects equilibrium population densities and could play a role in determining the outcome of interspecific competition. As such, mutation load may be subject to species-level selection. Selection among species is the stock-in-trade of community ecology, but it has had a more mixed reception in evolutionary biology (Whitlock 1996). Nonetheless, this idea is at the root of theoretical models using load-based explanations to provide an advantage of sexual populations over obligately asexual populations (Kondrashov 1982, Charlesworth 1990, Agrawal 2001, Agrawal & Chasnov 2001, Siller 2001) but is equally applicable to ecological competition between related sexual taxa. For example, Kawecki (1994) suggested that the evolution of ecological specialization could be enhanced by the greater genetic load of generalists (see also Fry 1996, Holt 1996, Whitlock 1996). Mutation load can also be a conservation concern because load (*a*) lowers initial vital rates and (*b*) is the source of the segregating deleterious alleles that can drift to fixation (i.e., the conversion of mutation load to drift load), potentially leading to mutational meltdowns (Lynch et al. 1995).

Recently, the possibility of using mutation load as a means of treating bacterial and viral infection has been studied (Loeb et al. 1999, Eigen 2002, Anderson et al. 2004, Bull et al. 2007, Bull & Wilke 2008, Martin & Gandon 2010). In lethal mutagenesis, drugs target pathogen DNA or RNA replication systems, increasing the mutation rate and driving up the load. If the load becomes sufficiently high, pathogens can be driven to lower densities or to extinction. Both theoretical (Eigen 2002, Bull et al. 2007, Bull & Wilke 2008, Martin & Gandon 2010) and empirical (Loeb et al. 1999, Crotty et al. 2001, Anderson et al. 2004) studies support the potential utility of this idea. As with other disease treatments, the evolution of resistance is possible, though it may be selected less directly than in response to typical treatments (Freistadt et al. 2004, Martin & Gandon 2010).

If we use the term mutation load more liberally to refer to the consequences of mutation-selection balance, then mutation load is relevant to a whole host of other issues, including inbreeding depression and the evolution of outcrossing (Lande & Schemske 1985, Charlesworth et al. 1990), the maintenance of genetic variance in fitness (Charlesworth & Hughes 1999), the evolution of sexually selected male traits and female preference (Rowe & Houle 1996, Houle & Kondrashov 2002), and rates of adaptation under the influence of background selection (Charlesworth 2012).

CONCLUSIONS: THE SUSTAINABILITY OF LOAD

Given current estimates of U, classical theory would predict large loads for many multicellular eukaryotes. Haldane's theory applied with a recent estimate of $U = 2.2$ (Keightley 2012) would predict the average person to be ~90% less fit than a mutant-free competitor (i.e., $L \approx 0.9$). Some have argued (e.g., Kondrashov 1988, Crow 1997, Nachman & Crowell 2000) that the high load predicted by Haldane's calculation is incompatible with the continued persistence of a species such as ours, which has a relatively low reproductive capacity. The conclusion, some researchers claim, is that mutations must act nonmultiplicatively to reconcile the high genomic mutation rate with the continued success of the human population. Although we agree that the increased efficiency of selection under some models of epistasis and inbreeding can play an important role in determining the exact value of load, we also argue that there is no reason to say that the load predicted by Haldane's equation is not sustainable by humans.

First, remember that the low reproductive capacity of humans is for typical (loaded) individuals, not unloaded ones. It is a mistake to ask whether our current population could withstand a further 90% reduction in fitness because that would be applying the load twice. More importantly, we have emphasized that load has no direct relationship to population abundance or persistence. Instead, mutation load refers to the reduction in fitness of individuals, not populations, relative to a mutation-free reference genotype. To ask whether a given amount of load is plausible, we can ask what mutation load predicts the fitness of an unloaded individual to be.

Some have skeptically asked, "Is it reasonable that an unloaded genotype placed into the current population would produce $1/(1 - L) \approx 10$ times as many offspring as the average?" It is difficult to envision what a mutant-free hominid would be capable of given that it is unlikely that a hominid with fewer than 1,000 deleterious alleles has ever existed (see Introduction, above). Even limiting ourselves to data, it is clear that human reproductive capacity is greater than what happens on average. For example, to sustain a population, each female needs to produce on average only two adult offspring, whereas the world record is 67 (Glenday 2010), which is greater than predicted for the mean of an unloaded genotype, and males have the capacity for even greater reproductive success. (Remember, load can affect fitness through any fitness component, so the fitness deficit may be expressed through lowered fecundity and lower mating success, as well as through mortality.) Nevertheless, physiological constraints on reproductive capacity may limit the fitness of an ideal genotype so that it is only slightly better than a typical genotype when both are given unlimited resources. However, such constraints do not set an upper limit on load.

The question posed at the beginning of the previous paragraph is not quite correct because it focuses on putting an unloaded individual into a loaded population rather than the reverse. The load as calculated under Haldane's assumption more accurately reflects how much less fit the average individual would be than its neighbors if it were placed into a population consisting only of mutant-free genotypes. In competition for limited resources (e.g., food, shelter, mates) against a population of unloaded individuals, a typical, heavily loaded individual may do very poorly, being only one-tenth (or less) as fit. In other words, a high maximal absolute fitness of the ideal genotype is not needed to explain a high load. Rather, we need only that the ideal genotype be much more fit than the typical individual when in competition, which can result from the low absolute fitness of the latter rather than the high fitness of the former. In principle, a load of any magnitude is compatible with species persistence because heavily loaded individuals can have high absolute fitness when competing against one another, even though each would have negligible fitness if forced to compete in a population of unloaded individuals.

The idea that species cannot persist with high loads, independent of other assumptions, is incorrect. However, load can reduce population sizes (even with density-dependent regulation) and possibly cause extinction, but the magnitude of these effects depends heavily on other assumptions. As we argue above, we know that populations can persist with their current values of load, but this does not imply that populations with even greater loads could persist in the face of competition with other, possibly less loaded, species.

Because deleterious mutations happen, individuals must be less fit than they could be. In principle, we could infer how fit individuals might be if we knew how load worked, but we do not. We do not even know whether mutation rate affects fitness in the manner Haldane predicted 70 years ago. If it does not, then why? Regardless, large loads have the potential to have ecological consequences. Is load (or relative loads) an important aspect of ecology? These are some of the simplest, yet most pressing, questions regarding mutation load that have remained unanswered for decades. Hopefully, they will not remain so for decades more.

DISCLOSURE STATEMENT

The authors are not aware of any affiliations, memberships, funding, or financial holdings that might be perceived as affecting the objectivity of this review.

ACKNOWLEDGMENTS

This paper is dedicated to the fond memory of James F. Crow. This work was supported by the Natural Science and Engineering Research Council (Canada). Thanks to Peter Abrams, Charlie Baer, Doug Futuyma, Peter Keightley, Sally Otto, Nathaniel Sharp, and Adam Eyre-Walker for discussions and helpful comments on the manuscript.

LITERATURE CITED

1000 Genomes Project Consortium. 2010. A map of human genome variation from population-scale sequencing. *Nature* 467:1061–73

Abrams PA. 2003. Effects of altered resource consumption rates by one consumer species on a competitor. *Ecol. Lett.* 6:550–55

Abrams PA, Brassil CE, Holt RD. 2003. Dynamics and responses to mortality rates of competing predators undergoing predator-prey cycles. *Theor. Popul. Biol.* 64:163–76

Agrawal AF. 2001. Sexual selection and the maintenance of sexual reproduction. *Nature* 411:692–95

Agrawal AF. 2002. Genetic loads under fitness-dependent mutation rates. *J. Evol. Biol.* 15:1004–10

Agrawal AF. 2010. Ecological determinants of mutation load and inbreeding depression in subdivided populations. *Am. Nat.* 176:111–22

Agrawal AF, Chasnov JR. 2001. Recessive mutations and the maintenance of sex in structured populations. *Genetics* 158:913–17

Agrawal AF, Wang AD. 2008. Increased transmission of mutations by low-condition females: evidence for condition-dependent DNA repair. *PLoS Biol.* 6:e30

Agrawal AF, Whitlock MC. 2010. Environmental duress and epistasis: How does stress affect the strength of selection on new mutations? *Trends Ecol. Evol.* 25:450–58

Agrawal AF, Whitlock MC. 2011. Inferences about the distribution of dominance drawn from yeast gene knockout data. *Genetics* 187:553–66

Anderson JP, Daifuku R, Loeb LA. 2004. Viral error catastrophe by mutagenic nucleosides. *Annu. Rev. Microbiol.* 58:183–205

Baer CF, Miyamoto MM, Denver DR. 2007. Mutation rate variation in multicellular eukaryotes: causes and consequences. *Nat. Rev. Genet.* 8:619–31

Baer CF, Shaw F, Steding C, Baumgartner M, Hawkins A, et al. 2005. Comparative evolutionary genetics of spontaneous mutations affecting fitness in rhabditid nematodes. *Proc. Natl. Acad. Sci. USA* 102:5785–90

Bataillon T, Kirkpatrick M. 2000. Inbreeding depression due to mildly deleterious mutations in finite populations: Size does matter. *Genet. Res.* 75:75–81

Bégin M, Schoen DJ. 2006. Low impact of germline transposition on the rate of mildly deleterious mutation in *Caenorhabditis elegans*. *Genetics* 174:2129–36

Boyko AR, Williamson SH, Indap AR, Degenhardt JD, Hernandez RD, et al. 2008. Assessing the evolutionary impact of amino acid mutations in the human genome. *PLoS Genet.* 4:e1000083

Bull JJ, Sanjuán R, Wilke CO. 2007. Theory of lethal mutagenesis for viruses. *J. Virol.* 81:2930–39

Bull JJ, Wilke CO. 2008. Lethal mutagenesis of bacteria. *Genetics* 180:1061–70

Charlesworth B. 1990. Mutation-selection balance and the evolutionary advantage of sex and recombination. *Genet. Res.* 55:199–221

Charlesworth B. 2012. The effects of deleterious mutations on evolution at linked sites. *Genetics* 190:5–22

Charlesworth B, Charlesworth D. 2010. *Elements of Evolutionary Genetics*. Greenwood Village, CO: Roberts and Co.

Charlesworth B, Hughes KA. 1999. The maintenance of genetic variation in life-history traits. In *Evolutionary Genetics: From Molecules to Morphology*, ed. RS Singh, CB Krimbas, pp. 369–92. Cambridge, UK: Cambridge Univ. Press

Charlesworth D, Morgan MT, Charlesworth B. 1990. Inbreeding depression, genetic load, and the evolution of outcrossing rates in a multilocus system with no linkage. *Evolution* 44:1469–89

Clarke B. 1973a. Effect of mutation on population size. *Nature* 242:196–97

Clarke B. 1973b. Mutation and population size. *Heredity* 31:367–79

Conrad DF, Keebler JEM, DePristo MA, Lindsay SJ, Zhang YJ, et al. 2011. Variation in genome-wide mutation rates within and between human families. *Nat. Genet.* 43:712–14

Crotty S, Cameron CE, Andino R. 2001. RNA virus error catastrophe: direct molecular test by using ribavirin. *Proc. Natl. Acad. Sci. USA* 98:6895–900

Crow JF. 1970. Genetic loads and the cost of natural selection. In *Mathematical Topics in Population Genetics*, ed. K-I Kojima, pp. 128–77. Berlin: Springer-Verlag

Crow JF. 1993. Mutation, mean fitness, and genetic load. In *Oxford Surveys in Evolutionary Biology*, ed. D Futuyma, J Antonovics, pp. 3–42. Oxford, UK: Oxford Univ. Press

Crow JF. 1997. The high spontaneous mutation rate: Is it a health risk? *Proc. Natl. Acad. Sci. USA* 94:8380–86

Crow JF, Kimura M. 1970. *An Introduction to Population Genetics Theory*. Minneapolis: Burgess Publ.

Cutter AD. 2008. Divergence times in *Caenorhabditis* and *Drosophila* inferred from direct estimates of the neutral mutation rate. *Mol. Biol. Evol.* 25:778–86

Davies EK, Peters AD, Keightley PD. 1999. High frequency of cryptic deleterious mutations in *Caenorhabditis elegans*. *Science* 285:1748–51

de Visser JAGM, Hoekstra RF, van den Ende H. 1997a. An experimental test for synergistic epistasis and its application in *Chlamydomonas*. *Genetics* 145:815–19

de Visser JAGM, Hoekstra RF, van den Ende H. 1997b. Test of interaction between genetic markers that affect fitness in *Aspergillus niger*. *Evolution* 51:1499–505

Denver DR, Dolan PC, Wilhelm LJ, Sung W, Lucas-Lledo JI, et al. 2009. A genome-wide view of *Caenorhabditis elegans* base-substitution mutation processes. *Proc. Natl. Acad. Sci. USA* 106:16310–14

Denver DR, Morris K, Lynch M, Thomas WK. 2004. High mutation rate and predominance of insertions in the *Caenorhabditis elegans* nuclear genome. *Nature* 430:679–82

Drake JW, Charlesworth B, Charlesworth D, Crow JF. 1998. Rates of spontaneous mutation. *Genetics* 148:1667–86

Eigen M. 2002. Error catastrophe and antiviral strategy. *Proc. Natl. Acad. Sci. USA* 99:13374–76

Elena SF, Lenski RE. 1997. Test of synergistic interactions among deleterious mutations in bacteria. *Nature* 390:395–98

Eöry L, Halligan DL, Keightley PD. 2010. Distributions of selectively constrained sites and deleterious mutation rates in the hominid and murid genomes. *Mol. Biol. Evol.* 27:177–92

Estes S, Phillips PC, Denver DR, Thomas WK, Lynch M. 2004. Mutation accumulation in populations of varying size: the distribution of mutational effects for fitness correlates in *Caenorhabditis elegans*. *Genetics* 166:1269–79

Eyre-Walker A, Woolfit M, Phelps T. 2006. The distribution of fitness effects of new deleterious amino acid mutations in humans. *Genetics* 173:891–900

Fawcett TW, Johnstone RA. 2003. Mate choice in the face of costly competition. *Behav. Ecol.* 14:771–79

Freistadt MS, Meades GD, Cameron CE. 2004. Lethal mutagens: broad-spectrum antivirals with limited potential for development of resistance? *Drug Resist. Updat.* 7:19–24

Fry JD. 1996. The evolution of host specialization: Are trade-offs overrated? *Am. Nat.* 148:S84–107

Glémin S. 2003. How are deleterious mutations purged? Drift versus nonrandom mating. *Evolution* 57:2678–87

Glémin S, Ronfort J, Bataillon T. 2003. Patterns of inbreeding depression and architecture of the load in subdivided populations. *Genetics* 165:2193–212

Glenday C, ed. 2010. *Guinness World Records 2011*. New York: Bantam Books

Goho S, Bell G. 2000. Mild environmental stress elicits mutations affecting fitness in *Chlamydomonas*. *Proc. R. Soc. B* 267:123–29

Haag CR, Roze D. 2007. Genetic load in sexual and asexual diploids: segregation, dominance and genetic drift. *Genetics* 176:1663–78

Haag-Liautard C, Dorris M, Maside X, Macaskill S, Halligan DL, et al. 2007. Direct estimation of per nucleotide and genomic deleterious mutation rates in *Drosophila*. *Nature* 445:82–85

Haldane JBS. 1937. The effect of variation on fitness. *Am. Nat.* 71:337–49

Haldane JBS. 1957. The cost of natural selection. *J. Genet.* 55:511–24

Halligan DL, Keightley PD. 2009. Spontaneous mutation accumulation studies in evolutionary genetics. *Annu. Rev. Ecol. Evol. Syst.* 40:151–72

Hansen TF, Wagner GP. 2001. Epistasis and the mutation load: a measurement-theoretical approach. *Genetics* 158:477–85

Holsinger KE, Pacala SW. 1990. Multiple-niche polymorphisms in plant populations. *Am. Nat.* 135:301–9

Holt RD. 1996. Adaptive evolution in source-sink environments: direct and indirect effects of density-dependence on niche evolution. *Oikos* 75:182–92

Houle D, Kondrashov AS. 2002. Coevolution of costly mate choice and condition-dependent display of good genes. *Proc. R. Soc. B* 269:97–104

Howard RS, Lively CM. 1998. The maintenance of sex by parasitism and mutation accumulation under epistatic fitness functions. *Evolution* 52:604–10

Jasnos L, Korona R. 2007. Epistatic buffering of fitness loss in yeast double deletion strains. *Nat. Genet.* 39:550–54

Jasnos L, Tomala K, Paczesniak D, Korona R. 2008. Interactions between stressful environment and gene deletions alleviate the expected average loss of fitness in yeast. *Genetics* 178:2105–11

Kawecki TJ. 1994. Accumulation of deleterious mutations and the evolutionary cost of being a generalist. *Am. Nat.* 144:833–38

Kawecki TJ. 1995. Demography of source-sink populations and the evolution of ecological niches. *Evol. Ecol.* 9:38–44

Kawecki TJ, Barton NH, Fry JD. 1997. Mutational collapse of fitness in marginal habitats and the evolution of ecological specialisation. *J. Evol. Biol.* 10:407–29

Keightley PD. 1996. Metabolic models of selection response. *J. Theor. Biol.* 182:311–16

Keightley PD. 2012. Rates and fitness consequences of new mutations in humans. *Genetics* 190:295–304

Keightley PD, Eyre-Walker A. 2007. Joint inference of the distribution of fitness effects of deleterious mutations and population demography based on nucleotide polymorphism frequencies. *Genetics* 177:2251–61

Keightley PD, Halligan DL. 2009. Analysis and implications of mutational variation. *Genetica* 136:359–69

Keightley PD, Lynch M. 2003. Toward a realistic model of mutations affecting fitness. *Evolution* 57:683–85

Keightley PD, Otto SP. 2006. Interference among deleterious mutations favours sex and recombination in finite populations. *Nature* 443:89–92

Kimura M, Maruyama T. 1966. The mutational load with epistatic interactions in fitness. *Genetics* 54:1337–51

Kimura M, Maruyama T, Crow JF. 1963. Mutation load in small populations. *Genetics* 48:1303–12

King JL. 1966. The gene interaction component of genetic load. *Genetics* 53:403–13

King JL. 1967. Continuously distributed factors affecting fitness. *Genetics* 55:483–92

Kishony R, Leibler S. 2003. Environmental stresses can alleviate the average deleterious effect of mutations. *J. Biol.* 2:14

Kondrashov AS. 1982. Selection against harmful mutations in large sexual and asexual populations. *Genet. Res.* 40:325–32

Kondrashov AS. 1988. Deleterious mutations and the evolution of sexual reproduction. *Nature* 336:435–40

Kondrashov AS. 1995. Contamination of the genome by very slightly deleterious mutations: Why have we not died 100 times over? *J. Theor. Biol.* 175:583–94

Kondrashov AS, Crow JF. 1988. King's formula for the mutation load with epistasis. *Genetics* 120:853–56

Kondrashov AS, Crow JF. 1993. A molecular approach to estimating the human deleterious mutation rate. *Hum. Mutat.* 2:229–34

Laffafian A, King JD, Agrawal AF. 2010. Variation in the strength and softness of selection on deleterious mutations. *Evolution* 64:3232–41

Lande R, Schemske DW. 1985. The evolution of self-fertilization and inbreeding depression in plants. I. Genetic models. *Evolution* 39:24–40

Loeb LA, Essigmann JM, Kazazi F, Zhang J, Rose KD, Mullins JI. 1999. Lethal mutagenesis of HIV with mutagenic nucleoside analogs. *Proc. Natl. Acad. Sci. USA* 96:1492–97

Lynch M. 2007. *The Origins of Genome Architecture*. Sunderland, MA: Sinauer Assoc.

Lynch M. 2010. Rate, molecular spectrum, and consequences of human mutation. *Proc. Natl. Acad. Sci. USA* 107:961–68

Lynch M, Conery JS. 2003. The origins of genome complexity. *Science* 302:1401–4

Lynch M, Conery J, Bürger R. 1995. Mutation accumulation and the extinction of small populations. *Am. Nat.* 146:489–518

MacArthur RH. 1972. *Geographical Ecology*. New York: Harper and Row

Mack PD, Lester VK, Promislow DEL. 2000. Age-specific effects of novel mutations in *Drosophila melanogaster*. II. Fecundity and male mating ability. *Genetica* 110:31–41

Mallet MA, Bouchard JM, Kimber CM, Chippindale AK. 2011. Experimental mutation-accumulation on the X chromosome of *Drosophila melanogaster* reveals stronger selection on males than females. *BMC Evol. Biol.* 11:156

Manna F, Martin G, Lenormand T. 2012. Fitness landscapes: an alternative theory for the dominance of a mutation. *Genetics* 189:923–37

Martin G, Elena SF, Lenormand T. 2007. Distributions of epistasis in microbes fit predictions from a fitness landscape model. *Nat. Genet.* 39:555–60

Martin G, Gandon S. 2010. Lethal mutagenesis and evolutionary epidemiology. *Philos. Trans. R. Soc. B* 365:1953–63

Martin G, Lenormand T. 2006. The fitness effect of mutations across environments: a survey in light of fitness landscape models. *Evolution* 60:2413–27

Matessi C, Gatto M. 1984. Does *K*-selection imply prudent predation? *Theor. Popul. Biol.* 25:347–63

Mather K. 1969. Selection through competition. *Heredity* 24:529–49

Mukai T. 1969. Genetic structure of natural populations of *Drosophila melanogaster*. VII. Synergistic interaction of spontaneous mutant polygenes controlling viability. *Genetics* 61:749–61

Muller HJ. 1950. Our load of mutations. *Am. J. Hum. Genet.* 2:111–76

Nachman MW, Crowell SL. 2000. Estimate of the mutation rate per nucleotide in humans. *Genetics* 156:297–304

Nothel H. 1987. Adaptation of *Drosophila melanogaster* populations to high mutation pressure: evolutionary adjustment of mutation rates. *Proc. Natl. Acad. Sci. USA* 84:1045–49

Otto SP, Goldstein DB. 1992. Recombination and the evolution of diploidy. *Genetics* 131:745–51

Peck JR, Waxman D. 2000. Mutation and sex in a competitive world. *Nature* 406:399–404

Peters AD, Keightley PD. 2000. A test for epistasis among induced mutations in *Caenorhabditis elegans*. *Genetics* 156:1635–47

Phadnis N, Fry JD. 2005. Widespread correlations between dominance and homozygous effects of mutations: implications for theories of dominance. *Genetics* 171:385–92

Phillips N, Salomon M, Custer A, Ostrow D, Baer CF. 2009. Spontaneous mutational and standing genetic (co)variation at dinucleotide microsatellites in *Caenorhabditis briggsae* and *Caenorhabditis elegans*. *Mol. Biol. Evol.* 26:659–69

Reed FA, Aquadro CF. 2006. Mutation, selection and the future of human evolution. *Trends Genet.* 22:479–84

Rice WR. 1998. Requisite mutational load, pathway epistasis and deterministic mutation accumulation in sexual versus asexual populations. *Genetica* 102–103:71–81

Rowe L, Houle D. 1996. The lek paradox and the capture of genetic variance by condition dependent traits. *Proc. R. Soc. B* 263:1415–21

Roze D. 2012. Spatial heterogeneity in the strength of selection against deleterious alleles may strongly reduce the mutation load. *Heredity* 109:137–45

Roze D, Rousset FO. 2004. Joint effects of self-fertilization and population structure on mutation load, inbreeding depression and heterosis. *Genetics* 167:1001–15

Sanjuán R, Elena SF. 2006. Epistasis correlates to genomic complexity. *Proc. Natl. Acad. Sci. USA* 103:14402–5

Sanjuán R, Moya A, Elena SF. 2004. The contribution of epistasis to the architecture of fitness in an RNA virus. *Proc. Natl. Acad. Sci. USA* 101:15376–79

Sanjuán R, Nebot MR. 2008. A network model for the correlation between epistasis and genomic complexity. *PLoS ONE* 3:e2663

Sankaranarayanan K. 1964. Genetic loads in irradiated experimental populations of *Drosophila melanogaster*. *Genetics* 50:131–50

Sankaranarayanan K. 1965. Further data on the genetic loads in irradiated experimental populations of *Drosophila melanogaster*. *Genetics* 52:153–64

Schoener TW. 1973. Population growth regulated by intraspecific competition for energy or time: some simple representations. *Theor. Popul. Biol.* 4:56–84

Schoener TW. 1976. Alternatives to Lotka-Volterra competition: models of intermediate complexity. *Theor. Popul. Biol.* 10:309–33

Schreiber S, Rudolf VHW. 2008. Crossing habitat boundaries: coupling dynamics of ecosystems through complex life cycles. *Ecol. Lett.* 11:576–87

Segrè D, DeLuna A, Church GM, Kishony R. 2005. Modular epistasis in yeast metabolism. *Nat. Genet.* 37:77–83

Sharp NP, Agrawal AF. 2009. Sexual selection and the random union of gametes: testing for a correlation in fitness between mates in *Drosophila melanogaster*. *Am. Nat.* 174:613–22

Sharp NP, Agrawal AF. 2012. Evidence for elevated mutation rates in low-quality genotypes. *Proc. Natl. Acad. Sci. USA* 109(16):6142–46

Shaw FH, Baer CF. 2011. Fitness-dependent mutation rates in finite populations. *J. Evol. Biol.* 24:1677–84

Siller S. 2001. Sexual selection and the maintenance of sex. *Nature* 411:689–92

Simmons MJ, Crow JF. 1977. Mutations affecting fitness in *Drosophila* populations. *Annu. Rev. Genet.* 11:49–78

Springman R, Keller T, Molineux IJ, Bull JJ. 2010. Evolution at a high imposed mutation rate: adaptation obscures the load in phage T7. *Genetics* 184:221–32

Szathmáry E. 1993. Do deleterious mutations act synergistically? Metabolic control theory provides a partial answer. *Genetics* 133:127–32

Tobari I, Murata M. 1970a. Changes of genetic loads in experimental populations of *Drosophila melanogaster* with radiation histories. *Jpn. J. Genet.* 45:387–97

Tobari I, Murata M. 1970b. Effects of X-rays on genetic loads in a cage population of *Drosophila melanogaster*. *Genetics* 65:107–19

Turner JRG, Williamson MH. 1968. Population size, natural selection and genetic load. *Nature* 218:700

Wallace B. 1956. Studies on irradiated populations of *Drosophila melanogaster*. *J. Genet.* 54:280–93

Wallace B. 1959. Studies of the relative fitnesses of experimental populations of *Drosophila melanogaster*. *Am. Nat.* 93:295–314

Wallace B. 1968. Polymorphism, population size, and genetic load. In *Population Biology and Evolution*, ed. RC Lewontin, pp. 87–108. Syracuse, NY: Syracuse Univ. Press

Wallace B. 1970. *Genetic Load. Its Biological and Conceptual Aspects*. Englewood Cliffs, NJ: Prentice-Hall

Wallace B. 1975. Hard and soft selection revisited. *Evolution* 29:465–73

Wallace B. 1991. *Fifty Years of Genetic Load: An Odyssey*. Ithaca, NY: Cornell Univ. Press

Wang AD, Sharp NP, Spencer CC, Tedman-Aucoin K, Agrawal AF. 2009. Selection, epistasis, and parent-of-origin effects on deleterious mutations across environments in *Drosophila melanogaster*. *Am. Nat.* 174:863–74

West SA, Peters AD, Barton NH. 1998. Testing for epistasis between deleterious mutations. *Genetics* 149:435–44

Whitlock MC. 1996. The Red Queen beats the Jack-of-All-Trades: the limitations on the evolution of phenotypic plasticity and niche breadth. *Am. Nat.* 148:S65–77

Whitlock MC. 2002. Selection, load, and inbreeding depression in a large metapopulation. *Genetics* 160:1191–202

Whitlock MC, Agrawal AF. 2009. Purging the genome with sexual selection: reducing mutational load through selection on males. *Evolution* 63:569–82

Whitlock MC, Bourguet D. 2000. Factors affecting the genetic load in *Drosophila*: synergistic epistasis and correlations among fitness components. *Evolution* 54:1654–60

From Animalcules to an Ecosystem: Application of Ecological Concepts to the Human Microbiome

Noah Fierer,[1,2] Scott Ferrenberg,[1] Gilberto E. Flores,[2]
Antonio González,[3] Jordan Kueneman,[1] Teresa Legg,[1]
Ryan C. Lynch,[1] Daniel McDonald,[4]
Joseph R. Mihaljevic,[1] Sean P. O'Neill,[1,5]
Matthew E. Rhodes,[1] Se Jin Song,[1]
and William A. Walters[6]

[1]Department of Ecology and Evolutionary Biology, [2]Cooperative Institute for Research in Environmental Sciences, [3]Department of Computer Science, [4]Biofrontiers Institute, [5]Institute of Arctic and Alpine Research, and [6]Department of Molecular, Cellular, and Developmental Biology, University of Colorado, Boulder, Colorado 80309; email: Noah.Fierer@colorado.edu

Annu. Rev. Ecol. Evol. Syst. 2012. 43:137–55

First published online as a Review in Advance on August 28, 2012

The *Annual Review of Ecology, Evolution, and Systematics* is online at ecolsys.annualreviews.org

This article's doi:
10.1146/annurev-ecolsys-110411-160307

1543-592X/12/1201-0137$20.00

Keywords

bacterial diversity, host-associated bacterial communities, microbial diversity, microbial ecology

Abstract

The human body is inhabited by billions of microbial cells and these microbial symbionts play critical roles in human health. Human-associated microbial communities are diverse, and the structure of these communities is variable across body habitats, through time, and between individuals. We can apply concepts developed by plant and animal ecologists to better understand and predict the spatial and temporal patterns in these communities. Due to methodological limitations and the largely unknown natural history of most microbial taxa, this integration of ecology into research on the human microbiome is still in its infancy. However, such integration will yield a deeper understanding of the role of the microbiome in human health and an improved ability to test ecological concepts that are more difficult to test in plant and animal systems.

1. INTRODUCTION

Like nearly all plants and animals, humans host a large number of microorganisms, both on and in our bodies. As we go about our daily lives, we are continually in the process of acquiring and shedding microbes, exchanging microbes directly and indirectly with friends, family members, strangers, and any environment with which we come into contact. However, the resulting structure of our microbial communities is not simply a product of this immigration and emigration of microbes. Current and past conditions of our body habitats, conditions that are largely a product of our anatomy, physiology, behavior, and immune system function, also affect the structure of our microbial communities. The converse is also true; the microbial communities living on and in our bodies can shape the characteristics of the human body in a myriad of ways.

Collectively, the human microbiome, which we define here as those microorganisms associated with the human body, is represented by 10^{14} to 10^{15} microbial cells and the majority of these cells are found in the large intestine (Bäckhed et al. 2005). In terms of numerical abundance, bacterial cells likely outnumber human cells by at least an order of magnitude (Savage 1977). The bacterial contribution to the total genetic diversity found within the human body is perhaps a more important consideration as the number of bacterial genes clearly outnumbers the number of genes in the human genome by several orders of magnitude (Qin et al. 2010). Although many of these bacterial genes may have no direct relevance to the health of the human host, the human microbiome has the potential to endow us with a large number of traits or characteristics that are not encoded in our own genome. In other words, our health is not simply a product of the genetic or epigenetic potential of human cells, but also a function of the structure and activities of our associated microbial communities.

Understanding the structure and function of the human microbiome requires knowledge from a wide range of disciplines—from immunology to genomics to microbial metabolism. Here we examine the human microbiome from an ecological perspective, exploring how ecological concepts derived largely from research on plant and animal communities may help us understand the structure and function of the human microbiome. We recognize that ours is not the first attempt to delve into the ecology of the human microbiome; there are a number of excellent reviews on the topic (e.g., Dethlefsen et al. 2007, Gonzalez et al. 2011, Robinson et al. 2010). We also recognize that the ecology of the human microbiome is an impossibly large topic to summarize in this space. Essentially every topic in the field of ecology, from autecology to biogeography, is relevant to understanding the spatial and temporal patterns exhibited by human-associated microbes. We focus on selected topics that represent key knowledge gaps in our understanding of the ecology of the human microbiome. Although we primarily focus on bacteria, we recognize that the human body is also home to other microbial taxa, including fungi, microeukaryotes, and archaea, which can have important effects on the health of the human host. Likewise, we do not devote a lot of attention to viruses as researchers are only now beginning to document the diversity of viruses found in the human body and their role in the human microbiome (e.g., Minot et al. 2011, Reyes et al. 2010).

2. RELEVANCE OF THE MICROBIOME TO HUMAN HEALTH

Research on the human microbiome is as old as the field of microbiology. The first person to observe and formally describe microbes was Antonie van Leeuwenhoek, who in the late seventeenth century looked at his own saliva through a rudimentary microscope and noted: "I then most always saw, with great wonder, that in the said matter there were many very little living animalcules, very prettily a-moving." In subsequent centuries, research on the human microbiome expanded as microbiologists began to discover that specific bacterial taxa caused diseases, with many of these pathogens fulfilling Koch's postulates. Although the limitations of Koch's postulates are well known, they remain a cornerstone of medical microbiology because they provide a useful framework for the identification of new pathogens. However, the reliance on Koch's postulates (or variants thereof, see Fredericks & Relman 1996) may have directed the focus of medical microbiology toward diseases that could be linked to a specific taxon and away from diseases associated with changes in multiple taxa within a microbial community. The recent launch of the Human Microbiome Project (**http://commonfund.nih.gov/hmp/**), MetaHIT (**http://www.metahit.eu/**), and other related large-scale research efforts signals a shift of focus as both medical professionals and research microbiologists are increasingly recognizing the importance of the underlying structure of the entire human microbiome for human health.

The human microbiome can affect human health in many ways, and new functions of the human microbiome are being discovered on a regular basis. For example, we know that the human microbiome can alter host susceptibility to microbial pathogens, aid in the digestion of complex polysaccharides, produce metabolites required by the host (e.g., vitamins or specific amino acids), modulate and educate the immune system, regulate environmental conditions within body habitats, and influence tissue development (Gill et al. 2006, Pflughoeft & Versalovic 2011, Robinson et al. 2010). Host-microbiome interactions span the spectrum of being beneficial to the host, having no detectable influence on host health (a commensal relationship), or having a net negative effect on the health of the host. For example, the pathogenic bacterium *Clostridium difficile* can, in some cases, become more abundant and permanent members of the human microbiome (McFarland 2008). Likewise, there is a wide range of diseases that have been linked to dysbioses, causing changes in resident microbial communities that are associated with negative effects on host health. Such diseases include bacterial vaginosis, Crohn's disease, psoriasis, gingivitis, obesity, antibiotic-associated diarrhea, and irritable bowel syndrome (Frank et al. 2011, Pflughoeft & Versalovic 2011). However, causality for dysbiosis is often difficult to determine; the microbiome could be causing the disease or the resident microbial communities could be reflecting changes in the host brought about by the disease state. In either case, ecological investigations of human-associated microbial communities have the potential to directly affect the way in which we prevent diseases, identify onset of disease, design treatments, and track disease recovery.

3. DESCRIBING DIVERSITY IN THE HUMAN MICROBIOME

Only in the past few decades have microbiologists begun to fully appreciate the extent of microbial diversity found within the human body. Although we have known for centuries that the mouth, for example, harbors large numbers of bacteria (van Leeuwenhoek's animalcules), until recently, our understanding of their taxonomy and function was drawn almost entirely from pure culture studies of individual bacterial isolates grown in vitro. Most microbes are difficult to culture in isolation, leading to biases in culture-dependent surveys of microbial diversity (Pace 1997). Moreover, although culture-based investigations are vital to understanding the ecology, physiology, and genetics of individual taxa, microorganisms may exhibit different characteristics in vitro than in vivo. For example, streptococci, which are often associated with dental caries, rely on a suite of other bacteria in order to effectively colonize and reproduce on tooth surfaces (Jenkinson 2011). Thus, streptococcal ecology cannot be understood by solely studying these organisms in pure culture.

The development of culture-independent tools, particularly 16S rRNA gene sequence analysis, has greatly expanded our understanding of the microbial diversity in the human microbiome. We now know that the human mouth, for example, contains hundreds of microbial taxa, most of which have yet to be cultured (Jenkinson 2011). This revolution in our ability to describe human-associated microbial communities continues unabated. New tools that enable detailed examination of the human microbiome across large numbers of individuals are introduced nearly every month, including both sequencing platforms and data analysis approaches. Reviews of the methodological advances in human microbiome research (Hamady & Knight 2009, Kuczynski et al. 2012) point out that it is increasingly feasible for researchers to rapidly characterize microbial communities in thousands of samples. These advances enable us to move research on the human microbiome into the realm of ecology, describing and predicting the inter- and intraindividual variability in microbial communities and their functional capabilities.

Many of the approaches microbiologists use to describe the variability in microbial communities are conceptually similar to those approaches long used by plant and animal ecologists. However, there are important distinctions between microbial and "macrobial" (i.e., plant and animal) diversity surveys. Studies of plant and animal systems often rely on visible observations of taxa and their responses to biotic and abiotic stimuli. However, modern microbiome studies typically rely on DNA or RNA sequencing to infer the presence or relative abundance of organisms because visual observation of microbial taxa in situ is often difficult. Also, though microbiologists do have extensive knowledge about a limited number of well-studied, cultured organisms, we lack basic knowledge about the natural histories of most microbial taxa, even those taxa commonly found in the human body. Furthermore, because the average sizes and generation times of plant and animal species are orders of magnitude larger or longer than those of bacterial taxa, microbial ecologists need to address a unique set of spatial and temporal issues. These limitations make it difficult to directly compare those ecological phenomena observed in macrobial versus microbial systems. Nevertheless, they should not prevent us from applying ecological concepts (which were largely derived from plant and animal systems) to the study of the human microbiome.

4. ALPHA DIVERSITY IN THE HUMAN MICROBIOME AND HOST HEALTH

Alpha diversity is a key metric used by community ecologists and may provide important insight into community assembly patterns and the interactions between the microbiome and its human host. Alpha diversity is generally defined as richness (the number of taxa or lineages in a given sample), evenness (the relative abundance of taxa present within a given sample), or a metric that combines these two parameters (e.g., Shannon-Weiner diversity index). Although ecologists are increasingly describing alpha diversity patterns using metrics that incorporate phylogenetic (Faith & Baker 2006) and functional or trait-based information (Petchey & Gaston 2007), we focus here on the taxonomic diversity of human-associated microbial communities as this is the alpha diversity metric most commonly applied in microbial community analyses.

To an ecologist more familiar with plant and animal communities, the alpha diversity observed in human-associated bacterial communities is immense. Individual body habitats typically harbor dozens of bacterial phyla and hundreds, if not thousands, of individual bacterial phylotypes, canonically referred to as operational taxonomic units (OTUs). The notion of an OTU is essential as there is no consensus definition of what constitutes a bacterial species (Zhi et al. 2012). However, most of these OTUs are rare, and the majority of body habitats are dominated by just a few bacterial phyla (**Figure 1**). We also know that alpha diversity levels can vary dramatically between individuals, across body habitats, and within an individual body habitat over time (Caporaso et al. 2011, Costello et al. 2009). From a long history of research on the alpha diversity patterns exhibited by plant and animal communities, we know that a wide range of processes could generate the alpha diversity patterns observed within the human microbiome. Such

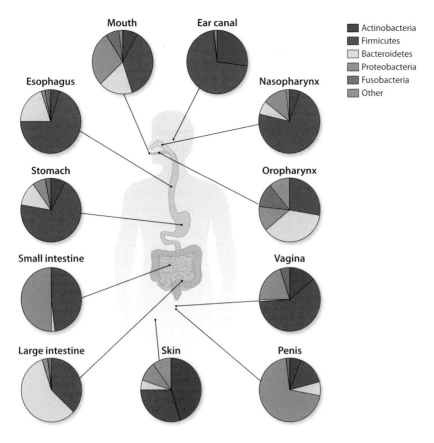

Figure 1

General patterns in the composition of bacterial communities in various body habitats. Pie charts illustrate percentage of 16S rRNA gene sequences representing the dominant bacterial phyla. Data compiled from Charlson et al. 2010, Costello et al. 2009, Dicksved et al. 2009, Frank et al. 2003, Hayashi et al. 2005, Kim et al. 2009, Pei et al. 2004, Price et al. 2010.

processes include environmental gradients, disturbance regimes, resource competition, predator-prey interactions, and niche differentiation (Ricklefs & Schluter 1993, Rosenzweig 1995), in addition to neutral processes (Hubbell 2001). Unfortunately, there are surprisingly few studies of the human microbiome that explicitly examine how these factors can influence alpha diversity patterns even though understanding the relationship between the alpha diversity of microbial communities and a disease state is one of the key questions in human microbiome research (Frank et al. 2011).

We know that certain diseases are associated with pronounced changes in alpha diversity within affected body habitats. However, the directional change associated with the disease state is not consistent across all diseases. Cystic fibrosis appears to cause an increase in the diversity of bacteria in patients' lungs (Harrison 2007), whereas Crohn's disease often results in a decrease in the microbial diversity of the intestine (Manichanh et al. 2006). Furthermore, it is often difficult to determine whether changes in alpha diversity are the cause or consequence of the disease state. For example, a study of human gut microbial communities found that there were differences in bacterial diversity between healthy individuals and those with inflammatory bowel disease (IBD; Frank et al. 2007); yet it was not clear if the healthy individuals were less susceptible to IBD because they had a diverse gut microbiota to begin with or if IBD caused a lower alpha diversity. For these reasons, alpha diversity may be a poor predictor of disease status (or susceptibility to disease). Nevertheless, changes in alpha diversity could provide a useful indicator of health status just as changes in plant diversity (Balvanera et al. 2006, Loreau et al. 2001) or animal diversity (Hudson et al. 2006) can sometimes be used as indicators of ecosystem health or function. Work on bacterial vaginosis lends support to this idea. In many cases, the onset of vaginosis is preceded by a large increase in bacterial diversity levels within the vagina (Lamont et al. 2011).

Although it is difficult to determine how shifts in alpha diversity within human-associated microbial communities may impact human health, we can apply concepts derived from research on plant and animal communities to speculate on possible linkages. In particular, we hypothesize that changes in alpha diversity may impact the stability of human-associated microbial communities and their susceptibility to invasion from microbial pathogens. Ecologists have long hypothesized that biodiversity might relate to these community-level properties of stability and invasibility (Elton 1958, MacArthur 1955). Despite some debate on this topic, there is increasing evidence from research in both terrestrial and aquatic systems that increases in alpha diversity may promote community stability and reduce invasibility. For example, higher alpha diversity tends to decrease the susceptibility of a wide range of ecosystem-level processes to disturbance events or environmental stressors (Cadotte et al. 2008, Cardinale et al. 2011, Hooper et al. 2005). Likewise, more diverse communities may be more resistant to invasion (Ives & Carpenter 2007). However, this is not always the case and ecologists have hypothesized that diverse communities may, in some cases, facilitate invasion by creating more niches than the endemic taxa can fill (Fridley et al. 2007). In addition, environmental conditions that favor highly diverse communities could also favor colonization by invasive taxa (Stohlgren et al. 2003). Still, a review of studies investigating links between plant community diversity and invasion found that higher alpha diversity was typically correlated with the reduced invasibility of communities at more local scales (Stohlgren et al. 2003).

There is some evidence that, within the human microbiome, more diverse communities may be less prone to microbial invaders. Blaser & Falkow (2009) have suggested that decreases in microbial diversity could create niches for microbial invaders including human pathogens. For example, antibiotic treatments typically reduce bacterial species richness, which may open up niches for pathogens such as *C. difficile* in the human gut (McFarland 2008) and *Pseudomonas aeruginosa* in the pulmonary tract (Flanagan et al. 2007). Unfortunately, we are not aware of any empirical studies that directly demonstrate a link between an increase, or decrease, in microbiome diversity and the onset of a disease in humans. Similarly, there is a dearth of empirical data indicating whether people with high or low diversity microbiomes are more prone or resistant to disease, although there is some evidence suggesting that such relationships may exist (Chang et al. 2008, Mazmanian et al. 2008, Packey & Sartor 2009).

5. BETA DIVERSITY IN THE HUMAN MICROBIOME

The human microbiome is frequently characterized as having high beta diversity, defined as the taxonomic or phylogenetic difference in community composition between samples [see Anderson et al. (2011) and Graham & Fine (2008) for an overview of the various beta diversity metrics]. Recent high-throughput studies reveal that even within a given body habitat, any pair of individuals share remarkably few bacterial taxa. For example, though the composition of gut microbial communities is more similar between family members than unrelated individuals, the relative abundances of the taxa found within the communities of family members can still vary by two orders of magnitude (Turnbaugh et al. 2009a). In addition, less than 30% of the microbial taxa sampled daily from specific habitats in a person's body were consistent community members in that body habitat over one month; after 120 days, less than 10% were consistent members (Caporaso et al. 2011). Although it is not a surprise that different body habitats harbor distinct bacterial communities (Costello et al. 2009), it is worth noting that the variability in bacterial communities across skin sites on one person, for example, exceeds the variability in community composition for a given skin location between individuals (Costello et al. 2009, Grice et al. 2009). All of this evidence suggests that at the finer levels of taxonomic resolution, a "core microbiome" (sensu Turnbaugh et al. 2007) is not likely to exist. Thus, one of the central questions in human microbiome research is: What mechanisms are responsible for these differences in the composition of human-associated microbial communities? As noted above, there is a long list of factors that can influence species diversity and cause high levels of beta diversity. For simplicity, we focus on niche and neutral theories that provide opposing explanations for beta diversity within and between individuals.

5.1. Niche Processes in the Human Microbiome

The majority of human microbiome studies conducted to date have focused on the role of niche processes in shaping beta diversity patterns. The niche perspective stresses the roles of environmental variables in driving patterns in community assembly and diversity. In fact, many human microbiome studies have found evidence supporting the role of niche-based processes in explaining beta diversity patterns. The significant differences between the microbial community structures of different body habitats provide a compelling example of niche differentiation in human-associated microbial communities (Costello et al. 2009). Within the skin body habitat, differences in microbial community composition are related to moisture content and pH of the skin (Grice et al. 2009). In addition, dietary habits (Ley et al. 2006) and obesity (Turnbaugh et al. 2006) have been shown to be correlated with differences in gut microbial community composition and phylogenetic structure.

However, there are limitations to niche-based approaches in human microbiome research. In most internal body habitats, it is difficult to accurately measure all of the environmental parameters, such as oxygen concentrations or moisture levels, which could be essential for defining microbial niches. Methodological constraints aside, one can imagine the difficulties associated with accurately characterizing the niches within the five to eight meters of the small intestine (for example) given that individual microbes are typically only a few microns in size. In particular, consider the environmental heterogeneity created by changes in the mucosal layers from the duodenum to the ileum, as well as the complexity created by villi and the continual breakdown of contents. For this reason, most studies of the gut rely on fecal samples, which are assumed to be representative of the average microbial community across the myriad of niches in the entire human gut. Consequently, it is not only difficult to determine the strength of niche processes in body habitats, it is also difficult to understand how niches influence the observed levels of beta diversity.

5.2. Neutral Processes in the Human Microbiome

It is possible that stochastic processes may also strongly influence beta diversity patterns in the human microbiome because microorganisms have high dispersal rates onto/into the human body and microbial taxa are subject to rapid evolutionary changes owing to relatively short generation times and horizontal gene transfer. In the neutral theory of biodiversity, interspecific trait differences are assumed irrelevant, and community structures arise via primarily stochastic processes (Hubbell 2001). In brief, neutral theory posits that stochastic extinction, immigration, and

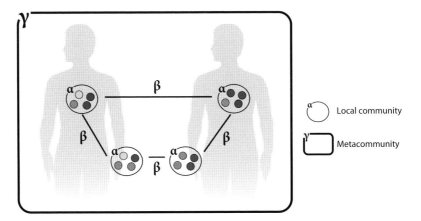

Figure 2

Three levels of diversity as defined by Whittaker et al. (2001) and applied to the human microbiome. Alpha diversity (α) describes the biodiversity found in one sample or a community (e.g., the microbes sampled from the skin on one's palm); beta diversity (β) describes the dissimilarity in communities or samples (e.g., the phylogenetic diversity difference between the group of microbes found on the skin of one's palm and chest, or between the palms of individuals); gamma diversity (γ) is the collective diversity of all samples, i.e., the grouped alpha diversity, (e.g., the total biodiversity found in all skin samples of one or more individuals).

speciation events can explain community composition within or between sites without knowledge of species-level traits. The appeal of these simple models is their parsimony or ability to accurately predict biodiversity while including few parameters or assumptions (e.g., Hubbell 2001). In studies of plants and animals, neutral models, though not always providing the best fit, have been able to predict species abundance curves, species area relationships, and distance-decay patterns across a variety of systems with similar accuracy as niche models that incorporate species-level traits and variability (McGill et al. 2006, Rosindell et al. 2011).

Neutral models pose unique challenges and opportunities in the study of the human microbiome. For example, defining the size and composition of the bacterial source pool (metacommunity; **Figure 2**), as well as determining the dispersal rates within the metacommunity, is conceptually and empirically challenging. Although initial studies have proposed elegant mathematical solutions to these challenges in model-fitting studies (Ofiteru et al. 2010, Sloan et al. 2006), empirical testing and validation of neutral model assumptions in bacterial communities remains difficult. Still, neutral models can be used to generate testable hypotheses of community patterns, which can then help us understand the relative roles of niche versus neutral processes in structuring the human microbiome [see discussions by Ofiteru et al. (2010) and Rosindell et al. (2011)].

Although the exploration of neutral processes in the human microbiome is relatively nascent, stochastic processes do seem to be significant predictors of the bacterial community composition within certain body habitats. For example, dispersal rates and the size of the bacterial source pool have been shown to be important in shaping the relative proportions of bacteria observed in human lungs, but not necessarily of communities found in fecal samples (Sloan et al. 2006). The same processes can also be important in structuring bacterial communities of lake and sewage treatment waters, as shown by theoretical and empirical studies (Lindström & Ostman 2011, Ofiteru et al. 2010). These results show that the influx of bacteria to certain habitats can be high, and stochastic departures and arrivals of bacterial community members can influence the variation observed between human body cavities and between human hosts.

Both niche and neutral processes are likely important for assembling the bacterial communities of humans, a pattern already reported for some plant and animal communities (Chase 2007, Chase & Myers 2011) and microbial communities (Cadotte 2007, Fukami et al. 2007, Zhang et al. 2009). Nevertheless, determining which process is most important under which conditions, or if these processes work simultaneously or conditionally, is imperative for the effective design of more targeted studies. For example, if neutral processes dictate community structure in the microbiome, it is important to empirically test the effect of dispersal rates among communities, as well as to determine the size and diversity of the relevant metacommunity. Alternatively, where niche processes are most important, research should focus on measuring community structure in response to variation in local environments and processes, as well as focusing on the role of interspecific and intraspecific trait variation in coexistence and community structuring (e.g., Clark et al. 2010). Finally, determining the relative strength of niche and neutral processes across spatial and temporal scales in the human microbiome is necessary in order to disentangle the effects of deterministic and stochastic processes on beta diversity. Considering beta diversity in the context of niche and neutral processes is particularly important for understanding how the microbiome responds to, and recovers from, disturbances. Also, we can use both niche and neutral approaches to refine our understanding of how community assembly in the microbiome is influenced by environmental heterogeneity and competition/facilitation, factors long considered by community ecologists and covered in greater detail below.

6. SUCCESSION AND DISTURBANCE

As with all biological communities, the composition and structure of the human microbiome is not static. The types of changes observed in the human microbiome can vary from small fluctuations around a relatively stable microbial community to complete shifts in community membership to alternate stable states. Furthermore, these variations in microbial community structure can occur over timescales ranging from hours to decades. Although some of the changes observed in the human microbiome occur during specific life stages and are thus predictable, other shifts are seemingly random or are triggered by specific disturbance events. In these regards, concepts borrowed from community ecology, such as succession and the temporal stability in community structure, may offer insights into the factors driving the temporal variability within the human microbiome.

Microbial colonization of the human body begins at birth, and delivery mode largely determines the pioneer colonizers (Dominguez-Bello et al. 2010). For example, the different body habitats of babies born vaginally are first colonized by *Lactobacillus* and *Prevotella* species originating from the mother's vagina. In contrast, babies born via Cesarean-section (C-section) are first colonized by *Staphylococcus*, *Corynebacterium*, and *Propionibacterium* species that originate from human skin. Interestingly, the pioneer microbiota of C-section babies differ from the skin communities of their mother, suggesting that these skin-associated taxa came from other people (nurses, doctors, the father) or from surfaces that the babies contact shortly after being born (Dominguez-Bello et al. 2010). These differences in initial colonization have a lasting effect on community composition as the intestinal microbiota of C-section babies remains distinct from vaginally delivered babies several months after birth (reviewed by Dominguez-Bello et al. 2011). This phenomenon is conceptually similar to the "priority effect" observed in other types of communities (Fukami & Nakajima 2011) and may be partially responsible for the higher occurrence of certain atopic diseases later in life. For example, the human microbiome can train the immune system to respond to various pathogens (Björkstén 2004), possibly contributing to the increased susceptibility of babies born via C-section to the development of allergies and asthma later in life (Salam et al. 2006).

Following initial colonization, our microbial communities continue to develop and diversify through the first couple of years of life (reviewed by Dominguez-Bello et al. 2011). During this life stage, bacterial communities change rapidly in response to our environment, health, and diet (Spor et al. 2011). For example, the intestinal community of an infant can switch from an *Actinobacteria*- and *Proteobacteria*-dominated community to an adult-like state dominated by *Firmicutes* and *Bacteroidetes* upon the introduction of plant-derived foods like peas (Koenig et al. 2011). Adolescents have distinct distal colon microbiota from adults (Agans et al. 2011), and shifts in the composition of the oral microbiota have been associated with the development of secondary sexual characteristics during puberty (Gusberti et al. 1990). This evidence indicates that some microbial communities undergo succession in concert with the developmental stages of the human host.

Until recently, there was a general consensus that healthy adults harbored a relatively stable, climax microbiome. Stability, however, is a relative term and means different things depending on the body habitat in question and the age of the individual. Furthermore, there is little empirical data to support this consensus, as only a few studies have examined temporal variability in the microbiome associated with healthy adults (e.g., Caporaso et al. 2011, Costello et al. 2009, Grice et al. 2009). From these studies, it is apparent that the adult microbiome is in a constant state of flux as microbial community composition on and in an individual varies substantially over time. Moreover, our body habitats exhibit differing degrees of variability; skin is the most variable, whereas the mouth appears to be the least variable (Caporaso et al. 2011). The differing degrees of variability observed across body habitats is likely dependent on a wide range of factors including the stability of environmental conditions, the turnover rate of the community, and the immigration rate from external sources. However, intraindividual variation is nearly always less than interindividual variation, and body habitats harbor distinct microbial communities (Caporaso et al. 2011, Costello et al. 2009). So, though there are variations in microbial community composition over time, each body habitat appears to exist in its own stability domain.

Although the healthy human microbiome is temporally variable and animal-associated microbiomes may transition between alternate stable states (Costello et al. 2010, Ravel et al. 2011), variation due to stressors may have detrimental effects on the function of our native microbiome. Stressors such as pathogenic invasions and exposure to broad-spectrum antibiotics (Dethlefsen et al. 2008, Hoffmann et al. 2009) can rapidly and dramatically alter the structure of the human microbiome. The recovery time after removal of the stress can be weeks to months, although evidence suggests that the microbial communities in some body habitats may never return to their prior state (Antonopoulos et al. 2009). In these cases, certain treatment measures, such as cohabitation with healthy individuals or intentional inoculation, may expedite the recovery process (Khoruts et al. 2010). In contrast, there are a wide range of stressors, such as hand washing, teeth brushing, and dieting, that have more transient effects on the skin, mouth, and gut microbiomes, respectively, and the residual populations rapidly recover after removal of the stressor (Fierer et al. 2008, Turnbaugh et al. 2009b).

Generally, the response of microbial communities to perturbations can be characterized by their resistance, resilience, and whether their stable states are changed (Allison & Martiny 2008) (**Figure 3**). Communities that are resistant change composition comparatively little in response to a disturbance. Although specific bacterial populations are resistant to certain disturbances (e.g., antibiotic-resistant taxa), it is likely that community-level resistance is dependent on the severity of the disturbance and is not a defining feature of the human microbiome. Instead, human-associated microbial communities appear to be highly resilient as there are numerous examples of communities either rapidly returning to a state that resembles the predisturbance community or moving to an entirely different stable state following a disturbance. For example, the gut communities of three individuals returned to their initial state after a single treatment with the antibiotic ciprofloxacin

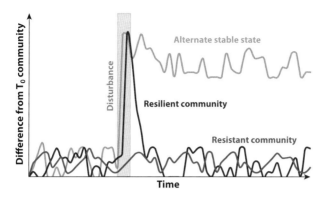

Figure 3

Biological communities respond to disturbances in at least three distinct ways. Resistant communities (*blue*) do not deviate considerably from an initial state following a disturbance. Resilient communities (*red*) deviate considerably but then return to resemble the initial community in time. Finally, communities may respond to a disturbance by moving to an alternate stable state (*green*), where community composition and structure stabilizes in a regime distinct from the initial stable state.

(Dethlefsen & Relman 2011). However, after a second treatment with the antibiotic, the communities stabilized around states different from their preantibiotic state. In this example, the impacts on the health of the hosts after the state change were unknown. However, other studies indicate that alternate states are not always healthy for the host, as illustrated by polymicrobial infections like bacterial vaginosis (Lamont et al. 2011). Moreover, care is needed in determining whether or not a community has entered an alternative stable state. This is because what appears to be an alternative stable state may actually represent an alternative transient state, and the conditions that promote alternative transient states can be different from those that promote alternative stable states (Fukami & Nakajima 2011).

7. MICROBE-MICROBE INTERACTIONS

Microbial diversity and function are driven to a large extent by biotic interactions occurring at the local scale (Lindström & Langenheder 2011). These include interactions between microbial taxa and, in the human microbiome, between microbes and the host. Host-microbe interactions play a significant role in regulating microbial community structure and function, and these relationships are reviewed in detail elsewhere (e.g., Bäckhed et al. 2005). Here we focus on the types of antagonistic and cooperative microbe-microbe interactions that are likely to play an important role in shaping both our health and the structure of our microbiome.

7.1. Antagonistic Interactions

Microbe-microbe competition for limited resources is likely a common occurrence in human-associated body habitats, and such competition may be an important driver of community structure and overall diversity through niche diversification. In the human gut, for example, many microbes are specialists in terms of nutrient sources despite the availability of a wide array of substrates, suggesting that competition for resources may be a strong driver of niche diversification in the gut environment. Studies show that different *Bacteroides* species exhibit corresponding changes in population size with the availability of specific substrates (Sonnenburg et al. 2010), and one study of two *Lactobacillus* strains in the mouse gut showed that resource specialization can allow for coexistence despite apparent overlap in resource requirements (Tannock et al. 2011). In a unique example, one strain of *Salmonella enterica* has been shown to have evolved a unique capacity to utilize ethanolamine, a byproduct of intestinal inflammation induced by virulence factors secreted by *S. enterica*. By stimulating an inflammation response, *S. enterica* is able to indirectly create a novel niche and thereby avoid competition for more commonly used redox combinations (Thiennimitr et al. 2011).

Microbial diversity within a given community may result from other types of antagonistic interactions besides the direct competition for resources. In the vagina, *Lactobacillus* spp. can lower the pH of the environment, thereby inhibiting the growth of potentially competitive microbes (Lamont et al. 2011). In addition, gut microbiota release molecules that trigger the mounting of antibacterial defenses such as defensins, mucin, and secretory IgA by host cells (Salzman 2011). These types of interactions between microbes and host can benefit both the resident microbiota and the host by protecting against invasive microbes that may compete for resources and/or function as host pathogens.

Predator-prey interactions, which are traditionally a major focus in population and community ecology, are also likely to be important in structuring the human microbiome. Gut microbial studies of predator-prey interactions have typically focused on bacteriophages (i.e., viruses) and their effects on bacterial population structure (Reyes et al. 2010). Viruses fulfill a unique and important role in the microbial trophic system because they can cause high levels of microbial mortality and also may promote horizontal gene transfer between bacterial lineages. New work detailing the effects of the predatory bacteria, such as *Bdellovibrio*-and-like organisms (BALO), in the gut adds yet another layer to our understanding of trophic interactions in the human microbiome. These studies suggest that predatory bacteria could be used in medical treatments to control pathogens in the human body (Schwudke et al. 2001, Van Essche et al. 2011).

7.2. Mutualistic Interactions

Cooperation between microbes is also important in structuring communities in the human microbiome. This is particularly true in the gut where low O_2 conditions promote syntrophic interactions—groups of microorganisms combining metabolic reactions to improve total energy yield (Stams & Plugge 2009). For example, the human gut archaeon *Methanobrevibacter smithii* reduces CO_2 to CH_4 with electrons from H_2 produced by bacteria such as *Bacteroides thetaiotaomicron* (Hansen et al. 2011). Removal of H_2 by *M. smithii* can accelerate *B. thetaiotaomicron* respiration, thereby accelerating the growth rates of both microbes (Samuel & Gordon 2006). Additionally, the augmented *B. thetaiotaomicron* levels lead to increased production of short-chain fatty acids, which can be used as source of nutrition by the human host. Alternatively, gut communities that lack methanogens can instead process excess H_2 through acetogenesis (Rey et al. 2010). In this case the acetogen *Bacteroides hydrogenotrophica* fills the role of H_2 consumer, with similar benefits to both the microbes involved in the syntrophic interaction and the host. More generally, these types of interactions are evidence that certain taxa, even rare taxa, can have effects on the host and on community assembly patterns that may be larger than their relative abundances might suggest.

Microbe-microbe signaling, such as quorum-sensing (QS), also demonstrates the advantage of cooperative behavior. QS is a form of chemical communication between and within species that allows for the coordination of activities according to changes in population density. Through QS, microbes such as *Vibrio cholera*, the bacterium that causes cholera, produce virulence factors and form biofilms at low cell densities, facilitating successful infection of hosts. In contrast, at higher cell densities *V. cholera* show a reduced rate of biofilm formation possibly to improve dispersal following diarrheal events (Hammer & Bassler 2003). Between microbial taxa, QS allows the peridontal pathogen *Poryphyromonas gingivalis* to colonize the biofilm created by a commensal bacteria, *Streptococcus gordonii* (McNab et al. 2003). In addition, evidence suggests that QS is used to create biofilms between two pathogens, *Pseudomonas aeruginosa* and *Burkholderia cepacia*, that are known to cause complications in cystic fibrosis patients (Riedel et al. 2001).

Overall, bacteria appear to exhibit numerous types of cooperative behaviors, many with significant manifestations at the community level. For example, many bacterial taxa release metabolites into the environment, which help other community members access essential resources such as complex organic compounds and iron (e.g., siderophores) or protect community members against toxic substances. In particular, horizontal gene transfer (HGT) generates unique opportunities for indirect community-level cooperation, including the transfer of novel traits between microbial taxa that may be unrelated. Recent research indicates that the gut bacteria unique to Japanese populations may have acquired novel enzymes for the degradation of porphyran, a chemical in seaweed, from marine bacteria associated with edible seaweed (Hehemann et al. 2010).

As the catalog of uncultured human microbial diversity continues to explode, we are faced with the difficult task of untangling the various interactions between microbes, and between microbes and host physiology, which could involve thousands of microbial taxa. Medical advances, such as the development of antibiotic and therapeutic drugs, have relied on our growing understanding of these relationships (Firn & Jones 2003). However, given the sheer scale and variability of the human microbiome, more research is required to fully understand the underlying interactions that define the human microbiome. Currently, we have the necessary tools for describing taxonomic co-occurrence networks, and these offer starting points for identifying relationships among taxa (e.g., Barberán et al. 2012, Freilich et al. 2010). However, we are only just beginning to resolve the functional components of these networks in the human microbiome, such as pathways of metabolites and trophic levels (levels that are more easily delineated in plant and animal systems). Furthermore, these interactions are difficult to study because they are dynamic; it is likely that multiple interactions occur simultaneously and that the characteristics of the interactions change depending on biotic and abiotic conditions. Also, interaction networks may be challenging to study in microorganisms because characteristics such as QS allow for the ecological attributes of a population (including functional capabilities and gene expression) to change with population density. Given the dynamism of microbial systems, detailed time course analyses will be particularly important to unraveling these webs of interactions. Likewise, integrating culture-based experiments, metabolic modeling, and the experimental manipulation of gnotobiotic communities should help address the knowledge gap between our cataloged microbial diversity and the respective functions of these complex communities.

8. CONCLUSIONS AND FUTURE DIRECTIONS

Our understanding of the ecology of the human microbiome clearly lags behind our understanding of plant and animal ecology. This discrepancy in the maturity of macrobial versus microbial ecology is not surprising—basic surveys of microbial diversity were, until recently, difficult to conduct; the functional attributes of many microbial taxa remain unknown; and the small size of microorganisms makes it inherently problematic to understand microbe-environment and microbe-microbe interactions at the spatial scales relevant to the size of the organisms. Fortunately, these barriers are dissolving as methodologies continue to advance at a rapid pace and as interdisciplinary research groups continue to focus on the human microbiome as a study system.

However, a number of ecological concepts are actually easier to study in microbial systems, like the human microbiome, than in plant or animal communities. Microbial communities are more amenable to experimental manipulations than plant and animal communities, where generation times are longer and logistical concerns prevent experimentation with large numbers of individuals in well-replicated studies. In addition, just as plant and animal ecologists increasingly incorporate phylogenetic information into models of community assembly and diversity patterns, microbiologists nearly always have such phylogenetic information as a product of the sequence-based diversity surveys, making such analyses even more straightforward. Likewise, with the short generation times of most microbial taxa, evolutionary processes can be directly observed during the course of studies lasting weeks to months, making it possible to assess how evolutionary processes may influence ecological processes and vice versa. Comparable studies in plant or animal communities could take decades or millennia, lengths of time far beyond the duration of most research programs.

As research on the human microbiome is advancing rapidly, it is our hope that this review will soon be rendered out of date. Thus, in lieu of a more formal conclusion, we conclude by highlighting a handful of topics that represent key knowledge gaps. This is not meant to be a comprehensive list, but rather a wish list of future research directions in the field.

- *Interactions between microbial taxa within the human body.* Bacteria, archaea, fungi, microeukaryotes, and viruses all share space within human body habitats, yet most studies focus on individual microbial groups in isolation (often just bacteria). Integrative studies that seek to understand the role of these cross-taxon interactions in shaping the structure and function of the human microbiome will be critical.

- *Moving beyond phylogenetic and taxonomic descriptions of the human microbiome.* Despite the explosion of information on the phylogenetic and taxonomic composition of the human microbiome, we often lack key information on the functional attributes of these communities and the traits of individual community members. In some cases, the phylogenetic structure of microbial communities may predict the functional characteristics of those communities, but this is clearly not true for all functional attributes. The wider application of "-omics"-based approaches (e.g., metabolomics, proteomics), together with more detailed analyses of individual taxa, may help resolve these knowledge gaps and improve our understanding of how the functional characteristics of the human microbiome shift across space and time.

- *Changes in microbial biomass within and between individuals.* Two of the fundamental metrics in community ecology are biomass (or productivity) and diversity. Researchers routinely characterize microbial diversity and changes in the relative abundances of specific microbial taxa, but it is surprisingly difficult to find robust estimates of total microbial abundances across human body habitats and how those abundances vary between individuals and over time. This is of particular importance as some disease states may not necessarily be associated with changes in diversity, but rather with changes in the absolute abundances of specific taxa or total microbial biomass levels.

- *Experimental manipulations of the microbiome.* Controlled experiments are critical for testing concepts in microbial ecology and building a more mechanistic understanding of community assembly within the human microbiome. For example, experimental manipulations could be used to determine how changes in biotic or abiotic conditions alter community composition, the role of niche versus neutral processes in governing community assembly patterns, and the importance of priority effects in diversity changes over time. Although controlled experiments and experimental treatments are logistically difficult to conduct with humans, such experiments could be conducted in model animal systems (e.g., mice) or in vitro using communities and environmental conditions that mimic those found within the human body.

- *Reconciling the disparity in scales.* We typically measure temporal and spatial patterns in microbial communities at scales that are far different from the scales at which microorganisms actually operate. For example, we assess bacterial community composition on skin regions that are often many square centimeters, yet bacteria are typically less than a few microns in diameter and they often interact with the human host and other microorganisms at spatial scales that are orders of magnitude smaller than the sampled area. Likewise, published studies of how microbial communities change over time have typically looked at patterns across temporal scales (weeks to months to years) that are far longer than the generation time of many microorganisms (hours to days). Although methodologically difficult, characterizing microbial communities and the environmental conditions of microbial habitats at scales that more closely approximate the scales at which they operate will undoubtedly improve our ability to describe and understand the ecology of the human microbiome.

DISCLOSURE STATEMENT

The authors are not aware of any affiliations, memberships, funding, or financial holdings that might be perceived as affecting the objectivity of this review.

LITERATURE CITED

Agans R, Rigsbee L, Kenche H, Michail S, Khamis HJ, Paliy O. 2011. Distal gut microbiota of adolescent children is different from that of adults. *FEMS Microbiol. Ecol.* 77:404–12

Allison SD, Martiny JB. 2008. Resistance, resilience, and redundancy in microbial communities. *Proc. Natl. Acad. Sci. USA* 105:11512–19

Anderson MJ, Crist TO, Chase JM, Vellend M, Inouye BD, et al. 2011. Navigating the multiple meanings of beta diversity: a roadmap for the practicing ecologist. *Ecol. Lett.* 14:19–28

Antonopoulos DA, Huse SM, Morrison HG, Schmidt TM, Sogin ML, Young VB. 2009. Reproducible community dynamics of the gastrointestinal microbiota following antibiotic perturbation. *Infect. Immun.* 77:2367–75

Bäckhed F, Ley RE, Sonnenburg JL, Peterson DA, Gordon JI. 2005. Host-bacterial mutualism in the human intestine. *Science* 307:1915–20

Balvanera P, Pfisterer AB, Buchmann N, He JS, Nakashizuka T, et al. 2006. Quantifying the evidence for biodiversity effects on ecosystem functioning and services. *Ecol. Lett.* 9:1146–56

Barberán A, Bates ST, Casamayor EO, Fierer N. 2012. Using network analysis to explore co-occurrence patterns in soil microbial communities. *Int. Soc. Microb. Ecol. J.* 6:343–51

Björkstén B. 2004. Effects of intestinal microflora and the environment on the development of asthma and allergy. *Semin. Immunopathol.* 25:257–70

Blaser MJ, Falkow S. 2009. What are the consequences of the disappearing human microbiota? *Nat. Rev. Microbiol.* 7:887–94

Cadotte MW. 2007. Concurrent niche and neutral processes in the competition-colonization model of species coexistence. *Proc. R. Soc. Lond. Ser. B* 274:2739–44

Cadotte MW, Cardinale BJ, Oakley TH. 2008. Evolutionary history and the effect of biodiversity on plant productivity. *Proc. Natl. Acad. Sci. USA* 105:17012–17

Caporaso JG, Lauber CL, Costello EK, Berg-Lyons D, Gonzalez A, et al. 2011. Moving pictures of the human microbiome. *Genome Biol.* 12:R50

Cardinale BJ, Matulich KL, Hooper DU, Byrnes JE, Duffy E, et al. 2011. The functional role of producer diversity in ecosystems. *Am. J. Bot.* 98:572–92

Chang JY, Antonopoulos DA, Kalra A, Tonelli A, Khalife WT, et al. 2008. Decreased diversity of the fecal microbiome in recurrent *Clostridium difficile*-associated diarrhea. *J. Infect. Dis.* 197:435–38

Charlson ES, Chen J, Custers-Allen R, Bittinger K, Li H, et al. 2010. Disordered microbial communities in the upper respiratory tract of cigarette smokers. *PLoS One* 5:e15216

Chase JM. 2007. Drought mediates the importance of stochastic community assembly. *Proc. Natl. Acad. Sci. USA* 104:17430–34

Chase JM, Myers JA. 2011. Disentangling the importance of ecological niches from stochastic processes across scales. *Proc. R. Soc. Lond. Ser. B* 366:2351–63

Clark JS, Bell D, Chu CJ, Courbaud B, Dietze M, et al. 2010. High-dimensional coexistence based on individual variation: a synthesis of evidence. *Ecol. Monogr.* 80:569–608

Costello EK, Gordon JI, Secor SM, Knight R. 2010. Postprandial remodeling of the gut microbiota in Burmese pythons. *Int. Soc. Microb. Ecol. J.* 4:1375–85

Costello EK, Lauber C, Hamady M, Fierer N, Gordon J, Knight R. 2009. Bacterial community variation in human body habitats across space and time. *Science* 326:1694–97

Dethlefsen L, Huse S, Sogin ML, Relman DA. 2008. The pervasive effects of an antibiotic on the human gut microbiota, as revealed by deep 16S rRNA sequencing. *PLoS Biol.* 6:e280

Dethlefsen L, McFall-Ngai M, Relman DA. 2007. An ecological and evolutionary perspective on human-microbe mutualism and disease. *Nature* 449:811–18

Dethlefsen L, Relman DA. 2011. Incomplete recovery and individualized responses of the human distal gut microbiota to repeated antibiotic perturbation. *Proc. Natl. Acad. Sci. USA* 108(Suppl 1):4554–61

Dicksved J, Lindberg M, Rosenquist M, Enroth H, Jansson JK, Engstrand L. 2009. Molecular characterization of the stomach microbiota in patients with gastric cancer and in controls. *J. Med. Microbiol.* 58:509–16

Dominguez-Bello MG, Blaser MJ, Ley RE, Knight R. 2011. Development of the human gastrointestinal microbiota and insights from high-throughput sequencing. *Gastroenterology* 140:1713–19

Dominguez-Bello MG, Costello EK, Contreras M, Magris M, Hidalgo G, et al. 2010. Delivery mode shapes the acquisition and structure of the initial microbiota across multiple body habitats in newborns. *Proc. Natl. Acad. Sci. USA* 107:11971–75

Elton C. 1958. *The Ecology of Invasions by Animals and Plants*. London: Methuen

Faith DP, Baker AM. 2006. Phylogenetic diversity (PD) and biodiversity conservation: some bioinformatics challenges. *Evol. Bioinform.* 2:121–28

Fierer N, Hamady M, Lauber C, Knight R. 2008. The influence of sex, handedness, and washing on the diversity of hand surface bacteria. *Proc. Natl. Acad. Sci. USA* 105:17994–99

Firn RD, Jones CG. 2003. Natural products—a simple model to explain chemical diversity. *Nat. Prod. Rep.* 20:382–91

Flanagan JL, Brodie EL, Weng L, Lynch SV, Garcia O, et al. 2007. Loss of bacterial diversity during antibiotic treatment of intubated patients colonized with *Pseudomonas aeruginosa*. *J. Clin. Microbiol.* 45:1954–62

Frank DN, Spiegelman GB, Davis W, Wagner E, Lyons F, Pace NR. 2003. Culture-independent molecular analysis of microbial constituents of the healthy human outer ear. *J. Clin. Microbiol.* 41:295–303

Frank DN, St Amand AL, Feldman RA, Boedeker EC, Harpaz N, Pace NR. 2007. Molecular-phylogenetic characterization of microbial community imbalances in human inflammatory bowel diseases. *Proc. Natl. Acad. Sci. USA* 104:13780–85

Frank DN, Zhu W, Sartor RB, Li E. 2011. Investigating the biological and clinical significance of human dysbioses. *Trends Microbiol.* 19:427–34

Fredericks DN, Relman DA. 1996. Sequence-based identification of microbial pathogens: a reconsideration of Koch's postulates. *Clin. Microbiol. Rev.* 9:18–33

Freilich S, Kreimer A, Meilijson I, Gophna U, Sharan R, Ruppin E. 2010. The large-scale organization of the bacterial network of ecological co-occurrence interactions. *Nucleic Acids Res.* 38:3857–68

Fridley JD, Stachowicz JJ, Naeem S, Sax DF, Seabloom EW, et al. 2007. The invasion paradox: reconciling pattern and process in species invasions. *Ecology* 88:3–17

Fukami T, Beaumont HJE, Zhang XX, Rainey PB. 2007. Immigration history controls diversification in experimental adaptive radiation. *Nature* 446:436–439

Fukami T, Nakajima M. 2011. Community assembly: alternative stable states or alternative transient states? *Ecol. Lett.* 14:973–84

Gill SR, Pop M, Deboy RT, Eckburg PB, Turnbaugh PJ, et al. 2006. Metagenomic analysis of the human distal gut microbiome. *Science* 312:1355–59

Gonzalez A, Clemente JC, Shade A, Metcalf JL, Song S, et al. 2011. Our microbial selves: what ecology can teach us. *EMBO Rep.* 12:775–84

Graham CH, Fine PV. 2008. Phylogenetic beta diversity: linking ecological and evolutionary processes across space in time. *Ecol. Lett.* 11:1265–77

Grice EA, Kong HH, Conlan S, Deming CB, Davis J, et al. 2009. Topographical and temporal diversity of the human skin microbiome. *Science* 324:1190–92

Gusberti FA, Mombelli A, Lang NP, Minder CE. 1990. Changes in subgingival microbiota during puberty. A 4-year longitudinal study. *J. Clin. Periodontol.* 17:685–92

Hamady M, Knight R. 2009. Microbial community profiling for human microbiome projects: tools, techniques, and challenges. *Genome Res.* 19:1141–52

Hammer BK, Bassler BL. 2003. Quorum sensing controls biofilm formation in *Vibrio cholerae*. *Mol. Microbiol.* 50:101–4

Hansen EE, Lozupone CA, Rey FE, Wu M, Guruge JL, et al. 2011. Pan-genome of the dominant human gut-associated archaeon, *Methanobrevibacter smithii*, studied in twins. *Proc. Natl. Acad. Sci. USA* 108:4599–606

Harrison F. 2007. Microbial ecology of the cystic fibrosis lung. *Microbiology* 153:917–23

Hayashi H, Takahashi R, Nishi T, Sakamoto M, Benno Y. 2005. Molecular analysis of jejunal, ileal, caecal and recto-sigmoidal human colonic microbiota using 16S rRNA gene libraries and terminal restriction fragment length polymorphism. *J. Med. Microbiol.* 54:1093–101

Hehemann JH, Correc G, Barbeyron T, Helbert W, Czjzek M, Michel G. 2010. Transfer of carbohydrate-active enzymes from marine bacteria to Japanese gut microbiota. *Nature* 464:908–12

Hoffmann C, Hill DA, Minkah N, Kirn T, Troy A, et al. 2009. Community-wide response of the gut microbiota to enteropathogenic *Citrobacter rodentium* infection revealed by deep sequencing. *Infect. Immunol.* 77:4668–78

Hooper DU, Chapin FS, Ewel JJ, Hector A, Inchausti P, et al. 2005. Effects of biodiversity on ecosystem functioning: a consensus of current knowledge. *Ecol. Monogr.* 75:3–35

Hubbell S. 2001. *The Unified Neutral Theory of Biodiversity and Biogeography*. Princeton, NJ: Princeton Univ. Press

Hudson PJ, Dobson AP, Lafferty KD. 2006. Is a healthy ecosystem one that is rich in parasites? *Trends Ecol. Evol.* 21:381–85

Ives AR, Carpenter SR. 2007. Stability and diversity of ecosystems. *Science* 317:58–62

Jenkinson H. 2011. Beyond the oral microbiome. *Environ. Microbiol.* 13:3077–87

Khoruts A, Dicksved J, Jansson JK, Sadowsky MJ. 2010. Changes in the composition of the human fecal microbiome after bacteriotherapy for recurrent *Clostridium difficile*-associated diarrhea. *J. Clin. Gastroenterol.* 44:354–60

Kim TK, Thomas SM, Ho M, Sharma S, Reich CI, et al. 2009. Heterogeneity of vaginal microbial communities within individuals. *J. Clin. Microbiol.* 47:1181–89

Koenig JE, Spor A, Scalfone N, Fricker AD, Stombaugh J, et al. 2011. Succession of microbial consortia in the developing infant gut microbiome. *Proc. Natl. Acad. Sci. USA* 108:4578–85

Kuczynski J, Lauber CL, Walters WA, Parfrey LW, Clemente J, et al. 2012. Experimental and analytical tools for studying the human microbiome. *Nat. Rev. Genet.* 13:47–58

Lamont RF, Sobel JD, Akins RA, Hassan SS, Chaiworapongsa T, et al. 2011. The vaginal microbiome: new information about genital tract flora using molecular based techniques. *Br. J. Obstet. Gynaecolog.* 118:533–49

Ley RE, Turnbaugh PJ, Klein S, Gordon JI. 2006. Microbial ecology: human gut microbes associated with obesity. *Nature* 444:1022–23

Lindström ES, Langenheder S. 2011. Local and regional factors influencing bacterial community assembly. *Environ. Microbiol. Rep.* 4:1–9

Lindström ES, Ostman O. 2011. The importance of dispersal for bacterial community composition and functioning. *PLoS One* 6:e25883

Loreau M, Naeem S, Inchausti P, Bengtsson J, Grime JP, et al. 2001. Biodiversity and ecosystem functioning: current knowledge and future challenges. *Science* 294:804–8

Macarthur R. 1955. Fluctuations of animal populations, and a measure of community stability. *Ecology* 36:533–36

Manichanh C, Rigottier-Gois L, Bonnaud E, Gloux K, Pelletier E, et al. 2006. Reduced diversity of faecal microbiota in Crohn's disease revealed by a metagenomic approach. *Gut* 55:205–11

Mazmanian SK, Round JL, Kasper DL. 2008. A microbial symbiosis factor prevents intestinal inflammatory disease. *Nature* 453:620–25

McFarland LV. 2008. Update on the changing epidemiology of *Clostridium difficile*-associated disease. *Nat. Clin. Pract. Gastroenterol. Hepatol.* 5:40–48

McGill BJ, Maurer BA, Weiser MD. 2006. Empirical evaluation of neutral theory. *Ecology* 87:1411–23

McNab R, Ford SK, El-Sabaeny A, Barbieri B, Cook GS, Lamont RJ. 2003. LuxS-based signaling in *Streptococcus gordonii*: autoinducer 2 controls carbohydrate metabolism and biofilm formation with *Porphyromonas gingivalis*. *J. Bacteriol.* 185:274–84

Minot S, Sinha R, Chen J, Li H, Keilbaugh SA, et al. 2011. The human gut virome: inter-individual variation and dynamic response to diet. *Genome Res.* 21:1616–25

Ofiteru ID, Lunn M, Curtis TP, Wells GF, Criddle CS, et al. 2010. Combined niche and neutral effects in a microbial wastewater treatment community. *Proc. Natl. Acad. Sci. USA* 107:15345–50

Pace NR. 1997. A molecular view of microbial diversity and the biosphere. *Science* 276:734–39

Packey CD, Sartor RB. 2009. Commensal bacteria, traditional and opportunistic pathogens, dysbiosis and bacterial killing in inflammatory bowel diseases. *Curr. Opin. Infect. Dis.* 22:292–301

Pei Z, Bini EJ, Yang L, Zhou M, Francois F, Blaser MJ. 2004. Bacterial biota in the human distal esophagus. *Proc. Natl. Acad. Sci. USA* 101:4250–55

Petchey OL, Gaston KJ. 2007. Dendrograms and measuring functional diversity. *Oikos* 116:1422–26

Pflughoeft KJ, Versalovic J. 2011. Human microbiome in health and disease. *Annu. Rev. Pathol. Mech. Dis.* 7:99–122

Price LB, Liu CM, Johnson KE, Aziz M, Lau MK, et al. 2010. The effects of circumcision on the penis microbiome. *PLoS One* 5:e8422

Qin J, Li R, Raes J, Arumugam M, Burgdorf KS, et al. 2010. A human gut microbial gene catalogue established by metagenomic sequencing. *Nature* 464:59–65

Ravel J, Gajer P, Abdo Z, Schneider GM, Koenig SS, et al. 2011. Vaginal microbiome of reproductive-age women. *Proc. Natl. Acad. Sci. USA* 108(Suppl 1):4680–87

Rey FE, Faith JJ, Bain J, Muehlbauer MJ, Stevens RD, et al. 2010. Dissecting the in vivo metabolic potential of two human gut acetogens. *J. Biol. Chem.* 285:22082–90

Reyes A, Haynes M, Hanson N, Angly FE, Heath AC, et al. 2010. Viruses in the faecal microbiota of monozygotic twins and their mothers. *Nature* 466:334–38

Ricklefs R, Schluter D. 1993. *Species Diversity in Space and Time*. Cambridge, UK: Cambridge Univ. Press

Riedel K, Hentzer M, Geisenberger O, Huber B, Steidle A, et al. 2001. N-acylhomoserine-lactone-mediated communication between *Pseudomonas aeruginosa* and *Burkholderia cepacia* in mixed biofilms. *Microbiology* 147:3249–62

Robinson CJ, Bohannan BJ, Young VB. 2010. From structure to function: the ecology of host-associated microbial communities. *Microbiol. Mol. Biol. Rev.* 74:453–76

Rosenzweig M. 1995. *Species Diversity in Space and Time*. Cambridge, UK: Cambridge Univ. Press

Rosindell J, Hubbell SP, Etienne RS. 2011. The unified neutral theory of biodiversity and biogeography at age ten. *Trends Ecol. Evol.* 26:340–48

Salam MT, Margolis HG, McConnell R, McGregor JA, Avol EL, Gilliland FD. 2006. Mode of delivery is associated with asthma and allergy occurrences in children. *Ann. Epidemiol.* 16:341–46

Salzman NH. 2011. Microbiota-immune system interaction: an uneasy alliance. *Curr. Opin. Microbiol.* 14:99–105

Samuel BS, Gordon JI. 2006. A humanized gnotobiotic mouse model of host-archaeal-bacterial mutualism. *Proc. Natl. Acad. Sci. USA* 103:10011–16

Savage DC. 1977. Microbial ecology of the gastrointestinal tract. *Annu. Rev. Microbiol.* 31:107–33

Schwudke D, Strauch E, Krueger M, Appel B. 2001. Taxonomic studies of predatory bdellovibrios based on 16S rRNA analysis, ribotyping and the hit locus and characterization of isolates from the gut of animals. *Syst. Appl. Microbiol.* 24:385–94

Sloan WT, Lunn M, Woodcock S, Head IM, Nee S, Curtis TP. 2006. Quantifying the roles of immigration and chance in shaping prokaryote community structure. *Environ. Microbiol.* 8:732–40

Sonnenburg ED, Zheng H, Joglekar P, Higginbottom SK, Firbank SJ, et al. 2010. Specificity of polysaccharide use in intestinal bacteroides species determines diet-induced microbiota alterations. *Cell* 141:1241–52

Spor A, Koren O, Ley R. 2011. Unravelling the effects of the environment and host genotype on the gut microbiome. *Nat. Rev. Microbiol.* 9:279–90

Stams AJ, Plugge CM. 2009. Electron transfer in syntrophic communities of anaerobic bacteria and archaea. *Nat. Rev. Microbiol.* 7:568–77

Stohlgren TJ, Barnett DT, Kartesz J. 2003. The rich get richer: patterns of plant invasions in the United States. *Front. Ecol. Environ.* 1:11–14

Tannock GW, Wilson CM, Loach D, Cook GM, Eason J, et al. 2011. Resource partitioning in relation to cohabitation of *Lactobacillus* species in the mouse forestomach. *Int. Soc. Microb. Ecol. J.* 6:927–38

Thiennimitr P, Winter SE, Winter MG, Xavier MN, Tolstikov V, et al. 2011. Intestinal inflammation allows *Salmonella* to use ethanolamine to compete with the microbiota. *Proc. Natl. Acad. Sci. USA* 108:17480–85

Turnbaugh PJ, Hamady M, Yatsunenko T, Cantarel BL, Duncan A, et al. 2009a. A core gut microbiome in obese and lean twins. *Nature* 457:480–84

Turnbaugh PJ, Ley RE, Hamady M, Fraser-Liggett CM, Knight R, Gordon JI. 2007. The human microbiome project. *Nature* 449:804–10

Turnbaugh PJ, Ley RE, Mahowald MA, Magrini V, Mardis ER, Gordon JI. 2006. An obesity-associated gut microbiome with increased capacity for energy harvest. *Nature* 444:1027–31

Turnbaugh PJ, Ridaura VK, Faith JJ, Rey FE, Knight R, Gordon JI. 2009b. The effect of diet on the human gut microbiome: a metagenomic analysis in humanized gnotobiotic mice. *Sci. Transl. Med.* 1:6ra14

Van Essche M, Quirynen M, Sliepen I, Loozen G, Boon N, et al. 2011. Killing of anaerobic pathogens by predatory bacteria. *Mol. Oral Microbiol.* 26:52–61

Whittaker RJ, Willis KJ, Field R. 2001. Scale and species richness: towards a general, hierarchical theory of species diversity. *J. Biogeogr.* 28:453–70

Zhang QG, Buckling A, Godfray HCJ. 2009. Quantifying the relative importance of niches and neutrality for coexistence in a model microbial system. *Funct. Ecol.* 23:1139–47

Zhi XY, Zhao W, Li WJ, Zhao GP. 2012. Prokaryotic systematics in the genomics era. *Antonie Van Leeuwenhoek* 101:21–34

Effects of Host Diversity on Infectious Disease

Richard S. Ostfeld[1] and Felicia Keesing[1,2]

[1]Cary Institute of Ecosystem Studies, Millbrook, New York 12545;
email: rostfeld@caryinstitute.org

[2]Biology Program, Bard College, Annandale-on-Hudson, New York 12504

Annu. Rev. Ecol. Evol. Syst. 2012. 43:157–82

First published online as a Review in Advance on August 28, 2012

The *Annual Review of Ecology, Evolution, and Systematics* is online at ecolsys.annualreviews.org

This article's doi:
10.1146/annurev-ecolsys-102710-145022

Keywords

biodiversity, dilution effect, disease ecology, ecosystem services, hantavirus, Lyme disease, West Nile virus

Abstract

The dynamics of infectious diseases can be affected by genetic diversity within host populations, species diversity within host communities, and diversity among communities. In principle, diversity can either increase or decrease pathogen transmission and disease risk. Theoretical models and laboratory experiments have demonstrated that a dilution effect (decreased disease risk with increasing diversity) can occur under a wide range of conditions. Field studies of plants, aquatic invertebrates, amphibians, birds, and mammals demonstrate that the phenomenon indeed does occur in many natural systems. A dilution effect is expected when (*a*) hosts differ in quality for pathogens or vectors; (*b*) higher quality hosts tend to occur in species-poor communities, whereas lower quality hosts tend to occur in more diverse communities; and (*c*) lower quality hosts regulate abundance of high-quality hosts or of vectors, or reduce encounter rates between these hosts and pathogens or vectors. Although these conditions characterize many disease systems, our ability to predict when and where the dilution effect occurs remains poor. The life-history traits that cause some hosts to be widespread and resilient might be correlated with those that promote infection and transmission by some pathogens, supporting the notion that the dilution effect might be widespread among disease systems. Criticisms of the dilution effect have focused on whether species richness or species composition (both being metrics of biodiversity) drives disease risk. It is well established, however, that changes in species composition correlate with changes in species richness, and this correlation could explain why the dilution effect appears to be a general phenomenon.

INTRODUCTION

At their simplest, infectious diseases involve only two species—a specialist pathogen and its sole host. The simplicity of these systems allows the development of ecological theory to understand the effects of variable host abundance, physiology, and behavior on pathogen transmission and disease dynamics within this two-species system (Anderson & May 1979, 1981). Models consisting of one host and one pathogen have been useful in allowing ecologists to isolate the effects on disease dynamics of, for instance, the length of the infectious period, the probability that an infectious and a susceptible individual make contact, and the probability that a pathogen is transmitted given contact (Anderson & May 1991). But these insights have required considerable simplifications of real host-pathogen systems. First, the models assume that host populations are homogeneous with respect to susceptibility to pathogen transmission, maintenance, and proliferation, but this is rarely the case. Most host populations are genetically heterogeneous in ways that affect their interactions with pathogens. Second, pathogens and parasites are allowed to infect only one host species in these models, but even many pathogens and parasites that are considered host specialists typically infect multiple host species. Some specialize on a particular host species but can spill over and temporarily invade others, whereas others readily proliferate within several or even many host species within a community. As a consequence, host species can differ strongly with respect to susceptibility to pathogen transmission, maintenance, and proliferation. Third, simple models assume that host species are affected only by their interaction with the specialist pathogen, but real host species occur within ecological communities. Host interactions with pathogens can be affected strongly by the presence of interacting species, including predators, competitors, and other parasites, within these communities (Holt & Roy 2007, Ostfeld & Holt 2004, Packer et al. 2003).

This review examines recent scientific studies that explore how disease dynamics are affected by variation in genetic diversity within a host species, variation among host species within communities, and variation among ecological communities. These three levels of variation—of diversity—correspond to the three levels invoked in leading definitions of the term biodiversity (Redford & Richter 1999). Indeed, current usage of the term biodiversity generally invokes genetic diversity, species diversity, and community diversity (we replace the more commonly used term ecosystem with community to reflect the notion that the abiotic component of ecosystems is not explicitly considered part of biodiversity (see Corvalan et al. 2005, Harper & Hawksworth 1994).

Within each of the three levels comprising biodiversity (genotypes, species, communities), diversity can be characterized in three ways: (*a*) by the number of different entities (e.g., the number of species in a community, or species richness), (*b*) by the relative abundances of the different entities (e.g., the evenness by which species are represented in a community, or species evenness), and (*c*) by the specific identities of the different entities (e.g., the species composition of a community). Quantifying biodiversity is fairly straightforward for *a* and *b*, but more challenging for *c*. This challenge is sometimes addressed by grouping particular genes or species into functional or taxonomic units. In other cases, it is addressed by attempting to correlate species identity with a quantitative measure of biodiversity. For example, when species losses from a community occur in a more or less predictable sequence, then changes in community composition are correlated with changes in species richness, a quantitative metric. In some cases, the same change in a community can increase biodiversity at one level while decreasing it at another. For example, the addition of an exotic species to a community increases species richness by one species but decreases community diversity by homogenizing the donor and recipient communities.

Metrics of biodiversity can be used to characterize either the long-term, static characteristics of a biota or the dynamic changes that accompany ecological and anthropogenic changes.

Within any species, community, or metacommunity (a group of communities), native biodiversity is the result of biogeographic processes (e.g., speciation, community diversification) occurring over evolutionary time. In contrast, dynamic changes in biodiversity are dominated by anthropogenic processes acting locally and quickly. Overwhelmingly, these changes consist of biodiversity losses rather than gains. This review addresses the relationship between infectious diseases and both static biodiversity and dynamic changes in biodiversity. Because of the importance of applying our scientific understanding to the management of diseases, we focus largely on how dynamic changes in biodiversity affect pathogen transmission and disease risk.

GENETIC DIVERSITY WITHIN HOSTS

Recently, Lively (2010) provided a simple model of the establishment and spread of an infectious disease in a host species with different numbers of genotypes. This model assumed that each host genotype was susceptible to one of many pathogen genotypes and resistant to the rest, reflecting the matching alleles model of infection (Lively 2010). Lively focused on R_0, the number of secondary infections produced by the initial infection in a susceptible population. He found that the ability of an invading pathogen to establish in a host population, as measured by R_0, was inversely proportional to the number of genotypes in the host population. He also found that infections that are able to spread initially die out more rapidly when the host population is more genetically diverse. In a separate study, the severity of modeled livestock epidemics was also inversely proportional to host genetic diversity under the assumption of a single pathogen genotype and variation in the genetic resistance of host individuals to this pathogen (Springbett et al. 2003).

These theoretical explorations support empirical observations from a variety of natural and laboratory systems. Monocultures of genetically similar or identical plants are notoriously susceptible to disease spread, whereas genetic mixtures of the same plants are more resistant (Elton 1958, Mundt 2002). For example, experimentally varying the frequency of a susceptible genotype of wheat (*Triticum aestivum*) from 1.0 (low diversity) to 0.25 (high diversity) in a field experiment led to dramatically reduced prevalence of disease caused by striped rust, *Puccinia striiformis* (Mundt et al. 2011). On an even larger scale, a field experiment of enormous scope addressed the effects of planting diverse genotypes of rice (*Oryza sativa*) on the incidence of disease caused by rice blast (*Magnaporthe grisea*) in Yunnan Province, China (Zhu et al. 2000). Traditionally, rice farmers throughout Yunnan Province plant vast monocultures of single rice genotypes, and they often require extensive applications of fungicides to prevent blast epidemics. Glutinous rice is particularly susceptible to rice blast. By planting glutinous rice interspersed with other rice varieties in thousands of fields, farmers reduced fungal rice blast disease by 94% and increased glutinous rice yields by 89% compared with monocultures (Zhu et al. 2000) (**Figure 1**).

Dennehy et al. (2007) experimentally assessed the effects of bacterial (host) genetic diversity on attack rates by a pathogenic bacteriophage. Phage Φ6 attacks and kills its bacterial host, *Pseudomonas phaseolicola*, by attaching to pili (small protuberances the bacteria use to attach to plants on which they feed). When the bacteria withdraw their pili, attached phages enter the cell, replicate, and kill the bacterium. In addition to wild-type *P. phaseolicola* with these pili, two naturally occurring mutants also exist. One, called superpiliated, has many pili but never retracts them, so that it attracts phages but the phages cannot enter the bacterial cell. The other mutant, called neutral, has no pili and so Φ6 phages are unable to attach. Dennehy et al. subjected monocultures of the wild-type *Pseudomonas* to its Φ6 pathogen and then added each of the naturally occurring *Pseudomonas* mutant genotypes, increasing genotypic diversity. Compared to the wild-type monoculture, a mixed culture of 50% wild type and 50% neutral reduced the abundance of Φ6 phage almost

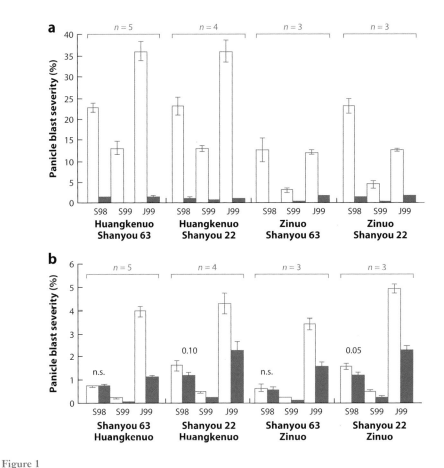

Figure 1

Severity of panicle rice blast disease (caused by *Magnaporthe grisea*) in monocultures (*light blue bars*) and polycultures (*dark blue bars*) of rice in China. (*a*) Susceptible, glutinous varieties of rice Huangkenuo and Zinuo; (*b*) more resistant, hybrid varieties Shanyuo 22 and Shanyuo 63. *n* = the number of plots included in the count for each bar. Error bars are 1 standard error. All differences between pairs of monoculture and polyculture bars are significant at P < 0.01 based on a one-tailed t-test, unless indicated by 0.05 (significant at P < 0.05), 0.10 (significant at P < 0.10), or n.s., (not significant at P = 0.10). Reproduced with permission from Zhu et al. (2000).

tenfold. When they presented the phage with a mixture of 50% wild type and 50% superpiliated *P. phaseolicola*, the phages declined in abundance by about 700-fold.

Similar experiments performed with animals give similar results. Using outdoor mesocosms, Altermatt & Ebert (2008) exposed populations of *Daphnia magna* water fleas to *Octosporea bayeri* microsporidian parasites. *Daphnia* populations were established by allowing clonal reproduction of genotypes taken from natural lakes and seeding experimental mesocosms with low (1 genotype) and high (10 genotypes) genetic diversity. Hosts in these mesocosms were exposed to parasite populations with different numbers of genotypes. Parasite prevalence (proportion of hosts infected) was consistently higher in low- than in high-diversity host populations; this result held irrespective of parasite genetic variation.

Several mechanisms appear capable of causing observed inverse relationships between host genetic diversity and disease severity or spread (Keesing et al. 2006). The data from all of the examples above are consistent with the encounter reduction mechanism of Keesing et al. in which host diversity (genetic diversity, in these cases) reduces rates of encounter between susceptible hosts and pathogens. For striped rust and rice blast, less susceptible genotypes appear to intercept dispersing fungal pathogens that might otherwise contact susceptible individuals. Superpiliated *P. phaseolicola* bacteria absorb but do not contribute phages, reducing phage encounters with wild-type hosts. Similarly, less susceptible *Daphnia* genotypes apparently act as dead-end hosts when they filter from the water column parasites that might otherwise infect susceptible genotypes. Apparently, encounter reduction can occur in systems in which genetic variation is spatially structured, as for plants in crop fields, and in which it is not, as for *Daphnia* in the water column.

Another mechanism by which genetic diversity might reduce pathogen transmission consists of the regulation of the population density of susceptible hosts by other competing genotypes [susceptible host regulation sensu Keesing et al. (2006)]. In the *Daphnia* study (Altermatt & Ebert 2008), this mechanism was rejected by the observation that genetically diverse host populations did not contain lower densities of susceptible hosts. In the bacterium-phage example, regulation of wild-type bacteria by competition with the neutral genotype appeared to reduce phage abundance (Dennehy et al. 2007). No evidence for susceptible host regulation has been detected in the plant-fungus studies described above, but an additional mechanism, which might be called resistance enhancement, might operate in the rice blast system. Zhu et al. (2000) postulated that the diverse mixtures of rice genotypes supported a diverse assemblage of fungal pathogens, none of which became very abundant. Each of these pathogens appeared to vaccinate rice plants and increase their resistance to multiple pathogens. Mechanisms underlying decreased disease prevalence with increasing genetic diversity require more attention.

In some of the examples, the reduction in pathogen transmission accompanying increased genetic diversity arose from specific traits of the genotypes used in field or lab experiments. For instance, in the experiments on fungal pathogens of plants, the monoculture consisted of a susceptible genotype and the polyculture included less susceptible ones. In other cases, however, reduced disease under increased genetic diversity occurred without regard to the specific genotypes used. Genetic diversity per se was responsible for reduced disease in the models of Lively (2010) and Springbett et al. (2003) and in the *Daphnia* experiments of Altermatt & Ebert (2008). In these cases, when faced with a genetically diverse group of pathogens, any host monoculture was more susceptible than any host polyculture.

SPECIES DIVERSITY AMONG HOSTS

In a sense, the effect of species diversity on pathogen transmission is simply an extension of the effect of genetic diversity within species. Just as with different host genotypes, different host species are expected to vary in their susceptibility to, and ability to support replication of, specific pathogens. But differences between species are expected to be larger than those within a species.

Both theoretical and empirical treatments of the effects of changes in species diversity on disease dynamics have tended to explore dynamics in a single-host system and then ask how disease dynamics change when species are added. An early example is transmission of the malaria parasite (*Plasmodium* spp.) by mosquitoes to humans. Investigators proposed that the presence of wild or domestic animals that did not support *Plasmodium* might divert blood-seeking mosquito vectors away from potential human hosts, an idea that has been called zooprophylaxis (Service 1991, World Health Organization 1982). This concept, following an earlier treatment by Brumpt (1944–45), implies two specific conditions: (*a*) that the added host does not amplify the pathogen and (*b*) that the added host reduces disease risk by diverting the vector from biting humans. Later, Matuschka & Spielman (1992) noticed parallels between malaria zooprophylaxis and another vector-borne disease, Lyme disease. Here the vector is an *Ixodes* tick and the pathogen is a spirochete, *Borrelia burgdorferi*. Ticks acquire pathogens readily by feeding on certain species of small mammals, but they acquire pathogens much less efficiently when feeding on other species of host, including some songbirds, ungulates, and carnivores (LoGiudice et al. 2003, Matuschka et al. 1993, Richter & Matuschka 2006, Richter et al. 2000). When these latter hosts, which are considered incompetent reservoirs for the Lyme pathogen, are present in the host community, they can potentially divert tick meals away from the competent reservoirs and reduce tick infection prevalence. Norman et al. (1999) used the term dilution effect to refer to the potential for increases in host diversity to drive another tick-borne pathogen, Louping ill virus, to extinction. They modeled the Louping ill system with both a reservoir-competent host (red grouse, *Lagopus lagopus scoticus*) and a reservoir-incompetent host (mountain hares, *Lepus timidus*); diversity was increased by increasing the relative abundance of the latter. Hence, their model did not manipulate species richness, but rather the evenness of the two hosts. As the ratio of incompetent to competent reservoirs increased from a low level, more tick bites were received by the incompetent host, reducing virus transmission until it was eliminated from the system.

Initial development of the concept that increased diversity decreases disease risk, under the aegis of either the term zooprophylaxis or dilution effect, was mechanistic and specific to a particular disease system. To generalize the concept and distinguish the phenomenon from the underlying mechanisms, Keesing et al. (2006) redefined the term dilution effect to refer to the inverse relationship between diversity and disease risk. They then characterized a set of potential mechanisms that can cause the relationship. Considerable recent research on the dilution effect (hereafter DE) has taken one of two general approaches. One approach asks whether a specific manipulation of host species diversity produces a DE—i.e., can it occur? The second asks, is the DE frequently observed in nature—i.e., does it occur?

Can the Dilution Effect Occur?

Addressing this question involves manipulating host species diversity either in laboratory experiments or in mathematical models and then measuring indicators of pathogen transmission or disease risk. Dobson (2004) and Rudolf & Antonovics (2005) used epidemiological models to explore the conditions under which increases in host diversity dilute, versus amplify, pathogen transmission. Both studies considered disease systems in which rates of pathogen transmission are proportional to the frequency of infected individuals in the host population (frequency-dependent transmission), or alternatively to the density of infected individuals in the host population (density-dependent transmission). Frequency-dependent transmission is thought to characterize most vector-borne diseases because a vector's biting rate (and thus its opportunities to become infected) is assumed to be determined by intrinsic physiological or behavioral features of the vector rather than by host density. Consequently, the probability of the vector becoming infected is determined by the frequency of infection in the host. Frequency dependence also pertains to sexually transmitted diseases. Density-dependent transmission is thought to characterize systems with free-living infective stages, aerial plumes of pathogens, or direct transmission in an unstructured (randomly mixing) host population. When transmission is frequency-dependent, increases in biodiversity suppress pathogen transmission, and this DE occurs regardless of the identities or changes in abundances of individual species (Dobson 2004, Rudolf & Antonovics 2005). When transmission is density dependent, increases in biodiversity tend to amplify pathogen transmission when the added species simply increase the total abundance of all hosts in the system, that is, when there are no compensatory reductions in host abundance with increased species richness. A DE occurs in systems with density-dependent transmission when compensatory

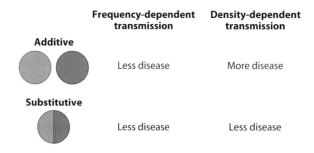

	Frequency-dependent transmission	Density-dependent transmission
Additive	Less disease	More disease
Substitutive	Less disease	Less disease

Figure 2

Effect of transmission mode on disease levels in a target (*blue*) population following the addition of a second (*red*) species, i.e., an increase in diversity. The bottom right-hand outcome panel represents density-dependent dilution, and outcomes in the two left-hand outcome panels represent frequency-dependent dilution. Reproduced with permission from Rudolf & Antonovics (2005).

reductions in abundance accompany the addition of new host species (Rudolf & Antonovics 2005) (**Figure 2**).

In addition to these general models, specific models of particular disease systems have been built to ask whether a DE can occur. In an epidemic model developed for hantavirus pulmonary syndrome, with a deer mouse host and a competing rodent nonhost, Peixoto & Abramson (2006) found that competition from the nonhost reduced prevalence of hantavirus infection in the mouse host and could, when competition was intense, drive the pathogen locally extinct. A detailed analytical model of Lyme disease showed that a DE occurred when abundances of inefficient reservoir hosts exceeded those of efficient reservoirs, and could also occur at any combined abundances of competent and incompetent reservoirs as long as tick abundance increases nonmonotonically with total host abundance (Rosa & Pugliese 2007). In a highly mechanistic simulation model of Lyme disease, a DE occurred under three conditions: (*a*) when high diversity reduced density of the main reservoir host, the white-footed mouse (*Peromyscus leucopus*) by any form of competition—direct, indirect or apparent; (*b*) when the nonmouse hosts in the community fed proportionally more nymphal than larval ticks compared to *P. leucopus*; and/or (*c*) when tick mortality was higher on nonmouse than mouse hosts (Ogden & Tsao 2009). When these conditions are not met, high diversity can amplify the number of infected ticks, and hence disease risk. Little direct evidence can be brought to bear on the first condition, although circumstantial evidence is supportive (Nupp & Swihart 2000). Supporting evidence for the second condition is provided by Ogden & Tsao's observation that in 12 of 18 studies nonmice had higher nymph:larva ratios than mice. The third condition is strongly supported by the dramatically higher mortality rates of larval ticks on nonmouse than mouse hosts observed by Keesing et al. (2009, 2010).

Manipulating host diversity in field experiments has also led to insights about whether the DE can occur. For example, Mitchell et al. (2002) varied the diversity of prairie plants from 1 to 24 species per plot and monitored the severity of disease caused by foliar fungal pathogens on all host plants. Plant species were selected by random draw, so species composition varied randomly in all field plots and did not confound variation in species richness. For 11 fungal pathogens, severity of disease decreased significantly with increasing plant species richness, whereas for only one pathogen did disease increase with increasing host richness. In about half of the pathogens showing a DE, reduced disease severity was associated with reduced host density, providing examples of susceptible host regulation (Keesing et al. 2006). Roscher et al. (2007) also studied disease severity in field plots planted with variable numbers of herbaceous plant species (from 0 to 59 additional species). In this study, the focal diseases were caused by crown rust (*Puccinia coronatum*) and stem rust (*P. graminis*), and the focal host plant was perennial ryegrass (*Lolium perenne*). Although different cultivars of ryegrass varied in their susceptibility to both pathogens, for all cultivars disease severity and infection prevalence declined dramatically with plant species richness. Although initial ryegrass density was constant in all treatments, both densities and individual plant sizes decreased with increasing diversity. Therefore, both these field experiments with plants strongly suggest that, even if pathogen transmission is density dependent, a DE occurs when increased diversity causes compensatory declines in host abundance or biomass, supporting a main conclusion of the Rudolf & Antonovics (2005) model. In contrast to these results that support the model, Borer et al. (2010) found that plant species richness did not predict prevalence of aphid-transmitted barley-cereal yellow dwarf virus in experimental plant communities. Instead, virus prevalence was predicted by local variation in percent cover of perennial grasses and phosphorus enrichment.

Experiments varying animal host diversity have demonstrated a DE for diseases of amphibians and humans. Johnson et al. (2008) examined infestation of American toads [*Anaxyrus* (= *Bufo*) *americanus*] with the trematode parasite *Ribeiroia ondatrae*, which causes limb malformations and increases mortality. They raised toads alone, in conspecific pairs (increased density), and in heterospecific pairs with either a treefrog [*Pseudacris* (= *Hyla*) *versicolor*] or the frog *Rana clamitans* or both. Adding a treefrog significantly reduced trematode burden in toads, and this effect of increased diversity was independent of either toad or total amphibian density. Adding *R. clamitans*, however, did not reduce trematode burden in toads. Using a similar experimental design, Searle et al. (2011) varied species richness of amphibians and total amphibian density and measured infection of a particularly susceptible species, the western toad (*Anaxyrus boreas*) with the fungal pathogen *Batrachochytrium dendrobatidis* (Bd), which is causing amphibian declines worldwide. Nontoad species added to manipulate species diversity were treefrogs (*Pseudacris regilla*) and the frog *Rana cascadae*. Increased host species richness consistently reduced Bd infection prevalence and infection severity both in toads alone and in all species combined (**Figure 3**). In contrast with the *Ribeiroia* study, both nontoad species reduced Bd infection, and the effect of adding *Rana* was somewhat stronger than that of adding *Pseudacris*.

To test for a DE in human schistosomiasis, Johnson et al. (2009) created monocultures of the snail *Biomphalaria glabrata*, which is an important intermediate host for *Schistosoma mansoni*. They also created polycultures containing *B. glabrata* and naturally co-occurring snail species. They added identical concentrations of *S. mansoni* miracidia to containers with either *B. glabrata* snails alone, *B. glabrata* with *Helisoma trivolvis* or *Lymnaea stagnalis* snails, or both (neither of these snail species supports replication by *S. mansoni* miracidia) and monitored cercarial release rates, which are strong predictors of human risk. In containers with two or three species of snail (*Biomphalaria* plus one or both of the other snails), both the proportion of snails infected and the cercarial release rates were strongly reduced. *Biomphalaria* snails raised alone released, on average, between two and five times as many schistosome cercariae as did the snails raised together with one or two other species. Several other experimental investigations of a DE in helminth parasites are reviewed by Johnson & Thieltges (2010) and Keesing et al. (2010).

From these modeling and experimental studies we can conclude that (*a*) a DE can occur under a wide range of conditions, including frequency- and density-dependent transmission of the pathogen, in pathogens with simple or complex life cycles, in terrestrial and aquatic systems, and in plants and animals; (*b*) in some cases, reduced pathogen transmission with increased diversity depends on the specific identity of the species added to the system, but in other cases the DE occurs irrespective of species identity; and (*c*) a DE can occur in host communities confronted with both single and multiple pathogens.

The potential for the DE to occur under a variety of conditions does not demonstrate that it does occur regularly in nature. A DE is not expected to occur, and indeed, an amplification

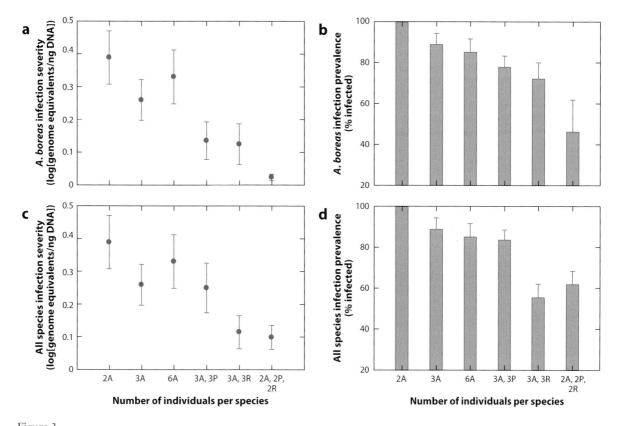

Figure 3

Average infection severity and prevalence of *Batrachochytrium dendrobatidis* (Bd) with varying host density and species richness. Treatments labeled on the horizontal axis represent the number of individuals of each species, with "A" for *Anaxyrus boreas*, "P" for *Pseudacris regilla*, and "R" for *Rana cascadae*. (*a,c*) Points show the average Bd infection severity (log + 1-transformed; ± standard error) for *A. boreas* (*a*) and for all species present (*c*). (*b,d*) Bars show the mean percentage (+standard error) of *A. boreas* testing positive for infection (*b*) and the percentage infection for all species (*d*). Reproduced with permission from Searle et al. (2011).

effect (Keesing et al. 2006) can occur when (*a*) the species added to foster diversity increases are the most competent reservoirs for the pathogen (Ostfeld & Keesing 2000, Schmidt & Ostfeld 2001) or (*b*) pathogen transmission is density dependent and more diverse host communities show no compensatory reductions in abundance of reservoir-competent hosts (Dobson 2004, Rudolf & Antonovics 2005). In addition, (*c*) if the pathogen is vector borne, host diversity might be relatively unimportant if the vector is a strict host specialist (i.e., changes in host diversity do not influence vector feeding patterns) and compensatory reductions in the most competent reservoir(s) do not accompany increased diversity (Loss et al. 2009, Ogden & Tsao 2009, Ostfeld & Keesing 2000, Schmidt & Ostfeld 2001).

Does the Dilution Effect Occur in Nature?: Plants and Aquatic Organisms

Correlative studies assessing the relationship between diversity and disease risk are accumulating, some of which use statistical approaches to infer, if not demonstrate, causality. Perhaps the most comprehensive individual field study to date concerns infection of woody plants by the causative agent of sudden oak death (SOD), *Phytophthora ramorum*, in central coastal California. Haas et al. (2011) analyzed plant communities and pathogen prevalence in 280 500-m² plots within a 79,356-ha study area over two years. They used Bayesian hierarchical modeling to assess the effects of host species diversity (richness and Shannon diversity) on pathogen prevalence in local plant communities, with models accounting for an abundance of plots with no pathogen (zero inflated models) and for spatial autocorrelation at a variety of scales. All of their Bayesian models indicated a negative relationship between species diversity (both richness and Shannon diversity) and pathogen incidence. This strong DE remained after statistically accounting for local density of the most competent host plants [bay laurel (*Umbellularia californica*) and tanoak (*Notholithocarpus densiflorus*)]. Detection of a DE was consistent with prior experimental studies on diseases of herbaceous plants (Knops et al. 1999, Mitchell et al. 2002).

Hall et al. (2009) examined the prevalence of fungal (*Metschnikowia bicuspidate*) infection in *Daphnia* communities in a series of lakes in Michigan, USA, that varied in host diversity. Some lakes were virtual monocultures of *Daphnia dentifera*, which is readily infected by the pathogen, whereas other lakes were more diverse, with *D. dentifera* comprising between about 25% and 80% of the *Daphnia* community. The relative host index, defined as the population density of *D. dentifera* divided by the summed densities of the three other *Daphnia* species, in a lake measured before fungal epidemics was a significant predictor of the size of the epidemic. Lower diversity communities were more dominated by *D. dentifera* and had larger epidemics.

In other freshwater systems, the introduction of an exotic species has been associated with strong declines in disease within native hosts. Introductions of a non-native freshwater snail (*L. stagnalis*) and non-native brown trout (*Salmo trutta*) reduced prevalence of infection of native snails with trematode parasites (Kopp & Jokela 2007) and of native fishes with several helminth parasites (Kelly et al. 2009a), respectively. Although these species introductions increased species richness by one species, it is important to note that they reduced community diversity by homogenizing the native and adoptive communities. In addition, in some cases, the addition of non-native species with their associated pathogens to susceptible host communities can increase pathogen transmission in the native community (Gurnell et al. 2006, Tompkins et al. 2003).

Does the Dilution Effect Occur in Nature?: Nonvector-Borne Zoonoses

Correlative studies of several zoonotic disease systems that lack vectors demonstrate the frequent occurrence of the DE in nature. Perhaps the most widely studied of these are the hantaviruses (family Bunyaviridae), common rodent-associated viruses widespread in Europe, Asia, North America, and South America. Virus can be transmitted among individual small mammals during fighting and other contacts, and humans can be exposed if they inhale airborne particles of small-mammal excreta that contain virus particles. Risk of human exposure is a function of both the population density and infection prevalence of the rodent reservoir (Yates et al. 2002). Each type (species) of hantavirus tends to be associated with only one species, or a group of closely related species, of reservoir host, although spillover to other species is possible. The list of rodent species that act as key reservoir hosts worldwide includes the Norway rat (*Rattus norvegicus*), black rat (*R. rattus*), bank vole (*Myodes glareolus*), deer mouse (*Peromyscus maniculatus*), white-footed mouse (*P. leucopus*), hispid cotton rat (*Sigmodon hispidus*), rice rat (*Oryzomys palustris*), and small vesper mouse (*Calomys laucha*). It is noteworthy that each of these species is a habitat generalist, geographically widespread, locally highly abundant, and resilient to anthropogenic disturbance. Consequently, these reservoir species tend to occur and even thrive in low-diversity small-mammal communities, potentially leading to higher transmission risk when biodiversity is low.

Studying Laguna Negra hantavirus in the Paraguayan Chaco, Yahnke et al. (2001) analyzed small mammal communities and virus distributions in four habitat types over 15 months. The most abundant and widespread mammal they trapped, *C. laucha*, was the only demonstrated host for the virus. Infection prevalence in host populations was positively correlated with the proportion of *C. laucha* in the rodent community, with highest prevalence found in sites where >30% of the rodent fauna consisted of *C. laucha*, but was not correlated with host density per se. Piudo et al. (2011) assessed the causes of variation in infection prevalence of Andes hantavirus in its reservoir host, *Oligoryzomys longicaudatus*, in both peridomestic and forest habitats of Patagonia. They found that *O. longicaudatus* was the numerically dominant small-mammal species in peridomestic habitats where small-mammal diversity was low but not in sylvan habitats where diversity was higher. They found that the number of infected *O. longicaudatus* was negatively correlated with small-mammal diversity during the same month and in the prior two months; however, they did not find a significant correlation between mammal diversity and the proportion of reservoir hosts infected. A similar result was obtained by Ruedas et al. (2004) studying Choclo and Calabazo hantaviruses in 9 sites in Panama's Azuero Peninsula, where the rodent reservoirs are *O. fulvescens* and *Zygodontomys brevicauda*. These researchers compared small-mammal communities at sites where human hantavirus disease was detected (case sites) with the fauna at a reference site with no human disease; case sites had significantly lower diversity (Shannon index) than reference sites. In a later live-trapping study in the same region, Suzan et al. (2009) permanently removed all small-mammal species except the community dominants *O. fulvescens* and *Z. brevicauda* on 16 sites, which created an experimental reduction of species diversity that mimicked natural patterns (Ruedas et al. 2004; Suzan et al. 2008a,b). On the remaining 8 sites—the control sites—the investigators marked and released all animals captured, but did not manipulate abundances. In the 16 removal plots, population densities of the two reservoir species were significantly increased relative to their densities on the control plots, indicating that high small-mammal diversity suppresses abundance of reservoir hosts. More strikingly, the removal of nonreservoir species was associated with a dramatic increase in the proportion of the reservoir populations that were infected with hantavirus, from 8% to 12% on control plots to about 35% on removal plots (**Figure 4**). Studying a European hantavirus (Puumala virus) in 14 sites in northern Belgium, Tersago et al. (2008) found that

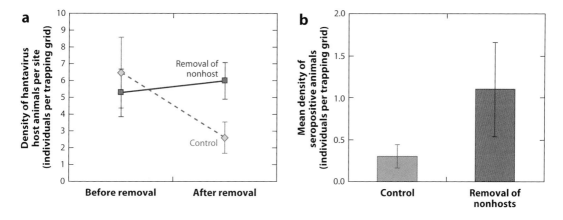

Figure 4

Effects of experimental removal of hantavirus reservoir species, *Oligoryzomys fulvescens* and *Zygodontomys brevicauda* on density of these hosts (*a*) and density of seropositive hosts (*b*). Shown in panel *a* are means (+ standard error) from field plots before and after nonhost species had been removed (*solid line*) and from unmanipulated controls (*dashed line*). Hosts on control plots underwent a strong seasonal decline in abundance, whereas those on plots where nonhosts were experimentally removed did not. Shown in panel *b* are means (+ standard error) of seropositive (currently or previously infected) animals on plots from which nonhosts had been removed and on control plots. Reproduced with permission from Keesing et al. (2010) based on data from Suzan et al. (2009).

hantavirus infection prevalence in *M. glareolus* reservoirs was strongly, negatively correlated with the relative abundance of a nonreservoir host, the wood mouse (*Apodemus sylvaticus*), which was the only other common rodent at these sites.

Several recent studies have assessed the relationship between mammalian diversity and hantavirus risk in North American Sin Nombre virus, for which the deer mouse *P. maniculatus* is the primary reservoir host. Dizney & Ruedas (2009) captured about 5,000 individuals of 21 species of small- and medium-sized mammals in five sites around Portland, OR, USA, and found that all sites that maintained moderate to high mammalian diversity were characterized by low (~2%) hantavirus infection prevalence in deer mice, whereas the one site with low mammalian diversity had 14% infection prevalence, a result that was highly statistically significant. In Montana, Carver et al. (2011) analyzed 15 years of live-trapping data on three field plots on which deer mice were always present but voles (nonreservoirs) were only occasionally trapped. They found that the presence of voles was associated, on average, with a 54% to 64% reduction in infection prevalence of deer mice. Deer mouse infection prevalence did not vary with vole population density; the mere presence of voles, even in low numbers, apparently was sufficient to suppress mouse infection rates. On the Channel Islands of southern California, Orrock et al. (2011) found that the species diversity of predators on small mammals (carnivores and raptors) was strongly negatively correlated with hantavirus infection prevalence in deer mice.

Attention has turned to the mechanisms that might underlie the negative correlation between mammalian biodiversity and hantavirus transmission. The Dizney & Ruedas (2009) study did not find that higher mammalian diversity was correlated with lower abundance of deer mouse reservoirs, nor that hantavirus infection prevalence was positively correlated with deer mouse abundance. They therefore conclude that, rather than regulating populations of reservoir hosts, high diversity reduced encounters among deer mice that could result in virus transmission. Studying Sin Nombre virus dynamics in Utah, Clay et al. (2009a,b) and Lehmer et al. (2008) uncovered a somewhat more complex set of factors influencing infection prevalence in deer mice. Clay et al. found a negative correlation between nocturnal mammal diversity and infection prevalence of deer mice with Sin Nombre virus across 16 field sites, supporting the occurrence of a DE. Although they found that both the population density of deer mice and the average survival (persistence) of individual mice were lower where diversity was higher, infection prevalence was negatively correlated only with deer mouse survival and not with deer mouse density (Clay et al. 2009c; Lehmer et al. 2008). By monitoring encounter rates between mice in foraging plots, Clay et al. (2009b) found that contacts between deer mice were less frequent in areas with high mammalian diversity. The weight of evidence for the Sin Nombre hantavirus system in western North America suggests that high diversity of mammals causes deer mice to increase their frequency of encounters with heterospecifics at the cost of encounters with conspecifics. Heterospecific encounters typically do not result in hantavirus transmission, whereas hantavirus transmission rates correlate positively with the frequency of intraspecific encounters (Dearing & Dizney 2010). In addition to reduced biodiversity per se, several studies suggest that land conversion to agriculture and cattle ranching (Goodin et al. 2006; Yan et al. 2007; Zhang et al. 2009a,b), human habitation (Kuenzi et al. 2001, Langlois et al. 2001), and habitat fragmentation (Mackelprang et al. 2001; Suzan et al. 2008a,b) can increase hantavirus infection prevalence in reservoir hosts (Dearing & Dizney 2010).

The DE has also been observed in other zoonoses that lack vectors. Derne et al. (2011) examined the relationship between mammalian species richness and per capita incidence of leptospirosis—the most common bacterial zoonosis worldwide—on the world's island nations. Rats (*R. rattus* and *R. norvegicus*) are the most competent reservoirs for this pathogen and occur throughout the world's islands mostly as a result of dispersal on ships. Human cases of leptospirosis (cases per 100,000 per year) across island nations were significantly, negatively correlated with mammalian species richness. The DE in leptospirosis persisted after statistically accounting for variation in land mass among islands.

In contrast to the experimental demonstration of a DE in North American chytridiomycosis discussed above (Searle et al. 2011), a field study on two common, widely distributed frogs in Costa Rica and Australia (Rain frog, *Craugastor fitzingeri*, and Stony Creek frog, *Litoria lesueuri*, respectively) found support for an amplification effect (Becker & Zamudio 2011). For both species, higher occurrence or prevalence of Bd at a site was correlated with both lower levels of habitat destruction and higher amphibian species richness. Becker & Zamudio postulated that habitat destruction creates less favorable abiotic conditions for Bd, and also that amphibian species that amplify the pathogen might be more likely to occur in species-rich communities. The qualitative differences between the experimental lab study (Searle et al. 2011) and nonexperimental field study (Becker & Zamudio 2011) might result from differences in the characteristics of both the focal species chosen and the nonfocal species providing variation in host biodiversity.

Does the Dilution Effect Occur in Nature?: Vector-Borne Zoonoses

The involvement of an arthropod vector in pathogen transmission adds at least two additional mechanisms by which species diversity can influence risk. High diversity can potentially influence the abundance of the vector, by increasing or decreasing total feeding opportunities or vector survival, and it can influence encounter rates between the vector and the most competent reservoir host (Keesing et al. 2006). The DE has been evaluated extensively in Lyme disease, a tick-borne bacterial zoonosis, and West Nile fever, a mosquito-borne viral zoonosis.

The hard (Ixodid) ticks are vectors of many human, livestock, and wildlife pathogens worldwide. Risk of human exposure to tick-borne pathogens is a function of both the population density of ticks and their infection prevalence (Ostfeld 2011). Each active life stage—larva, nymph, and adult—typically requires a single blood meal from a host before dropping off the host, undergoing diapause, and molting into the next stage, or reproducing and dying. Ticks can acquire an infection with the causative agent, *Borrelia burgdorferi*, only by feeding on a reservoir host; that is, they acquire the bacteria horizontally rather than vertically. Ticks are weakly motile and require a host to approach closely in order to embark and attempt a blood meal. The tick species responsible for transmitting most zoonotic pathogens are strong host generalists, typically feeding from dozens of different host species.

Because of the reliance of ticks on access to vertebrate hosts and high mortality when they fail to find a host, tick abundance is widely thought to be correlated with that of hosts. Models of tick-borne zoonoses therefore often predict that disease risk is positively correlated with host diversity, i.e., an amplification effect (e.g., Dobson 2004), as long as high diversity leads to high total abundance of hosts (i.e., hosts are additive). However, when positive correlations between tick abundance and host abundance are detected, the correlation is specific to one species of host. For example, some studies have found positive correlations between an abundance of white-tailed deer (*Odocoileus virginianus*) and that of ticks (e.g., Deblinger et al. 1993, Rand et al. 2004, Stafford et al. 2003; but see Ostfeld 2011), and others have detected positive correlations between the abundance of white-footed mice (*P. leucopus*) and that of ticks (Ostfeld et al. 2006). Deer and mice are important hosts for the adult and immature stages, respectively, of the blacklegged tick, *Ixodes scapularis*. To our knowledge, no studies have assessed the relationship between total host abundance and tick abundance. Consequently, whether tick-borne disease is accurately portrayed with density-dependent rather than frequency-dependent transmission models is unknown.

A DE is expected to operate in the Lyme disease system if increases in the species diversity of host communities (*a*) tend to add species that are relatively poor reservoirs for the bacterial agent

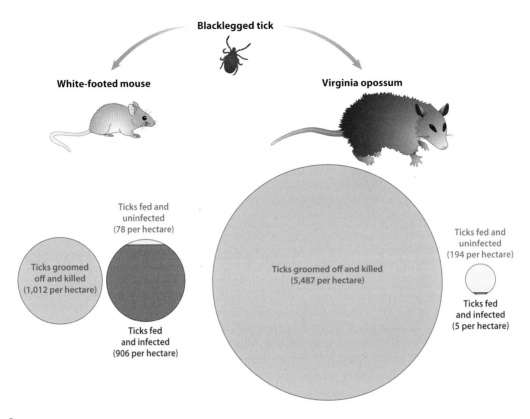

Figure 5

Roles of host species in the transmission of Lyme disease in the northeastern United States. Lyme disease is transmitted to humans by the bite of an infected blacklegged tick (*Ixodes scapularis*). Immature ticks can acquire the infection if they feed on an infected host and can become infectious to humans if they subsequently survive to the next life stage. White-footed mice are abundant in northeastern forests and feed many ticks. Ticks that attempt to feed on Virginia opossums are likely to be groomed off and killed. Red and yellow circles show the mean number of ticks per hectare fed by mice or opossums; yellow shading shows the proportion of ticks infected after feeding. Blue circles show the mean number of ticks per hectare groomed off and killed. Ticks that feed on mice are highly likely to become infected with the bacterium that causes Lyme disease, whereas those that feed on opossums are not. Reprinted with permission from Keesing et al. (2010).

and/or poor hosts for the tick vector, (*b*) regulate population densities of the most competent reservoir hosts, (*c*) regulate densities of the highest quality hosts for the tick vector, or (*d*) deflect tick meals away from the most competent pathogen reservoir and/or the highest quality tick host.

Extensive field and laboratory studies in the northeastern United States have documented that white-footed mice are the most competent reservoir for *B. burgdorferi* (Donahue et al. 1987, Lane et al. 1991, LoGiudice et al. 2003, Mather et al. 1989), with eastern chipmunks (*Tamias striatus*) and short-tailed shrews (*Blarina brevicauda*) being secondarily competent reservoirs and most other vertebrate species having low reservoir competence (Brisson et al. 2008, LoGiudice et al. 2003). Of four commonly parasitized mammals and two birds, white-footed mice were by far the most permissive host for blacklegged ticks, with tick survival rates dramatically higher on mice than on any other host (Brunner et al. 2011, Keesing et al. 2009) (**Figure 5**). Although the impact of mammalian biodiversity in regulating mouse populations is poorly studied, some evidence suggests that mouse density is reduced where species richness or abundance of nonmouse species is higher (LoGiudice et al. 2008; Nupp & Swihart 1998, 2000). A long-term study of the determinants of tick burdens on white-footed mice revealed that an increased abundance of chipmunks was the strongest factor reducing numbers of larval ticks on mice (Brunner & Ostfeld 2008). These data suggest that, at least for this pair of host species, increasing availability of nonmouse hosts strongly deflects tick meals away from mice.

Several studies in the eastern United States demonstrate that white-footed mice are among the most widespread vertebrate species within heterogeneous, fragmented landscapes (Nupp & Swihart 2000, Rosenblatt et al. 1999). White-footed mice were the only species to occupy all forest patches sampled for vertebrate occupancy in Connecticut, Indiana, New Jersey, and New York (LoGiudice et al. 2008, Nupp & Swihart 2000, Swihart et al. 2003b). These studies also show that eastern chipmunks were nearly ubiquitous. In Indiana, habitat occupancy was positively correlated with diet and niche breadth, providing the basis for the nested, rather than random, pattern of species occupancy (Swihart et al. 2003a,b). Therefore, low-diversity communities are highly likely to contain white-footed mice and eastern chipmunks.

Models parameterized with data on host attributes relevant to tick survival and infection indicate that, as species are added to low-diversity communities (mice only), both tick infection prevalence and population density of infected ticks decline (Keesing & Ostfeld 2012; Keesing et al. 2009; LoGiudice et al. 2003, 2008; Ostfeld & LoGiudice 2003; Schmidt & Ostfeld 2001). These studies all treat habitat occupancy to be nested, such that the sequence of species additions as diversity increases is nonrandom, reflecting empirical observations (Ostfeld 2011). Whenever habitat occupancy patterns are nested, species diversity and species composition are correlated. In a study testing these models with field data, LoGiudice et al. (2008) found that the prevalence of infection of nymphal ticks with Lyme spirochetes in a forest patch was negatively correlated with species richness, but that the relationship was weak. Including information on the specific identities of the species added as diversity increased dramatically strengthened their power to predict nymphal infection prevalence both across space and through time.

Species area curves observed for vertebrate hosts for ticks, combined with the nested occupancy patterns (LoGiudice et al. 2008, Rosenblatt et al. 1999, Swihart et al. 2003b), suggest that larger forest patches should contain tick populations that are less abundant, have lower infection prevalence, or both. Allan et al. (2003) found strong negative correlations between forest fragment size and both tick (nymph) density and infection prevalence in New York State. Brownstein et al. (2005) found that more highly fragmented landscapes around Lyme, CT, had higher densities of infected blacklegged ticks. Jackson et al. (2006a,b) found that landscapes in Maryland characterized by high percentages of edge between forest and herbaceous habitat, a measure of fragmentation, tended to have high incidences of Lyme disease in human residents. In contrast, Wilder & Meikle (2004) found that a lower proportion of mice in small forest fragments rather than large fragments in Ohio were infested with blacklegged ticks. And, despite a higher density of infected ticks in their fragmented sites in Connecticut, Brownstein et al. (2005) found lower incidence rates in human populations within these sites, suggesting that in some cases high ecological risk does not predict high disease incidence (Ostfeld 2011).

West Nile virus is a flavivirus (relative of the yellow fever virus) amplified in certain species of birds and transmitted among wildlife hosts by mosquitoes, predominantly *Culex* and *Aedes*. The virus was restricted to parts of Africa, Asia, and Europe until 1999 when it was inadvertently introduced into the New York City area, from which it has dispersed throughout much of North America, where it remains a serious health threat to both people and native avifauna (Allan et al. 2009; Kilpatrick et al. 2006a; LaDeau et al. 2007, 2008).

Like the Lyme disease spirochete, West Nile virus relies on horizontal (host to vector to host) transmission rather than on vertical (transovarial) transmission across generations of mosquitoes. Therefore, the prevalence of infection in the vector depends on which hosts are bitten and how likely those hosts are to transmit the pathogen to the feeding mosquito. Mosquitoes in the *Culex pipiens* complex often feed disproportionately on birds, and American robins in particular seem to be favored (Apperson et al. 2004; Hamer et al. 2011; Kilpatrick et al. 2006a,c). Nevertheless, within West Nile endemic zones, ample evidence indicates that infection is widespread among birds, mammals, and some reptiles (LaDeau et al. 2007, 2008; Marra et al. 2004), and birds other than robins often are disproportionately represented in blood-meal analyses (Loss et al. 2009), confirming that, despite their feeding preferences, these mosquitoes readily bite many different species of hosts (Kilpatrick et al. 2007, Komar et al. 2003, LaDeau et al. 2008, Marra et al. 2004). The different blood-meal hosts for mosquitoes vary dramatically in their ability to support replication by the pathogen, the length of time they remain infected, and their probability of transmitting virus to feeding mosquitoes (Kilpatrick et al. 2006b, Komar et al. 2003). Although studies at different North American localities often implicate different bird species as reservoirs, in general American robins, house sparrows, common grackles, American crows, blue jays, and house finches are the most competent reservoirs for West Nile virus. Many other passerines (songbirds), nonpasserine birds, mammals, and reptiles, either fail to permit virus infection or amplification, producing very low viremias, or they permit a high viremia but for only a very brief period, rendering them inefficient reservoirs.

Ezenwa et al. (2006) postulated that, because nonpasserine birds were generally incapable of producing high viremias, the presence of a high diversity of these bird species should cause mosquitoes to feed predominantly on poor reservoir hosts, reducing transmission rates of West Nile virus from host to vector. In their Louisiana, USA, study sites, they found that the proportion of mosquitoes infected with West Nile virus was significantly negatively correlated with species richness of nonpasserines. No correlation existed for passerines. Population density of infected *Culex* mosquitoes was also negatively correlated with nonpasserine diversity, suggesting that these birds might be injuring or killing mosquitoes that attempt to feed (Day & Edman 1984, Edman & Kale 1971, Edman et al. 1972). In addition, examining county-specific data throughout Louisiana, Ezenwa et al. found that the per capita incidence of West Nile disease in humans was negatively correlated with species richness of nonpasserines in that county.

Three later studies also detected a DE in West Nile virus, although in all these studies, virus prevalence was negatively correlated with passerine, rather than nonpasserine, diversity. Swaddle & Calos (2008) selected 65 adjacent pairs of counties in the eastern United States, with each pair consisting of one county with and an adjacent county without human cases of West Nile encephalitis. They tested the hypothesis that the West Nile–negative county in each pair would contain higher bird species diversity, as measured by the Breeding Bird Survey data from the United States Geological Survey. They found that the pairs of counties with the largest difference in bird diversity also had the largest difference in prevalence of West Nile encephalitis, with higher bird diversity being correlated with lower disease in humans.

Allan et al. (2009) examined human incidence of West Nile disease and bird diversity (Shannon Index, calculated from the Breeding Bird Survey) from 742 US counties within 38 states during three years (2002, 2003, and 2004) of rapid, cross-country spread. They calculated the total community competence of birds by multiplying the abundance of each species by its reservoir competence and summing the species-specific totals. Community competence provides an estimate of the likelihood that a particular bird community will produce infected mosquitoes. They found that bird assemblages with the highest community competence were also those with the lowest species diversity, whereas high-diversity communities tended to include poorer reservoirs. Consequently, as with observations on Lyme disease, host species most resilient to forces causing biodiversity loss tend to be the most competent reservoirs for the pathogen (see also Keesing et al. 2010). By comparing different models for explaining the variation in West

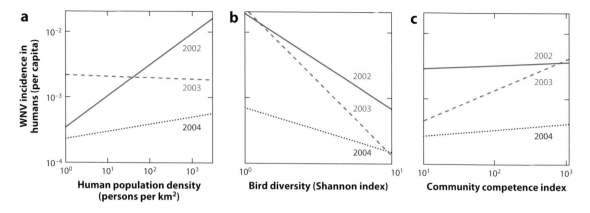

Figure 6

Relationship of human per capita incidence of West Nile virus (WNV) illness in the US counties with (*a*) human population density, (*b*) bird diversity (Shannon index), and (*c*) community competence index in 2002 (*solid line*), 2003 (*dashed line*) and 2004 (*dotted line*) as shown by multiple regression conducted separately for each year. Slopes reflect partial regression coefficients after statistically controlling for the other factors in the model. After controlling for spatial autocorrelation, relationships in panel *b* remained statistically significant, whereas many of the other relationships did not. Reprinted with permission from Allan et al. (2009).

Nile disease among counties, Allan et al. found that bird diversity was the strongest and most consistent factor—counties with low avian diversity had, on average, much higher rates of West Nile disease than did those with high diversity (**Figure 6**).

Koenig et al. (2010) examined the role of bird diversity in population declines of American crows during the initial five-year sweep of West Nile virus across the United States. Crows can experience rapid mortality after infection, resulting in severe population declines as West Nile virus invades new areas (LaDeau et al. 2007); therefore, crow population declines are a reasonable proxy for transmission rates within wildlife communities. Koenig et al. quantified the severity of crow declines using the Breeding Bird Survey data for individual survey routes conducted before and after invasion by the virus. They tested the ability of several factors to explain the severity of crow declines, including the year West Nile virus was first detected in the state, population density of crows, bird diversity (Shannon index), urbanization (estimated by human population density), total number of individual birds, mean rainfall, mean maximum temperature, and spatial autocorrelation. The causal variables best supported by the data were bird diversity (crows were more likely to decline where bird diversity was low), urbanization (crow declines were more likely in more urbanized areas), and crow density (declines were more likely where initial crow density was high).

Hamer et al. (2011) examined the relationship between bird diversity and West Nile virus infection in *Culex pipiens* mosquitoes in Chicago, IL. They assessed mosquito abundance and infection prevalence, total bird diversity (richness and Shannon index), and the diversity of the birds from which the mosquitoes fed based on molecular diagnostic tests of blood in captured mosquitoes. These data allowed them to test the hypothesis that selective feeding by the mosquitoes was strong enough to weaken the effect of the diversity of available bird hosts. Hamer et al. estimated each bird species' contribution to the pool of infected mosquitoes by calculating host-specific "force of infection" (the squared fraction of total blood meals taken from a host species times that species' reservoir competence), which also allowed them to calculate "community force of infection" by summing across all host species present at a site. Although they found strong apparent host preferences overall, these preferences differed among sites, suggesting that preferences are not fixed but might vary with local host availability. Diversity of birds available and diversity of birds represented in the blood-fed mosquitoes were not correlated. Despite apparently strong host preferences by mosquitoes, the best-supported model invoked an interaction between bird diversity and the community force of infection. The majority of infected mosquitoes had fed on American robins, house sparrows, and house finches, species that tend to reach high abundance in low-diversity communities (Allan et al. 2009). A plausible interpretation of these results is that selective feeding by mosquito vectors causes a multiplicative effect with diversity, whereby bird species with high reservoir competence are both preferred by mosquitoes and more likely to be abundant in low-diversity communities.

One recent study found no support for a DE in West Nile virus dynamics. In 2005 and 2006, Loss et al. (2009) sampled mosquito and bird communities in 13 sites within the Chicago metropolitan area to assess the relationship between bird diversity and the proportion of *Culex* mosquitoes infected with the virus. Avian species richness in a site was not a significant predictor of mosquito infection prevalence. Loss et al. argued that a DE did not occur because the most competent reservoirs were not disproportionately well represented in species-poor communities. Their data supported only a weak, positive relationship between reservoir competence of a bird host and its relative abundance (ubiquity) across sites in this urban environment ($r = 0.11$).

COMMUNITY DIVERSITY

The possible effects of diversity of community types on the dynamics of infectious diseases have received scant attention. The invasion of a community by nonindigenous species reduces community-level diversity by homogenizing donor and recipient communities. When communities are invaded by nonindigenous species that are infected with pathogens, these pathogens often spill over into the indigenous species and increase disease risk, incidence, and mortality (Parrish et al. 2008, Vandegrift et al. 2010). Even when nonindigenous species invade and do not import their pathogens, they can often become competent reservoirs for pathogens in the indigenous community and increase disease risk via spillback (Kelly et al. 2009b, O'Brien et al. 2011, Poulin et al. 2011). In other cases, nonindigenous species invading without their pathogens can reduce disease risk in indigenous hosts (reviewed above) by absorbing but not releasing pathogens. Recent reviews of these processes in both freshwater and terrestrial wildlife communities suggest that how reductions in community diversity via invasion by nonindigenous species affect pathogen transmission depends on (*a*) the ability of the invading species to support proliferation of pathogens native to the habitats it invades, (*b*) whether the nonindigenous species bring their pathogens into these habitats, and (*c*) the extent to which the nonindigenous species regulate the abundance of indigenous reservoir hosts (Johnson & Thieltges 2010; Poulin et al. 2011; Tompkins et al. 2003, 2011).

GLOBAL-SCALE PATTERNS OF DIVERSITY AND HUMAN INFECTIOUS DISEASES

Although high host diversity often reduces pathogen transmission for existing diseases, some researchers have postulated that high host diversity might increase the probability of disease emergence by providing a large source of potential zoonotic pathogens. Both vertebrate and parasite diversity tend to be higher at low rather than high latitudes (Guernier et al. 2004, Pianka 1966), leading to the hypothesis that emergence rates of human infectious diseases will be positively correlated with vertebrate diversity and negatively correlated with latitude (Dunn et al. 2010). Evidence pertaining to this hypothesis is equivocal. Using structural equation modeling, Dunn et al. concluded that bird and mammal species richness drives richness of human pathogens, which in turn drives the prevalence of 22 diseases in their database.

In contrast, Jones et al. (2008) compiled a data set consisting of 335 infectious diseases of humans that emerged after 1940 and found as follows:

> The highest concentration of EID [emerging infectious disease] events per million square kilometers of land was found between 30 and 60° north and between 30 and 40° south, with the main hotspots in the northeastern United States, western Europe, Japan and southeastern Australia.
>
> (Jones et al. 2008, p. 991)

To account statistically for the possibility that the geographic pattern of EIDs could be biased by geographic variation in the abundance or activity of scientists likely to detect and report these diseases, Jones et al. included as an independent variable the frequency of each country listed as the address for each author in all papers published in the *Journal of Infectious Diseases* since 1973. Using this added variable, Jones et al. found that the emergences of zoonotic pathogens from wildlife and of vector-borne pathogens were not significantly associated with latitude, but that emergence events of zoonotic pathogens from wildlife hosts were significantly, positively correlated with wildlife species richness. However, this correlation was quite weak, with wildlife species richness explaining only between 0.8% and 1.3% of the spatial variation in emergence. In contrast, human population density, a primary driver of biodiversity loss, was positively correlated with disease emergence, explaining 54% of the spatial variation in EIDs. Consequently, whether high host diversity is an important risk factor for the emergence of zoonotic diseases from wildlife remains an open question.

CONCLUDING COMMENTS

Neither the potential for the DE to occur nor its actual occurrence in a variety of natural and constructed systems is controversial, as the studies reviewed above indicate. For many but not all disease systems, host diversity explains much of the variation in disease risk. What remains unresolved is the degree to which the DE is general. A recent review of the consequences of biodiversity loss (Cardinale et al. 2012) found that 80% of statistical tests of associations between biodiversity and disease transmission showed a statistically significant negative relationship (DE), whereas 12% showed a significant positive relationship (amplification effect), and 8% were not significant (**Table 1**). These results provisionally indicate that the dilution effect is indeed widespread in a diversity of disease systems.

A DE is expected to occur when the following conditions are met: (*a*) hosts differ in their quality as hosts for a pathogen and/or its vector; (*b*) the taxa most likely to remain when diversity

Table 1   Number of statistical tests of the relationship between diversity and disease transmission in which the association was significantly ($P < 0.05$) negative (dilution), positive (amplification), or not significant (neither)[a]

Taxon (N)	Dilution	Amplification	Neither
Plants (107)	91	16	0
Animals (45)	30	2	13
Total (152)	121	18	13

[a]Data taken from Cardinale et al. (2012).

is lost tend to support greater abundance of the pathogen or vector, whereas those most likely to be added as diversity increases tend to be poorer hosts; and (*c*) the taxa most likely to be added as diversity increases reduce either encounter rates between high-quality hosts and pathogens or abundance of high-quality hosts. The generality of the first condition is abundantly demonstrated by the literature reviewed above. Although some support exists for the second and third conditions (Johnson & Thieltges 2010, Keesing et al. 2010, Ostfeld & Keesing 2000, Pongsiri et al. 2009), their generality remains to be evaluated comprehensively. The second condition can be further subdivided into the following two patterns: (*i*) communities have nested patterns of species composition across gradients in diversity—i.e., as species diversity varies among communities, some species tend to be ubiquitous and others are more likely to be absent when diversity is lower; and (*ii*) a positive correlation exists between resilience (defined here as the tendency to remain when diversity is lost) and host quality. Some supporting evidence for these subconditions exists for plants (Cronin et al. 2010), amphibians (Johnson et al. 2012), and rodents (Previtali et al. 2012). Furthermore, it has been postulated that certain life-history traits might underlie both resilience to anthropogenic forces that reduce biodiversity and permissiveness to pathogens and vectors. Specifically, species with a fast pace of life (short lifespan, early and rapid reproduction, high dispersal ability) might often adopt pathogen defense strategies that are innate (as opposed to induced), energetically inexpensive, and potentially less effective against some pathogens (Cronin et al. 2010; Johnson et al. 2012; Martin 2009; Martin et al. 2007, 2008; Palacios et al. 2011; Previtali et al. 2012).

Even where a DE might occur, it is expected to be weakened by strong specialization by either the pathogen or the vector on one species of host. Under these conditions, variation in host diversity will not cause similar variation in the distribution of pathogens or vectors among those hosts. Such an effect is thought to underlie the lack of a DE in West Nile virus in Chicago, IL (Loss et al. 2009). But it is important to note that the degree to which pathogens and vectors specialize is difficult to assess. For vectors, host specialization is typically assessed by analyzing the distribution of blood meals across a community of potential hosts (for mosquitoes) or by counting ectoparasites attached to hosts (e.g., ticks). But such analyses consider only those vector bites that were successful (i.e., resulted in a blood meal) and neglect the potential for hosts to kill or injure vectors attempting a blood meal (Edman & Kale 1971, Edman et al. 1972, Keesing et al. 2009). Blood-meal analyses therefore might not represent either feeding preferences or the effects of various host species on vector populations. Similarly, a DE is expected to be absent, weak, or complex where no single host is sufficient for pathogen maintenance and transmission. For instance, for a vector with a complex life cycle requiring multiple hosts, a low-diversity community might not provide sufficient host diversity for vector or pathogen persistence. One might expect a unimodal relationship between diversity and disease risk for such systems, whereby intermediate levels of diversity maximize pathogen transmission, below which an amplification effect occurs and above which a DE occurs (Ostfeld & Keesing 2000, Van Buskirk & Ostfeld 1995).

The possibility remains that high host diversity generally reduces the transmission and risk of existing diseases while it increases the probability of new diseases emerging. However, any positive correlation between wildlife diversity and the probability of emergence appears to be weak, explaining only ~1% of the variation in emergence. Nevertheless, it will be important to distinguish strategies for focusing our surveillance of potential emergence events from those for mitigating human-caused increases in risk of existing diseases. For anticipated emergence events, surveillance possibly should be concentrated in biogeographic areas where diversity of potential pathogens is known or expected to be high. But for many existing diseases, mitigation strategies should be considered in situations where human activities decrease host diversity.

DISCLOSURE STATEMENT

The authors are not aware of any affiliations, memberships, funding, or financial holdings that might be perceived as affecting the objectivity of this review.

ACKNOWLEDGMENTS

The authors thank the National Science Foundation, the National Institutes of Health (NIAID), the US Environmental Protection Agency, and Dutchess County, NY, for supporting original research and syntheses.

LITERATURE CITED

Allan BF, Keesing F, Ostfeld RS. 2003. Effect of forest fragmentation on Lyme disease risk. *Conserv. Biol.* 17:267–72

Allan BF, Langerhans RB, Ryberg WA, Landesman WJ, Griffin NW, et al. 2009. Ecological correlates of risk and incidence of West Nile virus in the United States. *Oecologia* 158:699–708

Altermatt F, Ebert D. 2008. Genetic diversity of *Daphnia magna* populations enhances resistance to parasites. *Ecol. Lett.* 11:918–28

Anderson RM, May RM. 1979. Population biology of infectious diseases: part 1. *Nature* 280:361–67

Anderson RM, May RM. 1981. The population dynamics of microparasites and their invertebrate hosts. *Philos. Trans. R. Soc. B* 291:451–524

Anderson RM, May RM. 1991. *Infectious Diseases of Humans*. Oxford, UK: Oxford Univ. Press. 757 pp.

Apperson CS, Hassan HK, Harrison BA, Savage HM, Aspen SE, et al. 2004. Host feeding patterns of established and potential mosquito vectors of West Nile virus in the eastern United States. *Vector-Borne Zoonotic Dis.* 4:71–82

Becker CG, Zamudio KR. 2011. Tropical amphibian populations experience higher disease risk in natural habitats. *Proc. Natl. Acad. Sci. USA* 108:9893–98

Borer ET, Seabloom EW, Mitchell CE, Power AG. 2010. Local context drives infection of grasses by vector-borne generalist viruses. *Ecol. Lett.* 13:810–18

Brisson D, Dykhuizen DE, Ostfeld RS. 2008. Conspicuous impacts of inconspicuous hosts on the Lyme disease epidemic. *Proc. R. Soc. B: Biol. Sci.* 275:227–35

Brownstein JS, Skelly DK, Holford TR, Fish D. 2005. Forest fragmentation predicts local scale heterogeneity of Lyme disease risk. *Oecologia* 146:469–75

Brumpt E. 1944–45. Anophélisme sans paludisme et régression spontanée du paludisme. *Ann. Parasitol.* 20:67–91

Brunner JL, Cheney L, Keesing F, Killilea M, Logiudice K, et al. 2011. Molting success of *Ixodes scapularis* varies among individual blood meal hosts and species. *J. Med. Entomol.* 48:860–66

Brunner JL, Ostfeld RS. 2008. Multiple causes of variable tick burdens on small-mammal hosts. *Ecology* 89:2259–72

Cardinale BJ, Duffy JE, Gonzalez A, Hooper DU, Perrings C, et al. 2012. Biodiversity loss and its impact on humanity. *Nature* 486:59–67

Carver S, Kuenzi A, Bagamian KH, Mills JN, Rollin PE, et al. 2011. A temporal dilution effect: hantavirus infection in deer mice and the intermittent presence of voles in Montana. *Oecologia* 166:713–21

Clay CA, Lehmer EM, Previtali A, St Jeor S, Dearing MD. 2009a. Contact heterogeneity in deer mice: implications for Sin Nombre virus transmission. *Proc. R. Soc. B: Biol. Sci.* 276:1305–12

Clay CA, Lehmer EM, St Jeor S, Dearing MD. 2009b. Testing mechanisms of the dilution effect: deer mice encounter rates, Sin Nombre virus prevalence and species diversity. *Ecohealth* 6:250–59

Clay CA, Lehmer EM, St Jeor S, Dearing MD. 2009c. Sin Nombre virus and rodent species diversity: a test of the dilution and amplification hypotheses. *PLoS ONE* 4:e6467

Corvalan CF, Hales S, McMichael AJ. 2005. *Ecosystems and health well-being: health synthesis: a report of the Millenium Ecosystem Assessment*. Geneva, Switzerland: World Health Organization. 53 pp.

Cronin JP, Welsh ME, Dekkers MG, Abercrombie ST, Mitchell CE. 2010. Host physiological phenotype explains pathogen reservoir potential. *Ecol. Lett.* 13:1221–32

Day JF, Edman JD. 1984. Mosquito engorgement on normally defensive hosts depends on host activity patterns. *J. Med. Entomol.* 21:732–40

Dearing MD, Dizney L. 2010. Ecology of hantavirus in a changing world. *Ann. NY Acad. Sci.* 1195:99–112

Deblinger RD, Wilson ML, Rimmer DW, Spielman A. 1993. Reduced abundance of immature *Ixodes dammini* (Acari:Ixodidae) following incremental removal of deer. *J. Med. Entomol.* 30:144–50

Dennehy JJ, Friedenberg NA, Yang YW, Turner PE. 2007. Virus population extinction via ecological traps. *Ecol. Lett.* 10:230–40

Derne BT, Fearnley EJ, Lau CL, Paynter S, Weinstein P. 2011. Biodiversity and leptospirosis risk: a case of pathogen regulation? *Med. Hypotheses* 77:339–44

Dizney LJ, Ruedas LA. 2009. Increased host species diversity and decreased prevalence of Sin Nombre virus. *Emerg. Infect. Dis.* 15:1012–18

Dobson A. 2004. Population dynamics of pathogens with multiple host species. *Am. Nat.* 164:S64–78

Donahue JG, Piesman J, Spielman A. 1987. Reservoir competence of white-footed mice for Lyme disease spirochetes. *Am. J. Trop. Med. Hyg.* 36:92–96

Dunn RR, Davies TJ, Harris NC, Gavin MC. 2010. Global drivers of human pathogen richness and prevalence. *Proc. R. Soc. B: Biol. Sci.* 277:2587–95

Edman JD, Kale HW. 1971. Host behavior: its influence on the feeding success of mosquitoes. *Ann. Entomol. Soc. Am.* 64:513–16

Edman JD, Webber LA, Kale HW. 1972. Effect of mosquito density on the interrelationship of host behavior and mosquito feeding success. *Am. J. Trop. Med. Hyg.* 21:487–92

Elton CS. 1958. *The Ecology of Invasions by Animals and Plants.* London: Methuen & Co.

Ezenwa VO, Godsey MS, King RJ, Guptill SC. 2006. Avian diversity and West Nile virus: testing associations between biodiversity and infectious disease risk. *Proc. R. Soc. B: Biol. Sci.* 273:109–17

Goodin DG, Koch DE, Owen RD, Chu YK, Hutchinson JMS, Jonsson CB. 2006. Land cover associated with hantavirus presence in Paraguay. *Glob. Ecol. Biogeogr.* 15:519–27

Guernier V, Hochberg ME, Guegan JFO. 2004. Ecology drives the worldwide distribution of human diseases. *PLoS Biol.* 2:740–46

Gurnell J, Rushton SP, Lurz PWW, Sainsbury AW, Nettleton P, et al. 2006. Squirrel poxvirus: landscape scale strategies for managing disease threat. *Biol. Conserv.* 131:287–95

Haas SE, Hooten MB, Rizzo DM, Meentemeyer RK. 2011. Forest species diversity reduces disease risk in a generalist plant pathogen invasion. *Ecol. Lett.* 14:1108–16

Hall SR, Becker CR, Simonis JL, Duffy MA, Tessier AJ, Caaceres CE. 2009. Friendly competition: evidence for a dilution effect among competitors in a planktonic host-parasite system. *Ecology* 90:791–801

Hamer GL, Chaves LF, Anderson TK, Kitron UD, Brawn JD, et al. 2011. Fine-scale variation in vector host use and force of infection drive localized patterns of West Nile virus transmission. *PLoS ONE* 6:e23767

Harper JL, Hawksworth DL. 1994. Biodiversity: measurement and estimation. Preface. *Philos. Trans. R. Soc. B* 345:5–12

Holt RD, Roy M. 2007. Predation can increase the prevalence of infectious disease. *Am. Nat.* 169:690–99

Jackson LE, Hilborn ED, Thomas JC. 2006a. Towards landscape design guidelines for reducing Lyme disease risk. *Int. J. Epidemiol.* 35:315–22

Jackson LE, Levine JF, Hilborn ED. 2006b. A comparison of analysis units for associating Lyme disease with forest-edge habitat. *Community Ecol.* 7:189–97

Johnson PTJ, Hartson RB, Larson DJ, Sutherland DR. 2008. Diversity and disease: community structure drives parasite transmission and host fitness. *Ecol. Lett.* 11:1017–26

Johnson PTJ, Lund PJ, Hartson RB, Yoshino TP. 2009. Community diversity reduces *Schistosoma mansoni* transmission, host pathology and human infection risk. *Proc. R. Soc. B: Biol. Sci.* 276:1657–63

Johnson PTJ, Rohr JR, Hoverman JT, Kellermans E, Bowerman J, Lunde KG. 2012. Living fast and dying of infection: host life history drives interspecific variation in infection and disease risk. *Ecol. Lett.* 15:235–42

Johnson PTJ, Thieltges DW. 2010. Diversity, decoys and the dilution effect: how ecological communities affect disease risk. *J. Exp. Biol.* 213:961–70

Jones KE, Patel NG, Levy MA, Storeygard A, Balk D, et al. 2008. Global trends in emerging infectious diseases. *Nature* 451:990–93

Keesing F, Belden LK, Daszak P, Dobson A, Harvell CD, et al. 2010. Impacts of biodiversity on the emergence and transmission of infectious diseases. *Nature* 468:647–52

Keesing F, Brunner J, Duerr S, Killilea M, LoGiudice K, et al. 2009. Hosts as ecological traps for the vector of Lyme disease. *Proc. R. Soc. B: Biol. Sci.* 276:3911–19

Keesing F, Holt RD, Ostfeld RS. 2006. Effects of species diversity on disease risk. *Ecol. Lett.* 9:485–98

Keesing F, Ostfeld RS. 2012. An ecosystem service of biodiversity—the protection of human health against infectious disease. In *New Directions in Conservation Medicine*, ed. AA Aguirre, RS Ostfeld, P Daszak. pp. 56–66. New York, NY: Oxford Univ. Press

Kelly DW, Paterson RA, Townsend CR, Poulin R, Tompkins DM. 2009a. Has the introduction of brown trout altered disease patterns in native New Zealand fish? *Freshw. Biol.* 54:1805–18

Kelly DW, Paterson RA, Townsend CR, Poulin R, Tompkins DM. 2009b. Parasite spillback: a neglected concept in invasion ecology? *Ecology* 90:2047–56

Kilpatrick AM, Daszak P, Jones MJ, Marra PP, Kramer LD. 2006a. Host heterogeneity dominates West Nile virus transmission. *Proc. R. Soc. B: Biol. Sci.* 273:2327–33

Kilpatrick AM, Daszak P, Kramer L, Marra PP, Dobson AP, et al. 2006b. Predicting the transmission of West Nile virus. *Am. J. Trop. Med. Hyg.* 75:139–39

Kilpatrick AM, Kramer LD, Jones MJ, Marra PP, Daszak P. 2006c. West Nile virus epidemics in North America are driven by shifts in mosquito feeding behavior. *PLoS Biol.* 4:606–10

Kilpatrick AM, LaDeau SL, Marra PP. 2007. Ecology of West Nile virus transmission and its impact on birds in the western hemisphere. *Auk* 124:1121–36

Knops JMH, Tilman D, Haddad NM, Naeem S, Mitchell CE, et al. 1999. Effects of plant species richness on invasion dynamics, disease outbreaks, insect abundances and diversity. *Ecol. Lett.* 2:286–93

Koenig WD, Hochachka WM, Zuckerberg B, Dickinson JL. 2010. Ecological determinants of American crow mortality due to West Nile virus during its North American sweep. *Oecologia* 163:903–9

Komar N, Langevin S, Hinten S, Nemeth N, Edwards E, et al. 2003. Experimental infection of North American birds with the New York 1999 strain of West Nile virus. *Emerg. Infect. Dis.* 9:311–22

Kopp K, Jokela J. 2007. Resistant invaders can convey benefits to native species. *Oikos* 116:295–301

Kuenzi AJ, Douglass RJ, White D, Bond CW, Mills JN. 2001. Antibody to Sin Nombre virus in rodents associated with peridomestic habitats in west central Montana. *Am. J. Trop. Med. Hyg.* 64:137–46

LaDeau SL, Kilpatrick AM, Marra PP. 2007. West Nile virus emergence and large-scale declines of North American bird populations. *Nature* 447:710–13

LaDeau SL, Marra PP, Kilpatrick AM, Calder CA. 2008. West Nile virus revisited: consequences for North American ecology. *Bioscience* 58:937–46

Lane RS, Piesman J, Burgdorfer W. 1991. Lyme borreliosis: relation of its causative agent to its vectors and hosts in North America and Europe. *Annu. Rev. Entomol.* 36:587–609

Langlois JP, Fahrig L, Merriam G, Artsob H. 2001. Landscape structure influences continental distribution of hantavirus in deer mice. *Landsc. Ecol.* 16:255–66

Lehmer EM, Clay CA, Pearce-Duvet J, St Jeor S, Dearing MD. 2008. Differential regulation of pathogens: the role of habitat disturbance in predicting prevalence of Sin Nombre virus. *Oecologia* 155:429–39

Lively CM. 2010. The effect of host genetic diversity on disease spread. *Am. Nat.* 175:E149–52

LoGiudice K, Duerr STK, Newhouse MJ, Schmidt KA, Killilea ME, Ostfeld RS. 2008. Impact of host community composition on Lyme disease risk. *Ecology* 89:2841–49

LoGiudice K, Ostfeld RS, Schmidt KA, Keesing F. 2003. The ecology of infectious disease: effects of host diversity and community composition on Lyme disease risk. *Proc. Natl. Acad. Sci. USA* 100:567–71

Loss SR, Hamer GL, Walker ED, Ruiz MO, Goldberg TL, et al. 2009. Avian host community structure and prevalence of West Nile virus in Chicago, Illinois. *Oecologia* 159:415–24

Mackelprang R, Dearing MD, St Jeor S. 2001. High prevalence of Sin Nombre virus in rodent populations, central Utah: a consequence of human disturbance? *Emerg. Infect. Dis.* 7:480–82

Marra PP, Griffing S, Caffrey C, Kilpatrick AM, McLean R, et al. 2004. West Nile virus and wildlife. *Bioscience* 54:393–402

Martin LB. 2009. Stress and immunity in wild vertebrates: timing is everything. *Gen. Comp. Endocr.* 163:70–76

Martin LB, Weil ZM, Nelson RJ. 2007. Immune defense and reproductive pace of life in *Peromyscus* mice. *Ecology* 88:2516–28

Martin LB, Weil ZM, Nelson RJ. 2008. Fever and sickness behaviour vary among congeneric rodents. *Funct. Ecol.* 22:68–77

Mather TN, Wilson ML, Moore SI, Ribeiro JMC, Spielman A. 1989. Comparing the relative potential of rodents as reservoirs of the Lyme disease spirochete (*Borrelia burgdorferi*). *Am. J. Epidemiol.* 130:143–50

Matuschka FR, Heiler M, Eiffert H, Fischer P, Lotter H, Spielman A. 1993. Diversionary role of hoofed game in the transmission of Lyme disease spirochetes. *Am. J. Trop. Med. Hyg.* 48:693–99

Matuschka FR, Spielman A. 1992. Loss of Lyme disease spirochetes from *Ixodes ricinus* ticks feeding on European blackbirds. *Exp. Parasitol.* 74:151–58

Mitchell CE, Tilman D, Groth JV. 2002. Effects of grassland plant species diversity, abundance, and composition on foliar fungal disease. *Ecology* 83:1713–26

Mundt CC. 2002. Use of multiline cultivars and cultivar mixtures for disease management. *Annu. Rev. Phytopathol.* 40:381–410

Mundt CC, Sackett KE, Wallace LD. 2011. Landscape heterogeneity and disease spread: experimental approaches with a plant pathogen. *Ecol. Appl.* 21:321–28

Norman R, Bowers RG, Begon M, Hudson PJ. 1999. Persistence of tick-horne virus in the presence of multiple host species: tick reservoirs and parasite mediated competition. *J. Theor. Biol.* 200:111–18

Nupp TE, Swihart RK. 1998. Effects of forest fragmentation on population attributes of white-footed mice and eastern chipmunks. *J. Mammal.* 79:1234–43

Nupp TE, Swihart RK. 2000. Landscape-level correlates of small-mammal assemblages in forest fragments of farmland. *J. Mammal.* 81:512–26

O'Brien VA, Moore AT, Young GR, Komar N, Reisen WK, Brown CR. 2011. An enzootic vector-borne virus is amplified at epizootic levels by an invasive avian host. *Proc. R. Soc. B: Biol. Sci.* 278:239–46

Ogden NH, Tsao JI. 2009. Biodiversity and Lyme disease: dilution or amplification? *Epidemics* 1:196–206

Orrock JL, Allan BF, Drost CA. 2011. Biogeographic and ecological regulation of disease: prevalence of Sin Nombre virus in island mice is related to island area, precipitation, and predator richness. *Am. Nat.* 177:691–97

Ostfeld R, Keesing F. 2000. The function of biodiversity in the ecology of vector-borne zoonotic diseases. *Can. J. Zool.* 78:2061–78

Ostfeld RS. 2011. *Lyme Disease: the Ecology of a Complex System*. New York: Oxford Univ. Press

Ostfeld RS, Canham CD, Oggenfuss K, Winchcombe RJ, Keesing F. 2006. Climate, deer, rodents, and acorns as determinants of variation in Lyme-disease risk. *PLoS Biol.* 4:1058–68

Ostfeld RS, Holt RD. 2004. Are predators good for your health? Evaluating evidence for top-down regulation of zoonotic disease reservoirs. *Front. Ecol. Environ.* 2:13–20

Ostfeld RS, LoGiudice K. 2003. Community disassembly, biodiversity loss, and the erosion of an ecosystem service. *Ecology* 84:1421–27

Packer C, Holt RD, Hudson PJ, Lafferty KD, Dobson AP. 2003. Keeping the herds healthy and alert: implications of predator control for infectious disease. *Ecol. Lett.* 6:797–802

Palacios MG, Sparkman AM, Bronikowski AM. 2011. Developmental plasticity of immune defence in two life-history ecotypes of the garter snake, *Thamnophis elegans*—a common-environment experiment. *J. Anim. Ecol.* 80:431–37

Parrish CR, Holmes EC, Morens DM, Park EC, Burke DS, et al. 2008. Cross-species virus transmission and the emergence of new epidemic diseases. *Microbiol. Mol. Biol. R.* 72:457–70

Peixoto ID, Abramson G. 2006. The effect of biodiversity on the hantavirus epizootic. *Ecology* 87:873–79

Pianka ER. 1966. Latitudinal gradients in species diversity: a review of concepts. *Am. Nat.* 100:33–46

Piudo L, Monteverde MJ, Walker RS, Douglass RJ. 2011. Rodent community structure and Andes virus infection in sylvan and peridomestic habitats in northwestern Patagonia, Argentina. *Vector-Borne Zoonotic Dis.* 11:315–24

Pongsiri MJ, Roman J, Ezenwa VO, Goldberg TL, Koren HS, et al. 2009. Biodiversity loss affects global disease ecology. *Bioscience* 59:945–54

Poulin R, Paterson RA, Townsend CR, Tompkins DM, Kelly DW. 2011. Biological invasions and the dynamics of endemic diseases in freshwater ecosystems. *Freshw. Biol.* 56:676–88

Previtali MA, Ostfeld RS, Keesing F, Jolles A, Hanselmann R, Martin LB. 2012. Relationship between pace of life and immune responses of wild rodents. *Oikos* In press (DOI:10.1111/j.1600-0706.2012.020215.x)

Rand PW, Lubelczyk C, Holman MS, Lacombe EH, Smith RP. 2004. Abundance of *Ixodes scapularis* (Acari: Ixodidae) after the complete removal of deer from an isolated offshore island, endemic for Lyme disease. *J. Med. Entomol.* 41:779–84

Redford KH, Richter BD. 1999. Conservation of biodiversity in a world of use. *Conserv. Biol.* 13:1246–56

Richter D, Matuschka FR. 2006. Modulatory effect of cattle on risk for Lyme disease. *Emerg. Infect. Dis.* 12:1919–23

Richter D, Spielman A, Komar N, Matuschka FR. 2000. Competence of American robins as reservoir hosts for Lyme disease spirochetes. *Emerg. Infect. Dis.* 6:133–38

Rosa R, Pugliese A. 2007. Effects of tick population dynamics and host densities on the persistence of tick-borne infections. *Math. Biosci.* 208:216–40

Roscher C, Schumacher J, Foitzik O, Schulze ED. 2007. Resistance to rust fungi in *Lolium perenne* depends on within-species variation and performance of the host species in grasslands of different plant diversity. *Oecologia* 153:173–83

Rosenblatt DL, Heske EJ, Nelson SL, Barber DH, Miller MA, MacAllister B. 1999. Forest fragments in east-central Illinois: islands or habitat patches for mammals? *Am. Midl. Nat.* 141:115–23

Rudolf VHW, Antonovics J. 2005. Species coexistence and pathogens with frequency-dependent transmission. *Am. Nat.* 166:112–18

Ruedas LA, Salazar-Bravo J, Tinnin DS, Armien B, Caceres L, et al. 2004. Community ecology of small mammal populations in Panama following an outbreak of hantavirus pulmonary syndrome. *J. Vector Ecol.* 29:177–91

Schmidt KA, Ostfeld RS. 2001. Biodiversity and the dilution effect in disease ecology. *Ecology* 82:609–19

Searle CL, Biga LM, Spatafora JW, Blaustein AR. 2011. A dilution effect in the emerging amphibian pathogen *Batrachochytrium dendrobatidis*. *Proc. Natl. Acad. Sci. USA* 108:16322–26

Service MW. 1991. Agricultural-development and arthropod-borne diseases: a review. *Rev. Saude Publ.* 25:165–78

Springbett AJ, MacKenzie K, Woolliams JA, Bishop SC. 2003. The contribution of genetic diversity to the spread of infectious diseases in livestock populations. *Genetics* 165:1465–74

Stafford KC, Denicola AJ, Kilpatrick HJ. 2003. Reduced abundance of *Ixodes scapularis* (Acari: Ixodidae) and the tick parasitoid *Ixodiphagus hookeri* (Hymenoptera: Encyrtidae) with reduction of white-tailed deer. *J. Med. Entomol.* 40:642–52

Suzan G, Armien A, Mills JN, Marce E, Ceballos G, et al. 2008a. Epidemiological considerations of rodent community composition in fragmented landscapes in Panama. *J. Mammal.* 89:684–90

Suzan G, Marce E, Giermakowski JT, Armien B, Pascale J, et al. 2008b. The effect of habitat fragmentation and species diversity loss on hantavirus prevalence in Panama. *Ann. N.Y. Acad. Sci.* 1149:80–83

Suzan G, Marce E, Giermakowski JT, Mills J, Ceballos G, et al. 2009. Experimental evidence for reduced rodent diversity causing increased hantavirus prevalence. *PLoS ONE* 4:e5461

Swaddle JP, Calos SE. 2008. Increased avian diversity is associated with lower incidence of human West Nile infection: observation of the dilution effect. *PLoS ONE* 3:e2488

Swihart RK, Atwood TC, Goheen JR, Scheiman DM, Munroe KE, Gehring TM. 2003a. Patch occupancy of North American mammals: Is patchiness in the eye of the beholder? *J. Biogeogr.* 30:1259–79

Swihart RK, Gehring TM, Kolozsvary MB, Nupp TE. 2003b. Responses of 'resistant' vertebrates to habitat loss and fragmentation: the importance of niche breadth and range boundaries. *Divers. Distrib.* 9:1–18

Tersago K, Schreurs A, Linard C, Verhagen R, Van Dongen S, Leirs H. 2008. Population, environmental, and community effects on local bank vole (*Myodes glareolus*) Puumala virus infection in an area with low human incidence. *Vector-Borne Zoonotic Dis.* 8:235–44

Tompkins DM, Dunn AM, Smith MJ, Telfer S. 2011. Wildlife diseases: from individuals to ecosystems. *J. Anim. Ecol.* 80:19–38

Tompkins DM, White AR, Boots M. 2003. Ecological replacement of native red squirrels by invasive greys driven by disease. *Ecol. Lett.* 6:189–96

Van Buskirk J, Ostfeld RS. 1995. Controlling Lyme disease by modifying the density and species composition of tick hosts. *Ecol. Appl.* 5:1133–40

Vandegrift KJ, Sokolow SH, Daszak P, Kilpatrick AM. 2010. Ecology of avian influenza viruses in a changing world. *Ann. N.Y. Acad. Sci.* 1195:113–28

World Health Organization. 1982. *Manual on Environmental Management for Mosquito Control, with Special Emphasis on Malaria Vectors.* WHO offset publication 66. Geneva: WHO. 283 pp.

Wilder SM, Meikle DB. 2004. Prevalence of deer ticks (*Ixodes scapularis*) on white-footed mice (*Peromyscus leucopus*) in forest fragments. *J. Mammal.* 85:1015–18

Yahnke CJ, Meserve PL, Ksiazek TG, Mills JN. 2001. Patterns of infection with Laguna Negra virus in wild populations of *Calomys laucha* in the central Paraguayan chaco. *Am. J. Trop. Med. Hyg.* 65:768–76

Yan L, Fang LQ, Huang HG, Zhang LQ, Feng D, et al. 2007. Landscape elements and hantaan virus-related hemorrhagic fever with renal syndrome, People's Republic of China. *Emerg. Infect. Dis.* 13:1301–6

Yates TL, Mills JN, Parmenter CA, Ksiazek TG, Parmenter RR, et al. 2002. The ecology and evolutionary history of an emergent disease: hantavirus pulmonary syndrome. *Bioscience* 52:989–98

Zhang WY, Fang LQ, Jiang JF, Hui FM, Glass GE, et al. 2009a. Predicting the risk of hantavirus infection in Beijing, People's Republic of China. *Am. J. Trop. Med. Hyg.* 80:678–83

Zhang YZ, Dong X, Li X, Ma C, Xiong HP, et al. 2009b. Seoul virus and hantavirus disease, Shenyang, People's Republic of China. *Emerg. Infect. Dis.* 15:200–6

Zhu YY, Chen HR, Fan JH, Wang YY, Li Y, et al. 2000. Genetic diversity and disease control in rice. *Nature* 406:718–22

Coextinction and Persistence of Dependent Species in a Changing World

Robert K. Colwell,[1,2] Robert R. Dunn,[3] and Nyeema C. Harris[4]

[1] Department of Ecology and Evolutionary Biology, University of Connecticut, Storrs, Connecticut 06269; email: colwell@uconn.edu

[2] University of Colorado Museum of Natural History, Boulder, Colorado 80309

[3] Department of Biology and [4] Department of Forestry and Environmental Resources, North Carolina State University, Raleigh, North Carolina 27607; email: rrdunn@ncsu.edu

Annu. Rev. Ecol. Evol. Syst. 2012. 43:183–203

First published online as a Review in Advance on August 28, 2012

The *Annual Review of Ecology, Evolution, and Systematics* is online at ecolsys.annualreviews.org

This article's doi:
10.1146/annurev-ecolsys-110411-160304

Keywords

affiliate species, commensalism, extinction cascade, extinction vortex, food web, host switching, interaction network, mutualism, parasitism, pollination, secondary extinction

Abstract

The extinction of a single species is rarely an isolated event. Instead, dependent parasites, commensals, and mutualist partners (affiliates) face the risk of coextinction as their hosts or partners decline and fail. Species interactions in ecological networks can transmit the effects of primary extinctions within and between trophic levels, causing secondary extinctions and extinction cascades. Documenting coextinctions is complicated by ignorance of host specificity, limitations of historical collections, incomplete systematics of affiliate taxa, and lack of experimental studies. Host shifts may reduce the rate of coextinctions, but they are poorly understood. In the absence of better empirical records of coextinctions, statistical models estimate the rates of past and future coextinctions, and based on primary extinctions and interactions among species, network models explore extinction cascades. Models predict and historical evidence reveals that the threat of coextinction is influenced by both host and affiliate traits and is exacerbated by other threats, including habitat loss, climate change, and invasive species.

INTRODUCTION

With the accelerating loss of species to extinction, much effort has been invested in identifying the life-history, morphological, and functional characteristics associated with extinction vulnerability for a wide range of taxa, ranging from tropical angiosperms (Sodhi et al. 2008) to coral reef fauna (Graham et al. 2011). Meanwhile, estimating the number of species already lost, in historical times, or predicted to go extinct in the future is complicated by our ignorance of the large proportion of living species that remain undescribed (Hubbell et al. 2008, Barnosky et al. 2011, Cardoso et al. 2011) and by weak support for most present estimates (Stork 2010).

Whatever their accuracy for primary extinctions, estimates of past and future extinction rates generally fail to take account of dependent species. Interred within the tombs of known and unknown species extinctions is the coterie of species dependent on the lost ones: parasites extinct with their hosts, specialist herbivores gone with their food plants, plants lost with their pollinators or seed dispersers. These coextinctions of affiliate species may exceed the number of primary extinctions (Koh et al. 2004a, Dunn et al. 2009). As a concept, coextinction in its simplest, binary form is straightforward. Two species can be linked by pure exploitation (one species benefits while the other is harmed), by mutualism (each partner bears a cost while reaping a net benefit), or by commensalism (one species benefits with negligible cost or benefit to the other). A species that obligately requires goods or services from another species faces coextinction when the other, its benefactor or codependant, meets its own demise (Stork & Lyal 1993).

The consequences of coextinctions extend beyond reduced biodiversity. Coextinctions have indirect consequences for the communities from which species have been lost. Interdependent species complexes are ubiquitous, and many ecological interaction networks are large and complex (May 2009). Some networks even connect species across biomes or link terrestrial and aquatic ecosystems. For example, fish abundance in ponds affects reproduction in a shoreline flowering plant (St John's wort, *Hypericum fasciculatum*) through consumption of dragonflies, which are important predators of bee pollinators of the plant (Knight et al. 2005). The threat of species extinction or population extirpation (local extinction, an indicator of species extinction risk) can propagate across such trophic and functional links. In a broader sense, then, secondary extinctions driven by trophic cascades (Estes et al. 2011) are also coextinctions. In the literature of ecological interaction networks, secondary extinctions of mutualists, consumers, and other affiliates are described as extinction cascades (e.g., Memmott et al. 2004, Eklöf & Ebenman 2006), and in a broad sense, we view these as further examples of coextinctions or indicators of coextinction risk.

Vulnerable ecological associations can simultaneously include multiple relationship modes, such as parasitism, commensalism, and mutualism. Ants in the genera *Atta* and *Acromyrmex* grow fungi to feed their larvae and in turn provide protection, nutrition, and dispersal for the fungi (Cafaro et al. 2011). But pathogens also depend on the fungi, whereas commensal mites and beetles (Navarrete-Heredia 2001), mutualist bacteria, and other species depend on the ants (Cafaro et al. 2011). In complex multitrophic associations like these, multiple coextinctions would likely follow the extinction of a key host species.

Sometimes we can anticipate the potential magnitude of coextinction risk from generalities. Roughly 78% of temperate and 94% of tropical plant species are pollinated by animals (Ollerton et al. 2011). Although many factors contribute to the vulnerability or resilience of such pollination networks in the face of potential coextinctions, the subset of those hundreds of thousands of species dependent on specialist pollinators are susceptible to coextinction should their pollinators fail, a troubling scenario given declines in the abundances of many native bees (Potts et al. 2010).

Despite longstanding interest in the numbers of primary extinctions, coextinctions remain a relatively unexplored but timely topic. In this review, which expands and updates earlier overviews

(Bronstein et al. 2004; Koh et al. 2004a; Dunn et al. 2009; Moir et al. 2010, 2012b), we provide theoretical constructs and empirical evidence for coextinctions and coendangerment following loss or decline of hosts or mutualist partners. We review the many impediments to documenting coextinction, discuss the implications of these limitations, and outline several approaches to modeling coextinction and extinction cascades. Because the risk of coextinction is exacerbated through interactions with other environmental stresses, we discuss coextinction in the contexts of habitat alterations, population declines, climate change, and invasive species. We evaluate circumstances, such as host-switching, that promote persistence of affiliate species despite the loss of hosts. We discuss the role of affiliates in ecological recovery and restoration and conclude by outlining challenges and suggesting ways forward.

Coendangerment:
the endangerment of an affiliate species as a result of the endangerment or extinction of its host(s) or mutualist partner(s)

HISTORICAL AND CONTEMPORARY COEXTINCTIONS AND COENDANGERMENT

Challenges in Documenting Coextinctions

Dunn (2009) searched widely for credible examples of historical coextinctions and came up with a short list. Among the most cited examples of coextinction are parasites (particularly lice; Stork & Lyal 1993) of the passenger pigeon (*Ectopistes migratorius*) and a louse restricted to the black-footed ferret (*Mustela nigripes*) (Gompper & Williams 1998). These cases have come to illustrate the challenges of studying coextinction empirically: determining host specificity, the limitations of historical collections, the importance of resampling hosts, and the taxonomic impediment.

Determining host specificity. Assessing host specificity is a crucial step in evaluating the risk of coextinction or the likelihood of historical coextinction of an affiliate (Koh et al. 2004a, Moir et al. 2010). Two examples epitomize how difficult this can be. The lice of the passenger pigeon, once the most abundant bird in North America (Webb 1986) but extinct since 1914, provided the first published example of alleged coextinction. Stork & Lyal (1993) reported that two of these louse species, *Columbicola extinctus* and *Campanulotes defectus*, were likely to be extinct, as they were known from no other hosts. Louse species often include two or more closely related bird or mammal species among their acceptable hosts (Bush & Clayton 2006), but at the time Stork and Lyal published, the phylogenetic placement of the passenger pigeon was not yet clear. Not until 2010 was the closest living relative of the passenger pigeon identified as likely to be the band-tailed pigeon (*Patagioenas fasciata*) (Johnson et al. 2010), on which one of the two "extinct" passenger pigeon lice, *C. extinctus*, had already been found, very much alive (Clayton & Price 1999).

Taylor & Moir (2009) recently described two new species of rare, herbivorous insects in the genus *Acizzia* (Psyllidae: Hemiptera), each from a different rare, threatened Australian host plant. Taylor & Moir (2009) suggested these species be considered coendangered as a precautionary measure. Powell et al. (2011) later discovered a few individuals of one of them, *Acizzia keithi*, in the same region on a common, widespread, congeneric host plant, thereby reducing concern for this species. The second species, *Acizzia veski*, is still believed to be host specific, and its conservation listing has been approved at the state (Western Australia) level (M. Moir, personal communication).

Even though detailed study commonly reveals affiliates to be less specialized than assumed given published specimen records (Shaw 1994, Tompkins & Clayton 1999), it is unwise to assume further that all host ranges are underestimated. Coextinction is a process, not an event, and affiliates may often go extinct before their declining hosts (Moir et al. 2010, 2012b; Powell 2011). High host specificity is always a risk factor for coextinction when hosts are threatened.

Limitations of historical collections. In most cases, documenting coextinction depends on comparison of extant forms of a lineage (for example, a louse genus) to historical collections from an extinct host or partner. Such comparisons are only as dependable as the quality of the historical specimens and the accuracy of their associated collection data. For many reasons, collection data are often not reliable. In this regard, the passenger pigeon louse, *C. defectus*, also offered as an example of a coextinction, is illustrative, though perhaps extreme. The single specimen of this species reported to have been collected from a passenger pigeon appears to have become mislabeled during a World War II bombing episode in Germany. "*Campanulotes defectus*" appears to be a specimen of *Campanulotes flavens*, a parasite of the common bronzewing (*Phaps chalcoptera*), an Australian pigeon that has never occurred in the Americas (Price et al. 2000).

The importance of resampling hosts. If the passenger pigeon lice are synonymous with coextinction, the black-footed ferret louse has become emblematic of the special case of threatened parasites under captive rearing. In a growing number of cases, when the last individuals of rare species are brought into captivity for breeding with the aim of subsequent reintroduction of the species, they are immediately treated with biocides (pesticides, endectocides, or antihelminthics) to eliminate parasites (see sidebar, Extinction by Eradication?). When the last wild black-footed ferrets were brought into captivity in 1987 (Biggins et al. 2011), they may have carried a host-specific, but undescribed, louse species of the genus *Neotrichodectes*, last collected decades earlier. After pesticide treatment (delousing), no lice could be found on any of the captive ferrets, raising the possibility that this louse may now be extinct, despite the successful re-establishment of the ferret in the wild (Gompper & Williams 1998). No one has yet resurveyed the re-established black-footed ferret populations to examine whether the louse has reappeared, other ectoparasites have colonized, or the black-footed ferret now harbors any specialist parasite. Resampling hosts after release or translocation is an important but often neglected step in assessing the status of coendangered affiliates and host-species recovery.

The taxonomic impediment. Many, perhaps most, eukaryotic affiliates belong to hyperdiverse taxa (arthropods, helminths, protists, and fungi) (Dobson et al. 2008) that are not only poorly studied but also often difficult to distinguish and describe using only morphological characters.

EXTINCTION BY ERADICATION?

Captive breeding and other ex situ conservation programs for threatened or endangered hosts offer excellent opportunities for coextinction research (Moir et al. 2012a,b). In cases such as the black-footed ferret (Gompper & Williams 1998), in which the extinction of affiliates is threatened by insecticide or antihelminthic treatment of hosts, the total number of hosts to be sampled is usually small, and care for the host species in captivity is well funded. Tests of the role of parasites in stimulating and moderating host immune response upon release suggest a positive effect for some parasites in the captive setting (e.g., Van Oosterhout et al. 2007). A key experiment would compare the fitness of individuals from captive-rearing programs that are released after having been treated to exterminate parasites to the fitness of matched individuals that have been released without having been treated. The objectives of conservation of hosts and conservation of affiliates have the greatest potential for conflict in this setting, but these conflicts have been poorly explored, despite frequent pleas in both the biological (Windsor 1995; Nichols & Gomez 2011; Colwell et al. 2009; Moir et al. 2012a,b) and veterinary (Adler et al. 2011) literature to consider the ecological, evolutionary, ethical, and societal consequences of intentional or incidental eradication of parasites and other affiliate species.

As a result, it is now common for molecular studies to reveal lineages of multiple specialist species within what were previously viewed as single generalist affiliate species (Poulin & Keeney 2008). Detection of potentially coendangered or coextinct species may be aided by the increasing use of DNA barcoding and related approaches (Janzen et al. 2005). For example, the initial concerns of Gompper & Williams (1998) about the extinction of a potentially host-specific black-footed ferret louse were based solely on a record from a published host-parasite checklist. Had the historical louse samples been preserved or had sampling occurred at the time when ferrets were brought into captivity, current DNA-barcoding techniques could illuminate whether the species on the black-footed ferret was unique and now extinct. Nevertheless, systematists, even when using all available tools, face daunting obstacles in documenting the existence, much less the coextinction, of many affiliates. For example, Hamilton et al. (2010) estimate that 70% of terrestrial arthropods (nearly 3 million species), a large proportion of them dependent species, remain undescribed.

Additional Examples of Contemporary Coextinction and Coendangerment

Similar to the case of the black-footed ferret, when the last 22 wild California condors (*Gymnogyps californianus*) were brought into captivity in 1987 for rearing, they were deloused (Snyder & Snyder 2000). Three species of louse had previously been known from wild California condors, and two of them have not, so far, been found elsewhere. But, apparently, no one has systematically searched for these louse species on other bird species, not even on the California condor's closest relative (Wink 1995), the historically co-occurring black vulture (*Coragyps atratus*). Nor do there appear to have been attempts to resample the California condor in populations re-established from released, captively bred individuals.

Mihalca et al. (2011) considered the evidence for the extinction of ticks with the loss of their hosts and highlighted three potential examples, including the tick *Ixodes nitens*, known only from a single endemic rat species, *Rattus maclaeri*, from Christmas Island. Ironically, the extinction of this endemic rat (sometime around 1900) appears to have been driven by the introduction of pathogenic trypanosomes vectored by fleas, themselves affiliates of the invasive black rat (*Rattus rattus*), as demonstrated by ancient-DNA evidence from museum specimens (Wyatt et al. 2008). Mey (2005) lists 12 bird lice species as probably coextinct, all but one from extinct island bird species. (The exception was the passenger pigeon louse, later rediscovered on band-tailed pigeons.)

Additional examples of coendangered species have steadily accumulated (e.g., Durden & Keirans 1996, Colwell et al. 2009, Daszak et al. 2011), although most tend to be based on literature review of host associations rather than field studies (e.g., Powell et al. 2012) and experiments (e.g., Moir et al. 2012a). An iconic example is the rhinoceros stomach bot, *Gyrostigma rhinocerontis* (Oestridae: Diptera), the largest fly in Africa, which is restricted to the critically endangered black (*Diceros bicornis*) and near-threatened white (*Ceratotherium simum*) rhinoceroses (**Figure 1**) (Colwell et al. 2009).

Experiments on Coendangerment and Coextinction

Although experiments on coendangerment and coextinction are not impossible, few have been reported. Experimental population translocation (or reintroduction) is increasingly discussed as a means of helping rare and poorly dispersing species to track climate change through "assisted migration" (McLachlan et al. 2007); experiments with affiliates could easily be integrated into such programs. Moir et al. (2012a) compared the arthropod affiliates associated with threatened plants in natural populations and affiliates of translocated populations of the same plants distant from the source populations. For some of the plant species studied, species composition of affiliates

Figure 1

The coendangered rhinoceros stomach bot fly (*Gyrostigma rhinocerontis*, Oestridae: Diptera), which is the largest fly in Africa (25–30 mm long, wingspan 50–57 mm), and one of its two host species, the critically endangered black rhinoceros (*Diceros bicornis*). The adult female fly attaches her eggs at the base of the horns or ears or on the neck or shoulders of the rhinoceros. The larvae enter the digestive tract of the rhinoceros and burrow into its stomach lining, where they feed until ready to emerge from the anus and pupate in rhinoceros dung piles. The other host is the near-threatened white rhinoceros (*Ceratotherium simum*) (Colwell et al. 2009; M. Hall, personal communication). Fly photograph by Harry Taylor, copyright Natural History Museum, London; rhinoceros photo from Ngorongoro N.P., Tanzania, copyright R.K. Colwell.

in translocated populations differed substantially from affiliates on hosts in natural populations, instead resembling the affiliate fauna of local, related plants. In some cases, host-specific affiliates appeared to be missing on the translocated plants.

A vast literature describes affiliates (both pests and beneficial species) of cultivated plants and domesticated animals introduced far from their place of origin, and an equally large literature covers invasive species and their affiliates. This work could be mined for quantitative patterns bearing on coendangerment and coextinction risk (Keane & Crawley 2002). For example, Torchin et al. (2003) showed that twice as many parasite species are associated with host species (molluscs, crustaceans, fishes, birds, mammals, amphibians, and reptiles) in the hosts' native range than on the same hosts living as exotics elsewhere, providing implicit evidence for extirpation of affiliate populations during range expansions. If source populations of the host were later lost from their native range (a common occurrence with cultivated and domesticated species), these missing affiliates would represent a special class of affiliate extinctions.

Anderson et al. (2011) considered the effects of a very different sort of natural experiment in New Zealand, involving a native bird-pollinated shrub (*Rhabdothamnus solandri*, Gesneriaceae) and its three endemic pollinators—bellbirds, tui, and stitchbirds. On small island sanctuaries where the birds are still abundant, pollination, fruit set, and plant density of the shrub were all greater than they were on the North Island of New Zealand, where two of the three pollinator species have been locally extinct since approximately 1870.

PERSISTENCE OF AFFILIATE SPECIES: HOST SHIFTS

Host shift or host switch: expansion of an affiliate's host repertoire to include an additional, alternative host or the abandonment of one host for another

In addition to the challenges of documenting coextinctions, some parasites and other affiliates may complicate matters by shifting or expanding affiliation from a rapidly declining host (or mutualist partner) to an alternative, more common host, even if such novel hosts are initially inferior in terms of the fitness benefits they offer (Dunn et al. 2009, Moir et al. 2010). Although we focus here on host shifts by parasites, analogous considerations apply to shifts in affiliation among mutualist partners (Bronstein et al. 2004). Ideally, in the context of coextinction, we must understand three aspects of host shifts: how common they are, the host attributes that promote them, and the consequences they imply for novel hosts.

Real-Time Studies of Host Shifts

The most unambiguous way to document changes in the host specificity of affiliates is to see them in action, an increasingly frequent occurrence as infectious or parasitic organisms exploit humans and our domesticates, either as an additional host (host-range or niche expansion) or as sole host (host switch). The majority of diseases that affect humans worldwide originated from a nonhuman animal reservoir (Taylor et al. 2001). Recent examples of host shifts of eukaryote human affiliates include *Cyptosporidium* protists (Guerrant 1997) and helminths (Hotez et al. 1997) as well as a relatively long list of emerging zoonotic diseases (reviewed by Jones et al. 2008). Emerging zoonotic diseases in humans and our domesticates generally involve host shifts, or expansion of the affiliate's host repertoire, from relatively rare to relatively more common hosts. (We prefer the term host repertoire instead of host range, reserving "range" for its more general, geographical sense.)

> **Host repertoire:** the set of alternative host species upon which an affiliate can survive and reproduce (also known as host range)

Are Affiliates More Likely To Switch When Their Hosts Become Rare?

If host rarity influences the probability of host switches, the endangerment of host species may drive a shift of parasites and other affiliates to more common hosts or mutualist partners, with consequences for those hosts or partners. A large literature considers the evolution of host preference among herbivores in light of host plant quality and density (e.g., Mayhew 1997). A few studies have examined experimentally how the local rarity and density of animal hosts influence the probability of host shift in their parasites. For example, in California, western fence lizards (*Sceloporus occidentalis*) serve as the blood-meal hosts for a large proportion of larval and nymphal western black-legged ticks (*Ixodes pacificus*). When Swei et al. (2011) removed lizards from enclosures and quantified ticks on alterative (mammal) hosts, the removal of lizards increased larval ticks on other hosts by only approximately 5%, leaving many larval ticks without any host.

Given the potential public health consequences (both for humans and our domesticated crops and animals) of host switches from rare to common hosts, a thorough literature review on host shifts as a function of decreasing host populations would prove useful, especially because both we and our domesticates tend to be among the most abundant potential hosts and the most likely to be in contact with dwindling species (particularly vertebrates) (Woolhouse et al. 2005, Wolfe et al. 2007). If host shifts occur, but do not tend to be directly influenced by the rarity or density of hosts, host shifts and coextinction are independent phenomena. By contrast, if host rarity influences the probability of a host shift, the two phenomena are coupled, with many potential implications.

Transitions Inferred from Phylogeny

Phylogenetic trees provide an additional means of assessing the propensity of affiliates to switch hosts, especially for taxa in which monitoring real-time host shifts is difficult (e.g., Reed et al. 2007). If host specificity is highly conserved, related affiliates would be expected to have the same or related hosts. Conversely, for affiliate taxa in which host specificity is a labile trait, closely related affiliate species often differ in their host preference, and host shifts may even be an important mode of speciation and radiation (e.g., Ziętara & Lumme 2002). The specificity of parasites is often, if not universally, well conserved among clades (Mouillot et al. 2006, Poulin et al. 2006) and in some cases in networks of mutualists (Rezende et al. 2007), but no comprehensive review appears to have considered how labile host preferences are in general and whether differences in lability among taxa or functional groups (e.g., parasites versus mutualists) are statistically consistent.

MODELING COEXTINCTION

Given the difficulty of demonstrating individual events of contemporary coextinction or documenting coextinctions from the historical or fossil record, estimating the number of coextinctions, for past or future periods, calls for a modeling approach. A spectrum of approaches to modeling coextinction events can be described. At one extreme lie purely statistical methods that rely on estimating rates of affiliate coextinction as a function of primary extinctions of hosts or partners. At the other extreme, secondary extinctions and extinction cascades are modeled for hypothetical or empirical interaction webs. Here we attempt to link the two extremes conceptually, compare the challenges they face, and sketch out a middle ground.

Statistical Host-Extinction Models: Discrete Models

A species that obligately requires goods or services from any of several other species faces unequivocal extinction only when the last of its benefactors becomes extinct or falls below some critical population size or density threshold (Anderson & May 1986, Altizer et al. 2007, Dobson et al. 2008, Moir et al. 2010). Koh et al. (2004a,b) and Dunn et al. (2009) applied a simple, probabilistic extinction model to this scenario, treating host species (or populations) as either extant or extinct, ignoring density-threshold effects. This model relies on published data yielding "affiliation matrices" between hosts or mutualist partners and their affiliates, recording the binary association of S affiliate species with each of H hosts as a set of ones and zeros. A generalist affiliate will have several hosts, whereas a maximally specialized affiliate has only one. This simple approach assumes that no affiliate requires more than one host species to survive and ignores the complex life cycles of many parasites (taking them into account would increase predicted rates of coextinction). The number of coextinctions (extinctions of affiliates) as a function of h host extinctions can be estimated by randomly eliminating $1, 2, \ldots h, \ldots H$ hosts from the matrix without replacement. We refer to this model as a uniform random coextinction model (**Figure 2a**).

Mathematically, when hosts are chosen randomly for extinction, this approach is precisely equivalent to sample-based rarefaction (Colwell et al. 2012), widely used to estimate the number of species expected for a subset of sampling units (e.g., quadrats, traps, culture plates) for which only the presence or absence of each species has been recorded for each sample in a reference sample of H sampling units. An analytical (combinatoric) solution and unconditional variance estimators are available for this problem, allowing estimation of a confidence interval around the expected number of affiliates surviving after extinction of a specified number of hosts. The model is analogous to using a species-area relation "backwards" to estimate the expected number of extinctions given a certain loss of habitat area (the endemics-area curve) (Kinzig & Harte 2000, He & Hubbell 2011). As demonstrated by Colwell et al. (2004), sample-based (incidence-based) rarefaction is not biased by aggregation (in this application, by the aggregation of affiliates on certain hosts).

Building on this simple, uniform random extinction model, it is straightforward to condition the extinction of each host species on its intrinsic organismal traits (e.g., size, vagility, clade position) or on its susceptibility to extinction based on population size, geographical range size, habitat fragmentation, or other risk factors for primary extinction (e.g., Bunker et al. 2005, Srinivasan et al. 2007, Carpaneto et al. 2011). In addition, the cell probabilities in the affiliation matrix [zero or one in the model of Koh et al. (2004a)] could be based on quantitative, empirical patterns as estimates of probabilities of affiliation with hosts so that the effects of host loss on affiliate

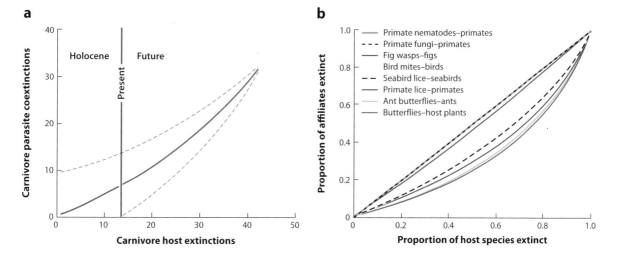

Figure 2

(*a*) Estimated number of coextinctions of louse, flea, and cestode (tapeworm) species as a function of the past and future extinctions of their hosts, 42 terrestrial North American Carnivora (NAC) species. Future coextinctions are based on the recorded affiliates of extant NAC, which are eliminated in random order in the model (Koh et al. 2004a). Dashed lines are estimated 95% confidence intervals (CI). Between 4 and 8 (mean 6.4, 4.3–8.5, 95% CI) NAC affiliate coextinctions are estimated to have occurred during the Holocene, on the basis of the known extinction of 13 NAC species since 11,000 years before present (Smith et al. 2003). Holocene coextinctions were estimated by extrapolating the future curve into the past using the statistical model of Colwell et al. (2012). The data considered here cover only metazoan ectoparasites and a small proportion of endoparasites (roughly one in 10 species), so the total number of coextinctions, past and future, would be greater. For the 10 NAC in the data set with ranges extending beyond North America, the losses should be considered coextirpations, unless the species is assumed extinct over its full range. Carnivora with ranges covering <10% of the Americas north of Panama were not considered. (*b*) The estimated proportion of affiliate species expected to face coextinction as a function of an increasing proportion of their hosts becoming extinct for eight affiliate-host systems: pollinating Agaonidae fig wasps—figs (*Ficus*), primate *Pneumocystis* fungi—primates, primate nematodes—primates, primate lice—primates, seabird lice—seabirds, bird mites—birds, butterflies—host plants, and Lycaenidae ant butterflies—ants. The curves were estimated on the basis of recorded affiliates for each host in each group, with hosts eliminated in random order in the model (figure from Koh et al. 2004a).

populations could be weighted (Vesk et al. 2010), akin to estimates of interaction strength in interaction-web models (e.g., Allesina & Pascual 2008).

Memmott et al. (2004) explored alternative host-elimination algorithms to estimate the expected number of plant coextinctions resulting from primary extinctions of their pollinators for two large pollination networks. Not surprisingly, when the deletion of pollinators from the affiliation matrix is ordered from the least-connected pollinators (specialists) to the most-connected pollinators (generalists), the number of plant coextinctions rises more slowly than in the uniform random coextinction model. In contrast, when deletions are ordered from generalist to specialist pollinators, the number of coextinctions rises more rapidly than they do in the uniform random model.

Using affiliation matrices for living species, Koh et al. (2004a) applied the uniform random co-extinction model to estimate the number of undocumented, historical coextinctions of affiliates as a result of recorded, historical extinctions of their hosts. They modeled two cases: the coextinction of butterflies, as a result of known historical plant extinctions (assumed to be larval host plants), and the number of bird mite extinctions, as a function of the number of historical avian extinctions. This approach assumes that affiliates that have already become extinct had the same quantitative pattern of association with extinct hosts as living affiliates have with extant hosts, when, in fact, past host extinction [or even current threat (Altizer et al. 2007)] is likely to have increased the host specificity of surviving affiliates through the extinction of some of their alternative hosts. A more rigorous approach would extrapolate the coextinction curve into the past [as Koh et al. (2004a) suggest in their supplemental online material].

With appropriate assumptions, recent advances in extrapolating sample-based species accumulation (rarefaction) curves beyond the empirical reference sample make this retrospective extrapolation feasible. Using sampling-theoretic, nonparametric predictors, Colwell et al. (2012) developed estimators (with unconditional confidence intervals) for the expected number of additional species to be found in h^* additional samples. By direct analogy, we can estimate the number of affiliate coextinctions that occurred as the result of the historical (or conjectured) extinction of h^* hosts. To demonstrate the method, **Figure 2a** shows an example for a subset of the parasites of North American Carnivora. The model estimates that between four and eight (unknown) species of fleas, lice, ticks, and cestodes were extirpated with the 13 known extinctions of North American Carnivora during the Holocene (Smith et al. 2003). Any affiliates that did not survive on non-Carnivora hosts or on non–North American Carnivora are now assumed globally extinct. Adding other parasitic taxa such as trematodes, nematodes, and protists would increase these estimates.

The critical statistics for all the above models are the numbers of hosts per affiliate (even if weighted for different hosts). Unfortunately, as for any measure of (statistical) range for finite samples, the observed host repertoire of an affiliate is a negatively biased estimator for its true host repertoire and, therefore, tends to overestimate coextinction risk, an issue of equal importance for interaction network models (Blüthgen 2010). Recognizing this problem, Vesk et al. (2010; see also Moir et al. 2011) have explored a promising approach toward reducing the bias in estimates of host breadth that is based on a Bayesian, zero-inflated Poisson regression approach.

Statistical Host-Extinction Models: Curve-Fitting Models

When normalized by percentages, as in **Figure 2b**, the shape of coextinction curves based on uniform random host-extinction models (**Figure 2a**) depends entirely on the frequency distribution of hosts per affiliate. If every affiliate has exactly one host and every host one affiliate, as in the three groups of primate parasites in **Figure 2b**, the curve is a straight line with slope 1. Called the linear approach by Moir et al. (2010), the assumption of such a one-to-one relationship underlies early estimates of coextinction risk (e.g., Stork & Lyal 1993, Poulin & Morand 2004) and estimates that are intentionally confined to monophagous species (e.g., Thacker et al. 2006). For groups in which some or all affiliates have more than a single host per affiliate, the curve is shallower, because a typical affiliate tends to survive early host extinctions, succumbing only when its last, alternative host goes extinct (Koh et al. 2004a, Dobson et al. 2008).

Koh et al. (2004a, see their supplemental online material) showed that the distributions of hosts per affiliate are similar in shape for the groups in **Figure 2b** and took advantage of the similarity to fit a regression model for the ratio of affiliate extinctions to host extinctions (the instantaneous slopes of the curves in **Figure 2b**) as a function of host-extinction percentage and the mean number of hosts per affiliate. Koh et al. (2004a) used this empirical "nomographic model" to estimate the expected number of affiliate extinctions for groups such as herbivorous beetles and bird lice, for which complete affiliate matrices are not available but mean number of hosts can be estimated. They applied the model to project future extinctions, in the event that all currently endangered hosts were to be lost, as well as to estimate historical coextinctions as a function of recorded historical host extinctions. Using the same fitted equation, Dunn (2009) extended the historical extinction results for bird and mammal ectoparasites to estimate the number of coextinctions in these groups for the entire Holocene on the basis of estimated Holocene bird and mammal extinctions. Because the relation is data dependent, having been fitted for a particular set of random host-extinction curves, these estimates of past and future extinctions depend closely on the assumption that affiliation matrices for all the groups to which they are applied are similar to the current distribution of number of hosts per affiliate in the "training" data sets. More work needs to be done comparing the host specificity distributions of different groups of affiliates, particularly to understand the extent to which taxa or classes of affiliates (e.g., parasites versus mutualists) differ systematically in the risk and rate of coextinction.

Linking Coextinction Matrix Models and Interaction Network Models

Coextinction models based on affiliation matrices, like those discussed in the previous sections, are conceptually a subset of ecological interaction network models (Bascompte & Jordano 2007), although this fact seems not to have previously been discussed in the literature. In both kinds of models, links between mutualist partners or between hosts and affiliates can be broken by extinctions (random or not), nodes can be unweighted or weighted, and the consequences interpreted as fragility or robustness.

A full account of food-web and interaction network studies would not be pertinent (or possible) here, and comprehensive reviews have recently appeared elsewhere (e.g., Bascompte & Jordano 2007, Bascompte 2009). From the point of view of coextinctions, the key consideration with food-web and interaction network studies is how the removal of species (primary extinctions) having identifiable characteristics leads to the secondary extinction of other species in the network. The characteristics of those consequential casualties are also of great interest (e.g., Petchey et al. 2008). Whereas the statistical coextinction models of Koh et al. (2004a) and Memmott et al. (2004) include only two trophic levels, many (but not all) food-web and interaction network models encompass multiple trophic levels and are thus capable of much more complex dynamic behavior. Here, we highlight the implications for coextinctions in a few pertinent network studies. We follow the literature of interaction networks in using "extinction" for local networks, recognizing that the species losses that the models predict are more accurately characterized as population extirpations.

Interaction Networks and Coextinctions

Examining three large empirical food webs, Solé & Montoya (2001) demonstrated that random deletion of species initially leads to few secondary extinctions, whereas [in accord with Memmott et al. (2004)] selective deletion of highly connected (i.e., keystone) species produces a rapid disintegration of the network and a high rate of secondary extinctions. Dunne et al. (2002) confirmed these conclusions for a larger set of empirical food webs, but they offered evidence that higher levels of connectance (more links per species independent of community size, the equivalent of decreasing host specificity in affiliation matrices) and increased species diversity (Dunne & Williams 2009) in model networks delay the nonlinear increase in secondary extinctions when highly connected species are targeted for primary extinction.

In a food-web study of 50 lakes in the Adirondack region of the United States, Srinivasan et al. (2007) found that deleting species according to prevalence (from least to most prevalent) among the lakes caused surprisingly few secondary extinctions until nearly all species were removed by primary extinction. Almost as few secondary extinctions resulted as when species were deleted according to increasing numbers of links. The reverse order of extinctions in both cases led to catastrophic cascades of extinction. For a very large collection of plant-pollinator and fruit-frugivore interaction networks, Rezende et al. (2007) showed that simulated secondary extinctions tend to eliminate species that are more closely related than would be expected at random, given the species in each network, presumably because of phylogenetic signaling in their niches.

Using dynamic population models (generalized Lotka-Volterra models) as well as topological analysis of empirical food webs, Petchey et al. (2008) showed that the primary extinction of species with unique trophic relationships is especially likely to drive secondary extinctions. Eklöf & Ebenman (2006) noted that topological analyses (unlike dynamic models) are blind to trophic cascades and other complex secondary interactions. Using dynamic models, they argued that trophic webs with higher connectance confer greater persistence (i.e., permanence) and that, within such high-connectance communities, species most vulnerable to secondary extinction tend to be in the middle of trophic webs.

Fundamental differences between mutualistic networks and trophic networks are under active exploration (Lewinsohn et al. 2006). On the basis of dynamic models and a meta-analysis of empirical pollination and herbivory networks, Thébault & Fontaine (2010) concluded that stability is promoted in mutualistic networks by highly connected and nested architectures, whereas stability in trophic networks is enhanced by weak connections and compartmentalization. By contrast, Allesina & Pascual (2008) argued on the basis of dynamic models that the stability of complex communities depends strongly on simple consumer-victim (including parasite-host) subsystems.

In summary, network studies indicate that the rate of secondary extinction is greater following the primary extinction of common species and highly connected species, whereas secondary extinction slows in networks with high average levels of connectance and high diversity. These results are closely in accord with conclusions from coextinction models, in which the extinction of hosts with many affiliates (such hosts tend to be widespread geographically; Lindenfors et al. 2007, Harris & Dunn 2010) drives faster affiliate coextinctions, whereas the presence of affiliates with many hosts slows the rate of coextinction. In both kinds of models, the survivors as well as the extinctions tend to be nonrandom subsets of the community.

FACTORS THAT INFLUENCE THE RISK OF COEXTINCTION

Host Specificity and Life Cycle

The factor most likely to influence the probability of coextinction directly is the host specificity of affiliates, including relative fitness on different hosts as well as the phylogenetic spectrum of hosts (Moir et al. 2010, Poulin et al. 2011) or mutualist partners (Bronstein et al. 2004). Host specificity is also both scale and context dependent (Krasnov et al. 2011). So far, coextinction models have ignored affiliates with complex life cycles (parasites, many free-living insects), which rely on different host species at different life stages. As hosts are lost, the risk of coextinction is assumed higher for such species than it is for affiliates with simple life cycles (Koh et al. 2004a, Poulin & Morand 2004, Lafferty 2012).

Coextinction is also a greater risk if specificity is evolutionarily inflexible. By evolving traits necessary for survival and reproduction on alternative or additional hosts, affiliates may escape extinction (Bronstein et al. 2004, Poulin et al. 2006). The characteristics of hosts that tend to have evolutionarily labile affiliates and of those affiliates could be better explored.

Affiliate and Host Traits

Affiliates predisposed to coextinction may share traits with species more generally vulnerable to extinction. If they are larger bodied, their total population size is likely to be smaller; larger-bodied affiliates also tend to depend on larger hosts (**Figure 1**) (Bush & Clayton 2006), which are also more vulnerable to extinction (e.g., Stork et al. 2009). Affiliates of hosts of higher trophic levels (**Figure 2**) share their host's vulnerability to extinction (Estes et al. 2011). Finally, just as for their hosts, affiliates with limited climatic tolerances will be at an increased risk of extinction if their climatic niches are narrow or if the geographic area corresponding to their climatic niches is small (Moir et al. 2010).

Host traits may influence coextinction through host rarity or abundance and through their influence on affiliate diversity. Hosts vary dramatically in their rarity: Some species consist of fewer than ten individuals in the wild, whereas others have hundreds of millions and perhaps even billions of individuals. Large, abundant, geographically widespread hosts tend to have the greatest diversity of parasites (Lindenfors et al. 2007, Harris & Dunn 2010), even though a high diversity of parasites on a single host species may increase interspecific competition among its affiliates (e.g., Moore & Simberloff 1990). Such competition notwithstanding, parasites and other affiliate species dependent on abundant, widespread, invasive, pest, and domestic species may be relatively unlikely to face coextinction under current regimes.

Although hosts with wide geographic ranges and large populations may offer affiliates a low probability of extinction, the consequences of their extinction in terms of the number of coextinctions may be disproportionately greater. This principle is encoded in the idea that the extinction of well-connected species is more likely to cause cascading extinctions (as discussed in Modeling Coextinction; see section above). *Eciton burchellii*, a wide-ranging Neotropical army ant species, hosts no fewer than 300 and perhaps even thousands of affiliate species (including mites, beetles, millipedes, and ant birds), many of which are likely to be host specific to *E. burchellii* (Rettenmeyer et al. 2011). *E. burchellii* is just one of more than 150 New World army ant species (Formicidae: Ecitoninae) and nearly as many species in the Old World army ant clade (Dorylinae), each of which likely hosts many unique species. The extinction of species such as *E. burchellii* has the potential to drive a remarkably large number of coextinctions. In all but a handful of cases, these coextinctions will be of unstudied species that, if known at all before their demise, will be found only in museum collections.

INTERACTIONS WITH OTHER THREATS AND CONDITIONS

Habitat Alterations

As with other threats to biodiversity, the risk and rate of coextinctions can be exacerbated by additional perturbations (Bronstein et al. 2004, Dunn et al. 2009, Moir et al. 2010). Such synergisms not only add complexity to any effort to reduce the risk of coextinction, but also complicate documenting its occurrence. Loss and alteration of habitat remain the predominant threat to species persistence and thus to ecological associations. On the island of Singapore, observed or inferred estimates of the historical "local extinction" rate for some taxa approach 90%, corresponding to rapid loss of tropical forest exceeding 95% (Brook et al. 2003). Koh et al. (2004a) estimated the historical local extinction of at least 56 butterfly species, which they suggested was a consequence of both direct impacts on their populations and the loss of 208 potential butterfly larval host plants.

Climate Change

Because shifts in geographic distributions of species in response to changing climates may often prove discordant among species (Gilman et al. 2010, Sheldon et al. 2011), shifts in the geographic range of hosts may drive spatial or temporal mismatches among previously co-occurring species, dramatically altering interactions between hosts and affiliate species (Traill et al. 2010). Although many affiliate species can be expected to survive by range shifts in step with their hosts or partners, host switches and conversion by affiliates to independent lifestyles may also allow survival. The abandonment of mutualisms has been documented, for example, when available pollinators are infrequent and plants switch from animal to wind dispersal (Kaiser-Bunbury et al. 2010).

In the larval stage, freshwater mussels (Unionoida) obligately parasitize fish. Spooner et al. (2011) modeled the effects of decreased river flow (expected under ongoing climate change) on the coextirpation of mussel species in rivers of the eastern United States, predicting that up to 43% of mussel populations may fail owing to loss of host-fish populations. Phenological mismatch of partner interactions, such as plants and their pollinators, may also alter key ecosystem processes under climate change scenarios (Yang & Rudolf 2010).

Exploitation

Given the continued reliance of human societies on natural resources, coextinction dynamics may be accelerated through harvest. Bushmeat harvest decimated vertebrate seed dispersers in Thailand, resulting in diminished seed dispersal, population growth, and heightened extinction risk of a canopy tree (*Choerospondias axillaris*, Anacardiaceae) (Brodie et al. 2009). Wood et al. (2010) showed that marine protected areas, which generally prohibit fishing and trolling activities, harbor greater parasite richness and host abundances than fished areas support (e.g., Loot et al. 2005). Not surprisingly, multiple threats may interact, as evident in the increased vulnerability of the affiliates of coral reef fishes to coextinction through fishing pressure and climate change (Graham et al. 2011).

Invasive Species

Invasions of exotic species continue to threaten biological diversity globally, and affiliate species are not spared. Native to Asia, emerald ash borer (*Agrilus planipennis*, Buprestidae: Coleoptera) is an epidemic pest on ash (*Fraxinus*) species of eastern North America that causes significant mortality and restructuring of forests. The native arthropod fauna, including an estimated 15% (43 species) of ash-dependent invertebrate species, has also suffered (Gandhi & Herms 2010). According to the IUCN Red List (Gandhi & Herms 2010), the extirpation of American chestnut by chestnut blight, an Asian fungus, drove the extinction of at least two specialist arthropods, American chestnut moth (*Ectodemia castaneae*, Nepticulidae: Lepidoptera) and phleophagan chestnut moth (*Ectodemia phleophaga*), and possibly others (Opler 1978; D.L. Wagner, personal communication). The coextinction of the endemic Christmas Island rat and its parasites (discussed above) after the introduction of the black rat with its flea-vectored trypanosome parasites (Wyatt et al. 2008) may represent an underappreciated scenario applicable to other settings.

Invasive species may disrupt mutualistic interactions among native species (Bronstein et al. 2004). In South Africa, invasive Argentine ants profoundly altered floral visitation by native arthropods (Lach 2008). A contrary, yet growing, literature documents the ability of invasive species to replace the function of lost native species, maintaining interactions to the benefit of affiliates (Bronstein et al. 2004). For example, following their recent arrivals to New Zealand, the black rat (*Rattus rattus*) and silvereye bird (*Zosterops lateralis*) have partially compensated for missing native vertebrate pollinators by pollinating some plant species, thereby sustaining species interactions and function (Pattemore & Wilcove 2011).

SPECIES CONSERVATION, RECOVERY, AND RESTORATION

Coextinction matters not only because of the intrinsic value of all species, but also because of the broader consequences of the loss of dependent species. Every class of affiliates, whether parasites (Hudson et al. 2006), commensals (Howells et al. 2011), or mutualists (Forup et al. 2008, Menz et al. 2011), provides ecosystem services (Cardoso et al. 2011). Thus the conservation or restoration of these services may often require attention to affiliate species and the dynamics of coextinction. For example, during rainforest regeneration in the Biological Dynamics of Forest Fragments Project (Laurance et al. 2011) in the Brazilian Amazon, processes linked to affiliate species lagged in their recovery: Research shows that decomposition of vertebrate feces by dung beetles required 20 years to recover fully (Quintero & Roslin 2005), whereas dispersal of large seeded trees dependent on large vertebrate dispersers has yet to recover fully after 30 years (Cramer et al. 2007).

The return of the ecological function of affiliates may be the ultimate measure of success in recovery or restoration projects (Huspeni & Lafferty 2004, Dixon 2009, Menz et al. 2011). Likewise, the continued functioning of affiliates may be a sensitive indicator of successful conservation programs; conserving affiliates in the first place will always be easier than restoring them once they are gone. As an obvious starting point, the fate of affiliate species of threatened hosts—whether the California condor (Snyder & Snyder 2000) or an endangered pitcher plant (Folkerts 1999)—deserves immediate attention (Moir et al. 2012a,b).

SUMMARY POINTS

1. The extinction of a single species that is large or conspicuous enough to be detected is rarely, if ever, an isolated loss. Parasites, commensals, and mutualists (affiliates), many of them hard to study and poorly known, face coextinction with the demise of their hosts or partners.

2. Coextinctions are difficult to document because of sampling difficulties, taxonomic uncertainty, the limitations of historical collections, and the potential for host shifts. Experimental studies of host-affiliate interactions are possible, but rare.

3. Statistical models based on patterns of association between hosts and affiliates can be used to estimate the rates of past coextinctions and the patterns and rates of future coextinctions.

4. Network models of species interactions explore secondary extinctions and extinction cascades driven by primary extinctions of specified categories of hosts and mutualist partners.

5. Coextinction risk and its management depend on the host specificity and evolutionary lability of affiliates and on the ecological traits of hosts or mutualist partners and of their affiliates. The risk of coextinction interacts with other threats, including habitat loss, climate change, and invasive species.

6. Restoration and recovery of ecosystems can depend on and be measured by the return of ecosystem services delivered by dependent species.

FUTURE ISSUES

1. Better understanding of the risk and rate of coextinction depends, unequivocally, on increased study and documentation of the natural history of affiliates and better support for the training and research of systematists working on understudied affiliate groups.

2. Intensified experimental study of hosts and affiliates in the laboratory and field would substantially increase understanding of coextinction dynamics.

3. Integration of statistical and network approaches to the modeling of coextinctions and extinction cascades, incorporating demographic and evolutionary dynamics, host switching, affiliate phylogeny, and risk factors for affiliate extinction, is likely to advance our ability to predict and prevent future coextinctions under conditions of ongoing global change.

DISCLOSURE STATEMENT

The authors are not aware of any affiliations, memberships, funding, or financial holdings that might be perceived as affecting the objectivity of this review.

ACKNOWLEDGMENTS

All the authors contributed equally to this review. We dedicate this review to the memory of our friend and colleague Navjot Sodhi. We thank L.P. Koh, M.L. Moir, and F.A. Powell for their thoughtful comments on a draft and M. Hall, K.L. Kuhn, and D.L. Wagner for fact checking. This work was supported by the US National Science Foundation (DEB-0639979 and DBI-0851245 to R.K.C.; NSF-0953390 to R.R.D.; DBI-1103661 to N.C.H.), the US National Aeronautics and Space Administration (ROSES-NNX09AK22G to R.R.D.), the US Department of Energy (DOE-PER DE-FG2-O8ER64510 to R.R.D.), and the Department of Forestry and Environmental Resources, North Carolina State University (Graduate Fellowship to N.C.H.). Present address for N.C.H.: Department of Environmental Science, Policy, and Management, University of California, Berkeley, California 94720; email: nyeema@berkeley.edu.

LITERATURE CITED

Adler PH, Tuten HC, Nelder MP. 2011. Arthropods of medicoveterinary importance in zoos. *Annu. Rev. Entomol.* 56:123–42

Allesina S, Pascual M. 2008. Network structure, predator-prey modules, and stability in large food webs. *Theor. Ecol.* 1:55–64

Altizer S, Nunn CL, Lindenfors P. 2007. Do threatened hosts have fewer parasites? A comparative study in primates. *J. Anim. Ecol.* 76:304–14

Anderson RM, May RM. 1986. The invasion, persistence and spread of infectious diseases within animal and plant communities [and discussion]. *Philos. Trans. R. Soc. Lond. Ser. B* 314:533–68

Anderson SH, Kelly D, Ladley JJ, Molloy S, Terry J. 2011. Cascading effects of bird functional extinction reduce pollination and plant density. *Science* 331:1068

Barnosky AD, Matzke N, Tomiya S, Wogan GOU, Swartz B, et al. 2011. Has the Earth's sixth mass extinction already arrived? *Nature* 471:51–57

Bascompte J. 2009. Disentangling the web of life. *Science* 325:416

Bascompte J, Jordano P. 2007. Plant-animal mutualistic networks: the architecture of biodiversity. *Annu. Rev. Ecol. Evol. Syst.* 38:567–93

Biggins D, Livieri T, Breck S. 2011. Interface between black-footed ferret research and operational conservation. *J. Mammal.* 9:699–704

Blüthgen N. 2010. Why network analysis is often disconnected from community ecology: a critique and an ecologist's guide. *Basic Appl. Ecol.* 11:185–95

Brodie JF, Helmy OE, Brockelman WY, Maron JL. 2009. Bushmeat poaching reduces the seed dispersal and population growth rate of a mammal-dispersed tree. *Ecol. Appl.* 19:854–63

Bronstein JL, Dieckmann U, Ferrièrre R. 2004. Coevolutionary dynamics and the conservation of mutualisms. In *Evolutionary Conservation Biology*, ed. R Ferrièrre, U Dieckmann, D Couvet, pp. 305–26. Cambridge, UK: Cambridge Univ. Press

Brook B, Sodhi N, Ng P. 2003. Catastrophic extinctions follow deforestation in Singapore. *Nature* 424:420–26

Bunker DE, DeClerck F, Bradford JC, Colwell RK, Perfecto I, et al. 2005. Species loss and aboveground carbon storage in a tropical forest. *Science* 310:1029–31

Bush SE, Clayton DH. 2006. The role of body size in host specificity: reciprocal transfer experiments with feather lice. *Evolution* 60:2158–67

Cafaro MJ, Poulsen M, Little AEF, Price SL, Gerardo NM, et al. 2011. Specificity in the symbiotic association between fungus-growing ants and protective *Pseudonocardia* bacteria. *Proc. R. Soc. Lond. Ser. B* 278:1814–22

Cardoso P, Erwin TL, Borges PAV, New TR. 2011. The seven impediments in invertebrate conservation and how to overcome them. *Biol. Conserv.* 144:2647–55

Carpaneto GM, Mazziotta A, Pittino R, Luiselli L. 2011. Exploring co-extinction correlates: the effects of habitat, biogeography and anthropogenic factors on ground squirrels-dung beetles associations. *Biodivers. Conserv.* 20:3059–76

Clayton DH, Price RD. 1999. Taxonomy of New World Columbicola (Phthiraptera: Philopteridae) from the Columbiformes (Aves), with descriptions of five new species. *Ann. Entomol. Soc. Am.* 92:675–85

Colwell DD, Otranto D, Stevens JR. 2009. Oestrid flies: eradication and extinction versus biodiversity. *Trends Parasitol.* 25:500–4

Colwell RK, Chao A, Gotelli NJ, Lin S-Y, Mao CX, et al. 2012. Models and estimators linking individual-based and sample-based rarefaction, extrapolation, and comparison of assemblages. *J. Plant Ecol.* 5:3–21

Colwell RK, Mao CX, Chang J. 2004. Interpolating, extrapolating, and comparing incidence-based species accumulation curves. *Ecology* 85:2717–27

Cramer JM, Mesquita RCG, Bentos TV, Moser B, Williamson GB. 2007. Forest fragmentation reduces seed dispersal of *Duckeodendron cestroides*, a Central Amazon endemic. *Biotropica* 39:709–18

Daszak P, Ball SJ, Streicker DG, Jones CG, Snow KR. 2011. A new species of *Caryospora* Léger, 1904 (Apicomplexa: Eimeriidae) from the endangered Round Island boa *Casarea dussumieri* (Schlegel) (Serpentes: Bolyeridae) of Round Island, Mauritius: an endangered parasite? *Syst. Parasitol.* 78:117–22

Dixon KW. 2009. Pollination and restoration. *Science* 325:571

Dobson A, Lafferty KD, Kuris AM, Hechinger RF, Jetz W. 2008. Homage to Linnaeus: how many parasites? How many hosts? *Proc. Natl. Acad. Sci. USA* 105:11482

Dunn RR. 2009. Coextinction: anecdotes, models, and speculation. In *Holocene Extinctions*, ed. S Turvey, pp. 167–80. Oxford: Oxford Univ. Press

Dunn RR, Harris NC, Colwell RK, Koh LP, Sodhi NS. 2009. The sixth mass coextinction: Are most endangered species parasites and mutualists? *Proc. R. Soc. Lond. Ser. B* 276:3037–45

Dunne JA, Williams RJ. 2009. Cascading extinctions and community collapse in model food webs. *Philos. Trans. R. Soc. Lond. Ser. B* 364:1711

Dunne JA, Williams RJ, Martinez ND. 2002. Network structure and biodiversity loss in food webs: Robustness increases with connectance. *Ecol. Lett.* 5:558–67

Durden LA, Keirans JE. 1996. Host-parasite coextinction and the plight of tick conservation. *Am. Entomol.* 42:87–91

Eklöf A, Ebenman B. 2006. Species loss and secondary extinctions in simple and complex model communities. *J. Anim. Ecol.* 75:239–46

Estes JA, Terborgh J, Brashares JS, Power ME, Berger J, et al. 2011. Trophic downgrading of planet Earth. *Science* 333:301–6

Folkerts D. 1999. Pitcher plant wetlands of the southeastern United States: arthropod associates. In *Invertebrates in Freshwater Wetlands of North America: Ecology and Management*, ed. D Batzer, R Radar, S Wissinger, pp. 247–75. New York: John Wiley & Sons

Forup ML, Henson KSE, Craze PG, Memmott J. 2008. The restoration of ecological interactions: plant-pollinator networks on ancient and restored heathlands. *J. Appl. Ecol.* 45:742–52

Gandhi KJK, Herms DA. 2010. North American arthropods at risk due to widespread *Fraxinus* mortality caused by the Alien Emerald ash borer. *Biol. Invasions* 12:1839–46

Gilman SE, Urban MC, Tewksbury J, Gilchrist GW, Holt RD. 2010. A framework for community interactions under climate change. *Trends Ecol. Evol.* 25:325–31

Gompper ME, Williams ES. 1998. Parasite conservation and the black-footed ferret recovery program. *Conserv. Biol.* 12:730–32

Graham NAJ, Chabanet P, Evans RD, Jennings S, Letourneur Y, et al. 2011. Extinction vulnerability of coral reef fishes. *Ecol. Lett.* 14:341–48

Guerrant RL. 1997. Cryptosporidiosis: an emerging, highly infectious threat. *Emerg. Infect. Dis.* 3:51

Hamilton AJ, Basset Y, Benke KK, Grimbacher PS, Miller SE, et al. 2010. Quantifying uncertainty in estimation of tropical arthropod species richness. *Am. Nat.* 176:90–95

Harris NC, Dunn RR. 2010. Using host associations to predict spatial patterns in the species richness of the parasites of North American carnivores. *Ecol. Lett.* 13:1411–18

He F, Hubbell SP. 2011. Species-area relationships always overestimate extinction rates from habitat loss. *Nature* 473:368–71

Hotez PJ, Zheng F, Long-qi X, Ming-gang C, Shu-hua X, et al. 1997. Emerging and reemerging helminthiases and the public health of China. *Emerg. Infect. Dis.* 3:303–10

Howells ME, Pruetz J, Gillespie TR. 2011. Patterns of gastro-intestinal parasites and commensals as an index of population and ecosystem health: the case of sympatric western chimpanzees (*Pan troglodytes verus*) and guinea baboons (*Papio hamadryas papio*) at Fongoli, Senegal. *Am. J. Primatol.* 73:173–79

Hubbell SP, He FL, Condit R, Borda-de-Agua L, Kellner J, ter Steege H. 2008. How many tree species and how many of them are there in the Amazon will go extinct? *Proc. Natl. Acad. Sci. USA* 105:11498–504

Hudson PJ, Dobson AP, Lafferty KD. 2006. Is a healthy ecosystem one that is rich in parasites? *Trends Ecol. Evol.* 21:381–85

Huspeni TC, Lafferty KD. 2004. Using larval trematodes that parasitize snails to evaluate a saltmarsh restoration project. *Ecol. Appl.* 14:795–804

Janzen DH, Hajibabaei M, Burns JM, Hallwachs W, Remigio E, Hebert PDN. 2005. Wedding biodiversity inventory of a large and complex Lepidoptera fauna with DNA barcoding. *Philos. Trans. R. Soc. Lond. Ser. B* 360:1835–45

Johnson KP, Clayton DH, Dumbacher JP, Fleischer RC. 2010. The flight of the passenger pigeon: phylogenetics and biogeographic history of an extinct species. *Mol. Phylogen. Evol.* 57:455–58

Jones KE, Patel NG, Levy MA, Storeygard A, Balk D, et al. 2008. Global trends in emerging infectious diseases. *Nature* 451:990–93

Kaiser-Bunbury CN, Muff S, Memmott J, Muller CB, Caflisch A. 2010. The robustness of pollination networks to the loss of species and interactions: a quantitative approach incorporating pollinator behaviour. *Ecol. Lett.* 13:442–52

Keane RM, Crawley MJ. 2002. Exotic plant invasions and the enemy release hypothesis. *Trends Ecol. Evol.* 17:164–70

Kinzig A, Harte J. 2000. Implications of endemics-area relationships for estimates of species extinctions. *Ecology* 81:3305–11

Knight TM, McCoy MW, Chase JM, McCoy KA, Holt RD. 2005. Trophic cascades across ecosystems. *Nature* 437:880–83

Koh LP, Dunn RR, Sodhi NS, Colwell RK, Proctor HC, Smith VS. 2004a. Species coextinctions and the biodiversity crisis. *Science* 305:1632–34

Koh LP, Sodhi NS, Brook BW. 2004b. Co-extinctions of tropical butterflies and their hostplants. *Biotropica* 36:272–74

Krasnov BR, Mouillot D, Shenbrot GI, Khokhlova IS, Poulin R. 2011. Beta-specificity: the turnover of host species in space and another way to measure host specificity. *Int. J. Parasitol.* 41:33–41

Lach L. 2008. Argentine ants displace floral arthropods in a biodiversity hotspot. *Divers. Distrib.* 14:281–90

Lafferty KD. 2012. Biodiversity loss decreases parasite diversity: theory and patterns. *Philos. Trans. R. Soc. Lond. Ser. B* 367:2814–27

Laurance WF, Camargo JLC, Luizão RCC, Laurance SG, Pimm SL, et al. 2011. The fate of Amazonian forest fragments: a 32-year investigation. *Biol. Conserv.* 144:56–67

Lewinsohn T, Prado I, Jordano P, Bascompte J, Olesen J. 2006. Structure in plant-animal interaction assemblages. *Oikos* 113:174

Lindenfors P, Nunn CL, Jones KE, Cunningham AA, Sechrest W, Gittleman JL. 2007. Parasite species richness in carnivores: effects of host body mass, latitude, geographical range and population density. *Glob. Ecol. Biogeogr.* 16:496–509

Loot G, Aldana M, Navarrete SA. 2005. Effects of human exclusion on parasitism in intertidal food webs of central Chile. *Conserv. Biol.* 19:203–12

May RM. 2009. Food-web assembly and collapse: mathematical models and implications for conservation. *Philos. Trans. R. Soc. Lond. Ser. B* 364:1643

Mayhew PJ. 1997. Adaptive patterns of host-plant selection by phytophagous insects. *Oikos* 79:417–28

McLachlan JS, Hellmann JJ, Schwartz MW. 2007. A framework for debate of assisted migration in an era of climate change. *Conserv. Biol.* 21:297–302

Memmott J, Waser NM, Price MV. 2004. Tolerance of pollination networks to species extinctions. *Proc. R. Soc. Lond. Ser. B* 271:2605

Menz MHM, Phillips RD, Winfree R, Kremen C, Aizen MA, et al. 2011. Reconnecting plants and pollinators: challenges in the restoration of pollination mutualisms. *Trends Plant Sci.* 16:4–12

Mey E. 2005. *Psittacobrosus bechsteini*: ein neuer ausgestorbener Federling (Insecta, Phthiraptera, Amblycera) vom Dreifarbenara *Ara tricolor* (Psittaciiformes), nebst einer annotierten Übersicht über fossile und rezent ausgestorbene Tierläuse. *Anz. Ver. Thüring Ornithol.* 5:201–17

Mihalca AD, Gherman CM, Cozma V. 2011. Coendangered hard-ticks: threatened or threatening? *Parasites Vectors* 4:71

Moir ML, Vesk PA, Brennan KEC, Keith DA, Hughes L, McCarthy MA. 2010. Current constraints and future directions in estimating coextinction. *Conserv. Biol.* 24:682–90

Moir ML, Vesk PA, Brennan KEC, Keith DA, McCarthy MA, Hughes L. 2011. Identifying and managing threatened invertebrates through assessment of coextinction risk. *Conserv. Biol.* 25:787–96

Moir ML, Vesk PA, Brennan KEC, McCarthy MA, Keith DA, et al. 2012a. A preliminary assessment of changes in plant-dwelling insects when threatened plants are translocated. *J. Insect Conserv.* 16:367–77

Moir ML, Vesk PA, Brennan KEC, Poulin R, McCarthy MA, et al. 2012b. Considering extinction of dependent species during translocation, ex situ conservation and assisted migration of threatened hosts. *Conserv. Biol.* 26:199–207

Moore J, Simberloff D. 1990. Gastrointestinal helminth communities of bobwhite quail. *Ecology* 71:344–59

Mouillot D, Krasnov BR, Shenbrot GI, Gaston KJ, Poulin R. 2006. Conservatism of host specificity in parasites. *Ecography* 29:596–602

Navarrete-Heredia JL. 2001. Beetles associated with *Atta* and *Acromyrmex* ants (Hymenoptera: Formicidae: Attini). *Trans. Am. Entomol. Soc.* 127:381–429

Nichols E, Gomez A. 2011. Conservation education needs more parasites. *Biol. Conserv.* 144:937–41

Ollerton J, Winfree R, Tarrant S. 2011. How many flowering plants are pollinated by animals? *Oikos* 120:321–26

Opler PA. 1978. Insects of American chestnut: possible importance and conservation concern. In *Proceedings of the American Chestnut Symposium*, ed. WL MacDonald, FC Cech, C Smith, pp. 83–85. Morganton: W. Va. Univ. Press

Pattemore DE, Wilcove DS. 2011. Invasive rats and recent colonist birds partially compensate for the loss of endemic New Zealand pollinators. *Proc. R. Soc. Lond. Ser. B* 279:1597–605

Petchey OL, Eklöf A, Borrvall C, Ebenman B. 2008. Trophically unique species are vulnerable to cascading extinction. *Am. Nat.* 17:568–79

Potts SG, Biesmeijer JC, Kremen C, Neumann P, Schweiger O, Kuhn WE. 2010. Global pollinator declines: trends, impacts and drivers. *Trends Ecol. Evol.* 25:345

Poulin R, Keeney DB. 2008. Host specificity under molecular and experimental scrutiny. *Trends Parasitol.* 24:24–28

Poulin R, Krasnov BR, Mouillot D. 2011. Host specificity in phylogenetic and geographic space. *Trends Parasitol.* 27:355–61

Poulin R, Krasnov BR, Shenbrot GI, Mouillot D, Khokhlova IS. 2006. Evolution of host specificity in fleas: Is it directional and irreversible? *Int. J. Parasitol.* 36:185–91

Poulin R, Morand S. 2004. *Parasite Biodiversity*. Washington, DC: Smithsonian

Powell FA. 2011. Can early loss of affiliates explain the coextinction paradox? An example from Acacia-inhabiting psyllids (Hemiptera: Psylloidea). *Biodivers. Conserv.* 20:1533–44

Powell FA, Hochuli DF, Cassis G. 2011. A new host and additional localities for the rare psyllid *Acizzia keithi* Taylor and Moir (Hemiptera: Psyllidae). *Aust. J. Entomol.* 50:441–44

Powell FA, Hochuli DF, Symonds CL, Cassis G. 2012. Are psyllids affiliated with the threatened plants *Acacia ausfeldii, A. dangarensis* and *A. gordonii* at risk of co-extinction? *Aust. Ecol.* 37:140–48

Price RD, Clayton DH, Adams RJ. 2000. Pigeon lice down under: taxonomy of Australian *Campanulotes* (Phthiraptera: Philopteridae), with a description of *C. durdeni* n. sp. *J. Parasitol.* 86:948–50

Quintero I, Roslin T. 2005. Rapid recovery of dung beetle communities following habitat fragmentation in central Amazonia. *Ecology* 86:3303–11

Reed D, Light J, Allen J, Kirchman J. 2007. Pair of lice lost or parasites regained: the evolutionary history of anthropoid primate lice. *BMC Biol.* 5:7

Rettenmeyer CW, Rettenmeyer ME, Joseph J, Berghoff SM. 2011. The largest animal association centered on one species: the army ant *Eciton burchellii* and its more than 300 associates. *Insectes Soc.* 58:281–92

Rezende EL, Lavabre JE, Guimarães PR, Jordano P, Bascompte J. 2007. Non-random coextinctions in phylogenetically structured mutualistic networks. *Nature* 448:925–28

Shaw MR. 1994. Paristoid host ranges. In *Parasitoid Community Ecology*, ed. BA Hawkins, W Sheehan, pp. 111–44. Oxford: Oxford Univ. Press

Sheldon KS, Yang S, Tewksbury JJ. 2011. Climate change and community disassembly: impacts of warming on tropical and temperate montane community structure. *Ecol. Lett.* 14:1191–200

Smith FA, Lyons SK, Ernest SKM, Jones KE, Kaufman DM, et al. 2003. Body mass of Late Quaternary mammals. *Ecology* 84:3403

Snyder N, Snyder H. 2000. *The California Condor: A Saga of Natural History and Conservation*. San Diego, CA: Academic

Sodhi NS, Koh LP, Peh KS-H, Tan HTW, Chazdon RL, et al. 2008. Correlates of extinction proneness in tropical angiosperms. *Divers. Distrib.* 14:1–10

Solé RV, Montoya M. 2001. Complexity and fragility in ecological networks. *Proc. R. Soc. Lond. Ser. B* 268:2039

Spooner DE, Xenopoulos MA, Schneider C, Woolnough DA. 2011. Coextirpation of host-affiliate relationships in rivers: the role of climate change, water withdrawal, and host-specificity. *Glob. Change Biol.* 17:1720–32

Srinivasan UT, Dunne JA, Harte J, Martinez ND. 2007. Response of complex food webs to realistic extinction sequences. *Ecology* 88:671–82

Stork NE. 2010. Re-assessing current extinction rates. *Biodivers. Conserv.* 19:357–71

Stork NE, Coddington JA, Colwell RK, Chazdon RL, Dick CW, et al. 2009. Vulnerability and resilience of tropical forest species to land-use change. *Conserv. Biol.* 23:1438–47

Stork NE, Lyal CHC. 1993. Extinction or 'co-extinction' rates? *Nature* 366:307

Swei A, Ostfeld RS, Lane RS, Briggs CJ. 2011. Impact of the experimental removal of lizards on Lyme disease risk. *Proc. R. Soc. Lond. Ser. B* 278:2970–78

Taylor GS, Moir ML. 2009. In threat of co-extinction: two new species of *Acizzia* Heslop-Harrison (Hemiptera: Psyllidae) from vulnerable species of *Acacia* and *Pultenaea*. *Zootaxa* 2249:20–32

Taylor LH, Latham SM, Mark E. 2001. Risk factors for human disease emergence. *Philos. Trans. R. Soc. Lond. Ser. B* 356:983–89

Thacker JI, Hopkins GW, Dixon AFG. 2006. Aphids and scale insects on threatened trees: Co-extinction is a minor threat. *Oryx* 40:233–36

Thébault E, Fontaine C. 2010. Stability of ecological communities and the architecture of mutualistic and trophic networks. *Science* 329:853

Tompkins DM, Clayton DH. 1999. Host resources govern the specificity of swiftlet lice: Size matters. *J. Anim. Ecol.* 68:489–500

Torchin ME, Lafferty KD, Dobson AP, McKenzie VJ, Kuris AM. 2003. Introduced species and their missing parasites. *Nature* 421:628–30

Traill LW, Lim MLM, Sodhi NS, Bradshaw CJA. 2010. Mechanisms driving change: altered species interactions and ecosystem function through global warming. *J. Anim. Ecol.* 79:937–47

Van Oosterhout C, Smith AM, Hänfling B, Ramnarine IW, Mohammed RS, Cable J. 2007. The guppy as a conservation model: implications of parasitism and inbreeding for reintroduction success. *Conserv. Biol.* 21:1573–83

Vesk PA, McCarthy MA, Moir ML. 2010. How many hosts? Modelling host breadth from field samples. *Methods Ecol. Evol.* 1:292–99

Webb SL. 1986. Potential role of passenger pigeons and other vertebrates in the rapid Holocene migrations of nut trees. *Quat. Res.* 26:367–75

Windsor DA. 1995. Equal rights for parasites. *Conserv. Biol.* 9:1–2

Wink M. 1995. Phylogeny of Old and New World vultures (Aves: Accipitridae and Cathartidae) inferred from nucleotide sequences of the mitochondrial cytochrome b gene. *Z. Naturforsch.* 50C:868–82

Wolfe ND, Dunavan CP, Diamond J. 2007. Origins of major human infectious diseases. *Nature* 447:279–83

Wood CL, Lafferty KD, Micheli F. 2010. Fishing out marine parasites? Impacts of fishing on rates of parasitism in the ocean. *Ecol. Lett.* 13:761–75

Woolhouse MEJ, Haydon DT, Antia R. 2005. Emerging pathogens: the epidemiology and evolution of species jumps. *Trends Ecol. Evol.* 20:238–44

Wyatt KB, Campos PF, Gilbert MTP, Kolokotronis SO, Hynes WH, et al. 2008. Historical mammal extinction on Christmas Island (Indian Ocean) correlates with introduced infectious disease. *PLoS ONE* 3:e3602

Yang LH, Rudolf VHW. 2010. Phenology, ontogeny and the effects of climate change on the timing of species interactions. *Ecol. Lett.* 13:1–10

Ziętara MS, Lumme J. 2002. Speciation by host switch and adaptive radiation in a fish parasite genus *Gyrodactylus* (Monogenea, Gyrodactylidae). *Evolution* 56:2445–58

Functional and Phylogenetic Approaches to Forecasting Species' Responses to Climate Change

Lauren B. Buckley and Joel G. Kingsolver

Department of Biology, University of North Carolina, Chapel Hill, North Carolina 27599;
email: buckley@bio.unc.edu, jgking@bio.unc.edu

Annu. Rev. Ecol. Evol. Syst. 2012. 43:205–26

First published online as a Review in Advance on August 28, 2012

The *Annual Review of Ecology, Evolution, and Systematics* is online at ecolsys.annualreviews.org

This article's doi:
10.1146/annurev-ecolsys-110411-160516

Keywords

phenology, phenotype, phylogeny, range shift, trait

Abstract

Shifts in phenology and distribution in response to both recent and paleontological climate changes vary markedly in both direction and extent among species. These individualistic shifts are inconsistent with common forecasting techniques based on environmental rather than biological niches. What biological details could enhance forecasts? Organismal characteristics such as thermal and hydric limits, seasonal timing and duration of the life cycle, ecological breadth and dispersal capacity, and fitness and evolutionary potential are expected to influence climate change impacts. We review statistical and mechanistic approaches for incorporating traits in predictive models as well as the potential to use phylogeny as a proxy for traits. Traits generally account for a significant but modest fraction of the variation in phenological and range shifts. Further assembly of phenotypic and phylogenetic data coupled with the development of mechanistic approaches is essential to improved forecasts of the ecological consequences of climate change.

1. INTRODUCTION

Predicting species responses to climate change is a challenge central to both maintaining biodiversity and assessing our understanding of constraints on species abundance and distribution. In response to past climate changes, species in a variety of taxa have shifted their phenology (Parmesan 2006), distribution (Davis & Shaw 2001), and abundances (Williams & Jackson 2007) in different directions and to different extents. These individualistic range shifts are inconsistent with the most common technique for predicting species' distribution responses to change—correlative species distribution models (SDMs). These models estimate a species' niche by correlating localities to environmental layers (e.g., temperature and precipitation) in order to define a climate envelope (or environmental niche). Range shifts are predicted by assuming that the species will follow its climate envelope through climate change.

Correlative SDMs require little understanding of organismal biology and are obtainable for a wide variety of organisms, but predictions are coarse and tend to get spatial details wrong (Helmuth et al. 2005, Kearney & Porter 2009, Buckley et al. 2010). Morphological and physiological traits can vary across the range of a species such that the environmental niche estimated by an SDM is an overestimate for any particular individual or population. Further, the associations between multiple abiotic variables may shift over time, leading to spurious estimates of environmental suitability.

In contrast to broad-scale modeling efforts that largely ignore biological details, small-scale experimental studies highlight how complex interactions between organismal traits and environmental conditions can be crucial to determining responses to environmental change (Jentsch et al. 2007). However, the generality and realism of these small-scale studies is uncertain. The studies tend to ignore stochastic environmental fluctuations and the complex context of actual ecosystems. Moreover, studying all species in all environmental conditions in detail is impractical.

Here, we ask what biological details are needed to more accurately predict how species will respond to climate change. Can traits such as individual and population growth rate and ecological generality predict climate change responses? How do predictor traits vary between taxa and regions? Can functional approaches successfully bridge between small-scale experimental studies where the biological details are thought to be crucial and broad-scale modeling efforts where the biological details are ignored? Can we use proxies for biology such as phylogenetic relatedness to inform predictions?

We start by reviewing aspects of the physical environment that can result in traits being central to an organism's response to environmental change. We next consider traits that may govern responses to environmental change. We evaluate whether these traits and evolutionary history can predict whether a species will move, acclimate, adapt, or go extinct in response to environmental change. We conclude by summarizing how the reviewed research should inform the next generation of models for forecasting ecological responses to environmental change.

2. PHYSICAL CONTEXT OF CLIMATE AND CLIMATE CHANGE

The challenge of addressing the particulars of organisms' interactions with the environment has led to increased attention to spatial and temporal patterns of climate and climate change. Characterizing these patterns can identify the aspects of the physical environment relevant to organisms without the uncertainty of accounting for biological details (Ackerly et al. 2010). Describing the geography of climate can address where disappearing and novel climates occur (Williams et al. 2007) and how fast and in what direction an organism would have to move to offset future climate change (Loarie et al. 2009). Climate metrics relevant to biological impacts include patterns of climate change across space and seasons and the incidence of extreme events.

2.1. Shifts in Mean Climate Conditions Vary Across Space and Seasons

The magnitude of climate change varies considerably across locations and seasons. One mid-range climate change scenario predicts that average annual global temperatures will increase by 2.45°C from 1950–1990 to 2070–2100. However, polar and temperate areas (>20°N and S) are expected to incur more pronounced changes than tropical areas (2.51°C versus 2.23°C) [based on climate scenario and data from Deutsch et al. (2008)]. Traditionally, it has been assumed that this differential warming will result in more severe impacts being experienced by polar organisms. However, the lesser climatic seasonality in the tropics may lead to greater thermal specialization. Thus, a lesser magnitude of temperature change may have a more severe impact for tropical organisms as they may easily exceed their narrow thermal safety margins (Tewksbury et al. 2008).

Climate novelty: combinations of environmental conditions, such as temperature and precipitation, without current analogs

Considering seasonal differences in climate change is also crucial for predicting biological impacts. For example, in boreal regions, winter warming is predicted to be more severe than summer warming, particularly at higher latitudes (Tebaldi et al. 2010). Seasonal climate manipulations in a subarctic peat land found that spring warming and winter snow addition had as much potential to impact plant phenotypes as did summer warming (Aerts et al. 2009). Furthermore, the frequency of cold extremes has been declining more rapidly than the frequency of heat extremes is increasing, and this trend is predicted to continue (Kharin et al. 2010). Although scenarios for future precipitation are less certain, it is likely that precipitation patterns will shift toward less frequent, more intense events (Kharin et al. 2010).

2.2. Climatic Extremes and Novelty

Most research concerning the link between environmental conditions and organismal performance and demography focuses on mean environmental conditions over time (Easterling et al. 2000, Stenseth et al. 2002). Yet, temperature variability and extreme events can substantially impact organismal stress and ultimately survival and fecundity (Helmuth et al. 2005). The incidence of extreme heat and precipitation events is expected to increase in response to climate change. Indeed, in Central Europe, heat waves have doubled in length, and the likelihood of extreme precipitation events has increased from 1.1% to 24.6% over the past century (Jentsch & Beierkuhnlein 2008).

Although extreme events are expected to substantially impact populations, communities, and ecosystems, calls are only now emerging for research to decipher the details of these impacts (Smith 2011). Paleorecords suggest that shifts in mean and extreme events will interact to produce complex biological responses (Jackson et al. 2009). We illustrate this interaction using a model of the responses of *Colias* butterflies in the Rocky Mountains to recent climate change (Buckley & Kingsolver 2012). The duration of flight, which is restricted to a narrow thermal window, is directly related to fecundity as the butterflies lay individual eggs on host plants. Although flight duration is largely a function of mean thermal conditions, egg viability is sensitive to heat extremes: Temperatures exceeding 40°C quarter egg viability (Kingsolver & Watt 1983). In a focal location in Colorado, recent increases in mean temperature are less dramatic than the increasing incidence of extremes (**Figure 1a,b**). Warm temperatures have led to decreases in estimates of available flight time, but the decline in population growth rates is more pronounced when decreases in egg viability due to extreme events are also considered (**Figure 1c,d**).

An additional challenge for predicting biological impacts is that much of Earth will experience novel climates by 2100 (Williams et al. 2007). Organisms are likely to respond to these new combinations of environmental variables such as temperature and precipitation in unexpected ways (Williams & Jackson 2007). Climate novelty presents a particular challenge to species distribution modeling techniques based on defining a multidimensional climatic niche. Assessing

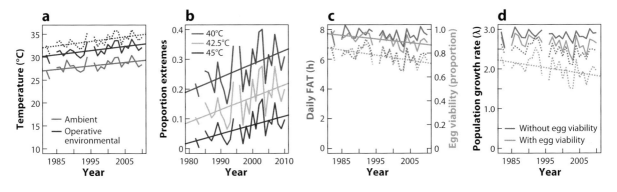

Figure 1

Responses to temperature change for two species of alpine butterflies, *Colias eriphyle* (*solid lines*) and *C. meadii* (*dashed lines*). The species display phenotypes adapted to lower and higher elevations, respectively. (*a*) Mean flight season (June 15–August 15) ambient and operative environmental temperatures have increased since 1980 in Grand Junction, CO (National Weather Service COOP Program #53489). These increases in mean conditions are less dramatic than (*b*) increases in the incidence of thermal extremes. Predicted (*c*) available flight time and (*d*) population growth rates (λ) have remained relatively constant in response to shifts in mean temperature for both species. Notably, decreases in population growth rates are predicted when impacts of extreme events on (*c*) egg viability are included. Adapted from Buckley & Kingsolver (2012). Lines are depicted for regressions significant at P < 0.05. Abbreviation: FAT, flight activity time.

potential niche locations of an organism following climate change requires extrapolation beyond the currently sampled environmental space (Veloz et al. 2012).

3. BIOLOGICAL DETERMINANTS OF CLIMATE CHANGE RESPONSES

The importance of measuring phenotypes for addressing climatic limits on species' distribution and abundance is increasingly being recognized (Helmuth et al. 2005, Gaston et al. 2009, Kingsolver 2009). Here, we review phenotypes governing responses to environmental change. How phenotypes mediate shifts in species interaction due to climate change is complex (Gilman et al. 2010), so we omit community responses despite their likely importance.

3.1. Thermal and Hydric Limits

Thermal performance curves (TPCs) describe the performance of a biological process as a function of temperature (Huey & Stevenson 1979). Processes that are frequently considered include aspects of whole-organism performance—such as rates of locomotion, growth, and feeding—and components of fitness—such as survival, reproductive rate, and generation time (**Figure 2**). The temperature dependence of performance and fitness is typically unimodal with a single optimal temperature and asymmetric such that performance declines more quickly at temperatures above rather than below the optimum (**Figure 2**). Classically, performance curves are thought to be constrained by trade-offs. For example, generalist-specialist trade-offs sometimes occur in which species or genotypes may either exhibit high performance over a narrow range of temperatures or low performance over a broad range of temperatures (Gilchrist 1995, Izem & Kingsolver 2005). However, a "hotter is better" phenomenon, in which maximum performance is greater as optimal temperature increases, has been observed for a variety of organisms (Angilletta 2009, Kingsolver 2009).

The temperature sensitivity of performance and fitness can vary considerably. For example, the thermal breadth for assimilation tends to be narrower than that for locomotion

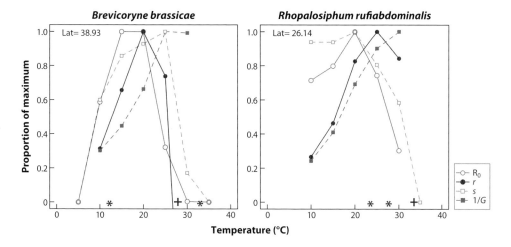

Figure 2

Components of fitness (proportion of maximum) as functions of mean rearing temperature for two aphid insect species: R_0, net reproductive rate; r, intrinsic rate of population increase; s, survival of juveniles (to adulthood or first reproduction); and $1/G$, generation time (reciprocal). Critical thermal maximum (CT_{max}) for each species is indicated as a plus sign. Mean annual temperature and the mean temperature for the hottest month at each site over the period 1960–2009 are also indicated as asterisks. Adapted from Kingsolver et al. (2011).

(Angilletta 2009). The processes often occur in sequence such that any one process can be a bottleneck for the overall energy balance of an organism. Similarly, different fitness components may differ in both thermal breadth and optimal temperature. For example, thermal breadths for survival probabilities are often wider than for net reproductive rates (**Figure 2**). Several recent data compilations have produced generalizations regarding geographic variation in TPCs. Thermal breadth tends to be narrower in the tropics, resulting in a smaller thermal safety margin (Huey et al. 2009, Sunday et al. 2010, Clusella-Trullas et al. 2011).

Thermal and water stresses can also have acute impacts on organisms. Water stress during heat extremes is predicted to result in mortality events for taxa including birds (McKechnie & Wolf 2009). A well-documented example of the interaction of acute and accumulated thermal impacts on organisms has been shown in intertidal mussels. Helmuth (2009) has developed a biophysical model for mussels that reveals the importance of the interaction between air temperature and tidal height for determining the incidence of heat mortality events. Due to this interaction, the incidence of thermal stress does not follow a consistent latitudinal gradient. Although the occurrence of thermal stress (confirmed by biophysical models and assessment of heat shock proteins) does successfully predict the distributions of mussels in some regions such as the east and west coasts of the United States, estimates of thermal stress overpredict distributions elsewhere. Historical records reveal that the southern range limit of some European intertidal species are more constrained by long-term energy budgets than by direct thermal stress (Wethey et al. 2011). How multiple environmental stressors will interact to impact organisms is an important and scarcely addressed question.

3.2. Seasonal Timing and the Life Cycle

At temperatures below the optimum, development rate declines with decreasing temperature and approaches zero at some lower developmental threshold (LDT) temperature. If the thermal sensitivity of the development rate is approximately linear within this temperature range (cf. $1/G$ in **Figure 2**), then a simple model for development time to adulthood applies: the accumulation of heat (temperature) over time above the LDT. The quantity is known as degree-days and has successfully predicted the development time required for a variety of organisms, including many plants and insects (McMaster & Wilhelm 1997). The acceleration of development, and thus phenological shifts, due to climate change has been successfully predicted for a variety of organisms [e.g., trees (Morin et al. 2009); and grasshoppers (Nufio et al. 2010)].

One complicating factor is that temperature changes have not occurred consistently throughout seasons (see Section 2). Thus, early- versus late-developing species may experience different degree-day augmentation from climate change. For Colorado grasshoppers, the predominance of late-summer warming since 1960 has led to late-season grasshoppers advancing their development more than early-season species (Nufio et al. 2010). Among both European butterflies and moths since 1980 (Altermatt 2010) and North American prairie flowers in a warming experiment (Sherry et al. 2007), early-season flyers and bloomers emerge earlier, whereas late-season flyers and bloomers emerge later in response to warming. Another complicating factor is that many organisms use photoperiodic cues for key developmental events such as diapause, hatching, and emergence. As thermal isoclines shift, the associations between temperature and photoperiod also shift. The resulting mismatches between thermal and photoperiod cues pose a major challenge to organisms (Bradshaw & Holzapfel 2010). Different developmental constraints between predators and prey can also lead to phenological mismatches with detrimental impacts to populations (Visser & Both 2005).

Life stages of a species can also differ in thermal sensitivity and sometimes in habitat requirements. This phenomenon is particularly apparent for insects (Kingsolver et al. 2011). For example, larvae of *Colias* butterflies are cryptically colored and exhibit only limited thermoregulation; optimal temperatures for larval feeding are 20–30°C and variable among species. In contrast, the wing absorptivity of adult *Colias* butterflies varies along elevation gradients and determines their ability to thermoregulate and achieve flight; optimal temperatures for flight are 33–38°C and similar among species (reviewed by Kingsolver et al. 2011). Salmon populations are tightly adapted to local conditions and face diverse selection pressures across their complex life cycles. Consequently, juvenile salmon survival is correlated with summer temperature in some populations and fall streamflow in others (Crozier et al. 2008).

Accelerated development time may also alter the number of generations organisms such as insects are able to complete in a year. For example, 44 of 263 species of butterflies have increased their annual number of generations (i.e., voltinism) and the frequency of second and subsequent generations has increased significantly for multivoltine species since 1980 (Altermatt 2009). The life stage at which organisms overwinter can influence how severely their seasonal timing is altered by climate change. For example, those insects that overwinter at more advanced stages may more readily shift their phenology in response to earlier snow melt and spring warming than those that overwinter as eggs (Altermatt 2010, Diamond et al. 2011). Longer growing seasons and an increased number of generations may augment population growth, but species may risk being exposed to more variable spring temperatures.

3.3. Ecological Breadth and Dispersal

The responses of organisms to climate change depend on their ability to evade shifts in their environmental niches via movement. A substantial number of organisms have demonstrated their ability to use movement to track recent climate shifts (e.g., Tingley et al. 2009). The dependence of niche tracking on organismal characteristics contributes to individualistic range shifts. Different dispersal abilities among similar species can explain the degree to which particular species have been impacted by recent climate change (Pearson 2006). The ability of an organism to use dispersal to alleviate thermal stress may depend on the relative timing of a dispersal-capable life stage and of climate conditions. For example, it may be the sessile adults that experience thermal stress rather than the mobile larvae, which have the potential to move to evade hot conditions. Development rate, which is tightly constrained by temperature, is a key determinant of dispersal distance for planktivorous marine larvae (O'Connor et al. 2007). An additional challenge is that distribution shifts tend to be influenced more by rare, long-distance dispersal events than by mean dispersal rates (Clark 1998), and shifts may also be impeded by habitat loss (Parmesan 2006, Loarie et al. 2009).

Diet and habitat breadth may be important determinants of the potential for phenological and distribution shifts, but the directionality of these effects is uncertain. A broader niche may facilitate the persistence of species in new environments (including a temporal shift in emergence). Species are also less likely to be constrained by the phenology of species with which they interact (Pelini et al. 2009). However, if the host or prey species of a specialist shifts, the specialist may be more likely to follow. Higher trophic levels may respond less readily to climate change due to being constrained by prey or host availability. However, mismatches in phenological shifts between interacting species are frequently observed (Visser & Both 2005). A number of readily quantified traits such as body and range size and historical locations offer only a limited functional basis for predicting climate change responses but are frequently used when attempting to explain observed responses (see below).

3.4. Fitness and Evolutionary Potential

The mean fitness (e.g., the intrinsic rate of increase, r) of a population is key to understanding the dynamics and trajectory of population change. Data for the thermal sensitivity of r (**Figure 2**) have been widely used to predict ecological responses to future climate change (Tewksbury et al. 2008, Huey et al. 2009). These studies have emphasized the negative ecological impacts of climate change for tropical insects and lizards, despite smaller predicted climate changes in tropical rather than in high-latitude regions (Deutsch et al. 2008, Tewksbury et al. 2008, Huey et al. 2009). However, because of the importance of temporal variation in weather and climate in temperate and high-latitude regions, more detailed models incorporating the thermal sensitivity of multiple fitness components may be required for realistic predictions of the ecological consequences of climate change in these regions (Kingsolver et al. 2011).

In these and similar models, ecologically important phenotypic traits are assumed to be constant and unvarying for a population. Phenotypic and genetic variation in such traits creates the potential for selection and evolutionary responses to climate change. A central issue is whether adaptive evolution can reduce the negative consequences of climate change for the mean fitness of a population, thereby reducing the likelihood of its extinction. Models for the evolution of thermally sensitive traits in response to sustained directional changes in climate yield several important insights (Huey & Kingsolver 1993, Lynch & Lande 1993, Bürger & Lynch 1995). First, there is a critical rate of environmental change beyond which the mean phenotype of the population lags too far behind the optimal phenotype, ensuring rapid extinction. Second, adaptive evolution is more likely to help maintain a population through methods of greater genetic variation, larger initial population size, shorter generation time, and higher maximum reproductive potential, and for slower and less variable rates of climate change. Third, the relationship between selection (reflecting variation in relative fitness) and mean absolute population fitness is critical: If selection is sufficiently strong to reduce mean absolute fitness below replacement, the conditions under which adaptive evolution can rescue the population from extinction are very restrictive (Holt et al.

2004). Understanding the significance of evolutionary responses to climate change for range shifts and population extinction remains a major challenge (Kingsolver 2009, Schoener 2011).

Niche conservatism: tendency for ecological traits to remain similar over time and, thus, be shared among closely related taxa

4. PREDICTING CLIMATE CHANGE RESPONSES

Species can effectively respond to climate change in three primary ways: moving in space or time to remain in a constant environmental niche, evolutionary adaptation, or phenotypic acclimation or plasticity. If these three response strategies fail, species face extinction. Can we use our knowledge of traits and phylogenetic relatedness to predict the relative importance of different responses to environmental change across different species? The potential for using species' characteristics to assess vulnerability to climate change has long been heralded, but only recently quantitatively assessed. For example, Williams et al. (2008) envisioned a vulnerability framework incorporating biotic vulnerability, exposure to climate change, potential evolutionary and acclimatory responses, and the potential efficacy of management strategies. One challenge is that it is often difficult to relate responses such as phenological shifts to fitness or population growth.

We review three approaches to ecological forecasting. First, traits have been incorporated in regressions attempting to explain recent distribution and phenological shifts. Second, traits have been used to parameterize process-based models that translate environmental conditions into energetics and demography. Third, phylogenetic relatedness may be used as a proxy for phenotypes to predict species responses. The rationale for this approach is that niche conservatism, the tendency for closely related species to share similar traits (Wiens et al. 2010), implies that closely related species will respond similarly to climate change (**Figure 3**).

4.1. Overview of Existing Studies

We searched the literature (through 2011) for studies examining the performance of traits or phylogenies as predictors of phenological or range shifts attributed to recent climate change. How numerous are such studies, and how much variation in climate change responses can they account for? We queried the Thomson Reuters (ISI) Web of Knowledge database and Google Scholar using all combinations of the following three search terms: climate change, phenology or range shift, and phylogeny or trait(s). We excluded studies with less than five species as we were interested in whether trait differences between species can predict differential climate change responses.

We summarize 11 studies of range shifts and 18 studies of phenological shifts in **Table 1** (see **Supplemental Table 1** for more detail; follow the **Supplemental Material link** from the Annual Reviews home page at **http://www.annualreviews.org**). The studies are primarily from Europe and North America, and each study includes between 9 and 566 species (mean = 167, median = 90). The average study interval is 58 years (median = 36 years). Most of the studies extracted trait information and phylogenies from published articles and atlases, suggesting that data availability is not a substantial barrier to including such analyses when reporting climate change responses. Traits were generally modeled using multivariate regressions. In cases in which a database was analyzed repeatedly, we included the study with the most thorough investigation of potential explanatory traits (e.g., Angert et al. 2011).

Additional studies have examined whether recent abundance changes can be predicted by traits and phylogeny. We excluded these studies as the link between the predictor variables and climate change responses may be more tenuous owing to other concurrent changes (e.g., land use). One interesting conclusion drawn from these studies is that the demography of shorter-lived species may be more responsive to increased climate variability associated with climate

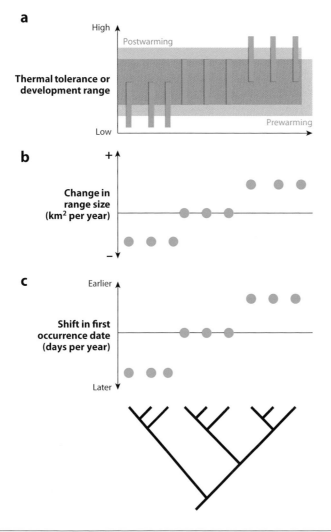

Figure 3

The conservatism of traits relevant to climate change responses may result in the biological impacts of climate change being phylogenetically biased. Climatically relevant traits (e.g., thermal tolerance and thermal limits on development, *green*) are conserved across a hypothetical phylogeny. (*a*) Shifts in environmental temperatures (from *blue* to *red shading*) may produce phylogenetic clustered responses such as range and phenological shifts. (*b*) The cool-adapted species may decrease their range size as an increased portion of habitat exceeds their thermal tolerance. The warm-adapted species may increase their range size as habitat warms to within their thermal tolerance range. (*c*) The cool-adapted species may emerge later as a greater proportion of temperatures exceed their thermal limits for development. The warm-adapted species may emerge earlier as a greater proportion of temperatures enter their thermal limits for development. After Davis et al. (2010).

change (Morris et al. 2008). Many of these studies employ long-term and broad-scale monitoring networks, particularly for European birds (e.g., Davis et al. 2010, Jiguet et al. 2010).

Our classification of utilized predictor traits (**Table 1**) does not correspond directly to the biological attributes that we present as potentially important predictors (Section 3). The discrepancies stem from the ease with which biological attributes resulting in climate sensitivity

Table 1 Studies examining the ability of phylogeny and traits to predict distribution and phenological shifts[a]

Taxa	Number of species (N)	Location	Time period	Phylogeny	↑Earlier season	↑Growth rate	↑Dispersal	↑Ecological generalization	↑Body or range size	↑Elevation or latitude	↓Trophic level	R^2	Reference
Distribution													
Plants	171	Western Europe	1905–2005			+		−					Lenoir et al. 2008
Plants (alpine)	133	Switzerland	1885–2004	1		o	o	o	o	−		0.18	Angert et al. 2011
Butterflies	51	United Kingdom	1970–1999				o	o		o			Hill et al. 2002
Butterflies	48	Finland	1992–2004	3	+	o+		×−o	+				Pöyry et al. 2009
Odonata	24	United Kingdom	1960–1995	1		o	o	+	o	o		0.24	Angert et al. 2011
Birds	254	North America	1775–2004	1		o	o	o	o−	o		0.07	Angert et al. 2011
Birds (passerine)	254	North America	1775–2004	1		o	o	+	o	o		0.07	Angert et al. 2011
Birds	55	Peru	1969–2010					o				0.03	Forero-Medina et al. 2011
Fish	90	North Sea	1977–2001		+				−			0.24[b]	Perry et al. 2005
Fish	28	North Sea	1980–2004							+			Dulvy et al. 2008
Mammals	28	Western North America	1914–2008	1		o		o	o	o−		0.33	Angert et al. 2011
Phenology													
Plants	557	United Kingdom	1954–2000		+	+	−					0.19[b]	Fitter & Fitter 2002
Plants	478	Eastern United States	1851–2007	5								NA	Willis et al. 2008
Plants	323	United Kingdom	1954–2000	5								NA	Davis et al. 2010
Butterflies/moths	566	Europe	1964–2008	4	+	+		+				0.47[b]	Altermatt 2010
Butterflies	51	United Kingdom	1976–2010	1	+	o	o	−	−	o		0.52	Diamond et al. 2011
Butterflies	17	Northwest Mediterranean	1988–2002	4	o	o	o	×					Stefanescu et al. 2003
Odonata	37	Netherlands	1995–2004	2								NA	Dingemanse & Kalkman 2008
Odonata	25	United Kingdom	1960–2004	3	+							0.31[b]	Hassall et al. 2007
Birds	307	Global	1995–2006				o						Gienapp et al. 2007
Birds	184	Europe	1960–2006	5				−				0.18	Rubolini et al. 2007
Birds	117	Eastern Hungary	1969–2007	1		o+	−	o+				0.11[b]	Vegvari et al. 2010
Birds	103	Eastern United States	1903–1993				−	×				NA	Butler 2003
Birds	100	Europe	1970–2000	1								NA	Davis et al. 2010
Birds	56	Northern Germany	1977–2006	3		o	−	o	o			0.11[b]	Rubolini et al. 2010
Birds	34	Scandinavia	1980–2004				+						Jonzén et al. 2006
Birds	18	United States	2000–2010									0.63	Hurlbert & Liang 2012
Birds	9	Northern Europe	1976–1997	3								0.64	Spottiswoode et al. 2006
Multiple	726	United Kingdom	1976–2005			o		o			+	0.06[b]	Thackeray et al. 2010
			Total +		5	6	2	4	1	1	1		
			Total o		1	12	7	9	6	6	0		
			Total –		0	0	6	3	3	2	0		

[a]The predictive ability of phylogeny is coded as follows from low to high (*light blue* to *dark blue*): 1, phylogenetic signal not significant or Pagel's $\lambda < 0.1$; 2, family or subfamily not a significant factor in a regression; 3, regression factors still significant when accounting for phylogeny; 4, family or subfamily is a significant factor in a regression; and 5, significant phylogenetic signal. The predictive ability of traits is coded as follows based on a significance level of $p < 0.05$ (from *light blue* to *dark blue*): o, no effect; ×, nondirectional difference between groups; −, slower/smaller response; +, accelerated/large response.
[b]Coefficient of variation (R^2) based on the single explanatory variable with the most explanatory power rather than the full model. Studies with NAs are those that only address phylogeny or that use nonparametric tests, and studies without R^2 values do not provide sufficient information to estimate R^2. Abbreviation: NA, not applicable.

can be quantified or grouped as a trait and the difficulty of measuring the trait. Notably, thermal tolerances and water requirements are absent from **Table 1**. Evolutionary potential is not directly assessed but is related to factors such as reproductive rate, generation time, and population size. We also include in **Table 1** some traits such as body size and historical locations that are frequently reported but whose predicted effects for responses to climate change are less clear (see below). In summary of Section 3, we expect that the following characteristics will yield more pronounced responses to climate change:

1. *Early season* species may have greater exposure to warming, which has been concentrated during spring seasons (Menzel et al. 2006), and may be selected to capitalize on an expanded

growing season (Pau et al. 2011). Species of arthropods with more advanced overwintering stages are more mobile and can respond quickly without waiting for further development (Dennis 1993).

2. *Higher individual and population growth rates* enable more rapid evolutionary and demographic responses to climate change (Perry et al. 2005).

3. *Greater dispersal* can enable more pronounced range shifts. However, this greater capacity to track climatic niches may dampen phenological shifts.

4. *Greater ecological generalization* (greater habitat or diet breadth) may release species from being constrained by the distribution or phenology of species they associate with.

5. *Larger body or range size* may correlate directly with dispersal ability, trophic level, environmental tolerance, ecological generalization, and life-cycle duration, and inversely with reproductive rates (Davies et al. 2009). Expectations based on range or body size are complicated by these correlations, but the traits may serve as informative and readily available proxies (Angert et al. 2011).

6. *Higher elevation or more poleward* species may be more responsive to climate change because they may be adapted to cooler temperatures, have experienced a greater magnitude of warming, or be more habitat restricted.

7. *Lower trophic levels* may respond more readily to climate change because they may be less constrained by prey or host availability (Thackeray et al. 2010).

4.2. Phenomenological Phenotypic Approaches

Attempts to relate traits to responses to recent climate change generally find that traits can explain a significant, but modest amount of the variation (**Table 1**). Traits tend to be better predictors of phenological shifts (mean $R^2 = 0.42$, $N = 18$) than of distribution shifts (mean $R^2 = 0.16$, $N = 11$). This may occur because phenology is typically measured as the date of first occurrence, whereas range shifts may be measured in different portions of the range (leading or trailing edge or center). Whether phenomenological approaches can account for a sufficient proportion of the variation to reliably forecast future responses and identify those species most likely to be impacted by climate change is questionable.

In a recent synthesis for multiple taxa, Angert et al. (2011) found that traits could predict a small, but often significant, percent of the variation in shifts of the leading range edge (**Table 1**). Although observed range shifts were generally consistent with their expectations (which parallel ours), models explained only a modest portion of the variation [worst for birds (4–7%) and best for mammals (22–31%)]. Few individual traits were significant predictors of range shifts. North American birds with smaller range sizes exhibited slightly but significantly larger range shifts. Range shifts among British Odonata were best, but modestly, predicted by egg habit. The duration of seed dispersal was a marginally significant predictor variable of elevation shifts in Swiss alpine plants. Small mammals in western North America from lower elevations shifted less.

Other studies have detected somewhat stronger effects of dispersal, life history, and ecological generalization [sometimes using the same data (Sorte & Thompson 2007, Holzinger et al. 2008, Moritz et al. 2008) but different metrics than Angert et al. (2011)]. Marine demersal fishes with smaller body sizes, faster maturation, and smaller sizes at maturity were more likely to shift their ranges (Perry et al. 2005). Grasses and those plant species restricted to mountain habitats experienced more pronounced range shifts in response to twentieth-century climate change (Lenoir et al. 2008). More mobile butterflies shifted their ranges further (Pöyry et al. 2009).

Attempts to predict phenological shifts have been somewhat more successful (Pau et al. 2011). Phenological shifts of butterflies are predicted, albeit sometimes weakly, by traits such as diet, generation time, overwintering stage, and dispersal ability (Stefanescu et al. 2003, Diamond et al. 2011). A larger study of phenological shifts for 566 European butterflies and moths over the past 150 years explained phenological changes using diet and life-cycle variables [with the strongest predictor variable (flight time) accounting for ~1/2 of the variation (Altermatt 2010)].

Trait-based expectations for phenology were a mix of consistent, inconsistent, and ambiguous responses compared with the observed responses to climate change (**Table 1**). The most consistent responses were that earlier season species (including those that overwinter in a more advanced stage) and species with a faster growth rate or more rapid life cycle had more accelerated or pronounced responses to climate change. Although some studies did not find a significant effect from earlier seasonality or faster growth rate, no observations reversed this expectation. Dispersal ability was a frequently utilized predictor of phenological shifts, particularly for birds. Notably, birds with smaller dispersal distances tended to exhibit more pronounced phenological shifts (**Table 1**). Long-distance migrants may be accustomed to heterogeneous climate conditions and their tropical wintering grounds may have experienced only limited warming (Rubolini et al. 2010). Additionally, they may have limited plasticity in their migration timing with which to respond to environmental variation (Pulido & Widmer 2005). These taxa-specific exceptions to common expectations highlight the challenges of generalizing phenomenological phenotypic models across taxa and regions.

4.3. Mechanistic Phenotypic Approaches

Other approaches explicitly incorporate traits through describing their influence on performance, energetic balances, and ultimately demography. Biophysical models use traits such as size and solar absorptivity to translate air temperatures (and other environmental variables such as radiation and wind speed) into body temperatures (Helmuth et al. 2005, Kearney & Porter 2009, Buckley et al. 2010). Predicted body temperatures can be compared to organisms' thermal limits to estimate activity time, energy balances, and thermal stress. Estimates of an organism's fundamental niche following climate change can be compared to its current realized niche to predict whether a species will need to move or adjust its phenotype to avoid extinction.

Ecosystem models based on plant functional types are frequently invoked to project ecosystems changes due to climate shifts (Van Bodegom et al. 2012). These models use empirically derived descriptions of the temperature dependence of processes such as survival, development, and photosynthesis to predict abundances and distributions (Moorcroft et al. 2001, Strigul et al. 2008). Although most vegetation models make fairly general predictions, models focused on particular species are emerging. For example, the PHENOFIT model incorporates thermal performance curves for numerous demographic constraints to produce estimates of a plant species' probability of persistence in a given region (Morin & Thuiller 2009). Demographic studies of forests have been used to construct probabilistic models based on individual sensitivities to the environment (Clark et al. 2012).

Analogous models for animals are less well established. As discussed above, Helmuth (2009) have used biophysical models to estimate thermal stress and energetics for intertidal organisms. Porter et al. (2000) have developed biophysical models to estimate body temperatures for a variety of terrestrial organisms. Information on the temperature dependence of development can be coupled with the biophysical model output to predict development rates and, thus, phenology (Kearney et al. 2010b). Body temperatures can be compared to thermal limits to predict activity time; knowledge of the amount of activity required to meet energetic demands can be used to estimate distributions (Kearney & Porter 2004). Recently, the biophysical models have been linked with dynamic energy budget models to produce demographic predictions based on rules of energy allocation (Kearney et al. 2010a, Kearney 2012). Activity time estimates can also be used to derive demographic predictions by examining energetic yield within an optimal foraging framework (Buckley 2008). Another approach involves documenting thermal performance curves for survival and fecundity to estimate population dynamics (Crozier & Dwyer 2006, Doak & Morris 2010). These mechanistic approaches have found that geographic trait variation can influence distributions substantially (Buckley 2008, Kolbe et al. 2010), highlighting the importance of species' traits for predicting climate change responses.

Phylogenetic signal: recognized when the traits of closely related taxa are more similar than expected under a specified model of trait evolution

Mechanistic models generally require detailed biological understanding for their parameterization. A less detailed but promising approach relies on the occurrence of niche conservatism and predictable broad-scale patterns in thermal tolerance to predict responses to climate change (Tewksbury et al. 2008; see also Section 3). For example, Huey et al. (2009) have documented that upper thermal tolerances are tightly constrained among temperate and tropical lizards. As a result, a Puerto Rican anole is predicted to have moved into the forest and to have displaced a forest anole through recent climate changes.

4.4. Phylogenetic Approaches

The conservatism of traits governing a species' environmental niche (Wiens et al. 2010) can result in responses to climate change being likewise phylogenetically conserved (Burns & Strauss 2011). If this is the case, studies of climate change impacts on a few species could be generalized to their relatives. The flip side of this is that if climate impacts (or predictor traits) are phylogenetically conserved, phylogeny must be controlled for in identifying predictor traits because the conservation of multiple traits can produce spurious correlations between particular traits and vulnerability to climate change. Phylogenetic signal can be assessed using metrics (e.g., Pagel's λ or Blomberg's K) or randomization tests (such as repeatedly swapping the tips of the phylogenetic tree). Evolutionary history can be controlled for in trait-based analyses by transforming the data into phylogenetically structured comparisons between sets of related species (independent contrasts). Another approach is to statistically account for expected covariance in trait values due to phylogenetic relatedness (phylogenetic generalized least squares, PGLS) (methods reviewed by Paradis 2006).

Of the analyses relating traits to climate change responses, half (54%) considered phylogeny (**Table 1**). Phylogeny was often controlled for in the trait regressions by using independent contrasts or PGLS. In other cases where phylogenies were unavailable, taxonomic groupings were included as an addition factor in the regression. We focus here on the several studies that examined phylogenetic signal in climate change responses directly using Pagel's λ ($\lambda = 0$, no phylogenetic signal; $\lambda = 1$, complete phylogenetic determination) or randomization tests (Davis et al. 2010). These studies suggest that climate change responses are constrained by phylogeny. Indeed, the degree to which flowering time in Thoreau's Woods (Concord, MA) tracked temperature was similar among closely related species and related to shifts in abundance (**Figure 4**) (Willis et al. 2008, Davis et al. 2010). The phylogenetic signal in flowering time tracking was confirmed for plants in Chinnor, United Kingdom (Davis et al. 2010). Flowering phenology itself tends to be phylogenetically conserved across broad geographic gradients (reviewed by Pau et al. 2011). Among European birds, population declines were phylogenetically clustered, but not fully attributable to phenological changes (Davis et al. 2010). Although the authors (Davis et al. 2010) identified traits (body size, overwintering location, and range latitude) exhibiting phylogenetic signal that might account for similar responses among related species, they did not find phylogenetic signal in phenological shifts, which is the best predictor of bird decline (Møller et al. 2008). Phenological shifts were not phylogenetically conserved among butterflies (Diamond et al. 2011) and some birds (Vegvari et al. 2010).

Less evidence exists for the phylogenetic conservatism of range shifts. This may be because range size shows only weak phylogenetic signal (Davies et al. 2009). All but one example we located that controlled for phylogeny in regressions with traits was by Angert et al. (2011). They found very low λ values (absence of significant phylogenetic signal) in their phylogenetic regressions. One other study (Pöyry et al. 2009) showed that the traits remained significant predictors of range shifts when accounting for phylogeny, suggesting that the traits may be robust predictor variables.

Do responses to climate change exhibit phylogenetic signal due to the conservation of traits governing a species' environmental niche? Analyses of phylogenetic signal in thermal tolerances offer a direct assessment of this issue. Similarities in thermal tolerance would provide a functional basis for predicting similar responses to climate shifts among closely related species. A global study of upper and lower thermal limits for a variety of terrestrial and marine taxa (reptiles, arthropods, amphibians, fish, and mollusks) found that including taxonomic nesting as a random effect significantly improved model performance (Sunday et al. 2010). More taxonomically focused analyses have enabled using phylogenies. Huey et al. (2009) found a high degree of phylogenetic signal in the field body temperatures, critical thermal minima and maxima, and optimal temperatures of 70 lizard species. The degree of niche conservatism varied between lineages. Lineages of forest-dwelling, nonbasking species have remained largely restricted to the tropics, whereas open-habitat, basking lineages have extended into temperate zones (Huey et al. 2009). Ants were found to exhibit a high degree of phylogenetic signal ($\lambda \sim 0.9$) in warming tolerance (Diamond et al. 2012). This is notable because ant species with higher critical thermal maxima were more abundant in experimental warming chambers (Diamond et al. 2012a).

Conservatism of realized climatic niches has been assessed for taxa lacking data on thermal tolerances. Phylogenetic signal has been detected in the realized climatic niches of amphibians (Hof et al. 2010) and mammals (Cooper et al. 2008). Environmental niches tend to be sufficiently conserved to justify predicting similar responses to climate change among closely related species. Yet, the potential to use phylogeny as a predictive tool has seldom been investigated. The tendency for environmental niches to be conserved also highlights the importance of controlling for phylogeny in trait-based analyses.

5. PROSPECTS FOR IMPROVED ECOLOGICAL FORECASTING

Researchers are still largely at a loss in explaining interspecific variation in responses to climate change. More pronounced biological responses do seem to correspond to exposure to greater magnitudes of climate warming (Chen et al. 2011). Attempts to explain the extent of range and phenological shifts have found that traits can account for a sometimes significant, but generally modest, proportion of the variation. In the few cases examined, climate change impacts have been observed to phylogenetically cluster. This highlights the need to control for phylogeny in searches for predictor traits. It also suggests that we may not have yet identified or appropriately quantified the most promising predictor traits given the modest success of trait-based analyses. Perhaps more likely, our often linear expectations for environment-trait interactions may be inadequate

Figure 4

Composite phylogenies depict the phylogenetic clustering of (*a*) phenological shifts since 1851 for plants from Concord, MA, and (*b*) changes in population size of European birds since 1970. Red and blue shading and dots at nodes indicate less and more responsive clades, respectively (*solid dots*, significant; *open dots*, marginally significant). Reproduced with permission from Davis et al. (2010).

a

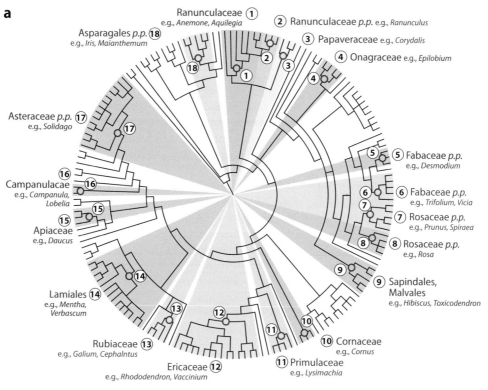

Ranunculaceae ① e.g., *Anemone, Aquilegia*

② Ranunculaceae *p.p.* e.g., *Ranunculus*

③ Papaveraceae e.g., *Corydalis*

④ Onagraceae e.g., *Epilobium*

Asparagales *p.p.* ⑱ e.g., *Iris, Maianthemum*

Asteraceae *p.p.* ⑰ e.g., *Solidago*

⑯ Campanulaceae e.g., *Campanula, Lobelia*

⑮ Apiaceae e.g., *Daucus*

Lamiales ⑭ e.g., *Mentha, Verbascum*

Rubiaceae ⑬ e.g., *Galium, Cephalntus*

Ericaceae ⑫ e.g., *Rhododendron, Vaccinium*

⑪ Primulaceae e.g., *Lysimachia*

⑩ Cornaceae e.g., *Cornus*

⑤ Fabaceae *p.p.* e.g., *Desmodium*

⑥ Fabaceae *p.p.* e.g., *Trifolium, Vicia*

⑦ Rosaceae *p.p.* e.g., *Prunus, Spiraea*

⑧ Rosaceae *p.p.* e.g., *Rosa*

⑨ Sapindales, Malvales e.g., *Hibiscus, Toxicodendron*

b

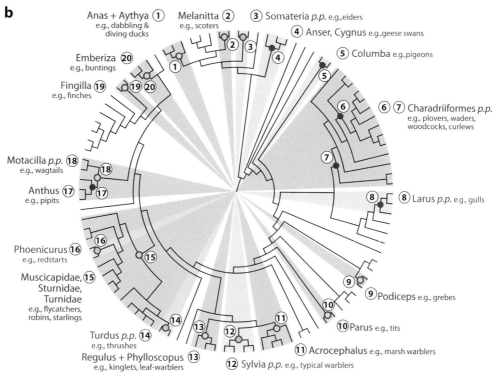

Anas + Aythya ① e.g., dabbling & diving ducks

Melanitta ② e.g., scoters

③ Somateria *p.p.* e.g.,eiders

④ Anser, Cygnus e.g.,geese swans

⑤ Columba e.g.,pigeons

⑥ ⑦ Charadriiformes *p.p.* e.g., plovers, waders, woodcocks, curlews

⑧ Larus *p.p.* e.g., gulls

⑨ Podiceps e.g., grebes

⑩ Parus e.g., tits

⑪ Acrocephalus e.g., marsh warblers

⑫ Sylvia *p.p.* e.g., typical warblers

Emberiza ⑳ e.g., buntings

Fingilla ⑲ e.g., finches

Motacilla *p.p.* ⑱ e.g., wagtails

Anthus ⑰ e.g., pipits

Phoenicurus ⑯ e.g., redstarts

Muscicapidae, ⑮ Sturnidae, Turnidae e.g., flycatchers, robins, starlings

Turdus *p.p.* ⑭ e.g., thrushes

Regulus + Phylloscopus ⑬ e.g., kinglets, leaf-warblers

to account for the complex interactions among suites of traits in heterogeneous and temporally varying environments. Additionally, behavior and acclimation can complicate environment-trait interactions. Evolution can lead to local adaptation that exceeds our coarse geographic knowledge of trait values.

The commonly implemented correlative SDMs are inherently limited in their ability to predict individualistic responses due to their assumption of constant traits across a range, fixed correlations between environmental variables, and persistent and generally simple environment-trait interactions. Extrapolating these models across time and space can be particularly problematic. Efforts to include more organismal biology (by including the output of mechanistic models as predictor variables) have met modest success (Elith et al. 2010, Buckley et al. 2011). Detailed mechanistic models for thoroughly studied species have succeeded in using the details of organismal biology to predict current distributions and climate change responses (Kearney & Porter 2009, Buckley et al. 2010), but how can these techniques be generalized to less studied species?

What aspect of the biology included in the models is responsible for their ability to predict individualistic range shifts? Applying functional and phylogenetic approaches to additional data sets may offer the most promise for addressing this question. Phylogenies and phylogenetic methods are becoming more accessible (e.g., R packages such as ape, CAIC, phylobase, phytools, and Picante; see Paradis 2006), and the assembly of trait data is accelerating (e.g., TRY initiative for plant traits). Most of the studies reviewed here used published phylogenies and trait data to investigate functional and phylogenetic predictors of climate change responses. It is thus our hope that many forthcoming reports of climate change responses will consider the traits and relatedness of species. Although initial functional and phylogenetic approaches demonstrate promise, many more case studies are needed to test our ability to predict responses to past climate change and, thus, provide a reliable basis for predicting future responses.

SUMMARY POINTS

1. Considering physical aspects of climate change beyond shifts in the mean and investigating the interaction of these physical characteristics with traits is central to forecasting impacts. Important considerations include variability in the magnitude of climate change across locations and seasons, increases in the incidences of climate extremes, and the occurrence of novel climates.

2. The complexity of an organism's life cycle along with its seasonal timing will determine its exposure to climate change. An organism's degree of ecological generalization is an important determinant of its ability to modify its timing or distribution. Thermal and hydric limits and evolutionary potential are conceptually important aspects of an organism's phenotype, but they are seldom empirically examined.

3. Traits can account for an often significant but generally modest amount of the variation in phenological and range shifts. Both traits and climate change responses tend to be phylogenetically conserved, suggesting that traits should be considered in an explicit phylogenetic framework. Further case studies and the development of more mechanistic techniques will enhance our ability to forecast the ecological consequences of climate change.

FUTURE ISSUES

1. Understanding and quantifying the phenotypic variation within and between species is becoming increasingly important in the context of climate change. Lab measurements using temperature-controlled environments are providing important insight into thermal limits and thermal constraints on development. Although basic data in constant conditions is crucial, experiments in fluctuating and stochastic environments are also important to understanding responses to climate extremes. Common garden experiments provide essential information on how locally adapted traits interact with the environment to determine performance and demography (e.g., Sexton et al. 2009). Manipulative field experiments directed at understanding responses to shifts in extremes in addition to mean conditions will provide important insight (Jentsch et al. 2007, Smith 2011).

2. Assembling phenotypic data and phylogenies is central to improved ecological forecasting. Investigating phenotypic shifts over time is also central to ecological forecasting. Museum specimens and data compilation are a valuable but largely untapped resource for examining temporal change in phenotypes (Johnson et al. 2011, Lister 2011). For example, recent studies have documented shifts in body size (Gardner et al. 2011) and coloration (linked to fitness as a function of snow cover; Karell et al. 2011) in response to recent climate change.

DISCLOSURE STATEMENT

The authors are not aware of any affiliations, memberships, funding, or financial holdings that might be perceived as affecting the objectivity of this review.

ACKNOWLEDGMENTS

This review stems from the Species Range Dynamics Working Group supported by the National Center for Ecological Analysis and Synthesis (NCEAS) (DEB-0553768) and the National Evolutionary Synthesis Center (EF-0423641). We thank participants of the working group and a related 2010 ESA Symposium for discussions that informed this review. Thanks to Amy Angert, Jonathan Davies, Sarah Diamond, Sharon Strauss, and members of our research groups for helpful comments on early versions of this review. This work was supported in part by NSF grants EF-1065638 to L.B.B., IOS-1120500 to J.G.K., and DEB-1120062 to L.B.B. and J.G.K.

LITERATURE CITED

Ackerly DD, Loarie SR, Cornwell WK, Weiss SB, Hamilton H, et al. 2010. The geography of climate change: implications for conservation biogeography. *Div. Distrib.* 16:476–87

Aerts R, Callaghan TV, Dorrepaal E, Van Logtestijn RS, Cornelissen JH. 2009. Seasonal climate manipulations result in species-specific changes in leaf nutrient levels and isotopic composition in a sub-arctic bog. *Funct. Ecol.* 23:680–88

Altermatt F. 2009. Climatic warming increases voltinism in European butterflies and moths. *Proc. R. Soc. B* 277:1281–87

Altermatt F. 2010. Tell me what you eat and I'll tell you when you fly: Diet can predict phenological changes in response to climate change. *Ecol. Lett.* 13:1475–84

Angert AL, Crozier LG, Rissler LJ, Gilman SE, Tewksbury JJ, Chunco AJ. 2011. Do species' traits predict recent shifts at expanding range edges? *Ecol. Lett.* 14:677–89

Provides a comprehensive study of life history correlates on phenological shifts in European butterflies and moths.

Synthesizes traits' and phylogeny's ability to predict leading-edge range shifts across taxa.

Angilletta MJ. 2009. *Thermal Adaptation: A Theoretical and Empirical Synthesis*. Oxford: Oxford Univ. Press

Bradshaw WE, Holzapfel CM. 2010. Light, time, and the physiology of biotic response to rapid climate change in animals. *Annu. Rev. Physiol.* 72:147–66

Buckley LB. 2008. Linking traits to energetics and population dynamics to predict lizard ranges in changing environments. *Am. Nat.* 171:E1–19

Buckley LB, Kingsolver JG. 2012. The demographic impacts of shifts in climate means and extremes on alpine butterflies. *Funct. Ecol.* In press

Buckley LB, Urban MC, Angilletta MJ, Crozier LG, Rissler LJ, Sears MW. 2010. Can mechanism inform species' distribution models? *Ecol. Lett.* 13:1041–54

Buckley LB, Waaser SA, MacLean HJ, Fox R. 2011. Does including physiology improve species distribution model predictions of responses to recent climate change? *Ecology* 92:2214–21

Bürger R, Lynch M. 1995. Evolution and extinction in a changing environment: a quantitative-genetic analysis. *Evolution* 49:151–63

Burns JH, Strauss SY. 2011. More closely related species are more ecologically similar in an experimental test. *Proc. Natl. Acad. Sci. USA* 108:5302–7

Butler CJ. 2003. The disproportionate effect of global warming on the arrival dates of short-distance migratory birds in North America. *Ibis* 145:484–95

Chen IC, Hill JK, Ohlemüller R, Roy DB, Thomas CD. 2011. Rapid range shifts of species associated with high levels of climate warming. *Science* 333:1024–26

Clark JS. 1998. Why trees migrate so fast: confronting theory with dispersal biology and the paleorecord. *Am. Nat.* 152:204–24

Clark JS, Bell DM, Kwit M, Stine A, Vierra B, Zhu K. 2012. Individual-scale inference to anticipate climate-change vulnerability of biodiversity. *Philos. Trans. R. Soc. B* 367:236–46

Clusella-Trullas S, Blackburn TM, Chown SL. 2011. Climatic predictors of temperature performance curve parameters in ectotherms imply complex responses to climate change. *Am. Nat.* 177:738–51

Cooper N, Bielby J, Thomas GH, Purvis A. 2008. Macroecology and extinction risk correlates of frogs. *Glob. Ecol. Biogeogr.* 17:211–21

Crozier L, Dwyer G. 2006. Combining population-dynamic and ecophysiological models to predict climate-induced insect range shifts. *Am. Nat.* 167:853–66

Crozier LG, Hendry AP, Lawson PW, Quinn TP, Mantua NJ, et al. 2008. Potential responses to climate change in organisms with complex life histories: evolution and plasticity in Pacific salmon. *Evol. Appl.* 1:252–70

Davies TJ, Purvis A, Gittleman JL. 2009. Quaternary climate change and the geographic ranges of mammals. *Am. Nat.* 174:297–307

Davis CC, Willis CG, Primack RB, Miller-Rushing AJ. 2010. The importance of phylogeny to the study of phenological response to global climate change. *Philos. Trans. R. Soc. B* 365:3201–13

Davis MB, Shaw RG. 2001. Range shifts and adaptive responses to Quaternary climate change. *Science* 292:673–79

Dennis RLH. 1993. *Butterflies and Climate Change*. Manchester, UK: Manchester Univ. Press

Deutsch CA, Tewksbury JJ, Huey RB, Sheldon KS, Ghalambor CK, et al. 2008. Impacts of climate warming on terrestrial ectotherms across latitude. *Proc. Natl. Acad. Sci. USA* 105:6668–72

Diamond SE, Frame AM, Martin RA, Buckley LB. 2011. Species' traits predict phenological responses to climate change in butterflies. *Ecology* 92:1005–12

Diamond SE, Nichols LM, McCoy N, Hirsch C, Pelini SL, et al. 2012a. A physiological trait-based approach to predicting the responses of species to experimental climatic warming. *Ecology*. In press

Diamond SE, Sorger DM, Hulcr J, Pelini SL, Toro ID, et al. 2012b. Who likes it hot? A global analysis of the climatic, ecological, and evolutionary determinants of warming tolerance in ants. *Glob. Change Biol.* 18:448–56

Dingemanse NJ, Kalkman VJ. 2008. Changing temperature regimes have advanced the phenology of Odonata in the Netherlands. *Ecol. Entomol.* 33:394–402

Doak DF, Morris WF. 2010. Demographic compensation and tipping points in climate-induced range shifts. *Nature* 467:959–62

Analyzes phylogenetic signal in phenological shifts of plants and birds.

Dulvy NK, Rogers SI, Jennings S, Stelzenmüller V, Dye SR, Skjoldal HR. 2008. Climate change and deepening of the North Sea fish assemblage: a biotic indicator of warming seas. *J. Appl. Ecol.* 45:1029–39

Easterling DR, Meehl GA, Parmesan C, Changnon SA, Karl TR, Mearns LO. 2000. Climate extremes: observations, modeling, and impacts. *Science* 289:2068–74

Elith J, Kearney M, Phillips S. 2010. The art of modelling range shifting species. *Methods Ecol. Evol.* 1:330–42

Fitter AH, Fitter RSR. 2002. Rapid changes in flowering time in British plants. *Science* 296:1689–91

Forero-Medina G, Terborgh J, Socolar SJ, Pimm SL. 2011. Elevational ranges of birds on a tropical montane gradient lag behind warming temperatures. *PLoS ONE* 6:e28535

Gardner JL, Peters A, Kearney MR, Joseph L, Heinsohn R. 2011. Declining body size: a third universal response to warming? *Trends Ecol. Evol.* 26:285–91

Gaston KJ, Chown SL, Calosi P, Bernardo J, Bilton DT, et al. 2009. Macrophysiology: a conceptual reunification. *Am. Nat.* 174:595–612

Gienapp P, Leimu R, Merila J. 2007. Responses to climate change in avian migration time: microevolution versus phenotypic plasticity. *Clim. Res.* 35:25–35

Gilchrist GW. 1995. Specialists and generalists in changing environments. 1. Fitness landscapes of thermal sensitivity. *Am. Nat.* 146:252–70

Gilman SE, Urban MC, Tewksbury J, Gilchrist GW, Holt RD. 2010. A framework for community interactions under climate change. *Trends Ecol. Evol.* 25:325–31

Hassall C, Thompson DJ, French GC, Harvey IF. 2007. Historical changes in the phenology of British Odonata are related to climate. *Glob. Change Biol.* 13:933–41

Helmuth B. 2009. From cells to coastlines: How can we use physiology to forecast the impacts of climate change? *J. Exp. Biol.* 212:753–60

Helmuth B, Kingsolver JG, Carrington E. 2005. Biophysics, physiological ecology, and climate change: Does mechanism matter? *Annu. Rev. Physiol.* 67:177–201

Hill JK, Thomas CD, Fox R, Telfer MG, Willis SG, et al. 2002. Responses of butterflies to twentieth century climate warming: implications for future ranges. *Proc. R. Soc. B* 269:2163–71

Hof C, Rahbek C, Araújo MB. 2010. Phylogenetic signals in the climatic niches of the world's amphibians. *Ecography* 33:242–50

Holt RD, Barfield M, Gomulkiewicz R. 2004. Temporal variation can facilitate niche evolution in harsh sink environments. *Am. Nat.* 164:187–200

Holzinger B, Hülber K, Camenisch M, Grabherr G. 2008. Changes in plant species richness over the last century in the eastern Swiss Alps: elevational gradient, bedrock effects and migration rates. *Plant Ecol.* 195:179–96

Huey RB, Deutsch CA, Tewksbury JJ, Vitt LJ, Hertz PE, et al. 2009. Why tropical forest lizards are vulnerable to climate warming. *Proc. R. Soc. B* 276:1939–48

Huey RB, Kingsolver JG. 1993. Evolution of resistance to high-temperature in ectotherms. *Am. Nat.* 142:S21–46

Huey RB, Stevenson RD. 1979. Integrating thermal physiology and ecology of ectotherms: a discussion of approaches. *Am. Zool.* 19:357–66

Hurlbert AH, Liang Z. 2012. Spatiotemporal variation in avian migration phenology: citizen science reveals effects of climate change. *PLoS ONE* 7:e31662

Izem R, Kingsolver JG. 2005. Variation in continuous reaction norms: quantifying directions of biological interest. *Am. Nat.* 166:277–89

Jackson ST, Betancourt JL, Booth RK, Gray ST. 2009. Ecology and the ratchet of events: climate variability, niche dimensions, and species distributions. *Proc. Natl. Acad. Sci. USA* 106:19685–92

Jentsch A, Beierkuhnlein C. 2008. Research frontiers in climate change: effects of extreme meteorological events on ecosystems. *C. R. Geosci.* 340:621–28

Jentsch A, Kreyling J, Beierkuhnlein C. 2007. A new generation of climate-change experiments: events, not trends. *Front. Ecol. Environ.* 5:365–74

Jiguet F, Gregory RD, DeVictor V, Green RE, Vorisek P, et al. 2010. Population trends of European common birds are predicted by characteristics of their climatic niche. *Glob. Change Biol.* 16:497–505

Johnson KG, Brooks SJ, Fenberg PB, Glover AG, James KE, et al. 2011. Climate change and biosphere response: unlocking the collections vault. *BioScience* 61:147–53

Jonzén N, Lindén A, Ergon T, Knudsen E, Vik JO, et al. 2006. Rapid advance of spring arrival dates in long-distance migratory birds. *Science* 312:1959–61

Karell P, Ahola K, Karstinen T, Valkama J, Brommer JE. 2011. Climate change drives microevolution in a wild bird. *Nat. Commun.* 2:208

Kearney M. 2012. Metabolic theory, life history and the distribution of a terrestrial ectotherm. *Funct. Ecol.* 26:167–79

Kearney M, Porter W. 2009. Mechanistic niche modelling: combining physiological and spatial data to predict species' ranges. *Ecol. Lett.* 12:334–50

Kearney M, Porter WP. 2004. Mapping the fundamental niche: physiology, climate, and the distribution of a nocturnal lizard. *Ecology* 85:3119–31

Kearney M, Simpson SJ, Raubenheimer D, Helmuth B. 2010a. Modelling the ecological niche from functional traits. *Philos. Trans. R. Soc. B* 365:3469–83

Kearney MR, Briscoe NJ, Karoly DJ, Porter WP, Norgate M, Sunnucks P. 2010b. Early emergence in a butterfly causally linked to anthropogenic warming. *Biol. Lett.* 6:674–77

Kharin VV, Zwiers FW, Zhang X, Hegerl GC. 2010. Changes in temperature and precipitation extremes in the IPCC ensemble of global coupled model simulations. *J. Clim.* 20:1419–44

Kingsolver JG. 2009. The well-temperatured biologist. *Am. Nat.* 174:755–68

Kingsolver JG, Watt WB. 1983. Thermoregulatory strategies in Colias butterflies: thermal-stress and the limits to adaptation in temporally varying environments. *Am. Nat.* 121:32–55

Kingsolver JG, Woods HA, Buckley LB, Potter KA, MacLean HJ, Higgins JK. 2011. Complex life cycles and the responses of insects to climate change. *Integ. Comp. Biol.* 51:719–32

Kolbe JJ, Kearney M, Shine R. 2010. Modeling the consequences of thermal trait variation for the cane toad invasion of Australia. *Ecol. Appl.* 20:2273–85

Lenoir J, Gegout JC, Marquet PA, De Ruffray P, Brisse H. 2008. A significant upward shift in plant species optimum elevation during the 20th century. *Science* 320:1768–71

Lister AM. 2011. Natural history collections as sources of long-term datasets. *Trends Ecol. Evol.* 26:153–54

Loarie SR, Duffy PB, Hamilton H, Asner GP, Field CB, Ackerly DD. 2009. The velocity of climate change. *Nature* 462:1052–55

Lynch M, Lande R. 1993. Evolution and extinction in response to environmental change. In *Biotic Interactions and Global Change*, ed. PM Kareiva, JG Kingsolver, RB Huey, pp. 234–50. Sunderland, MA: Sinauer Assoc.

McKechnie AE, Wolf BO. 2009. Climate change increases the likelihood of catastrophic avian mortality events during extreme heat waves. *Biol. Lett.* 6:253–56

McMaster GS, Wilhelm WW. 1997. Growing degree-days: one equation, two interpretations. *Agric. For. Meteorol.* 87:291–300

Menzel A, Sparks TH, Estrella N, Koch E, Aasa A, et al. 2006. European phenological response to climate change matches the warming pattern. *Glob. Change Biol.* 12:1969–76

Møller AP, Rubolini D, Lehikoinen E. 2008. Populations of migratory bird species that did not show a phenological response to climate change are declining. *Proc. Natl. Acad. Sci. USA* 105:16195–200

Moorcroft PR, Hurtt GC, Pacala SW. 2001. A method for scaling vegetation dynamics: the ecosystem demography model (ED). *Ecol. Monogr.* 71:557–86

Morin X, Lechowicz MJ, Augspurger C, O'Keefe J, Viner D, Chuine I. 2009. Leaf phenology in 22 North American tree species during the 21st century. *Glob. Change Biol.* 15:961–75

Morin X, Thuiller W. 2009. Comparing niche-and process-based models to reduce prediction uncertainty in species range shifts under climate change. *Ecology* 90:1301–13

Moritz C, Patton JL, Conroy CJ, Parra JL, White GC, Beissinger SR. 2008. Impact of a century of climate change on small-mammal communities in Yosemite National Park, USA. *Science* 322:261–64

Morris WF, Pfister CA, Tuljapurkar S, Haridas CV, Boggs CL, et al. 2008. Longevity can buffer plant and animal populations against changing climatic variability. *Ecology* 89:19–25

Nufio CR, McGuire CR, Bowers MD, Guralnick RP, Moen J. 2010. Grasshopper community response to climatic change: variation along an elevational gradient. *PLoS ONE* 5:1969–76

O'Connor MI, Bruno JF, Gaines SD, Halpern BS, Lester SE, et al. 2007. Temperature control of larval dispersal and the implications for marine ecology, evolution, and conservation. *Proc. Natl. Acad. Sci. USA* 104:1266–71

Paradis E. 2006. *Analysis of Phylogenetics and Evolution with R*. Berlin: Springer-Verlag

Parmesan C. 2006. Ecological and evolutionary responses to recent climate change. *Annu. Rev. Ecol. Evol. Syst.* 37:637–69

Pau S, Wolkovich EM, Cook BI, Davies TJ, Kraft NJB, et al. 2011. Predicting phenology by integrating ecology, evolution and climate science. *Glob. Change Biol.* 17:3633–43

Pearson RG. 2006. Climate change and the migration capacity of species. *Trends Ecol. Evol.* 21:111–13

Pelini SL, Dzurisin JDK, Prior KM, Williams CM, Marsico TD, et al. 2009. Translocation experiments with butterflies reveal limits to enhancement of poleward populations under climate change. *Proc. Natl. Acad. Sci. USA* 106:11160–65

Perry AL, Low PJ, Ellis JR, Reynolds JD. 2005. Climate change and distribution shifts in marine fishes. *Science* 308:1912–15

Porter WP, Budaraju S, Stewart WE, Ramankutty N. 2000. Physiology on a landscape scale: applications in ecological theory and conservation practice. *Am. Zool.* 40:1175–76

Pöyry J, Luoto M, Heikkinen RK, Kuussaari M, Saarinen K. 2009. Species traits explain recent range shifts of Finnish butterflies. *Glob. Change Biol.* 15:732–43

Pulido F, Widmer M. 2005. Are long-distance migrants constrained in their evolutionary response to environmental change? Causes of variation in the timing of autumn migration in a blackcap (*S. atricapilla*) and two garden warbler (*Sylvia borin*) populations. *Ann. N.Y. Acad. Sci.* 1046:228–41

Rubolini D, Møller AP, Rainio K, Lehikoinen E. 2007. Intraspecific consistency and geographic variability in temporal trends of spring migration phenology among European bird species. *Clim. Res.* 35:135–46

Rubolini D, Saino N, Møller AP. 2010. Migratory behaviour constrains the phenological response of birds to climate change. *Clim. Res.* 42:45–55

Schoener TW. 2011. The newest synthesis: understanding the interplay of evolutionary and ecological dynamics. *Science* 331:426–29

Sexton JP, McIntyre PJ, Angert AL, Rice KJ. 2009. Evolution and ecology of species range limits. *Annu. Rev. Ecol. Evol. Syst.* 40:415–36

Sherry RA, Zhou X, Gu S, Arnone JA, Schimel DS, et al. 2007. Divergence of reproductive phenology under climate warming. *Proc. Natl. Acad. Sci. USA* 104:198–202

Smith MD. 2011. An ecological perspective on extreme climatic events: a synthetic definition and framework to guide future research. *J. Ecol.* 99:656–63

Sorte FA, Thompson FR. 2007. Poleward shifts in winter ranges of North American birds. *Ecology* 88:1803–12

Spottiswoode CN, Tøttrup AP, Coppack T. 2006. Sexual selection predicts advancement of avian spring migration in response to climate change. *Proc. R. Soc. B* 273:3023–29

Stefanescu C, Peñuelas J, Filella I. 2003. Effects of climatic change on the phenology of butterflies in the northwest Mediterranean Basin. *Glob. Change Biol.* 9:1494–506

Stenseth NC, Mysterud A, Ottersen G, Hurrell JW, Chan KS, Lima M. 2002. Ecological effects of climate fluctuations. *Science* 297:1292–96

Strigul N, Pristinski D, Purves D, Dushoff J, Pacala S. 2008. Scaling from trees to forests: tractable macroscopic equations for forest dynamics. *Ecol. Monogr.* 78:523–45

Sunday JM, Bates AE, Dulvy NK. 2010. Global analysis of thermal tolerance and latitude in ectotherms. *Proc. R. Soc. B* 278:1823–30

Tebaldi C, Smith RL, Nychka D, Mearns LO. 2010. Quantifying uncertainty in projections of regional climate change: a Bayesian approach to the analysis of multimodel ensembles. *J. Clim.* 18:1524–40

Tewksbury JJ, Huey RB, Deutsch CA. 2008. Putting the heat on tropical animals. *Science* 320:1296–97

Thackeray SJ, Sparks TH, Frederiksen M, Burthe S, Bacon PJ, et al. 2010. Trophic level asynchrony in rates of phenological change for marine, freshwater and terrestrial environments. *Glob. Change Biol.* 16:3304–13

Tingley MW, Monahan WB, Beissinger SR, Moritz C. 2009. Birds track their Grinnellian niche through a century of climate change. *Proc. Natl. Acad. Sci. USA* 106:19637–43

Provides a framework for forecasting plant phenological shifts.

Documents phenological shifts for 726 UK taxa across environments and trophic levels.

Van Bodegom PM, Douma JC, Witte JPM, Ordoñez JC, Bartholomeus RP, Aerts R. 2012. Going beyond limitations of plant functional types when predicting global ecosystem-atmosphere fluxes: exploring the merits of traits-based approaches. *Glob. Ecol. Biogeogr.* 21:625–36

Vegvari Z, Bokony V, Barta Z, Kovacs G. 2010. Life history predicts advancement of avian spring migration in response to climate change. *Glob. Change Biol.* 16:1–11

Veloz SD, Williams JW, Blois JL, He F, Otto-Bliesner B, Liu Z. 2012. No-analog climates and shifting realized niches during the late quaternary: implications for 21st-century predictions by species distribution models. *Glob. Change Biol.* 18:1698–713

Visser ME, Both C. 2005. Shifts in phenology due to global climate change: the need for a yardstick. *Proc. R. Soc. B* 272:2561–69

Wethey DS, Woodin SA, Hilbish TJ, Jones SJ, Lima FP, Brannock PM. 2011. Response of intertidal populations to climate: effects of extreme events versus long term change. *J. Exp. Mar. Biol. Ecol.* 400:132–44

Reviews the relevance of niche conservatism for ecology and conservation.

Wiens JJ, Ackerly DD, Allen AP, Anacker B, Buckley LB, et al. 2010. Niche conservatism as an emerging principle in ecology and conservation biology. *Ecol. Lett.* 13:1310–24

Williams JW, Jackson ST. 2007. Novel climates, no-analog communities, and ecological surprises. *Front. Ecol. Environ.* 5:475–85

Projects novel climates and their relevance to individualistic responses to climate change.

Williams JW, Jackson ST, Kutzbach JE. 2007. Projected distributions of novel and disappearing climates by 2100 AD. *Proc. Natl. Acad. Sci. USA* 104:5738–42

Williams SE, Shoo LP, Isaac JL, Hoffmann AA, Langham G. 2008. Towards an integrated framework for assessing the vulnerability of species to climate change. *PLoS Biol.* 6:e325

Willis CG, Ruhfel B, Primack RB, Miller-Rushing AJ, Davis CC. 2008. Phylogenetic patterns of species loss in Thoreau's woods are driven by climate change. *Proc. Natl. Acad. Sci. USA* 105:17029–33

Rethinking Community Assembly through the Lens of Coexistence Theory

J. HilleRisLambers,[1] P.B. Adler,[2] W.S. Harpole,[3] J.M. Levine,[4] and M.M. Mayfield[5]

[1] Biology Department, University of Washington, Seattle, Washington 98195-1800; email: jhrl@u.washington.edu

[2] Department of Wildland Resources and the Ecology Center, Utah State University, Logan, Utah 84322; email: peter.adler@usu.edu

[3] Ecology, Evolution and Organismal Biology, Iowa State University, Ames, Iowa 50011; email: harpole@iastate.edu

[4] Institute of Integrative Biology, ETH Zurich, Zurich 8092, Switzerland, and Department of Ecology, Evolution, and Marine Biology, University of California, Santa Barbara, California 93106; email: jlevine@ethz.ch

[5] The University of Queensland, School of Biological Sciences, Brisbane, 4072 Queensland, Australia; email: m.mayfield@uq.edu.au

Annu. Rev. Ecol. Evol. Syst. 2012. 43:227–48

First published online as a Review in Advance on August 29, 2012

The *Annual Review of Ecology, Evolution, and Systematics* is online at ecolsys.annualreviews.org

This article's doi:
10.1146/annurev-ecolsys-110411-160411

1543-592X/12/1201-0227$20.00

Keywords

biotic filters, clustering, environmental filters, relative fitness differences, stabilizing niche differences, overdispersion

Abstract

Although research on the role of competitive interactions during community assembly began decades ago, a recent revival of interest has led to new discoveries and research opportunities. Using contemporary coexistence theory that emphasizes stabilizing niche differences and relative fitness differences, we evaluate three empirical approaches for studying community assembly. We show that experimental manipulations of the abiotic or biotic environment, assessments of trait-phylogeny-environment relationships, and investigations of frequency-dependent population growth all suggest strong influences of stabilizing niche differences and fitness differences on the outcome of plant community assembly. Nonetheless, due to the limitations of these approaches applied in isolation, we still have a poor understanding of which niche axes and which traits determine the outcome of competition and community structure. Combining current approaches represents our best chance of achieving this goal, which is fundamental to conceptual ecology and to the management of plant communities under global change.

1. INTRODUCTION

What drives the assembly of communities? By community assembly, we mean the process by which species from a regional pool colonize and interact to form local communities. Though there is still extensive debate about the details of community assembly, processes operating at a diverse range of spatiotemporal scales are thought to be important (**Figure 1**). For example, environmental drivers generate large-scale biogeographic patterns in diversity (Wiens & Donoghue 2004), whereas competitive interactions occurring in a small neighborhood contribute to local coexistence (Chesson 2000). The composition of local communities is constrained by the evolutionary history of the regional species pool (Ricklefs 2004), but also influenced on short timescales by demographic stochasticity (Tilman 2004). In short, the study of community assembly unites disciplines as diverse as evolutionary biology, biogeography, and community ecology.

Community assembly has not always been so broadly defined. From the 1970s through the 1980s, studies on community assembly primarily asked whether competitive interactions between species generated predictable patterns of species co-occurrence in communities (i.e., assembly

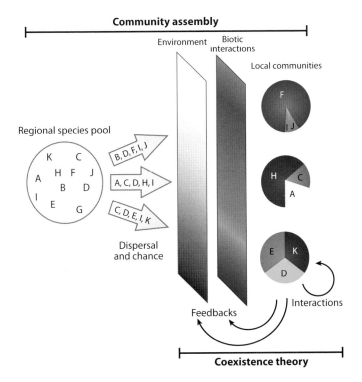

Figure 1

Community assembly is influenced by processes operating at a wide range of spatiotemporal scales. Species belong to a regional species pool that is constrained by historical processes (including evolution). A subset of the regional species pool (influenced by chance and dispersal limitation) is available for colonization of a particular site. The most common modern metaphor of community assembly then describes this subpool of species passing through an environmental (abiotic) filter and a biotic filter (e.g., Belyea & Lancaster 1999, Chase 2003, Götzenberger et al. 2012). Local communities are thus assumed to reflect the cumulative effects of these processes. In this review, we argue that contemporary coexistence theory, by highlighting the role of relative fitness differences and stabilizing niche differences, provides a more nuanced perspective on the role interactions between species and their environment (both abiotic and biotic) play during community assembly.

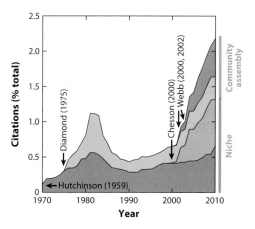

Figure 2

Relative interest in community assembly and the niche over the past 50 years, as reflected by the citation histories of influential articles about the niche (Hutchinson 1959, Chesson 2000) and community assembly (Diamond 1975, Webb 2000, Webb et al. 2002); their publication dates are indicated by the black arrows. Citation effort was standardized by the total number of articles published by journals in which these citations were published. We do not plot the overlaps because there are few (<5% of articles published since 1975 that cite Hutchinson and/or Chesson also cite Diamond and/or Webb).

rules; Diamond 1975, Weiher et al. 2011). The resurgence of interest in community assembly in the past decade, and arrival of a broader perspective, was fostered by two important developments (**Figure 2**). First, ecologists began integrating newly available phylogenetic data with community data, introducing an evolutionary perspective to community assembly (Webb 2000, Webb et al. 2002, Cavender-Bares et al. 2004a). Second, Hubbell's unified neutral theory, which explains high diversity with slow stochastic drift when species are equivalent in their competitive ability (i.e., are neutral), reinvigorated debate that the processes influencing diversity include both local and short-term mechanisms as well as regional processes occurring over longer timescales (Sale 1977, Hubbell 2001).

Recent theoretical advances in coexistence theory suggest, however, that there is still much to learn about how competitive interactions at a local scale influence community assembly (Chesson 2000) (**Figure 1**). In this review, we therefore do not focus on forces operating over longer temporal or larger spatial scales, like speciation or historical constraints to regional species pools, which are covered by other studies and reviews (e.g., Emerson & Gillespie 2008, Pavoine & Bonsall 2011), nor do we address how demographic drift or dispersal influences local communities. We do not mean to imply that these processes do not influence community assembly, but simply that they operate in addition to, not instead of, the environmental factors and biotic interactions that influence the composition and structure of communities at local scales. Contemporary coexistence theory can help link evidence from different empirical approaches, and in some cases refines our understanding of their central assumptions (Mayfield & Levine 2010). Our goal with this review is therefore to explore empirical studies relevant to community assembly through the lens of contemporary coexistence theory, and by doing so, identify important ways forward. We do so in three steps:

1. We redefine community assembly from the perspective of contemporary coexistence theory, distinguishing between stabilizing niche differences and relative fitness differences (Chesson 2000). (For expanded definitions of these terms, see sidebar, Coexistence Theory Terms Relevant to Community Assembly.)

2. Using this framework, we review studies that come from three empirical approaches: (*a*) experimental manipulations of niche differences and/or fitness differences; (*b*) quantification of relationships between community composition, traits and/or phylogenetic relatedness, and the environment; and (*c*) quantification or manipulation of frequency-dependent population growth (the signature of niche differences; see sidebar, Coexistence Theory Terms Relevant to Community Assembly).

3. Finally, we describe future directions that take advantage of developments in coexistence theory as well as the complementary strengths of the three approaches described to better understand the forces driving community assembly at local scales.

2. CONTEMPORARY COEXISTENCE THEORY AND COMMUNITY ASSEMBLY

Although the often-used metaphor of independent environmental and biotic filters is intuitively appealing (**Figure 1**), the role of the niche in determining community membership is more

COEXISTENCE THEORY TERMS RELEVANT TO COMMUNITY ASSEMBLY

- *Frequency-dependent population growth rates* occur when the per capita population growth of a species is determined by its frequency (relative abundance) within the community. Negative frequency-dependent population growth rates are the hallmark of stabilizing niche differences and arise when a focal species suppresses itself more than it does the resident species with which it competes, which can serve to "stabilize" coexistence (see definition of stable coexistence below). Negative frequency-dependent population growth can thus be used to assess whether community composition during community assembly is stabilized by niche differences. Frequency-dependent population growth rates can be measured directly by following population sizes of co-occurring species over time (Clark & McLachlan 2003) or from population dynamic models parameterized with field-based vital rates (e.g., germination, seed production) in communities where focal species differ in their frequency (Adler et al. 2006, Levine & HilleRisLambers 2009).

- *Stabilizing niche differences* are those differences that cause species to more strongly limit themselves than others through, for example, resource partitioning, host-specific natural enemies, or storage effects. When these stabilizing niche differences are greater than relative fitness differences, they foster diversity during community assembly by preventing competitive exclusion of inferior competitors by superior competitors. Stabilizing niche differences are challenging to quantify because they depend on all the interactions unique to the environment and the species composition of the community. However, they can be derived from phenomenological population dynamic models parameterized with field-based vital rates and interaction coefficients (Adler et al. 2010).

- *Relative fitness differences* are those differences between species that predict the outcome of competition in the absence of stabilizing niche differences. They have also been called fitness inequalities (Chesson 2000, Adler et al. 2007). Note that fitness is used in an ecological, not evolutionary, context—species are the unit of comparison for fitness differences in coexistence theory, not individuals (as in evolutionary studies). As with stabilizing niche differences, these fitness differences can arise through many mechanisms, including environmentally mediated differences in fecundity or differences in the ability to take up limiting resources and/or tolerate herbivores. Relative fitness differences influence the relative abundance of species (i.e., species composition) during community assembly. Similar to stabilizing niche differences, relative fitness differences depend on the specific environmental conditions and species composition unique to the community. They can be quantified by parameterizing population dynamic models with field-based estimates of vital rates (see Adler et al. 2007, 2010; Levine & HilleRisLambers 2009). Practically, relative fitness differences are difficult to disentangle from stabilizing niche differences.

- *Competitive exclusion* occurs when the presence of one competitor causes population growth rates of another to go from positive to negative, thus driving the extinction of the competitively inferior species. This occurs when stabilizing niche differences are smaller than is needed to overcome relative fitness differences. Competitive exclusion is often represented by the "biotic filter" (or the Hutchinsonian realized niche) in community assembly studies. Competitive exclusion may be observed following an experimental manipulation (by measuring population sizes of co-occurring species; e.g. Suding et al. 2005, Suttle et al. 2007) or can be inferred as having occurred by comparing trait distributions and phylogenetic relatedness in communities (e.g., Slingsby & Verboom 2006, Cornwell & Ackerly 2009).
- *Stable coexistence* refers to a community of species that stably co-occur within communities over long periods of time, with members of the community buffered from extinction. This occurs when stabilizing niche differences of species are greater than their relative fitness differences. Because stabilizing niche differences and relative fitness differences depend on both the environmental conditions and biotic interactions unique to that community, stable coexistence is sensitive to any perturbation of the environment and species composition of communities.

complicated and dynamic. For one, the niche of a species includes both its response to and impact on the abiotic and biotic environment (as emphasized by Hutchinson and Elton, respectively; Hutchinson 1957, Chase & Leibold 2003). Thus, while an environmental filter may reasonably describe how abiotic factors like climate prevent species without certain physiological traits from occurring in local communities (e.g., species without frost tolerance may not occur in alpine communities), it does not adequately describe the dynamic response to or impact of plant species on limiting resources and consumers (Tilman 1982). Second, species interactions with co-occurring competitors, consumers, mutualists, and natural enemies (the biotic environment) will not only depend on the environment, but can feed back to influence the environment (Tilman 1982). Our understanding of how these feedbacks influence the outcome of species interactions has greatly benefited from recent advances in coexistence theory (Chesson 2000).

Contemporary theory emphasizes that coexistence depends on both niche differences and fitness differences (see sidebar, Coexistence Theory Terms Relevant to Community Assembly). That niche differences are essential for long-term coexistence has long been recognized. After all, the competitive exclusion principle (see sidebar, Coexistence Theory Terms Relevant to Community Assembly), which states that no two species with the same niche can stably coexist, was first formulated by Gause in the 1930s (Gause 1934). In contemporary coexistence theory, the niche differences that underlie stable coexistence are termed stabilizing niche differences because they cause species to have higher population growth rates when the species is rare than when common, buffering them from extinction (see sidebar, Coexistence Theory Terms Relevant to Community Assembly) (Chesson 2000, Adler et al. 2007). The niche differences (and species' traits) driving this frequency-dependent population growth can arise from differences among species in their effect on and response to limiting factors like shared resources, consumers, and mutualists. These stabilizing niche differences can be extracted from mechanistic coexistence models, including those involving resource partitioning, storage effects, and density-dependent natural enemies (**Figure 3**) (Chesson 2000, Chase & Leibold 2003).

Coexistence theory illustrates, however, that not all differences between species are stabilizing niche differences. Rather, some differences drive competitive dominance and are termed relative fitness differences, as in Chesson's framework (see sidebar, Coexistence Theory Terms Relevant to Community Assembly) (Chesson 2000). For example, consider two plant species, both with growth limited by the same resources, but one has lower minimum requirements for those resources

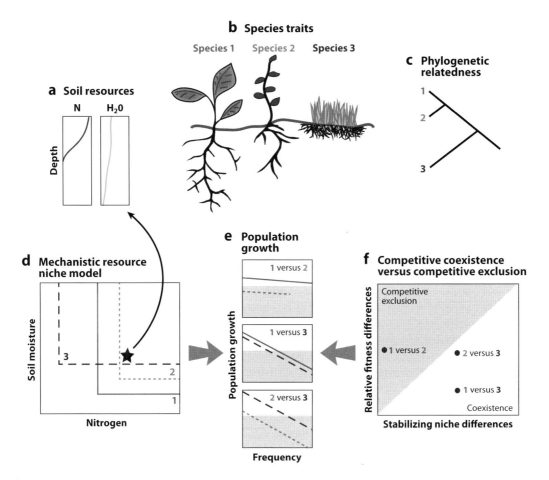

Figure 3

Relationship between the environment (*a*), species traits (*b*), phylogenetic relatedness (*c*), resource uptake (*d*), population growth rates (*e*), and coexistence (*f*). Species 1 is the superior competitor for water because its deeper roots (*b*) can access pools of soil water not available to species 2 and 3 (*a*). Species 3 is the superior competitor for nitrogen because it has greater root density at shallow soils (*b*), where nitrogen concentrations are high (*a*). Species 1 is closely related to species 2 (*c*), and therefore also more similar in rooting traits (*b*). In a mechanistic resource niche model, species 1 and 2 are similar in their resource uptake for the two resources, and differ from species 3 [in *d*, *lines* indicate the lowest levels at which each species can survive; see Tilman (1982), Chase & Leibold (2003)]. Species 1 and 3 or 2 and 3 can coexist at resource supply rates indicated by the star, but species 1 competitively excludes 2 because their resource uptake patterns are too similar (*d*). This is reflected in frequency-dependent population growth rates of each species pair (in *e*, *gray* represents population growth rates below zero), and the balance between stabilizing niche differences and the magnitude of fitness differences (in *f*, *gray* indicates competitive exclusion of one member of the species pair; *white* indicates coexistence; see Adler et al. (2007)].

(**Figure 3**). In theoretical terms, these species differ in their R^* for those resources (Tilman 1982, Chase & Leibold 2003), and one species will competitively exclude the other. Such relative fitness differences are what stabilizing niche differences must overcome to generate coexistence (Chesson 2000), and they can arise from some of the same traits and limiting factors that drive stabilizing niche differences (**Figure 3**).

The main message of Chesson's framework is that stabilizing niche differences facilitate coexistence, whereas relative fitness differences drive competitive exclusion (**Figure 3**) (see sidebar, Coexistence Theory Terms Relevant to Community Assembly). But how is this relevant to community assembly? First, competitive exclusion can either preferentially eliminate taxa that are too functionally similar when trait differences function as stabilizing niche differences or preferentially eliminate all taxa that do not possess the near optimal trait when such trait differences translate into fitness differences. Second, both stabilizing niche differences and relative fitness differences are influenced by abiotic and biotic factors. For both reasons, patterns of trait dissimilarity or similarity cannot easily be used to infer the relative importance of environmental versus biotic (competitive) filters, which is an important goal of community assembly studies (**Figure 1**). Third, frequency-dependent population growth rates are the signature of all stabilizing niche differences (Adler et al. 2007) and can thus be used to infer niche differences, but frequency-dependence alone cannot identify the mechanisms allowing for stable coexistence, which is another central goal of community assembly studies. Armed with these insights from contemporary coexistence theory, we next review three empirical approaches to understanding the role of competitive interactions and niches during community assembly.

3. A REVIEW OF EMPIRICAL APPROACHES

The literature exploring how species differences relate to the outcome of competitive interactions is enormous; for this reason, we do not review all empirical approaches used to test the niche concept or its various definitions. Instead, we focus on three approaches that provide relevant and contemporary (but not always complete) insights into our understanding of the interplay between stabilizing niche and relative fitness differences during community assembly. First, we review experimental manipulations of stabilizing niche differences and relative fitness differences, studies designed to directly test specific coexistence mechanisms (e.g., Turnbull et al. 1999). Second, we review studies relating trait and phylogenetic distributions within communities to the environment; these are the studies that are driving the recent resurgence of interest in community assembly (**Figure 2**) (Webb et al. 2002, Ackerly & Cornwell 2007). Finally we review studies that derive stabilizing niche differences and relative fitness differences from demographic data, studies that were directly inspired by recent breakthroughs in coexistence theory (Chesson 2000, Adler et al. 2007). Although community assembly processes are clearly relevant to all organisms, we primarily review literature associated with plant communities because this reflects our expertise and the predominant focus of recent studies. However, we emphasize that the conceptual topics reviewed here are relevant to all communities of interacting and competing organisms (e.g., Helmus et al. 2007, Horner-Devine et al. 2007, Chase 2010, Fukami et al. 2010).

3.1. Experimental Manipulations

To understand how community assembly is regulated by stabilizing niche differences and relative fitness differences, one can experimentally manipulate the environmental and biotic factors that control these forces (e.g., through nitrogen addition, herbivore removal, or competitor removal). On the plus side, such experimental manipulations can provide strong evidence that specific environmental factors or biotic interactions are critical for community assembly. However, a major challenge is the sheer amount of information needed to rigorously relate the results of such treatments to coexistence theory. For example, relating the impacts of nitrogen addition to predictions from a mechanistic resource niche model requires information on the differential abilities of species to compete for nitrogen and an understanding of what drives stabilizing niche differences (in a simple two-resource model, for example, the identity of the other limiting resource and the trade-offs involved). Perhaps this explains why, despite the large number of experiments manipulating niche axes or competitive interactions within plant communities, surprisingly few explicitly test coexistence theory (Siepielski & McPeek 2010).

Virtually every experiment in the history of community ecology manipulates relative fitness and stabilizing niche differences whether the investigators intend to or not. We thus narrow our review to those studies intentionally aiming to manipulate these determinants of coexistence. We highlight the following three types of studies: those that manipulate factors thought to control relative fitness differences (see Section 3.1.1), those that reduce stabilizing niche differences (see Section 3.1.2), and those that manipulate aspects of community assembly (see Section 3.1.3). We identify the circumstances under which each type of experiment can be linked to specific coexistence mechanisms and, thus, highlight the relative importance of fitness differences and stabilizing niche differences for community assembly.

3.1.1. Experimental manipulations of relative fitness differences.

Direct manipulation of limiting factors can reveal their role in generating relative fitness differences among co-occurring species. For example, resource competition models predict that the composition of plant communities limited by two resources depends on the relationship between resource uptake and the supply rate of those resources (**Figure 3**). If so, an increase in the supply rate of one of those resources should predictably lead to an increase in the abundance of the species that is most limited by that resource (Wedin & Tilman 1993). Similarly, community responses to experimental manipulations of rainfall (Suttle et al. 2007), consumers (HilleRisLambers et al. 2010), and pathogens (Allan et al. 2010) can provide information on the factors driving relative fitness differences and the species traits that underlie them.

For example, nitrogen (N) is the resource most often manipulated in terrestrial resource addition experiments because it often limits productivity, and the impacts of N deposition on ecosystem and community dynamics are of major concern in many ecosystems (Suding et al. 2005 and references therein; Clark et al. 2007 and references therein). N addition experiments generally lead to changes in species composition that are broadly consistent with predictions from resource competition theory. For example, N-fixing forbs and C4 grasses (considered good N competitors) generally decreased in abundance with N addition, whereas C3 grasses (less competitive for N) increased in abundance (Suding et al. 2005). A recent study provides even stronger mechanistic links: It demonstrates that changes in abundance following N addition are predictable from species-specific indices of competitive ability for N (R^* for N), with good competitors for N losing ground to species that are worse competitors for this limiting resource (Harpole & Tilman 2006).

However, these experiments also illustrate the complexity of experiments aiming to manipulate relative fitness differences. N addition is often accompanied by diversity loss, suggesting that the magnitude of stabilizing niche differences is also influenced by the addition of N (Clark et al. 2007). The mechanism for this is uncertain; one hypothesis holds that nutrient addition shifts the limiting resource to light, a resource for which relative fitness differences are more asymmetric (Hautier et al. 2009). Consistent with this hypothesis, Hautier et al. (2009) found that the addition of light reversed the decline in species diversity seen with fertilization alone. However, N addition can also reduce resource heterogeneity (Harpole & Tilman 2007), potentially affecting stabilizing niche differences and, thus, influencing diversity. In all, these results suggest that the traits that influence relative fitness differences can also influence stabilizing niche differences (e.g., resource uptake), complicating inference.

3.1.2. Experimental reductions in stabilizing niche differences.

A number of experimental studies aim to manipulate stabilizing niche differences by removing the environmental or biotic variables that provide species with their competitive advantages when rare. These manipulations can therefore take a wide range of forms, depending on the factors underlying stabilizing niche differences. For example, Turnbull et al. (1999) tested whether a competition-colonization trade-off maintained plant diversity in limestone grassland in England by experimentally eliminating colonization limitation. They found that even when seeds of large-seeded species were added in high numbers, high diversity remained, suggesting that the competition-colonization trade-off was not necessary for coexistence (Turnbull et al. 1999). By contrast, Dornbush & Wilsey (2010) reduced soil depth in an effort to eliminate resource partitioning and did find a decrease in species richness, suggesting impacts on stabilizing niche differences. Similarly, Carson & Root (2000) and Allan et al. (2010) excluded insects and foliar fungi from plant communities to remove the frequency-dependent advantage host-specific natural enemies might provide, and they also found reductions in plant diversity.

Just as with experimental studies aiming to manipulate relative fitness differences (see Section 3.1.1 above), a complicating factor is that it is almost impossible to manipulate a stabilizing niche difference without also affecting relative fitness differences. For example, the loss of diversity following the addition of multiple limiting resources to grasslands can result from the increased fitness differences that accompany competition for light (Dybzinski & Tilman 2007, Hautier et al. 2009), the reduction in the number of limiting resources (i.e., niche dimension; Harpole & Tilman 2007), or the removal of spatial heterogeneity in the limiting resources (a source of stabilizing niche differences; Tilman 1982). Similarly, forcing plants into a shallow rooting zone may eliminate niche differences, but will also alter the competitive balance between species by favoring shallow-rooted species (Dornbush & Wilsey 2010). Finally, eliminating natural enemies (Carson & Root 2000, Allan et al. 2010) may reduce the stabilizing effects of host-specific natural enemies, but will also alter frequency-independent performance of co-occurring species. It can therefore be difficult to distinguish which effect dominates using experimental manipulations of limiting factors alone.

3.1.3. Invasion/assembly experiments. A complementary approach to perturbing stabilizing niche differences or relative fitness differences is to experimentally manipulate the community assembly process itself. For example, several studies have documented a negative relationship between the relative abundance of a functional group in established communities and the probability that a new member of that functional group can invade (Fargione et al. 2003, Mwangi et al. 2007, Roscher et al. 2009, Hooper & Dukes 2010, Petermann et al. 2010). Specifically, Fargione et al. (2003) and Hooper & Dukes (2010) showed that invasion success was greatest when phenological differences between the invader and resident species were maximized. These results suggest that competitive interactions are more intense with other members of the same functional group, which is consistent with functional group differences translating into stabilizing niche differences (e.g., Burns & Strauss 2011). Other studies have documented a negative relationship between the functional diversity of resident species and invasibility (Levine 2000, Kennedy et al. 2002, Fargione et al. 2003). These experiments support the idea that more diverse systems should leave less vacant niche space for colonization, which also supports a role for stabilizing niche differences. Finally, the convergence in functional composition of experimentally assembled communities, despite dissimilar initial compositional starting points, suggests that strong relative fitness differences, rather than stochasticity, drive community composition toward species with similar functional traits (Seabloom et al. 2003, Fukami et al. 2005, but see Koerner et al. 2008). In all, invasion and experimental assembly experiments can provide strong but indirect evidence of stabilizing niche differences and relative fitness differences.

3.2. Trait-Phylogeny-Environment Relationships

A second approach to studying community assembly is to relate observed patterns of species presence/absence or abundance in communities to null expectations. Most of these approaches

a Traits and phylogenetic relatedness clustered   **b** Traits and phylogenetic relatedness overdispersed

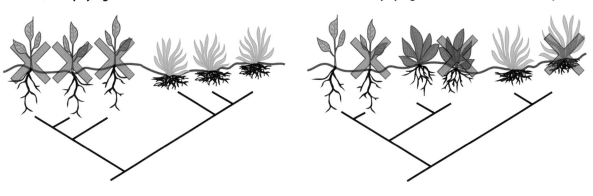

Figure 4

An example of (*a*) trait (and phylogenetic) clustering and (*b*) overdispersion, where species excluded from local sites (but present in the regional species pool) are marked by a red X. In panel *a*, closely related species with high root densities co-occur in communities, leading to trait and phylogenetic clustering, whereas in panel *b*, species competitively exclude closely related species with similar traits, leading to trait and phylogenetic overdispersion (see Mayfield & Levine, 2010).

are descendants of Diamond's assembly rules and earlier studies that used genus-to-species ratios in communities to test the competition relatedness hypothesis (**Figure 2**) (Diamond 1975, Webb 2000, Cavender-Bares et al. 2004b, Ackerly & Cornwell 2007). In the past decade, these approaches have been expanded in two important ways. First, the development of statistical methods to compare trait distributions within and between communities has increased our understanding of how traits are distributed among interacting species at various spatial scales (Ackerly & Cornwell 2007); and second, the insight that phylogenetic relatedness can be used as a proxy for trait similarity has expanded our ability to understand these interactions, even in the absence of trait data (Webb 2000, Webb et al. 2002, Cavender-Bares et al. 2004a). Contemporary studies using these approaches now relate trait distributions (see Section 3.2.1), sometimes with phylogenetic information (see Section 3.2.2) to null expectations to determine the overall importance of environmental factors and biotic filters for community assembly (**Figure 1**). Studies that incorporate trait, phylogenetic, and environmental data can further identify links between function and process (Mayfield et al. 2009, Pavoine et al. 2011).

There are clear advantages to trait- and/or phylogeny-based studies. First, community composition reflects the cumulative effects of stabilizing niche differences and relative fitness differences (as well as long-term evolutionary processes not considered here), as mediated by the environment and species interactions, during community assembly (**Figure 1**). Second, temporal replication or logistically difficult treatments are not required, and community composition data, trait data, and phylogenetic data are already available for many plant communities (e.g., Rees et al. 2001, Kattge et al. 2011). However, there are also disadvantages to these approaches. Contemporary coexistence theory demonstrates that large fitness differences and competitive exclusion can lead to trait clustering (**Figure 4**) (Mayfield & Levine 2010), calling into question the assumption that trait (or phylogenetic) clustering is solely the outcome of environmental filters. The interpretation of phylogenetic overdispersion as reflecting limiting similarity is also complicated because this pattern may reflect a lack of stabilizing niche differences between closely related species (if traits are conserved) or environmental filtering of species with similar traits (if traits are convergent; Cavender-Bares et al. 2004b, 2009). Finally, phylogenetic or trait distribution patterns that do not deviate from null expectations are difficult to interpret as these could reflect a combination or cancelling out of environmental filters, relative fitness differences, or stabilizing niche differences (Mayfield et al. 2005). Despite these disadvantages, advanced statistical techniques are continually improving our ability to infer processes from patterns (Pillar & Duarte 2010, Chase & Myers 2011, Ives & Helmus 2011, Pavoine et al. 2011).

3.2.1. Distributions of traits among and within communities. Many studies incorporating environmental data into their analyses find evidence of trait clustering (underdispersion), suggesting that local environments can exclude species without the necessary physiological trait values (**Table 1**) (Cavender-Bares et al. 2004a, Kraft et al. 2008, Cornwell & Ackerly 2009, Swenson & Enquist 2009, Kluge & Kessler 2011). For example, oak species with fire-resistant traits (thick bark, resprouting) occurred in fire-prone scrub and sandhill communities, whereas species with lower specific leaf area (SLA) and smaller leaves (correlated with drought tolerance) occurred on dry ridgetops in Amazonian forests (Cavender-Bares et al. 2004b, Kraft et al. 2008). In cases like these, where there is a clear relationship between an environmental gradient and traits, concluding that an environmental filter is important for assembly is reasonable. However, in the absence of such clear and measured environment/trait relationships, distinguishing environmental filters from competitive differences (i.e., large relative fitness differences) can be difficult (**Figure 4**). For example, are species with high requirements for N absent from infertile soils—assuming no dispersal limitation—because they cannot tolerate such low resource conditions or because they have been competitively excluded? Either scenario could lead to communities with species that have narrower trait distributions than expected (e.g., in rooting depth).

Trait-based studies have found evidence of overdispersion or limiting similarity that can be linked to stabilizing niche differences (**Table 1**). For example, several studies have found that seed size is overdispersed within local communities (Kraft et al. 2008, Cornwell & Ackerly 2009, Swenson & Enquist 2009), a finding consistent with the operation of competition-colonization trade-offs (Tilman 1994). Kursar et al. (2009) found that *Inga* species co-occurring within local communities differed more than expected in their defensive chemistry, suggesting that specialized herbivores may contribute to stabilizing niche differences between *Inga* species (Kursar et al. 2009). In other cases, trait overdispersion is more difficult to interpret. In part, this is because specific coexistence mechanisms are difficult to infer from trait differences assumed to reflect stabilizing niche differences. For example, SLA values in central Californian plant communities were more evenly spaced than expected (Cornwell & Ackerly 2009), potentially reflecting competitive interactions between species with similar SLA values. However, because SLA can reflect differences among species in water use efficiency, competitive abilities for N (Suding et al. 2005, Angert et al. 2009), or phenotypic plasticity, the exact mechanism responsible is difficult to determine.

3.2.2. Phylogenetic approaches. Traditionally, phylogenetic clustering and overdispersion are interpreted as manifestations of environmental and biotic filters, respectively, with phylogenetic relatedness assumed to be a proxy for trait and niche similarity (**Figure 4**) (Webb 2000). With contemporary coexistence theory, we can reinterpret these patterns as reflecting relative fitness differences or environmental filters in the case of phylogenetic clustering (underdispersion), and stabilizing niche differences in the case of phylogenetic overdispersion (Mayfield & Levine 2010). Critically, interpretations of these phylogenetic patterns rest on the assumption that niches are conserved (Cavender-Bares et al. 2009).

Just as in purely trait-based studies, phylogenetic studies provide evidence of both clustering (Webb 2000, Cavender-Bares et al. 2004a, Kembel & Hubbell 2006, Kraft & Ackerly 2010, Anderson et al. 2011, Fine & Kembel 2011) and overdispersion (Kembel & Hubbell 2006, Slingsby & Verboom 2006, Swenson et al. 2007, Letcher 2010, Anderson et al. 2011). Studies that combine

Table 1 Examples of studies documenting trait and/or phylogenetic "overdispersion" and "underdispersion" (clustering) in plant communities

Type of study	Habitat	Dispersion (o, over; u, under)	Source
Traits	California grasslands[a]	o, u	Ackerly & Cornwell 2007, Cornwell & Ackerly 2009
	Spanish pastures[a]	o, u	de Bello et al. 2009
	Amazonian palms	u	Anderson et al. 2011
	North American trees	u	Swenson & Weiser 2010
	Australian subtropical forests	u	Kooyman et al. 2010
	Amazonian tropical forests[a]	o, u	Kraft et al. 2008
	Tropical successional communities (Mexico)	u	Lebrija-Trejos et al. 2010
	Tropical cloud forests (China)	u	Long et al. 2011
	Tropical rain forest (French Guyana)	o, u	Paine et al. 2011
	Neotropical dry forest[a]	o, u	Swenson & Enquist 2009
	Costa Rican nontree communities	u	Mayfield et al. 2005
Phylogeny	Disturbed old fields (Canada)	u	Dinnage 2009
	California plant communities[a]	o, u	Cadotte et al. 2010
	Brazilian cerrado	u	Silva & Batalha 2009
	Bornean rainforest	u	Webb 2000
	Amazonian forests[a] (Peru)	o, u	Fine & Kembel 2011
	Neotropical forests[a] (Panama)	o, u	Kembel & Hubbell 2006
	Subtropical forests[a] (China)	o, u	Pei et al. 2011
	Tropical forests (Panama, Puerto Rico, Costa Rica)	o, u	Swenson et al. 2007
	Costa Rican secondary forests	o	Letcher 2010
Traits and phylogeny	Serengeti grasslands[a]	o, u	Anderson et al. 2011
	Algerian Xeric communities	u	Pavoine et al. 2011
	Cape Floristic region	o	Slingsby & Verboom 2006
	Dutch plant communities[a]	o, u	Prinzing et al. 2008
	Mediterranean communities[a] (Spain)	o, u	Verdu & Pausas 2007, Ojeda et al. 2010
	Minnesotan oak savannahs	u	Willis et al. 2010
	Floridian oaks[a]	o, u	Cavender-Bares et al. 2004a,b
	Floridian forests	u	Cavender-Bares et al. 2006
	Tropical lnga trees	o, u	Kursar et al. 2009
	Tropical fern communities[a] (Costa Rica)	o, u	Kluge & Kessler 2011
	Amazonian forests[a] (Ecuador)	o, u	Kraft & Ackerly 2010
	Costa Rican nontree communities	u	Mayfield et al. 2009

[a]Opposite patterns at different spatial scales, in different habitats, for different groups of species, or for different traits.

phylogenetic and trait analyses provide for a more nuanced view of assembly processes because they can test for niche conservatism versus convergence and because information on traits can help generate hypotheses as to the fitness differences and niche differences that lead to clustering or overdispersion (Cavender-Bares et al. 2004a, Mayfield et al. 2009, Kraft & Ackerly 2010). For example, phylogenetic clustering in Spanish Mediterranean communities has been linked to fire-resistance traits and, presumably, reflects differences in fire regimes across habitats (Ojeda et al. 2010), whereas phylogenetic and trait clustering in Amazonian forests was found to be related to growth traits (e.g., wood density), which likely reflects differences in the resource acquisition strategies best suited for ridgetop versus valley bottoms (Kraft et al. 2008). Similarly, phylogenetic overdispersion in co-occurring Floridian oaks likely reflects environmental filtering because these species share moisture preferences (i.e., convergent evolution), whereas phylogenetic overdispersion in co-occurring South African Sedges paired with niche conservatism in functional traits (e.g., leaf height, leaf area) is more consistent with competitive exclusion (Cavender-Bares et al. 2004b, Slingsby & Verboom 2006).

3.3. Demographic Analyses

A third research approach looks for signatures of niche differences not in the composition of communities, but rather in the demographic rates of component species. This approach relies on the premise that all stabilizing niche differences influence coexistence by causing species to have greater per capita growth rates when they are rare versus common (Chesson 2000, Adler et al. 2007). A major advantage of demographic approaches is that they do not require an a priori understanding of the environmental variables, biotic interactions, and species traits that collectively determine how stabilizing niche differences and relative fitness differences combine to determine community structure. Of course, the phenomenological nature of this approach is also its disadvantage because it prevents determination of the particular mechanisms operating that influence coexistence. Moreover, extensive demographic information is needed to quantify frequency-dependent population growth rates (rather than individual fitness components), information that is not always available (see sidebar, Coexistence Theory Terms Relevant to Community Assembly). We therefore first review studies that document negative frequency-dependent performance in plant communities, some of which are able to provide indirect evidence for the underlying mechanisms responsible (see Section 3.3.1). Because the importance of such stabilizing niche differences for coexistence also depends on the magnitude of relative fitness differences, we also review studies that quantify (or manipulate) the importance of both of these factors (see Section 3.3.2).

3.3.1. Frequency-dependent performance (the signature of stabilizing niche differences).
Numerous studies have documented frequency- or density-dependent performance in diverse communities, including Mediterranean grasslands, temperate woodlands, and tropical forests (**Table 2**) (Wills et al. 1997, Webb & Peart 1999, Harms et al. 2000, HilleRisLambers et al. 2002, Webb et al. 2006, Harpole & Suding 2007, Yamazaki et al. 2009, Clark 2010, Comita et al. 2010, Metz et al. 2010). For example, seedling germination and survival for many tree species are lower in high conspecific-density neighborhoods than in locations where species are rare (e.g., Webb & Peart 1999, HilleRisLambers et al. 2002, Comita et al. 2010). These studies assume that strong (negative) density-dependent effects mediated by conspecific densities translate to population growth advantages when species are rare. Inspired by the Janzen-Connell hypothesis, many of these studies also assume that host-specific natural enemies are responsible for these patterns (Janzen 1970, Connell 1971). Modern coexistence theory, however, illustrates that other stabilizing niche differences (e.g., resource partitioning) also result in frequency-dependent plant

Table 2 Examples of demographic studies documenting or quantifying frequency-dependent performance, the signature of stabilizing niche differences

Type of study	Habitat	Source(s)
Frequency-dependent performance	Tropical rainforests	Wills et al. 1997, Webb & Peart 1999, Harms et al. 2000, Hubbell et al. 2001, Peters 2003, Webb et al. 2006, Comita et al. 2010, Metz et al. 2010, Kobe & Vriesendorp 2011
	Temperate deciduous forests	HilleRisLambers et al. 2002, Clark 2010
	California annual grasslands	Harpole & Suding 2007
Frequency-dependent performance; mechanism	Prairie/climate variability	Adler et al. 2006
	Temperate forests/natural enemies	Yamazaki et al. 2009
	Temperate grasslands/soil microbiota	Klironomos 2002, Petermann et al. 2008
	Tropical forests/natural enemies	Webb et al. 2006, Mangan et al. 2010, Metz et al. 2010, Swamy & Terborgh 2010
	Tropical forests/allelochemicals	McCarthy-Neumann & Kobe 2010
Frequency-dependent performance + fitness difference	Serpentine grasslands	Levine & HilleRisLambers 2009
	Sagebrush steppe	Adler et al. 2010

performance. Regardless of mechanism, these studies are collectively consistent with the idea that niche differences stabilize community structure in many plant communities.

Which mechanisms are responsible for these stabilizing niche differences? A limited number of studies have provided insights into the processes responsible (**Table 2**). Much of the work has focused on host-specific pathogens or natural enemies. Greenhouse and field studies, for example, find evidence of microbially mediated plant-soil feedbacks that generate advantages when rare (Bever 1994, Klironomos 2002, Petermann et al. 2008, Mangan et al. 2010). Careful documentation of the agents of mortality in one study also pointed to the importance of fungal pathogens (Yamazaki et al. 2009). By contrast, McCarthy-Neumann & Kobe (2010) found that density dependence was mediated by soil chemistry; and Adler et al. (2006, 2009) combined demographic analyses with simulation modeling to demonstrate that a "storage effect" (likely mediated through climate) operated to foster diversity in a prairie community but not in a sagebrush steppe (Adler et al. 2006, 2009; McCarthy-Neumann & Kobe 2010).

3.3.2. Comparing stabilizing niche differences with fitness differences. Although the studies highlighted in the previous paragraphs demonstrate the pervasive effects of stabilizing niche differences as manifested through negative frequency-dependent population growth rates, a full understanding of the importance of niche differences also hinges on quantifying relative fitness differences (Chesson 2000, Adler et al. 2007). For example, several studies find large frequency-independent increases in seedling performance when natural enemies, which in some cases are larger than frequency-dependent effects, are excluded (McCarthy-Neumann & Kobe 2010, Swamy & Terborgh 2010, Kobe & Vriesendorp 2011). This result suggests that other agents of mortality (e.g., light, allelochemicals, generalist natural enemies) influence relative fitness differences, even while specialized natural enemies may influence stabilizing niche differences. Do the stabilizing effects of host-specific natural enemies, if present, outweigh relative fitness differences (from impacts of generalist natural enemies as well as other factors) and result in stable coexistence? Demographic analyses incorporating frequency-dependent and -independent performance at all life-history stages and of all individuals (e.g., Clark 2010) may be able to answer this question.

Unfortunately, few studies (to our knowledge) have attempted to quantify both the stabilizing effects of niche differences and relative fitness differences because of the wealth of data needed. However, an experiment with serpentine annuals suggests that the stabilizing effects of niche differences are required to overcome large fitness differences (Levine & HilleRisLambers 2009), whereas a study of sagebrush coexisting with three perennial bunchgrasses documented stabilizing effects of niche differences that were much greater than required to overcome fitness differences (Adler et al. 2010). Both studies demonstrate that removing the demographic influence of stabilizing niche differences causes reductions in diversity (Levine & HilleRisLambers 2009, Adler et al. 2010). Analyses with other long-term demographic sets (e.g., Clark 2010, Kraft & Ackerly 2010) may allow ecologists to determine whether these findings are generally applicable across ecosystems and communities.

4. PARTING THOUGHTS

We have argued that recent theoretical developments in coexistence theory allow for a greater understanding of the importance of competitive interactions in driving community assembly. With that in mind, we have provided numerous empirical examples of how stabilizing niche differences and relative fitness differences influence community assembly by influencing, for example, the identity of species that are excluded from communities through competition, the relative abundance of species within those communities, whether coexistence in those communities is stabilized, and their invasibility. However, the overall importance of stabilizing niche differences and relative fitness differences, as well as the mechanisms underlying these differences during plant community assembly, is still poorly understood. Fortunately, our review of empirical approaches suggests a way forward.

4.1. Future Directions

We believe that combining experimental manipulations or trait-based approaches with demographic models will allow ecologists to distinguish between the traits that tend to stabilize coexistence versus those that tend to drive competitive exclusion. This approach will allow empirical approaches to catch up to theoretical predictions, which have been significantly clarified over the past decade or so (Section 2). To their credit, current manipulative experiments and empirical trait-based approaches (Sections 3.1 and 3.2) both aim to identify the specific traits and mechanisms determining the outcome of competition (coexistence versus exclusion). However, both are limited by the investigators' ability to manipulate or measure all relevant niche axes and by the difficulty of distinguishing fitness differences from stabilizing niche differences. Meanwhile, demographic approaches (Section 3.3) can quantify the net effect of stabilizing niche and fitness differences among interacting species (e.g., Levine & HilleRisLambers 2009, Adler et al. 2010), but they have generally not been used to identify the traits or mechanisms that drive the outcomes of competition (but see Angert et al. 2009). By combining these empirical approaches (**Figure 3**), we believe that ecologists will be better able to link functional traits, phylogenetic relationships, and competitive interactions to community dynamics, long a goal of community assembly studies.

For example, rather than correlating the trait differences between competitors with their presence or absence in a community, we might instead correlate these differences with stabilizing niche and relative fitness differences from phenomenological demographic models, which better capture the actual drivers of competitive outcomes. This approach would require a diverse system, with species whose growth and interactions could be reasonably described by population dynamic models parameterized from data (e.g., Adler et al. 2006, 2010; Clark 2010). From such a model, pairwise stabilizing niche differences and relative fitness differences could be quantified (Chesson 2000, Adler et al. 2007) and correlated with their functional trait differences, including all traits thought to influence competitive outcomes.

One might find, for example, that disparities in rooting depth are excellent predictors of the stabilizing niche differences between competitors, whereas disparities in relative growth rates predict fitness differences. Given the interrelated nature of many plant traits, multivariate approaches will likely be necessary to properly disentangle the contributions of various traits to both stabilizing niche differences and relative fitness differences (many will influence both). Results such as these can not only help identify the mechanisms of coexistence but also predict the influence of competition on community assembly, i.e., overdispersion along the rooting depth axis and clustering along a relative growth rate axis. The potential power of combining demographic and trait-based approaches is exemplified by the work of Angert et al. (2009). They found that species position along a trade-off between water use efficiency and relative growth rates was predictive of how species fecundity fluctuated through time, a key ingredient of coexistence via the storage effect (Angert et al. 2009).

Finally, experiments can provide powerful empirical tests of the coexistence mechanisms inferred by such demographic and trait-based compositional analyses. In an idealized example, imagine that plant stoichiometry suggests that different species are limited by different nutrients, and that differences in species' stoichiometry correlate with the strength of stabilizing niche differences. Fertilizing with one of the limiting resources should not only erode species diversity but predictably result in the loss of species that are superior competitors for those resources, providing an experimental test of the hypothesized mechanism. It is possible that some of the many existing manipulative studies of community composition (e.g., Rees et al. 2001, Suding et al. 2005, Suttle et al. 2007, HilleRisLambers et al. 2010) can be reanalyzed using contemporary coexistence theory, leading to important insights into community assembly (e.g., Fukami et al. 2005, Cadotte & Strauss 2011). For example, using the results of a recent study manipulating the seasonality and timing of precipitation (Suttle et al. 2007) to parameterize population dynamic models (e.g., Adler et al. 2012) could provide insight into whether declining diversity with spring precipitation results from a reduction in stabilizing niche differences or an increase in relative fitness differences.

4.2. Conclusions

An understanding of how stabilizing niche differences and relative fitness differences contribute to plant community assembly is not simply an academic question. Clarifying the processes that influence the composition, diversity, and relative abundance of co-occurring species in local communities has long been a goal of community ecologists, but it holds particular urgency with the increasing impacts of humans on the environment. Climate change, habitat fragmentation, eutrophication, and invasive species have large impacts on local plant communities by altering both the environment and biotic milieu that operate during community assembly (Hobbs et al. 2009). Understanding how plant communities disassemble following these perturbations, or how we can drive reassembly to a desired outcome with restoration, can be achieved with studies that link coexistence theory (determining how species are influenced by the environment and each other) to the outcome of community assembly (see sidebar, Coexistence Theory and Global Change).

COEXISTENCE THEORY AND GLOBAL CHANGE

Global change alters the composition and relative abundance of species within communities (i.e., community disassembly), but negative impacts can be reversed through restoration (i.e., community reassembly). Contemporary coexistence theory suggests that a valuable way to predict impacts and optimize restoration is to determine how global change alters stabilizing niche differences and relative fitness differences. For example:

- Climate change impacts on plant communities range from losses of weak competitors as relative fitness differences increase (e.g., as population growth rates of warm-adapted species increase) to increased persistence of rare species as stabilizing niche differences increase (through a storage effect mediated by climate variability; Adler & Drake 2008).
- In restoration, a functionally diverse seed mix can minimize invasibility through limiting similarity (Holmes 2001, Fargione et al. 2003), and the abiotic environment can be altered to favor desirable species by increasing relative fitness differences (e.g., by decreasing N or P levels; Harpole 2006, Jeppesen et al. 2007), or these processes can be combined for the best outcomes (Funk et al. 2008).

DISCLOSURE STATEMENT

The authors are not aware of any affiliations, memberships, funding, or financial holdings that might be perceived as affecting the objectivity of this review.

ACKNOWLEDGMENTS

We thank Kevin Ford, Ailene Ettinger, Sharon Strauss, and Susan Waters for comments that substantially improved earlier drafts.

LITERATURE CITED

Ackerly DD, Cornwell WK. 2007. A trait-based approach to community assembly: partitioning of species trait values into within- and among-community components. *Ecol. Lett.* 10:135–45

Adler PB, Dalgleish HJ, Ellner SP. 2012. Forecasting plant community impacts of climate variability and change: When do competitive interactions matter? *J. Ecol.* 100:478–87

Adler PB, Drake JM. 2008. Environmental variation, stochastic extinction, and competitive coexistence. *Am. Nat.* 172:E186–95

Adler PB, Ellner SP, Levine JM. 2010. Coexistence of perennial plants: an embarrassment of niches. *Ecol. Lett.* 13:1019–29

Adler PB, HilleRisLambers J, Kyriakidis PC, Guan Q, Levine JM. 2006. Climate variability has a stabilizing effect on coexistence of prairie grasses. *Proc. Natl. Acad. Sci. USA* 103:12793–98

Adler PB, HilleRisLambers J, Levine JM. 2007. A niche for neutrality. *Ecol. Lett.* 10:95–104

Adler PB, HilleRisLambers J, Levine JM. 2009. Weak effect of climate variability on coexistence in a sagebrush steppe community. *Ecology* 90:3303–12

Allan E, van Ruijven J, Crawley MJ. 2010. Foliar fungal pathogens and grassland biodiversity. *Ecology* 91:2572–82

Anderson TM, Shaw J, Olff H. 2011. Ecology's cruel dilemma, phylogenetic trait evolution and the assembly of Serengeti plant communities. *J. Ecol.* 99:797–806

Angert AL, Huxman TE, Chesson P, Venable DL. 2009. Functional tradeoffs determine species coexistence via the storage effect. *Proc. Natl. Acad. Sci. USA* 106:11641–45

Belyea LR, Lancaster J. 1999. Assembly rules within a contingent ecology. *Oikos* 86:402–16

Bever JD. 1994. Feedback between plants and their soil communities in an old field community. *Ecology* 75:1965–77

Burns JH, Strauss SY. 2011. More closely related species are more ecologically similar in an experimental test. *Proc. Natl. Acad. Sci. USA* 108:5302–7

Cadotte MW, Borer ET, Seabloom EW, Cavender-Bares J, Harpole WS, et al. 2010. Phylogenetic patterns differ for native and exotic plant communities across a richness gradient in Northern California. *Divers. Distrib.* 16:892–901

Cadotte MW, Strauss SY. 2011. Phylogenetic patterns of colonization and extinction in experimentally assembled plant communities. *PLoS ONE* 6:e19363

Carson W, Root R. 2000. Herbivory and plant species coexistence: community regulation by an outbreaking phytophagous insect. *Ecol. Monogr.* 70:73–99

Cavender-Bares J, Ackerly DD, Baum DA, Bazzaz FA. 2004a. Phylogenetic overdispersion in Floridian oak communities. *Am. Nat.* 163:823–43

Cavender-Bares J, Keen A, Miles B. 2006. Phylogenetic structure of Floridian plant communities depends on taxonomic and spatial scale. *Ecology* 87:S109–22

Cavender-Bares J, Kitajima K, Bazzaz FA. 2004b. Multiple trait associations in relation to habitat differentiation among 17 Floridian oak species. *Ecol. Monogr.* 74:635–62

Cavender-Bares J, Kozak KH, Fine PVA, Kembel SW. 2009. The merging of community ecology and phylogenetic biology. *Ecol. Lett.* 12:693–715

Chase JM. 2010. Stochastic community assembly causes higher biodiversity in more productive environments. *Science* 328:1388–91

Chase JM. 2003. Community assembly: When should history matter? *Oecologia* 136: 489–98

Chase JM, Leibold MA. 2003. *Ecological Niches: Linking Classical and Contemporary Approaches.* London: Univ. Chicago Press

Chase JM, Myers JA. 2011. Disentangling the importance of ecological niches from stochastic processes across scales. *Philos. Trans. R. Soc. B Biol. Sci.* 366:2351–63

Chesson P. 2000. Mechanisms of maintenance of species diversity. *Annu. Rev. Ecol. Syst.* 31:343–66

Clark CM, Cleland EE, Collins SL, Fargione JE, Gough L, et al. 2007. Environmental and plant community determinants of species loss following nitrogen enrichment. *Ecol. Lett.* 10:596–607

Clark JS. 2010. Individuals and the variation needed for high species diversity in forest trees. *Science* 327:1129–32

Clark JS, McLachlan JS. 2003. Stability of forest biodiversity. *Nature* 423:635–38

Comita LS, Muller-Landau HC, Aguilar S, Hubbell SP. 2010. Asymmetric density dependence shapes species abundances in a tropical tree community. *Science* 329:330–32

Connell JH. 1971. On the role of natural enemies in preventing competitive exclusion in some marine animals and in rain forest trees. In *Dynamics of Populations*, ed. PJ den Boer, GR Gradwell, pp. 298–310. Wageningen, The Neth.: Cent. Agric. Publ. Doc.

Cornwell WK, Ackerly DD. 2009. Community assembly and shifts in plant trait distributions across an environmental gradient in coastal California. *Ecol. Monogr.* 79:109–26

de Bello F, Thuiller W, Leps J, Choler P, Clement J, et al. 2009. Partitioning of functional diversity reveals the scale and extent of trait convergence and divergence. *J. Vegetation Sci.* 20:475–86

Diamond JM. 1975. Assembly of species communities. In *Ecology and Evolution of Communities*, ed. ML Cody, JM Diamond, pp. 342–444. Cambridge, MA: Harvard University Press

Dinnage R. 2009. Disturbance alters the phylogenetic composition and structure of plant communities in an old field system. *PLoS ONE* 4:e7071

Dornbush ME, Wilsey BJ. 2010. Experimental manipulation of soil depth alters species richness and co-occurrence in restored tallgrass prairie. *J. Ecol.* 98:117–25

Dybzinski R, Tilman D. 2007. Resource use patterns predict long-term outcomes of plant competition for nutrients and light. *Am. Nat.* 170:305–18

Emerson BC, Gillespie RG. 2008. Phylogenetic analysis of community assembly and structure over space and time. *Trends Ecol. Evol.* 23:619–30

Fargione J, Brown CS, Tilman D. 2003. Community assembly and invasion: an experimental test of neutral versus niche processes. *Proc. Natl. Acad. Sci. USA* 100:8916–20

Fine PVA, Kembel SW. 2011. Phylogenetic community structure and phylogenetic turnover across space and edaphic gradients in western Amazonian tree communities. *Ecography* 34:552–65

Fukami T, Bezemer TM, Mortimer SR, van der Putten WH. 2005. Species divergence and trait convergence in experimental plant community assembly. *Ecol. Lett.* 8:1283–90

Fukami T, Dickie IA, Wilkie JP, Paulus BC, Park D, et al. 2010. Assembly history dictates ecosystem functioning: evidence from wood decomposer communities. *Ecol. Lett.* 13:675–84

Funk JL, Cleland EE, Suding KN, Zavaleta ES. 2008. Restoration through reassembly: plant traits and invasion resistance. *Trends Ecol. Evol.* 23:695–703

Götzenberger L, de Bello F, Bråthen KA, Davison J, Dubuis A, et al. 2012. Ecological assembly rules in plant communities—approaches, patterns and prospects. *Biol. Rev.* 87:111–27

Gause GF. 1934. *The Struggle for Existence*. New York: Hafner Publ. Co.

Harms KE, Wright SJ, Calderon O, Hernandez A, Herre EA. 2000. Pervasive density-dependent recruitment enhances seedling diversity in a tropical forest. *Nature* 404:493–95

Harpole WS. 2006. Resource-ratio theory and the control of invasive plants. *Plant Soil* 280:23–27

Harpole WS, Suding KN. 2007. Frequency-dependence stabilizes competitive interactions among four annual plants. *Ecol. Lett.* 10:1164–69

Harpole WS, Tilman D. 2006. Non-neutral patterns of species abundance in grassland communities. *Ecol. Lett.* 9:15–23

Harpole WS, Tilman D. 2007. Grassland species loss resulting from reduced niche dimension. *Nature* 446:791–93

Hautier Y, Niklaus PA, Hector A. 2009. Competition for light causes plant biodiversity loss after eutrophication. *Science* 324:636–38

Helmus MR, Savage K, Diebel MW, Maxted JT, Ives AR. 2007. Separating the determinants of phylogenetic community structure. *Ecol. Lett.* 10:917–25

HilleRisLambers J, Clark JS, Beckage B. 2002. Density-dependent mortality and the latitudinal gradient in species diversity. *Nature* 417:732–35

HilleRisLambers J, Yelenik SG, Colman BP, Levine JM. 2010. California annual grass invaders: the drivers or passengers of change? *J. Ecol.* 98:1147–56

Hobbs RJ, Higgs E, Harris JA. 2009. Novel ecosystems: implications for conservation and restoration. *Trends Ecol. Evol.* 24:599–605

Holmes P. 2001. Shrubland restoration following woody alien invasion and mining: effects of topsoil depth, seed source, and fertilizer addition. *Restor. Ecol.* 9:71–84

Hooper DU, Dukes JS. 2010. Functional composition controls invasion success in a California serpentine grassland. *J. Ecol.* 98:764–77

Horner-Devine MC, Silver JM, Leibold MA, Bohannan BJM, Colwell RK, et al. 2007. A comparison of taxon co-occurrence patterns for macro- and microorganisms. *Ecology* 88:1345–53

Hubbell SP. 2001. *The Unified Neutral Theory of Biodiversity and Biogeography*. Princeton, NJ: Princeton Univ. Press

Hubbell SP, Ahumada J, Condit R, Foster R. 2001. Local neighborhood effects on long-term survival of individual trees in a neotropical forest. *Ecol. Res.* 16:859–75

Hutchinson GE. 1957. Population studies: animal ecology and demography: concluding remarks. *Cold Spring Harb. Symp. Quant. Biol.* 22:415–27

Ives AR, Helmus MR. 2011. Generalized linear mixed models for phylogenetic analyses of community structure. *Ecol. Monogr.* 81:511–25

Janzen DH. 1970. Herbivores and number of tree species in tropical forests. *Am. Nat.* 104:501–28

Jeppesen E, Sondergaard M, Meerhoff M, Lauridsen TL, Jensen JP. 2007. Shallow lake restoration by nutrient loading reduction—some recent findings and challenges ahead. *Hydrobiologia* 584:239–52

Kattge J, Diaz S, Lavorel S, Prentice C, Leadley P, et al. 2011. TRY—a global database of plant traits. *Glob. Change Biol.* 17:2905–35

Kembel SW, Hubbell SP. 2006. The phylogenetic structure of a neotropical forest tree community. *Ecology* 87:S86–99

Kennedy TA, Naeem S, Howe KM, Knops JMH, Tilman D, Reich P. 2002. Biodiversity as a barrier to ecological invasion. *Nature* 417:636–38

Klironomos JN. 2002. Feedback with soil biota contributes to plant rarity and invasiveness in communities. *Nature* 417:67–70

Kluge J, Kessler M. 2011. Phylogenetic diversity, trait diversity and niches: species assembly of ferns along a tropical elevational gradient. *J. Biogeogr.* 38:394–405

Kobe RK, Vriesendorp CF. 2011. Conspecific density dependence in seedlings varies with species shade tolerance in a wet tropical forest. *Ecol. Lett.* 14:503–10

Koerner C, Stoecklin J, Reuther-Thiebaud L, Pelaez-Riedl S. 2008. Small differences in arrival time influence composition and productivity of plant communities. *New Phytol.* 177:698–705

Kooyman R, Cornwell WK, Westoby M. 2010. Plant functional traits in Australian subtropical rain forest: partitioning within-community from cross-landscape variation. *J. Ecol.* 98:517–25

Kraft NJB, Ackerly DD. 2010. Functional trait and phylogenetic tests of community assembly across spatial scales in an Amazonian forest. *Ecol. Monogr.* 80:401–22

Kraft NJB, Valencia R, Ackerly DD. 2008. Functional traits and niche-based tree community assembly in an Amazonian forest. *Science* 322:580–82

Kursar TA, Dexter KG, Lokvam J, Pennington RT, Richardson JE, et al. 2009. The evolution of antiherbivore defenses and their contribution to species coexistence in the tropical tree genus *Inga*. *Proc. Natl. Acad. Sci. USA* 106:18073–78

Lebrija-Trejos E, Pérez-García EA, Meave JA, Bongers F, Poorter L. 2010. Functional traits and environmental filtering drive community assembly in a species-rich tropical system. *Ecology* 91:386–98

Letcher SG. 2010. Phylogenetic structure of angiosperm communities during tropical forest succession. *Proc. R. Soc. B Biol. Sci.* 277:97–104

Levine JM. 2000. Species diversity and biological invasions: relating local process to community pattern. *Science* 288:852–54

Levine JM, HilleRisLambers J. 2009. The importance of niches for the maintenance of species diversity. *Nature* 461:254–57

Long W, Zang R, Schamp BS, Ding Y. 2011. Within- and among-species variation in specific leaf area drive community assembly in a tropical cloud forest. *Oecologia* 167:1103–13

Mangan SA, Schnitzer SA, Herre EA, Mack KML, Valencia MC, et al. 2010. Negative plant-soil feedback predicts tree-species relative abundance in a tropical forest. *Nature* 466:752–55

Mayfield MM, Boni MF, Ackerly DD. 2009. Traits, habitats, and clades: identifying traits of potential importance to environmental filtering. *Am. Nat.* 174:E1–22

Mayfield MM, Boni M, Daily G, Ackerly D. 2005. Species and functional diversity of native and human-dominated plant communities. *Ecology* 86:2365–72

Mayfield MM, Levine JM. 2010. Opposing effects of competitive exclusion on the phylogenetic structure of communities. *Ecol. Lett.* 13:1085–93

McCarthy-Neumann S, Kobe RK. 2010. Conspecific plant-soil feedbacks reduce survivorship and growth of tropical tree seedlings. *J. Ecol.* 98:396–407

Metz MR, Sousa WP, Valencia R. 2010. Widespread density-dependent seedling mortality promotes species coexistence in a highly diverse Amazonian rain forest. *Ecology* 91:3675–85

Mwangi PN, Schmitz M, Scherber C, Roscher C, Schumacher J, et al. 2007. Niche pre-emption increases with species richness in experimental plant communities. *J. Ecol.* 95:65–78

Ojeda F, Pausas JG, Verdu M. 2010. Soil shapes community structure through fire. *Oecologia* 163:729–35

Paine CET, Baraloto C, Chave J, Herault B. 2011. Functional traits of individual trees reveal ecological constraints on community assembly in tropical rain forests. *Oikos* 120:720–27

Pavoine S, Bonsall MB. 2011. Measuring biodiversity to explain community assembly: a unified approach. *Biol. Rev.* 86:792–812

Pavoine S, Vela E, Gachet S, de Belair G, Bonsall MB. 2011. Linking patterns in phylogeny, traits, abiotic variables and space: a novel approach to linking environmental filtering and plant community assembly. *J. Ecol.* 99:165–75

Pei N, Lian J, Erickson DL, Swenson NG, Kress WJ, et al. 2011. Exploring tree-habitat associations in a Chinese subtropical forest plot using a molecular phylogeny generated from DNA barcode loci. *PLoS ONE* 6:e21273

Petermann JS, Fergus AJF, Roscher C, Turnbull LA, Weigelt A, Schmid B. 2010. Biology, chance, or history? The predictable reassembly of temperate grassland communities. *Ecology* 91:408–21

Petermann JS, Fergus AJF, Turnbull LA, Schmid B. 2008. Janzen-Connell effects are widespread and strong enough to maintain diversity in grasslands. *Ecology* 89:2399–406

Peters H. 2003. Neighbour-regulated mortality: the influence of positive and negative density dependence on tree populations in species-rich tropical forests. *Ecol. Lett.* 6:757–65

Pillar VD, Duarte LDS. 2010. A framework for metacommunity analysis of phylogenetic structure. *Ecol. Lett.* 13:587–96

Prinzing A, Reiffers R, Braakhekke WG, Hennekens SM, Tackenberg O, et al. 2008. Less lineages—more trait variation: phylogenetically clustered plant communities are functionally more diverse. *Ecol. Lett.* 11:809–19

Rees M, Condit R, Crawley M, Pacala S, Tilman D. 2001. Long-term studies of vegetation dynamics. *Science* 293:650–55

Ricklefs R. 2004. A comprehensive framework for global patterns in biodiversity. *Ecol. Lett.* 7:1–15

Roscher C, Schmid B, Schulze E. 2009. Non-random recruitment of invader species in experimental grasslands. *Oikos* 118:1524–40

Sale P. 1977. Maintenance of high diversity in coral-reef fish communities. *Am. Nat.* 111:337–59

Seabloom EW, Harpole WS, Reichman OJ, Tilman D. 2003. Invasion, competitive dominance, and resource use by exotic and native California grassland species. *Proc. Natl. Acad. Sci. USA* 100:13384–89

Siepielski AM, McPeek MA. 2010. On the evidence for species coexistence: a critique of the coexistence program. *Ecology* 91:3153–64

Silva IA, Batalha MA. 2009. Phylogenetic overdispersion of plant species in southern Brazilian savannas. *Braz. J. Biol.* 69:843–49

Slingsby JA, Verboom GA. 2006. Phylogenetic relatedness limits co-occurrence at fine spatial scales: evidence from the schoenid seges (Cyperaceae: Schoeneae) of the Cape Floristic Region, South Africa. *Am. Nat.* 168:14–27

Suding KN, Collins SL, Gough L, Clark C, Cleland EE, et al. 2005. Functional- and abundance-based mechanisms explain diversity loss due to N fertilization. *Proc. Natl. Acad. Sci. USA* 102:4387–92

Suttle KB, Thomsen MA, Power ME. 2007. Species interactions reverse grassland responses to changing climate. *Science* 315:640–42

Swamy V, Terborgh JW. 2010. Distance-responsive natural enemies strongly influence seedling establishment patterns of multiple species in an Amazonian rain forest. *J. Ecol.* 98:1096–107

Swenson NG, Enquist BJ. 2009. Opposing assembly mechanisms in a Neotropical dry forest: implications for phylogenetic and functional community ecology. *Ecology* 90:2161–70

Swenson NG, Enquist BJ, Thompson J, Zimmerman JK. 2007. The influence of spatial and size scale on phylogenetic relatedness in tropical forest communities. *Ecology* 88:1770–80

Swenson NG, Weiser MD. 2010. Plant geography upon the basis of functional traits: an example from eastern North American trees. *Ecology* 91:2234–41

Tilman D. 1982. *Resource Competition and Community Structure*. New Jersey: Princeton Univ. Press

Tilman D. 1994. Competition and biodiversity in spatially structured habitats. *Ecology* 75:2–16

Tilman D. 2004. Niche tradeoffs, neutrality, and community structure: a stochastic theory of resource competition, invasion, and community assembly. *Proc. Natl. Acad. Sci. USA* 101:10854–61

Turnbull LA, Rees M, Crawley MJ. 1999. Seed mass and the competition/colonization trade-off: a sowing experiment. *J. Ecol.* 87:899–912

Verdu M, Pausas JG. 2007. Fire drives phylogenetic clustering in Mediterranean basin woody plant communities. *J. Ecol.* 95:1316–23

Webb CO. 2000. Exploring the phylogenetic structure of ecological communities: an example for rain forest trees. *Am. Nat.* 156:145–55

Webb CO, Gilbert GS, Donoghue MJ. 2006. Phylodiversity-dependent seedling mortality, size structure, and disease in a Bornean rain forest. *Ecology* 87:S123–31

Webb CO, Peart DR. 1999. Seedling density dependence promotes coexistence of Bornean rain forest trees. *Ecology* 80:2006–17

Webb CO, Ackerly D, McPeek M, Donoghue M. 2002. Phylogenies and community ecology. *Annu. Rev. Ecol. Syst.* 33:475–505

Wedin D, Tilman GD. 1993. Competition among grasses along a nitrogen gradient: initial conditions and mechanisms of competition. *Ecol. Monogr.* 63:199–229

Weiher E, Freund D, Bunton T, Stefanski A, Lee T, Bentivenga S. 2011. Advances, challenges and a developing synthesis of ecological community assembly theory. *Philos. Trans. R. Soc. B Biol. Sci.* 366:2403–13

Wiens J, Donoghue M. 2004. Historical biogeography, ecology and species richness. *Trends Ecol. Evol.* 19:639–44

Willis CG, Halina M, Lehman C, Reich PB, Keen A, et al. 2010. Phylogenetic community structure in Minnesota oak savanna is influenced by spatial extent and environmental variation. *Ecography* 33:565–77

Wills C, Condit R, Foster R, Hubbell S. 1997. Strong density- and diversity-related effects help to maintain tree species diversity in a neotropical forest. *Proc. Natl. Acad. Sci. USA* 94:1252–57

Yamazaki M, Iwamoto S, Seiwa K. 2009. Distance- and density-dependent seedling mortality caused by several diseases in eight tree species co-occurring in a temperate forest. *Plant Ecol.* 201:181–96

The Role of Mountain Ranges in the Diversification of Birds

Jon Fjeldså,[1] Rauri C.K. Bowie,[2] and Carsten Rahbek[3]

[1]Center for Macroecology, Evolution, and Climate, Natural History Museum of Denmark, University of Copenhagen, DK-2100 Copenhagen, Denmark; email: jfjeldsaa@snm.ku.dk

[2]Museum of Vertebrate Zoology & Department of Integrative Biology, University of California, Berkeley, California 94720; email: bowie@berkeley.edu

[3]Center for Macroecology, Evolution, and Climate, Department of Biology, University of Copenhagen, DK-2100 Copenhagen, Denmark; email: crahbek@bio.ku.dk

Annu. Rev. Ecol. Evol. Syst. 2012. 43:249–65

First published online as a Review in Advance on September 4, 2012

The *Annual Review of Ecology, Evolution, and Systematics* is online at ecolsys.annualreviews.org

This article's doi:
10.1146/annurev-ecolsys-102710-145113

Keywords

birds, global, marine impacts, mountains, persistence, speciation

Abstract

Avian faunas vary greatly among montane areas; those at high latitudes are biologically impoverished, whereas those of some low-latitude mountains are biologically very complex. Their high level of species richness is caused by the aggregation of many small-ranged species, which has been difficult to explain from purely macroecological models focusing on contemporary ecological processes. Because the individual mountain tracts harbor species that represent different evolutionary trajectories, it seems plausible to relate these species assemblages to high persistence (or absence of extinction) in addition to high levels of speciation. The distribution of small-ranged species is concentrated near tropical coasts, where moderation of the climate in topographically complex areas creates cloud forests and stable local conditions. The stability underpins specialization and resilience of local populations, and thereby the role of these places as cradles of biodiversity.

1. INTRODUCTION

Montane areas represent rugged landscapes that are uplifted to an extent that affects local climate. Mountains are therefore often viewed as bleak and biologically impoverished environments (e.g., Martin & Wiebe 2004). Mountains at high latitudes are essentially arctic in terms of climate and biota, and many mountains at moderate latitudes harbor relict populations of arctic species, which in most instances have persisted since the Pleistocene ice ages (e.g., Hughes & Eastwood 2006). Even some speciose components of the tropical montane biota are thought to have their origins in temperate environments (see examples in Vuilleumier & Monasterio 1986). Regardless of whether mountains were colonized from higher latitudes or from adjacent lowlands, it has long been acknowledged that some speciation must have taken place by isolation in allopatry or parapatry within montane systems (Vuilleumier & Monasterio 1986, Moritz et al. 2000). However, only recently has the role that mountains play as cradles of biodiversity become fully realized. Mountains contain half of the currently defined biodiversity hot spots (Kohler & Maselli 2009), although they cover only 16.5–27% of the land area (depending on how montane areas are defined). The traditionally delineated montane hot spots are widely recognized as areas of high priority for conservation, primarily as a consequence of the large number of endemic and threatened species they encompass (Stattersfield et al. 1998).

Macroecological analyses on continental distributions of birds have revealed that models based exclusively on contemporary climate fail to explain overall patterns of richness. Thus, it is necessary to incorporate topography relief as a feature to obtain statistical power to account for the unusually high diversity in tropical mountain regions (Rahbek & Graves 2000, 2001). Subsequent analyses have emphasized that models based on contemporary environmental variables explain well only the regional variation in richness of the most wide-ranging species (Jetz & Rahbek 2002, Rahbek et al. 2007). The corollary that models incorporating key variables such as contemporary water and energy availability fail to explain the aggregated occurrence of species with small distributions in montane regions has attracted renewed research effort. The spatial positioning of aggregations of small-ranged species also exceeds what can be predicted by including effects of topography, geometric constraints, and random draws from the total species pool in a given area (Jetz et al. 2004, Fjeldså & Rahbek 2006, Rahbek et al. 2007). As a consequence, effort has centered on better integrating contemporary, evolutionary, and historical variables in an attempt to understand the variation in diversity and composition of avifaunas among montane areas and between mountains and adjacent lowlands.

One influential step toward understanding the difference in biodiversity between high- and low-latitude mountains was formulated by Janzen (1967). He pointed out that high-latitude environments are characterized by seasonal temperature amplitude exceeding that of the elevational temperature gradient, making topographic barriers less important in temperate regions than in the tropics. In the tropics, species could evolve narrower thermal tolerances and thereby be able to permanently reside within distinct elevational zones. His argument of higher species turnover on tropical elevational gradients has been supported by recent studies (e.g., Ghalambor et al. 2006, McCain 2009). Furthermore, Cadena et al. (2012) established that tropical vertebrates tend to have greater evolutionary conservatism in their thermal niches, with sister species generally inhabiting very similar thermal niches.

Most studies of speciation in montane areas have focused on the physical barriers that arose during mountain building, but this approach may not adequately explain the observed variation among montane areas. In northern montane regions, the dispersal barriers are low (in Janzen's sense), and vicariant patterns may be rapidly erased by high levels of climate-driven range dynamics (Jansson & Dynesius 2002, Hawkins & Diniz-Filho 2006). In tropical mountains, congruent geographic patterns, when placed in a temporal framework, may represent distinct area cladograms, rejecting the idea of generalized vicariance patterns (Fjeldså & Bowie 2008). The problem is exemplified by the pattern of species richness across the Eastern Arc Mountains, a 600-km chain of 13 discrete sky islands in East Africa. Although species richness roughly follows the species-area relationship among the different sky islands, endemism does not, with three sky islands having excessive levels of endemism for their given area (Burgess et al. 2007). Although speciation probably proceeds through divergence in allopatry, the divergences among multiple lineages are not coincident in time (Fjeldså & Bowie 2008, Lawson 2010, Tolley et al. 2011). Most likely, adaptive redistribution and varying rates of extinction among these sky islands have played an important role in generating the observed pattern of endemism. Thus, in order to explain the aggregations of small-range species, we need to consider the whole process of lineage diversification (speciation and extinction). This is particularly relevant from a conservation perspective because the focus on areas where lineages persist (because of low extinction rates) will be more meaningful than the traditional focus on barriers between areas.

Our background for writing this review builds on decades of exploratory field studies in montane regions, notably in the Andes (e.g., Fjeldså & Krabbe 1990) and in the mountains of eastern Africa (e.g., Fjeldså & Bowie 2008, Fjeldså et al. 2010, Voelker et al. 2010), which revealed a much more localized pattern of aggregation of small-ranged species than is apparent from compiling coarse-scale geographical ranges from the literature. We have interpreted the local aggregation of young and old (relictual?) species as signs of local stability, resulting from local climate moderation as prevailing winds and atmospheric stratification interact with complex topography (Fjeldså 1995, Fjeldså & Lovett 1997). This was supported by weather satellite data sampled over a few years (Fjeldså et al. 1999). Furthermore, some descriptions of biodiversity hot spots in tropical coastal mountains (Lovett 1993, Best & Kessler 1995) provided compelling evidence of how the climatic influence from the nearest ocean could provide predictable conditions over longer periods of time.

Nevertheless, this represents an ad hoc explanation, which needs to be supplemented by a search for general patterns supported by quantitative analyses. This was not possible at the time of the first formulation of the above hypotheses, as adequate distributional data had not yet been compiled in digital form. With the recent development of such databases, several thorough continent-wide analyses of variation in species richness now allow a new class of studies that can reveal significant deviation from the expectations of general macroecological models. With the rapid development of phylogenetic data and tools for paleoclimatic modeling, we may now begin to analyze the complexity of the montane biota across the globe.

In this review, we use distributional and phylogenetic data, primarily for passerine birds (order Passeriformes) to illustrate general patterns, and we emphasize some environmental characteristics of the montane biodiversity hot spots. Addressing the possible underlying mechanisms remains a work in progress, although we do broadly outline some interesting patterns and trends. Our passerine focal group is suitable as a model group, as the order is well studied in terms of molecular phylogeny and represents the largest avian radiation with about 6,000 species. Furthermore, passerines are fairly homogeneous in terms of size and gross morphology and yet are highly diverse in their use of terrestrial habitats and in their diet.

2. MATERIALS AND METHODS

Detailed accounts of the montane regions of the world and their biodiversity are described in other published or emerging papers. We therefore provide only a very general outline based on our more detailed continental and especially regional studies from some of the ornithologically most outstanding parts of the world, with the following data primarily derived from our own extensive fieldwork:

1. The tropical Andes region of South America, a semicontinuous but narrow band of montane habitat covering a large latitudinal range (Fjeldså et al. 1999, Rahbek & Graves 2001, Rahbek et al. 2007, Fjeldså & Irestedt 2009),

2. The Afromontane region, mainly spatially discrete sky islands (Jetz & Rahbek 2002, Fjeldså & Bowie 2008), and

3. The Indo-Pacific region, a complex tropical archipelago that arose during the Neogene Australian-Asian collision, with many montane areas (e.g., Jønsson et al. 2011, Fritz et al. 2012).

4. In addition, significant new data are available for the Sino-Himalayan Mountains, a long mountain chain extending outside the tropics (Johansson et al. 2007a, Price et al. 2011, Päckert et al. 2012).

We defined mountain areas of the world primarily from a GIS (geographic information systems) model based on the range in local elevation and slope developed by the Mountain Research Initiative (Kapos et al. 2000). Because of the fine spatial resolution of this model, large areas appear as mosaics of montane and nonmontane pixels, with hundreds of isolated pixels that marginally qualify as montane scattered far outside the main mountain tracts. Thus, we developed a set of rules to circumscribe broader montane regions, subdivided these according to recognized biogeographic systems, and removed isolated rugged pixels. The biodiversity content (species richness and number of small-ranged species) per region could then be analyzed in relation to latitude, area, placement within continents, and environmental parameters extracted from global environmental models.

For all passerine birds, we compiled a global distribution database (presence-absence data for all species at the spatial resolution of a 1×1 latitudinal-longitudinal degree grid following the approach outlined by Rahbek & Graves 2001). From this we extracted species lists for the defined montane regions. In order to illustrate general patterns of diversification history, we have compiled divergence data for mitochondrial DNA from >150 publications to assess time since most recent common ancestor (TMRCA) for densely sampled groups. Reference to basal and terminal species generally refers to species with short or long root-paths (number of nodes from the base of a phylogeny), unless the TMRCA [in million years ago (Mya)] is specifically mentioned. To highlight different patterns in this review, we used a detailed phylogenetic framework for the endemic radiations of South American suboscine birds (Derryberry et al. 2011; J.I. Ohlson, M. Irestedt, P.G.P. Ericson, and J. Fjeldså, unpublished data) and for some African bird groups: the nonpasserine Galliformes (Cohen et al. 2012, Mandiwana-Neudani 2012), Malaconotidae (notably Fuchs et al. 2004, 2012; Njabo et al. 2008), and Pycnonotidae (Johansson et al. 2007b). Although we can never know the past distribution of a clade, we estimated the paleodistributions of some clades by merging the distributions of all constituent species, assuming diversification by vicariance. Because terminal subclades that obviously represent recent dispersal out of the ancestral area are irrelevant for the reconstruction of paleodistributions, these were removed (Fjeldså & Bowie 2008, Fjeldså 2012).

3. THE GLOBAL VARIATION OF PASSERINE DIVERSITY

Some aspects of the variation in the global diversity pattern of passerine birds are illustrated in **Figure 1**, where panel 1*a* expresses species richness as relative brightness: old species by purple hues, and recently evolved species by green hues (categories defined in the figure legend). To fully interpret this, one needs to bear in mind the austral origin (in Australia, South America, and almost certainly Antarctica) of passerine birds (Ericson et al. 2003). Whereas the suboscines were moderately successful in colonizing areas outside the Austral area of origin, the oscine passerines

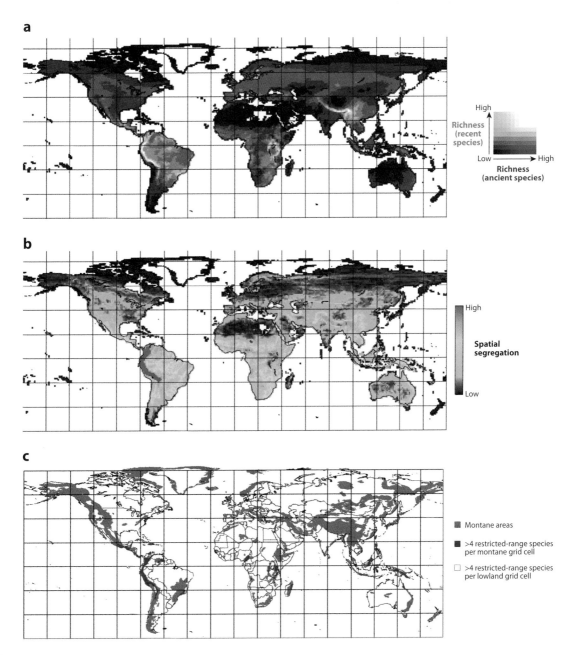

Figure 1

Global patterns of diversification. (*a*) Diversity of 2,291 species (brightness reflecting species richness; maximum value is 160 species) representing groups where time of diversification is verified by molecular data. Green hues show species with time since most recent common ancestor <3 million years, purple hues show ancient species (small clades originating >15 Mya; such clades may comprise a single species or up to three recently diverged allospecies), and gray hues reflect similar representation of both groups. (*b*) Spatial segregation of all passerine birds calculated for each grid cell as the species replacement rate in relation to the adjacent 8 grid cells; red represents maximum values. (*c*) The mountain areas of the world. Dark blue marks areas with >4 restricted-range species occurring together in a montane grid cell, yellow shows similar aggregations in the lowlands (tiny areas near the coasts of Ecuador and Brazil, and on several tropical islands).

(songbirds) underwent a worldwide phylogenetic expansion, initiated in the Oligocene by rapid island radiations in the proto-Papuan archipelago (Jønsson et al. 2011) and multidirectional dispersal across the oceans and the Wallacean archipelago to the Old World tropics. Further expansions went across the Palearctic region to North America and then onward to South America as part of the Great American Interchange (Weir et al. 2009). The historical biogeography of passerine birds explains the predominance of basal lineages in the Indo-Pacific island archipelago, with many relictual forms in the mountains of New Guinea, Sundaland, and the Indo-Burmese foothill mountains. Terminal radiations and high species turnover at northern latitudes (Weir & Schluter 2004), together with range dynamics in the Sino-Himalayan Mountains, would have provided ample opportunity for exchange between high- and low-latitude mountains during the climatically unstable Neogene period (Päckert et al. 2012).

3.1. Species Replacements and Endemism

High levels of species replacement (**Figure 1b**) characterize biome boundaries and are not particularly characteristic of montane regions (**Figure 1c**; see also Hawkins & Diniz-Filho 2006) at high latitudes (in agreement with Janzen's hypothesis). Presumably, the range dynamics at high latitudes and even in some tropical inland regions erased much of the population structure and local adaptations that took place intermittently and locally, leading to a moderate diversity and a predominance of widespread species.

The situation is quite different at low latitudes, where the tropical Andes region and some parts of the Afromontane and Sino-Himalayan montane regions stand out prominently (**Figure 1b**). These latter areas have particularly high aggregates of small-ranged species (**Figure 1c**) (Stattersfield et al. 1998, Davies et al. 2007, Rahbek et al. 2007). Phylogeographic data suggest a global tendency for a marked reduction in speciation events in the tropical lowlands during the climatically most unstable upper Pleistocene, while speciation continued in the tropical montane areas (Fjeldså & Bowie 2008, Päckert et al. 2012, Fjeldså 2012).

Small-range species apparently survived over time only where they could track their optimal ecological or climatic window by moving short distances (e.g., Tingley et al. 2009), which is typically the case in montane areas. It has been suggested and advocated that the distribution of small-ranged land vertebrates in montane regions may be related to low climate-change velocity (Sandel et al. 2011) and that this metric is a useful reflection of how fast species must move to keep track of climate change since the last Pleistocene glaciation, as well as in the future (Loarie et al. 2009). These are an approach and metric that so far rely exclusively on modeled data and, thus, have yet to be validated with historical empirical data.

As many as 1,116 (73.7%) of the 1,514 small-ranged passerine bird species (species inhabiting 1–20 grid cells, representing the fourth quartile of the 25% of species with the smallest range sizes among all passerine birds) live entirely or partly within the montane regions outlined in **Figure 1c**. In studies of species richness patterns of Africa and South America, the highest residual values of species diversity that could not be explained by the employed ecoclimatic models, notably of small-range species, were found in mountains and near coasts (Rahbek & Graves 2001, Jetz & Rahbek 2002, Jetz et al. 2004, Rahbek et al. 2007). Thus, because rugged landscapes are often found along continental margins, we need to consider what is most important: topography or proximity to the ocean. As many as 1,136 (75.0%) of the small-ranged passerine bird species live (wholly or partially) less than 300 km from sea coasts within 20° latitude (N and S). Of these, 862 inhabit montane regions or their coastal foothills, 39 are found in coastal lowlands, and a further 236 on islands, most of which are mountainous but too small for inclusion in our defined montane regions. The peak aggregations of small-ranged passerines occur in Costa Rica, the Panamanian highlands, the northern Andes, the Cameroon Mountains, mountains around the northern Indian Ocean, and mountainous islands along the southeast Asian gateway to the Pacific (**Figure 1c**). Globally, the highest density of small-ranged species is near warm coasts with year-round precipitation, but some monsoon coasts and mist-impacted coasts near the cool (upwelling) eastern boundary currents are also important. Farther inland, significant numbers of small-ranged species are found only in the Albertine Rift Mountains of central Africa, along the eastern Andean slope of southern Peru and Bolivia and locally in the Sino-Himalayan Mountains.

It has long been acknowledged that oceanic islands can act as refuges for ancient relictual species because of their thermal stability and slow biotic turnover (Cronk 1997). However, the remarkable aggregation of small-ranged birds outlined above points to a more general trend for such species to be found in rugged landscapes near tropical coasts. The existence of some ancient (relictual) species in the hot spots for small-ranged birds (such as *Sapayoa aenigma* in the South American Chocó region and the Modulatricidae in some African mountains) points to low rates of extinction (Fjeldså & Lovett 1997). There is now a need for a strong statistical approach to determine to what extent the diversification in the terrestrial biota is controlled by thermal inertia and the slow rhythms of variation in oceanic systems (Steele 1985), as well as by the stability of oceanic circulation patterns near the equator (e.g., von der Heydt & Dijkstra 2011).

3.2. Gardens of Eden in the Mist

The most outstanding places of montane endemism all share high air humidity (and thus gain heat from condensation) and high precipitation (at least seasonally). Cloud forests are with few exceptions confined to within 300 km of tropical coasts (Bruijnzeel et al. 2010), but there is a gradual transition to other humid montane forest types in the major mountain ranges extending deeper into the continents. In addition, mist-dependent vegetation types may exist in drier regions, notably on escarpments near coasts affected by cold eastern boundary currents (Bruijnzeel et al. 2010, p. 34). These vegetation formations are highly affected by sea temperatures and relative humidity, which determine where stratified clouds hit the mountain slopes. Persistent mist conditions are mainly found at 2,000–3,000-m elevation, but it is important to note that such habitats can exist below 1,000 m in small coastal mountains (also known as the telescoping effect; e.g., Monteverde near the Pacific coast of Costa Rica; see Foster 2001). Local drainage patterns of cold air may even lead to local development of cloud forest near sea level (Bruijnzeel et al. 2010, pp. 130–133). In these environments, water from wind-driven fog is added to the incident rainfall. This may not add much in places with heavy rainfall, but it is important outside the primary rainy periods and in the mist zones of arid regions. Thus, because of the dense condensation on the complex foliage of tiny, leathery leaves, a significant stemflow is observed even when the rainfall occurs as a fine drizzle or when the forest is cloud enshrouded (see figure 50.2 in Bruijnzeel et al. 2010). Whereas the annual mean precipitation declines slightly with elevation in large highlands, it increases with elevation in small mountains (see figure 3.6 in Bruijnzeel et al. 2010). These mountains are characterized by low evapotranspiration, strong infiltration, and constantly high soil moisture, leading to a large water storage capacity and constant flow of water to the adjacent lowlands.

Thus, cloud forests may represent a distinct habitat from that of the adjacent lowlands at early stages of mountain building, offering early opportunities for the evolution of a montane avifauna. Particular habitat characteristics that are likely to have been present at this stage include stunted/gnarled trees with tiny, leathery leaves and large amounts of epiphytes and mosses (which require constant air humidity). The leached and nutrient-poor soils would provide favorable conditions for nectarivorous birds (Rebelo 1991). This is a hypothesis that invites intensive research.

The high relative dominance of species representing mid-Tertiary radiations in many small mountains near tropical coasts (admittedly hard to see in **Figure 1a** because of the spatial

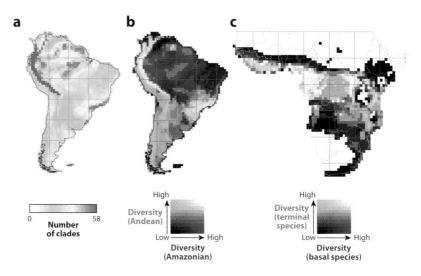

Figure 2

Diversification of South American and African birds. The two left maps show diversity of suboscines (mid-Miocene lineages, at 15 Mya, based on J.I. Ohlson, M. Irestedt, P.G.P. Ericson, and J. Fjeldså, unpublished data). (*a*) Shows all 76 clades (*darkest red* representing 58 clades present in one grid cell). (*b*) 35 clades rooted in the Andean region shown in green, 36 clades rooted in the Amazonian region in purple. (*c*) Diversity of greenbuls (Pycnonotidae). Basal species in the phylogeny (50% with shortest root paths, e.g., lowest numbers of nodes from the root of the phylogeny) are shown in purple, and the terminal species (longest root paths) are shown in green.

resolution) suggests that this is where the colonization road into the modern highlands started (see Bates et al. 1998 and Sedano & Burns 2010 for the Chocó region) and then extended deeper into the continents. Because of the way the climate is moderated by topography, atmospheric inversions and mist zones, the transition between mountain valleys and large highlands may persist in spite of climate fluctuations. Lush vegetation can under certain conditions exist even in the immediate proximity of glaciers, and some inland montane areas may have maintained refuge habitats for birds during the Pleistocene (see Qu et al. 2010 for the eastern outliers of the Tibet Plateau and Allen et al. 2010 for the Pamir-Tian-Altai Mountains).

4. THE HISTORY OF DIVERSIFICATION

4.1. The Long-Term History of Tropical Montane Avifaunas

Some elements of the long-term diversification history of montane birds is outlined below, starting with South America and the comprehensive molecular phylogenies that are available for suboscine birds (J.I. Ohlson, M. Irestedt, P.G.P. Ericson, and J. Fjeldså, unpublished data; Fjeldså 2012). **Figure 2** illustrates the diversity of suboscine clades in South America at 15 Mya, with a comparison of clades with distinct geographical origins in **Figure 2*b***. It is evident that the maximum diversity (*red cells* in **Figure 2*a***, *maximum brightness* in **Figure 2*b***) corresponds to grid cells located at the interface between the Andean and Amazonian biomes. Some (near) endemic clades of the Andean region date back to the Oligocene and early Miocene (Rhinocryptidae, *Chamaeza*, *Geositta*, and *Grallaria/Grallaricula*), and others appeared toward the mid-Miocene. These lineages are best represented in the old parts of the orocline, in Patagonia and in the submontane habitats of western Ecuador and Colombia. Some of them are small clades, suggesting a reduced net diversification rate since the initial adaptation to montane habitats. Other lineages maintained high rates of speciation throughout the period of orogeny.

Very few taxa among the Amazonian clades (e.g., *Thripadectes*) underwent marked upslope colonization into the Andes, but such an indication of constraining niche conservatism applies also to the Andean clades, which rarely colonized the tropical lowlands. Instead, the Andean avifauna is more strongly connected with the harsh lowland biomes of the Southern Cone of the continent (and with temperate North America). **Figure 2b** excludes some species-rich clades for which the geographical origins could not be precisely determined; however, they were probably south of the Amazon area and involved recent expansions into the tropical Andes region. Here the most intensive recent diversification took place along the *cis*-Andean tree-line zone (Fjeldså & Irestedt 2009). The furnariids (Derryberry et al. 2011) and fluvicoline flycatchers (Ohlson et al. 2008) were constantly able to maintain a high net speciation rate as the geographic center of diversification shifted over time from tropical lowland forests to the new savanna habitats in the south and then onward along the Andes.

At the same time, northern groups, notably the nine-primaried oscines, colonized South America during the Great American Interchange and initially diversified in the moderately high cloud forest ridges of Central America and the northern Andes, from where they proceeded, in a more dynamic way than the endemic South American groups, with several adaptive shifts and colonization of the highest mountain ranges as well as tropical and southern subtropical lowlands (Fjeldså & Rahbek 2006, Sedano & Burns 2010). Similarly, Santos et al. (2011) provide evidence that the Amazonian amphibian diversity arose by multiple colonizations out of the Andes.

Päckert et al. (2012) describe a similar progression for the Asian songbird radiation as outlined above for the Neotropical suboscines. Ancient groups inhabited the tropical lowlands and Indo-Burmese mountain foothills. These clades gave rise to a northward expansion through the interior mountain ranges of China (a connection that has been partly erased by Pleistocene aridification and diversity loss in China's inland mountains; see **Figure 1a**) to the extensive Palearctic forest region. Back colonization took place during the Pleistocene, along the Tian-Pamir-Hindu Kush Mountains to the subalpine Sino-Himalayan *Rhododendron*-coniferous forest zone, leading to the buildup of considerable species diversity in southern China. Thus, gray hues at the Chinese-Indochinese border zone in **Figure 1a** reflect the mix of species of Pleistocene age along the high ridges and ancient fauna in the warm valleys (see also López-Pujol et al. 2011).

Africa was once extensively forested, but savanna habitats expanded during the Miocene, with a tipping point around 5 Mya occurring in response to the formation of the west Antarctic ice sheet (Zachos et al. 2001). The result was severed connectivity of forest habitat across Africa as well as of the forest corridor across the Middle East that once connected Africa with India. Thus, the African montane avifauna has largely evolved in situ, except for rare long-distance dispersal events (e.g., *Hemitesia*; Irestedt et al. 2010) and some putative migratory drop-off from Palearctic radiations whose members winter in Africa (e.g., *Sylvia*; Voelker & Light 2011).

Greenbuls (Pycnonotidae) exemplify the large-scale pattern of lineage divergence in Africa. Here, the 50% most basal species (shortest root paths; *purple* in **Figure 2c**) occur widely across the lowland rainforest biomes, whereas terminal species (*green* in **Figure 2c**) occur primarily outside it, in the East African mosaic of highland habitats (Albertine Rift, Kenyan Highlands, and Eastern Arc Mountains) and savanna thickets (Fjeldså et al. 2007).

Most divergence times between sister taxa of African montane birds center on the Pliocene-Miocene boundary (see **Figure 3c**) (Voelker et al. 2010) and are not clustered in the Pleistocene as suggested in the past by several researchers (e.g., Diamond & Hamilton 1980). This clustering of divergence events in time suggests that most montane speciation events resulted from the rapid isolation of populations in separate sky islands, rather than through immigration from other areas

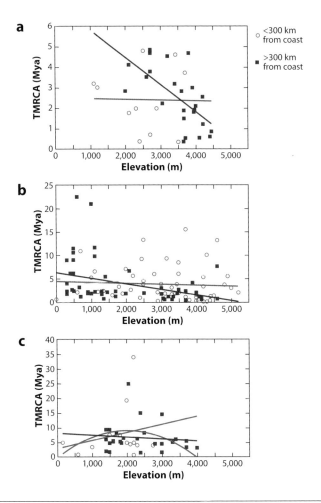

Figure 3

Time since most recent common ancestor (TMRCA) plotted against the upper elevational range of each species; red squares represent species living more than 300 km from the coast, blue open circles are species restricted to the zone less than 300 km from a warm coast. (*a*) Data for Sino-Himalayan mountains and Taiwan (redrawn from Päckert et al. 2012); (*b*) furnariid species of Peru and Bolivia, using the chronology developed by Derryberry et al. (2011); (*c*) data from eastern Africa (see the section on Materials and Methods).

as seen in the Andean and Sino-Himalayan avifaunas (Fjeldså & Bowie 2008, Voelker et al. 2010). This difference may also be a consequence of the isolation of the African montane fauna from that of the Palearctic region due to desert barriers, as well as from temperate environments in the south of the continent (e.g., Bowie et al. 2005).

4.2. Speciation and Species Turnover within Mountain Regions

A recent study of New World vertebrates documented that tropical sister taxa tend to have thermal niches that are both narrower and more evolutionarily conserved than those of temperate sister taxa (Cadena et al. 2012). Therefore, one may expect greater species packing in the tropics as a consequence of increased opportunity for isolation and allopatric divergence across elevational thermal gradients relative to temperate montane regions. However, it remains controversial whether disruptive selection across an elevational gradient can lead to parapatric speciation, as it requires selection to be sufficiently strong to counter the effects of recurrent gene flow (e.g., Fuchs et al. 2011). For birds in both the Andes and Sino-Himalayan montane systems, the dominant mode of divergence occurs via allopatry rather than parapatry, with sister taxa tending to occupy the same elevation on adjacent slopes across valleys (Weir 2009, Cadena et al. 2012, Päckert et al. 2012) and only tending to co-occur on the same slope as a consequence of secondary contact (García-Moreno & Fjeldså 2000).

Furthermore, the spatial connectivity with high-latitude avifaunas also appears to have played a major role in the diversification of the Andean and Sino-Himalayan birds. In both montane systems, the influx from higher latitudes is likely to have facilitated in situ diversification via isolation from temperate ancestral lineages preadapted to the harshness characterizing the highest elevations (Graves 1988, Sedano & Burns 2010, Price et al. 2011, Päckert et al. 2012), as revealed by the mean age estimate for lineage splits (**Figure 3a,b**). The corresponding plot for eastern Africa recovers a different pattern, and most lineage splits occur at mid-elevation (**Figure 3c**). This contrasting pattern perhaps reflects the fundamentally different spatial structure of the African montane biome, where montane systems typically consist of a series of spatially isolated sky islands and have very little area available above 4,000 m.

Among montane areas of endemism in Africa, most work to date has been conducted on the Eastern Arc Mountains of East Africa (e.g., Fjeldså & Bowie 2008, Lawson 2010, Tolley et al. 2011). Although some spatial concordance is present among area cladograms for several bird species, there is little temporal congruence (Fjeldså & Bowie 2008). For instance, although most lineages of montane birds exhibit a pronounced genetic break between the sky islands of the northern and central Eastern Arc, divergence estimates from mitochondrial DNA data vary from 10% to 1.5% (Fjeldså & Bowie 2008, Fuchs et al. 2011). This extensive variation among sister clades (most of them formally ranked as subspecies!) in adjacent areas questions the validity of a purely vicariance mechanism of diversification and emphasizes instead the role of persistence (lack of extinction).

Although a general tendency for more recent lineage divergence at high elevation seems to exist in the interior of the continents (*closed squares* in **Figure 3**), there appears to be no clear trend in mountains located near warm seas (*open circles* in **Figure 3**). This emphasizes the high lineage persistence in these mountains, as species diversity apparently accumulated over geological time along the entire elevational gradient. In the case of the Eastern Arc, it is remarkable that only three widely separated mountains (Usambara, Uluguru, and Udzungwa) among the 13 sky islands stand out as having particularly high numbers of small-ranged species, including some ancient relictual forms (Fjeldså & Bowie 2008). Among these, the Usambara mountains are located right at the coast and have cloud forest habitat that extends to the coastal foothills. The Uluguru and Udzungwa mountains are located farther inland but have steep and high elevational gradients that effectively capture the incoming precipitation. A climatic explanation is emerging for the Eastern Arc based on a circulation model for the Indian Ocean and is supported by the palynological record demonstrating an absence of vegetation change in the montane forest during the last glacial cycle (Marchant et al. 2007). Similar areas of stability have also been documented in the Andes (Fjeldså et al. 1999) and the montane wet tropics of Australia (Graham et al. 2006). Furthermore, paleoclimatic modeling can now be used to develop a predictive stability (habitat persistence) surface for any system of interest that can then be evaluated by using multilocus DNA sequence data together with phylogenetic and statistical phylogeographic analytical approaches (e.g., Carnaval et al. 2009).

Instead of always searching for barriers to explain speciation, we need to also consider the nature of some habitat configurations, such as the extremely long and narrow band of Andean tree-line habitat and the potential role that stochasticity and local weather phenomena can play in generating observed patterns of fragmentation (Graves 1988, Fjeldså et al. 1999). It has been suggested that small-ranged montane species could provide recruitment to the regional fauna (Roy et al. 1998). An analysis of range size versus TMRCA for furnariid clades that diversified in the Andes and in the tropical lowlands, respectively (*green* and *purple* in **Figure 2b**) recovered quite similar slopes ($y = 12.215 \times ^{0.3099}$, $R^2 = 0.0472$ for Andean groups; $y = 55.219 \times ^{0.3679}$, $R^2 = 0.0383$ for lowland groups) but different regression levels, with the range-size intercept at 1 Mya being 12.2 grid cells for Andean groups and 55.2 grid cells for lowland groups. The increase in range size over time could reflect higher extinction risks among small-ranged species. However, studies of historical population structures of Afromontane species, at finer temporal and spatial scales (e.g., Bowie et al. 2004, 2006; Voelker et al. 2010), suggest that species, which are now widespread, were fragmented and locally distributed during arid periods in the early Pleistocene. For instance, the Olive Sunbird, *Nectarinia olivacea*, which apparently originated locally in East Africa, underwent range expansion across the entire tropical region of Africa.

5. ADAPTATIONS TO LOCAL CONDITIONS

The geographical range of species is assumed to correlate with their abundance (Brown 1984, Gaston et al. 2000, Borregaard & Rahbek 2010). This occupancy-abundance relationship is considered one of the best-documented relationships in macroecology. Nevertheless, researchers with considerable field experience from tropical hot spots of endemism know that many species with tiny geographical distributions are abundant where they occur and are sometimes the most locally abundant taxon. This is at present poorly documented, probably because few biologists have spent sufficient time in the remaining patches of impenetrable virgin forest on steep, mist-enshrouded slopes to obtain adequate abundance data. However, some documentation has been provided for the Cameroon Mountains (Reif et al. 2006) and Eastern Arc Mountains (Romdal & Rahbek 2009, Fjeldså et al. 2010). Most convincingly, Williams et al. (2009) demonstrated for the montane forests of the Australian wet tropics that species with narrow environmental niches and small geographic ranges have high and uniform local abundance. Many of the narrow endemics are phylogenetically old and are probably locally specialized, and this may enhance the resilience of such species by maintaining high demographic connectivity throughout their ranges.

The analysis of plant-hummingbird mutualistic networks sampled at 31 localities spanning a wide range of climatic regimes across the Americas (Dalsgaard et al. 2011) established that the highest degree of mutualistic specialization among the sampled localities occurs in tropical montane forests (Costa Rica, Colombia, Brazilian Atlantic Forests). Across all 31 sites, the degree of mutualistic connectedness was well explained by the spatial variation of climate change velocity metrics (see above). Thus, such conditions not only benefit individual species but may also extend across communities.

5.1. Safe Havens for Montane Biodiversity?

Many researchers have expressed concern that montane species adapted to narrow elevational zones would be threatened under impending climate change. Montane species should be able to track their climatic niche by moving short distances up or down slope (Loarie et al. 2009, Tingley et al. 2009), but mountaintop inhabitants may have nowhere to go, as suggested by local empirical studies (Foster 2001, Moritz et al. 2008, Colwell & Rangel 2010), as well as by global modeling with different dispersal scenarios (La Sorte & Jetz 2010). The simulations by Williams et al. (2009) and Ohlemüller et al. (2011) project extensive loss of habitat, including within montane regions.

Similarly, the simulations conducted by McCain & Colwell (2011) find high risks of mountaintop extinctions during interglacials and lowland extinction thresholds during glacials, favoring mid-elevation lineages.

We agree that the biota of isolated mountains (conical-shaped, like the classical study sites of Volcan Barva in Costa Rica and Mount Kinabalu on Borneo) may be at risk as climate zones move up, but this may not be universally true. For realistic predictions of the fate of montane avifaunas, we need to know more about the local factors that have allowed small-ranged species to persist in spite of the instability of the global climate.

First, top-ridge habitats, with their distinct, low, gnarled and drought-resistant vegetation may be a consequence of exposure and seasonal dryness rather than elevation as such. Many birds may use this habitat to take advantage of special resources such as seasonal blooming. The recorded elevational distribution of such birds does not necessarily represent a climatic window.

Second, we need to consider whether the lapse-rate-based statistical climate models based on interpolation between widely scattered weather stations can account for the complex topography-driven patterns of temperature and humidity of larger montane regions (Nogués-Bravo et al. 2007). In particular, the models fail to predict the interaction between flows of humid air and topography and the position of significant cloud-affected zones within montane areas (Foster 2001; Bruijnzeel et al. 2010, p. 34). Topography may govern the wind systems in the mountain valleys and cause more or less stable air stratification and local cold air ponding. In the alpine environment, the vegetation is governed by the length of the growth season, temperature and, most importantly, the night-time soil temperatures during the growth season. In steep and complex landscapes, this leads to habitat mosaics, where sites a few meters apart may encompass strikingly different abiotic regimes (Scherrer & Körner 2011). For instance, Archaux (2004) found that elevational distributions of breeding birds in the Alps were related to site-specific factors rather than climatic warming.

5.2. Implications for Conservation

Small-ranged species in tropical montane hot-spot areas have been described in some conservation strategies as the living dead and, hence, deemed to be of low conservation value. However, some of these species are in fact common and well adapted to their local environment, and they represent lineages that have persisted for millions of years in spite of dramatic shifts in the global climate. Some of these birds may represent the last stage in a taxon cycle, as species that were once widespread in the end survive as local remnant populations. We suggest that such birds occur exactly where the chances for long-term survival are best and that they therefore represent viable components of the regional fauna. Whether they may expand again at some point in time and start a new taxon cycle, we cannot know.

DISCLOSURE STATEMENT

The authors are not aware of any affiliations, memberships, funding, or financial holdings that might be perceived as affecting the objectivity of this review.

ACKNOWLEDGMENTS

We thank the Danish National Research Foundation for funding the Center for Macroecology, Evolution, and Climate and thank the Hellman Foundation for support of Rauri Bowie's African montane biogeographic research. Louis A. Hansen is thanked for his painstaking effort to compile the global databases of bird distributions. Martin Päckert kindly provided primary data for drawing **Figure 3a**.

LITERATURE CITED

Allen JRM, Hickler T, Singarayer JS, Sykes MT, Valdes PJ, Huntley B. 2010. Last glacial vegetation of northern Eurasia. *Quat. Sci. Rev.* 29:2604–18

Archaux F. 2004. Breeding upwards when climate is becoming warmer: no bird responses in the French Alps. *Ibis* 148:138–44

Bates JM, Hackett SJ, Cracraft J. 1998. Area-relationships in the Neotropical lowlands: an hypothesis based on raw distributions of passerine birds. *J. Biogeogr.* 25:783–93

Best BJ, Kessler M. 1995. *Biodiversity and Conservation of Tumbesian Ecuador and Peru*. Cambridge, UK: BirdLife Int.

Borregaard MK, Rahbek C. 2010. Causality of the relationship between geographic distribution and species abundance. *Q. Rev. Biol.* 85:3–25

Bowie RCK, Fjeldså J, Hackett SJ, Bates JM, Crowe TM. 2006. Coalescent models reveal the relative roles of ancestral polymorphism, vicariance and dispersal in shaping phylogeographical structure of an African montane forest robin. *Mol. Phylogenet. Evol.* 38:171–88

Bowie RCK, Fjeldså J, Hackett SJ, Crowe TM. 2004. Molecular evolution in space and through time: mtDNA phylogeography of the Olive Sunbird (*Nectarinia olivacea/obscura*) throughout continental Africa. *Mol. Phylogenet. Evol.* 33:56–76

Bowie RCK, Voelker G, Fjeldså J, Lens L, Hackett SJ, Crowe TM. 2005. Systematics of the Olive Thrush *Turdus olivaceus* species complex with reference to the taxonomic status of the endangered Taita Thrush *T. helleri*. *J. Avian Biol.* 36:391–404

Brown JH. 1984. On the relationship between abundance and distribution of species. *Am. Nat.* 124:255–79

Bruijnzeel LA, Scatena FN, Hamilton LS. 2010. *Tropical Montane Cloud Forests*. Cambridge, UK: Cambridge Univ. Press

Burgess ND, Butynski TM, Cordeiro NJ, Doggart NH, Fjeldså J, et al. 2007. The biological importance of the Eastern Arc Mountains of Tanzania and Kenya. *Biol. Conserv.* 134:209–31

Cadena CD, Kozak KH, Gómez JP, Parra JL, McCain CM, et al. 2012. Latitude, elevational climatic zonation and speciation in New World vertebrates. *Proc. R. Soc. B* 279:194–201

Carnaval AC, Hickerson MJ, Haddad CFB, Rodrigues MT, Moritz C. 2009. Stability predicts genetic diversity in the Brazilian Atlantic Forest hotspot. *Science* 323:785–89

Cohen C, Wakeling JL, Mandiwana-Neudani TG, Dranzoa C, Crowe TM, Bowie RCK. 2012. Phylogenetic affinities of evolutionary enigmatic African galliformes: the Stone Partridge *Ptilopachus petrosus* and Nahan's Francolin *Francolinus nahani*, and support for their sister relationship with New World Quails. *Ibis* 154:768–80

Colwell RK, Rangel TF. 2010. A stochastic, evolutionary model for range shifts and richness on tropical elevational gradients under Quaternary glacial cycles. *Philos. Trans. R. Soc. B* 365:3695–707

Cronk Q. 1997. Islands: stability, diversity, conservation. *Biodiv. Conserv.* 6:477–93

Dalsgaard B, Magård E, Fjeldså J, Rahbek C, Olesen JM, et al. 2011. Specialization in hummingbird-plant networks is tightly linked with endemism. *PLoS ONE* 6(10):e25891

Davies RG, Orme CDL, Storch D, Olson VA, Thomas GH, et al. 2007. Topography, energy and the global distribution of bird species richness. *Proc. R. Soc. B* 274:1189–97

Derryberry EP, Claramunt S, Derryberry G, Chesser RT, Cracraft J, et al. 2011. Lineage diversification and morphological evolution in a large-scale continental radiation: the Neotropical ovenbirds and woodcreepers (Aves: Furnariidae). *Evolution* 65:2973–85

Diamond AW, Hamilton AC. 1980. The distribution of forest passerine birds and quaternary climate change in Africa. *J. Zool. Lond.* 191:379–402

Ericson PGP, Irestedt J, Johansson US. 2003. Evolution, biogeography, and patterns of diversification in passerine birds. *J. Avian Biol.* 34:3–15

Fjeldså J. 1995. Geographical patterns of neoendemic and relict species of Andean forest birds: the significance of ecologically stable areas. In *Biodiversity and Conservation of Neotropical Montane Forests*, ed. SP Churchill, H Balslev, E Forero, JL Luteyn, pp. 89–109. New York: New York Bot. Gard.

Fjeldså J. 2012. Diversification of the Neotropical avifauna: disentangling the geographical patterns of persisting ancient taxa and phylogenetic expansions. *Orn. Neotrop.* 23(Suppl.): In press

Fjeldså J, Bowie RCK. 2008. New perspectives on Africa's ancient forest avifauna. *Afr. J. Ecol.* 46:235–47

Fjeldså J, Irestedt M. 2009. Diversification of the South American avifauna: patterns and implications for conservation in the Andes. *Ann. Mo. Bot. Gard.* 96:398–409

Fjeldså J, Johansson US, Lokugalappatti LGS, Bowie RCK. 2007. Diversification of African greenbuls in space and time: linking ecological and historical processes. *J. Ornithol.* 148S:359–67

Fjeldså J, Kiure J, Doggart N, Hansen LA, Perkin AW. 2010. Distribution of highland forest birds across a potential dispersal barrier in the Eastern Arc Mountains of Tanzania. *Steenstrupia* 32:1–43

Fjeldså J, Krabbe N. 1990. *Birds of the High Andes.* Copenhagen: Zoological Museum

Fjeldså J, Lambin E, Mertens B. 1999. Correlation between endemism and local ecoclimatic stability documented by comparing Andean bird distributions and remotely sensed land surface data. *Ecography* 22:63–78

Fjeldså J, Lovett JC. 1997. Geographical patterns of phylogenetic relicts and phylogenetically subordinate species in tropical African forest biota. *Biodiv. Conserv.* 6:325–46

Fjeldså J, Rahbek C. 2006. Diversification of tanagers, a species rich bird group, from lowlands to montane regions in South America. *Integr. Comp. Biol.* 46:72–81

Foster P. 2001. The potential negative impacts of global climate change on tropical montane cloud forests. *Earth-Sci. Rev.* 55:73–106

Fritz SA, Jønsson KA, Fjeldså J, Rahbek C. 2012. Out of New Guinea: diversification and biogeographical patterns in island radiations of passerine birds. *Evolution* 66:179–90

Fuchs J, Bowie RCK, Fjeldså J. 2011. Diversification across an altitudinal gradient in the Tiny Greenbul (*Phyllastrephus debilis*) from the Eastern Arc Mountains of Africa. *BMC Evol. Biol.* 17:117

Fuchs J, Bowie RCK, Fjeldså J, Pasquet E. 2004. Phylogenetic relationships of the African bush-shrikes and helmet-shrikes (Passeriformes: Malaconotidae). *Mol. Phylogenet. Evol.* 33:428–39

Fuchs J, Irestedt M, Fjeldså J, Couloux A, Pasquet P, Bowie RCK. 2012. Molecular phylogeny of African bush-shrikes and allies: tracing the biogeographic history of an explosive radiation of corvoid birds. *Mol. Phylogenet. Evol.* 64:93–105

García-Moreno J, Fjeldså J. 2000. Chronology and mode of speciation in the Andean avifauna. *Bonn. Zool. Monogr.* 46:25–46

Gaston KJ, Greenwood JJD, Gregory RD, Quinn RM, Lawton JH. 2000. Abundance-occupancy relationships. *J. Appl. Ecol.* 37:39–59

Ghalambor CK, Huey RB, Martin PR, Tewksbury JJ, Wang G. 2006. Are mountain passes higher in the tropics? Janzen's hypothesis revisited. *Integr. Comp. Biol.* 46:5–17

Graham CH, Moritz C, Williams SE. 2006. Habitat history improves prediction of biodiversity in a rainforest fauna. *Proc. Natl. Acad. Sci. USA* 103:632–36

Graves GR. 1988. Linearity of geographic range and its possible effect on the population structure of Andean birds. *Auk* 105:47–52

Hawkins BA, Diniz-Filho JAF. 2006. Beyond Rapoport's rule: evaluating range size patterns of New World birds in a two-dimensional framework. *Glob. Ecol. Biogeogr.* 15:461–69

Hughes C, Eastwood R. 2006. Island radiation on a continental scale: exceptional rates of plant diversification after uplift of the Andes. *Proc. Natl. Acad. Sci. USA* 103:20334–39

Irestedt M, Gelang M, Sangster G, Olsson U, Ericson PGP, Alström P. 2010. *Hemitesia neumanni* (Aves: Sylvioidea): a relict member of a Paleotropic Miocene avifauna? *Ibis* 153:78–86

Jansson R, Dynesius M. 2002. The fate of clades in a world of recurrent climatic change: Milankovitch oscillations and evolution. *Annu. Rev. Ecol. Syst.* 33:741–77

Janzen DH. 1967. Why mountain passes are higher in the tropics. *Am. Nat.* 101:233–49

Jetz W, Rahbek C. 2002. Geographic range size and determinants of avian species richness. *Science* 297:1548–51

Jetz W, Rahbek C, Colwell RK. 2004. The coincidence of rarity and richness and the potential signature of history in centres of endemism. *Ecol. Lett.* 7:1180–91

Johansson US, Alström P, Olsson U, Ericson PCP, Sundberg P, Price TD. 2007a. Build-up of the Himalayan avifauna through immigration: a biogeographical analysis of the *Phylloscopus* and *Seicercus* warblers. *Evolution* 61:324–33

Johansson US, Fjeldså J, Lokugalappatti LGS, Bowie RCK. 2007b. A nuclear DNA phylogeny and proposed taxonomic revision of African greenbuls (Aves, Passeriformes, Pycnonotidae). *Zool. Scripta* 36:417–27

Jønsson KA, Fabre PH, Ricklefs RE, Fjeldså J. 2011. Major global radiation of corvoid birds originated in the proto-Papuan archipelago. *Proc. Natl. Acad. Sci. USA* 108:2328–33

Kapos V, Rhind J, Edwards M, Price MF. 2000. Developing a map of the world's mountain forests. In *Forest in Sustainable Mountain Development*, ed. MT Price, N Butts, pp. 4–9. Wallingford, UK: CAB Int.

Kohler T, Maselli D. 2009. *Mountains and Climate Change. From Understanding to Action*. Berne: Geogr. Bernesia

La Sorte FA, Jetz W. 2010. Projected range contractions of montane biodiversity under global warming. *Proc. R. Soc. B* 277:3401–10

Lawson LP. 2010. The discordance of diversification: evolution in the tropical-montane frogs of the Eastern Arc Mountains of Tanzania. *Mol. Ecol.* 19:4046–60

Loarie SR, Duffy PB, Hamilton H, Asner GP, Field CB, Ackerly DD. 2009. The velocity of climate change. *Nature* 462:052–55

López-Pujol J, Zhang F-M, Sun H-Q, Ying T-S, Ge S. 2011. Centres of plant endemism in China: places for survival or for speciation? *J. Biogeogr.* 38:1267–80

Lovett JC. 1993. Temperate and tropical floras in the mountains of eastern Tanzania. *Opera Bot.* 121:217–27

Mandiwana-Neudani TG. 2012. *Taxonomy, phylogeny and biogeography of francolins ('Francolinus' spp)*. Ph.D. Thesis, Univ. Cape Town, South Africa. 422 pp.

Marchant R, Mumbi C, Behera S, Yamagata T. 2007. The Indian Ocean dipole—the unsung driver of climatic variability in East Africa. *Afr. J. Ecol.* 45:4–16

Martin K, Wiebe KL. 2004. Coping mechanisms of alpine and arctic breeding birds: extreme weather and limitations to reproductive resilience. *Integr. Comp. Biol.* 44:177–85

McCain CM. 2009. Global analysis of bird elevational diversity. *Glob. Ecol. Biogeogr.* 19:346–60

McCain CM, Colwell RK. 2011. Assessing the threat to montane biodiversity from discordant shifts in temperature and precipitation in a changing climate. *Ecol. Lett.* 14:1236–45

Moritz C, Patton JL, Conroy CJ, Parra JL, White GC, Beissinger SR. 2008. Impact of a century of climate change on small-mammal communities in Yosemite National Park, USA. *Science* 322:261–64

Moritz C, Patton JL, Schneider CJ, Smith TB. 2000. Diversification of rainforest faunas: an integrated molecular approach. *Annu. Rev. Ecol. Syst.* 31:533–63

Njabo NY, Bowie RCK, Sorenson MD. 2008. Phylogeny, biogeography and taxonomy of the African wattle-eyes (Aves: Passeriformes: Platysteiridae). *Mol. Phylogenet. Evol.* 48:136–49

Nogués-Bravo D, Araújo MB, Errea MP, Martínez-Rica JP. 2007. Exposure of global mountain systems to climate warming during the 21st Century. *Glob. Envir. Chang.* 17:420–28

Ohlemüller R, Anderson BJ, Araújo MB, Butchart SHM, Kudrna O, et al. 2011. The coincidence of climatic and species rarity: high risk to small-range species from climate change. *Biol. Lett.* 4:568–72

Ohlson JI, Fjeldså J, Ericson PGP. 2008. Tyrant flycatchers coming out in the open: phylogeny and ecological radiations in Tyrannidae (Aves: Passeriformes). *Zool. Scripta* 37:315–35

Päckert M, Martens J, Sun Y-H, Severinghaus LL, Nazarenko AA, et al. 2012. Horizontal and elevational phylogeographic patterns of Himalayan and Southeast Asian forest patterines (Aves: Passeriformes). *J. Biogeogr.* 39:556–73

Price TD, Mohan D, Tietze DT, Hooper DM, Orme CDL, Rasmussen PC. 2011. Determinants of northerly range limits along the Himalayan bird diversity gradient. *Am. Nat.* 178:97–108

Qu Y, Lei F, Zhang R, Lu X. 2010. Comparative phylogeography of five avian species: implications for Pleistocene evolutionary history in the Qinghai-Tibetan plateau. *Mol. Ecol.* 19:338–51

Rahbek C, Gotelli NJ, Colwell RK, Entsminger GL, Rangel TFLVB, Graves GR. 2007. Predicting continental-scale patterns of bird species richness with spatially explicit models. *Proc. R. Soc. B* 274:165–74

Rahbek C, Graves GR. 2000. Detection of macro-ecological patterns in South American hummingbirds is affected by spatial scale. *Proc. R. Soc. B* 267:2259–65

Rahbek C, Graves GR. 2001. Multiscale assessment of patterns of avian species richness. *Proc. Natl. Acad. Sci. USA* 98:4534–39

Rebelo AG. 1991. Community organization of sunbirds in the Afro-tropical region. *Acta Congr. Int. Orn.* 20:1180–87

Reif J, Hořák D, Sedláček O, Riegert J, Pešata L, et al. 2006. Unusual abundance–range size relationship in an Afromontane bird community: the effect of geographical isolation. *J. Biogeogr.* 33:1959–68

Romdal TS, Rahbek C. 2009. Elevational zonation of afrotropical forest bird communities along a homogeneous forest gradient. *J. Biogeogr.* 36:327–36

Roy MS, Arctander P, Fjeldså J. 1998. Speciation and taxonomy of montane greenbuls of the genus *Andropadus* (Aves: Pycnonotidaae). *Steenstrupia* 24:51–66

Sandel B, Arge L, Dalsgaard B, Davies RG, Gaston KJ, et al. 2011. The influence of late quaternary climate-change velocity on species endemism. *Science* 334:660–64

Santos JC, Coloma LA, Summers K, Caldwell JP, Ree R, Cannatella DC. 2011. Amazonian amphibian diversity is primarily derived from late Miocene Andean lineages. *PLoS Biol.* 7(3):e1000056

Scherrer D, Körner C. 2011. Topographically controlled thermal-habitat differentiation buffers alpine plant diversity against climate warming. *J. Biogeogr.* 38:406–16

Sedano RE, Burns KJ. 2010. Are the Northern Andes a species pump for Neotropical birds? Phylogenetics and biogeography of a clade of Neotropical tanagers (Aves: Thraupini). *J. Biogeogr.* 37:325–43

Stattersfield AJ, Crosby MJ, Long AJ, Wege DC. 1998. *Endemic Bird Areas of the World*. Cambridge, UK: BirdLife Int.

Steele JH. 1985. A comparison of terrestrial and marine ecological systems. *Nature* 313:355–58

Tingley MW, Monahan WB, Beissinger SR, Moritz C. 2009. Birds track their Grinnellian niche through a century of climate change. *Proc. Natl. Acad. Sci. USA* 106:19637–43

Tolley KA, Tilbory CR, Measey GJ, Menegon M, Branch WR, Matthee CA. 2011. Ancient forest fragmentation or recent radiation? Testing refugial speciation models in chameleons within an African biodiversity hotspot. *J. Biogeogr.* 38:1748–60

Voelker G, Light JE. 2011. Palaeoclimatic events, dispersal and migratory losses along the Afro-European axis as drivers of biogeographic distributions in Sylvia warblers. *BMC Evol. Biol.* 11:163

Voelker G, Outlaw RK, Bowie RCK. 2010. Pliocene forest dynamics as a primary driver of African bird speciation. *Glob. Ecol. Biogeogr.* 19:111–21

von der Heydt AS, Dijkstra HA. 2011. The impact of ocean gateways on the ENSO variability in the Miocene. In *The SE Asian Gateway: History and Tectonics of the Australia-Asia Collision*, ed. R Hall, MA Cottam, MEJ Wilson, pp. 305–18. London: Geol. Soc.

Vuilleumier F, Monasterio M. 1986. *High Altitude Tropical Biogeography*. Oxford: Oxford Univ. Press

Weir JT. 2009. Implications of genetic differentiation in Neotropical montane forest birds. *Ann. Mo. Bot. Gard.* 96:410–33

Weir JT, Bermingham E, Schluter D. 2009. The Great American Biotic Interchange in birds. *Proc. Natl. Acad. Sci. USA* 106:21737–42

Weir JT, Schluter D. 2004. Ice sheets promote speciation in boreal birds. *Proc. R. Soc. B* 271:1881–87

Williams JW, Jackson ST, Kutzbach JE. 2007. Projected distributions of novel and disappearing climates by 2100 AD. *Proc. Natl. Acad. Sci. USA* 104:5738–42

Williams SE, Williams YM, VanDerWal J, Isaac JL, Shoo LP, Johnson CN. 2009. Ecological specialization and population size in a biodiversity hotspot: how rare species avoid extinction. *Proc. Natl. Acad. Sci. USA* 106:19737–41

Zachos J, Pagani M, Sloan L, Thomas E, Billups K. 2001. Trends, rhythms, and aberrations in global climate 65 Ma to present. *Science* 292:686–93

Evolutionary Inferences from Phylogenies: A Review of Methods

Brian C. O'Meara

Department of Ecology and Evolutionary Biology, University of Tennessee, Knoxville, Tennessee 37996; email: bomeara@utk.edu, http://www.brianomeara.info

Annu. Rev. Ecol. Evol. Syst. 2012. 43:267–85

First published online as a Review in Advance on September 4, 2012

The *Annual Review of Ecology, Evolution, and Systematics* is online at ecolsys.annualreviews.org

This article's doi:
10.1146/annurev-ecolsys-110411-160331

1543-592X/12/1201-0267$20.00

Keywords

methods, continuous-time Markov Chain, multivariate normal, birth-death, tree stretching

Abstract

There are many methods for making evolutionary inferences from phylogenetic trees. Many of these can be divided into three main classes of models: continuous-time Markov chain models with finite state space (CTMC-FSS), multivariate normal models, and birth-death models. Numerous approaches are just restrictions of more general models to focus on particular questions or kinds of data. Methods can be further modified with the addition of tree-stretching algorithms. The recent realization of the effect of correlated trait evolution with diversification rates represents an important advance that is slowly revolutionizing the field. Increased attention to model adequacy may lead to future methodological improvements.

INTRODUCTION

A phylogeny is an inferred history of species splitting through time. It may just show the relationships of the species, but it may also have branch lengths (in units of time, number of generations, or amount of change) and, more rarely, widths (population size through time). As species evolve up a phylogeny, various processes lead to changes in trait values within populations. For example, the optimal body size may change as climate warms, as competitors invade a habitat, or as new resources become available, so the average body size of individuals in a species can change through time as the population evolves toward the moving optimum.

Information about a phylogeny and trait values at the tips of the tree can be combined with an inference method to learn about evolutionary processes. One frequently applied example of this is methods for inferring character states for ancestral species. These methods have been used to reconstruct what calls extinct frogs may have made and to evaluate the reaction of extant frogs to these reconstructions to examine whether frogs evolve to sound unlike their neighbors (Ryan & Rand 1995). Similarly, the history of changes can be inferred from a tree, as was done recently with swine flu viruses to examine how certain strains lead to epidemics (Vijaykrishna et al. 2011). Methods that investigate the correlation of gene presence over evolutionary time can be used to predict functional links between genes better than methods that ignore a tree (Barker & Pagel 2005).

Some questions that can be addressed using this sort of approach include what trait the ancestor to a particular group of species had, how the possession of trait W influences the evolution of trait X, whether trait Y leads to a higher diversification rate, which group is evolving phenotypically the fastest, which group is diversifying the slowest, whether the rate of evolution has decreased through time, whether two particular clades are evolving toward different evolutionary optima, what the relative influences of past and present optima are on the observed value of a trait, whether trait Z is rarely observed because species with this trait go extinct rapidly, whether population size has changed through time, whether there has been gene flow between particular putative species, when a species dispersed between particular continents, and many more. However, though there is a dizzying array of methods for using phylogenies to make evolutionary inferences, certain themes keep arising. Rather than list all the methods, this review describes the connections between many methods. For example, the same basic model can be used for estimating whether an ancestral species had wings (ancestral state reconstruction), whether evolution of woody stems leads to evolution of bat-dispersed fruits (character correlation), and which tree best fits the data (DNA evolution). Many of the most widely used methods rely upon one of the three following classes of processes (which could themselves be grouped at deeper levels): continuous-time Markov chains with a finite state space (CTMC-FSS), multivariate normal distributions, and birth-death branching processes. Some important methods, however, do not fall neatly into these three bins. A set of approaches we can call tree stretching, the transformation of some or all branches of a tree, is a frequently employed approach used to build new methods from these common classes. New approaches jointly examine character transitions and diversification. This review covers these approaches at a general, and hopefully accessible, level.

CONTINUOUS-TIME MARKOV CHAINS WITH A FINITE STATE SPACE

Many methods are based on this one general process, ranging from estimating DNA substitution rates on a tree to looking at the effect of foraging time on coat color. Continuous-time means that time can be broken up into arbitrarily small chunks (rather than being in discrete units like

Q matrix

	A	B	C	D	E
A	-	r_{AB}	r_{AC}	r_{AD}	r_{AE}
B	r_{BA}	-	r_{BC}	r_{BD}	r_{BE}
C	r_{CA}	r_{CB}	-	r_{CD}	r_{CE}
D	r_{DA}	r_{DB}	r_{DC}	-	r_{DE}
E	r_{EA}	r_{EB}	r_{EC}	r_{ED}	-

Frequency vector

A	f_A
B	f_B
C	f_C
D	f_D
E	f_E

Figure 1

Basic structure of a continuous-time Markov chain finite state space (CTMC-FSS) model. The two main elements are an instantaneous rate matrix, **Q**, and a vector of state frequencies.

generations). A Markov chain is a random process in which the next step depends only on the present step, not steps further in the past. The probability of a nucleotide mutating from T to G when hit by a gamma ray does not depend on how long the position has been in state T or what state it was in before T. One common confusion is that while Markov chain Monte Carlo often appears in the context of Bayesian approaches in phylogenetics, there is nothing necessarily Bayesian about a Markov process. A finite state space means there are only certain states possible, such as A, T, G, or C for a particular site in DNA.

Figure 1 shows the basic elements of the model. When at a particular state, A, there is a probability that in a tiny time interval A becomes B. Call this rate r_{AB}. Such rates occur between every pair of states. They are often put into a matrix with the "from" state on the rows and the "to" state on the columns. This matrix is known as the instantaneous transition rate matrix, often given the symbol **Q**. The diagonal element r_{AA} is just the negative of the sum of the other elements, $-(r_{AB} + r_{AC} + r_{AD} + r_{AE})$. For convenience, this is usually just indicated as "$-$". There is also a vector of state frequencies. These can be optimized given a model set to fixed values (such as equal), set to empirical (observed) distributions, or set to equilibrium values. There is extensive work on the properties of continuous-time Markov chains, but to understand the links between methods, there are two things to remember. One is that while **Q** only has instantaneous rates, it is straightforward to get a matrix of probabilities of change between each state over a time interval t. This matrix is known as the transition-probability matrix and is frequently symbolized as $P(t)$ (P for probability, t because different length intervals have different transition probabilities). It is equal to $e^{\mathbf{Q}t}$. Take, for example, the Jukes-Cantor model of DNA evolution, where all transition rates are equal (all off-diagonal entries of **Q** are the same). On a short branch, the probability of starting at T and ending at A would be much smaller than the probability of starting at T and ending at T (not much time for change). But on a very long branch, the two probabilities would approach equality. Second, the vector of state probabilities at the end of a branch is the vector of probabilities at the beginning of a branch multiplied by $P(t)$. Pioneering work (Felsenstein 1981, Janson 1992, Pagel 1994, Yang 1994a) brought this model to tree inference and later to post-tree analysis of trait evolution.

The beauty of this model is that given a tree, transition rates, and frequencies at the root, one can calculate the probabilities of observed tip states. One could also examine different transition rates, root state frequencies, and even states elsewhere internally on the tree and see how they affect the probabilities of the tip states (the likelihood). A tree search involves optimizing some or all of

a GTR nucleotides

	A	G	C	T
A	-	r_{AG}	r_{AC}	r_{AT}
G	r_{AG}	-	r_{GC}	r_{GT}
C	r_{AC}	r_{GC}	-	r_{CT}
T	r_{AT}	r_{GT}	r_{CT}	-

b Binary correlation

	00	01	11	10
00	-	r_A	0	r_B
01	r_C	-	r_D	0
11	0	r_E	-	r_F
10	r_G	0	r_H	-

c Covarion

	A+	T+	A-	T-
A+	-	r_{AT}	δ	0
T+	r_{TA}	-	0	δ
A-	$k\delta$	0	-	0
T-	0	$k\delta$	0	-

d Ordered transitions

	0	1	2	3
0	-	r_{01}	0	0
1	r_{10}	-	r_{12}	0
2	0	r_{21}	-	r_{23}
3	0	0	r_{32}	-

Figure 2

Similarities between methods based on continuous-time Markov chain with a finite state space (CTMC-FSS) models. (*a*) General time reversible (GTR) nucleotide model; (*b*) model of correlated binary characters (Pagel 1994); (*c*) covarion model using variables from Penny et al. (2001), showing only two of four nucleotides for compactness; and (*d*) a model of ordered transitions (for example, state 1 can only go to its neighboring states, 0 or 2). Note that despite these models being intended for different kinds of characters (nucleotides, binary characters, and multistate characters) and for different purposes (tree inference, character correlation, examination of the process of hidden rate changes, and looking at transitions of a phenotypic character), they are restrictions of the same general **Q** matrix, with different rates set to equal each other or set to zero, and with the state names themselves varying. Not shown are the frequency vectors, which are also quite similar. If the frequencies are based on equilibrium values, they are specified by the **Q** matrix alone.

the above parameters as well as the tree itself: The basic model is the same (note that this review largely ignores methods intended just to infer the tree alone). Many Bayesian and likelihood-based models for discrete characters are just this general model with particular restrictions on the **Q** matrix. For example (**Figure 2**), when one infers a tree using a general time reversible (GTR) model, it is just a model where the **Q** matrix has four rows and columns (one for each nucleotide) and where rates are constrained such that the rate from state *i* to state *j* is forced to be equal to the rate from state *j* to state *i*, for any *i* and *j*. Pagel's model of binary character correlation (Pagel 1994) also has a four-by-four **Q** matrix: Each possible pair of binary states just becomes a single state in a four-state character, and the **Q** matrix is restricted so that changes in two binary states cannot happen simultaneously (the instantaneous rate of going from 00 to 11 is set to zero, so this transition must be accomplished by going through 01 or 10 intermediates). Seen this way, extensions for correlations of multiple, multistate characters is a straightforward change of mapping each possible combination to a different discrete state and making sure that instantaneous changes of multiple characters at once are prevented (making sure that there is enough data to fit the rates involved is a later empirical problem). A covarion model (Galtier 2001, Penny et al. 2001, Tuffley & Steel 1998), a model with characters invisibly converting between changeable and invariant modes, is also a restriction of the **Q** matrix (and if it had just two observed states, say A and T, as in **Figure 2**, it is also a four-by-four matrix; with four observed states, it would be an eight-by-eight matrix). Models that allow for different transition rates between different states for a single character, perhaps even treating characters as ordered (the character states form a chain), are restricted **Q** matrices, as are models of protein evolution that use a fixed rate matrix, such as a JTT (Jones et al. 1992) matrix. Statistical models for biogeography that allow species to occur in multiple discrete localities (Ree & Smith 2008) are also based on this basic model. The instantaneous rate parameters may be constrained in various ways (e.g., set to zero or set to equal one another) and may have different names (e.g, the rate of going from diurnal to nocturnal, the rate of changing from valine to lysine, the rate of entering into a variable covarion site, etc.), but the basic model still applies. Even maximum parsimony for discrete characters is an extension of the CTMC-FSS method that allows different rates on different branches for different characters [the no common mechanism model (Tuffley & Steel 1997)], though the basis for its appeal is not that it fits a complex model but rather that it assumes change is rare. Some of the methods involving parsimony, such as the concentrated changes test (Maddison 1990), may be outside the CTMC-FSS framework, but this has yet to be formally evaluated. Sometimes even models that appear to be unbounded birth-death models (see below) are actually CTMC-FSS models. For example, Hahn et al. (2005) look at the evolution of gene family size up a tree. They treat gene family size as an ordered, discrete character that can only attain a particular maximum size (100 genes, in their initial work). There are then transitions on the tree from one number of genes to a neighboring number of genes, with the same gain and loss rate between neighboring states (same chance of going from 2 to 3 genes as from 82 to 83 genes), except with no gain rate from state 0 (lost forever) and state 100 (higher numbers of genes than 100 are prohibited).

Ancestral states can be estimated at the root of the tree or at internal nodes by using this same framework. This can be done for joint estimates (Yang et al. 1995) or marginal estimates. Joint estimates find the single best reconstruction across all nodes that maximizes the likelihood; this is like inferring someone's favorite meal (a joint collection of several food items). Marginal estimates find the single best reconstruction at each node examined alone, integrating over the possibilities at all other nodes; this is like inferring someone's favorite food item, averaging across all the meals in which it could be eaten. If a question involves a complete set of reconstructions, the joint approach is better; if the question involves a reconstruction at one particular node, a marginal approach is preferred. Stochastic character mapping (Huelsenbeck et al. 2003, Nielsen 2002) can use the **Q** matrix to evolve data up a tree, subject to the constraint that the data at the tips must match the observed data. Each sampled history is a single instance of a character evolving up a tree; one can summarize across these histories the number of times a character changes on a branch or which parts of a branch are in a particular character state. This is the only currently available method to reconstruct changes at points within branches, though evolutionary pathway likelihood (Steel & Penny 2000) as a reconstruction method of traits along branches of a fixed tree remains relatively unexplored.

Other statistical ideas can be added to this basic model. For example, using priors and estimating posterior probabilities converts it into a Bayesian approach (note that if other parameters are being estimated, such as tree topology, those must also be embedded in a Bayesian model). There are several examples of this (Pagel et al. 2004, Schultz & Churchill 1999). If there is a biased sample of characters such that invariant characters are not sampled, the bias can be corrected for in the model (Felsenstein 1992, Lewis 2001). Uncertainty in observations can also be directly dealt with in this model (Felsenstein 2004).

Heterogeneity can be incorporated through additions to this model (**Figure 3**). Heterogeneity can mean different processes at different characters (such as the difference in rates and state frequencies between nuclear and mitochondrial genes), different processes at different parts of the tree (early versus late in evolution, or woody versus herbaceous plants), or a combination of these. One approach to dealing with different processes at different characters is to use a partitioned model (Yang 1996). One set of characters gets assigned to one model, and a different set gets assigned to a different model. If the only difference is rates at different sites, a common method is to use a discrete gamma distribution (Yang 1994b). This divides a gamma distribution into k rates, and the overall likelihood for each site is the weighted sum of the site likelihoods with the same **Q** matrix but scaled by one of the k overall rates. The phylogenetic mixture model (Pagel & Meade 2004) works in a similar way, but rather than calculating the likelihood at a site using a single **Q** matrix linearly transformed by several rates to form a set of **Q** matrices, it calculates the likelihood at a site using multiple **Q** matrices (and also, potentially, different overall rates). For a given **Q** matrix it is the same as the basic CTMC-FSS; the innovation is using the sum

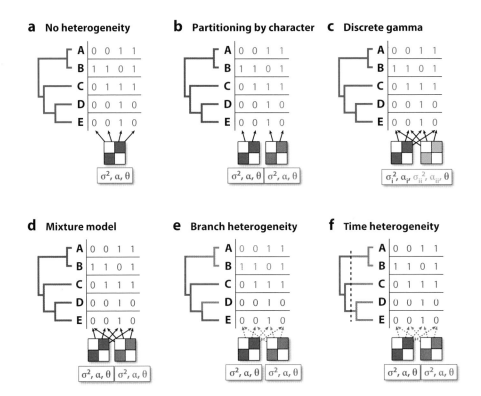

Figure 3

Dealing with heterogeneity on a tree. In each case, different models are represented in different colors (*red* and *blue*). Elements used with multiple models are in purple. Icons for continuous-time Markov chains with a finite state space (CTMC-FSS) models (**Q** matrix) and multivariate normal models (Ornstein-Uhlenbeck parameters σ^2, α, and θ) are shown for each to suggest how the same approach to heterogeneity can be used for each type of model. (*a*) The typical homogenous model. (*b*) Partitioning by character. Characters 1 and 2 (*red*) use one model, whereas characters 3 and 4 (*blue*) use another, but all share the same tree. (*c*) A discrete gamma model (Yang 1994b) with only two rate categories. With this approach the likelihood at each site is a weighted sum of likelihoods under different **Q** matrices, which are all linear transformations of the same basic matrix (thus showing the two models as very similar in *red* and *pink*). This has not yet been applied to continuous characters but could work in the same way. (*d*) A phylogenetic mixture model (Pagel & Meade 2004). This is like the discrete gamma model but allows for **Q** matrices that are not linear transformations of each other. (*e*) A branch heterogeneity model. Different branches may be assigned different models. This, and the next panel, can be done with birth-death models as well. (*f*) A time heterogeneity model. Different time periods are assigned to different models. Note that these various approaches to dealing with heterogeneity can be mixed with each other and with tree-stretching approaches.

of the likelihoods for the data under different **Q** matrices as the likelihood rather than using the likelihood under one **Q** matrix to account for differences between sites. Other approaches for dealing with rate heterogeneity across sites allow for rates to be correlated between neighboring sites (Felsenstein & Churchill 1996, Yang 1995). Heterogeneity across a tree has received less attention. Some have investigated allowing different state frequencies at different parts of the tree (Galtier & Gouy 1998, Yang & Roberts 1995). There has been some work on allowing different **Q** matrices on different, prespecified branches in order to evaluate different amounts of selection on different branches (Yang & Nielsen 2002, Zhang et al. 2005). It is important to realize that though many methods for dealing with heterogeneity envision using them for nucleotide data, there is no necessary requirement that these are the only sorts of data that can be used because underneath them is the same CTMC-FSS model. For heterogeneity across character models, anything other than full partitioning may not make sense unless there is some biological reason to think that different characters would have similar transition matrices (which may occur if they are multiple characters of the same type such as gene presence/absence or secondary compound presence/absence). For heterogeneity across the tree, which can be used for even a single character, characters of many types could be examined. For example, one analysis could investigate whether the ratios of transition rates between herbivory, omnivory, and carnivory in mammals changed after the KT extinction by fitting different **Q** matrices to branches (and parts of branches) before and after the KT event.

MULTIVARIATE NORMAL DISTRIBUTION

Normal distributions are common in biology. By the central limit theorem of statistics, the sum of a set of independent random variables (each with finite mean and variance, but no stronger requirements) is normally distributed. The log of a product of positive independent random variables is also normally distributed. For example, if body mass in a species continues to be multiplied by various factors (cooler climate leads to a 10% increase in size, lower food availability leads to a 5% decrease in size, competition with other species leads to a 7% increase in size) across many replicates of this sort of evolutionary process, the log(body mass) should be normally distributed. This process is often called Brownian motion. Many evolutionary processes, ranging from tracking a selective optimum to genetic drift, can lead to this pattern (Edwards & Cavalli-Sforza 1964, Felsenstein 1988, Hansen & Martins 1996).

If looking at a single trait for a single species, that trait would be normally distributed under this model; that is, rerunning evolution many times would lead to a normal distribution of trait values. This distribution could be characterized by its mean and variance. If considering a pair of species, or a pair of traits for a single taxon, it becomes a bivariate normal distribution: Each element has its own mean and variance but there may also be covariance between the two elements. For multiple species and/or multiple traits, it becomes a multivariate normal distribution. Each element has its own mean and variance, but may also have covariance with the other elements (**Figure 4**). For simple Brownian motion of a single character, the variances are the root-to-tip length for each taxon (in units of time) multiplied by the rate of Brownian motion. The covariances are the shared time from the root of the tree to the most recent common ancestor of the pair of taxa multiplied by the Brownian motion rate. The character means are equal to the state at the root.

Some approaches allow particular branches of a tree to be assigned a priori to different rate categories for continuous characters (though the same approaches would work for discrete characters) and then allow the rates in each category to be estimated (O'Meara et al. 2006, Thomas et al. 2006), which is equivalent to stretching particular branches [in the noncensored approach of O'Meara et al. (2006)]. Recent advancements allow automated detection of the optimal locations for these rate parameters (Eastman et al. 2011, Thomas & Freckleton 2012). Ornstein-Uhlenbeck models are similar in that they stretch branch lengths but may also transform expectations. Brownian motion allows traits to wiggle, with the same expectation of variance in each unit of time. Ornstein-Uhlenbeck models are sometimes described as rubber band models: A species trait value wiggles through time but is drawn back to a mean value with some force. Hansen (1997) developed a model that allowed the mean value to change over the tree while keeping the wiggle and attraction parameters constant. Butler & King (2004) developed a likelihood implementation of this model. Beaulieu et al. (2012) developed a more general model that allowed mean, attraction, and wiggle to all vary either together or at different parts of the tree. A different use of this model

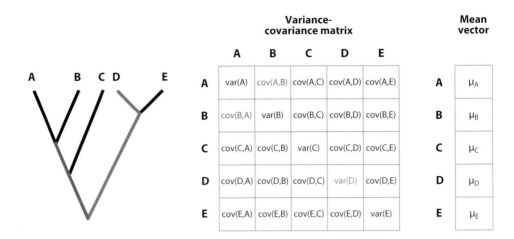

Figure 4

Multivariate normal distribution. The figure shows a tree, the tree's variance-covariance matrix, and the vector of means (which, under Brownian motion, would equal the root state). Highlighted are the branches leading to covariance between taxa A and B (*red*) and the branches leading to variance in D (*blue*).

is to look at correlations between characters. Revell & Collar (2009) developed an approach that estimates variances of and covariances between traits. Huelsenbeck & Rannala (2003) developed a model that can do that while also integrating uncertainty in the trees. Hansen et al. (2008) extend correlation approaches to allow Ornstein-Uhlenbeck processes. Ancestral states may be estimated under the multivariate normal process (Schluter et al. 1997) as they are with CTMC-FSS; in both cases, uncertainty in reconstructions can be substantial (Cunningham et al. 1998). The squared-change parsimony approach for continuous characters (Maddison 1991) recovers the same ancestral state estimates as Brownian motion models using a multivariate normal.

An interesting use of the multivariate normal process for discrete (and sometimes continuous) characters comes from Felsenstein's adoption of Wright's threshold model (Felsenstein 2005, 2012). A continuous trait, called the liability, evolves on the tree under a multivariate normal process. If this liability is above a certain threshold, a discrete character has one state; otherwise it has a different state. This can be used for multiple discrete and continuous characters (Felsenstein 2012). For the discrete character, the threshold model generates a different evolutionary pattern than would be expected under CTMC-FSS (for example, change in the discrete character is most likely immediately after a previous state change, where the liability is near the threshold, rather than constant through time), but the underlying process is multivariate normal.

BIRTH-DEATH BRANCHING PROCESSES

Finally, there are birth-death branching processes (Kendall 1948, Nee 2006). Each individual in a population has a certain probability of a birth (leading to two offspring) or a death (the individual ends). In the phylogenetic context, the individual is usually taken to be a species. In a pure birth model, also known as a Yule model (Yule 1924), the death rate is set to zero and the birth rate is constant through time. In other models, both rates are constant but allowed to be nonzero. It is also possible to vary the rates. Logistic growth, for example, is a pure birth model where the birth rate is a function of the number of individuals. There are many similarities between a birth-death branching process and a CTMC-FSS model, as they are both continuous-time Markov chains with discrete states. The main difference is that in a birth-death branching process there is no maximum number of states. State zero is also an absorbing state in birth-death models (no speciation follows after all the species are extinct), which introduces an ascertainment bias, at least when working with neontological data: One only studies groups where some members have survived. It is the same problem as studying whether dolphins actually rescue drowning sailors. The reputed pattern may be due entirely to ascertainment bias: Dolphins might push drowning sailors in random directions, but only those sailors pushed shoreward survive to tell of the dolphins' influence. Such ascertainment bias needs to be considered by methods dealing with only surviving species. CTMC-FSS methods may also have absorbing states, but a more likely ascertainment bias in those methods as used in phylogenetics is looking at only variable characters (see discussion above).

One major focus of evolutionary investigations using trees is examining what diversification (speciation minus extinction) rates are on a tree. Unlike trait evolution, a main focus here has been investigating heterogeneity across a tree. Point estimates of a net diversification rate can be calculated given just the age of a clade and the number of taxa [reviewed by Magallon & Sanderson (2001)]. There are likelihood estimators for birth and death rates given a tree with branch lengths in units of time (Bokma 2008a, Nee et al. 1994), though death rates may be quite uncertain, especially if estimated based on just extant taxa (Rabosky 2010).

One question often addressed is how the diversification rate has changed over time. A common quantitative test uses the gamma test statistic (Pybus & Harvey 2000), though its performance has been criticized (Cusimano & Renner 2010, Liow et al. 2010). This is intended to measure speedups or slowdowns in diversification rates relative to a constant birth-death null. There are also approaches to fitting other models to trees, such as a density-dependent model of the number of species in a group (Rabosky 2006), a slowdown in speciation rate immediately after a speciation event (Losos & Adler 1995), or the use of survival models to allow flexible estimation of diversification rates (Paradis 1997).

Rather than comparing early versus later parts in the tree, other questions relate to comparing different clades in the tree. One simple such test is the Slowinski-Guyer test (Slowinski & Guyer 1989), which can determine whether two sister clades are of unusually unequal diversity. There have been several later modifications or improvements to this test (McConway & Sims 2004, Paradis 2012). Sanderson & Donoghue (1994) developed an early approach that allowed for different diversification rates on different branches. The SymmeTREE approach was designed to identify heterogeneity in rates across a whole tree (Chan & Moore 2002) and in particular parts of a tree (Moore et al. 2004). The MEDUSA approach (Alfaro et al. 2009) can find areas on a tree where clades are diversifying unusually slowly or quickly.

TREE STRETCHING

Using any of the three general frameworks above, especially the first two, one common approach is to look at heterogeneity in rates of evolution by doing tree stretching, which allows one to algorithmically transform a set of branch lengths on the tree. Many of the most common stretching models come from work by Pagel (Pagel 1997, 1999). A lambda transform affects the ratio of internal to terminal branch lengths. With $\lambda = 0$, change only happens on terminal branches. This can measure how much of evolutionary change is due to evolution happening at a constant rate through the tree and how much is due to only very recent changes. For example, if species were evolving a continuous trait that rapidly moved between fixed bounds, the influence of phylogenetic history would be erased and lambda would thus be close to zero. A kappa transform raises all branch lengths by the power κ. A value of 0 creates equal lengths for all branches, which matches a model where changes only happen after speciation events and where there are no missing speciation events on the tree (a completely sampled tree with no extinctions). A delta transform stretches the tree to make changes more likely early (if $\delta < 1$) or later (if $\delta > 1$). Other transforms that may be applied include modifying branch lengths so that changes accelerate or decelerate through time [the ACDC model of Blomberg et al. (2003)] or modifying branch lengths so that branches after some threshold are a different length than branches before some threshold (Harmon et al. 2008). Some of the approaches in the Multivariate Normal Distribution section above (O'Meara et al. 2006, Thomas et al. 2006) also amount to simple stretching of prespecified parts of a tree. Note that any method that uses information on tree branch lengths can have tree stretching added to it, even if the particular stretching technique was developed for a different kind of data.

HYBRID METHODS

Historically, if a question were solely about trait evolution, differential diversification could be ignored. Many methods focused on the effect of traits on diversification itself would first reconstruct the tree, then reconstruct the traits, and only then examine the effect on diversification. However, in a key paper, Maddison (2006) showed that the distribution of a binary trait on a tree is affected by the effect of the trait's states both on diversification rates and on the transition rate between those states [note that this concern was raised fourteen years earlier by Janson (1992) but was largely ignored]. The traditional approach of separately looking at the transition and diversification aspects of the trait could provide the wrong answer. For example, imagine a trait where 0 to 1 transitions happen twenty times faster than 1 to 0 transitions. The number of taxa in state 1 should equilibrate at twenty times greater than the number of taxa in state 0. This could lead several methods for looking at diversification rates to estimate that the diversification rate in state 1 is greater than that in state 0 (thus leading to the increased number of taxa with state 1). In contrast, if state 1 does lead to higher diversification rates, this might be misinterpreted by a method looking at transition rates as a higher 0 to 1 than 1 to 0 rate. Maddison's (2006) work represents a change on the order of the change created by Felsenstein's independent contrasts paper (Felsenstein 1985), which showed that just conducting raw species contrasts might lead to misleading results because phylogeny is not taken into account. Community standards now generally expect a researcher looking into correlations between traits across many species to consider whether there is a phylogenetic effect that needs to be taken into account. We now know there could be an effect of differential diversification on estimates of trait evolution or an effect of unequal trait evolution on estimates of trait-associated diversification. Investigation of these effects is of growing importance. Note that this process need not only occur with key innovations or rapidly radiating groups: It could happen any time different states of a trait may be correlated (even via trait correlation with a second, unmeasured trait) with different extinction and/or speciation rates. For example, if carnivores tend to have shorter intestinal tracts than herbivores, and carnivory tends to lead to higher extinction rates (as carnivores are higher up the trophic pyramid), then a reconstruction of intestinal length ignoring potential correlation with speciation and extinction rates may be misleading. This happens even though there is no credible causal link between intestines and speciation or extinction: Lack of causation does not imply lack of correlation.

Fortunately, there are methods to deal with this. The first such method, BiSSE (Maddison et al. 2007), uses a model for binary traits that allows different states to have different transition rates, birth rates, and death rates. This requires fully resolved and sampled phylogenies. A later extension relaxed this requirement (FitzJohn et al. 2009), and other work by FitzJohn has extended it to multistate characters (FitzJohn 2012). There is now an approach for continuous characters (QuaSSE) that can deal with a single trait that evolves under a diffusion process (like Brownian motion or the Ornstein-Uhlenbeck process) but that also may affect diversification rate (FitzJohn 2010).

There is a model, GeoSSE, for dealing with geography, where a species may occur in multiple localities and where diversification rate may be tied to locality (Goldberg et al. 2011). There are not, yet, methods that deal with multiple characters (though the multistate character approach can deal with this if characters are recoded) or with heterogeneity over the tree, though such methods will undoubtedly be developed. Bokma (2002, 2008b) independently developed methods that examine the joint effect of diversification, missing species, and change at nodes and along branches.

OTHER KEY METHODS

Although many methods fit in the above three broad categories, many methods, including some of the most important ones, do not. For example, there is a rich array of regression-based methods that correct for phylogeny as an error term or as a correlation term. These include methods such as generalized linear models (Martins & Hansen 1997), generalized estimating equations (Paradis & Claude 2002), the phylogenetic mixed method (Housworth et al. 2004, Lynch 1991), phylogenetic generalized least squares, and phylogenetic autocorrelation (Cheverud et al. 1985, Gittleman & Kot 1990) among others (e.g., Ives & Garland 2010). They are too flexible to put into one category, but they are often essential tools, especially when examining data of mixed types. Independent contrasts (Felsenstein 1985), a method to correct for nonindependence of data on a tree, is consistent with the multivariate normal framework but works slightly differently (using estimates only incorporating information from descendant taxa, for example). There is a growing array of methods to use for phylogeographic inferences (history of population subdivision, history of population size, and so forth) that use richer models than any of the above (Knowles 2009). Some biogeographic models (Ronquist & Sanmartín 2011) and other coevolutionary models (Page 1994) also use methods beyond what was discussed here. There is also a wide array of randomization techniques for things like community phylogenetics (Webb et al. 2002) or phylogenetic signal (Blomberg et al. 2003). Methods that simulate based on actual mechanisms are growing in importance, as well. For example, Pigot et al. (2010) developed a model that simulates range changes and vicariance events explicitly, which generates distributions different from those seen under birth-death models. The growth of approximate Bayesian computation (ABC) methods (Beaumont et al. 2002, Huang et al. 2011, Rabosky 2009) is making the use of explicit models much more feasible.

PARAMETER ESTIMATION, HYPOTHESIS REJECTION, AND MODEL SELECTION

Practitioners of evolutionary inference using phylogenies differ in their statistical approaches. There are of course the differences between those who use explicit models and those who do not as well as between Bayesians and frequentists. However, given that many methods are not implemented in software (see the sidebar, Software) adopting all these perspectives, many empiricists care more about using a method that provides answers to their questions than about fealty to a particular statistical worldview. However, there is a more basic argument that affects which methods are developed and how they are applied. It is the ongoing debate about using methods to answer biological questions by estimating parameters versus using methods to reject null hypotheses (Stephens et al. 2007). A simple example of the latter is the Slowinski & Guyer (1989) test. It returns whether the distribution of diversity in two sister clades is more uneven than expected under a basic null model. There is utility to this: Learning that a clade of 1,000 taxa is not significantly more diverse than its sister clade having 30 taxa helps correct our perceptions of

SOFTWARE

Methods for evolutionary inferences using phylogenetics need to be implemented in software in order to be used. There are two notable recent trends in software for evolutionary inferences. The first is the move to open source: Rather than being a black box of compiled computer code, the human-readable code underlying the program is distributed freely for inspection and modification. Most new evolutionary methods are now released in open-source software.

The other major trend is the growth of R (R Dev. Core Team 2011) as a framework for methods for evolutionary inference. One of its main advantages is its use of packages to add functionality. For example, the APE package (Paradis et al. 2004) can load, traverse, and display trees, so someone developing a new method can use those existing functions. There are over thirty packages devoted to phylogenetics alone (O'Meara 2012), as well as a book (Paradis 2011). R is certainly not the only option. For example, Mesquite (Maddison & Maddison 2007) is a Java application that implements an abundance of methods and that can be extended with modules. There are also projects to build bioinformatics libraries in languages such as Python, Perl, and C++.

what is expected under null models. Likelihood ratio tests can be used with approaches like tree stretching to show that simple Brownian motion is rejected. Rejection of a null model is based on two factors, how well the data fit the null model and data set size. As data sets grow larger, it will become increasingly easy to reject basic null models: the chances of a clade of 10,000 taxa evolving traits exactly under simple Brownian motion, or having a completely constant diversification rate through time, are low. Rather than just inferring whether a parameter is statistically significantly different from zero, one must determine whether it is statistically and biologically significantly different from zero.

One response to criticisms of null hypothesis testing has been to adopt information theoretic approaches, such as the Akaike Information Criterion (Akaike 1973, Burnham & Anderson 2002). These approaches attempt to identify the model that minimizes information lost about nature. A model that is too simple may have missed important factors that should be included, whereas a model that is too complex may have too many parameters to fit well. This approach, like approaches that involve Bayesian model evaluations, lends itself well to multimodel inference: drawing conclusions based on a set of weighted models, rather than putting all faith in a single best model. The risk, however, is in interpreting model selection as hypothesis testing: treating a simple model with low weight as a rejected model and making that a focus.

Another recent advance has been the reintroduction of the idea of model adequacy. Existing approaches to model fitting [likelihood ratio tests, information theoretic measures, Bayes factors, reversible jump MCMC (Huelsenbeck et al. 2004, Pagel & Meade 2006), and so forth] estimate which model is least bad. There is little information from such approaches about whether a model is actually a good fit to the data [with the noteworthy exception of methods that are based on regression ideas (Martins & Hansen 1997) or various approaches to make sure contrasts are well-standardized (Garland et al. 1992)]. An approach to looking at whether the model adequately reflects the data is to simulate data under the inferred model and see if the simulated data are indistinguishable from the observed data by some measures. This has been done recently for comparative methods by Boettiger et al. (2011) but was done earlier for substitution models (Goldman 1993) and for looking at conflicting phylogenetic signal (Huelsenbeck & Bull 1996). In a Bayesian context, a posterior predictive distribution can be used in much the same way (Huelsenbeck et al. 2001). Doing this sort of test of adequacy will be important in developing more realistic models: comparing the fits of models will only tell which is better, but by showing that in some situations all are inadequate, researchers will know where to begin work on better models.

TREE SIZE AND HETEROGENEITY

In less than twenty years, large tree data sets have gone from 500 taxa (Chase et al. 1993) to over 70,000 taxa (Goloboff et al. 2009). The number of free parameters that can be accurately estimated has also increased, though the exact bounds of this remain to be explored. For example, if a reasonable rule of thumb of 100 observations per parameter is used, this suggests that a model with 700 free parameters could be estimated from a 70,000-taxon data set. In contrast, the commonly used Brownian motion model has just two parameters, the rate and the root state. It is an empirical question, but it seems likely that complex models with perhaps as many as dozens of free parameters may be appropriate at this scale. More importantly, as greater sections of the tree of life are used in analyses, the assumptions of the homogeneity of processes used in many models begin to break down. A phylogenetically corrected regression of brain size on body size may work well in New World monkeys, but as one expands the set of taxa to include all primates, then all mammals, then all vertebrates, the reasonableness of the model no longer holds. The factors linking brain and body size in arboreal mammals may be very different from those acting in teleost fishes or birds, as traits like aquatic environment, flight constraints, social structure, diet, and other factors may vary over the tree. Doing this analysis phylogenetically remains important, but a more complex model than a single fixed regression may be justified. More than just a complicating factor, this promises new discoveries about biology as the power to perform analyses with more realistic models grows. There have been initial approaches to try different models on different parts of trees using a priori assignment (Beaulieu et al. 2012, Butler & King 2004, O'Meara et al. 2006, Thomas et al. 2006) as well as automatic placement of models (Alfaro et al. 2009, Eastman et al. 2011, Thomas & Freckleton 2012), but this will represent an important area of growth in the future.

CONCLUSION

As species evolve through time, information about their traits and changes is passed on to their descendants. Although much of this information is erased by subsequent changes, much still persists: The ancestor of bats, pterosaurs, and cows had four limbs; the squid, human, and fly lineages have maintained a homologous gene for initiating a light detector; and angiosperms have a higher diversification rate than many other plant groups. The combination of trait data, a phylogeny, and various methods can be used to make inferences about past and ongoing evolutionary processes. Although there are scores of different methods for making inferences about evolutionary processes using phylogenies, many can be reduced to a few essential models. This does two important things. First, it allows models to be more readily understood: Once a GTR DNA model is grasped, one also then understands a model for evolution of flight as a discrete trait. Second, it allows advances in approaches where one set of questions can be easily ported to a different set of questions. For example, the discrete gamma approach to rate heterogeneity (Yang 1994b) has been applied to nucleotide data for tree inference, but it could be applied to data on expression/nonexpression of various genes in developing flowers in a more realistic model for ancestral state reconstruction of flower development. The only real difference between the models is whether the states are called A, T, G, and C at a site or presence/absence of expression for a gene.

We are in a golden age of phylogenetics. The number of taxa and data set size for phylogenetic trees are expanding, yet there are still discoveries about relationships to be made. Computational power on individual machines continues to grow as does the availability of free (to the community) supercomputing for phylogenetics (Goff et al. 2011, Miller et al. 2010). Data sets and trees are stored in accessible databases allowing reuse. This all points to more data allowing for more models, perhaps with more complexity and realism. As model adequacy is used more often, biologists will learn where existing methods are insufficient and will create new, more powerful approaches.

SUMMARY POINTS

1. There are a wide variety of methods for drawing evolutionary inferences given phylogenies, but despite this diversity, a few major models recur in many available methods.

2. Continuous-time Markov chain models with finite states underlie many diverse models, from DNA evolution to gene family size or biogeography.

3. The multivariate normal distribution underlies many methods for continuous data and at least one for discrete data.

4. Birth-death processes play major roles in inferences about speciation and extinction dynamics.

5. Tree stretching is a broad category of methods that allow tests of rate heterogeneity across time or taxa.

6. Methods that jointly estimate transition and diversification rates represent an important advance in comparative methods and may be of growing influence.

FUTURE ISSUES

1. Model adequacy will become more important as a way to measure how well models describe biological data.

2. As the sizes of trees grow, methods will increasingly have to deal with heterogeneity in processes at different parts of the tree.

3. Current models often describe patterns that may be caused by multiple processes. Mechanistic models may grow in importance, especially as simulation-based approaches become more accessible.

4. Most current methods assume species have a single trait value per character, with some important exceptions (Felsenstein 2005, 2008; Ives et al. 2007). Just as the realization of the importance of variation in gene histories is affecting tree reconstruction, appreciation of the variation in trait values within species will generate a need for further advances.

5. In many groups the true evolutionary history is reticulate, not fully tree-like. Methods for making evolutionary inferences on networks still need to be developed.

DISCLOSURE STATEMENT

The author is not aware of any affiliations, memberships, funding, or financial holdings that might be perceived as affecting the objectivity of this review.

ACKNOWLEDGMENTS

Support for some of the research leading to this review came from a fellowship to B.C.O. from the National Evolutionary Synthesis Center (NESCent), NSF #EF-0905606, and working group support from the iPlant Collaborative, NSF #DBI-0735191. I thank Barb Banbury, Jeremy Beaulieu, Martin Ryberg, Stacey Smith, and Peter Wainwright for feedback and guidance. Conversations with Joe Felsenstein contributed to an appreciation of the importance of dealing with heterogeneity of processes over a tree.

LITERATURE CITED

Akaike H. 1973. Information theory as an extension of the maximum likelihood principle. In *Second International Symposium on Information Theory*, ed. BN Petrov, F Csaki, pp. 267–81. Budapest: Akad. Kiado

Alfaro ME, Santini F, Brock C, Alamillo H, Dornburg A, et al. 2009. Nine exceptional radiations plus high turnover explain species diversity in jawed vertebrates. *Proc. Natl. Acad. Sci. USA* 106:13410–14

Barker D, Pagel M. 2005. Predicting functional gene links from phylogenetic-statistical analyses of whole genomes. *PLoS Comput. Biol.* 1:24–31

Beaulieu JM, Jhwueng D-C, Boettiger C, O'Meara BC. 2012. Modeling stabilizing selection: expanding the Ornstein-Uhlenbeck model of adaptive evolution. *Evolution* 66:2369–83

Beaumont MA, Zhang W, Balding DJ. 2002. Approximate Bayesian computation in population genetics. *Genetics* 162:2025–35

Blomberg SP, Garland T, Ives AR. 2003. Testing for phylogenetic signal in comparative data: Behavioral traits are more labile. *Evolution* 57:717–45

Boettiger C, Coop G, Ralph P. 2011. Is your phylogeny informative? Measuring the power of comparative methods. *Evolution* 66:2240–51

Bokma F. 2002. Detection of punctuated equilibrium from molecular phylogenies. *J. Evol. Biol.* 15:1048–56

Bokma F. 2008a. Bayesian estimation of speciation and extinction probabilities from (in)complete phylogenies. *Evolution* 62:2441–45

Bokma F. 2008b. Detection of "punctuated equilibrium" by Bayesian estimation of speciation and extinction rates, ancestral character states, and rates of anagenetic and cladogenetic evolution on a molecular phylogeny. *Evolution* 62:2718–26

Burnham KP, Anderson DR. 2002. *Model Selection and Multimodel Inference: A Practical Information-Theoretic Approach*. New York: Springer

Butler MA, King AA. 2004. Phylogenetic comparative analysis: a modeling approach for adaptive evolution. *Am. Nat.* 164:683–95

Chan KMA, Moore BR. 2002. Whole-tree methods for detecting differential diversification rates. *Syst. Biol.* 51:855–65

Chase MW, Soltis DE, Olmstead RG, Morgan D, Les DH, et al. 1993. Phylogenetics of seed plants: an analysis of nucleotide sequences from the plastid gene *rbc*L. *Ann. Mo. Bot. Gard.* 80:528–80

Cheverud JM, Dow MM, Leutenegger W. 1985. The quantitative assessment of phylogenetic constraints in comparative analyses: sexual dimorphism in body weight among primates. *Evolution* 39:1335–51

Cunningham CW, Omland KE, Oakley TH. 1998. Reconstructing ancestral character states: a critical reappraisal. *Trends Ecol. Evol.* 13:361–66

Cusimano N, Renner SS. 2010. Slowdowns in diversification rates from real phylogenies may not be real. *Syst. Biol.* 59:458–64

Eastman JM, Alfaro ME, Joyce P, Hipp AL, Harmon LJ. 2011. A novel comparative method for identifying shifts in the rate of character evolution on trees. *Evolution* 65:3578–89

Edwards AWF, Cavalli-Sforza LL. 1964. Reconstruction of evolutionary trees. In *Phenetic and Phylogenetic Classification*, Vol. 6, ed. VH Heywood, J McNeill, pp. 67–76. London: Syst. Assoc. Publ.

Felsenstein J. 1981. Evolutionary trees from DNA sequences—a maximum-likelihood approach. *J. Mol. Evol.* 17:368–76

Felsenstein J. 1985. Phylogenies and the comparative method. *Am. Nat.* 125:1–15

Felsenstein J. 1988. Phylogenies and quantitative characters. *Annu. Rev. Ecol. Syst.* 19:445–71

Felsenstein J. 1992. Phylogenies from restriction sites—a maximum-likelihood approach. *Evolution* 46:159–73

Felsenstein J. 2004. *Inferring Phylogenies.* Sunderland, MA: Sinauer

Felsenstein J. 2005. Using the quantitative genetic threshold model for inferences between and within species. *Philos. Trans. R. Soc. B* 360:1427–34

Felsenstein J. 2008. Comparative methods with sampling error and within-species variation: contrasts revisited and revised. *Am. Nat.* 171:713–25

Felsenstein J. 2012. A comparative method for both discrete and continuous characters using the threshold model. *Am. Nat.* 179:145–56

Felsenstein J, Churchill GA. 1996. A Hidden Markov Model approach to variation among sites in rate of evolution. *Mol. Biol. Evol.* 13:93–104

FitzJohn RG. 2010. Quantitative traits and diversification. *Syst. Biol.* 59:619–33

FitzJohn RG. 2012. *Diversitree: comparative phylogenetic tests of diversification. R package version 0.9–1.* **http://www.zoology.ubc.ca/prog/diversitree/**

FitzJohn RG, Maddison WP, Otto SP. 2009. Estimating trait-dependent speciation and extinction rates from incompletely resolved phylogenies. *Syst. Biol.* 58:595–611

Galtier N. 2001. Maximum-likelihood phylogenetic analysis under a covarion-like model. *Mol. Biol. Evol.* 18:866–73

Galtier N, Gouy M. 1998. Inferring pattern and process: maximum-likelihood implementation of a nonhomogeneous model of DNA sequence evolution for phylogenetic analysis. *Mol. Biol. Evol.* 15:871–79

Garland T Jr, Harvey PH, Ives AR. 1992. Procedures for the analysis of comparative data using phylogenetically independent contrasts. *Syst. Biol.* 41:18–32

Gittleman JL, Kot M. 1990. Adaptation: statistics and a null model for estimating phylogenetic effects. *Syst. Biol.* 39:227–41

Goff SA, Vaughn M, McKay S, Lyons E, Stapleton AE, et al. 2011. Frontiers: The iPlant Collaborative: cyberinfrastructure for plant biology. *Front. Plant Sci.* 2:34

Goldberg EE, Lancaster LT, Ree RH. 2011. Phylogenetic inference of reciprocal effects between geographic range evolution and diversification. *Syst. Biol.* 60:451–65

Goldman N. 1993. Simple diagnostic statistical tests of models for DNA substitution. *J. Mol. Evol.* 37:650–61

Goloboff PA, Catalano SA, Marcos Mirande J, Szumik CA, Salvador Arias J, et al. 2009. Phylogenetic analysis of 73 060 taxa corroborates major eukaryotic groups. *Cladistics* 25:211–30

Hahn MW, De Bie T, Stajich JE, Nguyen C, Cristianini N. 2005. Estimating the tempo and mode of gene family evolution from comparative genomic data. *Genome Res.* 15:1153–60

Hansen TF. 1997. Stabilizing selection and the comparative analysis of adaptation. *Evolution* 51:1341–51

Hansen TF, Martins EP. 1996. Translating between microevolutionary process and macroevolutionary patterns: the correlation structure of interspecific data. *Evolution* 50:1404–17

Hansen TF, Pienaar J, Orzack SH. 2008. A comparative method for studying adaptation to a randomly evolving environment. *Evolution* 62:1965–77

Harmon LJ, Weir JT, Brock CD, Glor RE, Challenger W. 2008. GEIGER: investigating evolutionary radiations. *Bioinformatics* 24:129–31

Housworth EA, Martins EP, Lynch M. 2004. The phylogenetic mixed model. *Am. Nat.* 163:84–96

Huang W, Takebayashi N, Qi Y, Hickerson MJ. 2011. MTML-msBayes: approximate Bayesian comparative phylogeographic inference from multiple taxa and multiple loci with rate heterogeneity. *BMC Bioinformatics* 12:1

Huelsenbeck JP, Bull JJ. 1996. A likelihood ratio test to detect conflicting phylogenetic signal. *Syst. Biol.* 45:92–98

Huelsenbeck JP, Larget B, Alfaro ME. 2004. Bayesian phylogenetic model selection using reversible jump Markov chain Monte Carlo. *Mol. Biol. Evol.* 21:1123–33

Huelsenbeck JP, Nielsen R, Bollback JP. 2003. Stochastic mapping of morphological characters. *Syst. Biol.* 52:131–58

Huelsenbeck JP, Rannala B. 2003. Detecting correlation between characters in a comparative analysis with uncertain phylogeny. *Evolution* 57:1237–47

Huelsenbeck JP, Ronquist F, Nielsen R, Bollback JP. 2001. Bayesian inference of phylogeny and its impact on evolutionary biology. *Science* 294:2310–14

Ives AR, Garland T Jr. 2010. Phylogenetic logistic regression for binary dependent variables. *Syst. Biol.* 59:9–26

Ives AR, Midford PE, Garland T Jr. 2007. Within-species variation and measurement error in phylogenetic comparative methods. *Syst. Biol.* 56:252–70

Janson CH. 1992. Measuring evolutionary constraints: a Markov model for phylogenetic transitions among seed dispersal syndromes. *Evolution* 46:136–58

Jones DT, Taylor WR, Thornton JM. 1992. The rapid generation of mutation data matrices from protein sequences. *Comput. Appl. Biosci.* 8:275–82

Kendall DG. 1948. On the generalized "birth-and-death" process. *Ann. Math. Stat.* 19:1–15

Knowles LL. 2009. Statistical phylogeography. *Annu. Rev. Ecol. Evol. Syst.* 40:593–612

Lewis PO. 2001. A likelihood approach to estimating phylogeny from discrete morphological character data. *Syst. Biol.* 50:913–25

Liow LH, Quental TB, Marshall CR. 2010. When can decreasing diversification rates be detected with molecular phylogenies and the fossil record? *Syst. Biol.* 59:646–59

Losos JB, Adler FR. 1995. Stumped by trees? A generalized null model for patterns of organismal diversity. *Am. Nat.* 145:329–42

Lynch M. 1991. Methods for the analysis of comparative data in evolutionary biology. *Evolution* 45:1065–80

Maddison WP. 1990. A method for testing the correlated evolution of two binary characters: Are gains or losses concentrated on certain branches of a phylogenetic tree? *Evolution* 44:539–57

Maddison WP. 1991. Squared-change parsimony reconstructions of ancestral states for continuous-valued characters on a phylogenetic tree. *Syst. Zool.* 40:304–14

Maddison WP. 2006. Confounding asymmetries in evolutionary diversification and character change. *Evolution* 60:1743–46

Maddison WP, Maddison DR. 2007. *Mesquite: a modular system for evolutionary analysis. Version 2.0.* **http://mesquiteproject.org**

Maddison WP, Midford PE, Otto SP. 2007. Estimating a binary character's effect on speciation and extinction. *Syst. Biol.* 56:701–10

Magallon S, Sanderson MJ. 2001. Absolute diversification rates in angiosperm clades. *Evolution* 55:1762–80

Martins EP, Hansen TF. 1997. Phylogenies and the comparative method: a general approach to incorporating phylogenetic information into the analysis of interspecific data. *Am. Nat.* 149:646–67

McConway KJ, Sims HJ. 2004. A likelihood-based method for testing for nonstochastic variation of diversification rates in phylogenies. *Evolution* 58:12–23

Miller MA, Pfeiffer W, Schwartz T. 2010. Creating the CIPRES Science Gateway for inference of large phylogenetic trees. Presented at *Gateway Comput. Environ. Workshop (GCE) Nov. 14, New Orleans, LA*, pp. 1–8

Moore BR, Chan KMA, Donoghue MJ. 2004. Detecting diversification rate variation in supertrees. In *Phylogenetic Supertrees: Combining Information to Reveal the Tree of Life*, ed. ORP Binida-Emonds, pp. 487–533. Dordrecht: Kluwer Acad.

Nee S. 2006. Birth-death models in macroevolution. *Annu. Rev. Ecol. Evol. Syst.* 37:1–17

Nee S, Holmes EC, May RM, Harvey PH. 1994. Extinction rates can be estimated from molecular phylogenies. *Philos. Trans. R. Soc. Lond. Ser. B* 344:77–82

Nielsen R. 2002. Mapping mutations on phylogenies. *Syst. Biol.* 51:729–39

O'Meara BC. 2012. *CRAN task view: phylogenetics. Version 2012-02-02.* **http://cran.r-project.org/web/views/ Phylogenetics.html**

O'Meara BC, Ane C, Sanderson MJ, Wainwright PC. 2006. Testing for different rates of continuous trait evolution using likelihood. *Evolution* 60:922–33

Page RDM. 1994. Parallel phylogenies: reconstructing the history of host-parasite assemblages. *Cladistics* 10:155–73

Pagel M. 1994. Detecting correlated evolution on phylogenies—a general method for the comparative analysis of discrete characters. *Proc. R. Soc. Lond. Ser. B* 255:37–45

Pagel M. 1997. Inferring evolutionary processes from phylogenies. *Zool. Scr.* 26:331–48

Pagel M. 1999. Inferring the historical patterns of biological evolution. *Nature* 401:877–84

Pagel M, Meade A. 2004. A phylogenetic mixture model for detecting pattern-heterogeneity in gene sequence or character-state data. *Syst. Biol.* 53:571–81

Pagel M, Meade A. 2006. Bayesian analysis of correlated evolution of discrete characters by reversible-jump Markov chain Monte Carlo. *Am. Nat.* 167:808–25

Pagel M, Meade A, Barker D. 2004. Bayesian estimation of ancestral character states on phylogenies. *Syst. Biol.* 53:673–84

Paradis E. 1997. Assessing temporal variations in diversification rates from phylogenies: estimation and hypothesis testing. *Proc. R. Soc. Lond. Ser. B* 264:1141–47

Paradis E. 2011. *Analysis of Phylogenetics and Evolution with R*. Berlin: Springer-Verlag. 2nd ed.

Paradis E. 2012. Shift in diversification in sister-clade comparisons: a more powerful test. *Evolution* 66:288–95

Paradis E, Claude J. 2002. Analysis of comparative data using generalized estimating equations. *J. Theor. Biol.* 218:175–85

Paradis E, Claude J, Strimmer K. 2004. APE: analyses of phylogenetics and evolution in R language. *Bioinformatics* 20: 289–90

Penny D, McComish BJ, Charleston MA, Hendy MD. 2001. Mathematical elegance with biochemical realism: the covarion model of molecular evolution. *J. Mol. Evol.* 53:711–23

Pigot AL, Phillimore AB, Owens IPF, Orme CDL. 2010. The shape and temporal dynamics of phylogenetic trees arising from geographic speciation. *Syst. Biol.* 59:660–73

Pybus OG, Harvey PH. 2000. Testing macroevolutionary models using incomplete molecular phylogenies. *Proc. R. Soc. Lond. Ser. B* 267:2267–72

R Dev. Core Team. 2011. *R: A Language and Environment for Statistical Computing*, Vol. 1. Vienna, Austria: R Found. Stat. Comput.

Rabosky DL. 2006. Likelihood methods for detecting temporal shifts in diversification rates. *Evolution* 60:1152–64

Rabosky DL. 2009. Heritability of extinction rates links diversification patterns in molecular phylogenies and fossils. *Syst. Biol.* 58:629–40

Rabosky DL. 2010. Extinction rates should not be estimated from molecular phylogenies. *Evolution* 64:1816–24

Ree RH, Smith SA. 2008. Maximum likelihood inference of geographic range evolution by dispersal, local extinction, and cladogenesis. *Syst. Biol.* 57:4–14

Revell LJ, Collar DC. 2009. Phylogenetic analysis of the evolutionary correlation using likelihood. *Evolution* 63:1090–100

Ronquist F, Sanmartín I. 2011. Phylogenetic methods in biogeography. *Annu. Rev. Ecol. Evol. Syst.* 42: 441–64

Ryan MJ, Rand AS. 1995. Female responses to ancestral advertisement calls in túngara frogs. *Science* 269:390–92

Sanderson MJ, Donoghue MJ. 1994. Shifts in diversification rate with the origin of angiosperms. *Science* 264:1590–93

Schluter D, Price T, Mooers AO, Ludwig D. 1997. Likelihood of ancestor states in adaptive radiation. *Evolution* 51:1699–711

Schultz TR, Churchill GA. 1999. The role of subjectivity in reconstructing ancestral character states: a Bayesian approach to unknown rates, states, and transformation asymmetries. *Syst. Biol.* 48:651–64

Slowinski JB, Guyer C. 1989. Testing the stochasticity of patterns of organismal diversity: an improved null model. *Am. Nat.* 907–21

Steel M, Penny D. 2000. Parsimony, likelihood, and the role of models in molecular phylogenetics. *Mol. Biol. Evol.* 17:839–50

Stephens PA, Buskirk SW, del Rio CM. 2007. Inference in ecology and evolution. *Trends Ecol. Evol.* 22:192–97

Thomas GH, Freckleton RP. 2012. MOTMOT: models of trait macroevolution on trees. *Methods Ecol. Evol.* 3:145–51

Thomas GH, Freckleton RP, Székely T. 2006. Comparative analyses of the influence of developmental mode on phenotypic diversification rates in shorebirds. *Proc. R. Soc. B* 273:1619–24

Tuffley C, Steel M. 1997. Links between maximum likelihood and maximum parsimony under a simple model of site substitution. *Bull. Math. Biol.* 59:581–607

Tuffley C, Steel M. 1998. Modeling the covarion hypothesis of nucleotide substitution. *Math. Biosci.* 147:63–91

Vijaykrishna D, Smith GJD, Pybus OG, Zhu H, Bhatt S, et al. 2011. Long-term evolution and transmission dynamics of swine influenza A virus. *Nature* 473:519–22

Webb CO, Ackerly DD, McPeek MA, Donoghue MJ. 2002. Phylogenies and community ecology. *Annu. Rev. Ecol. Syst.* 33:475–505

Yang ZH. 1994a. Statistical properties of the maximum likelihood method of phylogenetic estimation and comparison with distance matrix methods. *Syst. Biol.* 43:329–42

Yang ZH. 1994b. Maximum-likelihood phylogenetic estimation from DNA sequences with variable rates over sites—approximate methods. *J. Mol. Evol.* 39:306–14

Yang ZH. 1995. A space-time process model for the evolution of DNA sequences. *Genetics* 139:993–1005

Yang ZH. 1996. Maximum-likelihood models for combined analyses of multiple sequence data. *J. Mol. Evol.* 42:587–96

Yang ZH, Kumar S, Nei M. 1995. A new method of inference of ancestral nucleotide and amino acid sequences. *Genetics* 141:1641–50

Yang ZH, Nielsen R. 2002. Codon-substitution models for detecting molecular adaptation at individual sites along specific lineages. *Mol. Biol. Evol.* 19:908–17

Yang ZH, Roberts D. 1995. On the use of nucleic acid sequences to infer early branchings in the tree of life. *Mol. Biol. Evol.* 12:451–58

Yule GV. 1924. A mathematical theory of evolution, based on the conclusions of Dr. J. C. Willis, F.R.S. *Philos. Trans. R. Soc. Lond. B* 213:21–87

Zhang JZ, Nielsen R, Yang ZH. 2005. Evaluation of an improved branch-site likelihood method for detecting positive selection at the molecular level. *Mol. Biol. Evol.* 22:2472–79

A Guide to Sexual Selection Theory

Bram Kuijper,[1,2] Ido Pen,[1] and Franz J. Weissing[1]

[1]Theoretical Biology Group, Center for Ecological and Evolutionary Studies, University of Groningen, 9747 AG Groningen, The Netherlands; email: bk319@cam.ac.uk, i.r.pen@rug.nl, f.j.weissing@rug.nl

[2]Behavior and Evolution Group, Department of Zoology, University of Cambridge, CB2 3EJ Cambridge, United Kingdom

Annu. Rev. Ecol. Evol. Syst. 2012. 43:287–311

First published online as a Review in Advance on September 4, 2012

The *Annual Review of Ecology, Evolution, and Systematics* is online at ecolsys.annualreviews.org

This article's doi:
10.1146/annurev-ecolsys-110411-160245

1543-592X/12/1201-0287$20.00

Keywords

mate choice, population genetics, quantitative genetics, evolutionary game theory, adaptive dynamics, individual-based simulations, benefits of choice

Abstract

Mathematical models have played an important role in the development of sexual selection theory. These models come in different flavors and they differ in their assumptions, often in a subtle way. Similar questions can be addressed by modeling frameworks from population genetics, quantitative genetics, evolutionary game theory, or adaptive dynamics, or by individual-based simulations. Confronted with such diversity, nonspecialists may have difficulties judging the scope and limitations of the various approaches. Here we review the major modeling frameworks, highlighting their pros and cons when applied to different research questions. We also discuss recent developments, where classical models are enriched by including more detail regarding genetics, behavior, demography, and population dynamics. It turns out that some seemingly well-established conclusions of sexual selection theory are less general than previously thought. Linking sexual selection to other processes such as sex-ratio evolution or speciation also reveals that enriching the theory can lead to surprising new insights.

1. INTRODUCTION

Sexual selection is the process by which individuals compete for access to mates and fertilization opportunities (Andersson 1994, Jones & Ratterman 2009). Darwin (1871) developed the concept of sexual selection to explain the evolution of exaggerated and flamboyant characters such as calls, odors, ornaments, and conspicuous behaviors that are present in one sex only and cannot be easily explained as adaptations to the ecological conditions of a species. Darwin was well aware of the complex nature of sexual selection, "depending as it does, on the ardour of love, the courage, and the rivalry of the males, as well as on the powers of perception, the taste, and will of the female" (Darwin 1871, p. 296). Due to this interdependence of coevolving male and female traits, the essential features of sexual selection are inherently difficult to capture in verbal theories. Yet more than a century would have to pass after Darwin's seminal work before students of sexual selection started to develop mathematical models to capture the complexity of sexual selection in a rigorous fashion. Driven by these models, the empirical study of sexual selection has matured into one of the most active fields in evolutionary biology (Andersson & Simmons 2006).

There are numerous reasons why sexual selection models tend to be more complicated than "standard" models of natural selection. First, whereas models of natural selection often make the simplifying assumption of asexual reproduction or random mating, sexual reproduction and nonrandom mating lie at the heart of sexual selection. Second, natural selection models tend to avoid the intricacies of multilocus genetics. In contrast, sexual selection models are intrinsically multivariate because they reflect the coevolution of mating preferences, ornaments, and, in the case of the "good genes" process, variation in genetic quality. Moreover, the associations (linkage disequilibria) between traits or between preferences and traits are often crucial to understanding the evolutionary outcome (Lande 1981, Iwasa et al. 1991). Third, natural selection models tend to be based on a single fitness component, whereas sexual selection reflects the interplay of viability selection (e.g., costs of ornaments, costs of choosiness), fecundity selection (e.g., trade-offs between parental care and mating opportunities), and selection on mating and fertilization rates. Fourth, sexual selection models have to address sex differences, such as the sex-limited expression of traits and differences in the strength and direction of selection between the sexes. Fifth, as a consequence of sex-differential selection, genetic details may play a more prominent role than they do in other evolutionary models. In particular, autosomal versus sex chromosomal inheritance may strongly affect the outcome of evolution. Sixth, the coevolution between the sexes often takes the form of an evolutionary arms race, resulting in ongoing oscillations or even more complex nonequilibrium dynamics. Accordingly, the analysis of sexual selection often necessitates more refined dynamical approaches than those used in classical equilibrium-oriented methods (Gavrilets & Hayashi 2005, Van Doorn & Weissing 2006). Seventh, sexual selection is intrinsically linked to other processes such as sex-ratio evolution (Trivers & Willard 1973), the evolution of parental care (Trivers 1972), and speciation (Ritchie 2007, Weissing et al. 2011). It is becoming increasingly clear that robust conclusions on the outcome of sexual selection can be obtained only if such processes are explicitly included in the models (e.g., Kokko & Jennions 2008, Fawcett et al. 2011).

Given all these intricacies, it is no wonder that no single model has been able to capture all relevant aspects of sexual selection in a fully satisfactory way. Therefore, various modeling approaches have been developed, each of which has specific strengths and weaknesses. In Section 2, we briefly review these approaches, pointing out their scope and limitations. Section 3 provides an overview of the potential benefits (and costs) of mate choice behaviors. We discuss the relative importance of direct versus indirect benefits of choice and some recent insights, such as the realization that the classical Fisher model can exhibit ongoing oscillations of preferences and ornaments. In Section 4, we address the recent trend of adding mechanistic detail

Components of sexual selection models...

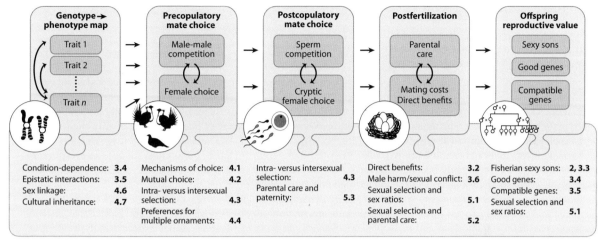

Figure 1

A global overview of the main components of sexual selection theory. Bold numbers refer to corresponding sections of this review.

to the classical models of sexual selection. Research shows that some seemingly well-established results are less robust than anticipated by "standard" theory. Finally, Section 5 discusses the implications of taking a more integrative approach and linking sexual selection to other evolutionary processes such as the evolution of sex, sex-ratio evolution, or speciation. **Figure 1** provides a summary of the various components of sexual selection theory that are discussed here. Throughout this review, the focus is on models that describe the evolution of female preferences for male ornaments, as this comprises the majority of work on the subject. Recent studies that focus on the evolution of male preferences are discussed in Section 4.2.

2. MODELING APPROACHES

The evolutionary process that underlies sexual selection can be mathematically described in a variety of ways (Dercole & Rinaldi 2008, ch. 2). Here we discuss the pros and cons of four of the most widely used descriptions of evolutionary change within the context of sexual selection: population genetics, quantitative genetics, invasion analysis, and individual-based simulations (Pen & Weissing 2000, Fawcett et al. 2007). The sidebar Four Implementations of the Fisher Process and **Figure 2** illustrate the application of the various modeling frameworks to the Fisher process, highlighting the congruencies and differences between the various approaches (for a more elaborate explanation, also see the **Supplemental Text**; for all **Supplemental Material**, follow the link from the Annual Reviews home page at **http://www.annualreviews.org**).

2.1. Population Genetics

Population genetics provides a description of evolution in terms of changes in genotype frequencies (see sidebar, Four Implementations of the Fisher Process). In principle, a population genetics framework is the most comprehensive approach to understanding sexual selection, as it directly models the evolutionary dynamics in terms of changing genotype frequencies. Whenever genetic

Fisher process: indirect selection on female preference caused by linkage disequilibrium with directly selected male ornament; leads to self-reinforced runaway selection

FOUR IMPLEMENTATIONS OF THE FISHER PROCESS

In a seminal contribution to sexual selection theory, Fisher (1915) predicted that female preferences could evolve through a self-reinforcing runaway process. Fisher argued that, once a female preference for a certain ornament has gained a foothold in a population (for whatever reason), both the preference and the ornament are subject to positive selection, but for different reasons. For the ornament, the argument is simple: Ornamented males will have a mating advantage if sufficiently many females mate preferentially with such males. For the preference, the argument is more sophisticated because selection on the preference is indirect. Because females with a strong preference tend to mate with males with a pronounced ornament, preference and ornament alleles often co-occur in the offspring of such matings, leading to a statistical association among these alleles. As a consequence, positive selection on the ornament will induce correlated positive selection on the preference. Hence, preferences induce the evolution of ornaments and subsequently become selected owing to their association with the ornament. Fisher realized that this self-reinforcing process could explain the huge exaggeration of sexual ornaments observed in many organisms. Interestingly, Fisher's arguments apply to arbitrary ornaments. In other words, ornaments that evolved through the so-called Fisher process do not necessarily indicate any inherent quality of their bearers. Notice that Fisher's argument on the statistical association between preference and ornament genes does not require the physical linkage of preference and ornament loci on the same chromosome.

For many decades, Fisher's ideas were greeted with skepticism. This verbal theory, deriving far-reaching and counterintuitive predictions from indirect processes and the emergence of statistical associations, was in need of a sound quantitative underpinning. Such would not be established until the 1980s, when models from quantitative genetics (Lande 1981) and population genetics (Kirkpatrick 1982) revealed the efficacy of the Fisher process.

Population genetics. Central to the population genetics approach is its ability to track genotype frequencies. Kirkpatrick's (1982) model of the Fisher process is a textbook example of a population genetics model of sexual selection. The model considers two haploid gene loci: (*a*) the female preference locus P with alleles P_0 (no preference; random mating) and P_1 (preference for mating with males carrying an ornament trait) and (*b*) the male trait locus T with alleles T_0 (no ornament) and T_1 (having a costly ornament). The relative allele frequencies of the preference and trait alleles are denoted by p and t, respectively, whereas the linkage disequilibrium parameter D describes the statistical association between alleles P_1 and T_1. As noted in the **Supplemental Text** (see also Bulmer 1994), the change in allele frequencies and genetic association from one generation to the next is described by equations of the form

$$\Delta t = \frac{1}{2}t(1 - t)A, \qquad\qquad\qquad 1.$$

$$\Delta p = \frac{1}{2}DA, \qquad\qquad\qquad 2.$$

and

$$\Delta D = \text{something complicated.} \qquad\qquad\qquad 3.$$

A is a function of p and t that describes the net effect of selection on the ornament (a balance between viability selection against the ornament and the mating advantage of ornamented males). Equation 2 shows that p changes only if $D \neq 0$, that is, if there is a statistical association between trait and preference alleles. A detailed analysis of the system represented by Equations 1, 2, and 3 reveals that, starting at zero, D becomes positive. As shown in **Figure 2a**, the system converges either to loss ($t = 0$) or fixation ($t = 1$) of the ornament or to a line of internal equilibria (given by $A = 0$). The line of internal equilibria corresponds to those combinations of t and p where the costs of carrying an ornament in terms of higher mortality are exactly balanced by the mating advantage of ornamented males. The whole set of equilibria is stable in the sense that selection prevents movements away from it, but each individual equilibrium is only neutrally stable, meaning that stochastic fluctuations can lead to shifts in p and t along the line of equilibria.

Quantitative genetics. Rather than tracking genotype frequencies, the quantitative genetics approach describes evolution in terms of changes of average phenotypic values. In the classical model by Lande (1981), the (phenotypic) values of a male ornament of size t and a female preference of intensity p are autosomally inherited, sex-limited, normally distributed traits with means $\bar{t}$ and $\bar{p}$. As shown in the **Supplemental Text** (see also Mead & Arnold 2004), the change in these means from one generation to the next can be described by the following equations:

$$\Delta \bar{t} = \tfrac{1}{2} G_t \beta_t; \qquad\qquad 4.$$

$$\Delta \bar{p} = \tfrac{1}{2} G_{tp} \beta_t. \qquad\qquad 5.$$

Here β_t is the total force of directional selection acting on the ornament (which is given by a combination of natural selection against and sexual selection in favor of large ornament size) (see Equation A7 in the **Supplemental Text**). The system does not include a corresponding term for the preference, because $\beta_p = 0$ in the absence of direct costs and benefits of choosiness. G_t is the additive genetic variance of the ornament, and G_{tp} is the additive genetic covariance between trait and preference. Equation 4 describes the evolution of the male trait under direct selection, whereas Equation 5 describes the correlated evolution of the female preference, which is mediated by the (positive) covariance between trait and preference. Like the system represented by Equations 1, 2, and 3, the system noted by Equations 4 and 5 has a line of equilibria. These equilibria correspond to the solutions of $\beta_t = 0$ (where natural and sexual selection are exactly balanced). **Figure 2b** shows when the line of equilibria is stable (which happens when the slope of the line of equilibria is larger than G_{tp}/G_t); if the covariance between trait and preference is very large, the line can also be unstable, leading to a never-ending runaway moving from the line with ever-increasing speed.

Invasion analysis. Here we briefly highlight a model of the Fisher process by Pen & Weissing (2000), which combines a reproductive-value approach with methods from adaptive dynamics theory. The model considers a class-structured population consisting of females and two types of males: nonornamented males (σ_0) and males expressing an ornament (σ_1) reducing their viability by a factor $1 - s$. Evolvable traits are the females' preference p for mating with ornamented males and the tendency t of a male to develop the ornament (i.e., the tendency to become a type-1 male). The aim is to find the evolutionarily stable values p^* and t^* of preference and ornament, respectively. As shown in the **Supplemental Text**, the fitness $W(t, p \mid t^*, p^*)$ of a rare mutant of type (t, p) in a resident population of type (t^*, p^*) can be derived systematically from life-history considerations. Evolutionarily stable strategies can be determined by inspecting the (total) derivatives of the fitness function W with respect to t and p at (t^*, p^*). These are of the form

$$dW/dt = (1 - s)v_{m1} - v_{m0}, \qquad\qquad 6.$$

$$dW/dp = b_{pt} \cdot ((1 - s)v_{m1} - v_{m0}), \qquad\qquad 7.$$

where v_{m1} and v_{m0} are the reproductive values of ornamented and nonornamented males in the resident population and b_{pt} is the slope of the regression of the female preference on the male trait, which describes the statistical association between trait and preference and is assumed to be positive. At an evolutionary equilibrium, both derivatives in Equations 6 and 7 have to be zero. This is the case when $(1 - s)v_{m1} = v_{m0}$, that is, when viability costs of the ornament are exactly balanced by the mating advantage provided by the ornament. As before, there is a line of equilibria (**Figure 2c**), and the approach to this line is governed by the canonical equation of adaptive dynamics.

Individual-based simulations. As indicated by **Supplemental Figure 2**, an individual-based simulation keeps track of a finite population of individuals, each of which has a set of properties (e.g., genotypes, sex, degree of preference, degree of ornamentation). Individuals interact, and owing to their properties and chance events, they differ in survival, mating success, and fecundity. During reproduction, individuals transmit (part of) their heritable properties to their offspring. Variation arises as a result of mutation. The simulation then tracks evolutionary change over the course of generations. **Figure 2d** shows the outcome of such a simulation resulting from the implementation of the Fisher process by Fawcett et al. (2007). In this simulation model, each individual harbors two loci (both with many alleles): one coding for an ornament of size t and one coding for a preference of intensity p. Making similar assumptions on the mortality costs of the ornament and the mating process as in Lande's (1981) model (see the **Supplemental Text**), the simulated population rapidly converges to a line of equilibria and subsequently drifts along this line. Interestingly, distinct behaviors (ongoing oscillations) can occur in the same model for different parameter settings (see Section 3.3 and **Figure 3d**).

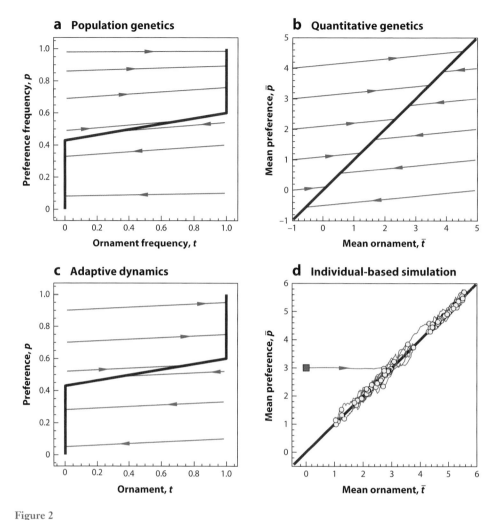

Figure 2

Four implementations of the Fisher process: (*a*) dynamics of Kirkpatrick's (1982) population genetics model, (*b*) dynamics of Lande's (1981) quantitative genetics model, (*c*) adaptive dynamics model by Pen & Weissing (2000), and (*d*) individual-based simulation based on the model by Fawcett et al. (2007). Red lines represent equilibria of the model (line of equilibrium), whereas blue lines with arrowheads indicate representative trajectories of (*a–c*) the dynamical systems as well as (*d*) a single simulation, with the square indicating the starting point of the simulation and circles indicating the state of the population at 50 generation intervals.

processes are crucial for a proper understanding of sexual selection, a population genetics model is typically the model of choice. A population genetics framework is often indispensable when studying the implications of a genetic architecture (such as sex linkage, recombination or epistatic gene interactions) on the course and outcome of sexual selection. Moreover, population genetics models are useful for delineating the scope and limitations of less comprehensive frameworks like quantitative genetics or adaptive dynamics.

Although population genetics methods have been very important for theory development, they are less popular in terms of practical applications for at least two reasons. First, despite enormous advances in unraveling the genetic underpinning of sexual characters (Chenoweth &

McGuigan 2010), the genetic basis of traits and preferences is generally not known. Instead of making hypothetical assumptions regarding the underlying genetics, students of sexual selection tend to find quantitative genetics (Section 2.2) and phenotypic approaches (Section 2.3) more appealing because these approaches do not specify the genetics and essentially treat it as a black box. Second, population genetics models quickly become mathematically intractable if several loci with genes of large effects are involved. Therefore, in practice, most population genetics models address only a small number of haploid loci in a setting of discrete, nonoverlapping generations. Perhaps more importantly, mathematical tractability necessitates that investigators make highly simplifying assumptions concerning the phenotypic level. As a result, addressing the mechanisms of behavioral interactions or complex trade-offs between fitness components becomes difficult.

However, the advent of the quasi-linkage-equilibrium (QLE) technique (Barton & Turelli 1991, Kirkpatrick et al. 2002; for a critical appraisal, see Pomiankowski & Bridle 2004) enabled researchers to overcome some of the disadvantages of the population genetics approach. Assuming that selection is weak relative to the rate of recombination, investigators avoid to some extent the intricacies of multilocus genetics by approximating the (high-dimensional) dynamics of genotype frequencies by the dynamics of allele frequencies and lower-level genetic associations (e.g., Kirkpatrick & Hall 2004, Servedio 2004, Greenspoon & Otto 2009). The conclusions obtained are often remarkably robust and not dependent on genetic detail or the mating system (e.g., Kirkpatrick & Barton 1997). However, due to the inherent assumption of weak selection, it is not always clear whether, and to what extent, the predictions of the QLE approach extend to scenarios involving strong selection (e.g., Servedio 2004).

Quasi-linkage-equilibrium (QLE) technique: technique to simplify population/quantitative genetics models by neglecting intricate interaction terms, assuming that selection is weak relative to recombination

2.2. Quantitative Genetics

Quantitative genetics is a widely used technique to model sexual selection (Mead & Arnold 2004). Quantitative genetics describes evolution at the phenotypic level but still takes account of genetics (to a certain extent), thus yielding plausible assumptions on the transmission of phenotypic traits from parents to their offspring. The latter are encapsulated in the so-called G-matrix, the collection of additive genetic variances and covariances of the phenotypic traits in question. As illustrated in the Four Implementations of the Fisher Process sidebar (and in more detail in the **Supplemental Text**), the change of (mean) traits and preferences from one generation to the next is characterized by a relatively simple equation that inspires considerable insights into the dynamics of sexual selection. However, the simplicity and elegance of the quantitative genetics approach comes at a cost, as many assumptions have to be made to justify this approach. Moreover, these assumptions are often implicit and not easily testable. One key assumption of the quantitative genetics approach is that breeding values have a multivariate normal distribution, which has been subject to considerable debate (Barton & Turelli 1991). Another common assumption is that viability costs are given by exponential functions (see the **Supplemental Text**). Assumptions like these are mathematically convenient because they assure that the distribution of phenotypes at the mating stage remains Gaussian. However, they are not always realistic and may be misleading. For example, the Fisher process has a stronger tendency to induce ongoing oscillations of traits and preferences if the costs of choosiness are related to the availability of the preferred males.

Quantitative genetics models often treat additive genetic variances and covariances as fixed parameters (e.g., Pomiankowski et al. 1991). However, selection affects the G-matrix entries (for a recent review, see Arnold et al. 2008) both indirectly by shaping the mutation rates and directly via the depletion of additive genetic variation or the buildup of additive genetic covariation (i.e., linkage disequilibria) due to assortative mating. Under certain assumptions regarding mutation, recombination, and the strength of selection, the evolution of the G-matrix can be studied within

Invasion fitness: the
exponential growth
rate of a rare mutant in
an environment
determined by the
common resident
strategy

Reproductive value:
the long-term
expected genetic
contribution of an
individual to the
population

the quantitative genetics framework by making use of the QLE approach (Barton & Turelli 1991, Pomiankowski & Iwasa 1993, Walsh & Lynch 2013). In most cases, however, individual-based simulations provide a more convenient and versatile tool to assess the evolution of the G-matrix (e.g., Arnold et al. 2008).

Studies of quantitative trait loci (QTL) have shown that quantitative traits are often affected by at least some major genes of large effect. In such cases in which one or more quantitative traits coevolve with large-effect modifiers, a hybrid approach can be taken (Lande 1983). The same type of approach can be used when studying the interplay of sexual selection and segregation distortion or sex-chromosome evolution.

2.3. Invasion Analysis

Whereas quantitative genetics is based on the assumption that a continuous distribution of phenotypes (and typically also genotypes) is available at all times, evolutionary game theory, adaptive dynamics, and other phenotypic approaches (Weissing 1996) consider the opposite extreme of a monomorphic resident population that is repeatedly challenged by the invasion attempts of rare mutants. The underlying idea is that evolution proceeds by a series of subsequent invasion and trait-substitution events. The dynamics of this process can be described by the canonical equation of adaptive dynamics (Dieckmann & Law 1996), where

$$\frac{d}{d\tau}\mathbf{x}^* = z\mathbf{M}\nabla W(\mathbf{x}|\mathbf{x}^*)$$

gives a description of change in the vector of characters $\mathbf{x}^*$ over evolutionary time τ through a successive series of invasion-substitution events. z reflects the variation in the rate of occurrence of mutations, whereas the mutational variance-covariance matrix $\mathbf{M}$ describes how a single mutation affects the different traits. $\mathbf{M}$ plays a role very similar to that of the additive genetic variance-covariance G-matrix in quantitative genetics. The direction of selection is given by the invasion-fitness gradient $\nabla W(\mathbf{x}|\mathbf{x}^*)$ of a rare mutant $\mathbf{x}$ invading in a population of $\mathbf{x}^*$ residents. Usually, the exact dynamics of how a mutant coexists with and replaces the resident are not modeled explicitly. Instead, population dynamical considerations are used to derive an expression for the invasion fitness of rare mutants (Metz et al. 1992). This fitness function is then systematically analyzed to identify evolutionarily stable strategies and other potential end points of the evolutionary process (Geritz et al. 1998, McGill & Brown 2007, Dercole & Rinaldi 2008). In a life-history context, it is convenient to frame invasion fitness in terms of reproductive values (Taylor 1996).

Because such phenotypic approaches neglect most genetic intricacies, they can address environmental feedbacks, frequency and density dependence, and age structure in more detail than can other approaches. As a result, recent models investigating the interaction of sexual selection with parental care (Kokko & Jennions 2008) or sex allocation (Fawcett et al. 2011) rely on a phenotypic approach. Nonetheless, the shortcomings of phenotypic models should not be ignored. One key assumption is that populations are nearly monomorphic. Because variation in male ornamentation is essential to the evolution of female choice, additional assumptions are necessary to maintain genetic variation (see the **Supplemental Text**), but the amount of variation that is maintained can alter the outcome in surprising ways (e.g., McNamara et al. 2008). Relaxing the assumption that mutants differ only slightly from the resident can also strongly affect the evolutionary dynamics (e.g., Wolf et al. 2008). Another disadvantage is that phenotypic models do not easily allow for a dynamical description of linkage disequilibria. Furthermore, most of these models assume that evolution proceeds at a much slower timescale than the ecological dynamics, even though it is known that sexually selected characters may evolve rapidly (e.g., Van Doorn et al. 2001, Swanson & Vacquier 2002, Shirangi et al. 2009).

2.4. Individual-Based Simulations

Individual-based simulations (see sidebar, Four Implementations of the Fisher Process) provide a flexible and easily extendable way of modeling complicated scenarios with a high degree of realism [e.g., environmental and demographic stochasticity, spatial population structure (Fromhage et al. 2009), complex genotype-phenotype maps (Ten Tusscher & Hogeweg 2009), a concrete representation of the sensory system (Fuller 2009)]. This is a clear advantage over all the simplifying assumptions made by analytical models. For example, the presence of stochasticity in individual-based simulations allows for a straightforward assessment of the importance of drift (Uyeda et al. 2009), which is much harder to assess in analytical models of sexual selection. On the downside, running complex simulations is often computationally demanding, limiting the number of parameter settings that can be investigated. If the model contains only 10 parameters (most have many more), then $3^{10} \approx 60,000$ simulations are needed to consider all the combinations of only three values for each parameter. Moreover, replicates of each simulation have to be run to cope with the stochasticity inherent in each simulation. Because this is not always feasible, it is often not clear whether and to what extent a given set of simulations is representative. Nonetheless, we believe that the disadvantages of simulation models are often overemphasized (e.g., McElreath & Boyd 2007, p. 8), especially given advantages such as relative ease of implementation and applicability to all kinds of situations. Perhaps most importantly, individual-based simulations can nicely complement an analytical approach. In fact, the theoretical justification of analytical approaches is often restricted to a narrow domain (e.g., weak selection), and simulations are useful for exploring the robustness and general applicability of analytical predictions beyond this domain.

2.5. A Plea for Pluralism

The famous quote that "each disadvantage has its advantage" (by the former Dutch football player Johan Cruijff) also applies to the various modeling frameworks considered thus far (**Table 1**). Accordingly, the choice of approach should depend mainly on the research question being tackled. But how should we deal with the often unrealistic assumptions made by virtually all approaches? Richard Levins's (1966, p. 423) statement that "our truth is the intersection of independent lies" provides an answer: If multiple modeling frameworks with varying underlying assumptions arrive at a similar outcome, we can be confident that this conclusion is robust and not just a result of some limiting assumptions. Hence, the use of multiple modeling frameworks in parallel helps us to delineate the scope and limitations of the predictions of sexual selection theory.

3. THE BENEFITS AND COSTS OF CHOICE

Much debate on sexual selection theory has focused on the benefits driving the evolution of mating preferences. Females can benefit directly from expressing a preference, if the resulting choice of mates gives them a higher viability or fecundity than they would have had without the preference (Møller & Jennions 2001). Alternatively, the particular choice of mating partners may lead choosy females to have offspring with a higher reproductive value (Kokko et al. 2002), for example, through attractive sons (Fisher 1915) or offspring of higher intrinsic quality (Zahavi 1975), in which case benefits are said to be indirect. In this context, offspring reproductive values often refer to offspring lifetime reproductive success, although there may be cases in which the reproductive success of grandoffspring or later generations should also be considered. Here, we highlight the major assumptions underlying models of direct and indirect benefits of sexual selection.

Table 1 Pros and cons of various modeling approaches[a]

	Population genetics	Quantitative genetics	Invasion approaches	Individual-based simulations
Pros	Most comprehensive description of sexual selection; explicit inclusion of genetic aspects (recombination, linkage disequilibrium); transparent model assumptions; provides justification of quantitative genetics approach	Focus on (measurable) phenotypic variation; method applicable to systems with limited information on genetics; model parameters (additive genetic variances and covariances) can be estimated	Minimal assumptions on genetics allow most comprehensive description of phenotypic level; based on transparent fitness concept; fitness function derived from first principles, allowing consideration of demography, class structure, frequency, and density dependence	Most versatile approach; few restrictions on model structure, allowing for complicated genetic architectures and intricate selection scenarios; natural inclusion of demographic and environmental stochasticity
Cons	Mainly suited for discrete variation; tractable only for highly simplified fitness scenarios; limited applicability because the genetic basis of most traits is unknown; analytical tractability limited to a small number of loci with few alleles	Assumptions (e.g., normal distribution of additive effects) often not met; G-matrix often assumed constant; difficult to include complex genetics, complex life histories, and complex fitness scenarios (age structure, trade-offs, etc.)	Assumption of monomorphic population clearly unrealistic; includes only simplistic genetics, although genetic associations may be crucial; multivariate adaptive dynamics theory not well developed	Danger of cherry-picking because only a small part of parameter space can be investigated; interpretation of results often subjective; difficult to judge the generality of conclusions; "coarse" description of simulation programs
Solutions	In the case of weak selection, powerful techniques (e.g., quasi-linkage equilibrium) are becoming available to address complex multilocus problems and fairly complex fitness scenarios	Methods for studying the evolution of the G-matrix are becoming more broadly available; hybrid models combine a quantitative genetic approach with population genetics and adaptive dynamics	Application of adaptive dynamics to genotypic level; combination of adaptive dynamics approach with genetics approaches	Application of simulations to systems in which analytical results are already available as a test case

[a]See Section 2 for a more detailed discussion of these modeling approaches.

3.1. No Benefits: Sensory By-Products

Female preferences can evolve in the absence of any benefits related to mate choice, for example, as a pleiotropic by-product of natural selection on the sensory system (Kirkpatrick 1987, Kirkpatrick & Ryan 1991). This is confirmed by models employing evolving neural networks (mimicking a simple sensory system), which often lead to mating preferences (e.g., in favor of symmetric mates) as a by-product (reviewed in Enquist & Ghirlanda 2005 and Phelps 2007). Whereas early models based their conclusions on highly simplified network architectures (Arak & Enquist 1993; for a critique, see Dawkins & Guildford 1995), more recent models are tailored to the sensory system of particular organisms. For example, Fuller's (2009) model based on the sensory system of guppies (*Poecilia reticulata*) reveals that details of the sensory architecture, such as the number of output neurons, can strongly affect the evolution of sensory biases. It remains to be seen which types of sensory architecture are most conducive to the evolution of preferences as sensory by-products, on which external conditions (i.e., the sensory environment) certainly have a large influence (Endler & Basolo 1998). In addition, there are few predictions on the long-term evolution of preferences that evolve as by-products of natural selection. Will pleiotropy in preferences always decay over time (when pleiotropic preferences lead to suboptimal female mating decisions) (Arnqvist 2006), or can we find situations in which the coevolutionary dynamics of both natural and sexual selection enhance the maintenance of pleiotropic preferences?

3.2. Direct Benefits

Direct selection on female choosiness occurs whenever the degree of choosiness is related to a female's survival and fecundity. Choosiness is often costly; for example, the search for mates can expose the female to predators, or females may risk ending up unfertilized (Kokko & Mappes 2005). As discussed below, even slight costs can override indirect benefits of choosiness, leading to the theoretical expectation that sexual selection driven only by indirect benefits of choosiness is rare in nature (Kirkpatrick & Barton 1997, Cameron et al. 2003).

The evolution of female preferences can most easily be explained if females gain direct benefits from being choosy. One likely mechanism is when females prefer males that advertise their quality to provide paternal care: Illuminating models on this good-parent process have been made by Price et al. (1993) and Iwasa & Pomiankowski (1999). Direct-benefit models have received relatively little attention in the theoretical literature presumably because the underlying mechanisms seem transparent and not too challenging (or sufficiently counterintuitive) to modelers. However, a number of recent results indicate that direct-benefit mechanisms are not as straightforward as previously thought. For example, males may differ not only in parental ability but also in genetic quality, and these aspects of quality are not necessarily related. Hence, it matters what aspect of quality is being signaled and how to interpret male signals (Kokko 1998, Alonzo 2012). Moreover, mating with a male of high parental quality may not assure a high level of paternal care if such males tend to mate with many females and therefore have to distribute their care over many offspring (Cotar et al. 2008, Tazzyman et al. 2012). In addition, females can be expected to change their own care level depending on their choice of mates (Ratikainen & Kokko 2010). All this leads us to conclude that the theory of direct benefits of sexual selection deserves more attention than currently is devoted to it.

3.3. Indirect Benefits: The Fisher Process

The Fisher process relates to the scenario in which female preferences are maintained as a result of self-reinforcing selection (see sidebar, Four Implementations of the Fisher Process). The key benefit associated with the Fisher process is a greater number of grandoffspring: According to this theory, choosy females will produce attractive sons, which in turn will have a higher mating rate. These benefits are relatively small: The slightest costs of choosiness break down the line of equilibria and costly choosiness disappears from models of the Fisher process (Kirkpatrick 1982, Pomiankowski 1987, Bulmer 1989) (see **Figure 3a**). However, the Fisher process can be rescued if additional mechanisms, such as a mutation bias (mutations that have mainly negative effects on male ornamentation) (Pomiankowski et al. 1991) (**Figure 3b**) or migration bias (influx of migrant males with smaller ornaments) (Day 2000) (**Figure 3c**), are included in the model. Even in the absence of such mutation or migration biases, the exaggeration of sexually selected traits beyond the naturally selected optimum is possible if the costs of choosiness and ornamentation are sufficiently weak (Hall et al. 2000). When this is the case, traits and preferences do not converge to equilibrium but oscillate forever on a limit cycle (B. Kuijper, L. Schärer, and I. Pen, unpublished manuscript)

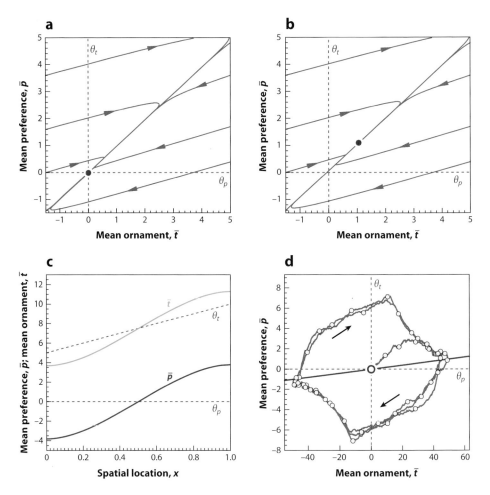

Figure 3

Costs of choice and the Fisher process. (*a*) In the presence of costly choice, the line of equilibria (see **Figure 2*b***) collapses to a single equilibrium point, coinciding with the naturally selected optima θ_t and θ_p of trait and preference, thus eliminating costly exaggeration of sexually selected traits (Pomiankowski 1987, Bulmer 1989). (*b*) Biased mutations tending to reduce ornamentation can "rescue" sexual selection, giving rise to an equilibrium point away from the naturally selected optimum (Pomiankowski et al. 1991). (*c*) Spatial variation and the influx of migrants with smaller ornaments can also lead to the exaggeration of trait *t* (*light blue line*) and preference *p* (*dark blue line*) beyond their naturally selected optima (*dotted lines*), which now vary along a spatial gradient (*x* axis) (Day 2000). (*d*) When the viability costs of preferences and traits are small, the equilibrium in panel *a* becomes unstable (indicated by the *red, open circle*) and the system converges to a limit cycle, corresponding to indefinite cycling of traits and preferences around the naturally selected optimum. Small circles indicate the state of the population in intervals of 200 generations.

(**Figure 3*d***). Similar cyclic dynamics were previously described in variants of the Fisher process with a curvilinear set of equilibria that ignored the costs of choice (Iwasa & Pomiankowski 1995).

3.4. Indirect Benefits: Good Genes

In good-genes scenarios of sexual selection, there is an evolution of female preferences for male indicators of heritable quality. Heritable "quality" is typically interpreted as enhanced offspring survival, but it should actually be interpreted to indicate enhanced offspring reproductive value (Kokko 2001, Kokko et al. 2002). Even if offspring inherit genes intrinsically favoring their viability, their survival may actually be lower than that of lower-quality offspring (for example, owing to intense signaling). This focus on reproductive values, in combination with the notion that all models of sexual selection, including the good-genes process, contain some element of the Fisher process, has led researchers to conclude that differences between the Fisher process and good-genes sexual selection are small and superficial (Kokko 2001, Kokko et al. 2002). However, good-genes sexual selection is conceptually different from the Fisher process in several ways. First, the dynamics of both processes are different: Whereas the Fisher process requires only coevolving preference and ornament genes, good-genes processes include at least one additional dimension (corresponding to heritable quality). Second, the evolution of a genetic correlation between trait and preference is a crucial aspect of the Fisher process (see sidebar, Four Implementations of the Fisher Process). In contrast, good-genes sexual selection still works (and, in fact, leads to a runaway process) even when such a correlation cannot be established (e.g., when preferences inherit exclusively through the matriline, in contrast to ornaments inherited exclusively through the patriline). In cases like this, preferences can evolve through a genetic correlation with alleles related to heritable quality. Third, in the Fisher process, ornaments and preferences are directly coupled; by definition, the ornament is the target of the preference. Discussions of whether the ornament can act as a reliable indicator of genetic quality (i.e., male attractiveness) are irrelevant for the Fisher process, even though they have played a dominant role in the context of good-genes processes. Hence, although it is important to point out the similarities between indirect-benefit models of sexual selection (Kokko et al. 2002), the conceptual differences between the Fisher process and good-genes processes should not be neglected.

Another debate regarding good-genes models of sexual selection has centered on the question of how females can reliably distinguish between males of different quality. According to Zahavi's (1975) handicap hypothesis, signals are reliable indicators of male heritable quality only if costs are associated with these signals. However, not all costly signals are reliable indicators of quality (Getty 2006). To separate the sheep from the goats, signals have been classified into different categories (Maynard Smith 1985; for formal definitions, see Van Doorn & Weissing 2006). Signals are condition dependent when the expression of a given signal is less costly for males of higher quality. Alternatively, signals are designated as revealing if for a given level of resource allocation into the development of an ornament high-quality males produce ornaments that are more pronounced than those of low-quality males. Whereas condition-dependent and revealing signals can evolve as reliable indicators of heritable quality (Andersson 1994), epistatic signals (which are neither condition dependent nor revealing) are usually considered unreliable and therefore cannot lead to the evolution of costly female preferences. However, several studies have reported the contrary. In a general model, Kirkpatrick & Barton (1997) show that—irrespective of the type of signal—the buildup of a correlation between male ornamentation and heritable quality can lead to indirect selection on the female preference (although this effect is weaker in the case of epistatic signals) (see also Siller 1998). Van Doorn & Weissing (2006) show that female preferences for epistatic signals can evolve if the coevolution of ornaments and preferences leads, not to a stable equilibrium, but to a limit cycle (as shown in **Figure 3d**).

Another problem for good-genes sexual selection is that female preference for high-quality males can lead to the depletion of genetic variation in quality, commonly referred to as the lek paradox (for a review, see Kotiaho et al. 2008). A potential resolution to the lek paradox is the genic capture hypothesis (Rowe & Houle 1996). The central tenet of this hypothesis is that an individual's condition is determined by a large number of genes, providing a large mutational target

Good-genes sexual selection: indirect selection on female preference caused by linkage disequilibrium with directly selected male ornament; leads to self-reinforced runaway selection so that some genetic variation in condition always exists. Indeed, sexual selection may cause the number of genes that underlie a signal to increase, giving rise to genic capture (Lorch et al. 2003).

Directional preferences: when all females prefer similar male phenotypes instead of basing their preference on their own genotype or state

3.5. Indirect Benefits: Compatible Genes

Whereas good-genes models assume that genetic quality is an intrinsic property of the genetic makeup of an organism, reality may not be that simple (Hunt et al. 2004, Puurtinen et al. 2009). For example, males may signal local adaptation (Proulx 2001, Reinhold 2004) or adaptation in contexts of frequency-dependent, disruptive selection (Van Doorn et al. 2009). In addition, off-spring performance may not directly reflect the genetic quality of the parents if it is affected by epistatic interactions between paternally and maternally inherited genes. Epistatic interactions are a complicating factor for sexual selection theory: Depending on their own genotype, different females may prefer different types of males. The existence of sexually antagonistic variation (Albert & Otto 2005, Arnqvist 2011) or selection on heterozygosity and other forms of epistatic variation are further examples of such compatible-allele effects (Puurtinen et al. 2009). In general, such effects weaken the selection for preference alleles. However, recent models show that directional preferences can nevertheless evolve if certain conditions, such as biased mutations (Lehmann et al. 2007) or spatial variation in finite populations (Fromhage et al. 2009), are met.

3.6. Avoiding Male-Induced Costs: Sexual Conflict

A large body of literature has revealed that females may have to endure mating-related costs, such as seminal toxins or damage by male genital spines (e.g., Rice 1996, Arnqvist & Rowe 2005). Thus, resistance to harmful mating may be thought of as a mating preference, because it can restrict the potential set of mating partners (Halliday 1983). Importantly, selection for female resistance is direct, because any female that evolves higher levels of resistance reaps the benefits in terms of increased fecundity or survival (Kokko et al. 2006).

The most influential coevolutionary models envisage female resistance as a threshold character that causes females to mate only with those males that have harm-trait values that surpass the female's threshold (Gavrilets et al. 2001). Females that accept too many males will incur a fecundity cost on top of the mortality cost of developing a resistance trait. Depending on these costs, the course of evolution varies, but exaggeration of male harm and female resistance is a common outcome. The situation is substantially different when female resistance evolves in the form of insensitivity, making a female reluctant to mate with any male phenotype (Rowe et al. 2005). In this case, male harm is effectively neutralized by female insensitivity, and an evolutionary standstill is a likely outcome, showing that mechanisms with which female resistance is realized can strongly change conclusions. Sexual conflict models have to account for the further complication that females mating with the most persistent males may accrue indirect benefits because they will tend to produce more persistent sons. Although these indirect benefits are unlikely to recoup the direct costs of harm for the reasons stated above, they may nonetheless alter the coevolutionary dynamics substantially (Härdling & Karlsson 2009).

3.7. Multiple Costs and Benefits

Whereas a single benefit or cost component has been the focus of most models, the co-occurrence of multiple costs and benefits has hardly been explored (but see Iwasa & Pomiankowski 1999, Van Doorn & Weissing 2004, Härdling & Karlsson 2009). One particular aspect that deserves attention is the potential for trade-offs between direct and indirect benefits. For example, in the good-parent process, females may incur more direct benefits (i.e., more care) by mating with unattractive males that have few mating opportunities and, hence, can focus their care on a small number of offspring (e.g., Cotar et al. 2008, Tazzyman et al. 2012). These direct benefits are associated with indirect costs, as the sons sired by these males will not be attractive and thus have a limited mating rate. Such trade-offs may lead to condition-dependent choices (Cotton et al. 2006) in which some females prefer males that provide direct benefits and others prefer males with indirect benefits.

4. ADDING MECHANISTIC DETAIL TO SEXUAL SELECTION MODELS

Traditionally, sexual selection models incorporated the processes under study in an abstract and seemingly general way. Recent models tend to include more mechanistic detail, thereby making the models more specific but also more easily applicable to real systems. Here we highlight how seemingly general insights of traditional models can change when the mechanisms underlying sexual selection are explicitly incorporated.

4.1. The Mechanisms of Mate Choice

Mate-choice behaviors refer to any behavior that limits an individual's set of potential partners (Halliday 1983). Although any review on mate choice is eager to point out the behavioral and social complexities involved (Cotton et al. 2006), little has been done to integrate these notions with formal models of sexual selection. In practice, most models assume that females sample from an effectively infinite pool of males in which attractive males are more "apparent" to choosy females and are thus more easily encountered (e.g., Lande 1981, Kirkpatrick 1982). Because slight changes to these assumptions can dramatically affect an outcome (e.g., Seger 1985), the question of how females should optimally sample males in the face of costs and constraints arises. Optimal-sampling models predict that females should evaluate males sequentially and stop sampling when they sample a male that matches a certain threshold (Real 1990, Wiegmann et al. 2010), such that the value of this threshold may depend on the particular way females acquire information about the male phenotype distribution (e.g., Dombrovsky & Perrin 1994, Luttbeg 1996). Houle & Kondrashov (2002) show that sequential sampling in a good-genes model enhances sexual selection. By contrast, in classical models, the costs of preferences increase with the level of exaggeration beyond an abstract survival optimum. Thus, sampling costs allow a female to have very strong preferences as long as she is likely to encounter a suitable male within a limited number of samples. In general, this finding by Houle & Kondrashov (2002) shows that there is much to say for a further integration of mechanistic aspects of mate choice within models of sexual selection.

4.2. Mutual Choice and Sex-Role Reversal

The focus on female choice within many models denies the observed variation in choosiness between the different sexes that exists across taxa; this varies from exclusive female choice to male or mutual choice (Clutton-Brock 2007). In general, mutual choice evolves when both sexes exhibit sufficient variation in quality, but it is hampered by the possibility that choice may be more costly for the sex that competes most intensely for matings (Johnstone et al. 1996). Moreover, the intensity of competition among members of one sex may be a function of the amount of choice exerted by the opposite sex, indicating that the evolution of sex differences in mate choice is the result of a complicated feedback between choice and competition (see Kokko & Johnstone 2002, Kokko & Jennions 2008). To disentangle this, recent models have taken a self-consistent approach whereby individuals of each sex are considered to be either breeding (time out) or busy acquiring matings (time in). The time and mortality costs of each activity feed back on the densities of individuals that breed or that compete for mates, which subsequently influences the evolution of mate choice in each sex (Kokko & Jennions 2008). Using a version of this framework, Kokko & Johnstone (2002) found that mutual choice evolves only under restrictive conditions because choosiness in one sex immediately reduces the mean mating rate of the other sex, which subsequently becomes more competitive and less likely to evolve choice (see also Servedio & Lande 2006). Only when both the cost of breeding and the mate-encounter rates are high for both sexes does it pay for both males and females to be choosy (Kokko & Johnstone 2002).

4.3. Intrasexual versus Intersexual Selection

In addition to ornamentation that has evolved in the context of mate attraction, males may also evolve weapons or signals (i.e., badges of status) in the context of male-male competition. Although male ornaments may be used for both mate attraction and male-male competition, they have been the subject of little formal attention to date despite considerable empirical support (Berglund et al. 1996). A recent model by Veen (2008), however, considers the coevolution of female preferences for male signals that signal both dominance in male-male competition and quality to a female. Interestingly, whereas mate choice and male-male competition in isolation may lead to the evolution of such signals only under particular conditions (see Sections 2 and 3), the interaction between both processes appears to be particularly conducive to the evolution of male signals.

Focusing on the postcopulatory stage, a sound body of theoretical predictions on intrasexual selection (i.e., sperm competition) exists (Parker & Pizzari 2010). In addition, recent efforts have started to consider trade-offs between male investment in traits that increase a male's mating rate and his success in sperm competition (e.g., Tazzyman et al. 2009). However, the role of female choice in these contexts is only starting to be assessed (Ball & Parker 2003, Fromhage et al. 2008), and we know of no formal studies on the coevolution between female choice and heritable male traits that are directly related to ejaculate investment.

4.4. Preferences for Multiple Ornaments

Even though the vast majority of sexual selection models typically focus on single, univariate display traits, sexual displays often involve many different components (for recent reviews, see Candolin 2003, Bro-Jørgensen 2010). Most models that have formally investigated multiple ornaments are based on the redundant-signal hypothesis (Møller & Pomiankowski 1993) in which preferences evolve for multiple indicator traits reflecting the same quality. The widely accepted view is that there is only a limited scope for preferences based on these backup signals because females should always favor the most honest and reliable ornaments; preferences for any additional ornament will be tolerated only when its costs are sufficiently low (Schluter & Price 1993, Iwasa & Pomiankowski 1994) (**Supplemental Figure 1a** versus **Supplemental Figure 1b**). Nonetheless, these analyses rely on equilibrium arguments, whereas more dynamical analysis shows that females can easily maintain preferences for multiple redundant ornaments through conflicts between males and females over the honesty of signaling (Van Doorn & Weissing 2006) (**Supplemental Figure 1c**). Moreover, preferences for multiple ornaments can also easily evolve when each ornament signals a different aspect of quality (multiple messages) (Johnstone 1995, Van Doorn & Weissing 2004) (**Supplemental Figure 1d**). Hence, in contrast to more classical models (e.g., Schluter & Price 1993), various studies note that the evolution of multiple indicators of quality does not appear as restrictive as previously thought. Yet, these recent studies generate as many new questions as they resolve: Can external factors, such as context dependence, also be responsible for the maintenance of multiple preferences? For multiple ornaments, how much more likely is it that reproductive isolation a result of drift (e.g., Pomiankowski & Iwasa 1998)? These and other questions await further investigation.

4.5. Individual Variation in Ornamentation

Along with the inheritance of sexually selected characters, the large developmental variation and plasticity observed in sexually selected characters are also poorly understood (Pomiankowski & Møller 1995). To date, how sexual selection affects developmental variation has been investigated only in the context of condition-dependent indicator traits. In particular, a number of theoretical studies have investigated how indicator traits develop over an individual's life span (Kokko 1997, Rands et al. 2011). Interestingly, these studies find that low-quality males may express larger ornaments than do high-quality males because their higher mortality rate induces them to make a terminal investment to achieve matings. It remains to be seen if such dishonest signals are more widespread and extend to contexts in which individual variation is expressed in aspects other than variation in life span.

4.6. Genetic Architecture: Sex Linkage

Based on the observation that organisms with Z-W sex chromosomes (e.g., birds, butterflies) tend to have more strongly exaggerated ornaments than organisms with X-Y sex chromosomes (e.g., mammals, flies) (Hastings 1994), the sex-linked inheritance of sexually selected traits is recently receiving much attention (Reinhold 1998, Reeve & Pfennig 2003). This research also takes into account a growing number of concrete examples of sex-linked sexually selected traits (Qvarnström & Bailey 2009). Although simple haploid inheritance models apply to cases of Y or W linkage, other patterns of sex linkage require a diploid locus in one sex (i.e., XX or ZZ), requiring a more complex model. Multilocus approximations such as QLE can considerably simplify such models. Using these techniques, Kirkpatrick & Hall (2004) confirmed that, compared with X-Y systems, Z-W systems are more conducive to sexual selection. For example, Z-linked preferences are favorable to Fisherian sexual selection: When present in males, such preferences endure indirect selection two-thirds of the time. By contrast, when present in females, such preferences are found one-third of the time, where they endure no or negative selection. Z-W systems are also more conducive to sexual selection in the presence of sexual antagonism: when females express a costly male ornament (i.e., owing to a lack of sex-limited expression) (Albert & Otto 2005). Z linkage of the ornament ensures that an ornament endures net positive sexual selection (i.e., is present in males two-thirds of the time), whereas costs due to expression in females are minimized (present in females one-third of the time). Besides ornaments and preferences, other traits such as genetic quality may be sex linked and thus can also affect sexual selection (Connallon 2010), illustrating the importance of the genetic architecture when making predictions about the strength of sexual selection.

4.7. Cultural Imprinting

Cultural inheritance may also play an important role in the evolution of sexually selected characters. Examples include song imitation in passerine birds and mate-choice copying in guppies (Dugatkin 1996). In principle, cultural evolution can be understood by the frameworks described in Section 2, with the modifications that transmission may occur horizontally and that genetic and culturally inherited traits may evolve independently on separate timescales (Lachlan & Feldman 2003). As a result, imitation often leads to positive frequency dependence, given that the most prevalent variant is also most likely to be imitated (Laland 1994). Thus, the cultural variants and genotypes that are initially present in a population have a large impact on the eventual outcome because any novel and rare variant is unlikely to invade in regimes of positive frequency dependence (e.g., Kirkpatrick & Dugatkin 1994, Lachlan & Feldman 2003). As a consequence, drift may play a crucial role in the eventual fixation of sexually selected characters, suggesting a large scope for population divergence when imitation is important (Lachlan & Servedio 2004). In general, aspects such as learning and imitation as well as the social context in which they occur (Vakirtzis 2011) should play a more central role in sexual selection theory.

5. SEXUAL SELECTION AND OTHER PROCESSES

5.1. Sexual Selection and Sex-Ratio Evolution

Trivers & Willard (1973) predicted that females mated to attractive males should bias their sex allocation toward sons. Formal models confirm this (Pen & Weissing 2000, Fawcett et al. 2007) but also find that females mated to unattractive males almost exclusively produce daughters. Moreover, a more inclusive coevolutionary model in which sex allocation feeds back on the evolution of ornaments and preferences shows that sex allocation undermines sexual selection (Fawcett et al. 2011): By producing only daughters (which are certain to reproduce) but no unattractive sons, females mated to unattractive males have a fitness level that approaches that of females mated to attractive males. As a result, mating with attractive males ceases to yield any advantage to choosy females, thereby eliminating sexual selection and sex allocation based on it (Fawcett et al. 2011). This example is a clear demonstration of how more inclusive coevolutionary models can change our insights. Research has yet to determine whether sex allocation based on male attractiveness can be maintained in the presence of mate choice, for example, in cases in which sex allocation is based on sexually antagonistic alleles (Alonzo & Sinervo 2007, Blackburn et al. 2010).

5.2. Sexual Selection and Parental Care

Conventional sexual selection models assume females provide care and males compete over females. Although parental care is more commonly provided by females than males, substantial taxonomic variation in sex biases in parental care exists and is poorly explained by conventional theory (Kokko & Jennions 2008). Modeling the evolution of parental care is complicated: The decision of one parent to provide care versus to compete for matings depends not only on the behavior of its current partner but also on the opportunity to gain future matings, which, in turn, is a function of the population-wide density of members of its sex that are competing for matings as opposed to caring for their young (Houston et al. 2005, Kokko & Jennions 2008). To understand this better, investigators need more inclusive modeling approaches that take into account both the evolutionary dynamics (evolution of ornaments, preferences, care decisions) and the ecological dynamics (acting at a much faster timescale) in which individual decisions feed back on the densities of caring and competing individuals.

McNamara et al. (2000) provide one of the first models to take such a dynamical approach. These authors found that high population-wide levels of care select for individuals to desert their brood more rapidly (because they are likely to have mated with a partner providing care) and pursue more matings. However, as desertion rates increase, individuals face more competition for future matings, making it less worthwhile to desert and more preferable to continue caring. The authors also showed that alternations in patterns of care may result, such that bouts of biparental care may alternate with uniparental or no care. The feedback between desertion and competition also cautions researchers against attributing sex differences in care to biases in the operational sex ratio (OSR): For example, male-biased OSRs are widely assumed to lead to increased male-male competition and, hence, less male care (Trivers 1972). As reviewed by Kokko & Jennions (2008), a male-biased OSR may actually select for increased parental care by males because males now face increased competition over matings, thereby increasing the relative value of parental care. Hence, the co-occurrence of male-biased OSRs and female-biased patterns of parental care requires more specific explanations that consider the intricacies of the sexual selection processes or species-specific differences within costs of competition.

Operational sex ratio (OSR): the ratio of male versus female individuals that are available for mating at any given time

5.3. Sexual Selection and the Evolution of Sex

Despite recent theoretical progress on the evolution of sex along various lines, we still have a limited understanding of the various factors that can overcome the costs of sexual reproduction (Lehtonen et al. 2012). Sexual selection, which requires sexual reproduction, may positively feed back on the maintenance of sex (for a recent review, see Whitlock & Agrawal 2009). Specifically, if females choose males with the fewest mutations, the load of deleterious mutations is reduced in comparison with asexual populations (Siller 2001, Agrawal 2001). The twofold cost of sex is also overcome for those individuals that bear high-fitness offspring (e.g., attractive sons), which will result in a larger number of grandoffspring (Hadany & Beker 2007).

However, sexual selection may also work against the evolution of sex (Whitlock & Agrawal 2009). For example, the evolution of male harm or costly competition can reduce the mean fitness of sexual populations. The expected population size of sexual, as opposed to asexual, organisms is further reduced by the increased variance in male reproductive success, which may lead to an increased load of deleterious mutations due to drift. Last, the presence of divergent selection pressures in the two sexes (i.e., sexually antagonistic selection) also disfavors sexual reproduction (Roze & Otto 2012). To determine whether sexual selection defies the evolution of asexual reproduction, a more inclusive approach is required.

5.4. Sexual Selection and Speciation

Closely related species often differ most dramatically in their mating traits, suggesting that sexual selection plays an important role in speciation. Indeed, sexual selection has been ascribed a prominent role in virtually all processes related to speciation. For example, Lande (1981) has described how the Fisher process contributes to the divergence of mating preferences in geographically isolated populations and, hence, to the evolution of reproductive isolation mechanisms (Uyeda et al. 2009). Such divergence can be strongly enhanced when the driving force is sexual conflict, leading to antagonistic coevolution of the two sexes (Hayashi et al. 2007). Sexual selection can contribute to the low fitness of hybrids (and, hence, postzygotic reproductive isolation) if hybrids are less attractive as mates. Mating preferences may be a potent mechanism underlying reinforcement (selection against the occurrence of deleterious hybridization), for example, when individuals tend to mate with genetically compatible partners (Servedio & Noor 2003, Servedio 2004) (Section 3.5). In fact, reinforcement can give rise to particularly strong selection of female preferences for signals that indicate low degrees of hybrid incompatibility (Kirkpatrick & Servedio 1999). Sexual selection can also lead to the evolution of prezygotic isolation, which may occur in sympatry, when two diverging Fisher processes co-occur in populations with a broad variation in female preferences (Higashi et al. 1999; Van Doorn et al. 2001, 2004). A more likely scenario, however, is the evolution of preferences for indicators of local adaptation (Proulx 2001, Reinhold 2004) (Section 3.5) that, under parapatric conditions, can strongly enhance disruptive natural selection (Van Doorn et al. 2009). For recent reviews of the role of sexual selection in speciation, readers are referred to Ritchie (2007), Weissing et al. (2011), and the Marie Curie Speciation Network (2012).

FUTURE ISSUES

Despite the hundreds of models on sexual selection presented over the past four decades, we feel that the field is only at the beginning of a more integrative theory of sexual selection. In particular, the following aspects should receive major attention in future models:

1. Research needs to apply a more robust approach to the study of mating traits, which are typically modeled in a rather simplistic manner. In nature, preferences as well as traits are often conditional strategies, depending on an individual's position within the overall mating market. Virtually no models consider the possibility that individuals reallocate resources from ornamentation to the provisioning of direct benefits like parental care (or vice versa) in a dynamic way, which may be of particular relevance for species with mates with mutual choosiness. The evolution of preferences when different types of benefits are at stake has also not received much theoretical attention. Current models (e.g., Kirkpatrick & Barton 1997) lead to the clear-cut prediction that direct benefits play a role more prominent than that of indirect benefits. However, if this is true, why do many females engage in extrapair copulations (yielding only indirect benefits), which may endanger their social mate's investment in their joint clutch (thereby risking the loss of direct benefits)? Questions like these have hardly been considered by sexual selection theory.

2. The interplay between natural and sexual selection is still poorly understood: Are mating preferences and natural selection typically antagonistic (as often envisaged), or do they more often act in concert (Proulx 2001, Van Doorn et al. 2009)? How do ecological conditions affect the perception (and evolution) of sexually selected signals (Endler & Basolo 1998)? Under which conditions do preferences evolve as a pleiotropic by-product of natural selection (Kirkpatrick & Ryan 1991), and will such preferences be maintained in the long term? How do more explicit formulations of ecological interactions (predation, host-parasite interactions) (Hamilton & Zuk 1982) and environmental dynamics influence male quality and the benefits of choice?

3. An integrative theory of sexual selection should highlight inclusive models in which ecological aspects (i.e., density-dependent feedbacks, resource dynamics) are modeled dynamically in combination with evolving preferences and ornaments. Such an approach also requires that the models explicitly account for the coevolutionary interaction of female preferences, male ornaments, and traits involved in other processes, such as parental care, sex allocation, and male-male competition.

4. Most current predictions regarding the evolution of sexually selected traits are based on equilibrium situations, whereas several lines of evidence indicate that such traits exhibit rapid turnovers and strong interpopulation divergence (Wiens 2001, Bro-Jørgensen 2010). More effort is needed to understand when nonequilibrium dynamics occur and how aspects such as the genetic architecture (e.g., Van Doorn & Weissing 2006) or environmental dynamics (Bro-Jørgensen 2010) affect the continuous evolution of sexually selected traits.

DISCLOSURE STATEMENT

The authors are not aware of any affiliations, memberships, funding, or financial holdings that might be perceived as affecting the objectivity of this review.

ACKNOWLEDGMENTS

Sander Van Doorn and Tim Fawcett gave helpful suggestions to improve this manuscript. We thank the editor, Mark Kirkpatrick, for his useful comments and stimulating discussion.

LITERATURE CITED

Agrawal AF. 2001. Sexual selection and the maintenance of sexual reproduction. *Nature* 411:692–95

Albert AYK, Otto SP. 2005. Sexual selection can resolve sex-linked sexual antagonism. *Science* 310:119–21

Alonzo S, Sinervo B. 2007. The effect of sexually antagonistic selection on adaptive sex ratio allocation. *Evol. Ecol. Res.* 9:1097–117

Alonzo SH. 2012. Sexual selection favours male parental care, when females can choose. *Proc. R. Soc. Lond. Ser. B.* 279:1784–90

Andersson M. 1994. *Sexual Selection*. Princeton: Princeton Univ. Press

Andersson M, Simmons LW. 2006. Sexual selection and mate choice. *Trends Ecol. Evol.* 21:296–302

Arak A, Enquist M. 1993. Hidden preferences and the evolution of signals. *Philos. Trans. R. Soc. Lond. Ser. B* 340:207–13

Arnold SJ, Bürger R, Hohenlohe PA, Ajie BC, Jones AG. 2008. Understanding the evolution and stability of the G-matrix. *Evolution* 62:2451–61

Arnqvist G. 2006. Sensory exploitation and sexual conflict. *Philos. Trans. R. Soc. Lond. Ser. B* 361:375–86

Arnqvist G. 2011. Assortative mating by fitness and sexually antagonistic genetic variation. *Evolution* 65:2111–16

Arnqvist G, Rowe L. 2005. *Sexual Conflict*. Princeton: Princeton Univ. Press

Ball MA, Parker GA. 2003. Sperm competition games: sperm selection by females. *J. Theor. Biol.* 224:27–42

Barton NH, Turelli M. 1991. Natural and sexual selection on many loci. *Genetics* 127:229–55

Berglund A, Bisazza A, Pilastro A. 1996. Armaments and ornaments: an evolutionary explanation of traits of dual utility. *Biol. J. Linn. Soc.* 58:385–99

Blackburn G, Albert A, Otto S. 2010. The evolution of sex ratio adjustment in the presence of sexually antagonistic selection. *Am. Nat.* 176:264–75

Bro-Jørgensen J. 2010. Dynamics of multiple signalling systems: animal communication in a world in flux. *Trends Ecol. Evol.* 25:292–300

Bulmer M. 1989. Structural instability of models of sexual selection. *Theor. Popul. Biol.* 35:195–206

Bulmer M. 1994. *Theoretical Evolutionary Ecology*. Sunderland, MA: Sinauer

Cameron E, Day T, Rowe L. 2003. Sexual conflict and indirect benefits. *J. Evol. Biol.* 16:1055–60

Candolin U. 2003. The use of multiple cues in mate choice. *Biol. Rev.* 78:575–95

Chenoweth SF, McGuigan K. 2010. The genetic basis of sexually selected variation. *Annu. Rev. Ecol. Evol. Syst.* 41:81–101

Clutton-Brock T. 2007. Sexual selection in males and females. *Science* 318:1882–85

Connallon T. 2010. Genic capture, sex linkage, and the heritability of fitness. *Am. Nat.* 175:564–76

Cotar C, McNamara JM, Collins E, Houston AI. 2008. Should females prefer to mate with low-quality males? *J. Theor. Biol.* 254:561–67

Cotton S, Small J, Pomiankowski A. 2006. Sexual selection and condition-dependent mate preferences. *Curr. Biol.* 16:R755–65

Darwin C. 1871. *The Descent of Man, and Selection in Relation to Sex*. London: John Murray

Day T. 2000. Sexual selection and the evolution of costly female preferences: spatial effects. *Evolution* 54:715–30

Dawkins MS, Guildford T. 1995. An exaggerated preference for simple neural network models of signal evolution? *Proc. R. Soc. Lond. Ser. B.* 261:357–60

Dercole F, Rinaldi S. 2008. *Analysis of Evolutionary Processes: The Adaptive Dynamics Approach and its Applications.* Princeton: Princeton Univ. Press

Dieckmann U, Law R. 1996. The dynamical theory of coevolution: a derivation from stochastic ecological processes. *J. Math. Biol.* 34:579–612

Dombrovsky Y, Perrin N. 1994. On adaptive search and optimal stopping in sequential mate choice. *Am. Nat.* 144:355–61

Dugatkin LA. 1996. Interface between culturally based preferences and genetic preferences: female mate choice in *Poecilia reticulata*. *Proc. Natl. Acad. Sci. USA* 93:2770–73

Endler JA, Basolo AL. 1998. Sensory ecology, receiver biases and sexual selection. *Trends Ecol. Evol.* 13:415–20

Enquist M, Ghirlanda S. 2005. *Neural Networks and Animal Behavior.* Princeton: Princeton Univ. Press

Fawcett TW, Kuijper B, Pen I, Weissing FJ. 2007. Should attractive males have more sons? *Behav. Ecol.* 18:71–80

Fawcett TW, Kuijper B, Weissing FJ, Pen I. 2011. Sex-ratio control erodes sexual selection, revealing evolutionary feedback from adaptive plasticity. *Proc. Natl. Acad. Sci. USA* 108:15925–30

Fisher R. 1915. The evolution of sexual preference. *Eugen. Rev.* 7:184–92

Fromhage L, Kokko H, Reid JM. 2009. Evolution of mate choice for genome-wide heterozygosity. *Evolution* 63:684–94

Fromhage L, McNamara J, Houston A. 2008. Sperm allocation strategies and female resistance: a unifying perspective. *Am. Nat.* 172:25–33

Fuller RC. 2009. A test of the critical assumption of the sensory bias model for the evolution of female mating preference using neural networks. *Evolution* 63:1697–711

Gavrilets S, Arnqvist G, Friberg U. 2001. The evolution of female mate choice by sexual conflict. *Proc. R. Soc. Lond. Ser. B* 268:531–39

Gavrilets S, Hayashi TI. 2005. Speciation and sexual conflict. *Evol. Ecol.* 19:167–98

Geritz S, Kisdi E, Meszéna G, Metz J. 1998. Evolutionarily singular strategies and the adaptive growth and branching of the evolutionary tree. *Evol. Ecol.* 12:35–57

Getty T. 2006. Sexually selected signals are not similar to sports handicaps. *Trends Ecol. Evol.* 21:83–88

Greenspoon PB, Otto SP. 2009. Evolution by Fisherian sexual selection in diploids. *Evolution* 63:1076–83

Hadany L, Beker T. 2007. Sexual selection and the evolution of obligatory sex. *BMC Evol. Biol.* 7:245

Hall DW, Kirkpatrick M, West B. 2000. Runaway sexual selection when female preferences are directly selected. *Evolution* 54:1862–69

Halliday T. 1983. The study of mate choice. In *Mate Choice*, ed. P Bateson, pp. 3–32. Cambridge, UK: Cambridge Univ. Press

Hamilton W, Zuk M. 1982. Heritable true fitness and bright birds: a role for parasites? *Science* 218:384–87

Härdling R, Karlsson K. 2009. The dynamics of sexually antagonistic coevolution and the complex influences of mating system and genetic correlation. *J. Theor. Biol.* 260:276–82

Hastings IM. 1994. Manifestations of sexual selection may depend on the genetic basis of sex determination. *Proc. R. Soc. Lond. Ser. B* 258:83–87

Hayashi TI, Vose M, Gavrilets S. 2007. Genetic differentiation by sexual conflict. *Evolution* 61:516–29

Higashi M, Takimoto G, Yamamura N. 1999. Sympatric speciation by sexual selection. *Nature* 402:523–26

Houle D, Kondrashov AS. 2002. Coevolution of costly mate choice and condition-dependent display of good genes. *Proc. R. Soc. Lond. Ser. B* 269:97–104

Houston AI, Szèkely T, McNamara JM. 2005. Conflict between parents over care. *Trends Ecol. Evol.* 20:33–38

Hunt J, Bussiere LF, Jennions MD, Brooks R. 2004. What is genetic quality? *Trends Ecol. Evol.* 19:329–33

Iwasa Y, Pomiankowski A. 1994. The evolution of mate preferences for multiple sexual ornaments. *Evolution* 48:853–67

Iwasa Y, Pomiankowski A. 1995. Continual change in mate preferences. *Nature* 377:420–22

Iwasa Y, Pomiankowski A. 1999. Good parent and good genes models of handicap evolution. *J. Theor. Biol.* 200:97–109

Iwasa Y, Pomiankowski A, Nee S. 1991. The evolution of costly mate preferences II. The "handicap" principle. *Evolution* 45:1431–42

Johnstone RA. 1995. Honest advertisement of multiple qualities using multiple signals. *J. Theor. Biol.* 177:87–94

Johnstone RA, Reynolds JD, Deutsch JC. 1996. Mutual mate choice and sex differences in choosiness. *Evolution* 50:1382–91

Jones AG, Ratterman NL. 2009. Mate choice and sexual selection: What have we learned since Darwin? *Proc. Natl. Acad. Sci. USA* 106:10001–8

Kirkpatrick M. 1982. Sexual selection and the evolution of female choice. *Evolution* 36:1–12

Kirkpatrick M. 1987. Sexual selection by female choice in polygynous animals. *Annu. Rev. Ecol. Syst.* 18:43–70

Kirkpatrick M, Barton N. 1997. The strength of indirect selection on female mating preferences. *Proc. Natl. Acad. Sci. USA* 94:1282–86

Kirkpatrick M, Dugatkin LA. 1994. Sexual selection and the evolutionary effects of copying mate choice. *Behav. Ecol. Sociobiol.* 34:443–49

Kirkpatrick M, Hall DW. 2004. Sexual selection and sex linkage. *Evolution* 58:683–91

Kirkpatrick M, Johnson T, Barton N. 2002. General models of multilocus evolution. *Genetics* 161:1727–50

Kirkpatrick M, Ryan MJ. 1991. The evolution of mating preferences and the paradox of the lek. *Nature* 350:33–38

Kirkpatrick M, Servedio MR. 1999. The reinforcement of mating preferences on an island. 151:865–84

Kokko H. 1997. Evolutionarily stable strategies of age-dependent sexual advertisement. *Behav. Ecol. Sociobiol.* 41:99–107

Kokko H. 1998. Should advertising parental care be honest? *Proc. R. Soc. Lond. Ser. B* 265:1871–78

Kokko H. 2001. Fisherian and "good genes" benefits of mate choice: how (not) to distinguish between them. *Ecol. Lett.* 4:322–26

Kokko H, Brooks R, McNamara JM, Houston AI. 2002. The sexual selection continuum. *Proc. R. Soc. Lond. Ser. B* 269:1331–40

Kokko H, Jennions MD. 2008. Parental investment, sexual selection and sex ratios. *J. Evol. Biol.* 21:919–48

Kokko H, Jennions MD, Brooks R. 2006. Unifying and testing models of sexual selection. *Annu. Rev. Ecol. Evol. Syst.* 37:43–66

Kokko H, Johnstone RA. 2002. Why is mutual mate choice not the norm? Operational sex ratios, sex roles and the evolution of sexually dimorphic and monomorphic signalling. *Philos. Trans. R. Soc. Lond. Ser. B* 357:319–30

Kokko H, Mappes J. 2005. Sexual selection when fertilization is not guaranteed. *Evolution* 59:1876–85

Kotiaho JS, LeBas NR, Puurtinen M, Tomkins JL. 2008. On the resolution of the lek paradox. *Trends Ecol. Evol.* 23:1–3

Lachlan RF, Feldman MW. 2003. Evolution of cultural communication systems: the coevolution of cultural signals and genes encoding learning preferences. *J. Evol. Biol.* 16:1084–95

Lachlan RF, Servedio MR. 2004. Song learning accelerates allopatric speciation. *Evolution* 58:2049–63

Laland KN. 1994. Sexual selection with a culturally transmitted mating preference. *Theor. Popul. Biol.* 45:1–15

Lande R. 1981. Models of speciation by sexual selection on polygenic traits. *Proc. Natl. Acad. Sci. USA* 78:3721–25

Lande R. 1983. The response to selection on major and minor mutations affecting a metrical trait. *Heredity* 50:47–65

Lehmann L, Keller LF, Kokko H. 2007. Mate choice evolution, dominance effects, and the maintenance of genetic variation. *J. Theor. Biol.* 244:282–95

Lehtonen J, Jennions MD, Kokko H. 2012. The many costs of sex. *Trends Ecol. Evol.* 27:172–78

Levins R. 1966. The strategy of model building in population biology. *Am. Sci.* 54:421–31

Lorch PD, Proulx S, Rowe L, Day T. 2003. Condition-dependent sexual selection can accelerate adaptation. *Evol. Ecol. Res.* 5:867–81

Luttbeg B. 1996. A comparative Bayes tactic for mate assessment and choice. *Behav. Ecol.* 7:451–60

Marie Curie Speciation Network. 2012. What do we need to know about speciation? *Trends Ecol. Evol.* 27:27–39

Maynard Smith J. 1985. Sexual selection, handicaps and true fitness. *J. Theor. Biol.* 115:1–8

McElreath R, Boyd R. 2007. *Mathematical Models of Social Evolution.* Chicago: Univ. Chic. Press

McGill BJ, Brown JS. 2007. Evolutionary game theory and adaptive dynamics of continuous traits. *Annu. Rev. Ecol. Evol. Syst.* 38:403–35

McNamara JM, Barta Z, Fromhage L, Houston AI. 2008. The coevolution of choosiness and cooperation. *Nature* 451:189–92

McNamara JM, Székely T, Webb JN, Houston AI. 2000. A dynamic game-theoretic model of parental care. *J. Theor. Biol.* 205:605–23

Mead LS, Arnold SJ. 2004. Quantitative genetic models of sexual selection. *Trends Ecol. Evol.* 19:264–71

Metz J, Nisbet R, Geritz S. 1992. How should we define 'fitness' for general ecological scenarios? *Trends Ecol. Evol.* 7:198–202

Møller A, Jennions M. 2001. How important are direct fitness benefits of sexual selection? *Naturwissenschaften* 88:401–15

Møller A, Pomiankowski A. 1993. Why have birds got multiple sexual ornaments? *Behav. Ecol. Sociobiol.* 32:167–76

Parker GA, Pizzari T. 2010. Sperm competition and ejaculate economics. *Biol. Rev.* 85:897–934

Pen I, Weissing FJ. 2000. Sexual selection and the sex ratio: an ESS analysis. *Selection* 1:111–21

Phelps S. 2007. Sensory ecology and perceptual allocation: new prospects for neural networks. *Philos. Trans. R. Soc. Lond. Ser. B* 362:355–67

Pomiankowski A. 1987. The costs of choice in sexual selection. *J. Theor. Biol.* 128:195–218

Pomiankowski A, Bridle J. 2004. Evolutionary genetics: no sex please we're at QLE (quasi-linkage equilibrium). *Heredity* 93:407

Pomiankowski A, Iwasa Y. 1993. Evolution of multiple sexual preferences by Fisher's runaway process of sexual selection. *Proc. R. Soc. Lond. Ser. B* 253:173–81

Pomiankowski A, Iwasa Y. 1998. Runaway ornament diversity caused by Fisherian sexual selection. *Proc. Natl. Acad. Sci. USA* 95:5106–11

Pomiankowski A, Iwasa Y, Nee S. 1991. The evolution of costly mate preferences I. Fisher and biased mutation. *Evolution* 45:1422–30

Pomiankowski A, Møller AP. 1995. A resolution of the lek paradox. *Proc. R. Soc. Lond. Ser. B* 260:21–29

Price T, Schluter D, Heckman NE. 1993. Sexual selection when the female directly benefits. *Biol. J. Linn. Soc.* 48:187–211

Proulx SR. 2001. Female choice via indicator traits easily evolves in the face of recombination and migration. *Evolution* 55:2401–11

Puurtinen M, Ketola T, Kotiaho J. 2009. The good-genes and compatible-genes benefits of mate choice. *Am. Nat.* 174:741–52

Qvarnström A, Bailey RI. 2009. Speciation through evolution of sex-linked genes. *Heredity* 102:4–15

Rands SA, Evans MR, Johnstone RA. 2011. The dynamics of honesty: modelling the growth of costly, sexually-selected ornaments. *PLoS ONE* 6:e27174

Ratikainen II, Kokko H. 2010. Differential allocation and compensation: Who deserves the silver spoon? *Behav. Ecol.* 21:195–200

Real L. 1990. Search theory and mate choice. I. Models of single-sex discrimination. *Am. Nat.* 136:376–405

Reeve HK, Pfennig DW. 2003. Genetic biases for showy males: Are some genetic systems especially conducive to sexual selection? *Proc. Natl. Acad. Sci. USA* 100:1089–94

Reinhold K. 1998. Sex linkage among genes controlling sexually selected traits. *Behav. Ecol. Sociobiol.* 44:1–7

Reinhold K. 2004. Modeling a version of the good-genes hypothesis: female choice of locally adapted males. *Org. Divers. Evol.* 4:157–63

Rice WR. 1996. Sexually antagonistic male adaptation triggered by experimental arrest of female evolution. *Nature* 381:232–34

Ritchie MG. 2007. Sexual selection and speciation. *Annu. Rev. Ecol. Evol. Syst.* 38:79–102

Rowe L, Cameron E, Day T. 2005. Escalation, retreat, and female indifference as alternative outcomes of sexually antagonistic coevolution. *Am. Nat.* 165:S5–18

Rowe L, Houle D. 1996. The lek paradox and the capture of genetic variance by condition dependent traits. *Proc. R. Soc. Lond. Ser. B* 263:1415–21

Roze D, Otto SP. 2012. Differential selection between the sexes and selection for sex. *Evolution* 66:558–74

Schluter D, Price T. 1993. Honesty, perception and population divergence in sexually selected traits. *Proc. R. Soc. Lond. Ser. B* 253:117–22

Seger J. 1985. Unifying genetic models for the evolution of female choice. *Evolution* 39:1185–93

Servedio MR. 2004. The evolution of premating isolation: local adaptation and natural and sexual selection against hybrids. *Evolution* 58:913–24

Servedio M, Lande R. 2006. Population genetic models of male and mutual mate choice. *Evolution* 60:674–85

Servedio MR, Noor MAF. 2003. The role of reinforcement in speciation: theory and data. *Annu. Rev. Ecol. Evol. Syst.* 34:339–64

Shirangi TR, Dufour HD, Williams TM, Carroll SB. 2009. Rapid evolution of sex pheromone-producing enzyme expression in *Drosophila*. *PLoS Biol.* 7:e1000168

Siller S. 1998. The epistatic handicap principle does work. *J. Theor. Biol.* 191:141–61

Siller S. 2001. Sexual selection and the maintenance of sex. *Nature* 411:689–92

Swanson WJ, Vacquier VD. 2002. The rapid evolution of reproductive proteins. *Nat. Rev. Genet.* 3:137–44

Taylor PD. 1996. Inclusive fitness arguments in genetic models of behaviour. *J. Math. Biol.* 34:654–74

Tazzyman SA, Pizzari T, Seymour R, Pomiankowski A. 2009. The evolution of continuous variation in ejaculate expenditure strategy. *Am. Nat.* 174:E71–82

Tazzyman SJ, Seymour RM, Pomiankowski A. 2012. Fixed and dilutable benefits: female choice for good genes or fertility. *Proc. R. Soc. Lond. Ser. B* 279:334–40

Ten Tusscher K, Hogeweg P. 2009. The role of genome and gene regulatory network canalization in the evolution of multi-trait polymorphisms and sympatric speciation. *BMC Evol. Biol.* 9:159

Trivers R. 1972. Parental investment and sexual selection. In *Sexual Selection and the Descent of Man*, ed. B Campbell, pp. 136–79. Chicago: Aldine

Trivers RL, Willard DE. 1973. Natural selection of parental ability to vary the sex ratio of offspring. *Science* 179:90–92

Uyeda JC, Arnold SJ, Hohenlohe PA, Mead LS. 2009. Drift promotes speciation by sexual selection. *Evolution* 63:583–94

Vakirtzis A. 2011. Mate choice copying and nonindependent mate choice: a critical review. *Ann. Zool. Fenn.* 48:91–107

Van Doorn GS, Dieckmann U, Weissing FJ. 2004. Sympatric speciation by sexual selection: a critical re-evaluation. *Am. Nat.* 163:709–25

Van Doorn GS, Edelaar P, Weissing FJ. 2009. On the origin of species by natural and sexual selection. *Science* 326:1704–7

Van Doorn GS, Luttikhuizen PC, Weissing FJ. 2001. Sexual selection at the protein level drives the extraordinary divergence of sex related genes during sympatric speciation. *Proc. R. Soc. Lond. Ser. B* 268:2155–61

Van Doorn GS, Weissing FJ. 2004. The evolution of female preferences for multiple indicators of quality. *Am. Nat.* 164:173–86

Van Doorn GS, Weissing FJ. 2006. Sexual conflict and the evolution of female preferences for indicators of male quality. *Am. Nat.* 168:742–57

Veen T. 2008. *Mating decisions in a hybrid zone*. PhD thesis. Univ. Groningen, Neth.

Walsh B, Lynch M. 2013. *Evolution and Selection of Quantitative Traits: I. Foundations.* In press **http://nitro. biosci.arizona.edu/zbook/NewVolume_2/newvol2.html**.

Weissing FJ. 1996. Genetic versus phenotypic models of selection: can genetics be neglected in a long-term perspective? *J. Math. Biol.* 34:533–55

Weissing FJ, Edelaar P, Van Doorn G. 2011. Adaptive speciation theory: a conceptual review. *Behav. Ecol. Sociobiol.* 65:461–80

Whitlock MC, Agrawal AF. 2009. Purging the genome with sexual selection: reducing mutation load through selection on males. *Evolution* 63:569–82

Wiegmann DD, Seubert SM, Wade GA. 2010. Mate choice and optimal search behavior: fitness returns under the fixed sample and sequential search strategies. *J. Theor. Biol.* 262:596–600

Wiens JJ. 2001. Widespread loss of sexually selected traits: how the peacock lost its spots. *Trends Ecol. Evol.* 16:517–23

Wolf M, Van Doorn GS, Leimar O, Weissing FJ. 2008. Do animal personalities emerge? Reply. *Nature* 451:E9–10

Zahavi A. 1975. Mate selection-a selection for a handicap. *J. Theor. Biol.* 53:205–14

Ecoenzymatic Stoichiometry and Ecological Theory

Robert L. Sinsabaugh[1] and Jennifer J. Follstad Shah[2]

[1]Biology Department, University of New Mexico, Albuquerque, New Mexico 87131;
email: rlsinsab@unm.edu

[2]Watershed Sciences Department, Utah State University, Logan, Utah 84322

Annu. Rev. Ecol. Evol. Syst. 2012. 43:313–43

First published online as a Review in Advance on
September 4, 2012

The *Annual Review of Ecology, Evolution, and Systematics* is online at ecolsys.annualreviews.org

This article's doi:
10.1146/annurev-ecolsys-071112-124414

Keywords

ecoenzyme, ecological stoichiometry, biogeochemistry, decomposition,
microbial growth, resource allocation model

Abstract

The net primary production of the biosphere is consumed largely by
microorganisms, whose metabolism creates the trophic base for detrital
foodwebs, drives element cycles, and mediates atmospheric composition.
Biogeochemical constraints on microbial catabolism, relative to primary
production, create reserves of detrital organic carbon in soils and sediments
that exceed the carbon content of the atmosphere and biomass. The production of organic matter is an intracellular process that generates thousands
of compounds from a small number of precursors drawn from intermediary
metabolism. Osmotrophs generate growth substrates from the products of
biosynthesis and diagenesis by enzyme-catalyzed reactions that occur largely
outside cells. These enzymes, which we define as ecoenzymes, enter the
environment by secretion and lysis. Enzyme expression is regulated by environmental signals, but once released from the cell, ecoenzymatic activity is
determined by environmental interactions, represented as a kinetic cascade,
that lead to multiphasic kinetics and large spatiotemporal variation. At the
ecosystem level, these interactions can be viewed as an energy landscape
that directs the availability and flow of resources. Ecoenzymatic activity
and microbial metabolism are integrated on the basis of resource demand
relative to environmental availability. Macroecological studies show that the
most widely measured ecoenzymatic activities have a similar stoichiometry
for all microbial communities. Ecoenzymatic stoichiometry connects the
elemental stoichiometry of microbial biomass and detrital organic matter
to microbial nutrient assimilation and growth. We present a model that
combines the kinetics of enzyme activity and community growth under
conditions of multiple resource limitation with elements of metabolic and
ecological stoichiometry theory. This biogeochemical equilibrium model
provides a framework for comparative studies of microbial community
metabolism, the principal driver of biogeochemical cycles.

1. INTRODUCTION

The net primary production (NPP) of the biosphere is consumed largely by microorganisms, whose metabolism creates the trophic base for detrital foodwebs, drives global carbon (C) and nutrient cycles, and mediates atmospheric composition. Biogeochemical constraints on these catabolic processes, relative to primary production, create reserves of detrital organic C in soils and sediments that exceed the C content of the atmosphere and biomass by a factor of two or more (Cole et al. 2007, Houghton 2007).

The production of organic molecules is an intracellular process, broadly similar across domains, that generates thousands of compounds from a small number of precursors drawn from intermediary metabolism, fueled by the consumption of C from the environment. For autotrophs, the C source is CO_2. For osmotrophic prokaryotes and fungi, the C sources are low–molecular mass compounds, with individual taxa limited to a small number ($\sim$1–20) of growth substrates. These growth substrates are generated from the myriad products of biosynthesis and diagenesis by enzyme-catalyzed reactions that occur largely outside cells (Burns 1978, Chróst 1991, Burns & Dick 2002, Shukla & Varma 2011, Trasar-Cepeda et al. 2011, Dick 2012). The production of these catabolic enzymes is directed by environmental signals in relation to cellular resources, but once released into the environment by secretion or lysis, their activity and turnover are determined by complex physicochemical and biochemical interactions. The extracellular catabolism of organic matter, i.e., decomposition, is considered a rate-controlling step in the global C cycle. For this reason, considerable research has focused on environmental enzyme reactions across molecular to biosphere scales.

Herein, we briefly review the history of environmental enzyme research with a focus on concepts and models that link these enzyme activities to biogeochemical processes, microbial metabolism, and ecological theory. We refer readers to previous reviews and texts for background on the biochemistry of particular enzymes and the methodology for measuring reaction rates. We conclude with a synthesis of empirical data and model consensus that describes the equilibrium relationships among environment resource availability, microbial metabolism, and enzymatic indices of catabolic potential, the three components of the successional loop that drives decomposition (Sinsabaugh et al. 2002).

2. TERMINOLOGY AND ENZYMES OF INTEREST

Several terms have been used to describe the distribution or origin of enzymes found outside of cells, including extracellular enzymes, ectoenzymes, exoenzymes, abiontic enzymes, and free enzymes, with varying definitions. In recent papers, we use the term ecoenzyme to broadly encompass all enzymes located outside the confines of intact cell membranes regardless of whether such enzymes enter the environment by secretion or lysis (Sinsabaugh et al. 2009). This definition provides the closest correspondence between environmental enzyme activity and organic matter decomposition.

The most studied ecoenzymes catalyze the degradation of the largest environmental sources of organic C, nitrogen (N), and phosphorus (P) (**Figure 1**). The largest organic C pool is structural polysaccharides that form the cell walls and matric glycolates of plants and microorganisms, followed by lignins and other secondary polyphenolic molecules, storage polysaccharides, and lipids. Polysaccharide degradation is primarily hydrolytic; lipid and phenolic degradation is primarily oxidative. The organic N pool includes polymers of amino acids and aminosaccharides, which are sources of C and N. The organic P pool includes labile nucleic acids and phospholipids and more recalcitrant P storage products, principally inositol phosphates.

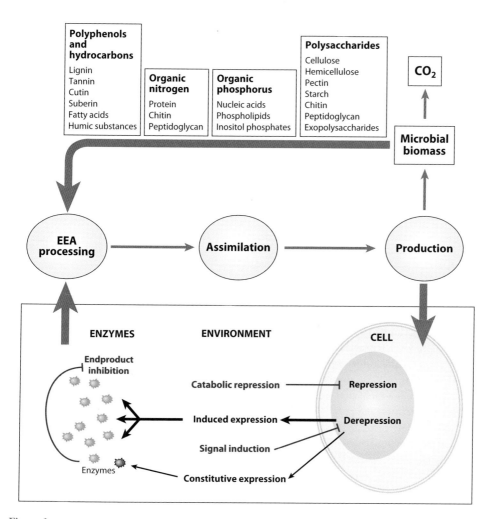

Figure 1

Expanded successional loop that links microbial production, detrital organic matter, and ecoenzyme activity (EEA). (*Upper loop*) Principal environmental sources of organic carbon, nitrogen, and phosphorus. (*Lower loop*) Expression of extracellular enzymes is controlled by signal pathways linked to environmental cues. Some enzymes are produced constitutively, usually at low levels. More generally, production is upregulated by induction-derepression pathways and downregulated by catabolic repression pathways controlled by environmental signals and cellular resources. Once released, the activity of extracellular enzymes is subject to environmental controls such as end-product inhibition. Intracellular enzymes released through cell lysis are subject to the same suite of controls. A broader description of environmental controls is presented in **Figure 2**.

Within these classes, the ecoenzymatic activities (EEAs) most commonly measured generate low–molecular mass products that can be directly consumed by microorganisms. These include α- and β-1,4-glucosidase, which catalyze the terminal reactions in the hydrolysis of storage and structural glucans; leucine and alanine aminopeptidase, which hydrolyze the two most abundant amino acids from the N-terminus of polypeptides; β-1,4-N-acetylglucosaminidase, which catalyzes the terminal reaction in the hydrolysis of chitin; and phosphatase, which hydrolyzes phosphate from phosphoesters. Phenol oxidases and peroxidases, which use molecular oxygen and peroxide, respectively, as electron acceptors, catalyze the oxidative degradation of lignin and the formation of humus. These choices are partially methodological; it is easier to assay activities using soluble substrates that yield soluble products than to study the degradation of insoluble polymers or humic complexes. This bias has value in that reactions that yield assimilable products are the ones most directly linked to microbial metabolism (Meyer-Reil 1987, Hoppe et al. 1988, Münster 1991). In the molecular sieve model (Burns 1978), such enzymes are intermediate agents in reaction pathways that connect the activities of polymer-degrading enzymes distributed throughout the environmental matrix and cell membrane permeases that transport substrates into the cell. In many cases, the enzymes that catalyze the terminal reactions in polymer degradation are localized on cell surfaces and periplasmic spaces.

3. HISTORY OF ECOENZYMATIC RESEARCH

Skujiņš (1978) summarized the history of soil enzyme research, beginning with the first report of catalase activity in soil in 1899. Through the 1970s, the major research topics were the source of soil enzymes (plant or microbe, intracellular or extracellular), the physicochemical interactions of enzymes with the soil matrix, correlating EEA with soil properties, and using EEA to classify soils. Phosphatase and urease (amidohydrolase) activities were of particular interest because of their role in generating P and N for plant growth.

Progress was limited by the lack of sensitive methods for measuring EEA and microbial dynamics. But there was a consensus that phosphatase activity increased in response to N fertilization, an indication that microbial communities allocate resources in relation to environmental nutrient availability (Skujiņš 1978). In the final synthesis chapter of *Soil Enzymes*, Burns (1978) described the soil system as a molecular sieve of stabilized matric enzymes linked to cell surface enzymes, periplasmic enzymes, and permeases.

Overbeck (1991) summarized the history of aquatic enzyme research, beginning with a 1906 paper on proteolytic activity in surface water. The importance of ecoenzymes in aquatic systems was highlighted by ZoBell (1943), who included ecoenzymes in his model of the organization of attached microbial communities. The text *Microbial Enzymes in Aquatic Environments* (Chróst 1991) contains several reviews and perspectives on the role of ecoenzymes in the organization of aquatic microbial communities. In particular, Hoppe (1991) added ecoenzymes as an integral component of the microbial loop model for the trophic organization of planktonic microbial communities, and Wetzel (1991) described the functional organization of aquatic ecosystems as a system of stored, immobilized enzymes.

The questions of interest to aquatic researchers differed from those of soil researchers. Many studies correlated phosphatase, aminopeptidase and glucosidase activities to the composition and turnover of dissolved organic and inorganic nutrient pools and rates of microbial production (Chróst 1991), establishing the fundamental relationships on which subsequent models are based.

By the time *Enzymes in the Environment* was published (Burns & Dick 2002), the perspectives of soil and aquatic researchers had largely converged through improvements in methodology that facilitated comparisons across systems and through conceptual advances such as the biofilm concept (Characklis & Marshall 1990) that established a common paradigm for the organization of attached microbial communities. During the past decade, EEA studies have merged with the biomics paradigm, which considers ecological communities as metagenomes and metaproteomes, while simulation models with increased mechanistic resolution connect EEA to ecological processes and theory.

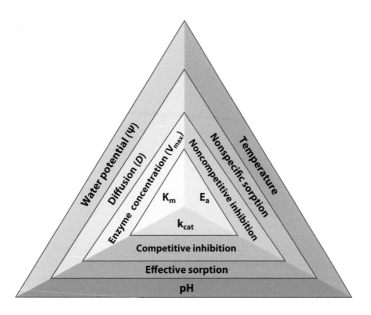

Figure 2

Environmental control of ecoenzymatic activity: a kinetic cascade. The kinetics of enzyme reactions can be characterized by the parameters k_{cat} (the rate of substrate conversion), K_m (the half-saturation constant), and E_a (activation energy), which are determined by enzyme structure. A cascade of enzyme-environment interactions affects the apparent values of these parameters. Enzyme interactions with reaction substrates and products (competitive inhibition), other reactive species (noncompetitive inhibition), and particle surfaces generally reduce k_{cat} and increase K_m and E_a values. Other rate-controlling matric effects include sorption-desorption reactions and diffusion rates of enzymes and substrates. These effects, in turn, are influenced by broader environmental variables such as pH, temperature, and water potential. As a result, kinetic parameters show multiphasic distributions and high spatiotemporal variation.

4. ENZYME EXPRESSION AND ENVIRONMENTAL CONTROL

4.1. Kinetic Cascade

Approximately 1–4% of the production of heterotrophic microbial communities is used to make enzymes for secretion into the environment; many more are released through lysis (Maire et al. 2012). Modeling studies suggest that production of extracellular enzymes has first priority on cell metabolism above the level of maintenance respiration (Schimel & Weintraub 2003, Moorhead & Sinsabaugh 2006, Moorhead et al. 2012, Wang et al. 2012a). Given the energy and material cost of producing enzymes whose control is lost to the cell, expression is closely regulated at the level of transcription. Specifics vary among enzymes and organisms, but the general model is that transcription is ultimately linked to environmental signals (**Figure 1**). These signals may be substrates consumed by the cell, indicators of toxicity, or quorum sensing molecules (DeAngelis et al. 2008). However, once released from the cell, whether by direct secretion or cell lysis, EEA is determined by a hierarchy of interactions that can be represented as a kinetic cascade (**Figure 2**).

In the kinetic cascade, ecoenzyme function, measured by the parameters k_{cat} (the rate of substrate conversion), K_m (the half-saturation constant), and E_a (activation energy), is determined by interactions with reaction substrates and products (competitive inhibition), other reactive species (noncompetitive inhibition), and matric interactions that control the sorption-desorption and diffusion of enzymes and substrates (Resat et al. 2012). These molecular processes are also

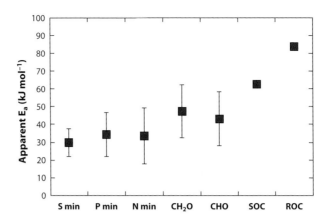

Figure 3

Mean apparent activation energy (E_a ± standard deviation) for enzyme catalyzed processes in litter and soils based on short-term temperature manipulations. Enzyme data for sulfur (S), phosphorus (P), and nitrogen mineralization (N min), carbohydrate hydrolysis (CH_2O), and polyphenol oxidation (CHO) come from **Table 1**. Values for mineralization (respiration) of soil organic carbon (SOC) and recalcitrant organic carbon (ROC) are taken from Gillooly et al. (2001) and Ramirez et al. (2012), respectively.

influenced by broader environmental variables such as pH, temperature, and water potential. As a result, EEA exhibits multiphasic kinetics (Overbeck 1975, 1991; McLaren 1978; Azam & Hodson 1981; Vrba et al. 2004) and high spatiotemporal variation (Sinsabaugh et al. 2008).

4.2. Ecoenzymatic Activities and Thermodynamics

In general, environmental interactions decrease the apparent V_{max} and increase the apparent K_m and E_a of EEAs. At a macroecological scale, apparent E_a receives the most attention because of its debatable utility for predicting changes in decomposition processes, microbial respiration, and C storage as a consequence of warming climate (Davidson & Janssens 2006, Sierra 2012). In general, the activation energy of a process increases with its complexity, i.e., the number of reactions (Sierra 2012). Short-term temperature manipulations of soils show that the apparent E_a of enzymes mediating sulfur, P, and N mineralization are lower on average (30–34 kJ mol^{-1}) than the apparent E_a of enzymes involved in lignocellulose degradation (44–47 kJ mol^{-1}) (**Figure 3**). An independent compilation by Wang et al. (2012b) yielded mean E_a estimates of 54 kJ mol^{-1} for phenol oxidase and peroxidase, 43 kJ mol^{-1} for β-glucosidase, 34 kJ mol^{-1} for β-endoglucanase, and 32 kJ mol^{-1} for cellobiohydrolase. The apparent E_a of C mineralization (65 kJ mol^{-1}), measured as respiration, is greater than the apparent E_a of ecoenzymatic reactions, and values for the mineralization of recalcitrant C range up to 85 kJ mol^{-1} (**Table 1**, **Figure 3**).

The totality of reactions that direct resource flow to microorganisms, which include sorption-desorption and diffusion as well as catalysis, can be represented as an activation energy landscape (Gfeller et al. 2007). As system complexity increases, mean activation energies increase (**Figure 4**). In laboratory trials, the temperature sensitivity of EEA is typically measured by adding substrate at a saturating concentration, which simplifies the energy landscape by overwhelming potential kinetic bottlenecks associated with sorption-desorption and diffusion processes. In contrast, respiration responses are often measured in situ without substrate amendment and the more complex landscape contributes to greater apparent E_a values.

In an ecosystem context, changes in the energy landscape translate to changes in the availability of substrates for EEA and microbial metabolism. A dynamic energy landscape complicates the

Table 1 Apparent activation energies (E_a) for ecoenzymes that catalyze sulfur, P, and N mineralization; carbohydrate hydrolysis; and polyphenol oxidation

Reference	System	Enzyme	E_a (kJ mol^{-1})
Trasar-Cepeda et al. 2007	Soil	Sulfatase	29.6
Elsgaard & Vinther 2004	Soil	Sulfatase	42.2
Tabatabai & Bremner 1970	Soil	Sulfatase	25.3
Oshrain & Wiebe 1976	Soil	Sulfatase	29.0
Beil et al. 1995	Free	Sulfatase	46.6
Ramirez-Martinez & McLaren 1966	Clay loam	Phosphatase	63.2
Kaziev 1975	Soil	Phosphatase	23.0
Menezes-Blackburn et al. 2011	Free	Phytase	35.8
Stone et al. 2012	Soils	N-acetylglucosamindase	44.1
Bremner & Mulvaney 1978	Soils	Urease	52.2
Ambus 1993	Riparian soil	Denitrification	64.9
Peterjohn 1991	Desert soil	Denitrification	41.0
Frankenberger & Tabatabai 1991b	Soils	L-glutaminase	32.4
Frankenberger & Tabatabai 1991a	Soils	L-asparaginase	26.6
Trasar-Cepeda et al. 2007	Grassland soil	Casein protease	38.0
Trasar-Cepeda et al. 2007	Grassland soil	L-argininase	23.3
Trasar-Cepeda et al. 2007	Grassland soil	Urease	29.5
McClaugherty & Linkins 1990		Chitinase	25.4
Trasar-Cepeda et al. 2007	Grassland soil	Cellulase	48.6
Stone et al. 2012	Soil	α-glucosidase	38.7
Stone et al. 2012	Soil	β-glucosidase	41.5
Stone et al. 2012	Soil	β-xylosidase	46.8
Stone et al. 2012	Soil	Cellobiohydrolase	52.8
Trasar-Cepeda et al. 2007	Grassland soil	β-glucosidase	28.6
McClaugherty & Linkins 1990	Forest soil	Exocellulase	44.8
McClaugherty & Linkins 1990	Forest soil	Endocellulase	50.4
Kahkonen et al. 2001	Pine soil	β-glucosidase	56.3
Davidson et al. 2012	Forest soil	β-glucosidase	61.8
Kocabas et al. 2008	Free	Phenol oxidase	42.3
Zhang et al. 2008	Soils	Laccase	44.8
Di Nardo et al. 2004	Oak litter	Laccase	55.0
Di Nardo et al. 2004	Oak litter	Peroxidase	60.0
McClaugherty & Linkins 1990	Forest soil	Laccase	54.4
McClaugherty & Linkins 1990	Forest soil	Peroxidase	39.6
Valtcheva et al. 2003	Wood pulp	Laccase	22.3
Lo et al. 2001	Free	Laccase	12.4
Aktaş et al. 2001	Free	Laccase	57.0
Acevedo et al. 2010	Clay bound	Manganese peroxidase	51.9
Acevedo et al. 2010	Free	Manganese peroxidase	34.4
Davidson et al. 2012	Soil	Phenol oxidase	32.5
Annuar et al. 2009	Bound	Laccase	23.0

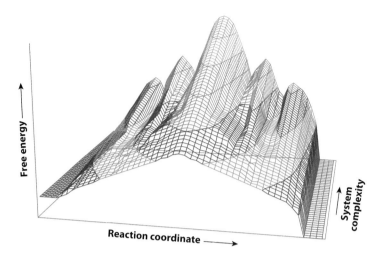

Figure 4

The effect of enzyme-environment interactions on the availability of resources can be represented as an activation energy landscape. In this conception, resource availability is related to free energy changes associated with catalytic, sorption-desorption and diffusion reactions involving enzymes, substrates, products, and inhibiters. As the matric and enzymatic complexity of the system increases, the energy landscape becomes more complex and apparent activation energies trend greater. Colors indicate increasing elevation on the vertical axis.

extrapolation of process rates to future conditions using Arrhenius models because altered resource flows affect microbial metabolism and enzyme expression. Consequently, the equilibrium between environmental resources, EEAs, and microbial metabolism is not easily linked to temperature.

5. ECOENZYMATIC ACTIVITY AND ECOLOGICAL STOICHIOMETRY

The metabolic basis of EEA stoichiometry is organismal control over enzyme expression based on environmental signals (**Figure 1**). The kinetic cascade (**Figure 2**) that controls the activity and turnover of ecoenzymes provides feedback by generating assimilable substrates and other signals of nutrient availability in relation to growth requirements. Studies of EEA stoichiometry, although not described by that term, have proceeded largely independently of ecological stoichiometry theory (Sterner & Elser 2002), which originated from comparisons of the elemental ratios of biomass composition.

5.1. Phosphorus

The most studied case of stoichiometric control of EEA is the generally inverse relationship between phosphatase activity and environmental P availability (Reichardt et al. 1967, Berman 1970, Jones 1972, Speir & Ross 1978, Wetzel 1981, Chróst & Overbeck 1987). Correlations and relative response magnitudes vary widely across aquatic and terrestrial systems in relation to measures of organic and inorganic P concentration or availability. But overall, the generalization is well supported by both small-scale and large-scale studies (e.g., Olander & Vitousek 2000; Sinsabaugh et al. 2008; Hill et al. 2010a,b, 2012; Marklein & Houlton 2012; Williams et al. 2012).

5.2. Nitrogen

Relating the activities of proteases, aminopeptidases, and amidohydrolases to measures of N availability is tenuous because N sources are more varied, and amino acids and amino sugars can be important sources of C as well as N. Proteins are the largest source of organic N. In soils, Abe & Watanabe (2004) found that peptide N accounted for 66–90% of the N content of humic acids. In planktonic systems, amino acids support a large fraction of bacterial production, ranging up to 100% with a median value of about 40% (Kirchman 2003). As a result, relationships between proteolytic activities and inorganic or organic N concentrations are variable. High-resolution planktonic studies (e.g., Hollibaugh & Azam 1983, Someville & Billen 1983, Meyer-Reil 1987, Cunningham & Wetzel 1989, Billen 1991, Hoppe 1991, Münster 1991) generally show a close linkage between the amino acid hydrolysis, amino acid uptake, and microbial metabolism with turnover rates for free amino acids ranging from minutes to a few hours. Amendment studies show that aminopeptidase activity can be depressed by additions of nitrate, ammonium, and amino acids and induced by protein addition (Chróst 1991, Foreman et al. 1998).

Jones et al. (2009) found that turnover rates for amino acids across a broad range of soils were similar to those in aquatic systems, i.e., 1–4 h, and suggested that enzymatic hydrolysis was the rate-limiting step for microbial metabolism. A study of 84 soils by Hofmockel et al. (2010) found that potential rates of proteolysis, measured as amino acid production, were similar to rates of N mineralization (amidohydrolase activity) in all ecosystems except semiarid grasslands, where proteolysis rates were six times greater than amidohydrolase. In addition, proteolytic activity was inversely related to extractable ammonium. In a study of leaf litter decomposition, Wanek et al. (2010) used ^{15}N isotopic dilution of labeled amino acids to show that rates of proteolysis exceeded N mineralization by eight fold.

After protein, the aminopolysaccharides of chitin and peptidoglycan are the most abundant sources of N. About 16% of the dry mass of filamentous fungi is chitin (Dahiya et al. 2006). Zhang & Amelung (1996) reported that hydrolysable muramic acid, glucosamine, mannosamine, and galactosamine concentration in soils was 6.2–9.5% of soil C. In a study of the Delaware estuary, Kirchman & White (1999) found that potential chitinase activity usually exceeded ^{14}C-chitin degradation rates, but both measures were of the same order of magnitude. From these data and other studies, they estimated that ~10% of bacterial production in marine systems is supported by chitin. Zeglin et al. (2012) found that N-acetylglucosamine and chitin addition to coniferous forest soils increased β-N-acetylglucosaminidase activity, N mineralization, and microbial respiration. Olander & Vitousek (2000) found that β-N-acetylglucosaminidase activity decreased as soil N increased along a Hawaiian chronosequence, but N addition depressed β-N-acetylglucosaminidase activity only in the youngest N-limited soil.

Experimental N amendment studies have been conducted in nearly every ecosystem type with no clear trend for either proteolytic or chitinolytic enzyme activities. Reported responses, if significant, are generally small. Enowashu et al. (2009) filtered N from rainwater in a spruce forest and measured the response of 15 N-acquiring enzyme activities. Some activities increased, e.g., urease (amidohydrolase), whereas, others decreased, including leucine aminopeptidase and β-N-acetylglucosaminidase. Zeglin et al. (2007) also found complementary responses in grassland ecosystems: where the leucine aminopeptidase to β-N-acetylglucosaminidase ratio was high, N amendment reduced leucine aminopeptidase activity and increased β-N-acetylglucosaminidase; where the ratio was low, N deposition depressed β-N-acetylglucosaminidase and increased leucine aminopeptidase. These examples highlight that there is no simple relationship between bulk measures of N availability and microbial N-acquiring EEAs at the ecosystem scale. A further complication is that N addition to soils depresses microbial biomass and respiration (Treseder 2008, Ramirez et al. 2012).

Because amidohydrolases are the proximate agents of N mineralization, analogous to phosphatase and P mineralization, these activities may be more directly linked to N availability and uptake than proteases or chitinases. But these activities are not commonly measured in ecological studies, with the exception of urease (Bremner & Mulvaney 1978) whose activity can be induced by urea fertilization and repressed by ammonium (Mobley et al. 1995, Hasan 2000). Blank (2002) found that the activities of four amidohydrolases were positively correlated with N mineralization and plant N uptake in riparian habitats colonized by an invasive crucifer. In a broad survey of soils, Hofmockel et al. (2010) found a 1:1 ratio of protease to amidohydrolase activities.

Cellular N metabolism is mediated by the activities of glutamate dehydrogenase (GDH) and glutamine synthetase (GS) (Marzluf 1997). GDH catalyzes a low-affinity NH_4^+ assimilation reaction, the interconversion of glutamate and α-ketoglutarate; GS catalyzes a high-affinity NH_4^+ assimilation reaction, the interconversion of glutamate and glutamine (Hoch et al. 2006). Jorgensen et al. (1999) found that dissolved free amino acids were the dominant N sources for estuarine and oceanic bacterioplankton and that greater cell-specific amino acid assimilation was associated with greater cell-specific leucine aminopeptidase activity, a higher ratio of GDH:GS activities, and a lower cell-specific respiration rate. Hoch & Bronk (2007) found that amendments of amino acids and ammonium to marine bacterioplankton increased growth and repressed GS and assimilatory nitrate reductase activities. Similarly, Geisseler et al. (2009) found that the GDH:GS activity ratio for soil microbial communities was inversely related to the C:N availability.

5.3. Carbon

P and N mineralization are predominantly hydrolytic reactions. The ecoenzymatic degradation of carbon molecules requires a large number of both hydrolytic and oxidative reactions, whose stoichiometry is linked to the composition and oxidation state of available organic matter.

5.3.1. Polysaccharides. Cellulose production accounts for about half of terrestrial NPP (Ericksson et al. 1990). Hemicelluloses, polymers of xylose, mannose, galactose, arabinose, and glucose, compose 20–30% of the mass of plant cell walls (Ericksson et al. 1990). In aquatic systems, the capsules and glycocalyces of microorganisms and biofilms are composed of complex polysaccharides whose mass can far exceed that of the microorganisms themselves (Decho 1990, Leppard 1995, Pereira et al. 2009, Bellinger et al. 2010). Given the range of monomers, linkages, and crystallinity, polysaccharide degradation requires the interaction of diverse enzymes (Warren 1996). In aquatic ecosystems, β- and α-glucosidase activities are correlated with the abundance of dissolved polysaccharides and the turnover and uptake of neutral monosaccharides (Hoppe 1983, Hoppe et al. 1988, Münster 1991, Arnosti 2011). Polysaccharides are estimated to support ~20% of bacterial production in marine systems (Kirchman 2003, Piontek et al. 2011). In terrestrial ecosystems, the activities of β-glucosidase and other cellulolytic enzymes are correlated with microbial metabolism and rates of mass loss from plant litter (Sinsabaugh et al. 1992, 1994; Jackson et al. 1995; Allison & Vitousek 2004; Snajdr et al. 2011).

5.3.2. Lignin and humus. Lignin, the most recalcitrant component of plant litter, accounts for about one-quarter of terrestrial NPP. The degradation of lignin and humus is mediated by an array of oxidases, peroxidases, dehydrogenases, and supporting enzymes that vary widely among taxa (Rabinovich et al. 2004, Baldrian 2006, Sinsabaugh 2010, Theuerl & Buscot 2010, Bugg et al. 2011, Strong & Claus 2011). Many of these enzymes are produced for purposes other than C acquisition including morphogenesis, response to oxidative stress, antimicrobial defense, and detoxification of reactive phenols. Many activities are nonspecific and may involve the generation of secondary organic and inorganic redox mediators.

Fungi, particularly Basidiomycota, are the most efficient lignin degraders in terrestrial systems, but bacteria are the ultimate consumers of most lignin-derived C. Lignin degradation is a rate-limiting process in litter decomposition and a lignocellulose index (ratio of lignin to lignin plus cellulose) of 0.7 is considered the limit of decomposition, i.e., the transition between plant litter and soil organic matter (Berg & McClaugherty 2010). In soils, the mining of humus for polysaccharide and polypeptide C and N may lead to a positive correlation between oxidative and hydrolytic enzyme activities, but more commonly these activities are uncorrelated (Sinsabaugh 2010, Sinsabaugh & Follstad Shah 2011, Sinsabaugh et al. 2011). Phenol oxidase and peroxidase activities in soils and sediments generally increase with the concentration of polyphenols, but other factors, including pH and oxygen levels and the availability of manganese and iron mediators, are also important. In soils, N saturation reduces phenol oxidase activity, decomposition rates, microbial respiration, and microbial biomass (Knorr et al. 2005, Treseder 2008, Sinsabaugh 2010, Ramirez et al. 2012). Oxidative activities are thought to play a major role in mediating C sequestration (Freeman et al. 2001, 2004; Collins et al. 2008; Sinsabaugh & Follstad Shah 2011).

6. MICROBIAL RESOURCE ALLOCATION

6.1. Conceptual Model

Correlating EEAs with the availability of target substrates provides empirical relationships for decomposition models based on the successional loop and corroborates the utility of EEA measurements for inferring microbial nutrient needs in relation to supply. At the community and ecosystem scale, the research focus shifts from the biochemistry of specific enzymes to microbial metabolism as the driver of decomposition processes. For terrestrial systems, early research involved correlating EEA with soil respiration. Correlating EEA to rates of litter decomposition began in the 1980s (Sinsabaugh et al. 1981), but it was problematic to correlate snapshots of highly dynamic activities with cumulative changes in litter mass or composition. Integrating EEA over time provided the first statistical models for mass-loss rates as a function of potential EEAs (Sinsabaugh & Linkins 1993). In aquatic systems, it is much easier to measure the turnover of dissolved organic matter pools and its relationship to EEA and microbial metabolism (Chróst 1991). The ^{3}H-thymidine and ^{3}H-leucine assays for bacterial productivity, which came into widespread use in the 1980s, facilitated empirical analyses of the microbial loop model (Fuhrman & Azam 1982).

The development of resource allocation models in the 1990s provided a conceptual paradigm for integrating EEA and microbial metabolism (Sinsabaugh & Moorhead 1994). This effort paralleled the development of the biofilm concept, which extended a common model of microbial community organization across disciplines (Characklis & Marshall 1990). The view of microbial communities as interacting consortia with 10^2–10^4 populations and multiple resource requirements moved microbial ecology beyond the single limiting-nutrient approach of the Monod (1949) and Droop (1977) models and moved EEA studies closer to the field of ecological stoichiometry (Zinn et al. 2004, Cherif & Loreau 2007, Danger et al. 2008).

6.2. Empirical Studies of Resource Allocation

In his history of soil enzyme research, Skujiņš (1978) noted that N fertilization of soils generally increased phosphatase activity. A meta-analysis of soil N and P fertilization studies by Marklein & Houlton (2012) confirms this effect for a wide variety of soils. In aquatic systems, several studies showed that the ratio of glucosidase:aminopeptidase activities shifts in response to nutrient availability (e.g., Hoppe 1991, Christian & Karl 1995).

Sinsabaugh et al. (1992, 1993) showed that differences in decomposition rate of birch sticks placed at six sites were directly related to cellulolytic enzyme activities and inversely related to the activities of enzymes involved in P and N acquisition. These studies provided the empirical basis for a model that linked EEA to mass-loss rates using ratios of C-, N-, and P-acquiring EEAs as indicators of microbial resource allocation (Sinsabaugh & Moorhead 1994).

Foreman et al. (1998) evaluated the resource allocation model by adding eight C and N amendments to bacterioplankton communities from a eutrophic river sampled on eight dates over an annual cycle. The responses of five enzymatic activities varied widely over the year with a mix of induction-depression, feedback inhibition, and resource allocation effects. On average, ammonium and leucine amendments increased bacterial production and growth efficiency and reduced enzymatic activity (feedback inhibition). Albumin had similar effects on metabolism, but increased peptidase and phosphatase activities (resource allocation). Glucose, cellobiose, and starch increased production, respiration, and glycosidase activities (induction-derepression). Vanillin and tannin increased respiration, but reduced growth efficiency and enzymatic activities. Across treatments, EEAs were closely correlated with productivity and weakly correlated with respiration. Response ratios were greatest for α- and β-glucosidase followed by phosphatase and leucine aminopeptidase. These ratios were inversely related to ambient V_{max} and apparent K_m. The results supported the resource allocation model, but illustrated that multiple mechanisms underlie shifts in ratios of EEAs.

Allison & Vitousek (2005) tested the resource allocation model in a low-nutrient tropical soil amended with simple and complex (insoluble) C, N, and P substrates, measuring respiration and the activities of β-glucosidase, glycine aminopeptidase, and acid phosphatase. The activities of enzymes directed at complex nutrients generally increased in response to the addition of complementary simple nutrients. However, ecoenzymatic responses were tempered by the activity of enzymes stabilized on soil particles, and therefore uncoupled from microbial responses. Hernandez & Hobbie (2010) conducted a similar experiment adding nine substrates singly or in varying combinations and quantities to a low-nutrient grassland soil. Ecoenzymatic responses included complementation consistent with resource allocation as well as substrate induction-derepression. Respiration rates were directly related to the sum of β-glucosidase, α-glucosidase, phosphatase, and leucine aminopeptidase activities.

Sinsabaugh & Follstad Shah (2010) compared the apparent activation energy of bacterioplankton production in two rivers sampled over an annual cycle to EEA kinetics. Bacterial consumption of carbohydrates and proteins varied by season due to changes in substrate concentration and the EEAs that mediated substrate turnover. Production was closely related to the ecoenzyme-mediated generation of assimilable substrates from dissolved carbohydrate, protein, and organic phosphate pools. The analyses demonstrated that EEA kinetics and stoichiometry can be used to resolve nutrient and temperature constraints on microbial community metabolism.

Collectively, these studies, and others, support resource allocation and multiple resource limitation models for the functional organization of microbial communities. Changes in substrate availability caused by altered inputs or shifts in the activation energy landscape affect the stoichiometry of EEA and nutrient supply, altering microbial metabolism (Allison et al. 2007).

6.3. Macroecological Patterns of Resource Allocation

Recently, EEA data sets extending to continental and global scales make it possible to compare EEA patterns to large-scale biogeochemical trends and evaluate models that link EEA stoichiometry to metabolic and stoichiometric theories of ecology.

Sinsabaugh et al. (2008) described the distribution of β-1,4-glucosidase (BG), leucine aminopeptidase (LAP), β-1,4-N-acetylglucosaminidase (NAG), acid (alkaline) phosphatase (AP), phenol oxidase and peroxidase activities in soils in relation to edaphic (soil pH, soil organic matter) and climatic (mean annual temperature, mean annual precipitation) variables. All activities correlated with soil pH. Ratios of C:N and C:P acquiring hydrolytic activities reflected latitudinal trends in P and N availability, with a global mean BG:(NAG+LAP):AP ratio of approximately 1:1:1. Sinsabaugh et al. (2009) extended these analyses, showing that terrestrial soils and freshwater sediments have similar stoichiometry with slopes near 1.0 for regressions of ln(BG) versus ln(AP) and ln(BG) versus ln(NAG+LAP). They proposed that EEAs connect the stoichiometric and metabolic theories of ecology by reflecting the equilibrium between the elemental composition of microbial biomass and detrital organic matter and the efficiencies of nutrient assimilation and growth. Sinsabaugh et al. (2010) evaluated one of these predictions by normalizing EEA to microbial productivity rates and showing that the regression slopes of ln(BG) versus ln(AP) and ln(BG) versus ln(NAG+LAP) for plankton and biofilm communities differ in proportion to the elemental C:P and C:N ratios of biomass.

Sinsabaugh & Follstad Shah (2011) analyzed the relationships between hydrolytic and oxidative activities in soils and presented a conceptual model that links organic matter recalcitrance and EEA stoichiometry. As N becomes increasingly concentrated in humus, N acquisition becomes linked to phenol oxidase activity, and N availability to microbial communities is lower than elemental C:N ratios suggest. The growth rate hypothesis (GRH) proposes that microbial growth rates are related to the cellular P quota (Frost et al. 2006, Allen & Gillooly 2009). As microbial growth slows with increasing organic matter recalcitrance, P demand should decrease relative to N. Normalizing hydrolytic activities to phenol oxidase activity reduces the slope of the ln(BG) versus ln(AP) regression and increases the slope of ln(BG) versus ln(NAG+LAP), reflecting the shift toward N acquisition expected as microbial growth rates decline with increasing organic matter recalcitrance.

Kelley et al. (2011) conducted a meta-analysis of 34 studies that measured soil enzyme responses to elevated atmospheric CO_2 treatments. β-N-acetylglucosaminidase activity increased significantly with CO_2 enrichment across studies, suggesting increased N demand. Activities directed toward lignocellulose degradation and P mineralization increased on average, but the responses were not statistically significant across all studies. Consequently, shifts in EEA stoichiometry varied with ecosystem type and duration of treatment.

Williams et al. (2012) sampled 50 streams, measuring planktonic EEA and bacterial abundance and production as well as taking several measures of land use. Abundance, productivity, and EEA were positively related to nutrient availability and anthropogenic land use. The ratio of β-glucosidase to alkaline phosphatase activity approached 1:1 with increasing anthropogenic land use and total dissolved N. The ratio of leucine aminopeptidase to alkaline phosphatase activity approached 1:1 with increased dissolved organic C and N. Ecoenzymatic C:N:P ratios moved closer to 1:1:1 as bacterial turnover increased.

Sinsabaugh et al. (2011) analyzed EEA data collected from 2,200 stream sites nationwide by the US Environmental Protection Agency. The data included nine hydrolytic activities; phenol oxidase and peroxidase activities; elemental analyses of sediment C, N, and P content; and dehydrogenase activity as a measure of microbial respiratory potential. On average, EEAs in stream sediments are two to five times greater per gram C than those of terrestrial soils. The mean ratios of BG:AP and BG:(NAG+LAP) were 1.64 ± 0.18 [95% CI (confidence interval)] and 1.83 ± 0.04, respectively, compared to 0.62 ± 0.04 and 1.43 ± 0.22 for soils, reflecting differences in the mean elemental C:P and C:N ratios of sediment and soil (57 versus 186 and 18.2 versus 14.3, respectively).

Here, we extend these stoichiometric analyses of soils and sediments by pooling data from studies of limnetic and marine plankton (**Table 2**). The slopes for the standardized major axis

Table 2 α-glucosidase (AG), β-glucosidase (BG), alkaline phosphatase (AP), leucine aminopeptidase (LAP), and β-N-acetylglucosaminidase (NAG) activities (in nanomoles per hour per liter) of planktonic microbial communities

Reference	System	N	AG	BG	AP	LAP	NAG	BG/AP	BG/LAP
Hendel & Marxsen 1997	Streams	8		4.11	37.9	54.1		0.108	0.076
Williams et al. 2012	Streams	37		12.9	71.0	138		0.182	0.093
Sinsabaugh et al. 1997	River	51	23.5	32.0	148	759		0.216	0.042
Findlay et al. 1998	River	18	10.0	19.4	179	179	14.1	0.108	0.108
Sinsabaugh & Foreman 2001	River	10	37.7	99.2	227	243		0.437	0.408
Boucher & Debroas 2009	Lake	25	3.64	3.71	3.74	69.6		0.992	0.053
Pamer et al. 2011	Lakes	10		61.0	852			0.072	
Vrba et al. 2004	Lakes	10	3.25	8.97			12.9		
Münster et al. 1992	Humic lake	32		53.9	134	14.1		0.402	3.823
Mudryk & Skórczewski 2004	Estuarine lake	5	96.4	166	1646	3223	56.8	0.101	0.052
Caruso et al. 2005	Littoral ponds	7		6.49	177	209		0.037	0.031
Taylor et al. 2003	Estuary	47		10.4		26.1	8.62		0.398
Celussi et al. 2009a	Ross Sea	14		1.67	2.20		1.54	0.759	
Celussi et al. 2009b	Ross Sea	23		0.61	1.54	22.3	0.60	0.396	0.027
Monticelli et al. 2003	Ross Sea	8		0.26	0.64	79.1		0.406	0.003
Hoppe 1983	Baltic Sea	5	4.38	3.63	15.3	45.7	4.32	0.237	0.079
Caruso 2010	Mediterranean Sea	10		0.70	34.8	16.0		0.020	0.044
Karner & Rassoulzadegan 1995	Mediterranean Sea	5	0.31	0.84		24.8			0.034
Grossart & Simon 2002	Red Sea	2		21.0		26.5			0.792
Baltar et al. 2009	Atlantic Ocean	2	0.04	0.04	0.75	5.5		0.053	0.007
Christian & Karl 1995	Pacific Ocean	6		0.06		8.0			0.008
Hoppe & Ullrich 1999	Indian Ocean	31		1.46	2.91	9.93		0.502	0.147
Mean		366	9.90	7.60	31.3	71.0	5.47	0.243	0.107

regressions of ln(BG) versus ln(AP) and ln(BG) versus ln(NAG+LAP) are similar to those for soils and sediments, showing that the ecoenzymatic scaling of planktonic communities does not differ substantially from that of attached microbial communities (**Figure 5**). However, the normalization constants and mean EEA ratios for plankton are much lower than those of soils and sediments [BG:AP and BG:(NAG+LAP) ratios are 0.26 and 0.12, respectively], reflecting

Figure 5

Global regressions for β-glucosidase (BG), acid (alkaline) phosphatase (AP), leucine aminopeptidase (LAP), and β-N-acetylglucosaminidase (NAG). For sediments and soils, activity units are in nanomoles per hour per gram of organic matter; for plankton, activity units are in nanomoles per hour per liter. (*a*) Standardized major axis (SMA) regressions for ln(BG) versus ln(AP). Freshwater sediments (*blue*): $b = 0.946 \pm 0.030$ (95% CI), $a = 1.02 \pm 0.29$, $R^2 = 0.43$, $n = 2{,}208$, geomean BG:AP ratio is 1.636 ± 0.179. Soils (*orange*): $b = 1.162 \pm 0.056$, $a = -1.715 \pm 0.481$, $R^2 = 0.40$, $n = 929$, geomean BG:AP ratio is 0.617 ± 0.045. Plankton (*gray*): $b = 0.855 \pm 0.506$, $a = -0.836 \pm 0.219$, $R^2 = 0.72$, $n = 292$, geomean BG:AP ratio is 0.243 ± 0.086. The slope of the soil regression is significantly greater than that of the sediment regression ($p < 0.05$). (*b*) SMA regressions for ln BG versus ln (NAG+LAP). Freshwater sediments (*blue*): $b = 1.098 \pm 0.028$, $a = -0.42 \pm 0.28$, $R^2 = 0.62$, $n = 2{,}208$, geomean BG:(NAG+LAP) ratio is 1.832 ± 0.036. Soils (*orange*): $b = 1.091 \pm 0.063$, $a = -0.59 \pm 0.503$, $R^2 = 0.16$, $n = 929$, geomean BG:(NAG+LAP) ratio is 1.434 ± 0.220. Plankton (*gray*): $b = 1.276 \pm 0.176$, $a = -3.300 \pm 0.808$, $R^2 = 0.50$, $n = 106$, geomean BG:(NAG+LAP) ratio is 0.123 ± 0.139. The slopes of the sediment, soil and plankton regressions are not significantly different. The BG versus LAP regression for plankton: $b = 1.086 \pm 0.104$, $a = -2.509 \pm 0.521$, $R^2 = 0.23$, $n = 292$, geomean BG/LAP ratio is 0.091 ± 0.068. Trend lines highlight the range of values and do not correspond with the SMA regressions. Freshwater sediment data from Sinsabaugh et al. (2011). Soil data from Sinsabaugh et al. (2009). Plankton data from **Table 2**

P limitation on growth and the importance of dissolved proteins and aminopolysaccharides as sources of both C and N.

Some of the residual variance associated with the global regressions for soils, sediments, and plankton (**Figure 5**) can be attributed to differences in nutrient availability, pH, and other parameters that differ at the ecoregional scale. For example, within the soil data, which have the weakest regressions, the alkaline soils of arid lands have high aminopeptidase and phenol oxidase activities (relative to β-glucosidase) compared to acidic soils; and highly weathered tropical soils have high phosphatase activities (relative to β-glucosidase) compared to higher latitude soils (Sinsabaugh et al. 2008). Further data collection may lead to specific stoichiometries for individual soil orders.

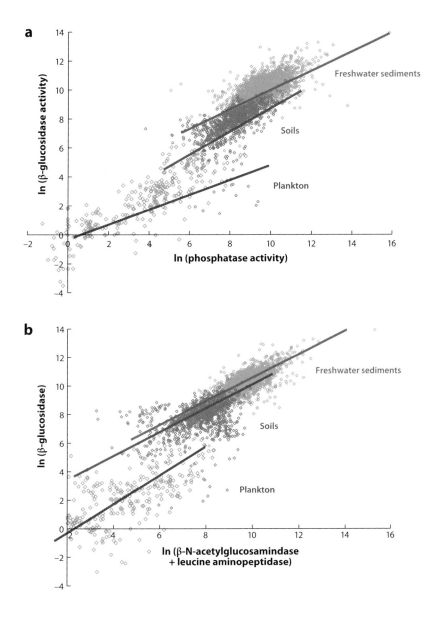

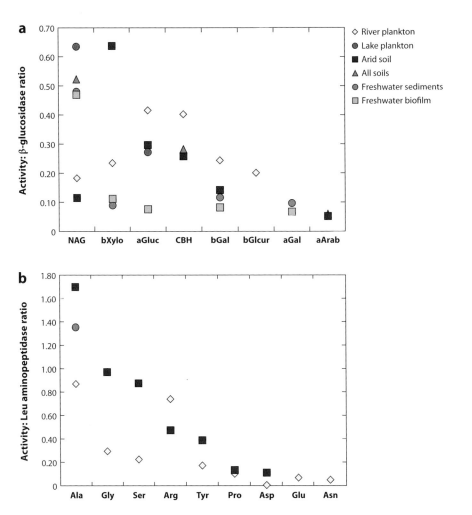

Figure 6

Mean glycosidase and aminopeptidase activities relative to β-glucosidase from representative studies. River bacterioplankton data from Sinsabaugh & Foreman (2001; $n = 10$); arid soil data from Stursova et al. (2006; $n = 24$–48); all soils data from Sinsabaugh et al. (2008; $n = 948$–1012); freshwater sediments from Hill et al. (2012; $n = 2,200$); freshwater biofilm data from L.L.P. Lehto and B.H. Hill (unpublished data; $n = 72$); lake plankton data from Vrba et al. (2004; $n = 360$). Abbreviations: NAG, β-N-acetylglucosaminidase; bXylo, β-xylosidase; aGluc, α-glucosidase; CBH, cellobiohydrolase; bGal, β-galactosidase; bGlcur, β-glucuronidase; aGal, α-galactosidase; aArab, α-arabinosidase; Ala, alanine; Gly, glycine; Ser, serine; Arg, arginine; Tyr, tyrosine; Pro, proline; Asp, aspartate; Glu, glutamate; Asn, asparagine aminopeptidase.

From the freshwater sediment data, Hill et al. (2012) were able to resolve EEA stoichiometries for nine ecoregions.

β-glucosidase and leucine aminopeptidase are the most widely measured glycosidase and aminopeptidase activities because they generally yield the greatest potential rates and hydrolyze the most abundant substrates. Some studies include other measures. The expectation is that activities will scale in proportion to the relative abundance of substrates; that appears to be the case (**Figure 6**), but there are few studies that combine high-resolution analyses of substrate composition with corresponding EEAs. We note that leucine and alanine, the two most abundant protein amino acids, are hydrolyzed by the same enzymes, but hydrolysis of coumarin-linked alanine is generally greater than coumarin-linked leucine even though the latter is more commonly used (and costs more).

7. MODELS

7.1. Ecological Theory

The metabolic and stoichiometric theories of ecological function are founded on the kinetics and structure of cellular components that mediate production and respiration (Sterner & Elser 2002, Brown et al. 2004, Gillooly et al. 2005). Each theory is based on invariance rules and allometric relationships (Allen & Gillooly 2009, Doi et al. 2010). The metabolic theory extrapolates the functions of terminal metabolic units involved in C fixation, respiration, and translation to organismal and ecological levels of organization based on thermodynamics and the limitations of material distribution, which differ for prokaryotes, protists, and metazoans (DeLong et al. 2010). The stoichiometric theory extends the elemental composition of cellular components to ecological processes and organization using cellular growth models and ratios of nutrient availability. These theories are interrelated through the threshold element ratio (TER) and the GRH.

The TER is the elemental C:N or C:P ratio of available growth substrates at which control of cellular or community metabolism switches from energy supply (C) to nutrient supply (N, P) (Sterner & Elser 2002, Elser et al. 2003, Frost et al. 2006, Doi et al. 2010). Frost et al. (2006) defined the TER as

$$\text{TER}_{C:P} = (A_P/GE) \cdot B_{C:P} \quad \text{and} \quad \text{TER}_{C:N} = (A_N/GE) \cdot B_{C:N}, \qquad 1.$$

where A_P and A_N are assimilation efficiencies for P and N, GE is microbial growth efficiency with respect to C, and $B_{C:P}$ and $B_{C:N}$ are the elemental C:P and C:N ratios of microbial biomass. Doi et al. (2010) presented a TER definition that replaces the A_P/GE and A_N/GE terms with GE_P^{max}/GE_C^{max} and GE_N^{max}/GE_C^{max}, which are maximum growth efficiencies with respect to C, N, and P. This definition emphasizes that the TER is the condition for optimal growth. For microbial communities and invertebrates, the ratio of GE_C^{max}/GE_N^{max} is approximately 0.77; the GE_C^{max}/GE_P^{max} is more variable with a mean of about 0.3 (Herron et al. 2009, Jones et al. 2009, Doi et al. 2010, Zeglin et al. 2012). Moorhead et al. (2012) developed a simulation model for microbial community growth on C and C+N substrates in which TER is defined as $B_{C/N}/CUE$ (carbon use efficiency). Because TER links biomass composition and growth efficiency, values vary by the extent to which elemental homeostasis is maintained in relation to external resource supply (Gusewell & Gessner 2009, Hladyz et al. 2009, Franklin et al. 2011, Hall et al. 2011).

Elemental homeostasis is itself dependent on growth rate. The GRH predicts that the cellular P quota increases with growth rate because growth rate is directly related to ribosome density (Elser et al. 2003, Doi et al. 2010), and ribosomes are the largest P stocks in many cells. Franklin et al. (2011) show that the GRH does not apply under conditions of N limitation, because cells have little capacity to alter their C:N stoichiometry by altering the relative abundance of biomolecules. Sinsabaugh & Follstad Shah (2011) showed that increased N limitation associated with the humification of organic matter reverses the GRH by slowing growth and resource allocation to P-acquiring enzymes.

7.2. Substrate Availability, Ecoenzymatic Activity, and Growth

The generation of an assimilable substrate from a single extracellular enzymatic reaction can be represented by the Michaelis-Menten function (Michaelis & Menten 1913):

$$V = V_{max} \cdot S/[K_m + S], \qquad 2.$$

where V is the reaction rate; V_{max} is the maximum reaction rate when available enzyme capacity is saturated; S is environmental substrate concentration; and K_m is the half-saturation constant for the enzyme, i.e., S at $V_{max}/2$. The Monod model uses the same formulation to represent the growth of single cell organisms as a function of the environmental concentration of a limiting substrate (Monod 1949):

$$\mu = \mu_{max} \cdot S/[K_s + S], \qquad 3.$$

where μ is growth rate, μ_{max} is the maximum growth rate, and K_S is the half-saturation constant. The presumption is that μ is controlled by the substrate affinity and active transport capacity of integral membrane proteins (Button 1993). In simulation models, growth is more complex because the energy and materials represented by substrate consumption must be allocated to a variety of maintenance processes, only resource consumption beyond these requirements contributes to growth (Moorhead et al. 2012, Wang et al. 2012a).

The Monod model is not broadly applicable to microbial communities because single substrate limitation is not the typical condition (Chen & Christensen 1985, Zinn et al. 2004, Cherif & Loreau 2007, Danger et al. 2008). Within microbial communities the thresholds for substrate limitation of growth vary across populations. At any given state, there are many resources that are limiting the growth of one or more of the 10^2–10^4 extant populations. In addition, (a) populations have some physiological capacity to adapt to changing nutrient availabilities and (b) the acquisition of various substrates may not be independent. The latter occurs because (a) the organic matter available for degradation by heterotrophic microorganisms has a limited range of biochemical and stoichiometric composition and (b) the acquisition of multiple resources is integrated through optimal resource allocation strategies selected to maximize growth. If environmental conditions remain stable for an extended period, community adaptation also includes succession of populations. As a result, the relationship between microbial community production rate and resource availability is less subject to discrete thresholds and sequential limitations than single populations. It is possible to posit growth models for specific types of resource interactions (Danger et al. 2008). But for microbial communities as a whole, we propose that the relationship among multiple resources and community production may be generally described as

$$\mu = \mu_{max} \cdot \{(S_1 \cdot S_2 \cdot \ldots \cdot S_n)/[(K_{S1} + S_1) \cdot (K_{S2} + S_2) \cdot (\ldots K_{Sn} + S_n)]\}^{1/n}. \qquad 4.$$

This community growth model relates production to the geometric mean of multiple resource availabilities (**Figure 7**). In this formulation, the acquisition of multiple resources is neither wholly independent nor fully integrated. In microbial communities, the assimilation of a limiting nutrient into the community is mediated by populations that have the greatest affinity for the substrate (lowest K_S). But once acquired by the community, nutrients can be internally recycled. Within biofilms, nutrient concentrations can be one or more orders of magnitude greater than the environmental concentration (Hall-Stoodley et al. 2004, Van Horn et al. 2011). As a result, community growth rates should exceed those predicted by the Monod model for a single population.

In the metabolic theory of ecology and ecological stoichiometry theory, microbial growth efficiency, rather than growth rate, is linked to the availability of limiting nutrients, commonly N and P relative to C, at both the organismal and community levels (Equation 1). The Michaelis-Menten and Monod models, and the metabolic theory of ecology and ecological stoichiometry theory models derived from them, can be related to the stoichiometry of the EEAs that mediate microbial nutrient acquisition from environmental organic matter (Sinsabaugh et al. 2009, 2010,

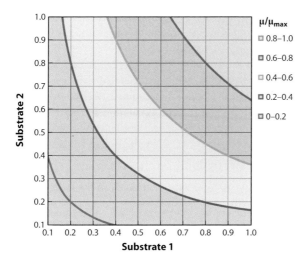

Figure 7

Microbial community growth rate as a fraction of maximal growth rate (μ/μ_{max}; see Equation 4 in text) based on availability of two substrates.

2011; Sinsabaugh & Follstad Shah 2011):

$$\text{EEA}_{C/N} \propto B_{C/N}/L_{C/N} \propto \text{TER}_{C/N}/B_{C/N} = A_N/\text{GE}_C, \qquad 5.$$

$$\text{EEA}_{C/P} \propto B_{C/P}/L_{C/P} \propto \text{TER}_{C/P}/B_{C/P} = A_P/\text{GE}_C, \qquad 6.$$

where $\text{EEA}_{C/N}$ and $\text{EEA}_{C/P}$ are ratios of ecoenzymatic C acquisition activity to ecoenzymatic N and P acquisition activity, and $L_{C/N}$ and $L_{C/P}$ are the C:N and C:P ratios of labile organic matter in the environment. Using the most widely measured activities, we define

$$\text{EEA}_{C/N} = \text{BG}/(\text{LAP} + \text{NAG}), \qquad 7.$$

$$\text{EEA}_{C/P} = \text{BG}/\text{AP}, \qquad 8.$$

where BG = β-1,4-glucosidase activity; LAP = leucine aminopeptidase activity; NAG = β-1,4-N-acetylglucosaminidase activity; and AP = acid (alkaline) phosphatase activity.

Following Equations 5 and 6, the generation of assimilable N and P substrates from environmental sources via EEA, expressed as a fraction relative to C, can be represented as

$$S_{C/N} = B_{C/N}/L_{C/N} \cdot 1/\text{EEA}_{C/N}, \qquad 9.$$

$$S_{C/P} = B_{C/P}/L_{C/P} \cdot 1/\text{EEA}_{C/P}. \qquad 10.$$

$S_{c/n}$ and $S_{C/P}$ are scalars for the relative availability of hydrolyzed N and P in relation to microbial community composition. Microbial growth efficiency as a function of the stoichiometry of nutrients generated by EEA can be represented as

$$\text{GE}_C = \text{GE}_C^{\text{max}} \cdot S_{C/N}/[K_{C/N} + S_{C/N}] = A_N \cdot B_{C/N}/\text{TER}_{C/N}, \qquad 11.$$

$$\text{GE}_C = \text{GE}_C^{\text{max}} \cdot S_{C/P}/[K_{C/P} + S_{C/P}] = A_P \cdot B_{C/P}/\text{TER}_{C/P}, \qquad 12.$$

where $K_{C/N}$ and $K_{C/P}$ are half-saturation constants for GE_C based on the stoichiometry of C:N and C:P availabilities. GE_C approaches GE_C^{max}, approximately 0.60, when $S_{C/N} \gg K_{C/N}$, $S_{C/P} \gg K_{C/P}$, and the ratios $B_{C/N}/\text{TER}_{C/N}$ and $B_{C/P}/\text{TER}_{C/P}$ are near 0.6 (assuming A_N and $A_P = 1$).

Equations 11 and 12 propose that community growth efficiency is a saturating function of N and P availability relative to C, linking substrate generation via EEA to microbial substrate assimilation and biomass composition.

If community metabolism is controlled by the resource in least supply, following the Liebig assumption, Equations 11 and 12 will generate independent estimates of GE_C. In that case, community growth efficiency is equal to the lower estimate. At the other extreme, the equations should yield identical estimates of GE_C if the generation and assimilation of C, N, and P are perfectly integrated. For reasons presented above, neither condition is likely to apply. Following the community growth model (Equation 4), we merge Equations 11 and 12 as

$$GE_C = GE_C^{max} \cdot \{(S_{C/N} \cdot S_{C/P})/[(K_{C/N} + S_{C/N}) \cdot (K_{C/P} + S_{C/P})]\}^{0.5}$$
$$= \{[A_N \cdot B_{C/N}/TER_{C/N}] \cdot [A_P \cdot B_{C/P}/TER_{C/P}]\}^{0.5}. \qquad 13.$$

This biogeochemical equilibrium model expresses growth efficiency at the community level as the geometric mean of the N and P supply, relative to C. The models for biogeochemical equilibrium (Equation 13) and community growth (Equation 4) can be merged using $GE_C = \mu/(\mu + R)$, where R is the community respiration rate. Equation 13 can be simplified to

$$GE_C = GE_C^{max} \cdot (N_{sat} \cdot P_{sat})^{0.5} = (N_{con} \cdot P_{con})^{0.5}. \qquad 14.$$

by defining N_{sat} and P_{sat} as stoichiometric indices of environmental N and P saturation relative to C,

$$N_{sat} = S_{C/N}/(K_{C/N} + S_{C/N}) \text{ and } P_{sat} = S_{C/P}/(K_{C/P} + S_{C/P}), \qquad 15.$$

and N_{con} and P_{con} as indices of microbial N and P consumption relative to C,

$$N_{con} = A_N \cdot B_{C/N}/TER_{C/N} \text{ and } P_{con} = A_P \cdot B_{C/P}/TER_{C/P}. \qquad 16.$$

7.3. Model Prediction

For most ecosystems, the mean GE_C for heterotrophic microbial communities is near $GE_C^{max}/2$ or 0.30 (Manzoni et al. 2012). In aquatic ecosystems, bacterial respiration (BR) generally increases sublinearly with production (BR $= 3.42BP^{0.61}$; del Giorgio & Cole 1998). Consequently, GE_C also generally increases with production, approaching an asymptote of approximately 0.50, with a global average of 0.26 (del Giorgio & Cole 1998). For decomposing plant litter in terrestrial ecosystems, microbial C use efficiency, an indirect measure of growth efficiency, averages about 0.30 based on C:N stoichiometry (Manzoni & Porporato 2009, Manzoni et al. 2010). There are fewer data for terrestrial soils, and most measurements involve calculating growth yield using labile substrate (usually glucose) amendments. In some cases, N and P additions are also included. Most of these estimates are in the range of 0.50–0.60 (Frey et al. 2001, Manzoni et al. 2012). Herron et al. (2009) reported a microbial growth efficiency of 0.46 for semiarid soils using ^{13}C-labeled acetic acid vapor. These growth efficiency estimates for labile substrates approach maximal values observed in culture. Community growth efficiency under ambient substrate and nutrient conditions is probably lower and comparable to values for other ecosystems.

In theory, at $GE_C = 0.5 \cdot GE_C^{max}$ balanced flows of C, N, and P would set $S_{C/N} = K_{C/N} = S_{C/P} = K_{C/P} = 0.5$ (Equations 11 and 12). The data set for terrestrial soils (**Figure 5**) lacks estimates of $L_{C/N}$, $L_{C/P}$, $B_{C/N}$, and $B_{C/P}$ for individual samples. Using mean values for $L_{C/N}$ (14.3), $L_{C/P}$ (186), $B_{C/N}$ (8.6), and $B_{C/P}$ (60) presented by Cleveland & Liptzin (2007), the estimated mean values of $S_{C/N}$ and $S_{C/P}$ are 0.42 and 0.52, respectively. Using these values and assuming $K_{C/N} = K_{C/P} = 0.5$, the estimated mean growth efficiency for soil microbial communities is 0.29 (Equation 13; **Figure 8**). The stream sediment data (**Figure 5**) include

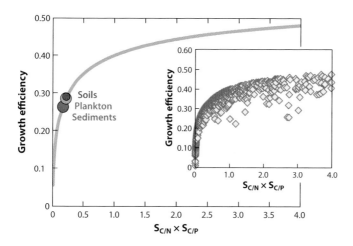

Figure 8

Biogeochemical equilibrium model (Equation 13) that predicts microbial community growth efficiency from the elemental C:N and C:P ratios of biomass ($B_{C/N}$, $B_{C/P}$) and environmental organic matter ($L_{C/N}$, $L_{C/P}$), and the ratios of ecoenzymatic activities (EEAs) that mediate C, N, and P acquisition ($EEA_{C/N}$, $EEA_{C/P}$). $S_{C/N} = B_{C/N}/L_{C/N} \cdot 1/ EEA_{C/N}$ and $S_{C/P} = B_{C/P}/L_{C/P} \cdot 1/ EEA_{C/P}$. Circles are mean values for the soil, sediment, and plankton data sets shown in **Figure 5**. (*Inset*) Predicted growth efficiencies for freshwater sediments; this data set includes $L_{C/N}$ and $L_{C/P}$ measures for each sample ($n = 2,096$; Hill et al. 2012).

$L_{C/N}$ and $L_{C/P}$ estimates for each sample, but not values for $B_{C/N}$ and $B_{C/P}$. Using mean $B_{C/N}$ and $B_{C/P}$ values of 8.6 and 60, respectively, for all samples, the mean values of $S_{C/N}$ and $S_{C/P}$ for stream sediments are 0.26 (95% CI = 0.02, $n = 2192$) and 0.64 (95% CI = 0.15, $n = 2192$), respectively. The differences in $S_{C/N}$ and $S_{C/P}$, relative to soils is driven by the low C:P ratio of sediments (57), which approximates $B_{C/P}$. Using $K_{C/N} = K_{C/P} = 0.5$, the mean GE_C for stream sediments is 0.27 (SD = 0.10, 95% CI = 0.004) (**Figure 8**), which is similar to the mean GE_C of 0.26 estimated by del Giorgio & Cole (1998).

Compared to soils and sediments, the mean BG/(LAP+NAG) ratio for plankton (0.123) is low (**Table 2**; **Figure 5**) and yields a $S_{C/N}$ value of 3.1 using a mean $B_{C/N}$ of 6.6 (Sinsaugh & Follstad Shah 2010) and a mean $L_{C/N}$ of 17.3 (Seitzinger et al. 2005), suggesting that BG/(LAP+NAG) is not a good indicator of $EEA_{C/N}$ for bacterioplankton. Analyses of dissolved organic matter composition and bacterial metabolism show that carbohydrates and proteins make roughly equal contributions to bacterioplankton production on a global basis (e.g., Kirchman & White 1999, Kirchman 2003, Rosenstock & Simon 2003, Simon & Rosenstock 2007). If proteins and aminopolysaccharides supply about half the C for production (Kirchman & White 1999, Kirchman 2003), a better indicator of $EEA_{C/N}$ for bacterioplankton is (BG+LAP+NAG)/(LAP+NAG), which has a mean value of 1.123 and yields a $S_{C/N}$ value of 0.34.

Using a mean $B_{C/P}$ of 106 (Sinsaugh & Follstad Shah 2010), a mean $L_{C/P}$ of 656 (Seitzinger et al. 2005), and a mean $EEA_{C/P}$ ratio of 0.257 (**Table 2**; **Figure 5**), the $S_{C/P}$ estimate for plankton is 0.63. This estimate is similar to that for stream sediments, but in this case $S_{C/P}$ is driven by high phosphatase activity in relation to environmental P availability, whereas sediments have high environmental P availability and low phosphatase activities. For plankton, phosphatase expression is presumably facilitated by the tight coupling of C and N acquisition that comes from reliance on protein and chitin for growth. Using $K_{C/N} = K_{C/P} = 0.5$ and values of 0.34 and 0.63 for $S_{C/N}$ and $S_{C/P}$, respectively, the mean GE_C for plankton is 0.28 (**Figure 8**), which approximates the mean GE_C of 0.26 estimated by del Giorgio & Cole (1998). Thus, Equation 13 produces similar estimates for the mean growth efficiency of microbial communities in soils, sediments, and plankton (0.29, 0.27, 0.28, respectively) despite large differences in resource availability and biomass composition.

7.4. Model Synthesis

The biogeochemical equilibrium model (Equation 13) yields microbial community GE_C values that converge on $GE_C^{max}/2$ because resource allocation to substrate acquisition, measured as potential EEA, generally scales with environmental substrate concentration such that enzymes operate at about one-half their maximum catalytic capacity, i.e., the Michaelis-Menten parameters V_{max}, S, and apparent K_m (Equation 2) are positively correlated (Williams 1973, Sinsabaugh & Follstad Shah 2010). The Monod parameters K_S and μ_{max} (Equation 3) are also positively correlated (Lobry et al. 1992). This condition sustains high substrate turnover while optimizing responsiveness to fluctuations in substrate availability. Expending resources to produce additional enzyme has marginal utility because additional enzyme depresses soluble substrate concentration, reducing the reaction velocity per enzyme. In addition, excess enzyme production depresses growth efficiency by increasing maintenance costs (Wang et al. 2012a).

An analogous condition applies to insoluble substrates because the optimal substrate:enzyme ratio is determined by the density of effective enzyme binding sites (Bünemann 2008, Lynd et al. 2002). This condition is modeled using adsorption isotherms or "reverse" Michaelis-Menten kinetics (Schimel & Weintraub 2003). In this case, expressing more enzyme has marginal value because substrate accessibility limits their effective use. This constraint can lead to C and nutrient limitation even in environments with high organic matter concentrations.

The ecoenzymatic controls on microbial growth become less important as environmental nutrient availability increases to the point of saturating microbial demand ($S_{C/N}$ and $S_{C/P} > 1$). In Equations 5 and 6 the terms $EEA_{C/N}$ and $B_{C/N}/L_{C/N}$ are linearly related to $A_N \cdot B_{C/N}/ TER_{C/N}$ and the terms $EEA_{C/P}$ and $B_{C/P}/L_{C/P}$ are linearly related to $A_P \cdot B_{C/P}/TER_{C/P}$. In Equations 11 and 12 these linearities are limited to values of $S_{C/N}$ and $S_{C/P} < 1$, corresponding to $GE_C < 0.30$ (**Figure 8**). As $S_{C/N}$ and $S_{C/P}$ increase beyond the half-saturation constants $K_{C/N}$ and $K_{C/P}$, control of growth efficiency progressively shifts from environmental resource availability to constraints imposed by thermodynamics and cellular organization as predicted by metabolic theory.

Equation 13 defines a biogeochemical equilibrium or attractor for environmental resource availability, EEA stoichiometry, biomass stoichiometry, and microbial community metabolism (**Figure 8**). At low resource availability, microbial growth is controlled by substrate generation via EEA. At high resource availability, microbial growth is controlled by membrane transport capacity and metabolic constraints. Throughout this range, growth efficiency is modulated by the stoichiometry of C, N, and P availability in relation to microbial growth requirements. This equilibrium attractor provides a framework for comparative studies of organic matter decomposition and microbial community metabolism.

8. FUTURE DIRECTIONS

Several simulation models that include ecoenzyme pools have been recently published (e.g., Davidson et al. 2012, Moorhead et al. 2012, Resat et al. 2012, Wang et al. 2012a). These models highlight the uncertainties in our understanding of biogeochemical processes at various scales of resolution. They also show that resource flows and microbial community metabolism depend on (*a*) rates and priorities of ecoenzyme expression and (*b*) the effective activity and turnover rate of ecoenzymes in the environment. Better estimates of these parameters, and their range of variation, are needed to translate these models into applications that predict process rates, e.g., respiration or production, from EEA monitoring data. The existence of a common EEA stoichiometry in dynamic equilibrium with resource availability, biomass composition and community metabolism suggests that such applications are feasible. One advantage of process models based on EEA stoichiometry is the potential to include multiple resource drivers. Microbial metabolism in soils, for example, is generally considered to be C limited. Some models include C and N, but EEA stoichiometry suggests that P acquisition, rather than N, is often a constraint on microbial metabolism, particularly for highly weathered soils.

Models that entrain EEA stoichiometry also facilitate monitoring of ecosystem responses to environmental changes because EEA is easily measured at high throughput and relatively low cost compared to most of the resource and metabolic variables to which it is dynamically linked. However, almost all empirical research is based on assays of potential activities most often conducted under substrate saturating conditions at ~20 C with pH buffering. Unfortunately, there are no commercially available systems for continuous monitoring of in situ EEA at present, due largely to lack of a market rather than technological impediments. Such systems could resolve a variety of mechanistic questions and provide data to parameterize process models. Questions of interest include the decomposition/nutrient mining activities of mycorrhizal fungi, signal transduction between plants and rhizosphere microbial communities, pulse-response dynamics to precipitation in arid land ecosystems or flooding in riparian ecosystems, and controls on soil organic C stores in relation to fluctuating nutrient availability and temperature regimes.

In the biomics paradigm, ecoenzymes link microbial community structure to biogeochemical process. As the technology for extracting and sequencing these enzymes advances, it will be increasingly possible to directly match enzymes to their producers. Eventually, comparisons of metatranscriptomes to metaproteomes will also show the relative turnover time of ecoenzymes produced by various populations within the community, resolving legacy effects. These analyses are the final link in the genes to ecosystems integration of biological organization.

DISCLOSURE STATEMENT

The authors are not aware of any affiliations, memberships, funding, or financial holdings that might be perceived as affecting the objectivity of this review.

ACKNOWLEDGMENTS

This work was supported by the NSF EaGER (DEB-0946288) and Ecosystem Studies programs (DEB-0918718) and the Sevilleta LTER. We thank Steve Allison, Christine Hawkes, Brian Hill, Kirsten Hofmockel, Rob Jackson, Daryl Moorhead, Eldor Paul, Kathleen Treseder, Bonnie Waring, and Michael Weintraub for providing comments on various drafts.

LITERATURE CITED

Abe T, Watanabe A. 2004. X-ray photoelectron spectroscopy of nitrogen functional groups in soil humic acids. *Soil Sci.* 169:35–43

Acevedo F, Pizzul L, Castillo MD, Gonzalez ME, Cea M, et al. 2010. Degradation of polycyclic aromatic hydrocarbons by free and nanoclay-immobilized manganese peroxidase from *Anthracophyllum discolor*. *Chemosphere* 80:271–78

Aktaş N, Çiçek H, Ünal AT, Kibarer G, Kolankaya N, Tanyolaç A. 2001. Reaction kinetics for laccase-catalyzed polymerization of 1-naphthol. *Bioresour. Technol.* 80:29–36

Allen AP, Gillooly JF. 2009. Towards an integration of ecological stoichiometry and the metabolic theory of ecology to better understand nutrient cycling. *Ecol. Lett.* 12:369–84

Allison SD, Gartner T, Holland K, Weintraub M, Sinsabaugh RL. 2007. Soil enzymes: linking proteomics and ecological process. In *Manual of Environmental Microbiology*, pp. 704–11. Washington, DC: ASM Press

Allison SD, Vitousek PM. 2004. Extracellular enzyme activities and carbon chemistry as drivers of tropical plant litter decomposition. *Biotropica* 36:285–96

Allison SD, Vitousek PM. 2005. Responses of extracellular enzymes to simple and complex nutrient inputs. *Soil Biol. Biochem.* 37:937–44

Ambus P. 1993. Control of denitrification enzyme-activity in a streamside soil. *FEMS Microbiol. Ecol.* 102:225–34

Annuar MSM, Adnan S, Vikineswary S, Chisti Y. 2009. Kinetics and energetics of azo dye decolorization by *Pycnoporus sanguineus*. *Water Air Soil Pollut.* 202:179–88

Arnosti C. 2011. Microbial extracellular enzymes and the marine carbon cycle. *Annu. Rev. Mar. Sci.* 3:401–25

Azam F, Hodson RE. 1981. Multiphasic kinetics for D-glucose uptake by assemblages of natural marine bacteria. *Mar. Ecol. Prog. Ser.* 6:213–22

Baldrian P. 2006. Fungal laccases—occurrence and properties. *FEMS Microbiol. Rev.* 30:215–42

Baltar F, Aristegui J, Sintes E, van Aken HM, Gasol JM, Herndl GJ. 2009. Prokaryotic extracellular enzymatic activity in relation to biomass production and respiration in the meso- and bathypelagic waters of the (sub)tropical Atlantic. *Environ. Microbiol.* 11:1998–2014

Beil S, Kehrli H, James P, Staudenmann W, Cook AM, et al. 1995. Purification and characterization of the arylsulfatase synthesized by *Pseudomonas aeruginosa* PAO during growth in sulfate-free medium and cloning of the arylsulfatase gene (*atsA*). *Eur. J. Biochem.* 29:385–94

Bellinger BJ, Gretz MR, Domozych DS, Kiemle SN, Hagerthey SE. 2010. Composition of extracellular polymeric substances from periphyton assemblages in the Florida Everglades. *J. Phycol.* 46:484–96

Berg G, McClaugherty CM. 2010. *Plant Litter: Decomposition, Humus Formation, Carbon Sequestration*. Berlin: Springer

Berman T. 1970. Alkaline phosphatases and phosphorus availability in Lake Kinneret. *Limnol. Oceanogr.* 15:663–74

Billen G. 1991. Protein degradation in aquatic environments. See Chróst 1991, pp. 123–43

Blank RR. 2002. Amidohydrolase activity, soil nitrogen status, and the invasive crucifer *Lepidium latifolium*. *Plant Soil* 239:155–63

Boucher D, Debroas D. 2009. Impact of environmental factors on couplings between bacterial community composition and ectoenzymatic activities in a lacustrine ecosystem. *FEMS Microbiol. Ecol.* 70:66–78

Bremner JM, Mulvaney RL. 1978. Urease activity in soils. See Burns 1978, pp. 149–96

Brown JH, Gillooly JF, Allen AP, Savage VM, West GB. 2004. Toward a metabolic theory of ecology. *Ecology* 85:1771–89

Bugg TDH, Ahmad M, Hardiman EM, Singh R. 2011. The emerging role for bacteria in lignin degradation and bio-product formation. *Curr. Opin. Biotechnol.* 22:394–400

Bünemann EK. 2008. Enzyme additions as a tool to assess the potential bioavailability of organically bound nutrients. *Soil Biol. Biochem.* 40:2116–29

Burns RG, ed. 1978. *Soil Enzymes*. New York: Academic

Burns RG, Dick RP, eds. 2002. *Enzymes in the Environment*. New York: Marcel Dekker

Button DK. 1993. Nutrient-limited microbial-growth kinetics—overview and recent advances. *Antonie van Leeuwenhoek Int. J. Gen. Mol. Microbiol.* 63:225–35

Caruso G, Monticelli L, Azzaro F, Azzaro M, Decembrini F, et al. 2005. Dynamics of extracellular enzymatic activities in a shallow Mediterranean ecosystem (Tindari ponds, Sicily). *Mar. Freshwater Res.* 56:173–88

Caruso G. 2010. Leucine aminopeptidase, beta-glucosidase and alkaline phosphatase activity rates and their significance in nutrient cycles in some coastal Mediterranean sites. *Mar. Drugs* 8:916–40

Celussi M, Cataletto B, Umani SF, Del Negro P. 2009a. Depth profiles of bacterioplankton assemblages and their activities in the Ross Sea. *Deep-Sea Res. Part I: Oceanogr. Res. Pap.* 56:2193–205

Celussi M, Paoli A, Crevatin E, Bergamasco A, Margiotta F, et al. 2009b. Short-term under-ice variability of prokaryotic plankton communities in coastal Antarctic waters (Cape Hallett, Ross Sea). *Estuar. Coast. Shelf Sci.* 81:491–500

Characklis WG, Marshall K. 1990. *Biofilms: A Basis for an Interdisciplinary Approach*. New York: Wiley-Intersci.

Chen CY, Christensen ER. 1985. A unified theory for microbial growth under multiple nutrient limitation. *Water Res.* 19:791–98

Cherif M, Loreau M. 2007. Stoichiometric constraints on resource use, competitive interactions, and elemental cycling in microbial decomposers. *Am. Nat.* 169:709–24

Christian JR, Karl DM.1995. Bacterial ectoenzymes in marine waters: activity ratios and temperature responses in three oceanographic provinces. *Limnol. Oceanogr.* 40:1042–49

Chróst RJ, ed. 1991. *Microbial Enzymes in Aquatic Environments*. New York: Springer-Verlag

Chróst RJ, Overbeck J. 1987. Kinetics of alkaline phosphatase and phosphorus availability for phytoplankton and bacterioplankton in Lake Plussee. *Microb. Ecol.* 13:229–48

Cleveland CC, Liptzin D. 2007. C:N:P stoichiometry in soil: Is there a "Redfield ratio" for the microbial biomass? *Biogeochemistry* 85:235–52

Cole JJ, Prairie YT, Caraco NF, McDowell WH, Tranvik LJ, et al. 2007. Plumbing the global carbon cycle: integrating inland waters into the terrestrial carbon budget. *Ecosystems* 10:171–84

Collins SL, Sinsabaugh RL, Crenshaw C, Green LE, Porras-Alfaro A, et al. 2008. Pulse dynamics and microbial processes in aridland ecosytems. *J. Ecol.* 96:413–20

Cunningham HW, Wetzel RG. 1989. Kinetic analysis of protein degradation by a freshwater wetland sediment community. *Appl. Environ. Microbiol.* 55:1963–67

Dahiya N, Tewari R, Hoondal GS. 2006. Biotechnological aspects of chitinolytic enzymes: a review. *Appl. Microbiol. Biotechnol.* 71:773–82

Danger M, Daufresne T, Lucas F, Pissard S, Lacroix G. 2008. Does Liebig's law of the minimum scale up from species to communities? *Oikos* 117:1741–51

Davidson EA, Jannsens IA. 2006. Temperature sensitivity of soil carbon decomposition and feedbacks to climate change. *Nature* 440:165–73

Davidson EA, Samanta S, Caramori SS, Savage K. 2012. The dual Arrhenius and Michaelis-Menten kinetics model for decomposition of soil organic matter at hourly to seasonal time scales. *Glob. Chang. Biol.* 18:371–84

DeAngelis KM, Lindow SW, Firestone MK. 2008. Bacterial quorum sensing and nitrogen cycling in rhizosphere soil. *FEMS Microbiol. Ecol.* 66:197–207

Decho AW. 1990. Microbial exopolymer secretions in ocean environments: their role(s) in food webs and marine processes. *Oceanogr. Mar. Biol. Annu. Rev.* 28:73–153

del Giorgio PA, Cole JJ. 1998. Bacterial growth efficiency in natural aquatic ecosystems. *Annu. Rev. Ecol. Syst.* 29:503–41

DeLong JP, Okie JG, Moses ME, Sibly RM, Brown JH. 2010. Shifts in metabolic scaling, production and efficiency across major evolutionary transitions of life. *Proc. Natl. Acad. Sci. USA* 107:12941–45

Di Nardo C, Cinquegrana A, Papa S, Fuggi A, Fioretto A. 2004. Laccase and peroxidase isoenzymes during leaf litter decomposition of *Quercus ilex* in a Mediterranean ecosystem. *Soil Biol. Biochem.* 36:1539–44

Dick R, ed. 2012. *Methods of Soil Enzymology*. Madison, WI: Soil Sci. Soc. Am.

Doi H, Cherif M, Iwabuchi T, Katano I, Stegen JC, Striebel M. 2010. Integrating elements and energy through the metabolic dependencies of gross growth efficiency and the threshold elemental ratio. *Oikos* 119:752–65

Droop MR. 1977. Approach to quantitative nutrition of phytoplankton. *J. Protozool.* 24:528–32

Elser JJ, Acharya K, Kyle M, Cotner J, Makino W, et al. 2003. Growth rate-stoichiometry couplings in diverse biota. *Ecol. Lett.* 6:936–43

Elsgaard L, Vinther FR. 2004. Modeling of the fine-scale temperature response of arylsulfatase activity in soil. *J. Plant Nutr. Soil Sci.* 167:196–201

Enowashu E, Poll C, Lamersdorf N, Kandeler E. 2009. Microbial biomass and enzyme activities under reduced nitrogen deposition in a spruce forest soil. *Appl. Soil Ecol.* 43:11–21

Ericksson K-E, Blanchette RA, Ander P. 1990. *Microbial and Enzymatic Degradation of Wood Components*. Berlin: Springer. 407 pp.

Findlay S, Sinsabaugh RL, Fischer DT, Franchini P. 1998. Sources of dissolved organic carbon supporting planktonic bacterial production in the tidal freshwater Hudson River. *Ecosystems* 1:227–39

Foreman CM, Franchini P, Sinsabaugh RL. 1998. The trophic dynamics of riverine bacterioplankton: relationships among substrate availability, ectoenzyme kinetics and growth. *Limnol. Oceanogr.* 43:1344–52

Frankenberger WT, Tabatabai MA. 1991a. L-asparaginase activity of soils. *Biol. Fertil. Soils* 11:6–12

Frankenberger WT, Tabatabai MA. 1991b. L-glutaminase activity of soils. *Soil Biol. Biochem.* 23:869–74

Franklin O, Hall EK, Kaiser C, Battin TJ, Richter A. 2011. Optimization of biomass composition explains microbial growth-stoichiometry relationships. *Am. Nat.* 177:E29–42

Freeman C, Ostle NJ, Fenner N, Kang H. 2004. A regulatory role for phenol oxidase during decomposition in peatlands. *Soil Biol. Biochem.* 36:1663–67

Freeman C, Ostle N, Kang H. 2001. An enzymic 'latch' on a global carbon store—a shortage of oxygen locks up carbon in peatlands by restraining a single enzyme. *Nature* 409:149–53

Frey SD, Gupta VVSR, Elliott ET, Paustian K. 2001. Protozoan grazing affects estimates of carbon utilization efficiency of the soil microbial community. *Soil Biol. Biochem.* 33:1759–68

Frost PC, Benstead JP, Cross WF, Hillebrand H, Larson JH, et al. 2006. Threshold elemental ratios of carbon and phosphorus in aquatic consumers. *Ecol. Lett.* 9:774–79

Fuhrman JA, Azam F. 1982. Thymidine incorporation as a measure of heterotrophic bacterial production in marine surface waters—evaluation and field results. *Mar. Biol.* 66:109–20

Geisseler D, Doane TA, Horwath WR. 2009. Determining potential glutamine synthetase and glutamate dehydrogenase activity in soil. *Soil Biol. Biochem.* 41:1741–49

Gfeller D, De Los Rios P, Caflisch A, Rao F. 2007. Complex network analysis of free-energy landscapes. *Proc. Natl. Acad. Sci. USA* 104:1817–22

Gillooly JF, Allen AP, Brown JH, Elser JJ, del Rio CM, et al. 2005. The metabolic basis of whole-organism RNA and phosphorus content. *Proc. Natl. Acad. Sci. USA* 102:11923–27

Gillooly JF, Brown JH, West GB, Savage VM, Charnov EL. 2001. Effects of size and temperature on metabolic rate. *Science* 293:2248–51

Grossart HP, Simon M. 2002. Bacterioplankton dynamics in the Gulf of Aqaba and the Northern Red Sea in early spring. *Mar. Ecol. Prog. Ser.* 239:263–76

Gusewell S, Gessner MO. 2009. N:P ratios influence litter decomposition and colonization by fungi and bacteria in microcosms. *Funct. Ecol.* 23:211–19

Hall EK, Maixner F, Franklin O, Daims H, Richter A, Battin T. 2011. Linking microbial and ecosystem ecology using ecological stoichiometry: a synthesis of conceptual and emprical approaches. *Ecosystems* 14:261–73

Hall-Stoodley L, Costerton JW, Stoodley P. 2004. Bacterial biofilms: from the natural environment to infectious diseases. *Nat. Rev. Microbiol.* 2:95–108

Hasan HAH. 2000. Ureolytic organisms and soil fertility: a review. *Comm. Soil Sci. Plant Anal.* 31:2565–89

Hendel B, Marxsen J. 1997. Measurement of low-level extracellular enzyme activity in natural waters using fluorigenic model substrates. *Acta Hydrochim. Hydrobiol.* 25:253–58

Hernandez DL, Hobbie SE. 2010. The effects of substrate composition, quantity and diversity on microbial activity. *Plant Soil* 335:397–411

Herron PM, Stark JM, Holt C, Hooker T, Cardon ZG. 2009. Microbial growth efficiencies across a soil moisture gradient assessed using ^{13}C-acetic acid vapor and ^{15}N-ammonia gas. *Soil Biol. Biochem.* 41:1262–69

Hill BH, Elonen CM, Jicha TM, Bolgrien DW, Moffett MF. 2010a. Sediment microbial enzyme activity as an indicator of nutrient limitation in the great rivers of the Upper Mississippi River basin. *Biogeochemistry* 97:195–209

Hill BH, Elonen CM, Seifert LR, May AA, Tarquinio E. 2012. Ecoenzymatic stoichiometry and nutrient limitation in US streams and rivers. *Ecol. Indic.* 18:540–55

Hill BH, McCormick FH, Harvey BC, Johnson SL, Warren ML, Elonen CM. 2010b. Microbial enzyme activity, nutrient uptake and nutrient limitation in forested streams. *Freshwater Biol.* 55:1005–19

Hladyz S, Gessner MO, Giller PS, Pozo J, Woodward G. 2009. Resource quality and stoichiometric constraints on stream ecosystem functioning. *Freshw. Biol.* 54:957–70

Hoch MP, Bronk DA. 2007. Bacterioplankton nutrient metabolism in the Eastern Tropical North Pacific. *J. Exp. Mar. Biol. Ecol.* 349:390–404

Hoch MP, Snyder RA, Jeffrey WH, Dillon KS, Coffin RB. 2006. Expression of glutamine synthetase and glutamate dehydrogenase by marine bacterioplankton: assay optimizations and efficacy for assessing N to carbon metabolic balance in situ. *Limnol. Oceanogr. Methods* 4:318–28

Hofmockel KS, Fierer N, Colman BP, Jackson RB. 2010. Amino acid abundance and proteolytic potential in North American soils. *Oecologia* 163:1069–78

Hollibaugh JT, Azam F. 1983. Microbial degradation of dissolved proteins in seawater. *Limnol. Oceanogr.* 28:1104–16

Hoppe HG. 1983. Significance of exoenzymatic activities in the ecology of brackish water: measurements by means of methylumbelliferyl substrates. *Mar. Ecol. Prog. Ser.* 11:299–308

Hoppe HG. 1991. Microbial extracellular enzyme activity: a new key parameter in aquatic ecology. See Chróst 1991, pp. 60–83

Hoppe HG, Kim SJ, Gocke K. 1988. Microbial decomposition in aquatic environments: combined processes of extracellular enzyme activity and substrate uptake. *Appl. Environ. Microbiol.* 54:784–90

Hoppe HG, Ullrich S. 1999. Profiles of ectoenzymes in the Indian Ocean: phenomena of phosphatase activity in the mesopelagic zone. *Aquat. Microb. Ecol.* 19:139–48

Houghton RA. 2007. Balancing the global carbon budget. *Annu. Rev. Earth Planet. Sci.* 35:313–47

Jackson C, Foreman C, Sinsabaugh RL. 1995. Microbial enzyme activities as indicators of organic matter processing rates in a Lake Erie coastal wetland. *Freshw. Biol.* 34:329–42

Jones DL, Kielland K, Sinclair FL, Dahlgren RA, Newsham KK, et al. 2009. Soil organic nitrogen mineralization across a global latitudinal gradient. *Glob. Biogeochem. Cycles* 23:GB1016 (DOI:10.1029/2008GB003250)

Jones JG. 1972. Studies on freshwater microorganisms: phosphatase activity in lakes of differing degrees of eutrophication. *J. Ecol.* 60:777–91

Jorgensen NOG, Kroer N, Coffin RB, Hoch MP. 1999. Relations between bacterial nitrogen metabolism and growth efficiency in an estuarine and an open-water ecosystem. *Aquat. Microb. Ecol.* 18:247–61

Kahkonen MA, Wittmann C, Kurola J, Ilvesniemi H, Salkinoja-Salonen MS. 2001. Microbial activity of boreal forest soil in a cold climate. *Boreal Environ. Res.* 6:19–28

Karner M, Rassoulzadegan F. 1995. Extracellular enzyme activity: indications for high short-term variability in a coastal marine ecosystem. *Microb. Ecol.* 30:143–56

Kaziev FKH. 1975. Thermodynamic characteristics of enymic reactions in soil. *Biol. Nauki* 10:121–27

Kelley AM, Fay PF, Polley HW, Gill RA, Jackson RB. 2011. Atmospheric CO_2 and soil extracellular enzyme activity: a meta-analysis and CO_2 gradient experiment. *Ecosphere* 2(8):art96 (DOI:10.1890/ES11-00117.1)

Kirchman DL. 2003. The contribution of monomers and other low-molecular weight compounds to the flux of dissolved organic material in aquatic ecosystems. In *Aquatic Ecosystems: The Interactivity of Dissolved Organic Matter*, ed. S Findlay, RL Sinsabaugh, pp. 218–39. San Diego: Academic

Kirchman DL, White J. 1999. Hydrolysis and mineralization of chitin in the Delaware estuary. *Aquat. Microb. Ecol.* 18:187–96

Knorr M, Frey SD, Curtis PS. 2005. Nitrogen additions and litter decomposition: a meta-analysis. *Ecology* 86:3252–57

Kocabas DS, Bakir U, Phillips SEV, McPherson MJ, Ogel ZB. 2008. Purification, characterization, and identification of a novel bifunctional catalase-phenol oxidase from *Scytalidium thermophilum*. *Appl. Microbiol. Biotechnol.* 79:407–15

Leppard GG. 1995. The characterization of algal and microbial mucilages and their aggregates in aquatic ecosystems. *Sci. Total Environ.* 165:103–31

Lo SC, Ho YS, Buswell JA. 2001. Effect of phenolic monomers on the production of laccases by the edible mushroom *Pleurotus sajor-caju*, and partial characterization of a major laccase component. *Mycologia* 93:413–21

Lobry JR, Flandrois JP, Carret G, Pave A. 1992. Monod's microbial growth model revisited. *Bull. Math. Biol.* 54:117–21

Lynd LR, Weimer PJ, van Zyl WH, Pretorius IS. 2002. Microbial cellulose utilization: fundamentals and biotechnology. *Microbiol. Mol. Biol. Rev.* 66:506–77

Maire V, Alvarez G, Colombet J, Comby A, Despinasse R, et al. 2012. An unknown respiration pathway substantially contributes to soil CO_2 emissions. *Biogeosci. Discuss.* 6:8663–91

Manzoni S, Porporato A. 2009. Soil carbon and nitrogen mineralization: theory and models across scales. *Soil Biol. Biochem.* 41:1355–79

Manzoni S, Taylor P, Richter A, Porporato A, Ågren GI. 2012. Environmental and stoichiometric controls on microbial carbon-use efficiency in soils. *New Phytol.* 196:79–91

Manzoni S, Trofymow JA, Jackson RB, Porporato A. 2010. Stoichiometric controls on carbon, nitrogen, and phosphorus dynamics in decomposing litter. *Ecol. Monogr.* 80:89–106

Marklein AR, Houlton BZ. 2012. Nitrogen inputs accelerate phosphorus cycling rates across a wide variety of terrestrial ecosystems. *New Phytol.* 193(3):696–704

Marzluf GA. 1997. Genetic regulation of nitrogen metabolism in the fungi. *Microbol. Mol. Biol. Rev.* 61:17–32

McClaugherty CA, Linkins AE. 1990. Temperature responses of enzymes in two forest soils. *Soil Biol. Biochem.* 22:29–33

McLaren AD. 1978. Kinetics and consecutive reactions of soil enzymes. See Burns 1978, pp. 97–116

Menezes-Blackburn D, Jorquera M, Gianfreda L, Rao M, Greiner R, et al. 2011. Activity stabilization of *Aspergillus niger* and *Escherichia coli* phytases immobilized on allophanic synthetic compounds and montmorillonite nanoclays. *Bioresour. Technol.* 102:9360–67

Meyer-Reil LA. 1987. Seasonal and spatial distribution of extracellular enzymatic activities and microbial incorporation of dissolved organic substrates in marine sediments. *Appl. Environ. Microbiol.* 53:1748–55

Michaelis L, Menten MI. 1913. Die Kinetik der Invertinwirkung. *Biochemische Z.* 49:333–69

Mobley HLT, Island MD, Hausinger RP. 1995. Molecular biology of microbial ureases. *Microbiol. Rev.* 59:451–80

Monod J. 1949. The growth of bacterial cultures. *Annu. Rev. Microbiol.* 3:371–94

Monticelli LS, La Ferla R, Maimone G. 2003. Dynamics of bacterioplankton activities after a summer phytoplankton bloom period in Terra Nova Bay. *Antarct. Sci.* 15:85–93

Moorhead DL, Lashermes G, Sinsabaugh RL. 2012. A theoretical model of C- and N-acquiring exoenzyme activities balancing microbial demands during decomposition. *Soil Biol. Biochem.* 53:133–41

Moorhead DL, Sinsabaugh RL. 2006. A theoretical model of litter decay and microbial interaction. *Ecol. Monogr.* 76:151–74

Mudryk ZJ, Skórczewski P. 2004. Extracellular enzyme activity at the air-water interface of an estuarine lake. *Estuar. Coast. Shelf Sci.* 59:59–67

Münster U. 1991. Extracellular enzyme activity in eutrophic and polyhumic lakes. See Chróst 1991, pp. 96–122

Münster U, Nurminen J, Einio P, Overbeck J. 1992. Extracellular enzymes in a small polyhumic lake—origin, distribution and activities. *Hydrobiologia* 243:47–59

Olander LP, Vitousek PM. 2000. Regulation of soil phosphatase and chitinase activity by N and P availability. *Biogeochemistry* 49:175–90

Oshrain RL, Wiebe WJ. 1976. Arylsulfatase activity of salt marsh soils. *Appl. Environ. Microbiol.* 38:337–40

Overbeck J. 1975. Distribution pattern of uptake kinetic responses in a stratified eutrophic lake. *Verh. Int. Vereinigung Theor. Angew. Limnol.* 19:2600–15

Overbeck J. 1991. Early studies on ecto- and extracellular enzymes in aquatic environments. See Chróst 1991, pp. 1–5

Pamer E, Vujovic G, Knezevic P, Kojic D, Prvulovic D, et al. 2011. Water quality assessment in lakes of Vojvodina. *Int. J. Environ. Res.* 5:891–900

Pereira S, Zille A, Micheletti E, Moradas-Ferreira P, De Philippis R, Tamagnini P. 2009. Complexity of cyanobacterial exopolysaccharides: composition, structures, inducing factors and putative genes involved in their biosynthesis and assembly. *FEMS Microbiol. Rev.* 33:917–41

Peterjohn WT. 1991. Denitrification—enzyme content and activity in desert soils. *Soil Biol. Biochem.* 23:845–55

Piontek J, Handel N, De Bodt C, Harlay J, Chou L, Engel A. 2011. The utilization of polysaccharides by heterotrophic bacterioplankton in the Bay of Biscay (North Atlantic Ocean). *J. Plankton Res.* 33:1719–35

Rabinovich ML, Bolobova AV, Vasilchenko LG. 2004. Fungal decomposition of natural aromatic structures and xenobiotics: a review. *Appl. Biochem. Microbiol.* 40:1–17

Ramirez KS, Craine JM, Fierer N. 2012. Consistent effects of nitrogen amendments on soil microbial communities and processes across biomes. *Glob. Change Biol.* 18:1918–27

Ramirez-Martinez JR, McClaren AD. 1966. Some factors influencing the determination of phosphatase activity in native soils and in soils sterilized br irradiation. *Enzymologia* 31:23–38

Reichardt W, Overbeck J, Steubing L. 1967. Free dissolved enzymes in lake water. *Nature* 216:1345–47

Resat H, Bailey V, McCue LA, Konopka A. 2012. Modeling microbial dynamics in heterogeneous environments: growth on soil carbon sources. *Microb. Ecol.* 63:883–97

Rosenstock B, Simon M. 2003. Consumption of dissolved amino acids and carbohydrates by limnetic bacterioplankton according to molecular weight fractions and proportions bound to humic matter. *Microb. Ecol.* 45:433–43

Schimel JP, Weintraub MN. 2003. The implications of exoenzyme activity on microbial carbon and nitrogen limitation in soil: a theoretical model. *Soil Biol. Biochem.* 35:549–63

Seitzinger SP, Harrison JA, Dumont E, Beusen AHW, Bouwman AF. 2005. Sources and delivery of carbon, nitrogen and phosphorus to the coastal zone: an overview of global nutrient export from watersheds (NEWS) models and their application. *Glob. Biogeochem. Cycles* 19:GB4S01 (DOI:10.1029/2005GB002606)

Shukla G, Varma A, eds. 2011. *Soil Enzymology*. Berlin: Springer-Verlag

Sierra CA. 2012. Temperature sensitivity of organic matter decomposition in the Arrhenius equation: some theoretical considerations. *Biogeochemistry* 108:1–15

Simon M, Rosenstock B. 2007. Different coupling of dissolved amino acid, protein, and carbohydrate turnover to heterotrophic picoplankton production in the southern ocean in austral summer and fall. *Limnol. Oceanogr.* 52:85–95

Sinsabaugh RL. 2010. Phenol oxidase, peroxidase and organic matter dynamics of soil. *Soil Biol. Biochem.* 42:391–404

Sinsabaugh RL, Antibus RK, Linkins AE, McClaugherty CA, Rayburn L, et al. 1992. Wood decomposition over a first-order watershed: mass loss as a function of lignocellulase activity. *Soil Biol. Biochem.* 24:743–49

Sinsabaugh RL, Antibus RK, Linkins AE, Rayburn L, Repert D, Weiland T. 1993. Wood decomposition: nitrogen and phosphorus dynamics in relation to extracellular enzyme activity. *Ecology* 74:1586–93

Sinsabaugh RL, Benfield EF, Linkins AE. 1981. Cellulase actvity associated with the decomposition of leaf litter in a woodland stream. *Oikos* 36:184–90

Sinsabaugh RL, Carreiro MM, Alvarez S. 2002. Enzyme and microbial dynamics during litter decomposition. See Burns 2002, pp. 249–66

Sinsabaugh RL, Findlay S, Franchini P, Fischer D. 1997. Enzymatic analysis of riverine bacterioplankton production. *Limnol. Oceanogr.* 42:29–38

Sinsabaugh RL, Follstad Shah JJ. 2010. Integrating resource utilization and temperature in metabolic scaling of riverine bacterial production. *Ecology* 91:1455–65

Sinsabaugh RL, Follstad Shah JJ. 2011. Ecoenzymatic stoichiometry of recalcitrant organic matter decomposition: the growth rate hypothesis in reverse. *Biogeochemistry* 102:31–43

Sinsabaugh RL, Follstad Shah JJ, Hill BH, Elonen CM. 2011. Ecoenzymatic stoichiometry of stream sediments with comparison to terrestrial soils. *Biogeochemistry*. Houten, The Neth.: Springer (DOI:10.1007/s10533-011-9676-x)

Sinsabaugh RL, Foreman CM. 2001. Activity profiles of bacterioplankton in a eutrophic river. *Freshw. Biol.* 46:1–12

Sinsabaugh RL, Hill BH, Follstad Shah JJ. 2009. Ecoenzymatic stoichiometry of microbial organic nutrient acquisition in soil and sediment. *Nature* 462:795–98

Sinsabaugh RL, Lauber CL, Weintraub MN, Ahmed B, Allison SD, et al. 2008. Stoichiometry of soil enzyme activity at global scale. *Ecol. Lett.* 11:1252–64

Sinsabaugh RL, Linkins AE. 1993. Statistical modeling of litter decomposition from integrated cellulase activity. *Ecology* 74:1594–97

Sinsabaugh RL, Moorhead DL. 1994. Resource allocation to extracellular enzyme production: a model for nitrogen and phosphorus control of litter decomposition. *Soil Biol. Biochem.* 26:1305–11

Sinsabaugh RL, Osgood MP, Findlay S. 1994. Enzymatic models for estimating decomposition rates of particulate detritus. *J. North Am. Benthol. Soc.* 13:160–69

Sinsabaugh RL, Van Horn DJ, Follstad Shah JJ, Findlay S. 2010. Ecoenzymatic stoichiometry in relation to productivity for freshwater biofilm and plankton communities. *Microb. Ecol.* 60:885–93

Skujiņš J. 1978. History of abiontic soil enzyme research. See Burns 1978, pp. 1–49

Snajdr J, Cajthaml T, Valaskova V, Merhautova V, Petrankova M, et al. 2011. Transformation of *Quercus petraea* litter: successive changes in litter chemistry are reflected in differential enzyme activity and changes in the microbial community composition. *FEMS Microbiol. Ecol.* 75:291–303

Someville M, Billen G. 1983. A method for determining exoproteolytic activity in natural waters. *Limnol. Oceanogr.* 28:190–93

Speir TW, Ross DJ. 1978. Soil phosphatases and sulphatase. See Burns 1978, pp. 197–250

Sterner RW, Elser JJ. 2002. *Ecological Stoichiometry: The Biology of Elements from Molecules to the Biosphere.* Princeton: Princeton Univ.

Stone MM, Weiss MS, Goodale CL, Adams MB, Fernandez IJ, et al. 2012. Temperature sensitivity of soil enzyme kinetics under N-fertilization in two temperate forests. *Glob. Chang. Biol.* 18(3):1173–84

Strong PJ, Claus H. 2011. Laccase: a review of its past and its future in bioremediation. *Crit. Rev. Environ. Sci.Technol.* 41:373–434

Stursova M, Crenshaw C, Sinsabaugh RL. 2006. Microbial responses to long-term N deposition in a semiarid grassland. *Microb. Ecol.* 51:90–98

Tabatabai MA, Bremner JM. 1970. Arylsulfatase activity of soils. *Soil Sci. Soc. Am. Proc.* 34:225–29

Taylor GT, Way J, Yu Y, Scranton MI. 2003. Ectohydrolase activity in surface waters of the Hudson River and western Long Island Sound estuaries. *Mar. Ecol. Prog. Series* 263:1–15

Theuerl S, Buscot F. 2010. Laccases: toward disentangling their diversity and functions in relation to soil organic matter cycling. *Biol. Fertil. Soils* 46:215–25

Trasar-Cepeda C, Gil-Sotres F, Leiros MC. 2007. Thermodynamic parameters of enzymes in grassland soils from Galicia, NW Spain. *Soil Biol. Biochem.* 39:311–19

Trasar-Cepeda C, Hernández T, Garcia C, Rad C. 2011. *Soil Enzymology in the Recycling of Organic Wastes and Environmental Restoration.* Berlin: Springer

Treseder KK. 2008. Nitrogen additions and microbial biomass: a meta-analysis of ecosystem studies. *Ecol. Lett.* 11:1111–20

Valtcheva E, Veleva S, Radeva G, Valtchev I. 2003. Enzyme action of the laccase-mediator system in the pulp delignification process. *React. Kinet. Catal. Lett.* 78:183–91

Van Horn DJ, Sinsabaugh RL, Takacs-Vesbach CD, Mitchell KR, Dahm CN. 2011. The response of heterotrophic stream biofilm communities to a resource gradient. *Aquat. Microb. Ecol.* 64:149–61

Vrba J, Callier C, Bittl T, Simek K, Bertoni R, et al. 2004. Are bacteria the major producers of extracellular glycolytic enzymes in aquatic environments? *Int. Rev. Hydrobiol.* 89:102–17

Wanek W, Mooshammer M, Blochl A, Hanreich A, Richter A. 2010. Determination of gross rates of amino acid production and immobilization in decomposing leaf litter by a novel N-15 isotope pool dilution technique. *Soil Biol. Biochem.* 42:1293–302

Wang G, Post WM, Mayes MA. 2012a. Development of microbial-enzyme-mediated decomposition model parameters through steady-state and dynamic analyses. *Ecol. Appl.* In press. (DOI:10.1890/12-0681.1)

Wang G, Post WM, Mayes MA, Frerichs JT, Sindhu J. 2012b. Parameter estimation for models of ligninolytic and cellulolytic enzyme kinetics. *Soil Biol. Biochem.* 48:28–38

Warren RAJ. 1996. Microbial hydrolysis of polysaccharides. *Annu. Rev. Microbiol.* 50:183–212

Wetzel RG. 1981. Longterm dissolved and particulate alkaline phosphatase activity in a hardwater lake in relation to lake stability and phosphorus enrichments. *Verhandlungen Int. Vereinigung Theor. Angwandte Limnol.* 21:337–39

Wetzel RG. 1991. Extracellular enzymatic interactions: storage, redistribution and interspecific communication. See Chróst 1991, pp. 6–28

Williams CJ, Scott AB, Wilson HF, Xenopoulos MA. 2012. Effects of land use on water column bacterial activity and enzyme stoichiometry in stream ecosystems. *Aquat. Sci.* 74(3):483–94

Williams PJL. 1973. The validity of the application of simple kinetic analysis to heterogeneous microbial populations. *Limnol. Oceanogr.* 18:159–65

Zeglin LH, Kluber LA, Myrold DD. 2012. The importance of amino sugar turnover to C and N cycling in organic horizons of old growth Douglas-fir forest soils colonized by ectomycorrhizal mats. *Biogeochemistry.* In press (DOI:10.1007/s10533-012-9746-8)

Zeglin LH, Stursova M, Sinsabaugh RL, Collins SL. 2007. Microbial responses to nitrogen addition in three contrasting grassland ecosystems. *Oecologia* 296:65–75

Zhang JB, Liu XP, Xu ZQ, Chen H, Yang YX. 2008. Degradation of chlorophenols catalyzed by laccase. *Int. Biodeterior. Biodegrad.* 61:351–56

Zhang XD, Amelung W. 1996. Gas chromatographic determination of muramic acid, glucosamine, mannosamine, and galactosamine in soils. *Soil Biol. Biochem.* 28:1201–6

Zinn M, Witholt B, Egli T. 2004. Dual nutrient limited growth: models, experimental obsevations, and applications. *J. Biotechnol.* 113:263–79

ZoBell CE. 1943. The effect of solid surfaces upon bacterial activity. *J. Bacteriol.* 46:39–56

Origins of New Genes and Evolution of Their Novel Functions

Yun Ding,[1,2] Qi Zhou,[3] and Wen Wang[1]

[1]State Key Laboratory of Genetic Resources and Evolution, Kunming Institute of Zoology, Chinese Academy of Sciences, Kunming 650223, China; email: wwang@mail.kiz.ac.cn

[2]Janelia Farm Research Campus, Howard Hughes Medical Institute, Ashburn, Virginia 20147; email: dingy@janelia.hhmi.org

[3]Department of Integrative Biology, University of California, Berkeley, California 94709; email: zhouqi@berkeley.edu

Annu. Rev. Ecol. Evol. Syst. 2012. 43:345–63

First published online as a Review in Advance on September 4, 2012

The *Annual Review of Ecology, Evolution, and Systematics* is online at ecolsys.annualreviews.org

This article's doi: 10.1146/annurev-ecolsys-110411-160513

Keywords

origin of new genes, evolution of novel functions, de novo origination

Abstract

The origination of novel genes is an important process during the evolution of organisms because it provides critical sources for evolutionary innovation. Addressing how novel genes emerged and acquired novel and adaptive functions is of fundamental importance. Here we summarize the newest advances in our understanding of the molecular mechanisms and genome-wide patterns of new gene origination and new gene functions. We pay special attention to the origins of noncoding RNA genes and de novo genes, whose processes had been previously overlooked but are gaining increasingly visible importance. We then introduce recent findings that have opened a path to the study of the evolution of novel functions and pathways via novel genes. We also discuss the important issues and potential developments in the field.

1. INTRODUCTION

The origination of novel functional genes is one of the principal processes contributing to evolutionary innovation. Its significance was first recognized in the 1930s by Haldane (1935) and Muller (1933) and then formulated in Ohno's (1970) seminal monograph, *Evolution by Gene Duplication*. Ohno emphasized gene duplication as the most important mechanism producing new genes. Although this view might have been refined by recent studies (Levine et al. 2006, Zhou et al. 2008, Knowles & McLysaght 2009, Tautz & Domazet-Loso 2011, Wu et al. 2011), his idea that gene duplication provides functionally redundant new copies as raw material for natural selection has set the tone for subsequent theoretical and experimental studies. Under this hypothesis, new gene copies survive in the host genome in three main ways: preservation as a functionally redundant copy (Clark 1994), transformation into a complementary copy (subfunctionalization) (Lynch & Force 2000), or evolution into novel functions (neofunctionalization) (Walsh 1995). These models, especially the latter two, have been extensively documented in case studies of new genes within diverse organisms.

However, these models have been developed only for new genes that have another gene as its ancestor. Not until the recent development of whole-genome scans has another important source of new genes been recognized: de novo genes originating from noncoding ancestors (Levine et al. 2006, Zhou et al. 2008, Knowles & McLysaght 2009, Wu et al. 2011). Besides the exciting discovery of a novel mechanism, a more comprehensive picture of the origin and evolution of new genes is emerging in the genomic era (Zhou & Wang 2008, Kaessmann 2010). The abundant genomic resources now available allow investigators to identify and reconstruct an unbiased history of young genes (usually <10 million years) whose sequence changes can be clearly traced back to their parental sequences. The time at which a species diverges sets the upper limit for the ages of lineage-specific young genes. As such, they can be grouped into different ages within a phylogenetic context, which allows for a survey of the dynamic changes that new genes undergo over time. For example, two recent studies using 12 *Drosophila* genomes have found the patterns of origination mechanisms and gene expressions differ among new *Drosophila* genes of different ages (Zhou et al. 2008, Zhang et al. 2010b).

Besides the published reviews on the origins of new genes (Long et al. 2003, Zhou & Wang 2008, Kaessmann 2010), additional summaries are needed for other important and fast-developing topics such as the recent advances in our understanding of the functional significance of new genes, the genome-wide patterns of new gene origination, the discovery of the importance of de novo origination, and functional evolution of new genes. In this review, we attempt to summarize the newest advances in our understanding of the origin of novel genes and their new functions. We conclude with a preview of the future directions of this field.

2. MOLECULAR MECHANISMS FOR ORIGINS OF NEW GENES

2.1. Gene Duplication

Various mechanisms can give rise to a new gene (Long et al. 2003, Zhou & Wang 2008, Kaessmann 2010). Among them, gene duplication is probably the most commonly and extensively studied. A preexisting gene can spawn another copy through small-scale events, such as duplication of the complete/partial gene regions or segmental duplication encompassing several genes. A new gene can also result from an event of much larger scale such as polyploidy or whole-genome duplication (WGD).

WGD is a special case of gene duplication (Semon & Wolfe 2007). Analysis of yeast species' genomes before and after WGD proposed that whether a WGD duplicate is retained as a new gene is largely determined by its dosage effects rather than by the novel functions it has acquired (Wapinski et al. 2007). For example, genes that encode cellular components or participate in essential growth processes tend to display haploinsufficiency and maintain uniform copy numbers after WGD, while genes that encode peripheral transporters or participate in stress responses tend to show both gene gain and loss through WGD (Wapinski et al. 2007). Recent studies in *Arabidopsis thaliana* and *Saccharomyces cerevisiae* have also revealed some distinctive features of new genes derived from WGDs and single-gene duplications: Whereas single-gene duplication is usually involved in distinctive functional categories (Maere et al. 2005), WGD duplicates tend to share more protein interactions and exhibit less profound phenotypic effects than single-gene duplicates (Hakes et al. 2007).

Smaller-scale gene duplication can be mediated by both DNA and RNA. DNA-based duplication, in a broader sense, occurs with genome segments regardless of whether genes are involved within the segment, whereas an RNA-mediated duplication event (also called retroposition; see below) occurs only in complete or partial gene regions. In detail, DNA-based duplication can emerge through the mediation of repetitive elements [nonallelic homologous recombination (NAHR)] (Roth et al. 1985, Bailey et al. 2003) or as a result of replication errors [nonhomologous end joining (NHEJ)] (Roth & Wilson 1988, Koszul et al. 2004). NHEJ seems to have a broader impact and may occur prior to NAHR (**Figure 1a**) (Hastings et al. 2009). For example, examination of the breakpoints of 53 telomeric duplications revealed that 92% of them are consistent with NHEJ (Linardopoulou et al. 2005). A recent whole-genome scan of DNA copy-number variation in *Drosophila melanogaster* and *Drosophila simulans* also found no association between the duplication hot spots and the repetitive elements, but it did find an enrichment of duplications in late-replicating regions, suggesting NHEJ rather than NAHR underlies the pattern (Cardoso-Moreira et al. 2011). However, these results do not preclude the importance of NAHR to the production of new duplicates. In a laborious search using fluorescence in situ hybridization for species-specific new genes in eight *Drosophila* species, Yang et al. (2008) identified 17 new genes dispersedly located from their parental genes. Most (82%) of them also have repetitive elements located at one or both of their duplication breakpoints. Further comparative genomics studies (Zhou et al. 2008) confirmed NAHR may play a more important role producing dispersed gene duplicates, whereas NHEJ dominates the generation of tandem duplication in *Drosophila*.

2.2. Retroposition

RNA-mediated duplication or retroposition (for a thorough review, see Kaessmann et al. 2009) is distinct from the above-mentioned molecular processes: The new retrocopy is usually intronless, most likely retaining only part of the parental gene and rarely inheriting the parental gene's promoters (**Figure 1b**). These defects led to the earlier thought that retrogenes are usually processed pseudogenes (Brosius 1991, Jeffs & Ashburner 1991, Petrov et al. 1996, Mighell et al. 2000). But since the late 1980s (McCarrey & Thomas 1987), abundant retrogenes with intriguing novel functions have been reported. They frequently evolve testis-biased gene expression patterns (Betran & Long 2003, Vinckenbosch et al. 2006, Bai et al. 2007) and may participate in spermatogenesis (Marques et al. 2005), brain function (Viale et al. 2000, Rosso et al. 2008), immune defense (Sayah et al. 2004), and courtship behavior (Dai et al. 2008). These exciting case studies from various species have shifted genome-wide inspections of the evolutionary pattern of retrogenes to a broader scale.

So far, retrogenes have been systematically characterized throughout the genomes of plants (Zhang et al. 2005, Wang et al. 2006), fruit flies (Bai et al. 2007, Zhou et al. 2008), and mammals (human, mouse, opossum) (Emerson et al. 2004, Vinckenbosch et al. 2006, Potrzebowski et al.

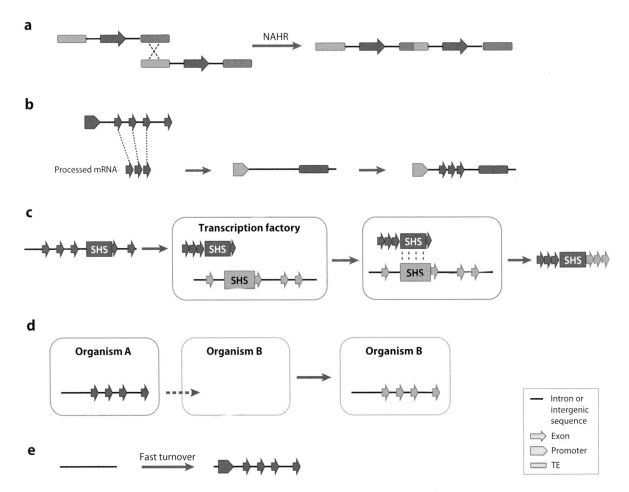

Figure 1

Molecular mechanisms for origins of new genes. (*a*) DNA-level gene duplication produced by nonallelic homologous recombination (NAHR). Orange and blue bars represent nonhomologous transposable elements (TEs). The recombination between them produces a tandem duplication of the focal gene (*red arrows*). (*b*) Retroposition followed by exon shuffling. The reverse-transcription products of the parental genes (*red arrows*) are inserted into a new locus and then recruit nearby sequences as promoters (*orange arrows*). Exon shuffling can happen via the evolution of chimeric gene structure formed by events such as the recruitment of TE sequences into the coding region (*red bar*). (*c*) RNA-level exon shuffling. Two DNA templates share the same transcription factory (TF) for processing mRNAs. Occasionally, the TF containing the processed RNA of one gene (*red arrow*) switches to another gene via the pairing of short homologous sequences (SHSs) without finishing the transcription of the first gene. The following transcription process on another gene creates a chimeric RNA. (*d*) Horizontal gene transfer. Genes of organism A (*red arrows*) may occasionally be transferred to organism B. (*e*) De novo origination. Some noncoding sequences undergo fast turnover and become a functional gene.

2008). General principles regarding the evolution of novel retrogenes, independent of the species being surveyed, have begun to emerge. First, the distinctive testis-biased expression patterns of novel retrogenes do not appear to be restricted to limited cases. For example, a search for species (group)-specific retrogenes using 12 *Drosophila* genomes found that 53% of retrogenes were expressed in the testis and that the retrogenes also tended to be expressed in fewer tissues than were their parental copies (Bai et al. 2007). Similarly, 50 out of 120 identified human retrogenes have evolved testis-related functions (Vinckenbosch et al. 2006). In particular, for both fruit flies and eutherian mammals, the testis-biased expression pattern for the retrogenes of X chromosome origin is more striking than that of autosomal origin (Betran et al. 2002, Emerson et al. 2004, Potrzebowski et al. 2008). In addition to their regulatory and expression innovations, retrogenes frequently recruit new exons from transposable elements or the sequences of another gene at the inserted site to form a chimeric structure. Investigators estimated that 42% (380 out of 898) of the rice retrogenes (Wang et al. 2006) and 31% (36 out of 117) of the human retrogenes (Vinckenbosch et al. 2006) have formed such a structure, suggesting the innovation of a protein function. These results together indicate that, because retrogenes will insert into a new genomic locus that has a distinctive chromatin status and adjacent sequence feature, the new genomic environment may endow the retrogenes with a new promoter, exon sequences, or chromatin-expression context to evolve novel functions. Interestingly, a recent scan of imprinted genes in the mouse genome has characterized four retrogenes derived from the X chromosome. They are all embedded within another host gene and evolved new methylation patterns during oogenesis (Wood et al. 2007). Although the molecular details of the evolution of imprinting on these retrogenes remain to be disclosed, this case represents a good example of how retrogenes acquire new epigenetic features following their insertion into different genomic loci (Wood et al. 2007).

2.3. Exon Shuffling and *Trans*-Splicing

The idea that exon shuffling would have great impact on protein evolution and new gene formation can be traced back to when the intron was discovered (Gilbert 1978). Intron-mediated recombination was hypothesized to reshuffle exons of different genes and thus create new genes (Patthy 1999). Broadly speaking, the shuffling process could also occur through DNA-level duplication and subsequent recruitment via processes other than intron recombination, such as formation of the chimeric structure after retroposition (**Figure 1b**). At the RNA level, chimeric RNAs of different gene sources can also form through *trans*-splicing or transcription slippage (**Figure 1c**) (Li et al. 2009).

The best new gene example of DNA-level exon shuffling may be the case of *jingwei* (Long & Langley 1993), which was the first characterized young gene. This gene was so labeled in reference to a princess in a Chinese myth who is reincarnated into a beautiful bird after death, so as to indicate the rebirth and reformation of the retrogene (Long & Langley 1993). This gene is found only in the *Drosophila yakuba* lineage and is only several million years old. Its emergence started from a retroposition inserted into another duplicated gene *yande*. The retrocopy then recruited three exons of *yande* and formed the new gene *jingwei*. It also evolved novel functions participating in hormone/pheromone metabolism different from its parental gene *Adh* (Zhang et al. 2004). DNA-level exon shuffling can also occur by recruiting intergenic sequences into new exons after gene duplication, such as the case of *Hun* (Arguello et al. 2006), or by recruiting both exon and intron sequences to form novel spliced RNA genes, as in the case of *sphinx* (Wang et al. 2002). The domestication of transposable elements can also result in DNA-level exon shuffling, as in the case of *SETMAR* (Cordaux et al. 2006). A systematic examination of *Drosophila* new genes confirmed a high frequency (30%) of chimeric genes among new genes and showed that various genomic sources can provide materials for exon shuffling (Zhou et al. 2008).

Chimeric genes can also form at the RNA level by *trans*-splicing, which fuses partial pre-RNAs of two distant genes during RNA processing (Gingeras 2009). Early examples of chimeric RNA and their protein products were detected in plants (Chapdelaine & Bonen 1991), insects (Horiuchi et al. 2003, Robertson et al. 2007), and normal (Li et al. 2008) and cancerous (Eychène et al. 2008) human cells, some of which were characterized with differential expression patterns (Li et al. 2008) or conserved across distant species (Gabler et al. 2005). Later genome-wide searches in model organisms using expressed sequence tags or next-generation sequencing illustrate that *trans*-splicing may be a universal phenomenon increasing transcriptome diversities (Li et al. 2009, McManus et al. 2010, Zhang et al. 2010a), although many of these chimeric RNAs may be derived from artificial chimeric cDNAs (McManus et al. 2010).

In addition, a novel mechanism producing chimeric RNAs starts to emerge (**Figure 1***c*). For example, a recent examination of junction sequences between two source genes within a large pool of chimeric RNAs did not produce canonical GU-AG splicing sites, inconsistent with the splicing model producing the chimeras. Instead, short homologous sequences appear frequently at the junctions and seem to mediate the production of the chimeric RNAs: Mutations in these sequences could abolish the corresponding chimeric RNAs in yeast (Li et al. 2009). Li et al. (2009) thus proposed a transcription slippage to account for the generation of these chimeric RNAs. Nevertheless, the functions of these chimeric RNAs remain largely uncharacterized, and some of the chimeric RNAs have recently been proposed to be experimental artifacts (Houseley & Tollervey 2010). Further studies are needed to clarify this new genetic phenomenon.

2.4. Horizontal Gene Transfer

Horizontal gene transfer (HGT) describes the nonsexual movement of genetic information between different species or between organelles and nuclei (**Figure 1***d*) (Keeling & Palmer 2008). It occurs frequently among prokaryotes (Beiko et al. 2005) and microbial eukaryotes (Gladyshev et al. 2008) and between host and parasitic plants (Yoshida et al. 2010) but rarely in other eukaryotic lineages, probably because the separation of germ and soma sets up a barrier to HGT in most eukaryotes. One of the most striking examples of bacteria-to-animal HGT may be the case of *Wolbachia*, which is an intracellular endosymbiont parasitic to a wide range of arthropods and filarial nematodes (Stouthamer et al. 1999). Its genome sequences have been identified in a wide variety of its host species, and in an extreme case, the entire *Wolbachia* genome (~1.4 Mb) was integrated into the *Drosophila ananassae* nuclear genome with at least 28 transferred genes transcribed (Hotopp et al. 2007). However, no solid case has yet shown the functionality of HGT genes in a multicellular eukaryote. Thus, its contribution to the origin of novel functional genes in multicellular eukaryotes remains to be confirmed.

2.5. De Novo Origination

Initially, the probability that noncoding sequences undergo fast turnover and become a functional new gene (**Figure 1***e*) was thought to be very low (Ohno 1970, Jacob 1977). However, a growing resource of genome sequences has enabled researchers to search for such de novo–originated genes and for their origination rate. The results indicate that de novo origination played an important role during new gene evolution, and case studies of de novo genes further confirmed their functional importance, despite their young ages. Below, we describe and discuss more about this long-underappreciated mechanism of gene origination.

3. GENOME-WIDE PATTERNS OF THE ORIGIN OF NEW GENES

3.1. The Rate of New Gene Origination

Because the majority of new genes are derived from gene duplication or retroposition, our current knowledge about the genome-wide tempo of new gene origination comes mainly from studies of gene duplication rates. As a main cause of spontaneous mutations, the rate of gene duplication

Table 1 Gene duplication rates of different species[a]

Species	Inference from recent duplicates	Inference from mutation accumulation experiments
Saccharomyces cerevisiae	$1 \times 10^{-5} - 8 \times 10^{-3}$	5.9×10^3
Caenorhabditis elegans	0.0208	10
Drosophila melanogaster	0.001–0.0023	NA
Homo sapiens	0.00369	NA

[a]Duplication rates are counted as duplicates per gene per million years. The conversion between units of different studies was made assuming 4.8 cell cycles per day for *S. cerevisiae*, and 4 days as the generation time for *C. elegans*. Data were collected or converted from Lipinski et al. (2011), Lynch & Conery (2000), Lynch et al. (2008), and Zhou et al. (2008). Abbreviation: NA, not available.

is of great interest and essential to understanding the impact of duplication on genetic variation. There are two main methods of calculating gene duplication rates: via inference from recent duplicates in sequenced genomes (Lynch & Conery 2000) or directly estimated from mutation-accumulating lines. The current estimates in yeast and worm show the latter will be several orders of magnitude higher than the former (**Table 1**) (Lynch et al. 2008, Lipinski et al. 2011). Such a drastic difference may be derived from differences in the quantification method; more importantly, inferences from recent genome duplicates may suffer from the effects of selection and gene conversion. Nevertheless, these data show the rate of gene duplication in different species is sufficiently large to spawn new genes.

A recent study in the *Drosophila* genome (Zhou et al. 2008) focused on the rate of origination of functional new genes. Because newly nascent genes may still be in the processes of accumulating deleterious/adaptive mutations and often show copy-number polymorphism within the population, this study investigated only functional new genes, which are of greater biological significance and are already fixed within a species. Zhou et al. (2008) studied the new genes that originated in a common ancestor of three *Drosophila* species. Such new genes have recent origins (aged 5–10 million years) and can be found intact in multiple species. The origination rate estimated from these genes is approximately three to five new genes per genome per million years, which is twofold lower than the rate from species-specific new genes.

3.2. A Dynamic View of New Gene Evolution over Time

Comparisons between new genes of different ages uncovered their dynamic patterns over time. For example, characterization of *Drosophila* species-specific new genes showed tandem duplication dominates the production of very nascent copies (~82%), but not of those "older" new genes that are shared by several species (whose proportion drastically decreased to ~34%). Instead, dispersed duplication that separates the new gene copy from its parental gene accounts for 44% of the new genes. These results indicate that individual mechanisms contribute differently to new genes of different ages and that dispersed duplication is more likely to allow for the functional diversification between the new gene and its parental gene (Zhou et al. 2008).

The chromosomal distribution of new genes also changed over time, particularly for male-biased new genes. Studies in mammalian (Zhang et al. 2010c) and *Drosophila* genomes (Zhang et al. 2010b) consistently found an excess of young male-biased protein-coding/microRNA genes that originated on the X chromosome, but this pattern reversed through time, i.e., older male-biased genes that originated earlier are enriched on autosomes and are deficient on the X chromosome. This striking switch is probably a result of several antagonistic factors simultaneously acting on the X chromosome: The fast-X effect tends to fix an excess of recessive male-beneficial new genes on the X chromosome, whereas meiotic sex chromosome inactivation or sexual antagonism demasculinize the X chromosome and relocate the male-biased genes to autosomes.

4. ORIGINS OF NOVEL NONCODING RNA GENES

Noncoding RNA (ncRNA) genes compose a specific group of genes that function without being translated into proteins; they include the well-documented transfer RNA (tRNA) and ribosomal RNA (rRNA), many kinds of small nonmessenger RNA (snmRNA) genes, as well as the long noncoding RNA (lncRNA) genes (Eddy 2001). They are widespread and play critical roles in regulating various biological processes (Mattick & Makunin 2006, Mercer et al. 2009). The origin of ncRNA had not received much attention until recently.

The snmRNA molecules consist of many different functional categories: small nuclear RNA (snRNA), small nucleolar RNA (snoRNA), small interfering RNA (siRNA), microRNA (miRNA), and piwi-interacting RNA (piRNA) (Eddy 2001). Among these, the repertoires of miRNA genes are in constant flux, with high birth and death rates during evolution (Fahlgren et al. 2007, Lu et al. 2008, Nozawa et al. 2012). In plants, the transcription of self-complementary RNA structures by inverted gene duplication is a very common mechanism for the origin of miRNA (Allen et al. 2004, Rajagopalan et al. 2006, Fahlgren et al. 2007, Fahlgren et al. 2010). In contrast, de novo emergence of RNA hairpins seems to be a major route for miRNA genesis in animals (Axtell et al. 2011). A study in *Drosophila* species found that none of the newly identified miRNAs were originated by inverted duplication of preexisting miRNA (Lu et al. 2008). Other mechanisms for miRNA genesis in plants and animals include evolution from transposable elements such as miniature inverted-repeat transposable elements (Smalheiser & Torvik 2005, Piriyapongsa et al. 2007, Piriyapongsa & Jordan 2008) and duplication of preexisting miRNA genes with subsequent divergence (Zhang et al. 2007, 2008; Nozawa et al. 2010, 2012). The evolution dynamics of other snmRNA categories has been much less explored but may involve processes similar to those of miRNA genes. For example, the expansion of piRNA in rodents was reported to occur via repetitive element-mediated duplication (Assis & Kondrashov 2009), a process similar to *Alu*-mediated expansion of miRNA in primates (Zhang et al. 2008). More studies of these snmRNAs will provide further insights into their origination patterns and mechanisms.

For lncRNA genes, our current knowledge of their origination processes is based mostly on case studies. The first characterized young lncRNA gene *sphinx* was identified specifically in *D. melanogaster* (Wang et al. 2002). This gene originated through the insertion of a retroposed copy of the ATP synthase chain F gene with subsequent recruitments of nearby exons and introns to form a chimeric gene structure. The newly formed *sphinx* gene experienced rapid sequence turnover and evolved sex-specific alternative splicing. It does not encode proteins, and, surprisingly, abolishment of splicing sites at its recruited intron in the male-specific transcript leads to male-male courtship behavior. This indicates a novel role for *sphinx* in enhancing the heterosexual courtship of *D. melanogaster* (Dai et al. 2008). A similar mechanism was also found for the well-known lncRNA gene *Xist*, which is the key initiator of X chromosome inactivation in eutherian mammals (Duret et al. 2006). Interestingly, in addition to protein-coding sequences, lncRNA genes can also evolve from previously unexpressed noncoding sequences. An example is the *Poldi* gene in house mice, which arose within the past 2.5–3.5 million years from a stretch of intergenic sequences with preexisting cryptic signals for transcript regulation and processing (Heinen et al. 2009). These cases indicate that lncRNA genes can originate from both protein-coding and "junk" DNA sequences.

The ongoing intensive investigation on lncRNAs may discover more novel lncRNA genes and the genome-wide pattern of their origination.

5. THE IMPORTANT ROLE OF DE NOVO GENE ORIGINATION

Complete de novo origination of a protein-coding gene from a noncoding sequence was thought to be nearly impossible: "Each new gene must have arisen from an already existing gene" (Ohno 1970, p. 72). "The probability that a functional protein would appear de novo by random association of amino acids is practically zero" (Jacob 1977, p. 1164). Nevertheless, the idea of de novo origination is particularly fascinating, as it is intrinsically related to the fundamental question of how the protein repertoire evolved to include such enormous diversity (Bornberg-Bauer et al. 2010). Mechanistically, the de novo route is assumed capable of generating proteins that are very different from those encoded by the existing genome. Therefore, it may have a special role in providing genetic materials for drastic and radical functional renovations. Recent characterizations of de novo genes, including both genome-wide analyses and case studies, have evoked an increasing appreciation of the importance of the de novo mechanism in various organisms (Tautz & Domazet-Loso 2011).

The early clues regarding de novo genes were derived from yeast with the completion of the first eukaryotic genome sequence, of which approximately one-third of the identified genes lacked homologues in other lineages (Dujon 1996). However, it was not until the recent availability of genomes of closely related species that the systematic identification of de novo genes has become a possibility, as the use of recently diverged species allows discrimination between de novo origination and rapid gene evolution as well as retrospection of the noncoding history of de novo genes. The first straightforward search for de novo genes at the genome level was performed by Levine et al. (2006) in *Drosophila*. Using the newly available *Drosophila* genomes, they identified five putative *D. melanogaster* and/or *D. simulans*–specific de novo genes that resemble no/poor BLAST hits to the genomes of *Drosophila yakuba*, *Drosophila erecta*, and *Drosophila ananassae*. All five candidates are expressed predominantly in testes. This finding stimulated the subsequent identification of several male-related de novo genes that emerged in *D. yakuba* or in the ancestor of *D. yakuba* and *D. erecta*, using the data of either accessory gland transcriptome (Begun et al. 2006) or testis-derived expressed sequence tags (Begun et al. 2007). For the genome-wide pattern, it was estimated that 5% of the primate orphan genes (Toll-Riera et al. 2009) and 12% of the *Drosophila* new genes originated from noncoding sequences (Zhou et al. 2008), thus casting doubt on the traditional view that de novo origination is extremely rare and indicating a considerable role for de novo mechanisms in creating novel genes. All the above-mentioned studies have adopted a similar criterion in defining de novo genes, i.e., they should not have recognizable homologues in the genomes of their closely related species. Other studies have employed a different criterion requiring the presence of non-protein-coding homologous sequences in the syntenic regions of the out-group species. This strategy led to the identification of 3 putative protein-coding de novo genes in human and 13 in the malaria parasite *Plasmodium vivax* (Yang & Huang 2011). Using both expressional and proteomic evidence, a recent study discovered 60 human-specific de novo genes (Wu et al. 2011).

Given these results together with other case studies, some common features of de novo genes are emerging. For example, these genes are relatively simple in intron/exon structure and tend to encode short and poorly structured proteins (Begun et al. 2006, Levine et al. 2006, Begun et al. 2007, Knowles & McLysaght 2009). Such features are consistent with speculation that de novo genes are nascent genes that evolved from random noncoding sequences and also raise the possibility of an underestimation of de novo origination owing to the potentially incomplete annotation of genes encoding short peptides; this is supported by the detection of translation signatures of hundreds of short species (group)-specific open reading frames located in nongenic sequences of the *S. cerevisiae* genome (Carvunis et al. 2012). Nevertheless, there are exceptions: Six of the 13 de novo genes identified in *Plasmodium vivax* possess an intron in the gene region (Yang & Huang 2011), and the yeast de novo gene *MDF1* encodes a three-helix-bearing protein (Li et al. 2010b).

Analysis of the yeast de novo gene *BSC4* by Cai et al. (2008) showed that its origination may be a two-step process. The orthologous, but non-protein-coding, loci of the *BSC4* gene in the out-group species are expressed; thus, an ncRNA state might exist between noncoding DNA and protein-coding genes. Similar processes may also occur within the human-specific de novo gene *FLJ33706*, an idea implied by the low expression of its orthologous locus in *Rhesus macaque* (Li et al. 2010a). Considering the abundance of lncRNAs in mammals and other eukaryotes (Mercer et al. 2009, Ponting et al. 2009), the above observations are particularly illuminative in suggesting the potential of lncRNAs to serve as a rich resource for de novo genes. However, de novo genes may also evolve directly from noncoding DNA, as supported by the example of *MDF1*, whose orthologous loci are not expressed in the out-group species (Li et al. 2010b). In its orthologous sequences, the number of "disabling" nucleotides that can abolish the coding of a fairly long open reading frame seems to decrease gradually toward the branch of *S. cerevisiae MDF1*, raising the possibility that a stepwise process could also exist at the protein-coding level.

As more newly evolved de novo genes are identified across various organisms, few receive comprehensive functional characterization. For the young de novo genes, only the case of *MDF1* (Li et al. 2010b) was narrowed down to the molecular pathway and its functional consequences. *MDF1* is a young de novo gene restricted only to the *S. cerevisiae* lineage. However, it regulates the canonical mating pathway and thereby enables the yeast to make the favorable choice between sexual and asexual reproduction under different nutritional environments (see details below). A series of other recent studies has provided clues regarding the functionality of de novo genes in various organisms. These findings include the discovery of the synthetic lethal gene *BSC4* in yeast, the bacterial infection–associated gene *OsDR1* in rice (Xiao et al. 2009), the spermatogenesis-related gene *hydra* (Chen et al. 2007), three pupae-lethal genes in fruit fly (Chen et al. 2010), and a human-specific gene *FLJ33706* that shows elevated expression in the brain samples of Alzheimer's patients (Li et al. 2010a). Despite the fact that these case studies show the involvements of de novo genes in multiple biological processes, they likely have a much more prominent role in male reproduction, as suggested by its strong testis-biased expression pattern (Levine et al. 2006, Zhou et al. 2008, Wu et al. 2011). Except for testes, de novo genes also show high expression in the human cerebral cortex (Wu et al. 2011), indicating that they may also be functionally important for the evolution of human-unique phenotypic traits such as our dramatically increased cognitive ability.

6. HOW NEW GENES EVOLVE NOVEL FUNCTIONS

Previous attempts to address the functional origin of new genes mainly focused on discussions of the functional fate of new genes, leading to the development of a theoretical framework consisting of various empirically supported models (for reviews, see Conant & Wolfe 2008, Innan & Kondrashov 2010), such as the neofunctionalization (Ohno 1970, Walsh 1995) and subfunctionalization (Force et al. 1999, Lynch & Force 2000) models. However, other fundamental issues remained: What are the functional roles of new genes, and how do they integrate into the preexisting pathways of the host organisms to execute functions? Studies of the functions of young genes are emerging to reveal a critical role of new genes in fundamental developmental processes, and several recent cases have started to shed light on how new genes may have acquired novel functions through pathway integration and their contributions to pathway evolution.

6.1. Essentiality of Newly Evolved Genes

Given the functional characterizations of newly originated genes, there does not seem to be a necessary association of new genes with species-specific traits. Research now indicates that new genes may play critical roles in essential biological processes that are responsible for reproduction and survival of the organisms. For example, the *Drosophila* new gene *K81* is a paternal-effect gene required for the first round of zygotic division (Loppin et al. 2005), *mojoless* is essential for the survival of male germ line (Kalamegham et al. 2007), and a very young duplicate gene in *D. melanogaster*, *nsr*, is essential for sperm maturation. For the genome-wide-level evidence, Chen et al. (2010) identified 566 young *D. melanogaster* genes that originated within 3–35 million years and systematically tested their phenotypic effects by RNA interference. Surprisingly, they found that 30% of these genes are essential for viability, which is comparable to that estimated for all genes in *D. melanogaster* ($\sim$25–35%). The functional essentiality of new genes revealed by these studies has challenged the traditional view that critical biological functions are always encoded by evolutionarily conserved genes. More such studies, especially in other organisms besides fruit fly, are needed to clarify the generality of this observation.

6.2. Pathway Integration of Novel Genes

Genes do not function alone. Upon entering the genome, a new gene has to establish interplay with preexisting gene pathways/networks. Even though a number of new genes have been demonstrated to be functionally important, an in-depth understanding regarding the functional origination of new genes at the pathway level has been lacking. Recently, through rigorous efforts, investigators have made available several cases showing how new genes acquired functions through pathway integration (Matsuno et al. 2009, Li et al. 2010b, Ding et al. 2010, Chen et al. 2012). These studies have also opened up the possibility that new genes are significant drivers of invention, adaptive evolution, and/or turnover of genetic pathways.

Gene duplication can provide enzymes with novel catalytic and recognition properties; this role in the evolution of biochemical pathways has been emphasized since the 1960s (Bryson & Vogel 1965, Jensen 1976). Matsuno et al. (2009) reported the first explicit example of how newly originated genes led to the invention of a novel metabolic pathway. They identified *CYP98A8* and *CYP98A9* as a pair of novel *Brassicaceae* P450 genes that arose via retroposition and duplication, and experienced neofunctionalization as a result of selective and local amino acid replacement. The two new genes then recruited novel catalytic substrates, leading to the synthesis of phenolamide and the formation of a novel phenolic pathway.

Li et al. (2010b) provided a detailed example of how a newly originated de novo gene integrated into the upstream position of the canonical mating pathway and facilitated the organism's adaptation (**Figure 2a**). The *MDF1* gene evolved from the noncoding antisense sequences of the *ADF1* locus in *S. cerevisiae* and encodes an open reading frame of 152 amino acids. It promotes vegetative growth and decreases mating efficiency in rich mediums. In yeast, the mating pathway is controlled by the *MAT* loci that encode the master regulators of cell types: the *MATa1* locus, present in cells and diploids, and the *MATαl/α2* locus, present in α cells and diploids. The mating behavior is triggered when α-specific genes are turned on by the MATα1 protein and α-specific genes are turned

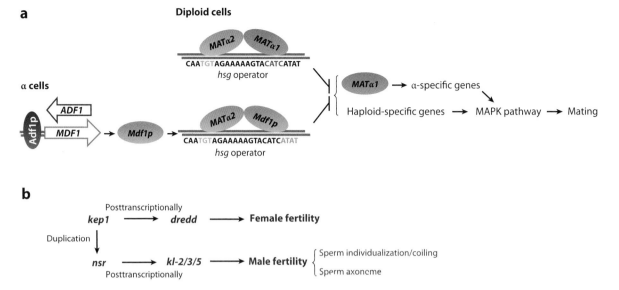

Figure 2

Examples of pathway integration of novel genes. (*a*) The molecular pathway of the de novo new gene *MDF1* in baker's yeast (Li et al. 2010b). In the diploid cell *(upper panel)*, *MATα1* interacts with *MATα2* and then suppresses the downstream α cell-specific and haploid-specific genes. But in the early growing α cell, *Mdf1p* can instead interact with *MATα2* and suppress the similar downstream pathways *(lower panel)*, making the haploid α cell look like a diploid cell and thus avoid mating. (*b*) Pathway integration of the young duplicate gene *nsr* in *Drosophila* (Ding et al. 2010). The parental gene *kep1* takes part in the splicing of *dredd* in females, while its young daughter gene *nsr* perhaps joins in the processing of transcripts of three male-specific genes (*kl-2/3/5*) in testis.

off by the MATα2 protein in α cells. At the early proliferation stage with sufficient nutritional supplement, the MDF1 protein, Mdf1p, binds the MATα2 protein and cooperatively targets the promoters of many haploid-specific genes (*hsg*) and the opener of α-specific genes, *MATα1*. It also downregulates their gene expression levels in α cells. Thus, the MAPK pathway responsible for switching on the physiological changes in preparation for mating is silenced, pushing the yeast toward mitotic cell growth. The mystery of Mdf1p's binding to MATα2 may lie in its structural similarity to MATa1, which binds MATα2 and suppresses mating in diploid cells. However, as a de novo originated protein, Mdf1p is not identical to MATa1: The position of its precise binding sites on the *hsg* operator is four nucleotides away from that of MATa1. Intriguingly, the inhibitory effect of *MDF1* on sexual reproduction is further regulated by its antisense counterpart *ADF1*. The ADF1 protein, Adf1p, can bind to the promoter of *MDF1* and act as a negative regulator. It lets yeast escape from the mitosis cell cycle under unfavorable conditions, such as nutritional limit, to enjoy the benefit of sexual reproduction. Corroborating this, Mdf1p is unable to promote growth in a nonfermentative medium. Therefore, the de novo emergence of the *MDF1* gene has added a new layer of regulatory control to the canonical mating pathway, enabling baker's yeast to adapt well to changing nutritional conditions. The example of *MDF1* shows that newly created genes can regulate a preexisting core pathway in an amazingly delicate manner and has elevated our understanding of the functional contribution of novel genes to organism adaptation.

Two case studies in *Drosophila* show that recently originated genes can rapidly evolve functions and gene networks different from those of their ancestral copy (Ding et al. 2010, Chen et al. 2012). The *D. melanogaster* new gene *nsr* emerged within the past 5.4–12.8 million years through duplication of the ancestral locus *kep1*. Both *nsr* and *kep1* encode RNA-binding proteins and likely act as splicing factors. The new gene *nsr* is essential for male fertility as it is required for the normal sperm individualization/coiling processes and the structural integrity of the sperm axoneme by regulating the Y chromosome fertility genes *kl-2*, *kl-3*, and *kl-5*, whereas the ancestral gene *kep1* is required for female fertility by regulating the apoptosis molecule *dredd* (**Figure 2b**) (Ding et al. 2010). *Zeus* is another young male fertility gene, and it originated through retroposition of a highly conserved housing keeping gene, *Caf40*. *Zeus* has a diverged genomic binding profile from *Caf40* and has recruited a new set of downstream genes, which might shape the evolution of gene expression in its germ line (Chen et al. 2012). These cases indicate that the newly derived gene could be quickly incorporated into a pathway that is not analogous to that of its parental copy. It is interesting to note that the sperm individualization process and the structure of the sperm axoneme have been evolutionarily conserved from fly to human. Thus, the case of *nsr* also indicates that the evolution of molecular pathways for essential and conserved biological aspects may be an ongoing process, and new genes may play a crucial role in the adaptive evolution of these molecular pathways. This understanding may help us to explain the "nonorthologous gene displacement" phenomena observed among different organisms (Koonin et al. 1996, Chothia et al. 2003) that a lot of apparently indispensable biological functions are encoded by nonorthologous, and sometimes unrelated, genes. Via this new perspective, the origin of new genes provides the raw material for evolution to play rock and roll.

FUTURE ISSUES

1. The current genome-wide pattern of the origin of new genes was elucidated in *Drosophila* using new genes aged 5–10 million years (Zhou et al. 2008). Does this pattern apply to other species or a different timescale? Beyond the origin of new genes, identifying novel genes in different taxa is also important to understand the evolution of specific traits in a particular taxon. With the rapid advancement of genome sequencing, more reference genomes within phylogenetic contexts are becoming available, and the study of novel genes may be extended to other taxa, such as ants in which the genomes of seven species were recently reported (Gadau et al. 2012). According to the BGI G10K Plant and Animal Reference Genomes Project, many reference genome sequences from a phylogenetic taxon will be deciphered in the near future (**http://ldl.genomics.cn/page/pa-research.jsp**). The Oryza Map Alignment Project will provide a series of reference genomes for rice species with AA genomes (**http://www.omap.org/resources.html**). When these projects are completed, the study of novel genes will proceed with unprecedented breadth and depth.

2. A fast-growing data set is starting to piece together the picture revealing that different genomic components, including noncoding DNA, ncRNA loci/genes, and protein-coding genes, have the potential to "transform" from one to the other. For instance, ncRNA genes can evolve from both noncoding DNA and protein-coding genes. To what extent and scale such events happened at the genome level awaits further elucidation. How our understanding progresses will depend largely on the development and application of advanced proteomics technology and the accumulation of transcriptome and proteomics data regarding closely related species.

3. Though great progress has been made in characterizing the novel functions of new genes, little is known about how they acquired these functions through pathway/network integration. Thus, rigorous investigation could focus on the functions of these new genes at the pathway level, and several recent cases could provide paradigm examples for such explorations. One unanticipated, but increasingly appreciated, finding is that a lot of new genes play key roles in very basic developmental processes, implying a significant contribution of new genes to the evolutionary turnover of fundamental pathways. Further efforts are needed to help researchers explain how closely related species without a certain new gene manage to fulfill the related developmental process and wherein lies the adaptive significance of the pathway turnover driven by the new genes.

4. In recent years, synthetic biology has opened a promising direction in biotechnology that may revolutionize many manufacturing sections with the potential of contributing significantly to human society in an environmentally friendly manner. The study of novel genes and the evolution of their functions can shed insight into the two main aspects of synthetic biology, i.e., constructing artificial biological systems and synthesizing genes or, perhaps, complete genomes (Nielsen & Keasling 2011). Analogous to exon shuffling in nature, artificially shuffling domains of different proteins has successfully rewired pathways, thereby changing the morphology and behavior of cells (Yeh et al. 2007, Peisajovich et al. 2010). After revealing many different kinds of exon/domain-shuffling of novel genes, we may get some ideas of how to design domain recombinants. Alien metabolic pathways have also been introduced into chassis host cells to produce drug precursors (Ro et al. 2006, Westfall et al. 2012). Understanding how novel gene products integrate into or potentially recruit pathways/networks in nature will provide clues valuable for the engineering of manufacture pathways. Nowadays large genomes can be chemically synthesized (Gibson et al. 2010); we believe comprehensive knowledge regarding the origins of novel genes and the evolution of their functions will help researchers design artificial genes, including ones that do not exist in nature, to fulfill human demands.

DISCLOSURE STATEMENT

The authors are not aware of any affiliations, memberships, funding, or financial holding that might be perceived as affecting the objectivity of this review.

ACKNOWLEDGMENTS

Y.D. and Q.Z. contributed equally to this review. All the authors thank Drs. Huifeng Jiang, Guojie Zhang, and Yang Dong for providing useful information. This work was supported by the National Natural Science Foundation of China (30990242, 30930056) and the Chinese 973 program (2007CB815700) to W.W.

LITERATURE CITED

Allen E, Xie Z, Gustafson AM, Sung GH, Spatafora JW, Carrington JC. 2004. Evolution of microRNA genes by inverted duplication of target gene sequences in *Arabidopsis thaliana*. *Nat. Genet.* 36:1282–90

Arguello JR, Chen Y, Yang S, Wang W, Long M. 2006. Origination of an X-linked testes chimeric gene by illegitimate recombination in *Drosophila*. *PLoS Genet.* 2:e77

Assis R, Kondrashov AS. 2009. Rapid repetitive element-mediated expansion of piRNA clusters in mammalian evolution. *Proc. Natl. Acad. Sci. USA* 106:7079–82

Axtell MJ, Westholm JO, Lai EC. 2011. Vive la différence: biogenesis and evolution of microRNAs in plants and animals. *Genome Biol.* 12:221

Bai Y, Casola C, Feschotte C, Betran E. 2007. Comparative genomics reveals a constant rate of origination and convergent acquisition of functional retrogenes in *Drosophila*. *Genome Biol.* 8:R11

Bailey JA, Liu G, Eichler EE. 2003. An *Alu* transposition model for the origin and expansion of human segmental duplications. *Am. J. Hum. Genet.* 73:823–34

Begun DJ, Lindfors HA, Kern AD, Jones CD. 2007. Evidence for de novo evolution of testis-expressed genes in the *Drosophila yakuba/Drosophila erecta* clade. *Genetics* 176:1131–37

Begun DJ, Lindfors HA, Thompson ME, Holloway AK. 2006. Recently evolved genes identified from *Drosophila yakuba* and *D. erecta* accessory gland expressed sequence tags. *Genetics* 172:1675–81

Beiko RG, Harlow TJ, Ragan MA. 2005. Highways of gene sharing in prokaryotes. *Proc. Natl. Acad. Sci. USA* 102:14332–37

Betran E, Long M. 2003. *Dntf-2r*, a young *Drosophila* retroposed gene with specific male expression under positive Darwinian selection. *Genetics* 164:977–88

Betran E, Thornton K, Long M. 2002. Retroposed new genes out of the X in *Drosophila*. *Genome Res.* 12:1854–59

Bornberg-Bauer E, Huylmans AK, Sikosek T. 2010. How do new proteins arise? *Curr. Opin. Struct. Biol.* 20:390–96

Brosius J. 1991. Retroposons—seeds of evolution. *Science* 251:753

Bryson V, Vogel HJ. 1965. Evolving genes and proteins. *Science* 147:68–71

Cai J, Zhao R, Jiang H, Wang W. 2008. De novo origination of a new protein-coding gene in *Saccharomyces cerevisiae*. *Genetics* 179:487–96

Cardoso-Moreira M, Emerson JJ, Clark AG, Long M. 2011. *Drosophila* duplication hotspots are associated with late-replicating regions of the genome. *PLoS Genet.* 7:e1002340

Carvunis A, Rolland T, Wapinski I, Calderwood MA, Yildirim MA, et al. 2012. Proto-genes and de novo gene birth. *Nature* 487:370–74

Chapdelaine Y, Bonen L. 1991. The wheat mitochondrial gene for subunit I of the NADH dehydrogenase complex: a *trans*-splicing model for this gene-in-pieces. *Cell* 65:465–72

Chen S, Ni X, Krinsky BH, Zhang YE, Vibranovski MD, et al. 2012. Reshaping of global gene expression networks and sex-biased gene expression by integration of a young gene. *EMBO J.* 31:2798–809

Chen S, Zhang YE, Long M. 2010. New genes in *Drosophila* quickly become essential. *Science* 330:1682–85

Chen ST, Cheng HC, Barbash DA, Yang HP. 2007. Evolution of hydra, a recently evolved testis-expressed gene with nine alternative first exons in *Drosophila melanogaster*. *PLoS Genet.* 3:e107

Chothia C, Gough J, Vogel C, Teichmann SA. 2003. Evolution of the protein repertoire. *Science* 300:1701–3

Clark AG. 1994. Invasion and maintenance of a gene duplication. *Proc. Natl. Acad. Sci. USA* 91:2950–54

Conant GC, Wolfe KH. 2008. Turning a hobby into a job: how duplicated genes find new functions. *Nat. Rev. Genet.* 9:938–50

Cordaux R, Udit S, Batzer MA, Feschotte C. 2006. Birth of a chimeric primate gene by capture of the transposase gene from a mobile element. *Proc. Natl. Acad. Sci. USA* 103:8101–6

Dai H, Chen Y, Chen S, Mao Q, Kennedy D, et al. 2008. The evolution of courtship behaviors through the origination of a new gene in *Drosophila*. *Proc. Natl. Acad. Sci. USA* 105:7478–83

Ding Y, Zhao L, Yang S, Jiang Y, Chen Y, et al. 2010. A young *Drosophila* duplicate gene plays essential roles in spermatogenesis by regulating several Y-linked male fertility genes. *PLoS Genet.* 6:e1001255

Dujon B. 1996. The yeast genome project: What did we learn? *Trends Genet.* 12:263–70

Duret L, Chureau C, Samain S, Weissenbach J, Avner P. 2006. The *Xist* RNA gene evolved in eutherians by pseudogenization of a protein-coding gene. *Science* 312:1653–55

Eddy SR. 2001. Non-coding RNA genes and the modern RNA world. *Nat. Rev. Genet.* 2:919–29

Emerson JJ, Kaessmann H, Betran E, Long M. 2004. Extensive gene traffic on the mammalian X chromosome. *Science* 303:537–40

Eychène A, Rocques N, Pouponnot C. 2008. A new *MAF*ia in cancer. *Nat. Rev. Cancer* 8:683–93

Fahlgren N, Howell MD, Kasschau KD, Chapman EJ, Sullivan CM, et al. 2007. High-throughput sequencing of *Arabidopsis* microRNAs: evidence for frequent birth and death of *MIRNA* genes. *PLoS One* 2:e219

Fahlgren N, Jogdeo S, Kasschau KD, Sullivan CM, Chapman EJ, et al. 2010. MicroRNA gene evolution in *Arabidopsis lyrata* and *Arabidopsis thaliana*. *Plant Cell* 22:1074–89

Force A, Lynch M, Pickett FB, Amores A, Yan YL, Postlethwait J. 1999. Preservation of duplicate genes by complementary, degenerative mutations. *Genetics* 151:1531–45

Gabler M, Volkmar M, Weinlich S, Herbst A, Dobberthien P, et al. 2005. *Trans*-splicing of the *mod(mdg4)* complex locus is conserved between the distantly related species *Drosophila melanogaster* and *D. virilis*. *Genetics* 169:723–36

Gadau J, Helmkampf M, Nygaard S, Roux J, Simola DF, et al. 2012. The genomic impact of 100 million years of social evolution in seven ant species. *Trends Genet.* 28:14–21

Gibson DG, Glass JI, Lartigue C, Noskov VN, Chuang RY, et al. 2010. Creation of a bacterial cell controlled by a chemically synthesized genome. *Science* 329:52–56

Gilbert W. 1978. Why genes in pieces? *Nature* 271:501

Gingeras TR. 2009. Implications of chimaeric non-co-linear transcripts. *Nature* 461:206–11

Gladyshev EA, Meselson M, Arkhipova IR. 2008. Massive horizontal gene transfer in bdelloid rotifers. *Science* 320:1210–13

Hakes L, Pinney JW, Lovell SC, Oliver SG, Robertson DL. 2007. All duplicates are not equal: the difference between small-scale and genome duplication. *Genome Biol.* 8:R209

Haldane J. 1935. *The Causes of Evolution*. London: Longmans & Green

Hastings PJ, Lupski JR, Rosenberg SM, Ira G. 2009. Mechanisms of change in gene copy number. *Nat. Rev. Genet.* 10:551–64

Heinen TJ, Staubach F, Haming D, Tautz D. 2009. Emergence of a new gene from an intergenic region. *Curr. Biol.* 19:1527–31

Horiuchi T, Giniger E, Aigaki T. 2003. Alternative *trans*-splicing of constant and variable exons of a *Drosophila* axon guidance gene, *lola*. *Genes Dev.* 17:2496–501

Hotopp JCD, Clark ME, Oliveira DCSG, Foster JM, Fischer P, et al. 2007. Widespread lateral gene transfer from intracellular bacteria to multicellular eukaryotes. *Science* 317:1753–56

Houseley J, Tollervey D. 2010. Apparent non-canonical trans-splicing is generated by reverse transcriptase in vitro. *PLoS One* 5:e12271

Innan H, Kondrashov F. 2010. The evolution of gene duplications: classifying and distinguishing between models. *Nat. Rev. Genet.* 11:97–108

Jacob F. 1977. Evolution and tinkering. *Science* 196:1161–66

Jeffs P, Ashburner M. 1991. Processed pseudogenes in *Drosophila*. *Proc. Biol. Sci.* 244:151–59

Jensen RA. 1976. Enzyme recruitment in evolution of new function. *Annu. Rev. Microbiol.* 30:409–25

Kaessmann H. 2010. Origins, evolution, and phenotypic impact of new genes. *Genome Res.* 20:1313–26

Kaessmann H, Vinckenbosch N, Long M. 2009. RNA-based gene duplication: mechanistic and evolutionary insights. *Nat. Rev. Genet.* 10:19–31

Kalamegham R, Sturgill D, Siegfried E, Oliver B. 2007. *Drosophila mojoless*, a retroposed GSK-3, has functionally diverged to acquire an essential role in male fertility. *Mol. Biol. Evol.* 24:732–42

Keeling PJ, Palmer JD. 2008. Horizontal gene transfer in eukaryotic evolution. *Nat. Rev. Genet.* 9:605–18

Knowles DG, McLysaght A. 2009. Recent de novo origin of human protein-coding genes. *Genome Res.* 19:1752–59

Koonin EV, Mushegian AR, Bork P. 1996. Non-orthologous gene displacement. *Trends Genet.* 12:334–36

Koszul R, Caburet S, Dujon B, Fischer G. 2004. Eucaryotic genome evolution through the spontaneous duplication of large chromosomal segments. *EMBO J.* 23:234–43

Levine MT, Jones CD, Kern AD, Lindfors HA, Begun DJ. 2006. Novel genes derived from noncoding DNA in *Drosophila melanogaster* are frequently X-linked and exhibit testis-biased expression. *Proc. Natl. Acad. Sci. USA* 103:9935–39

Li CY, Zhang Y, Wang Z, Cao C, Zhang PW, et al. 2010a. A human-specific de novo protein-coding gene associated with human brain functions. *PLoS Comput. Biol.* 6:e1000734

Li D, Dong Y, Jiang Y, Jiang H, Cai J, Wang W. 2010b. A de novo originated gene depresses budding yeast mating pathway and is repressed by the protein encoded by its antisense strand. *Cell Res.* 20:408–20

Li H, Wang J, Mor G, Sklar J. 2008. A neoplastic gene fusion mimics *trans*-splicing of RNAs in normal human cells. *Science* 321:1357–61

Li X, Zhao L, Jiang H, Wang W. 2009. Short homologous sequences are strongly associated with the generation of chimeric RNAs in eukaryotes. *J. Mol. Evol.* 68:56–65

Linardopoulou EV, Williams EM, Fan Y, Friedman C, Young JM, Trask BJ. 2005. Human subtelomeres are hot spots of interchromosomal recombination and segmental duplication. *Nature* 437:94–100

Lipinski KJ, Farslow JC, Fitzpatrick KA, Lynch M, Katju V, Bergthorsson U. 2011. High spontaneous rate of gene duplication in *Caenorhabditis elegans*. *Curr. Biol.* 21:306–10

Long M, Betran E, Thornton K, Wang W. 2003. The origin of new genes: glimpses from the young and old. *Nat. Rev. Genet.* 4:865–75

Long M, Langley CH. 1993. Natural selection and the origin of *jingwei*, a chimeric processed functional gene in *Drosophila*. *Science* 260:91–95

Loppin B, Lepetit D, Dorus S, Couble P, Karr TL. 2005. Origin and neofunctionalization of a *Drosophila* paternal effect gene essential for zygote viability. *Curr. Biol.* 15:87–93

Lu J, Shen Y, Wu Q, Kumar S, He B, et al. 2008. The birth and death of microRNA genes in *Drosophila*. *Nat. Genet.* 40:351–55

Lynch M, Conery JS. 2000. The evolutionary fate and consequences of duplicate genes. *Science* 290:1151–55

Lynch M, Force A. 2000. The probability of duplicate gene preservation by subfunctionalization. *Genetics* 154:459–73

Lynch M, Sung W, Morris K, Coffey N, Landry CR, et al. 2008. A genome-wide view of the spectrum of spontaneous mutations in yeast. *Proc. Natl. Acad. Sci. USA* 105:9272–77

Maere S, De Bodt S, Raes J, Casneuf T, Van Montagu M, et al. 2005. Modeling gene and genome duplications in eukaryotes. *Proc. Natl. Acad. Sci. USA* 102:5454–59

Marques AC, Dupanloup I, Vinckenbosch N, Reymond A, Kaessmann H. 2005. Emergence of young human genes after a burst of retroposition in primates. *PLoS Biol.* 3:e357

Matsuno M, Compagnon V, Schoch GA, Schmitt M, Debayle D, et al. 2009. Evolution of a novel phenolic pathway for pollen development. *Science* 325:1688–92

Mattick JS, Makunin IV. 2006. Non-coding RNA. *Hum. Mol. Genet.* 15:R17–29

McCarrey JR, Thomas K. 1987. Human testis-specific *PGK* gene lacks introns and possesses characteristics of a processed gene. *Nature* 326:501–5

McManus CJ, Duff MO, Eipper-Mains J, Graveley BR. 2010. Global analysis of *trans*-splicing in *Drosophila*. *Proc. Natl. Acad. Sci. USA* 107:12975–79

Mercer TR, Dinger ME, Mattick JS. 2009. Long non-coding RNAs: insights into functions. *Nat. Rev. Genet.* 10:155–59

Mighell AJ, Smith NR, Robinson PA, Markham AF. 2000. Vertebrate pseudogenes. *FEBS Lett.* 468:109–14

Muller HJ. 1935. The origination of chromatin deficiencies as minute deletions subject to insertion elsewhere. *Genetica* 17:237–52

Nielsen J, Keasling JD. 2011. Synergies between synthetic biology and metabolic engineering. *Nat. Biotechnol.* 29:693–95

Nozawa M, Miura S, Nei M. 2010. Origins and evolution of microRNA genes in *Drosophila* species. *Genome Biol. Evol.* 2:180–89

Nozawa M, Miura S, Nei M. 2012. Origins and evolution of microRNA genes in plant species. *Genome Biol. Evol.* 4:230–39

Ohno S. 1970. *Evolution by Gene Duplication*. New York: Springer

Patthy L. 1999. Genome evolution and the evolution of exon-shuffling: a review. *Gene* 238:103–14

Peisajovich SG, Garbarino JE, Wei P, Lim WA. 2010. Rapid diversification of cell signaling phenotypes by modular domain recombination. *Science* 328:368–72

Petrov DA, Lozovskaya ER, Hartl DL. 1996. High intrinsic rate of DNA loss in *Drosophila*. *Nature* 384:346–49

Piriyapongsa J, Jordan IK. 2008. Dual coding of siRNAs and miRNAs by plant transposable elements. *RNA* 14:814–21

Piriyapongsa J, Marino-Ramirez L, Jordan IK. 2007. Origin and evolution of human microRNAs from transposable elements. *Genetics* 176:1323–37

Ponting CP, Oliver PL, Reik W. 2009. Evolution and functions of long noncoding RNAs. *Cell* 136:629–41

Potrzebowski L, Vinckenbosch N, Marques AC, Chalmel F, Jegou B, Kaessmann H. 2008. Chromosomal gene movements reflect the recent origin and biology of therian sex chromosomes. *PLoS Biol.* 6:e80

Rajagopalan R, Vaucheret H, Trejo J, Bartel DP. 2006. A diverse and evolutionarily fluid set of microRNAs in *Arabidopsis thaliana. Genes Dev.* 20:3407–25

Ro DK, Paradise EM, Ouellet M, Fisher KJ, Newman KL, et al. 2006. Production of the antimalarial drug precursor artemisinic acid in engineered yeast. *Nature* 440:940–43

Robertson HM, Navik JA, Walden KK, Honegger HW. 2007. The bursicon gene in mosquitoes: an unusual example of mRNA *trans*-splicing. *Genetics* 176:1351–53

Rosso L, Marques AC, Weier M, Lambert N, Lambot MA, et al. 2008. Birth and rapid subcellular adaptation of a hominoid-specific CDC14 protein. *PLoS Biol.* 6:e140

Roth DB, Porter TN, Wilson JH. 1985. Mechanisms of nonhomologous recombination in mammalian cells. *Mol. Cell. Biol.* 5:2599–607

Roth DB, Wilson J. 1988. Illegitimate recombination in mammalian cells. In *Genetic Recombination*, ed. R Kucherlapati, GR Simth, pp. 621–53. Washington, DC: Am. Soc. Microbiol.

Sayah DM, Sokolskaja E, Berthoux L, Luban J. 2004. Cyclophilin A retrotransposition into TRIM5 explains owl monkey resistance to HIV-1. *Nature* 430:569–73

Semon M, Wolfe KH. 2007. Consequences of genome duplication. *Curr. Opin. Genet. Dev.* 17:505–12

Smalheiser NR, Torvik VI. 2005. Mammalian microRNAs derived from genomic repeats. *Trends Genet.* 21:322–26

Stouthamer R, Breeuwer J, Hurst G. 1999. *Wolbachia pipientis*: microbial manipulator of arthropod reproduction. *Annu. Rev. Microbiol.* 53:71–102

Tautz D, Domazet-Loso T. 2011. The evolutionary origin of orphan genes. *Nat. Rev. Genet.* 12:692–702

Toll-Riera M, Bosch N, Bellora N, Castelo R, Armengol L, et al. 2009. Origin of primate orphan genes: a comparative genomics approach. *Mol. Biol. Evol.* 26:603–12

Viale A, Courseaux A, Presse F, Ortola C, Breton C, et al. 2000. Structure and expression of the variant melanin-concentrating hormone genes: Only PMCHL1 is transcribed in the developing human brain and encodes a putative protein. *Mol. Biol. Evol.* 17:1626–40

Vinckenbosch N, Dupanloup I, Kaessmann H. 2006. Evolutionary fate of retroposed gene copies in the human genome. *Proc. Natl. Acad. Sci. USA* 103:3220–25

Walsh JB. 1995. How often do duplicated genes evolve new functions? *Genetics* 139:421–28

Wang W, Brunet FG, Nevo E, Long M. 2002. Origin of *sphinx*, a young chimeric RNA gene in *Drosophila melanogaster*. *Proc. Natl. Acad. Sci. USA* 99:4448–53

Wang W, Zheng H, Fan C, Li J, Shi J, et al. 2006. High rate of chimeric gene origination by retroposition in plant genomes. *Plant Cell* 18:1791–802

Wapinski I, Pfeffer A, Friedman N, Regev A. 2007. Natural history and evolutionary principles of gene duplication in fungi. *Nature* 449:54–61

Westfall PJ, Pitera DJ, Lenihan JR, Eng D, Woolard FX, et al. 2012. Production of amorphadiene in yeast, and its conversion to dihydroartemisinic acid, precursor to the antimalarial agent artemisinin. *Proc. Natl. Acad. Sci. USA* 109:655–56

Wood AJ, Roberts RG, Monk D, Moore GE, Schulz R, Oakey RJ. 2007. A screen for retrotransposed imprinted genes reveals an association between X chromosome homology and maternal germ-line methylation. *PLoS Genet.* 3:e20

Wu DD, Irwin DM, Zhang YP. 2011. De novo origin of human protein-coding genes. *PLoS Genet.* 7:e1002379

Xiao W, Liu H, Li Y, Li X, Xu C, et al. 2009. A rice gene of de novo origin negatively regulates pathogen-induced defense response. *PLoS One* 4:e4603

Yang S, Arguello JR, Li X, Ding Y, Zhou Q, et al. 2008. Repetitive element-mediated recombination as a mechanism for new gene origination in *Drosophila. PLoS Genet.* 4:e3

Yang Z, Huang J. 2011. De novo origin of new genes with introns in *Plasmodium vivax. FEBS Lett.* 585:641–44

Yeh BJ, Rutigliano RJ, Deb A, Bar-Sagi D, Lim WA. 2007. Rewiring cellular morphology pathways with synthetic guanine nucleotide exchange factors. *Nature* 447:596–600

Yoshida S, Maruyama S, Nozaki H, Shirasu K. 2010. Horizontal gene transfer by the parasitic plant *Striga hermonthica. Science* 328:1128

Zhang G, Guo G, Hu X, Zhang Y, Li Q, et al. 2010a. Deep RNA sequencing at single base-pair resolution reveals high complexity of the rice transcriptome. *Genome Res.* 20:646–54

Zhang J, Dean AM, Brunet F, Long M. 2004. Evolving protein functional diversity in new genes of *Drosophila*. *Proc. Natl. Acad. Sci. USA* 101:16246–50

Zhang R, Peng Y, Wang W, Su B. 2007. Rapid evolution of an X-linked microRNA cluster in primates. *Genome Res.* 17:612–17

Zhang R, Wang YQ, Su B. 2008. Molecular evolution of a primate-specific microRNA family. *Mol. Biol. Evol.* 25:1493–502

Zhang Y, Wu Y, Liu Y, Han B. 2005. Computational identification of 69 retroposons in *Arabidopsis*. *Plant Physiol.* 138:935–48

Zhang YE, Vibranovski MD, Krinsky BH, Long M. 2010b. Age-dependent chromosomal distribution of male-biased genes in *Drosophila*. *Genome Res.* 20:1526–33

Zhang YE, Vibranovski MD, Landback P, Marais GA, Long M. 2010c. Chromosomal redistribution of male-biased genes in mammalian evolution with two bursts of gene gain on the X chromosome. *PLoS Biol.* 8:e1000494

Zhou Q, Wang W. 2008. On the origin and evolution of new genes: a genomic and experimental perspective. *J. Genet. Genomics* 35:639–48

Zhou Q, Zhang G, Zhang Y, Xu S, Zhao R, et al. 2008. On the origin of new genes in *Drosophila*. *Genome Res.* 18:1446–55

Climate Change, Aboveground-Belowground Interactions, and Species' Range Shifts

Wim H. Van der Putten

Department of Terrestrial Ecology, Netherlands Institute of Ecology
(NIOO-KNAW)/Laboratory of Nematology, Wageningen University, 6700 ES, Wageningen,
The Netherlands; email: w.vanderputten@nioo.knaw.nl

Annu. Rev. Ecol. Evol. Syst. 2012. 43:365–83

First published online as a Review in Advance on
September 4, 2012

The *Annual Review of Ecology, Evolution, and
Systematics* is online at ecolsys.annualreviews.org

This article's doi:
10.1146/annurev-ecolsys-110411-160423

Keywords

climate warming, extinction, invasiveness, geographic range, multitrophic
interactions, no-analog communities

Abstract

Changes in climate, land use, fire incidence, and ecological connections all
may contribute to current species' range shifts. Species shift range individually, and not all species shift range at the same time and rate. This variation causes community reorganization in both the old and new ranges. In
terrestrial ecosystems, range shifts alter aboveground-belowground interactions, influencing species abundance, community composition, ecosystem
processes and services, and feedbacks within communities and ecosystems.
Thus, range shifts may result in no-analog communities where foundation species and community genetics play unprecedented roles, possibly
leading to novel ecosystems. Long-distance dispersal can enhance the disruption of aboveground-belowground interactions of plants, herbivores,
pathogens, symbiotic mutualists, and decomposer organisms. These effects
are most likely stronger for latitudinal than for altitudinal range shifts. Disrupted aboveground-belowground interactions may have influenced historical postglacial range shifts as well. Assisted migration without considering
aboveground-belowground interactions could enhance risks of such range
shift–induced invasions.

INTRODUCTION

A range, or distribution, is the geographical area where a species can be found. The range is determined by numerous environmental factors, including climate, soil type, and species interactions. Over geological timescales, adaptive radiation, speciation, and plate tectonics can also influence the range of a species. The range of a species can shift owing to one or more changes in environmental conditions, such as climate warming, land-use change, new ecological connections, or artificial introductions of the species to a new environment. Nevertheless, many reports on current massive range shifts of species toward higher altitudes and latitudes suggest that climate warming is a key driving factor (Grabherr et al. 1994, Walther et al. 2002, Parmesan & Yohe 2003, Parmesan 2006, Walther 2010). If land-use change were the main driver, species' range shifts would occur in more directions.

Compared with historical geographic range shifts, such as those that have taken place during glaciation-deglaciation cycles over the past two million years (Bush 2002), the rate of current climate warming is unprecedented (Walther et al. 2002). The earliest reports on species' adaptation to climate change suggested that many species were failing to shift range fast enough to keep up with climate warming (Warren et al. 2001, Thomas et al. 2004, Thuiller et al. 2005). But more recent studies suggest that at least some species might respond adequately to climate warming by shifting their ranges (Chen et al. 2011) and that a number of species can reach enhanced dominance in the new range (Walther et al. 2002, Tamis et al. 2005, Engelkes et al. 2008). Thus far, most predictions on range shifts have been made independent of species interactions, and the question is whether including species interactions may change the outcomes of the model predictions (Lavergne et al. 2010, Van der Putten et al. 2010).

Species abundance can be influenced by resource availability, predation, propagule availability, symbioses, competition, and facilitation. As all these factors may vary between the old and new ranges, species that can move may not necessarily encounter suitable circumstances for establishment, growth, and reproduction. Moreover, these factors may also vary after a species has been introduced to a new range, which can affect community composition in a dynamic way. Species interactions can drive evolution or be subject to it, as seen in highly specialized pollination or parasitism patterns or in other symbiotic mutualisms. Climate change may disrupt those evolutionary processes as well as initiate new processes (Lavergne et al. 2010).

Besides range shifts, species may also respond to climate warming and other environmental changes by adapting to them. For example, there is scope for genetic adaptation of plants to climate warming, but there are also limitations that may contribute to diversity loss (Jump & Peñuelas 2005). Climate warming is highly multidimensional. Local effects of climate warming may result from changes in temperature, precipitation, or length of the growing season. Species that shift range may also be exposed to different day length (Jump & Peñuelas 2005). Investigators have not yet determined how adaptation and migration interact during range shifts (Lavergne et al. 2010).

Terrestrial ecosystems are composed of aboveground and belowground subsystems, which have been examined separately for many years even though the different subcomponents clearly interact with each other (Wardle 2002). Plants connect the aboveground and belowground subsystems, and interactions belowground can, directly or indirectly, influence interactions aboveground (and vice versa). Species in aboveground and belowground subsystems are differently susceptible to climate warming (Berg et al. 2010), leading to—at least temporarily—new species combinations in the new range. As aboveground-belowground interactions have the potential to impose selection on plants (Schweitzer et al. 2008), range shifts may influence selection and adaptation. In spite of rapidly increasing interest in the subject of aboveground-belowground interactions, the effects of climate warming–induced range shifts have been poorly studied thus far (Bardgett & Wardle 2010). In this review, I combine reported knowledge on range shifts with information on the functional role of aboveground-belowground species interactions in community organization and ecosystem processes.

Belowground subsystems include biota that interact with plants directly (herbivores, pathogens, and symbionts) or indirectly (natural enemies of the directly interacting species and components of the decomposer subsystem). The direct and indirect interactions with plant roots can influence aboveground biota and can result in effects that feed back to the soil subsystem (Wardle et al. 2004). Expanding from a previous review that argued that trophic interactions need to be considered when predicting consequences of climate warming (Van der Putten et al. 2010), I focus here on how range shifts may influence community organization and ecosystem processes. I do not pretend to be complete in my review, and a part of my conclusions are speculative, but I hope to encourage thinking about species' range shifts from a more complex (and realistic) ecological perspective.

I discuss recent work on aboveground-belowground interactions in relation to climate warming–induced species' range shifts. I compare altitudinal gradients—where dispersal distances may not be a major limitation—with latitudinal gradients—where range shifts may disrupt aboveground-belowground interactions more severely, owing to larger dispersal distances and differences in dispersal rates. I also provide a brief paleoecological view and discuss how aboveground-belowground interactions in the past might have changed during deglaciation periods. In the next sections, community and ecosystem consequences of range shifts are reviewed from the perspective of aboveground-belowground interactions. I discuss community assembly processes, including species loss and species gain, from an aboveground-belowground perspective while discussing their roles in no-analog communities (and novel ecosystems), foundation species, and assisted migration.

SPECIES' RANGE SHIFTS

Patterns along Altitudinal Gradients

The earliest signals showing that the rapid climate warming of recent decades is leading to plant range shifts resulted from work along altitudinal gradients in alpine ecosystems (Grabherr et al. 1994, Walther et al. 2002, Parmesan & Yohe 2003). Alpine vegetation responses to climate warming may depend on plant type and altitude. For example, along an elevation gradient of 2,400 and 2,500 m above sea level, shrubs expanded 5.6% per decade, but above 2,500 m, unexpected patterns of regression occurred that were associated with increased precipitation and permafrost degradation (Cannone et al. 2007).

At lower altitudes in mountains, effects of climate warming are difficult to disentangle from those of changes unrelated to climate, such as land-use change. At high altitudes, where land use does not play a major role, effects of climate warming are clearer (Cannone et al. 2007). Nevertheless, even in low-altitude areas such as the Jura (France), effects of warming can be detected over a 20-year period (Lenoir et al. 2008, 2010). At a subarctic island, analyses of 40 years of species data revealed an average upward elevation shift of half the plant species (Le Roux & McGeoch 2008). Both here and in the Jura, only a subset of plant species responded to climate warming. Remarkably, although the species that determined the pattern of upslope expansion may be considered highly responsive, the response was still lower than expected based on the rate of warming (Le Roux & McGeoch 2008). Such species-specific range shift responses may result in no-analog communities at higher elevations, consisting of the original plant species and the range expanders. Downhill species shifts can also be observed, for example in California, where the water deficit at higher elevations increased over time (Crimmins et al. 2011).

Although much work has focused on patterns of altitudinal range shifts, less work has been done on the consequences of altered species interactions in relation to climate warming. In general, high-altitude plant communities may be structured more by facilitative interactions than by competitive ones (Callaway et al. 2002). However, plant facilitation could also be influenced by aboveground and belowground multitrophic interactions, which may need more attention for researchers to understand the consequences of climate warming in high-altitude habitats. Because range shift distances are relatively short in altitudinal gradients, dispersal is less limited than along latitudinal gradients, but aboveground-belowground interaction patterns may still be altered in highly complex ways. For example, the development of bare soil surface at higher altitudes (Walther et al. 2002) considerably influences belowground decomposition processes (Wardle et al. 1999). In contrast, ecosystem regression toward pioneer stages can affect the outcome of plant community interactions by a shift from symbiotic (arbuscular) mycorrhizal fungi toward soil-borne pathogens being the most important soil biota influencing plant community composition (Kardol et al. 2006). In general, global change effects on soil biota are relatively predictable (Blankinship et al. 2011), but interactive consequences of climate warming, such as altered frost incidence, rainfall patterns, plant types, and plant cover, may complicate predictions of soil biota responses and their feedback effects on plants and aboveground interactions.

Patterns along Latitudinal Gradients

Patterns of latitudinal range shifts have been predicted based on altitudinal shifts (Walther et al. 2002). Climate effects of 1 m in altitudinal range shift may be considered equal to 6.1 km in latitudinal shift (Parmesan & Yohe 2003). However, these conversion factors do not account for dispersal limitations that may arise from, for example, poor dispersal capacity, effects of habitat fragmentation, or limitations of vector organisms. In northwestern Europe, for example, there are clear patterns in seed dispersal limitations, as some vectors, especially large vertebrates, are much more limited in migration now than they were in the past (Ozinga et al. 2009). Such limitations may also apply to insect range shifts. A study in the United Kingdom showed that range expansion by habitat-specialist butterflies was constrained following climate warming because the specific habitats lacked connections. Only habitat generalists could keep up with climate warming because their dispersal was less limited by unsuitable corridors (Warren et al. 2001).

Poor dispersal capacities of certain soil biota, especially soil fauna, have been mentioned in several studies. For example, the highest nematode diversity occurs in temperate zones, where there are more root feeders of higher plants than exist in the tropics. Nematode diversity is lower in Antarctic than in Arctic zones, which suggests that dispersal limitations are, at least in part, causing the latitudinal zonation of nematodes (Procter 1984). There may also be gradients within latitudes, but these are related more to community similarity than to community richness. For example, in a comparison of nematodes and microbial assemblages among 30 chalk grasslands in the United Kingdom roughly scattered across a west-east gradient of 200 km, similarity in both nematodes and bacteria declined with distance (Monroy et al. 2012). Therefore, soil communities may vary with distance, irrespective of orientation (Fierer et al. 2009). Hence, range shifts in any direction can expose that plant species to novel soil biota and disconnect it from the usual biota with which it interacts.

Applications of findings from altitudinal shifts to range shift predictions in lowlands may also be complicated for other reasons. In a 44-year study (1965–2008) of climate warming in lowland and highland forests in France, latitudinal range shifts were expected in the lowland forests. However, in lowland forests, the responses of latitudinal range shifts were 3.1 times less strong than those of altitudinal range shifts in highland forests (Bertrand et al. 2011). There are several possible explanations: Lowland forests may have proportionally more species that are persistent in the face of warming, there may be fewer opportunities for short-distance escapes, or the greater habitat fragmentation in lowlands may prevent range shifting.

Range shifts can be limited by the availability of sites for establishment. This has been shown not only for butterflies (Warren et al. 2001) but also for plants. For example, Leithead et al. (2010) showed that range-shifting tree species from a temperate forest in Canada, such as red maple (*Acer rubrum*), can establish in a boreal red pine (*Pinus resinosa*) forest only if there are large tree-fall gaps. Native red pine forest species, in contrast, were not influenced by gap size or gap age. Interestingly, pine dominance in the red pine forest is maintained by wildfires, which selectively omit competitors and reset succession. Fire incidence can be altered by climate warming. Because southern tree species establish in tree-fall gaps too fast for the rate of wildfires to control, the combined effects may be enhanced colonization of northern forests by southern tree species. Tropical lowlands may be especially sensitive to climate warming for other reasons. The tropical climate now is warmer than at any time in the past two million years (Bush 2002). The spread of species from tropical forests to cooler areas may be constrained by long dispersal distances and poor colonization sites along the dispersal routes. Therefore, tropical regions may be sensitive to species loss owing to climate warming. Moreover, lowland tropics lack a species pool to provide new species that may favor the new climate conditions (Colwell et al. 2008). Range shifts of species from tropical lowlands to tropical highlands are possible, but they may result in depauperate lowland plant communities, which will be increasingly dominated by early successional species (Bush 2002, Colwell et al. 2008).

Researchers have investigated aboveground-belowground interactions in relation to latitudinal range shifts. A comparison of range-expanding plant species from Eurasia and other continents with species that are phylogenetically related to those from the invaded range showed that both types of range expanders develop less pathogenic activity in their soils than related natives do. Moreover, the range expanders on average were more tolerant of or were better defended against two polyphagous invertebrate aboveground herbivores. The pattern coincided with induced levels of phenolic compounds, which are general secondary metabolites used for plant defense (Engelkes et al. 2008). Therefore, successful range-expanding plant species may have invasive properties irrespective of their origin. Interestingly, although belowground and aboveground effect sizes were additive, there was no correlation between aboveground and belowground effect strengths (Morriën et al. 2011). Thus, plant species that resisted or tolerated belowground enemy effects in the new range were not necessarily well protected against generalist aboveground herbivores.

Analysis of soil samples along a latitudinal gradient of a range-expanding plant species (*Tragopogon dubius*) showed soil pathogen effects in several sites in the native range, but not in the range the species had shifted into recently (Van Grunsven et al. 2010). Thus, range shifts enabled the plants to escape their original soil pathogens, although successful range shifters defended themselves well against unknown and cosmopolitan aboveground polyphagous herbivorous insects (Van Grunsven et al. 2007, Engelkes et al. 2008). These results were based on growth trials in greenhouse mesocosms. The next step should be to determine the consequences of altered belowground and aboveground biotic interactions under field conditions.

Historical Patterns of Range Shifts

Species' range shifts have occurred throughout the Earth's history. For example, it is well documented that glacial cycles have caused species' range shifts (Jackson & Overpeck 2000, Williams et al. 2007, Willis et al. 2010). There have been approximately 20 cycles of glaciation and deglaciation during the Quaternary (the last 2.58 million years), especially in the Northern Hemisphere (Dawson et al. 2011). The last ice age occurred about 10,000 years ago. Based on pollen records from late Quaternary Europe, paleovegetation maps have been constructed at the level of formations. As these vegetation maps are not analogous with contemporary vegetation, Huntley (1990a) concluded that the macroclimate in the late Quaternary might have been completely different from the present one. But a complication of comparing paleobiology data with contemporary ecosystems is that current vegetation in Europe has been strongly influenced by human activities and the continent's heterogeneity (Huntley 1990a). In spite of these uncertainties, we can still surmise that communities have become reorganized over and over again during cycles of warming and cooling (Jackson & Overpeck 2000).

Historic range shift data still cast doubts on the rate of plant dispersal. The proposed average northward spread of 1 km per year during deglaciation periods is most likely 10 times as fast as the average dispersal capacity of individual plant species. This discrepancy in migration distances can be due to a hitherto undetected role of long-distance dispersal (Loarie et al. 2009). Long-distance dispersal likely played an important role in prehistoric times. In a modeling study (K.M. Meyer & M. van Oorschot, unpublished results), long-distance dispersal turned out to be crucial for enemy release, in their case from root-feeding nematodes. Long-distance dispersal of plants may also reduce their exposure to specialized aboveground enemies because these enemies may have difficulties reaching the new plant populations. Therefore, we can expect that during deglaciation range shifts, plant species might have become exposed to different aboveground-belowground interactions.

It is also possible that aboveground or belowground enemies have promoted tree range shifts (Moorcroft et al. 2006). In a modeling study, natural enemies were able to influence the spread of tree species into ecosystems where equally strong competitors were present. Adding host-specific pathogens to the model resulted in dispersal distances equal to the ones that have been reported by paleoecologists based on pollen patterns (Moorcroft et al. 2006). Obviously, research should place more emphasis on the issue of long-distance dispersal in relation to range shifts and relationships with aboveground and belowground natural enemies and their antagonists. This might also provide a different view on evolution during glaciation-deglaciation cycles.

In a review of postglacial range expansion effects on the evolution of insects, Hill et al. (2011) found that rapid evolution of dispersal may be promoted in the expansion zones. This suggests a positive feedback between range expansion and the evolution of traits (in this case dispersal) that accelerates range expansion capacity. Thus, the feedback between ecology and evolution is strongest at range boundaries where selection is assumed to be strongest and where population bottlenecks are common (Hill et al. 2011). But these data may not translate to present-day range shifts because of the unprecedented rate of the current warming. Moreover, modern landscapes are much more fragmented than the original postglacial landscapes, and this fragmentation may lead to loss of genetic variation rather than enable trait evolution (Hill et al. 2011).

Current insights on aboveground-belowground species interactions may be used to assess how they operated during prehistorical changes in vegetation types. For example, in a flood plain in Pakistan, isotope records reveal shifts from C3 to C4 grass-dominated ecosystems (Barry et al. 2002). There were also pulses in (vertebrate) fauna turnover, resulting in a loss of biodiversity and an accelerated pace of extinction in this region once C4 vegetation occurred on the flood plain. Overall, species composition was relatively steady, with brief, irregularly spaced temporal spikes of species turnover and ecological change. Time intervals of the assessment were at least 100,000 years (Barry et al. 2002). In contrast to these aboveground changes in vertebrate fauna, selective plant removal studies in New Zealand (Wardle et al. 1999) and sampling of C3 and C4 grasses in the United States (Symstad et al. 2000, Porazinska et al. 2003) suggest that conversion of C3 into C4 grasslands might have had very little effect on soil fauna or aboveground arthropod diversity. The C4 grass vegetation might have been a response to warming and drier conditions, which could have had a much stronger effect on soil community composition and the resulting ecosystem functioning (Blankinship et al. 2011).

Another example concerns the last postglacial period in Europe, during which mixed decid-uous forests received their current distribution around 8,000 years before present. Relative tree abundance changed in those forests over the past 13,000 years, as they were dominated first by *Pinus*, then by *Tilia*, and during the past few millennia by *Fagus* species (Huntley 1990b). How exactly these vegetation changes have taken place and at what rate are difficult issues to explain because these data, among others, are based on chord-distance maps that have intervals of 1,000 years (Huntley 1990a). Nevertheless, litter composition is known to influence decompo-sition (Hättenschwiler & Gasser 2005), and it also influences soil organisms, such as earthworms (Muys & Lust 1992) and microbes (Ayres et al. 2009, Strickland et al. 2009). These examples show that responses of plant communities to climate changes and consequences for ecosystem processes in the (late) Quaternary might have been quite dynamic. Over these long time periods, climate was the overarching driver. Belowground-aboveground interactions might have driven community responses at shorter spatial and temporal scales.

Other Drivers of Range Shifts

There are some, though not many, examples of range shifts caused by factors other than climate warming or cooling. For example, intensified grazing and fire regimes enabled range expansion of shrubs in Colorado (Archer et al. 1995), whereas the El Niño–Southern Oscillation influences the frequency and extent of wildfires, which in turn influence tree stand composition in the southern United States (Swetnam et al. 1999). Furthermore, there are examples of bird range expansion owing to land-use change. Improved feeding or nesting sites can drive such range shifts. For example, the Black-shouldered Kite (*Elanus caeruleus*) has shifted range northward into Spain because, during the last half of the previous century, cultivated Dehesa systems became more similar to African savannahs, where this species originated (Balbontín et al. 2008).

Habitat fragmentation, such as that caused by intensified land use, can limit the capacity of species' range shifts. Currently, this is considered one of the major constraints for species' responses to climate warming (Warren et al. 2001). Habitat fragmentation might also have limited range shifts in postglacial periods under specific conditions. In Finland, recolonization of former islands after land-ice retreat during the Holocene might have been hampered by poor connectedness to the surrounding mainland (Heikkilä & Seppä 2003). One possibility to determine if climate warming is the key factor leading to range shift is to determine if the pattern is one-directionally correlated with the warming gradient. But terrestrial range shifts often cause mosaiclike patterns rather than wavelike phenomena because the velocity of climate change on land is far more patchy than it is in the oceans (Burrows et al. 2011).

Conclusions on Species' Range Shifts

Patterns of individual species' range shifts in response to climate change are less uniform than general averages suggest because there are fast- and slow-responding species, time lags, downhill instead of uphill range shifts, and long-distance dispersal. Some range shifts are due to factors other than climate, such as changing land use or altered fire incidence. Uphill range shifts are better correlated with warming than are lowland range shifts toward the poles, probably due to shorter dispersal distances along altitudinal gradients and fewer constraints such as habitat fragmentation at high elevations. Lowland tropical systems may be highly sensitive to warming because temperatures are already higher than in the past two million years and dispersal distances to cooler areas are generally large, except in tropical lowland-mountain areas where uphill range shifts are possible. Range contractions are less well studied than range expansions, and in some cases downhill range shifts have been recorded (e.g., cases have been reported where water is more available at low elevation or where microclimate is cooler owing to forest regrowth downhill).

Aboveground, plants may also be released from their natural enemies, especially in the case of long-distance dispersal. This phenomenon is supposed to have played a role in recolonization during postglacial range shifts. Therefore, although little information exists on this subject, disassembly of aboveground-belowground interactions during range shifts may influence ecology and evolution during climate warming–induced range shifts. This may happen now, but it could also have played a role during prehistoric range shifts. Such disruptions of aboveground and belowground interactions have the potential of influencing community assemblage processes as well as the evolution of the species involved.

COMMUNITY CONSEQUENCES OF ALTERED ABOVEGROUND-BELOWGROUND INTERACTIONS DURING RANGE SHIFTS

Understanding species' range shifts requires addressing a key question in ecology: How will biodiversity and ecosystem functioning be influenced by the disappearance of existing species and the arrival of new species (Wardle et al. 2011)? The research on range shifts initially was dominated by reports on species extinctions due to climate warming (Warren et al. 2001, Thuiller et al. 2005), whereas later the emphasis also included consequences of climate warming for range shifts of exotic invaders (Walther et al. 2009). Other studies have shown that the number of species from warm climate regions in temperate areas is increasing (Tamis et al. 2005); thus, there is a group of species that may shift range to higher altitude or latitude in accordance with the rate of climate warming (Chen et al. 2011).

Patterns of Species Gains and Losses

Which species will be lost or gained following climate warming depends on a large number of aspects, including the tolerance of species to the environmental change (warming or an associated change, such as drought or extreme weather events), the time needed for species to disperse and the time needed by other species to be lost from communities, sensitivity to habitat fragmentation, habitat specialization, dispersal mode, etc. The net effect of species gains and species losses can be that total biodiversity remains constant, but biodiversity can also decrease, or even increase (Jackson & Sax 2010). As time proceeds, net effects of gains and losses of species may vary, and the total number of species in communities may temporarily go up or down. Although net effects of species gain and loss can be positive locally, worldwide biodiversity will decline, and communities across the world, in the same climatic zones, will appear more similar because of an increasing number of shared species.

The traits of the species coming in and going out will strongly influence that species's role in ecosystem processes. For example, novel chemistry may influence ecological relationships, as herbivores and decomposer organisms from the invaded range may not be capable of dealing with those compounds (Callaway & Ridenour 2004). Phylogenetic nonrelatedness with other species that are native in these communities can play an important role in predicting the success of species introductions (Strauss et al. 2006). Losses or gains of dominant species should have more impact on ecosystem processes (Grime 1998), although some low-abundant species may have disproportional effects. For example, microbial pathogens or endophytes have low abundance, but they can substantially influence plant community composition (Clay & Holah 1999) and therefore ecosystem functioning.

In general, new species most likely will have characteristics of early successionals because such species have good dispersal abilities. Long-distance dispersal may enable them to escape from natural enemies, to which early successional plant species can be sensitive (Kardol et al. 2006). For example, the range shift of *T. dubius* has not yet led to the establishment of specific soil-borne pathogens in the new range (Van Grunsven et al. 2010). Although not all species will respond to climate warming by range shift (Le Roux & McGeoch 2008), little is known about which species will stay behind, what traits they have, or what their fate will be in the long term.

Assessing Ecological Consequences

An increasing number of studies have assessed how aboveground and belowground interactions may change in relation to plant species gains (Maron & Vilà 2001, Agrawal et al. 2005, Parker & Gilbert 2007, Peltzer et al. 2010) and plant species losses (Wardle et al. 1999, Scherber et al. 2010). But few such studies have focused explicitly on plant range shifts (Engelkes et al. 2008, Morriën et al. 2010, Van Grunsven et al. 2010, Meisner et al. 2012). Interestingly, plant species that shift range and are successful in their new range have invasive properties with respect to aboveground and belowground enemy effects, that are similar to intercontinental exotic invaders (Engelkes et al. 2008). **Figure 1** presents different scenarios of aboveground-belowground range shifts and consequences for plant biomass. Depending on how fast plants, herbivores, and carnivores shift range, in the new range plants can produce more or less biomass than in the native range.

Further studies using aboveground and belowground surveys and manipulations along a range expansion gradient are needed to tease apart the ecological and evolutionary consequences of individual effects. Studies of natural enemy species on invasive plants have shown contrasting degrees of enemy exposure in the new range (Mitchell & Power 2003, Van Kleunen & Fischer 2009), whereas ecological responses are not necessarily in line with the assumed enemy release effects (Parker & Gilbert 2007). These results call into question whether enemy release may explain plant invasiveness in a new range. Long-term experiments and studies along latitudinal or elevation gradients (Sundqvist et al. 2011) are needed to determine extended effects of plant range shifts on decomposition, nutrient cycling, and plant performance under field conditions. Transplantation studies, for example, may reveal the extent to which specificity in litter decomposition exists along latitudinal or altitudinal gradients. This specificity has been described as a home-field advantage (Ayres et al. 2009, Strickland et al. 2009), as the soil communities of some plant species decompose their own litter faster than soil communities for other plant species. This home-field advantage is also specific to plant genotype (Madritch & Lindroth 2011).

It is important to include negative controls in experiments when testing species' responses to climate warming. For example, in aboveground-belowground interaction studies, successful range expanders may be compared with unsuccessful ones to test aboveground-belowground interaction effects (Morriën et al. 2011) and consequences for plant abundance (Klironomos 2002). Besides effects of species gains, consequences of species losses due to climate warming need to be tested experimentally. This will yield information on the traits of species that are under threat of extinction by climate warming, their ecological relationships, and the number of generalist and specialist relationships with other plants and multitrophic organisms. These integrated and field-based approaches may help to further conceptualizations of species loss and gain (Jackson & Sax 2010) from a multitrophic perspective. Ultimately, these approaches will show how food webs are being influenced by global changes (Tylianakis et al. 2008) and how trophic networks may function under dynamic restructuring.

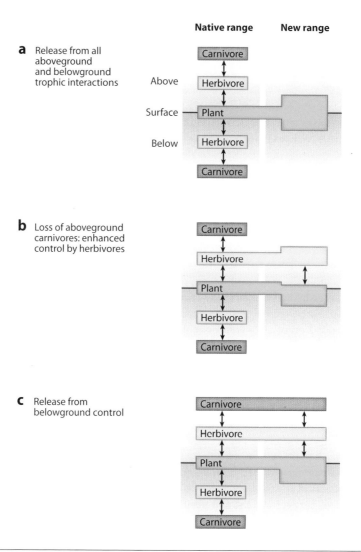

Figure 1

Scenarios for range shifts of plants, aboveground and belowground herbivores and their natural enemies, and consequences for plant size (or abundance). According to scenario (*a*), plants shift range faster than all aboveground and belowground biota and do not encounter biotic resistance in the new range. This leads to enhanced biomass in the new range both aboveground and belowground. In scenario (*b*), aboveground herbivores shift range as fast as plants and are released from their natural enemies. This leads to overexploitation of the plants aboveground (note that whether this also results in reduced belowground biomass that is due to a lack of photosynthesis products to support root and rhizome growth is still debated). In scenario (*c*), aboveground herbivores and carnivores shift range equally as fast as plants, resulting in unchanged aboveground biomass compared with the native range, whereas root biomass may be enhanced owing to lack of belowground herbivory (but see scenario *b*).

Long-Term Perspectives on Range Shifts

Aboveground and belowground interactions of range-shifting plants will not be static over time, as has been demonstrated for host-parasite interactions (Phillips et al. 2010). As time proceeds, the natural enemies, symbionts, and decomposer organisms and their antagonists may colonize the expanded range, but it is not yet known how fast this process may develop and how completely the original communities may become reassembled. Historical data from paleobiology do not provide such detailed information. Range-shifting plants that arrive without their naturally coevolved insects, microbes, and nematodes may or may not establish interactions with species from the new range. Provided that suitable conditions exist, natural selection may cause changes in the genetic structure of the range-shifted plants. For example, when exposure to natural enemies diminishes, selection against the production of costly defenses is to be expected (Müller-Schärer et al. 2004), which could lead to a trade-off between defense and growth (Blossey & Nötzold 1995). This process has been tested for cross-continental introductions of exotic plant species, although these costs are difficult to quantify and experimental tests sometimes show opposite results (Wolfe et al. 2004).

There are spectacular studies of introduced exotic species that lose their capability to produce high defense levels. For example, in a chronosequence representing over 50 years of *Alliaria petiolata* introduction to North America, phytotoxin production decreased as the time since introduction increased (Lankau et al. 2009). Variation in allelochemical concentrations also influenced soil microbes, including fungi that had mutualistic interactions with a native tree species (Lankau 2011). Over time, introduced plants may become less resistant or native biota may become more aggressive. For example, New Zealand plant species that varied in the amount of time since introduction (with 250 years as the maximum) were experimentally exposed to soil biota. This study showed that the longer the time since introduction, the stronger the pathogenic effects from the soil community (Diez et al. 2010). These studies suggest that introduced exotic species may become less invasive over time, owing to natural selection of the introduced species themselves or the belowground or aboveground species from the new habitat. These temporal processes may contribute to the sudden population crashes that have been observed for a number of introduced species (Simberloff & Gibbons 2004). A possible long-term scenario for such a boom-bust pattern has been worked out in **Figure 2**.

Conclusions on Altered Aboveground-Belowground Interactions

Thus far, most work on pattern analyses of range shifts has been dedicated to understanding the consequences of species loss due to climate warming. Effects of species introductions by range shifts from lower to higher latitudes and altitudes have not received much attention yet, and the ecological consequences, as well as temporary developments of range-shifting species, are only beginning to be explored. However, we can expect that successful range shifting involves a gradual response of the available plant species and that aboveground and belowground organisms expand their range subsequently, but at lower and variable rates. In the meantime, aboveground and belowground organisms from the native range will establish interactions with the new species and may adapt to the new plants by natural selection. The biotic interactions established in the new range will in return also impose natural selection on the range-shifting species, and this natural selection may reduce the invasive performance as the time since introduction increases. The question is what might happen when the former natural enemies become cointroduced as well: Will they recognize their original host (Menéndez et al. 2008), will they overexploit their former host, or will the novel biotic interactions completely alter priority effects (Lau 2006)? When we consider all these possible changes, one might reasonably conclude that the original host-consumer interactions are unlikely to be restored to their state in the original range. The outcome of this complex process may contribute to boom-bust patterns of abundance that have been observed for some introduced exotic species' novel community composition and functioning, or they may enable a soft landing for the range-shifting species in their novel habitats following restoration of the original species' interactions.

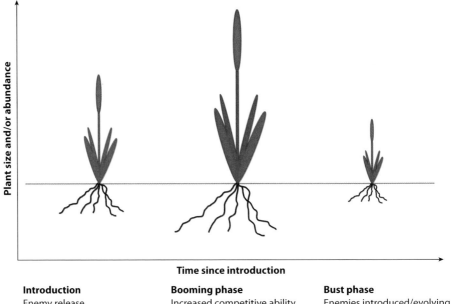

Introduction	Booming phase	Bust phase
Enemy release	Increased competitive ability	Enemies introduced/evolving
No home-field advantage	Home-field advantage	Home-field advantage
General symbionts available	General symbionts available	Specialized symbionts available

Figure 2

Hypothetical explanation for an introduction-boom-bust pattern of size (or abundance) of a range-shifting plant species. Following introduction, there may be benefits from (*a*) a release from native enemies, (*b*) the absence of biotic resistance in the new range, and (*c*) the presence of generalist mutualistic symbionts (pollinators, arbuscular mycorrhizal fungi) in the new range that outweigh poor home-field advantage. These benefits may further increase due to the evolution of increased competitive ability and the development of home-field advantage due to specialization of decomposer organisms from the new range. However, that benefit can turn into a major disadvantage when natural enemies from the native range migrate as well or when enemies from the new range break through plant resistance. In that case, plants are poorly defended, and despite their home-field advantage, the top-down control becomes so severe that plant size or abundance is strongly reduced, with an accompanying risk of extinction.

ECOSYSTEM CONSEQUENCES OF CHANGED ABOVEGROUND-BELOWGROUND INTERACTIONS DURING RANGE SHIFTS

Until this point, range shifts have been considered mainly from the perspectives of species' response patterns and community interactions. The questions now are whether and how these altered species' assemblages and community interactions translate into ecosystem consequences. These consequences may be expressed as altered ecosystem processes (nutrient cycles), resilience, and stability, and these may in turn influence the provisioning of ecosystem services (for example, primary production, control of greenhouse gas emissions, and control of pests and pathogens) (Naeem et al. 2009). Few analyses have been made of ecosystem consequences of range shifts in comparison to the numerous studies that have been conducted recently on how climate change might result in biodiversity loss and exotic species invasions (Wardle et al. 2011).

In a comparative study of range-shifted plant species and phylogenetically related natives from the new range, nutrient dynamics in the root zone (Meisner et al. 2011) and litter decomposition (Meisner et al. 2012) were affected by plant (genus-related) traits, rather than by plant origin.

This is analogous to work done on intercontinental invasive plant species, showing that some, but clearly not all exotics will enhance nutrient cycling (Ehrenfeld et al. 2005, Vilà et al. 2011).

Interestingly, plant origin affected sensitivity to aboveground polyphagous insects (Engelkes et al. 2008) and feedback effects from the soil community (Van Grunsven et al. 2007, Engelkes et al. 2008) in similar, phylogenetically controlled comparisons. Therefore, nutrient cycling–related ecosystem services may not be altered by range shifts as much as biocontrol-related services are. Failing top-down control in the new range can be due to enemy release of the range expanders (Van Grunsven et al. 2010), failing biotic resistance from the natural enemies present in the new range, or a combination of the two (Keane & Crawley 2002). In a survey of intercontinental invasive exotic plant species, exotic plants had fewer pathogen and virus species in the new range than expected (Mitchell & Power 2003). Little is known about whether this also applies to plant species that have shifted range intracontinentally.

Another ecosystem consequence of range shifts is related to the question of whether diversity begets diversity (Whittaker 1972, Janz et al. 2006). Some plant species can have a disproportional role in sustaining aboveground and belowground biodiversity. These so-called foundation species (Ellison et al. 2005) strongly influence aboveground and belowground community composition and species interactions, which can be considered extended phenotypes. Little is known about range shift potentials of such foundation species and whether species assemblages aboveground and belowground in the new range may be as extended as in the native range. Non-foundation species may have less far-reaching effects on aboveground and belowground communities. Nevertheless, many of those species may also have individual aboveground (Bukovinszky et al. 2008) and belowground (Bezemer et al. 2010) food webs that could be altered by differential range shift capacity (Berg et al. 2010). Therefore, ecosystem consequences of range shifts may be that foundation species, as well as non-foundation species, lose at least part of their extended phenotypes (**Figure 1**). Consequences for ecosystem processes, resilience, and stability are as yet unknown.

The altered community composition of range-shifted plant species potentially influences community genetics (Hersch-Green et al. 2011). When range-shifting plant species have fewer ecological interactions in the new range than in the original range, patterns of community genetics and evolutionary processes can be completely different. These changes at the genetic level may have consequences at the level of ecosystem processes and functioning (Whitham et al. 2006). Therefore, range shifts will provide interesting opportunities for community genetics approaches by testing how microevolutionary processes may play a role during disintegration and (re)assemblage of multitrophic interactions under climate warming. In these studies, abiotic stress conditions should also be included, as they can change during climate warming, in both the native and new range, and they can alter composition and functioning of entire food webs belowground (De Vries et al. 2012).

Investigators have proposed that assisted migration and colonization (Hoegh-Guldberg et al. 2008) may help solve problems of species that cannot shift range under climate warming. However, assisted migration may also involve risks with consequences for the composition as well as functioning of ecosystems of the new range. Successful range-expanding plant species have invasive properties similar to intercontinental invaders (Engelkes et al. 2008). Whether these invasive properties are already intrinsic in the original populations, are selected during range shift, or are due to rapid evolution in the new range is unknown. All these possibilities will be relevant when preparing for assisted migration: which genotypes to select for dispersal, how to test their ecological suitability to become established in the new range, and how to assess ecological consequences in case the assisted species does disproportionally well in its new range. There are already too many examples from intentional or unintentional cross-continental invasions in which taking species out of their original community context resulted in enemy release (Keane & Crawley 2002).

Therefore, before considering assisted migration and other climate warming–mitigation activities, community and ecosystem consequences of such actions need to be carefully assessed, including consequences of aboveground or belowground enemy release (Engelkes et al. 2008), symbiont availability (Hegland et al. 2009), and loss of the home-field advantage of decomposition (Ayres et al. 2009).

Researchers may need to consider this context when discussing emerging ecosystems (Milton 2003) and novel ecosystems (Hobbs et al. 2006). As in restoration ecology, where the role of soil communities and aboveground-belowground interactions are acknowledged (Harris 2009, Kardol & Wardle 2010), ecosystem-level consequences of aboveground-belowground interactions influenced by range shifts need to be considered as well. Most likely, the concept of novel ecosystems will require the consideration of species as related to aboveground-belowground interactions, rather than of the presence or absence of species in isolation. However, ecological novelty may change over time because of the temporal dynamics of the dispersal of associated species as well as the community genetics processes to which the new and the resident species will be exposed. Therefore, ecosystem consequences of (climate warming–induced) range shifts may be predicted better by including the interactions of aboveground and belowground species from a combined ecological and evolutionary perspective. This work could also help us better understand historical range shifts during glaciation-deglaciation cycles, the way those processes might have shaped current aboveground-belowground communities in terrestrial ecosystems, and the potential consequences of the current unprecedented rates of warming for future ecosystem functions and services.

SUMMARY POINTS

1. Terrestrial ecosystems consist of aboveground and belowground subsystems, and the species in these subsystems can all interact.

2. Range shifts of plant species may result in temporary release from natural enemies or symbionts, which may cause invasions or establishment failures in the new range.

3. Decomposition-related processes are supposed to be less specific, but recent work has pointed to considerable specificity in decomposer organisms, even down to the plant genetic level.

4. Latitudinal range shifts will be more sensitive to disruption of aboveground-belowground interactions than altitudinal range shifts.

5. No-analog communities have no-analog aboveground-belowground interactions, which may completely change patterns of community organization, species abundance, and biodiversity.

6. Landscape configuration may be important for range shifts, as it influences dispersal capacities of plants as well as aboveground and belowground biota.

7. Range shifts will be crucial for maintaining ecosystem functioning and ecosystem services.

8. The role of foundation species and community genetics may change substantially due to range shifts.

9. Assisted migration should be considered with care, as it may cause more problems than it solves.

DISCLOSURE STATEMENT

The author is not aware of any affiliations, memberships, funding, or financial holdings that might be perceived as affecting the objectivity of this review.

ACKNOWLEDGMENTS

I thank Tadashi Fukami for proposing that I write this review and for providing helpful comments; Pella Brinkman for helping me throughout the writing process; and Jennifer Krumins, Annelein Meisner, Elly Morriën, Pella Brinkman, and Martijn Bezemer for comments and suggestions on previous versions of the manuscript. This is NIOO publication 5266.

LITERATURE CITED

Agrawal AA, Kotanen PM, Mitchell CE, Power AG, Godsoe W, Klironomos J. 2005. Enemy release? An experiment with congeneric plant pairs and diverse above- and belowground enemies. *Ecology* 86:2979–89

Archer S, Schimel DS, Holland EA. 1995. Mechanisms of shrubland expansion: land use, climate or CO_2? *Clim. Chang.* 29:91–99

Ayres E, Steltzer H, Simmons BL, Simpson RT, Steinweg JM, et al. 2009. Home-field advantage accelerates leaf litter decomposition in forests. *Soil Biol. Biochem.* 41:606–10

Balbontín J, Negro JJ, Sarasola JH, Ferrero JJ, Rivera D. 2008. Land-use changes may explain the recent range expansion of the Black-shouldered Kite *Elanus caeruleus* in southern Europe. *Ibis* 150:707–16

Bardgett RD, Wardle DA. 2010. *Aboveground-Belowground Linkages: Biotic Interactions, Ecosystem Processes and Global Change.* New York: Oxford Univ. Press

Barry JC, Morgan MLE, Flynn LJ, Pilbeam D, Behrensmeyer AK, et al. 2002. Faunal and environmental change in the late Miocene Siwaliks of northern Pakistan. *Paleobiology* 28:1–71

Berg MP, Kiers ET, Driessen G, Van der Heijden M, Kooi BW, et al. 2010. Adapt or disperse: understanding species persistence in a changing world. *Glob. Chang. Biol.* 16:587–98

Bertrand R, Lenoir J, Piedallu C, Riofrío-Dillon G, de Ruffray P, et al. 2011. Changes in plant community composition lag behind climate warming in lowland forests. *Nature* 479:517–20

Bezemer TM, Fountain MT, Barea JM, Christensen S, Dekker SC, et al. 2010. Divergent composition but similar function of soil food webs of individual plants: plant species and community effects. *Ecology* 91:3027–36

Blankinship JC, Niklaus PA, Hungate BA. 2011. A meta-analysis of responses of soil biota to global change. *Oecologia* 165:553–65

Blossey B, Nötzold R. 1995. Evolution of increased competitive ability in invasive nonindigenous plants: a hypothesis. *J. Ecol.* 83:887–89

Bukovinszky T, van Veen FJF, Jongema Y, Dicke M. 2008. Direct and indirect effects of resource quality on food web structure. *Science* 319:804–7

Burrows MT, Schoeman DS, Buckley LB, Moore P, Poloczanska ES, et al. 2011. The pace of shifting climate in marine and terrestrial ecosystems. *Science* 334:652–55

Bush MB. 2002. Distributional change and conservation on the Andean flank: a palaeoecological perspective. *Glob. Ecol. Biogeogr.* 11:463–73

Callaway RM, Brooker RW, Choler P, Kikvidze Z, Lortie CJ, et al. 2002. Positive interactions among alpine plants increase with stress. *Nature* 417:844–48

Callaway RM, Ridenour WM. 2004. Novel weapons: invasive success and the evolution of increased competitive ability. *Front. Ecol. Environ.* 2:436–43

Cannone N, Sgorbati S, Guglielmin M. 2007. Unexpected impacts of climate change on alpine vegetation. *Front. Ecol. Environ.* 5:360–64

Chen I-C, Hill JK, Ohlemüller R, Roy DB, Thomas CD. 2011. Rapid range shifts of species associated with high levels of climate warming. *Science* 333:1024–26

Clay K, Holah J. 1999. Fungal endophyte symbiosis and plant diversity in successional fields. *Science* 285:1742–44

Colwell RK, Brehm G, Cardelús CL, Gilman AC, Longino JT. 2008. Global warming, elevational range shifts, and lowland biotic attrition in the wet tropics. *Science* 322:258–61

Crimmins SM, Dobrowski SZ, Greenberg JA, Abatzoglou JT, Mynsberge AR. 2011. Changes in climatic water balance drive downhill shifts in plant species' optimum elevations. *Science* 331:324–27

Dawson TP, Jackson ST, House JI, Prentice IC, Mace GM. 2011. Beyond predictions: biodiversity conservation in a changing climate. *Science* 332:53–58

De Vries F, Liiri M, Bjørnlund L, Bowker M, Christensen S, et al. 2012. Land use alters the resistance and resilience of soil food webs to drought. *Nat. Clim. Chang.* 2:276–80

Diez JM, Dickie I, Edwards G, Hulme PE, Sullivan JJ, Duncan RP. 2010. Negative soil feedbacks accumulate over time for non-native plant species. *Ecol. Lett.* 13:803–9

Ehrenfeld JG, Ravit B, Elgersma K. 2005. Feedback in the plant-soil system. *Annu. Rev. Environ. Resour.* 30:75–115

Ellison AM, Bank MS, Clinton BD, Colburn EA, Elliott K, et al. 2005. Loss of foundation species: consequences for the structure and dynamics of forested ecosystems. *Front. Ecol. Environ.* 3:479–86

Engelkes T, Morriën E, Verhoeven KJF, Bezemer TM, Biere A, et al. 2008. Successful range-expanding plants experience less above-ground and below-ground enemy impact. *Nature* 456:946–48

Fierer N, Strickland MS, Liptzin D, Bradford MA, Cleveland CC. 2009. Global patterns in belowground communities. *Ecol. Lett.* 12:1238–49

Grabherr G, Gottfried M, Pauli H. 1994. Climate effects on mountain plants. *Nature* 369:448

Grime JP. 1998. Benefits of plant diversity to ecosystems: immediate, filter and founder effects. *J. Ecol.* 86:902–10

Harris J. 2009. Soil microbial communities and restoration ecology: facilitators or followers? *Science* 325:573–74

Hättenschwiler S, Gasser P. 2005. Soil animals alter plant litter diversity effects on decomposition. *Proc. Natl. Acad. Sci. USA* 102:1519–24

Hegland SJ, Nielsen A, Lázaro A, Bjerknes AL, Totland O. 2009. How does climate warming affect plant-pollinator interactions? *Ecol. Lett.* 12:184–95

Heikkilä M, Seppä H. 2003. A 11,000 yr palaeotemperature reconstruction from the southern boreal zone in Finland. *Quat. Sci. Rev.* 22:541–54

Hersch-Green EI, Turley NE, Johnson MTJ. 2011. Community genetics: What have we accomplished and where should we be going? *Philos. Trans. R. Soc. B-Biol. Sci.* 366:1453–60

Hill JK, Griffiths HM, Thomas CD. 2011. Climate change and evolutionary adaptations at species' range margins. *Annu. Rev. Entomol.* 56:143–59

Hobbs RJ, Arico S, Aronson J, Baron JS, Bridgewater P, et al. 2006. Novel ecosystems: theoretical and management aspects of the new ecological world order. *Glob. Ecol. Biogeogr.* 15:1–7

Hoegh-Guldberg O, Hughes L, McIntyre S, Lindenmayer DB, Parmesan C, et al. 2008. Assisted colonization and rapid climate change. *Science* 321:345–46

Huntley B. 1990a. Dissimilarity mapping between fossil and contemporary pollen spectra in Europe for the past 13,000 years. *Quat. Res.* 33:360–76

Huntley B. 1990b. European postglacial forests: compositional changes in response to climatic change. *J. Veg. Sci.* 1:507–18

Jackson ST, Overpeck JT. 2000. Responses of plant populations and communities to environmental changes of the late Quaternary. *Paleobiology* 26:194–220

Jackson ST, Sax DF. 2010. Balancing biodiversity in a changing environment: extinction debt, immigration credit and species turnover. *Trends Ecol. Evol.* 25:153–60

Janz N, Nylin S, Wahlberg N. 2006. Diversity begets diversity: host expansions and the diversification of plant-feeding insects. *BMC Evol. Biol.* 6:4

Jump AS, Peñuelas J. 2005. Running to stand still: adaptation and the response of plants to rapid climate change. *Ecol. Lett.* 8:1010–20

Kardol P, Bezemer TM, Van der Putten WH. 2006. Temporal variation in plant-soil feedback controls succession. *Ecol. Lett.* 9:1080–88

Kardol P, Wardle DA. 2010. How understanding aboveground-belowground linkages can assist restoration ecology. *Trends Ecol. Evol.* 25:670–79

Keane RM, Crawley MJ. 2002. Exotic plant invasions and the enemy release hypothesis. *Trends Ecol. Evol.* 17:164–70

Klironomos JN. 2002. Feedback with soil biota contributes to plant rarity and invasiveness in communities. *Nature* 417:67–70

Lankau RA. 2011. Intraspecific variation in allelochemistry determines an invasive species' impact on soil microbial communities. *Oecologia* 165:453–63

Lankau RA, Nuzzo V, Spyreas G, Davis AS. 2009. Evolutionary limits ameliorate the negative impact of an invasive plant. *Proc. Natl. Acad. Sci. USA* 106:15362–67

Lau JA. 2006. Evolutionary responses of native plants to novel community members. *Evolution* 60:56–63

Lavergne S, Mouquet N, Thuiller W, Ronce O. 2010. Biodiversity and climate change: integrating evolutionary and ecological responses of species and communities. *Annu. Rev. Ecol. Evol. Syst.* 41:321–50

Le Roux PC, McGeoch MA. 2008. Rapid range expansion and community reorganization in response to warming. *Glob. Chang. Biol.* 14:2950–62

Leithead MD, Anand M, Silva LCR. 2010. Northward migrating trees establish in treefall gaps at the northern limit of the temperate-boreal ecotone, Ontario, Canada. *Oecologia* 164:1095–106

Lenoir J, Gégout JC, Dupouey JL, Bert D, Svenning J-C. 2010. Forest plant community changes during 1989–2007 in response to climate warming in the Jura Mountains (France and Switzerland). *J. Veg. Sci.* 21:949–64

Lenoir J, Gégout JC, Marquet PA, de Ruffray P, Brisse H. 2008. A significant upward shift in plant species optimum elevation during the 20th century. *Science* 320:1768–71

Loarie SR, Duffy PB, Hamilton H, Asner GP, Field CB, Ackerly DD. 2009. The velocity of climate change. *Nature* 462:1052–55

Madritch MD, Lindroth RL. 2011. Soil microbial communities adapt to genetic variation in leaf litter inputs. *Oikos* 120:1696–704

Maron JL, Vilà M. 2001. When do herbivores affect plant invasion? Evidence for the natural enemies and biotic resistance hypotheses. *Oikos* 95:361–73

Meisner A, De Boer W, Cornelissen JHC, Van der Putten WH. 2012. Reciprocal effects of litter from exotic and congeneric native plant species via soil nutrients. *PLoS One* 7:e31596

Meisner A, De Boer W, Verhoeven KJF, Boschker HTS, Van der Putten WH. 2011. Comparison of nutrient acquisition in exotic plant species and congeneric natives. *J. Ecol.* 99:1308–15

Menéndez R, González-Megías A, Lewis OT, Shaw MR, Thomas CD. 2008. Escape from natural enemies during climate-driven range expansion: a case study. *Ecol. Entomol.* 33:413–21

Milton SJ. 2003. 'Emerging ecosystems'—a washing-stone for ecologists, economists and sociologists? *S. Afr. J. Sci.* 99:404–6

Mitchell CE, Power AG. 2003. Release of invasive plants from fungal and viral pathogens. *Nature* 421:625–27

Monroy F, Van der Putten WH, Yergeau E, Duyts H, Mortimer SR, Bezemer TM. 2012. Structure of microbial, nematode and plant communities in relation to geographical distance. *Soil Biol. Biochem.* 45:1–7

Moorcroft PR, Pacala SW, Lewis MA. 2006. Potential role of natural enemies during tree range expansions following climate change. *J. Theor. Biol.* 241:601–16

Morriën E, Engelkes T, Macel M, Meisner A, Van der Putten WH. 2010. Climate change and invasion by intracontinental range-expanding exotic plants: the role of biotic interactions. *Ann. Bot.* 105:843–48

Morriën E, Engelkes T, Van der Putten WH. 2011. Additive effects of aboveground polyphagous herbivores and soil feedback in native and range-expanding exotic plants. *Ecology* 92:1344–52

Müller-Schärer H, Schaffner U, Steinger T. 2004. Evolution in invasive plants: implications for biological control. *Trends Ecol. Evol.* 19:417–22

Muys B, Lust N. 1992. Inventory of the earthworm communities and the state of litter decomposition in the forests of Flanders, Belgium, and its implications for forest management. *Soil Biol. Biochem.* 24:1677–81

Naeem S, Bunker DE, Hector A, Loreau M, Perrings C. 2009. *Biodiversity, Ecosystem Functioning, and Human Wellbeing: An Ecological and Economic Perspective*. New York: Oxford Univ. Press. 368 pp.

Ozinga WA, Römermann C, Bekker RM, Prinzing A, Tamis WLM, et al. 2009. Dispersal failure contributes to plant losses in NW Europe. *Ecol. Lett.* 12:66–74

Parker IM, Gilbert GS. 2007. When there is no escape: the effects of natural enemies on native, invasive, and noninvasive plants. *Ecology* 88:1210–24

Parmesan C. 2006. Ecological and evolutionary responses to recent climate change. *Annu. Rev. Ecol. Evol. Syst.* 37:637–69

Parmesan C, Yohe G. 2003. A globally coherent fingerprint of climate change impacts across natural systems. *Nature* 421:37–42

Peltzer DA, Allen RB, Lovett GM, Whitehead D, Wardle DA. 2010. Effects of biological invasions on forest carbon sequestration. *Glob. Chang. Biol.* 16:732–46

Phillips BL, Brown GP, Shine R. 2010. Life-history evolution in range-shifting populations. *Ecology* 91:1617–27

Porazinska DL, Bardgett RD, Blaauw MB, Hunt HW, Parsons AN, et al. 2003. Relationships at the aboveground-belowground interface: plants, soil biota, and soil processes. *Ecol. Monogr.* 73:377–95

Procter DLC. 1984. Towards a biogeography of free-living soil nematodes. I. Changing species richness, diversity and densities with changing latitude. *J. Biogeogr.* 11:103–17

Scherber C, Mwangi PN, Schmitz M, Scherer-Lorenzen M, Bessler H, et al. 2010. Biodiversity and belowground interactions mediate community invasion resistance against a tall herb invader. *J. Plant Ecol.* 3:99–108

Schweitzer JA, Bailey JK, Fischer DG, Leroy CJ, Lonsdorf EV, et al. 2008. Plant-soil-microorganism interactions: heritable relationship between plant genotype and associated soil microorganisms. *Ecology* 89:773–81

Simberloff D, Gibbons L. 2004. Now you see them, now you don't!—Population crashes of established introduced species. *Biol. Invasions* 6:161–72

Strauss SY, Webb CO, Salamin N. 2006. Exotic taxa less related to native species are more invasive. *Proc. Natl. Acad. Sci. USA* 103:5841–45

Strickland MS, Osburn E, Lauber C, Fierer N, Bradford MA. 2009. Litter quality is in the eye of the beholder: initial decomposition rates as a function of inoculum characteristics. *Funct. Ecol.* 23:627–36

Sundqvist MK, Giesler R, Graae BJ, Wallander H, Fogelberg E, Wardle DA. 2011. Interactive effects of vegetation type and elevation on aboveground and belowground properties in a subarctic tundra. *Oikos* 120:128–42

Swetnam TW, Allen CD, Betancourt JL. 1999. Applied historical ecology: using the past to manage for the future. *Ecol. Appl.* 9:1189–206

Symstad AJ, Siemann E, Haarstad J. 2000. An experimental test of the effect of plant functional group diversity on arthropod diversity. *Oikos* 89:243–53

Tamis WLM, Van't Zelfde M, Van Meijden R, Haes HAU. 2005. Changes in vascular plant biodiversity in the Netherlands in the 20th century explained by their climatic and other environmental characteristics. *Clim. Chang.* 72:37–56

Thomas CD, Cameron A, Green RE, Bakkenes M, Beaumont LJ, et al. 2004. Extinction risk from climate change. *Nature* 427:145–48

Thuiller W, Lavorel S, Araújo MB, Sykes MT, Prentice IC. 2005. Climate change threats to plant diversity in Europe. *Proc. Natl. Acad. Sci. USA* 102:8245–50

Tylianakis JM, Didham RK, Bascompte J, Wardle DA. 2008. Global change and species interactions in terrestrial ecosystems. *Ecol. Lett.* 11:1351–63

Van der Putten WH, Macel M, Visser ME. 2010. Predicting species distribution and abundance responses to climate change: why it is essential to include biotic interactions across trophic levels. *Philos. Trans. R. Soc. B-Biol. Sci.* 365:2025–34

Van Grunsven RHA, Van der Putten WH, Bezemer TM, Berendse F, Veenendaal EM. 2010. Plant-soil interactions in the expansion and native range of a poleward shifting plant species. *Glob. Chang. Biol.* 16:380–85

Van Grunsven RHA, Van der Putten WH, Bezemer TM, Tamis WLM, Berendse F, Veenendaal EM. 2007. Reduced plant–soil feedback of plant species expanding their range as compared to natives. *J. Ecol.* 95:1050–57

Van Kleunen M, Fischer M. 2009. Release from foliar and floral fungal pathogen species does not explain the geographic spread of naturalized North American plants in Europe. *J. Ecol.* 97:385–92

Vilà M, Espinar JL, Hejda M, Hulme PE, Jarošík V, et al. 2011. Ecological impacts of invasive alien plants: a meta-analysis of their effects on species, communities and ecosystems. *Ecol. Lett.* 14:702–8

Walther GR. 2010. Community and ecosystem responses to recent climate change. *Philos. Trans. R. Soc. B-Biol. Sci.* 365:2019–24

Walther GR, Post E, Convey P, Menzel A, Parmesan C, et al. 2002. Ecological responses to recent climate change. *Nature* 416:389–95

Walther GR, Roques A, Hulme PE, Sykes MT, Pyšek P, et al. 2009. Alien species in a warmer world: risks and opportunities. *Trends Ecol. Evol.* 24:686–93

Wardle DA. 2002. *Communities and Ecosystems: Linking the Aboveground and Belowground Components.* Princeton, NJ: Princeton Univ. Press

Wardle DA, Bardgett RD, Callaway RM, Van der Putten WH. 2011. Terrestrial ecosystem responses to species gains and losses. *Science* 332:1273–77

Wardle DA, Bardgett RD, Klironomos JN, Setälä H, Van der Putten WH, Wall DH. 2004. Ecological linkages between aboveground and belowground biota. *Science* 304:1629–33

Wardle DA, Bonner KI, Barker GM, Yeates GW, Nicholson KS, et al. 1999. Plant removals in perennial grassland: vegetation dynamics, decomposers, soil biodiversity, and ecosystem properties. *Ecol. Monogr.* 69:535–68

Warren MS, Hill JK, Thomas JA, Asher J, Fox R, et al. 2001. Rapid responses of British butterflies to opposing forces of climate and habitat change. *Nature* 414:65–69

Whitham TG, Bailey JK, Schweitzer JA, Shuster SM, Bangert RK, et al. 2006. A framework for community and ecosystem genetics: from genes to ecosystems. *Nat. Rev. Genet.* 7:510–23

Whittaker RH. 1972. Evolution and measurement of species diversity. *Taxon* 21:213–51

Williams JW, Jackson ST, Kutzbacht JE. 2007. Projected distributions of novel and disappearing climates by 2100 AD. *Proc. Natl. Acad. Sci. USA* 104:5738–42

Willis KJ, Bennett KD, Bhagwat SA, Birks HJB. 2010. 4 °C and beyond: What did this mean for biodiversity in the past? *Syst. Biodivers.* 8:3–9

Wolfe LM, Elzinga JA, Biere A. 2004. Increased susceptibility to enemies following introduction in the invasive plant *Silene latifolia*. *Ecol. Lett.* 7:813–20

Inflammation: Mechanisms, Costs, and Natural Variation

Noah T. Ashley,[1] Zachary M. Weil,[2]
and Randy J. Nelson[2]

[1] Department of Biology, Western Kentucky University, Bowling Green, Kentucky 42101;
email: noah.ashley@wku.edu

[2] Department of Neuroscience, Wexner College of Medicine, Ohio State University, Columbus,
Ohio 43210; email: Zachary.Weil@osumc.edu, Randy.Nelson@osumc.edu

Annu. Rev. Ecol. Evol. Syst. 2012. 43:385–406

First published online as a Review in Advance on
September 4, 2012

The *Annual Review of Ecology, Evolution, and
Systematics* is online at ecolsys.annualreviews.org

This article's doi:
10.1146/annurev-ecolsys-040212-092530

Keywords

host-commensal, immunity, life-history trade-off, pathogen, Th1/Th2
paradigm, virulence

Abstract

Inflammation is a pervasive phenomenon that operates during severe perturbations of homeostasis, such as infection, injury, and exposure to contami-
nants, and is triggered by innate immune receptors that recognize pathogens
and damaged cells. Among vertebrates, the inflammatory cascade is a complex network of immunological, physiological, and behavioral events that are
coordinated by cytokines, immune signaling molecules. Although the molecular basis of inflammation is well studied, its role in mediating the outcome
of host-parasite interactions has received minimal attention by ecologists.
This review provides a synopsis of vertebrate inflammation, its life-history
modulation, and its effects upon host-pathogen dynamics as well as host-
commensal microbiota interactions in the gut. What emerges is evidence
for phenotypic plasticity of inflammatory responses despite the apparently
invariant and redundant nature of the immunoregulatory networks that regulate them.

INTRODUCTION

Virulence: damage to a host arising from extraction of resources by a pathogenic infection or collateral damage produced by the host's immune response

Inflammation is a pervasive form of defense that is broadly defined as a nonspecific response to tissue malfunction and is employed by both innate and adaptive immune systems to combat pathogenic intruders. A distinctive feature of inflammatory responses in relation to other facets of antiparasite defenses is that damage to the self is unavoidable. Importantly, collateral damage from inflammation is not the same as immunopathology, which involves a specific immune-mediated attack on target tissue that is no longer recognized by the immune system as self. Autoimmune pathology reflects dysregulation of adaptive immune components, such as antibody and cell-mediated functions, and has both genetic and environmental influences (Graham et al. 2005, Råberg et al. 1998). Although inflammation-induced collateral damage can certainly contribute to immunopathology (e.g., rheumatoid arthritis, multiple sclerosis, diabetes), the damage invoked by inflammation represents a basic biological trade-off between damage control and self-maintenance and does not require the presence of self-antigens to become activated.

The emergence of the field of ecoimmunology has spurred a renewed interest in quantifying and understanding variation of immune function, which has traditionally been under the purview of cellular and molecular immunologists. However, the major difference of this relatively new field involves the assessment of immunity in nonmodel organisms in their natural environment, which has been challenging and biased toward measurement of the adaptive immune system. Ecoimmunologists are also interested in the fitness costs of immunity. A major hurdle of ecoimmunology has been integrating host immune function with host-pathogen dynamics and disease ecology (Graham et al. 2011, Hawley & Altizer 2010). In effect, studies have quantified immune function in a vacuum without assessing how these measures relate to disease resistance. Lastly, the importance of inflammation in regulating the outcome of host-pathogen interactions has received minimal attention by ecologists (Sears et al. 2011, Sorci & Faivre 2009).

This review provides an overview of inflammation and its role in mediating the ecology and evolution of host parasite and host commensal interactions. This review primarily focuses on inflammation in vertebrates, though we draw upon several studies among invertebrates. Although the molecular basis of inflammation is well described, we provide basic definitions of inflammation, a synopsis of inflammatory pathways, and the types of inflammatory response to provide ecologists and evolutionary biologists with proximate mechanisms that inform ultimate levels of analysis. We then present studies demonstrating life-history variation of inflammatory responses, in particular seasonal and latitudinal variation, as well as trade-offs with reproduction. Inflammation also affects both the host and pathogen, and we review the evidence that host-pathogen dynamics can be altered, such as virulence and transmission. Throughout this review, we refer to pathogen and parasite interchangeably, which we define as infectious agents that have the capability to harm hosts. The regulation of inflammation also shapes host-commensal interactions in the vertebrate gut. Such interactions benefit commensal organisms, whereas the host species neither benefits nor is harmed. The selective forces governing host-microbiota interactions in the gut is a rapidly developing field that has received minimal input from evolutionary ecologists.

WHAT IS INFLAMMATION?

Inflammation is a biological reaction to a disrupted tissue homeostasis (Medzhitov 2008). At its basic level, it is a tissue-destroying process that involves the recruitment of blood-derived products, such as plasma proteins, fluid, and leukocytes, into perturbed tissue. This migration is facilitated by alterations in the local vasculature that lead to vasodilation, increased vascular permeability, and increased blood flow.

Infection by microbial invaders is often implicated as the major culprit that promotes inflammatory responses (**Figure 1a**). However, injury or trauma (in the absence of parasitic infection) and exposure to foreign particles/irritants/pollutants are also potent activators of inflammation (Medzhitov 2008), suggesting that this response likely evolved as a general adaptation for coping with damaged or malfunctioning tissue (Matzinger 2002). A common explanation for why infection and trauma might evoke similar inflammatory responses is that infection often follows wounding, which implies that it would be advantageous to respond to trauma as if infection occurred (Nathan 2002). The more parsimonious explanation is that both pathogens and wounding cause damage to cells and tissue and trigger similar responses (Bianchi 2007).

The primary functions of inflammation are to rapidly destroy or isolate the underlying source of the disturbance, remove damaged tissue, and then restore tissue homeostasis (Medzhitov 2008, Soehnlein & Lindbon 2010). Inflammation, when regulated properly, is putatively adaptive. This statement is supported by the increased risk of serious infections in humans with genetic deficiencies in primary components of inflammation, such as neutropenia (abnormally low level of circulating neutrophils). Defects in the genes that encode proinflammatory cytokines and effectors of inflammation using mouse knock-out studies are also characterized by increased susceptibility to infection (Martinon et al. 2009). Conversely, there are several immune-relevant genes whose disruption leads to spontaneous inflammation, suggesting that the inflammatory response is actively suppressed by regulatory gene products to maintain health when inflammatory stimuli are not present (Nathan 2002). When not regulated properly, excessive inflammation can have devastating effects, resulting in excessive collateral damage and pathology.

On an evolutionary level, inflammation is a highly conserved phenomenon and appears to be an important first line of defense for both invertebrates and vertebrates. Many of the components associated with the inflammatory cascade, such as chemotaxis and phagocytosis, are readily employed by unicellular organisms and were later co-opted as defensive mechanisms to maintain the integrity of more complex multicellular organisms (Rowley 1996). Innate immunity in the form of phagocytosis and antimicrobial peptides is present in the earliest of invertebrates, whereas the adaptive immune system evolved later and is unique to jawed vertebrates (Flajnik & Du Pasquier 2004). Adaptive immunity is hypothesized to have evolved to recognize and manage the complex communities of microbes that reside in the vertebrate digestive tract, which harbors a greater diversity of microbial fauna than the invertebrate gut (McFall-Ngai 2007).

MECHANISMS OF INFLAMMATION

Inflammation consists of a tightly regulated cascade of immunological, physiological, and behavioral processes that are orchestrated by soluble immune signaling molecules called cytokines. The first step of the inflammatory cascade involves recognition of infection or damage (**Figure 1b**). This is typically achieved by the detection of pathogen-associated molecular patterns (PAMPs), which are specifically directed toward general motifs of molecules expressed by pathogens that are essential for pathogen survival. Alarmins, or damage-associated molecular patterns (DAMPs), are endogenous molecules that signal damage or necrosis and are also recognized by the innate immune system. An advantage of detecting these signals is that inadvertent targeting of host cells and tissues is minimized. Unlike adaptive immunity, the innate immune system lacks the ability to distinguish among different strains of pathogens and whether such strains are virulent (harmful to the host) (Janeway et al. 2005).

Many damage signals are recognized by germ-line encoded receptors, such as transmembrane Toll-like receptors (TLRs) and intracellular nucleotide binding domain and leucine-rich-repeat-containing receptors (NOD-like receptors or NLRs) (Lange et al. 2001, Proell et al. 2008, Roach

Cytokines: soluble proteins of low molecular weight that modulate the differentiation, proliferation, and function of immune cells, and coordinate inflammatory responses

DAMP: damage-associated molecular pattern

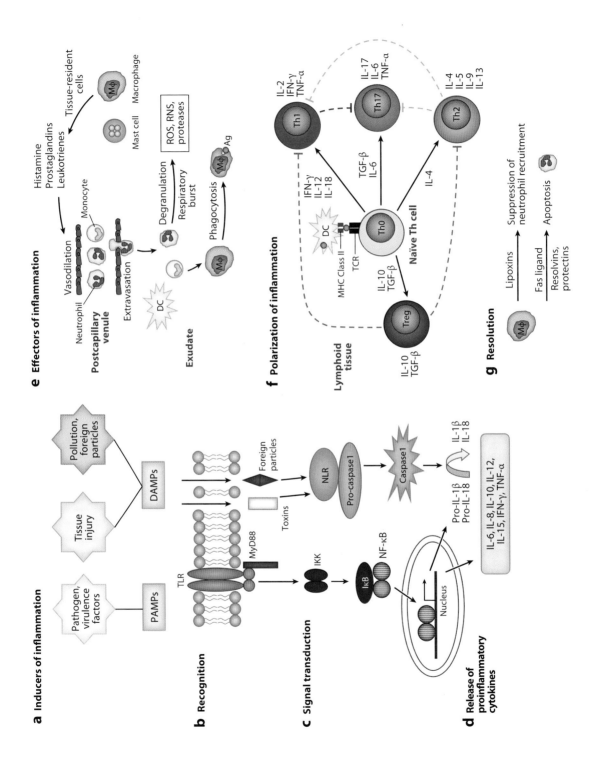

a Inducers of inflammation

Pathogen, virulence factors → PAMPs

Tissue injury → DAMPs

Pollution, foreign particles → DAMPs

b Recognition

TLR
MyD88
Toxins
Foreign particles
NLR
Pro-caspase1
Caspase1
Pro-IL-1β
Pro-IL-18
IL-1β
IL-18

c Signal transduction

IKK
IκB
NF-κB
Nucleus

d Release of proinflammatory cytokines

IL-6, IL-8, IL-10, IL-12, IL-15, IFN-γ, TNF-α

e Effectors of inflammation

Histamine
Prostaglandins
Leukotrienes

Tissue-resident cells
Mφ Macrophage
Mast cell

ROS, RNS, proteases

Vasodilation
Monocyte
Neutrophil
Postcapillary venule
Extravasation
DC
Exudate
Degranulation
Respiratory burst
Phagocytosis
Mφ Ag
Mφ

f Polarization of inflammation

IL-2
IFN-γ
TNF-α
Th1

IL-17
IL-6
TNF-α
Th17

IL-4
IL-5
IL-9
IL-13
Th2

IFN-γ
IL-12
IL-18

DC
MHC Class II
TCR
Th0
Naïve Th cell

TGF-β
IL-6

IL-4

IL-10
TGF-β

Treg

IL-10
TGF-β

Lymphoid tissue

g Resolution

Lipoxins → Suppression of neutrophil recruitment

Mφ

Fas ligand → Apoptosis

Resolvins, protectins

et al. 2005). Once recognition of ligands occurs, TLRs activate common signaling pathways that culminate in the activation of NF-κB (nuclear factor kappa-light-chain-enhancer of activated B cells; **Figure 1c**). This transcription factor is found in virtually all cell types and remains in an inactivated state bound to an inhibitor protein, IκB (Ghosh et al. 1998). Upon transduction of the signal, NF-κB is released from IκB and translocates to the nucleus, where transcription is upregulated through binding to target genes. Importantly, activation of NF-κB does not require new protein synthesis, which permits a rapid response. The NF-κB signaling system is ancient, but there is phylogenetic evidence that regulation of immune function by this pathway in vertebrates evolved independently from invertebrate immune mechanisms (Friedman & Hughes 2002). Intracellular NLRs respond to increasing numbers of DAMPs that alert the immune system to cell injury and provide a proximate pathway for sensing exposure to possible toxins or pollutants in the environment.

Transcription and translation of genes lead to the third stage of the inflammatory cascade, which is the inducible expression of proinflammatory cytokines, such as interleukin-1-beta (IL-1β), IL-6, tumor necrosis factor-alpha (TNF-α), and others (**Figure 1d**). In conjunction with chemokines (attractants) and various costimulatory molecules, these soluble proteins facilitate the recruitment of effector cells (**Figure 1e**), such as monocytes and neutrophils, to the site of disturbance. Neutrophils create a cytotoxic environment by releasing noxious chemicals from cytoplasmic granules (a process called degranulation). Rapid release of these chemicals requires consumption of both glucose and oxygen, known as the respiratory burst. Toxic chemicals released include highly reactive oxygen and nitrogen species (ROS and RNS, respectively) and various proteinases. These substances are destructive to both pathogens and hosts and essentially induce

---

Figure 1

A primer of the inflammatory cascade. (*a*) Pathogens, tissue injury, and foreign particles induce inflammation. (*b*) Transmembrane TLRs and intracellular NLRs bind to PAMPs or DAMPs, respectively. (*c*) TLRs activate a MyD88-dependent signal transduction pathway that involves the phosphorylation of the inhibitory IκB protein by IKK. NF-κB is released from IκB and translocates to the nucleus where transcription is upregulated through binding to target inflammatory genes. NLRs signal the inflammasome, which activates caspase-1 to convert cytokines into active forms (IL-1β and IL-18), which then elicit inflammation after being released from the cell. (*d*) A variety of proinflammatory cytokines and chemokines are produced and released to promote effector functions of inflammation. (*e*) Blood-borne neutrophils and monocytes migrate to the site of disturbance by chemotaxis and selectively pass through endothelial cells to reach target sites (extravasation). This influx of cells is accompanied by protein-rich fluid, known as the exudate, and promotes edema (swelling). Mast cells and tissue-resident macrophages promote this migration by releasing histamine, leukotrienes, and prostaglandins, which have rapid effects upon the vasculature, including vasodilation and increased vascular permeability. Neutrophils release toxic compounds, including ROS, RNS, and various proteases, which are nonspecific and harm both pathogen and host. Macrophages and dendritic cells participate in phagocytosis of Ag. (*f*) These cells migrate to lymphoid tissue and prime naïve T cells (Th0) to become polarized through stimulation of the TCR by antigen bound to MHC class II receptors. Th0 cells differentiate into several different types of effector and regulatory cells: Th1 cells (proinflammatory), Th2 cells (anti-inflammatory), Tregs (regulatory) and Th17 cells (proinflammatory). Depending upon the type of pathogen and other factors, the resulting Th population can be biased toward a proinflammatory, anti-inflammatory, or regulatory phenotype. Cytokines produced by polarized Th1 and Th2 are mutually inhibitory, whereas cytokines produced by Treg cells dampen both Th1 and Th2 responses. Th17 cells are highly proinflammatory and are regulated by the other Th subsets. Black arrowed and dashed lines represent stimulatory and inhibitory actions. (*g*) Resolution of inflammation occurs when neutrophils promote the switch of leukotrienes produced by macrophages and other cells to lipoxins, which initiates termination of inflammation. Fas ligand, resolvins, and protectins promote apoptosis of neutrophils. Macrophages phagocytose apoptotic neutrophils and cellular debris. Abbreviations: Ag, antigen; DC, dendritic cell; DAMP, damage-associated molecular pattern; IκB, nuclear factor of kappa light polypeptide gene enhancer in B-cells inhibitor; IKK, inhibitor of kappa B kinase; IFN-γ, interferon-gamma; IL, interleukin; Mφ, macrophage; MHC, major histocompatibility complex; MyD88, myeloid differentiation primary response gene (88); NF-κB, nuclear factor kappa-light-chain-enhancer of activated B cells; NLR, nucleotide binding domain and leucine-rich-repeat-containing receptors; PAMP, pathogen-associated molecular pattern; RNS, reactive nitrogen species; ROS, reactive oxygen species; TCR, T-cell receptor; TGF-β, transforming growth factor-beta; Th, T helper cell; TLR, Toll-like receptor; TNF-α, tumor necrosis factor-alpha; Treg, regulatory T cell. Adapted from Ghosh et al. 1998, Janeway et al. 2005, Anthony et al. 2007, and Soehnlein & Lindbom 2010.

liquefaction of surrounding tissue to stave off microbial metastasis (Nathan 2002). These effector mechanisms are thus major contributors to host collateral damage.

The net effect of these interactions culminates in the stereotypical cardinal signs of local inflammation: heat, swelling, redness, pain, and loss of function. The effector functions of inflammation are further regulated by the adaptive immune system (**Figure 1f**), which is discussed below.

The last phase of inflammation is its resolution (**Figure 1g**), which is critical for limiting collateral damage to the host (Serhan & Savill 2005). After the first few hours of inflammation, a coordinated program of resolution is set into motion by tissue-resident and recruited macrophages. During acute inflammation, these cells produce proinflammatory prostaglandins and leukotrienes, but rapidly switch to lipoxins, which block further neutrophil recruitment and instead favor enhanced infiltration of monocytes important for wound healing.

POLARIZATON OF INFLAMMATION

Over the course of evolution, animals have encountered a diverse array of parasites that range from microscopic bacteria and viruses to metazoan parasites, such as helminths and arthropods. As a general rule, proinflammatory responses are more effective against smaller pathogenic microbes (bacteria, viruses) than larger parasites. This is because multicellular parasites, as well as foreign bodies and indigestible particles, are too large to be phagocytosed by individual cells. In such cases, inflammation presents a less effective option for hosts as collateral damage increases substantially in responding to a larger, uncontained pathogen (Allen & Wynn 2011). In the early stages of parasite invasion, this method of defense attempts to heal wounded tissue through the formation of granulomas (fibrous connective tissue that replaces fibrin clots) to isolate and encapsulate the invader while preventing secondary bacterial infections. It is tempting to conclude that such a repair response would provide only benefits to the host. However, excessive granuloma formation can lead to fibrosis (scarring), which impedes normal functioning of host tissue and can lead to organ failure (Wynn 2004). Taken together, optimizing both types of defense appears to be critical for mitigating fitness costs of infection.

T-helper (Th) cells, a specific population of lymphocytes of the adaptive immune system, are largely responsible for activating and orchestrating these responses. When stimulated by cells presenting antigen, naïve Th cells (Th0; never exposed to antigen) can differentiate into several different types of effector and regulatory cells: Th1 cells (proinflammatory), Th2 cells (anti-inflammatory), regulatory T-cells (Tregs), and Th17 cells (Abbas et al. 1996, Anthony et al. 2007) (**Figure 1f**). Th1 cells regulate cellular immunity and proinflammatory responses against intracellular parasites through the release of interferon-gamma (IFN-γ), a cytokine that has potent antiviral and immunoregulatory properties and promotes further Th1 cell differentiation. Th1 cells also secrete IL-2 and TNF-α, which are important in mediating delayed type hypersensitivity responses and macrophage activation. By contrast, Th2 cells are important for humoral immunity, B-cell proliferation, regulation of allergic responses, and protection against infection from macroparasites (e.g., helminths). Th2 cells produce a different characteristic set of anti-inflammatory cytokines, such as IL-4, IL-5, IL-10, and IL-13, that stimulate further differentiation of the Th2 phenotype, promote alternative activation of macrophages, and induce B-cell antibody switching to IgE and eosinophil maturation, while downregulating the production of Th1 cytokines. In effect, Th1 and Th2 responses are mutually antagonistic and represent a balance between proinflammatory and anti-inflammatory mechanisms that may be selected upon to influence the outcome of infection (**Figure 1f**). Ideally, an optimization of Th1/Th2 phenotypes facilitates pathogen clearance with minimum damage to host tissues (**Figure 2**), although studies

Helminth: parasitic worms that include cestodes (tapeworms), trematodes (flukes), and nematodes (roundworms)

Granuloma: fibrous tissue formed by Th2 responses to encapsulate macroparasites or foreign particles

IL: interleukin

TNF: tumor necrosis factor

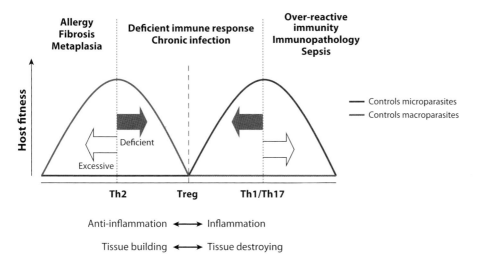

Figure 2

Host fitness is optimized relative to type of pathogen (micro- or macroparasite) and whether proinflammatory or anti-inflammatory responses are activated. A proinflammatory Th1 or Th17 response (*red line*) is more effective at controlling microparasites (viruses, bacteria, fungi, etc.), whereas an anti-inflammatory Th2 response (*blue line*) is more effective at controlling macroparasites (helminths, arthropods). To prevent excessive responses that can lead to various pathologies, these polarizing responses are tightly regulated by Treg cells. When regulation overrides effector mechanisms, chronic infection can result. Excessive and deficient responses are represented by white and gray arrows, respectively.

in captive animals demonstrate constrained ability to mount effective Th1 and Th2 responses simultaneously (Graham 2002, Mosmann & Coffman 1989).

The notion of a simplified bimodal response to infection that activates either the Th1 or Th2 phenotype has been eclipsed by recent discoveries of cross-regulation by additional T cells. Treg cells are important for suppressing the activation, proliferation, and effector functions of various immune cells that include T cells, NK cells, B cells, and antigen-presenting cells (Sakaguchi et al. 2010). Although the role of Treg cells in regulating homeostasis of the immune system is still being elucidated, there is evidence that these immune cells dampen both Th1 and Th2 responses by secreting IL-10 and TGF-β. These immunosuppressive cytokines inhibit the proliferation of both Th1 and Th2 responses to presumably minimize tissue damage (**Figure 2**). Treg cells also appear to play a role in mediating the outcome of chronic infection while preventing inflammation in immune-privileged organs. For example, removal of Treg cells during chronic infection with *Schistosoma mansoni* leads to increased liver damage in mice (Suvas et al. 2004). If Treg cells are inhibited, then tissue-damaging immunopathology results. However, an increased reliance upon Treg cells could lead to enhanced pathogen survival and, in some instances, long-term chronic infection (Belkaid & Tarbell 2009). Thus, a fine balance between regulatory and effector functions is established for hosts to effectively cope with pathogenic challenges (**Figure 2**). Th17 cells, a third subset of effector Th cells that was discovered in 2007, are largely involved in clearing extracellular microparasites that require a massive amount of inflammation and are not adequately handled by Th1 or Th2 responses (Korn et al. 2009).

Few coinfection studies have examined host Th1/Th2 polarization in free-living populations. One notable exception is in African buffalo (*Syncerus caffer*), where IFN-γ production and eosinophil counts in the blood were used to quantify Th1 (inflammatory) and Th2 (anti-inflammatory) phenotypes, respectively. African buffalo are one of the primary hosts of bovine

Treg cell: regulatory T cell

Th cell: T helper cell

tuberculosis (TB; caused by *Mycobacterium bovis*) and are also parasitized by a diverse assemblage of gastrointestinal nematodes. At the herd level and across the entire population, a negative correlation between TB prevalence and worm infection prevalence was documented (Jolles et al. 2008). These observed infection patterns were attributed to a combination of factors. First, coinfected buffalo suffered accelerated mortality compared to animals infected with either TB or worms. Second, trade-offs between the Th1 and Th2 responses affected TB transmission patterns. Buffalo that were worm-free had strong Th2 responses, and there was a significant negative correlation between Th1 and Th2 responses in TB-negative buffalo during the dry season (but not the wet season) (Jolles et al. 2008). These data suggest that animals whose immune systems were effective at fighting off worms were less able to simultaneously mount a strong Th1 response toward TB. This apparent cross-regulation of Th1 and Th2 responses is also more pronounced under conditions of seasonal resource limitation. Cross-regulation of Th1/Th2 responses is upheld in other naturally occurring populations (Jackson et al. 2011, Robinson et al. 2011), suggesting functional constraints.

Another example of the costs of polarizing inflammation comes from humans. The increased prevalence of allergic disease (asthma, rhinoconjuctivitis, and eczema) in developing countries over the past several decades has been attributed to a decline in childhood infections as a result of improved hygiene, vaccination, and use of antibiotics (Strachan 1989). Coined the hygiene hypothesis, this theory has an immunological explanation. If reduction in early-life microbial burden leads to insufficient stimulation of Th1 (proinflammatory) responses, then this effect could upset the Th1/Th2 balance, leading to expansion of Th2 (anti-inflammatory) cells. Excessive Th2 responses are characterized by increased IgE production to allergens, mastocytosis, and eosinophilia. This hypothesis is supported by the lack of allergic disease reported in developing countries where early-life infections are more common. However, this hypothesis fails to explain why there is a corresponding increase in Th1-autoimmune diseases, such as Type I diabetes, or why Th2-mediated helminth infections do not cause allergy. Instead, it has been hypothesized that persistent immune challenge in developing countries, with recurrent cycles of infection and inflammation, has resulted in a robust anti-inflammatory network that controls allergic disease (Yazdanbakhsh et al. 2002). This network would be weakly formed in children of industrialized countries, possibly leading to inappropriate allergic responses. Together, these studies support the concept that the apparent dichotomy between proinflammatory (Th1/Th17) and anti-inflammatory (Th2) defenses can significantly alter the course of infectious disease in vertebrate hosts. Thus, there is a pressing need to incorporate measures of Th1/Th2/Th17/Treg immunity (e.g., measurement of cytokines) into ecoimmunology studies to assess relationships with pathogen load and fitness parameters of the host (Graham et al. 2011).

TYPES OF INFLAMMATION DRIVE THEIR COSTS

Location

Distinguishing among different types of inflammation (**Table 1**, **Figure 3**) is critical for understanding relative fitness costs to the host. Importantly, inflammation and its sequelae vary both spatially and temporally (Medzhitov et al. 2012). Inflammation normally begins in a localized area, but depending upon the severity of the infection/wound, it can spread rapidly to the periphery. This systemic response is triggered by proinflammatory cytokines, particularly IL-1, IL-6, and TNF-α, which are released in the circulation and activate fever and sickness behaviors in the brain as well as acute phase protein secretion from the liver. Inflammation in this peripheral acute phase response is costlier to produce and maintain than local inflammation. In the absence of

Table 1 Types of inflammatory responses are categorized by intensity (low-grade versus high-grade) and duration (acute versus chronic)

Intensity	Duration	
	Acute	**Chronic**
Low-grade	Para-inflammation Metaplasia	Inflammatory diseases (diabetes mellitus, atherosclerosis) Autoimmune disorders Neurodegenerative diseases Tumor growth Tissue damage (fibrosis)
High-grade	Acute phase response Release of cytokines Neutrophil migration Recruitment of effector cells (neutrophils, macrophages) Localized tissue damage	Sepsis Cytokine storm Tissue destruction

infection, increased inflammation is clearly maladaptive and can lead to inappropriate patholo-gies, such as inflammatory diseases and autoimmunity (**Figure 3a**). However, the inflammatory response benefits hosts during a pathogenic challenge (**Figure 3b**). If the infection is cleared by inflammatory processes, then resolution of inflammation can occur (**Figure 3b**). Another scenario involves pathogen manipulation of the inflammatory response. Several different types of pathogens are known to trigger inflammatory defenses rather than evade them (reviewed by Sorci & Faivre 2009). For example, some intracellular pathogens have evolved to replicate within immune cells. During an inflammatory response, immune cells are recruited to the site of infection, which in-advertently increases transmission and spread of the pathogen (Sorci & Faivre 2009). For these

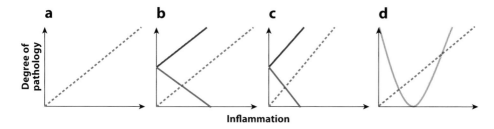

Figure 3

Relationship between inflammatory response and degree of pathology (morbidity). (*a*) In the absence of infection, inflammation increases pathology, as is the case for many autoinflammatory and autoimmune disorders. (*b*) During a pathogenic infection, the effectiveness of peripheral inflammation in clearing the pathogen is an important factor influencing host morbidity. If inflammation decreases pathogen load, then clearance of the pathogen will result when the morbidity costs of parasitism (*blue line*) equal the costs of the host inflammatory response in the absence of pathogen (*gray dashed line*). Once clearance occurs, inflammation resolves. However, if the pathogen evades or manipulates the host inflammatory response, then the combination of host inflammatory collateral damage and direct damage from the pathogen increases pathology (*red line*). (*c*) In the brain, the survival costs of exhibiting inflammation are greater than in the periphery, as represented by an increased slope relating morbidity to host inflammation (*gray dashed line*). Red and blue lines represent relationships as described above. (*d*) The diverse microbiota in the gut requires mild inflammation (*green line*) to control opportunistic pathogens while minimizing damage to the self and commensal flora.

special cases, the costs of pathology should increase owing to damage incurred from both host inflammation and pathogen exploitation (*red line*, **Figure 3b**).

Even more costly is the activation of inflammatory responses within tissues or organs that exhibit immunological privilege, a tissue-specific phenomenon that involves a lack of a response, or tolerance, to foreign antigens. In these organs, immunological tolerance of antigens appears to be favored over activation of the inflammatory response. It is hypothesized that immunological privilege is an adaptation to protect essential structures from the damaging effects of inflammation (Kaplan & Niederkorn 2007). When inflammation does occur, the resulting tissue damage can lead to permanent loss of function, morbidity, and mortality. For example, inflammation in the brain can have rapid and deleterious consequences for the host that are far greater than the costs of peripheral inflammation, as depicted by a steep morbidity curve in **Figure 3c**. Most of the neuronal damage from bacterial meningitis and cerebral malaria is attributed to over-reactive inflammatory responses. Also, a significant proportion of damage that occurs from traumatic brain injury is from inflammatory processes rather than from direct mechanical damage. This suggests that inflammation in the brain after trauma is maladaptive rather than beneficial, as many traumatic injuries to the central nervous system are fatal in humans (and presumably other species) without modern medical intervention (reviewed by Weil et al. 2008). Furthermore, unlike peripheral tissues, damage to the central nervous system can cause irreparable harm because regeneration potential is limited (Weil et al. 2008). A unique case of selection for mild, yet persistent, inflammation (rather than no inflammation) in the vertebrate gut is discussed in a later section.

Timing

Another important consideration is the time course of inflammation. Acute inflammation encompasses the immediate and early responses to an injurious agent and is quickly resolved. Once the disturbance is removed, there is strong selective pressure for the inflammatory response to end so that tissue repair can restore functionality. Chronic inflammation results when the disturbance persists. The beginning of chronic inflammation is often characterized by the replacement of neutrophils with macrophages and other immune cells, such as T cells (Medzhitov 2008). During a chronic inflammatory state, granulomas are typically formed in a final attempt to wall off the host from pathogens. Persistent inflammation leads to increased cellular turnover and provides selection pressure that results in the emergence of cells that are at high risk for cancer. Chronic inflammation is also associated with a variety of cardiovascular, metabolic, and neurodegenerative diseases (**Table 1**), as well as stroke and myocardial infarction (reviewed by Hotamisligil 2006). Many of these diseases are often tied to aging, suggesting that the costs of chronic inflammation are not realized until later in life. For example, it has been proposed that the historical increase in human life span is a result of reduced life-time exposure to inflammatory insults from infectious disease (Finch & Crimmins 2004). Age-related inflammatory disease, from the progressive buildup of oxidative damage from chronic inflammation, could be the cost paid for having a robust immune system during early life. Although there is a general lack of empirical evidence to support this hypothesis, a recent study in mealworm beetles (*Tenebrio molitor*) demonstrated that increased inflammatory response early in life accelerates aging (Pursall & Rolff 2011).

Magnitude

The magnitude of the inflammatory response is also important when assessing costs. A low response could result in an underreactive and deficient immune response, whereas excessive

Immunological privilege: lack of a response (tolerance) against an introduced antigen to protect against inflammatory damage exhibited by certain tissues, such as brain, eye, placenta, and testis

inflammation leads to costly collateral damage that, if unchecked, causes serious pathology. Thus, there is balancing selection for an intermediate level of inflammation to occur in most cases (**Figure 2**). Even moderate tissue stress, known as low-grade or para-inflammation, can trigger the inflammatory cascade. This response is graded and intermediate between baseline homeostasis and a classic inflammatory response (Medzhitov 2008). Para-inflammation is hypothesized to play a large role in the development of chronic inflammatory conditions and the eventual pathogenesis of several modern human diseases, as previously discussed.

When location, timing, and magnitude of inflammation are considered together, a hierarchy of costs is established. Chronic and high-grade inflammatory responses in immune-privileged organs lead to high fitness costs that include extensive damage, whereas acute, low-grade inflammation in localized tissue is the least expensive option. Thus, we predict that hosts should attempt to minimize these costs through the efficient delivery of immune cells to focal sites, the use of regulatory networks that quickly resolve inflammation, and the minimization of inflammation spreading into the periphery (such as the blood, which triggers cytokine storm and sepsis) or organs sensitive to inflammatory damage (brain, eye, testes, etc.).

NATURAL VARIATION OF INFLAMMATION

Hosts vary substantially in immune function and their ability to resist or tolerate infection (Medzhitov et al. 2012, Råberg et al. 2007). Ecoimmunology studies have demonstrated that activation of the immune system typically generates measurable costs through trade-offs with other life-history parameters (e.g., reproduction, growth, age) (Sadd & Schmid-Hempel 2009, Schmid-Hempel & Ebert 2003, Sheldon & Verhulst 1996). Some of these trade-offs are not apparent unless the host is exposed to stressful conditions (Hanssen et al. 2004, Moret & Schmid-Hempel 2000, Råberg et al. 1998). In contrast, the costs of inflammatory responses are normally overt. First, substantial energetic costs are involved. Mounting a febrile response involves a rise in the thermoregulatory set point that results in a 10–15% increase in metabolic rate for every 1°C increase in body temperature (Roe & Kinney 1965). Glucose use can also increase up to 68% during an acute phase response (Klasing 1998). Perhaps the most striking example involves sepsis, which is a whole-body inflammatory state that involves a 30–60% increase in metabolic rate (Lochmiller & Deerenberg 2000). Second, inflammation-induced collateral damage that is reparable requires time and energy to restore through tissue remodeling. Life-history costs of repair are largely ignored by ecoimmunology studies, but they could play a critical role during host recovery. Host damage that is irreparable obviously has deleterious effects upon normal physiological functioning and, ultimately, host fitness. Third, activation of sickness behavior involves a temporal cost to hosts. Most animals generally forego normal functions during behavioral symptoms of infections perhaps as an adaptation to conserve energy and minimize the risk of predation (Hart 1988). These symptoms of infection are regulated by inflammatory mediators, as blocking the effects of proinflammatory cytokines in the brain using receptor antagonists can diminish sickness behaviors (Dantzer 2001). Modification of host behavior can also affect disease dynamics, such as contact rates (Hawley & Altizer 2010).

Increasingly, experimental studies have identified life-history trade-offs where activation of the immune system leads to a corresponding decline in reproductive success or somatic growth (reviewed by Schmid-Hempel & Ebert 2003). Conversely, experimentally increasing reproductive effort (e.g., increasing brood size) can lead to a corresponding decrease in a host's immune response. There are significant well-described effects upon normal life-history functions during an inflammatory response. Local tissue inflammation leads to increased pain, which can deter animals from participating in normal activities. If inflammation spreads peripherally or centrally, then

Table 2 Examples of studies demonstrating life-history variation of lipopolysaccharide (LPS)-induced inflammation

Type of variation	Species	Study	Result
Age	House Mouse (*Mus musculus*)	Godbout et al. 2005	Neuroinflammation and sickness behavior increase in aged mice
	Norway Rat (*Rattus norvegicus*)	Foster et al. 1992	Increased duration of fever and elevated IL-6 and TNF in aged rats
	Tree Swallow (*Tachycineta bicolor*)	Palacios et al. 2011	Stronger anorexic responses and less parental care in older females[a]
Latitudinal	Song Sparrow (*Melospiza melodia*)	Adelman et al. 2010b	Fever diminished at higher latitude
	White-crowned Sparrow (*Zonotrichia leucophrys*)	Owen-Ashley et al. 2008	Anorexic responses vary by latitude, but are affected by body condition
Seasonal	Siberian Hamster (*Phodopus sungorus*)	Bilbo et al. 2002	Short day lengths attenuate fever and IL-1, IL-6
	White-crowned Sparrow (*Zonotrichia leucophrys*)	Owen-Ashley et al. 2006	Short day lengths diminish anorexic response
	Song Sparrow (*Melospiza melodia*)	Owen-Ashley & Wingfield 2006	Long days diminish sickness behavior in breeding males
Trade-offs with reproduction	House Sparrow (*Passer domesticus*)	Bonneaud et al. 2003	Increasing clutch size ameliorates feeding of nestlings in breeding females
	House Mouse (*Mus musculus*)	Aubert et al. 1997, Weil et al. 2006a	Parental care is resumed if survival of offspring is threatened; maternal aggression is unaffected
	Siberian Hamster (*Phodopus sungorus*)	Weil et al. 2006b	Short-day regression of gonads is delayed

[a]Effect observed only during the first year of the study.

proinflammatory cytokines can have direct inhibitory effects upon the hypothalamo-pituitary-gonadal axis that governs the release of sex steroids important for male and female reproduction. Behaviors associated with reproduction, such as sexual behavior, mating, aggression, and parental care, are typically suppressed in favor of sickness behaviors, which involve somnolence (sleepiness), reduced food and water intake, and depressive-like behaviors (reviewed by Ashley & Wingfield 2012). Although debated, such behaviors are thought to impart a selective advantage to hosts during an infection to conserve energy while reducing the intake of micronutrients that are important for pathogen growth (Hart 1988).

If inflammation carries substantial costs, then its activation should vary in relation to other life-history processes that have lower fitness costs. Accumulating evidence from both captive and field studies suggests that inflammatory responses are not fixed but display phenotypic plasticity (**Table 2**). To trigger inflammation, many studies have relied upon challenging animals with lipopolysaccharide (LPS; an immunogenic compound isolated from the cell wall of bacteria that mimics the onset of an infection in a dose-dependent fashion and is readily recognized by TLR4 on vertebrate immune cells), and then measuring physiological (cytokines, fever) or behavioral (anorexia, reductions in parental care or territorial aggression) responses.

Several studies have demonstrated seasonal modulation of inflammation (**Table 2**). Resources fluctuate dramatically and predictably in seasonal environments, especially in temperate and boreal regions, where low ambient temperatures and reduced food supply characterize winter. Seasonal

LPS:
lipopolysaccharide

environments can dramatically alter disease dynamics (Altizer et al. 2006), as well as the allocation of resources that hosts dedicate toward immunological defenses (Martin et al. 2008, Nelson 2004, Nelson & Demas 1996). Many seasonally breeding rodents prepare for the arrival of winter by enhancing various components of immune function, and this upregulation can be triggered experimentally by exposing captive rodents to short day lengths, a reliable environmental cue that forecasts winter (Nelson & Demas 1996). For example, in field voles (*Microtus agrestis*), both proinflammatory and anti-inflammatory cytokine mRNAs were generally upregulated during winter, and lowest during the breeding season (Jackson et al. 2011). Alternatively, seasonal changes in immune function could also be driven by intraseasonal trade-offs where immune system demands compete directly with reproduction or other costly life-history activities, such as migration and molting (Martin et al. 2008). In Siberian hamsters (*Phodopus sungorus*), exposure to short days increased several parameters of innate immunity (Martin et al. 2008), but attenuated LPS-induced fever and sickness behavior (Bilbo et al. 2002). It is hypothesized that prolonged inflammatory responses are selected against to presumably optimize energy expenditure with survival outcome when energy availability is low (Nelson 2004). A similar seasonal pattern has been observed in seasonally breeding songbirds (Owen-Ashley et al. 2006). Whether inflammation is suppressed during breeding or nonbreeding is highly contingent upon seasonal fluctuations in energy stores, with poor body condition and elevated glucocorticosteroids associated with diminished inflammation (Owen-Ashley & Wingfield 2007). It is unknown whether host inflammation is modulated according to seasonal parasite dynamics (e.g., prevalence, transmission); this requires further carefully designed experiments.

As latitude increases, seasonal fluctuations in energy availability become more pronounced. The summer breeding season is truncated at higher latitudes, and any disruption can delay reproduction until the following year. Coupled with decreased pathogen risk at higher latitudes (Piersma 1997), it is hypothesized that selection should minimize inflammatory responses that would otherwise interfere with limited reproductive opportunities. In support of this hypothesis, free-ranging song sparrows (*Melospiza melodia*) breeding in Washington State reduced febrile responses to LPS compared to conspecifics breeding in California, and these differences persisted in captivity, suggesting a role for genetic (or maternal) effects (Adelman et al. 2010a,b).

Reproduction is also costly and can potentially trade-off with the ability to mount inflammatory responses. LPS treatment reduces the nestling feeding rate of female house sparrows (*Passer domesticus*). However, when brood size is increased, feeding rates are ameliorated (Bonneaud et al. 2003). Female mice are less sensitive to LPS challenge when survival of pups is threatened (Aubert et al. 1997, Weil et al. 2006a), and LPS treatment delays gonadal regression upon exposure to short day lengths in Siberian hamsters (Weil et al. 2006b). These studies support the terminal investment hypothesis, which predicts that organisms should invest in current reproduction if the probability of the next reproductive event is low. The physiological mechanisms underlying terminal investment remain largely unexplored.

Other studies have reported increased proinflammatory responses with age (**Table 2**), which is consistent with human studies (Licastro et al. 2005). Genes regulating inflammation might provide a benefit to hosts early in life by controlling infection, but later become deleterious, as proposed by the antagonistic pleiotropy model (Williams 1957). Alternatively, the disposable soma hypothesis states that there is little incentive to invest in mechanisms of self-maintenance, such as Th2-type tissue repair from proinflammatory damage, because the payoff is minimal relative to the expected longevity of the individual (Kirkwood 1977). Differentiating between these two hypotheses represents a major challenge for evolutionary ecologists.

INFLAMMATION AND HOST-PARASITE INTERACTIONS

Effects Upon Host Fitness

At the individual level, there is ample evidence that inflammation is beneficial in the short term, but detrimental at chronic levels. For example, disabling inflammation through the use of antagonists or selective gene knock-outs can increase susceptibility to infection (Nathan 2002). These studies suggest that inflammation improves host resistance in the short term. There is also evidence that the inflammatory cascade is important for resistance against multicellular pathogens. For example, knock-out mice deficient in IL-1 were more susceptible to infection by a gastrointestinal nematode, *Trichuris muris*, than wild-type mice (Helmby & Grencis 2004). When scaling up to population-level effects, there is a lack of studies that have examined inflammatory markers in relation to host population dynamics.

Infection can negatively impact hosts through damage from the parasite and through tissue damage caused by inflammatory responses to the pathogen. **Table 3** provides examples of human pathologies from various infections where a proportion of the damage to the host is caused by inflammation. Such damage includes oxidative stress from ROS and RNS as well as degradation of proteins by proteases. Hosts can mitigate some of this damage and increase tolerance by producing antioxidants (Finkel & Holbrook 2000) or switching to a Th2 (anti-inflammatory) phenotype. There are very few examples where virulence is determined solely by direct damage delivered by the pathogen and its virulence factors (e.g., dental caries, paralysis from neurotoxins) (Margolis & Levin 2008). Instead, host inflammation significantly contributes to disease pathology, and in some cases, the damage from self-harm is greater than the damage caused by the pathogen itself (Graham et al. 2005).

In principle, any mechanism that involves self-damage should be selected against, yet inflammation is one of the most highly conserved biological processes and is routinely involved in the pathogenesis of many diseases. Why is this the case? The evolution of inflammatory collateral damage may depend upon two limitations of host defense: delayed activation of adaptive immunity and weak specificity of innate immunity (Sorci & Faivre 2009). First, whereas adaptive immunity requires days or weeks to fully develop, inflammatory responses are activated within minutes. Pathogens have much shorter generation times than their hosts and rapidly proliferate unless appropriate host defenses are in place. The onset of pathogenic infection represents a vulnerable period for the hosts because adaptive immune function is not fully developed. Hosts may only be left with the option of activating highly costly and damaging defenses to minimize pathogen load until more effective defenses can take over. As long as the benefit of mitigating or clearing infection is greater than the deleterious costs of collateral tissue damage, this response should prevail in the short term. Second, inflammation entails a nonspecific response to infection. Although pathogens are readily distinguished from the self to trigger inflammation, further refinement of immunity to specific injurious agents is constrained by nonspecific immune responses. It should be noted that collateral immunological damage is not unique to vertebrates, as self-damage from inflammation occurs in invertebrates as well (Sadd & Siva-Jothy 2006). Thus, the evolution of highly cytotoxic defenses likely represents a significant yet costly response for the host.

Effects Upon Parasite Fitness

The study of pathogen virulence has attempted to understand the morbidity and mortality of hosts caused by parasites and pathogens. Most models of virulence have arisen from the classic virulence-transmission trade-off model formulated by Anderson & May (1982) and

Table 3 Examples of inflammation contributing to pathogenesis in humans[a]

Pathology	Pathogen	Site of damage	Type of damage	Mechanism
Flu	1918 Spanish Influenza (reconstructed)	Lung	Alveolar damage Edema Hemorrhagic exudate	Alteration of host IFN-mediated antiviral response (Kobasa et al. 2007)
Duodenal Ulcer	*Helicobacter pylori*	Gastric and duodenal mucosa	Mucosal atrophy	Chronic inflammation
Toxic Shock Syndrome/Scarlet Fever	*Staphylococcus Streptococcus*	Circulatory system Systemic	Septic shock	Extreme inflammation Cytokine storm Oxidative damage
Pneumonia	Various species, e.g., *Streptococcus pneumoniae, Neisseria meningitides*	Lung	Alveolar damage Edema	Induction of proinflammatory cytokines Edema Fibrin deposition
Cerebral Malaria	*Plasmodium* spp.	Brain	Vascular leakage Brain hemorrhage	Release of proinflammatory cytokines
Cutaneous Anthrax	*Bacillus anthracis*	Skin	Tissue necrosis	Release of proinflammatory cytokines
Tuberculosis	*Mycobacterium tuberculosis*	Lung	Recruitment of fluid and cells into the air spaces of the lungs Necrosis	Release of proinflammatory cytokines, namely TNF-α
Meningitis	*S. pneumoniae N. meningitides*	Brain Spinal cord	Brain damage Increased blood-brain permeability Increased intracranial pressure	Release of proinflammatory cytokines

[a]Modified from Margolis & Levin (2008).

Ewald (1983):

$$R_0 = \frac{\beta S}{\mu + \alpha + \gamma}.$$

In this equation, R_0 is the basic reproductive number and is defined as the average number of secondary infections arising from one infected individual from a wholly susceptible host population. Pathogens can increase R_0 by enhancing transmission (β) or by decreasing the rate of parasite-induced mortality or virulence (α) or recovery rate of hosts (γ). α is influenced by the rate at which the parasite is cleared, which is a function of host resistance. S is the density of susceptible hosts in the population and μ is the natural mortality rate of the host, which can also affect pathogen fitness. If the host dies before transmission can occur, then the fitness of the pathogen effectively drops to zero (Anderson & May 1982, Frank 1996). Following this model, a fundamental trade-off between transmission and virulence is formed. Increased virulence should favor the production of more transmission forms per unit time (benefit) but limits the infectious period by host death (cost). Natural selection favors those pathogen strains that optimally balance these costs and benefits, leading to an intermediate level of virulence in most cases (reviewed in Alizon et al. 2008).

The trade-off model assumes that infection-induced mortality is solely a consequence of pathogen exploitation. However, a wealth of evidence suggests that host mortality from infection is often dependent upon the amount of collateral damage caused by host inflammation (Graham et al. 2005, Margolis & Levin 2008). Several testable predictions can be made when extending the trade-off model to incorporate inflammatory responses. First, it is predicted that inflammation directly enhances clearance of intracellular pathogens, albeit in a biased manner. Although inflammation is broadly effective at clearing infection, mathematical models predict that avirulent (or slow-growing) strains would be cleared faster than virulent (fast-growing) strains, selecting for more virulent strains to proliferate (Antia et al. 1994). This assumption, however, is entirely dependent upon the effectiveness of the inflammatory defense (magnitude, duration) in clearing the particular pathogen. Second, inflammatory responses cause damage to both host and pathogen and should thus directly contribute to increased virulence. Based upon these relationships, it is tempting to presume that inflammation leads to an obligatory increase in virulence and should, therefore, be selected against. However, before arriving at such a conclusion, it is necessary to investigate whether increased virulence from inflammation leads to a corresponding increase in parasite transmission. There are certainly cases where increased inflammation enhances the transmission of pathogens (e.g., diarrhea). However, if host inflammatory responses increase virulence without increasing transmission, then there would be selection for the opposite effect, which is decreased virulence (Graham et al. 2005). This is because any increase in self-damage would not provide an incremental advantage to the pathogen. In addition, whether or not damage from the parasite is dependent upon collateral damage is also important. If collateral damage is independent from pathogen-induced damage, then a high level of infection-induced mortality is expected because self-harm undermines the restrained exploitation of the parasite (Day et al. 2007). Conversely, reduced host mortality is predicted if a rise in collateral damage is accompanied by increased pathogen exploitation. In this case, natural selection would favor parasite strains that minimize collateral damage (Day et al. 2007). Thus, inflammatory responses have the potential to influence the outcome of host-parasite interactions, although empirical studies that manipulate levels of collateral host damage (without affecting pathogen-mediated virulence) are needed (Long & Boots 2011).

Because inflammation is nonspecific, there is ample evidence from the literature demonstrating the ability of microbes to manipulate or undermine the inflammatory cascade of hosts for their own benefit, thereby generating conditions that ensure their survival for a protracted period of time (reviewed in Sorci & Faivre 2009). A common mechanism employed is the induction of host anti-inflammatory cytokines, such as IL-10 and TGF-β, to temper the inflammatory cascade (Redpath et al. 2001). This can be achieved through the production of molecules that are similar to IL-10, such as viral IL-10, or through the direct induction of host immunosuppressive cytokines. Other mechanisms of evasion include interference with ROS and RNS production and resistance to phagocytic digestion (reviewed in Sorci & Faivre 2009). Although pathogens regularly employ antigenic variation to evade recognition by antigen-specific defenses, there are constraints in the ability of pathogens to modify PAMPs, which are molecules essential for pathogen survival and consequently recognized by the innate immune system. In other scenarios, within-host selection might favor pathogens triggering inflammation to exclude within-host competitors. A strategy of proactive invasion can allow a rare pathogen with enhanced immune resistance to invade and supplant competitors by promoting inflammation (Brown et al. 2008).

GUT INFLAMMATION AND HOST-COMMENSAL INTERACTIONS

The vertebrate gut is home to one of the most dense microbial populations on Earth, with up to 100 trillion (10^{14}) microbes inhabiting the human intestinal tract (Ley et al. 2006). Within this ecosystem, bacteria species dominate, especially the Firmicutes and the Bacteroidetes. Equally impressive is the ability of the intestinal epithelium (composed of a single layer of cells) to function as a selective barrier that must repel pathogens, tolerate commensals and food antigens, and rapidly incorporate nutrients essential for survival (MacDonald & Monteleone 2005). Besides producing microbial peptides, proinflammatory cytokines are also secreted by these cells in response to microbes. Gut-associated lymphoid tissue (GALT) is a collective term for lymphoid tissues located in the intestine that include Peyer's patches and isolated lymphoid follicles. These tissues play a large role in activating immune function that is restricted to the gut environment.

Gut inflammation can be triggered a number of different ways that include increasing exposure to pathogenic organisms, breaching the intestinal lumen, being exposed to irritants, or disabling the immunoregulatory network (Izcue et al. 2009). Intestinal homeostasis of immunity is achieved through the coordination of various regulatory mechanisms, including regulatory T cells (Tregs). These cells have been shown to significantly control gut inflammation by secreting immunosuppressive cytokines, such as IL-10 and TGF-β.

Under normal conditions, regulatory mechanisms override inflammatory signals reacting to stimuli produced by intestinal flora. The goal is to prevent immunological defenses from responding to harmless antigens found in the intestine while keeping opportunistic flora at bay, which requires low-grade, chronic inflammation. This statement is evidenced by studies in germ-free animals that demonstrate a mild state of inflammation when colonized by commensal flora (**Figure 2d**). The absence of flora results in underdeveloped immune organs, such as the spleen, and in GALT. These differences are no longer apparent once hosts are colonized with intestinal flora (Rakoff-Nahoum et al. 2004). If regulatory mechanisms are removed due to dysregulation, then the balance is tipped toward a heightened immune response, which leads to chronic inflammation and the development of inflammatory bowel disease. Disruption of the intestinal barrier also leads to chronic inflammation because the immune system is exposed to a greater amount of proinflammatory stimuli. A reduction in basal inflammation would presumably fail to control opportunistic pathogens, which would result in the accumulation of proinflammatory stimuli and trigger chronic inflammation despite an initial reduced inflammatory state (Izcue et al. 2009). Thus, the balance between host and flora is precarious, requiring finely tuned regulatory and effector mechanisms.

Does gut inflammation in turn alter the evolutionary dynamics of commensal microflora? Recent evidence suggests that acute inflammation can enhance the rate of horizontal gene transfer between pathogenic and commensal flora in the gut, which fosters the spread of virulence and antibiotic-resistant genes (Stecher et al. 2012). These findings suggest that inflammation can promote the coevolution of pathogenic microbes and commensal organisms in the gut, which paves the way for rapid evolution of emerging infectious diseases with novel phenotypes.

CONCLUSIONS

This review provides a framework for linking proximate mechanisms of inflammation with ecological and evolutionary perspectives to better understand the selective forces shaping natural variation of inflammation in vertebrates. Inflammation is a costly host defense mechanism that exhibits phenotypic plasticity despite the seemingly rigid and redundant immunoregulatory networks that govern it. However, these networks can be altered or dysregulated over the life span of the individual, as evidenced by an increase in chronic inflammation with age (Licastro et al. 2005). Furthermore, a bias in inflammatory status during early life can potentially reinforce the strength of the regulatory network during adulthood (Yazdanbakhsh et al. 2002). Whether such relationships apply to wild populations is untested. Examination of the potential organizational effects that alter immunoregulation of inflammation is a promising area of research to understand the trade-offs underlying inflammation.

Despite the tremendous progress that has been made toward understanding the proximate mechanisms of acute and chronic inflammation, very little is known about the role inflammation plays in regulating host population dynamics in the wild. Understanding some of the proximate mechanisms, such as signaling transduction and T-cell polarization, allows us to make predictions regarding how hosts respond to and control helminth parasites versus viral infections, for example. It is tempting to assume that chronic inflammation is rare in populations because individuals exhibiting such a state would presumably experience high fitness costs in the form of reduced fecundity and increased extrinsic mortality from predation or disease. However, we would also predict that chronic inflammation should increase in populations that have been exposed to rapidly emerging threats of the twenty-first century that include increased chemical and light pollution and the emergence of novel pathogens, which can trigger inflammatory states.

These possibilities raise several intriguing questions that could be experimentally tested: Are inflammatory states present in natural populations? If so, do individuals with high inflammation have the highest or lowest pathogen loads? Does this proportion vary in relation to sex, age, or season? Does inflammation resolve for some individuals, but not others? Does the activation and maintenance of host inflammation affect vital rates, such as fecundity and mortality? To begin answering these types of questions, immunological methods to accurately measure inflammatory states are required. Although there are several classic markers of inflammation, such as acute phase proteins (e.g., C-reactive protein, haptoglobin), cytokine levels measured in blood or other tissues using enzyme-linked immunosorbent assay (ELISA) or quantification of cytokine mRNA using rtPCR represent attractive options for assessing inflammatory states in wild populations. A panel of inflammatory (IL-1, IL-6, TNF, IFN, IL-17), anti-inflammatory (IL-4, IL-5), and regulatory (IL-10) cytokine activity could be ascertained in a single sample (Graham et al. 2007). A major challenge is developing such assays for nonmodel species because cytokine structure is highly variable between species, and antibody reagents used for measuring cytokines in murine and human studies do not necessarily cross-react with nonmodel species. However, the availability of high-throughput sequencing technologies and entire genome sequences makes it entirely possible to measure cytokine mRNA (Jackson et al. 2011). Coupling these measures with assessment of pathogen load and host vital rates would represent a powerful combination for assessing the role of inflammation in regulating host population dynamics. Lastly, monitoring the inflammatory state of populations could provide valuable insight for evaluating how animal populations respond to and cope with increasing environmental threats from pollution as well as emerging infectious disease from global climate change.

FUTURE ISSUES

1. Increasingly sophisticated molecular techniques need to be developed to explore inflammatory markers to elucidate the extent and natural variation of inflammation in free-living populations.

2. Although there are a number of studies demonstrating modulation of inflammation in a variety of life-history contexts, future studies should address how this modulation leads to alterations in life span, fecundity, survival, and resistance to disease.

3. Understanding the linkages between inflammation-induced collateral damage, tolerance to infection, and host recovery/repair will help address methods that measure the capacity of a host to cope with infection.

4. More examples that document polarized inflammatory states in hosts that are infected with an assemblage of different parasite species are needed.

5. Further insights in the role that inflammation plays in shaping the interactions between commensal and pathogenic bacteria in the vertebrate gut will increase our understanding of within-host competition, evolution of virulence, and emergence of novel pathogens.

DISCLOSURE STATEMENT

The authors are not aware of any affiliations, memberships, funding, or financial holdings that might be perceived as affecting the objectivity of this review.

ACKNOWLEDGMENTS

Support for writing this review comes from the US National Science Foundation (grant IOS-08-38098).

LITERATURE CITED

Abbas AK, Murphy KM, Sher A. 1996. Functional diversity of helper T lymphocytes. *Nature* 383:787–93

Adelman JS, Bentley GE, Wingfield JC, Martin LB, Hau M. 2010a. Population differences in fever and sickness behaviors in a wild passerine: a role for cytokines. *J. Exp. Biol.* 213:4099–109

Adelman JS, Córdoba-Córdoba K, Spoelstra K, Wikelski M, Hau M. 2010b. Radiotelemetry reveals variation in fever and sickness behaviors with latitude in a free-living passerine. *Funct. Ecol.* 24:813–23

Alizon S, Hurford A, Mideo N, van Baalen M. 2008. Virulence evolution and the trade-off hypothesis: history, current state of affairs and the future. *J. Evol. Biol.* 22:245–59

Allen JE, Wynn TA. 2011. Evolution of Th2 immunity: a rapid repair response to tissue destructive pathogens. *PLoS Pathog.* 7:1–4

Altizer SM, Dobson A, Hosseini P, Hudson P, Pascual M, Rohani P. 2006. Seasonality and the dynamics of infectious diseases. *Ecol. Lett.* 9:467–84

Anderson RM, May RM. 1982. Coevolution of hosts and parasites. *Parasitology* 85:411–26

Anthony RM, Rutitzky LI, Urban JF, Stadecker MJ, Gause WC. 2007. Protective immune mechanisms in helminth infection. *Nat. Rev. Immunol.* 7:975–87

Antia R, Levin BR, May RM. 1994. Within-host population dynamics and the evolution and maintenance of microparasite virulence. *Am. Nat.* 144:457–72

Ashley NT, Wingfield JC. 2012. Sickness behavior in vertebrates: allostasis, life-history modulation, and hormonal regulation. In *Ecoimmunology*, ed. GE Demas, RJ Nelson, pp. 45–91. Oxford: Oxford Univ. Press

Aubert A, Goodall G, Dantzer R, Gheusi G. 1997. Differential effects of lipopolysaccharide on pup retrieving and nest building in lactating mice. *Brain Behav. Immun.* 11:107–18

Belkaid Y, Tarbell K. 2009. Regulatory T cells in the control of host-microorganism interactions. *Annu. Rev. Immunol.* 2:551–89

Bianchi ME. 2007. DAMPs, PAMPs and alarmins: all we need to know about danger. *J. Leuk. Biol.* 81:1–5

Bilbo SD, Drazen DL, Quan N, He L, Nelson RJ. 2002. Short day lengths attenuate the symptoms of infection in Siberian hamsters. *Proc. R. Soc. Lond. B* 269:447–54

Bonneaud C, Mazuc J, Gonzalez G, Haussy C, Chastel O, et al. 2003. Assessing the cost of mounting an immune response. *Am. Nat.* 161:367–79

Brown SP, Le Chat L, Taddei F. 2008. Evolution of virulence: triggering host inflammation allows invading pathogens to exclude competitors. *Ecol. Lett.* 11:44–51

Dantzer R. 2001. Cytokine-induced sickness behavior: Where do we stand? *Brain Behav. Immun.* 15:7–24

Day T, Graham AL, Read AF. 2007. Evolution of parasite virulence when host responses cause disease. *Proc. R. Soc. Lond. B* 274:2685–92

Ewald PW. 1983. Host-parasite relations, vectors, and the evolution of disease severity. *Annu. Rev. Ecol. Syst.* 14:465–85

Finch CE, Crimmins EM. 2004. Inflammatory exposure and historical changes in human life spans. *Science* 305:1736–39

Finkel T, Holbrook NJ. 2000. Oxidants, oxidative stress and biology of ageing. *Nature* 408:239–47

Flajnik MF, Du Pasquier L. 2004. Evolution of innate and adaptive immunity: Can we draw a line? *Trends Immunol.* 25:640–44

Foster KD, Conn CA, Kluger MJ. 1992. Fever, tumor necrosis factor, and interleukin-6 in young, mature, and aged Fischer 344 rats. *Am. J. Physiol. Regul. Physiol.* 262:R211–15

Frank SA. 1996. Models of parasite virulence. *Q. Rev. Biol.* 71:37–78

Friedman R, Hughes AL. 2002. Molecular evolution of the NF-κβ signaling system. *Immunogenetics* 53:964–74

Ghosh S, May MJ, Kopp EB. 1998. NF-κβ and Rel proteins: evolutionarily conserved mediators of immune responses. *Annu. Rev. Immunol.* 16:225–60

Godbout JP, Chen J, Abraham J, Richwine AF, Berg BM, et al. 2005. Exaggerated neuroinflammation and sickness behavior in aged mice after activation of the peripheral innate immune system. *FASEB* 19:1329–31

Graham AL. 2002. When T-helper cells don't help: immunopathology during concomitant infection. *Q. Rev. Biol.* 77:409–34

Graham AL, Allen JE, Read AF. 2005. Evolutionary causes and consequences of immunopathology. *Annu. Rev. Ecol. Evol. Syst.* 36:373–93

Graham AL, Cattadori IM, Lloyd-Smith JO, Ferrari MJ, Bjørnstad ON. 2007. Transmission consequences of coinfection: cytokines writ large. *Trends Parasitol.* 23:284–91

Graham AL, Shuker DM, Pollitt LC, Auld SKJR, Wilson AJ, Little TJ. 2011. Fitness consequences of immune responses: strengthening the empirical framework for ecoimmunology. *Funct. Ecol.* 25:5–17

Hanssen SA, Hasselquist D, Folstad I, Erikstad KE. 2004. Costs of immunity: immune responsiveness reduces survival in a vertebrate. *Proc. R. Soc. Lond. B* 271:925–30

Hart BL. 1988. Biological basis of the behavior of sick animals. *Neurosci. Biobehav. Rev.* 12:123–37

Hawley DM, Altizer SM. 2010. Disease ecology meets ecological immunology: understanding the links between organismal immunity and infection dynamics in natural populations. *Funct. Ecol.* 25:48–60

Helmby H, Grencis RK. 2004. Interleukin 1 plays a major role in the development of Th2-mediated immunity. *Eur. J. Immunol.* 34:3674–81

Hotamisligil GS. 2006. Inflammation and metabolic disorders. *Nature* 444:860–67

Izcue A, Coombes JL, Powrie F. 2009. Regulatory lymphocytes and intestinal inflammation. *Annu. Rev. Immunol.* 27:313–38

Jackson JA, Begon M, Birtles R, Paterson S, Friberg IM, et al. 2011. The analysis of immunological profiles in wild animals: a case study of immunodynamics in the field vole. *Mol. Ecol.* 20:893–909

Janeway CA, Travers P, Walport M, Shlomchik MJ. 2005. *Immunobiology: The Immune System in Health and Disease*. New York: Garland Sci.

Jolles AE, Ezenwa V, Etienne RS, Turner WC, Olff H. 2008. Interactions between macroparasites and microparasites drive infection patterns in free-ranging African buffalo. *Ecology* 89:2239–50

Kaplan HJ, Niederkorn JY. 2007. Regional immunity and immune privilege. *Chem. Immunol. Allergy* 2007:11–26

Kirkwood TBL. 1977. Evolution of aging. *Nature* 270:301–4

Klasing KC. 1998. Nutritional modulation of resistance to infectious diseases. *Poult. Sci.* 77:1119–25

Kobasa D, Jones SM, Shinya K, Kash JC, Copps J, et al. 2007. Aberrant innate immune response in lethal infection of macaques with the 1918 influenza virus. *Nature* 445:319–23

Korn T, Bettelli E, Oukka M, Kuchroo VK. 2009. IL-17 and Th17 cells. *Annu. Rev. Immunol.* 27:485–517

Lange C, Hemmrich G, Klostermeier UC, López-Quintero JA, Miller DJ, et al. 2001. Defining the origins of the NOD-like receptor system at the base of animal evolution. *Mol. Biol. Evol.* 28:1687–702

Ley RE, Peterson DA, Gordon JI. 2006. Ecological and evolutionary forces shaping microbial diversity in the human intestine. *Cell* 24:837–48

Licastro F, Candore G, Lio D, Porcellini E, Colonna-Romano G, et al. 2005. Innate immunity and inflammation in ageing: a key for understanding age-related disease. *Immun. Ageing* 2:8–21

Lochmiller RL, Deerenberg C. 2000. Trade-offs in evolutionary immunology. Just what is the cost of immunity? *Oikos* 88:87–98

Long GH, Boots M. 2011. How can immunopathology shape the evolution of parasite virulence? *Trends Parasitol.* 27:300–5

MacDonald TT, Monteleone G. 2005. Immunity, inflammation, and allergy in the gut. *Science* 307:1920–25

Margolis E, Levin BR. 2008. The evolution of bacteria-host interactions: virulence and the immune over-response. In *Introduction to the Evolutionary Biology of Bacterial and Fungal Pathogens*, ed. F Baquero, C Nombela, GH Cassell, JA Gutiérrez, pp. 3–13. Washington, D.C.: ASM Press

Martin LB, Weil ZM, Nelson RJ. 2008. Seasonal changes in vertebrate immune activity: mediation by physiological trade-offs. *Philos. Trans. R. Soc. B* 363:321–39

Martinon F, Mayor A, Tschopp J. 2009. The inflammasomes: guardians of the body. *Annu. Rev. Immunol.* 27:229–65

Matzinger P. 2002. The danger model: a renewed sense of self. *Science* 296:301–5

McFall-Ngai M. 2007. Adaptive immunity: care for the community. *Nature* 445:153

Medzhitov R. 2008. Origin and physiological roles of inflammation. *Nature* 454:428–35

Medzhitov R, Schneider DS, Soares MP. 2012. Disease tolerance as a defense strategy. *Science* 335:936–41

Moret Y, Schmid-Hempel P. 2000. Survival for immunity: the price of immune system activation for bumblebee workers. *Science* 290:1166–70

Mosmann TR, Coffman RL. 1989. TH1 and TH2 cells: different patterns of lymphokine secretion lead to different functional properties. *Annu. Rev. Immunol.* 7:145–73

Nathan C. 2002. Points of control in inflammation. *Nature* 420:846–52

Nelson RJ. 2004. Seasonal immune function and sickness responses. *Trends Immunol.* 25:187–92

Nelson RJ, Demas GE. 1996. Seasonal changes in immune function. *Q. Rev. Biol.* 71:511–48

Owen-Ashley NT, Hasselquist D, Råberg L, Wingfield JC. 2008. Latitudinal variation of immune defense and sickness behavior in the white-crowned sparrow (*Zonotrichia leucophrys*). *Brain Behav. Immun.* 22:614–25

Owen-Ashley NT, Turner M, Hahn TP, Wingfield JC. 2006. Hormonal, behavioral, and thermoregulatory responses to bacterial lipopolysaccharide in captive and free-living white-crowned sparrows (*Zonotrichia leucophrys gambelii*). *Horm. Behav.* 49:15–29

Owen-Ashley NT, Wingfield JC. 2006. Seasonal modulation of sickness behavior in free-living northwestern song sparrows (*Melospiza melodia morphna*). *J. Exp. Biol.* 209:3062–70

Owen-Ashley NT, Wingfield JC. 2007. Acute phase responses of passerines: characterization and seasonal variation. *J. Ornithol.* 148(Suppl 2):S583–91

Palacios MG, Winkler DW, Klasing KC, Hasselquist D, Vleck CM. 2011. Consequences of immune system aging in nature: a study of immunosenescence costs in free-living Tree Swallows. *Ecology* 92:952–66

Piersma T. 1997. Do global patterns of habitat use and migration strategies co-evolve with relative investments in immunocompetence due to spatial variation in parasite pressure? *Oikos* 80:623–31

Proell M, Riedel SJ, Fritz JH, Rojas AM, Schwarzenbacher R. 2008. The Nod-like receptor (NLR) family: a tale of similarities and differences. *PLoS One* 3:e2199

Pursall ER, Rolff J. 2011. Immune responses accelerate aging: proof-in-principle in an insect model. *PLoS One* 6:e19972

Råberg L, Grahn M, Hasselquist D, Svensson E. 1998. On the adaptive significance of stress-induced immunosuppression. *Proc. R. Soc. Lond. B* 265:1637–41

Råberg L, Sim D, Read AF. 2007. Disentangling genetic variation for resistance and tolerance to infectious diseases in animals. *Science* 318:812–14

Rakoff-Nahoum S, Pagliono J, Eslami-Varzaneh F, Edberg S, Medzhitov R. 2004. Recognition of commensal microflora by Toll-like receptors is required for intestinal homeostasis. *Cell* 118:229–41

Redpath S, Ghazal P, Gasciogne NRJ. 2001. Hijacking and exploitation of IL-10 by intracellular parasites. *Trends Microbiol.* 9:86–92

Roach JC, Glusman G, Rowen L, Kaur A, Purcell MK, et al. 2005. The evolution of vertebrate Toll-like receptors. *Proc. Natl. Acad. Sci. USA* 102:9577–82

Robinson MW, O'Brien R, MacKintosh CG, Clark RG, Griffin JFT. 2011. Immunoregulatory cytokines are associated with protection from immunopathology following *Mycobacterium avium* subspecies *paratuberculosis* infection in red deer. *Infect. Immun.* 79:2089–97

Roe C, Kinney J. 1965. The caloric equivalent of fever. II. Influence of major trauma. *Ann. Surg.* 161:140–48

Rowley AF. 1996. The evolution of inflammatory mediators. *Med. Inflamm.* 5:3–13

Sadd BM, Schmid-Hempel P. 2009. Principles of ecological immunology. *Evol. Appl.* 2:113–21

Sadd BM, Siva-Jothy MT. 2006. Self-harm caused by an insect's innate immunity. *Proc. R. Soc. Lond. B* 273:2571–74

Sakaguchi S, Miyara M, Costantino CM, Hafler DA. 2010. FOXP3+ regulatory T cells in the human immune system. *Nat. Rev. Immunol.* 10:490–500

Schmid-Hempel P, Ebert D. 2003. On the evolutionary ecology of specific immune defence. *Trends Ecol. Evol.* 18:27–32

Sears BF, Rohr JR, Allen JE, Martin LB. 2011. The economy of inflammation: when is less more. *Trends Parasitol.* 27:382–87

Serhan CN, Savill J. 2005. Resolution of inflammation: the beginning programs the end. *Nat. Immunol.* 6:1191–97

Sheldon BC, Verhulst S. 1996. Ecological immunology: costly parasite defenses and trade-offs in evolutionary ecology. *Trends Ecol. Evol.* 11:317–21

Soehnlein O, Lindbon L. 2010. Phagocyte partnership during the onset and resolution of inflammation. *Nat. Rev. Immunol.* 10:427–39

Sorci G, Faivre B. 2009. Inflammation and oxidative stress in vertebrate host-parasite systems. *Philos. Trans. R. Soc. B* 364:71–83

Stecher B, Denzler R, Maier L, Bernet F, Sanders MJ, et al. 2012. Gut inflammation can boost horizontal gene transfer between pathogenic and commensal *Enterobacteriaceae*. *Proc. Natl. Acad. Sci. USA* 109:1269–74

Strachan DP. 1989. Hay fever, hygiene, and household size. *Br. Med. J.* 299:1259–60

Suvas S, Azkur AK, Kim BS, Kumaraguru U, Rouse BT. 2004. CD4+CD25+ regulatory T cells control the severity of viral immunoinflammatory lesions. *J. Immunol.* 172:4123–32

Weil ZM, Bowers SL, Dow ER, Nelson RJ. 2006a. Maternal aggression persists following lipopolysaccharide-induced activation of the immune system. *Physiol. Behav.* 87:694–99

Weil ZM, Martin LB, Workman JL, Nelson RJ. 2006b. Immune challenge retards seasonal reproductive regression in rodents: evidence for terminal investment. *Biol. Lett.* 2:393–96

Weil ZM, Norman GJ, Devries AC, Nelson RJ. 2008. The injured nervous system: a Darwinian perspective. *Prog. Neurobiol.* 86:48–59

Williams GC. 1957. Pleiotropy, natural selection, and the evolution of senescence. *Evolution* 11:398–411

Wynn TA. 2004. Fibrotic disease and the T_H1/T_H2 paradigm. *Nat. Rev. Immunol.* 4:583–94

Yazdanbakhsh M, Kremsner PG, van Ree R. 2002. Allergy, parasites, and the hygiene hypothesis. *Science* 296:490–94

New Pathways and Processes in the Global Nitrogen Cycle

Bo Thamdrup

Nordic Center for Earth Evolution, Institute of Biology, University of Southern Denmark, DK-5230 Odense M, Denmark; email: bot@biology.sdu.dk

Annu. Rev. Ecol. Evol. Syst. 2012. 43:407–28

First published online as a Review in Advance on September 4, 2012

The *Annual Review of Ecology, Evolution, and Systematics* is online at ecolsys.annualreviews.org

This article's doi:
10.1146/annurev-ecolsys-102710-145048

Keywords

denitrification, anammox, ammonium oxidation, nitrogen loss, bacteria, eukaryotes

Abstract

Our understanding of the players and pathways of the global nitrogen cycle has advanced substantially over recent years with discoveries of several new groups of organisms and new types of metabolism. This review focuses on recently discovered processes that add new functionality to the nitrogen cycle and on the organisms that perform these functions. The processes include denitrification and other dissimilatory nitrogen transformations in eukaryotes, anaerobic ammonium oxidation, and anaerobic methane oxidation with nitrite. Of these, anaerobic ammonium oxidation coupled to nitrite reduction by anammox bacteria has been well documented in natural environments and constitutes an important sink for fixed nitrogen. Benthic foraminifera also contribute substantially to denitrification in some sediments, in what potentially represents an ancestral eukaryotic metabolism. The ecophysiology of the novel organisms and their interactions with classical types of nitrogen metabolism are important for understanding the nitrogen cycle and its tight links to the cycling of carbon today, in the past, and in the future.

INTRODUCTION

In many ways, the nitrogen cycle is the most complex of Earth's biogeochemical cycles, constituted by an unusually diverse set of transformations, many of which are carried out only by distinct groups of specialized microorganisms. Furthermore, it involves reservoirs of different forms of nitrogen in the atmosphere, oceans, soils and sediments, the crust, and biota. Most of the major transformations were discovered more than a century ago, and by 1934, Bass Becking (1934) was discussing the fundamental concepts of the nitrogen cycle. Nonetheless, our understanding of the functional relationships within the nitrogen cycle has changed substantially during the past 10–15 years. Important discoveries include new types of organisms that are involved in the well-known processes as well as those that convey new types of processes. After a brief introduction to the nitrogen cycle, this review provides an overview of the novel pathways and attempts to evaluate their importance for our understanding of the nitrogen cycle via an ecological and evolutionary perspective.

The Biological Nitrogen Cycle

The roles that nitrogen plays for living organisms can be grouped into two general categories: assimilation, i.e., the acquisition of matter for the incorporation into biomass, and dissimilation, which designates processes that are associated with the extraction of energy from the environment. Nitrogen is an essential element required in large amounts by all life, mainly for the synthesis of amino acids and nucleotides, and it takes part in several different types of respiratory energy metabolism in which nitrogen compounds may serve as either oxidant or reductant. These dissimilatory transformations are largely carried out by specialized groups of prokaryotes, but new findings discussed below suggest that various eukaryotes also contribute.

The nitrogen cycle (**Figure 1**; also see examples of the reactions in **Table 1**) is driven by a combination of these assimilatory and dissimilatory biological transformations. Also included are biotic redox transformations, such as atmospheric nitrogen fixation associated with lightning, as well as chemodenitrification (e.g., Tai & Dempsey 2009), although these processes make only small contributions relative to those of the microbial transformations (Canfield et al. 2005, Gruber & Galloway 2008; but also see Samarkin et al. 2010). The geological nitrogen cycle, i.e., the flux of nitrogen through the Earth's crust driven by sediment burial and rock weathering, is another minor component (Berner 2006), although weathering may constitute a substantial source of nitrogen to terrestrial ecosystems on particularly nitrogen-rich bedrock (Morford et al. 2011).

The primary ecological and evolutionary significance of the nitrogen cycle lies in its ability to regulate the availability of fixed nitrogen to the biota. Free nitrogen, N_2, constitutes most of Earth's atmosphere but is accessible only to N_2-fixing bacteria and archaea, which reduce it to ammonium and incorporate it into biomass. By contrast, other prokaryotes as well as all eukaryotes require fixed nitrogen (also called reactive or combined nitrogen) in forms such as nitrate, ammonium, or organic nitrogen for assimilation. Fixed nitrogen in the biosphere corresponds to less than 0.1% of the N_2 pool and limits primary production in both terrestrial and marine ecosystems; this likely describes the case through most of Earth's history (Vitousek & Howarth 1991, Canfield et al. 2010). On a global scale, the availability of fixed nitrogen is controlled by the balance between nitrogen fixation and the recycling of fixed nitrogen to N_2 through dissimilatory transformations (**Figure 1**). Thus, the nitrogen cycle is closely coupled to the carbon cycle, and its processes and their regulation are of fundamental importance in both modern and ancient ecosystems.

During the past century, human activities have shifted the balance between N_2 fixation and recycling to the extreme. In terrestrial environments, more nitrogen is now fixed through industrial

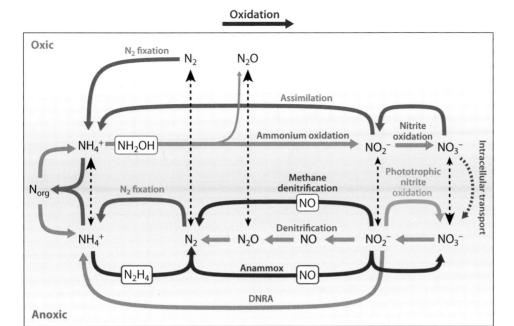

Oxidation →

Reduction ←

Figure 1

Schematic representation of the nitrogen cycle. Metabolic transformations are shown as thick arrows. Also shown are the classical processes of assimilation (*green*) and dissimilation (*gray*) as well as recently discovered metabolisms discussed in the text (*colored*). Aerobic and anaerobic processes are separated, and dashed vertical arrows indicate exchange or transport between oxic and anoxic environments, with the relative size of arrowheads indicating the dominant direction of transport. Abbreviation: DNRA, dissimilatory nitrate reduction to ammonium.

fertilizer production (140 Tg N year^{-1}) than through natural sources (110 Tg N year^{-1}) (Canfield et al. 2010). Human activities further contribute fixed nitrogen through the cultivation of legumes with associated nitrogen-fixing bacteria and by burning fossil fuels. Altogether, anthropogenic sources approach half of the estimated global nitrogen fixation (Gruber & Galloway 2008, Canfield et al. 2010). These sudden changes emphasize the need for a detailed understanding of the nitrogen cycle and the controls on its many different parts.

In the classical view, two dissimilatory microbial processes convey the recycling of ammonium, as generated by N_2 fixation or released during the degradation of organic matter, to N_2 (**Figure 1, Table 1**): (*a*) nitrification, the aerobic oxidation of ammonium to nitrite and then to nitrate, with each step performed by a specialized group of bacteria, and (*b*) denitrification, the respiratory anaerobic reduction of nitrate via nitrite, nitric oxide, and nitrous oxide to N_2, coupled to the oxidation of organic matter, hydrogen, or reduced iron or sulfur species. The coupling to carbon oxidation introduces a second important link between the nitrogen and carbon cycles in addition to the nitrogen limitation of primary production. Nitrifying bacteria are autotrophic, using some of the electrons from oxidation of ammonium and nitrite to reduce CO_2 and build biomass. Denitrifiers that consume organic substrates are heterotrophic, whereas those that utilize inorganic substrates can be autotrophic. Instead of performing denitrification, some microbes employ a different reductive pathway known as dissimilatory nitrate reduction to ammonium (DNRA),

Table 1 Stoichiometries normalized to reactant nitrogen species and standard free energy yields[a] of dissimilatory nitrogen transformations discussed in this review

Process	Reaction	$\Delta G^{\circ\prime}$ (kJ [mol reaction]$^{-1}$)
Lithotrophy		
Ammonium oxidation	$NH_4^+ + \frac{3}{2}O_2 \rightarrow NO_2^- + H_2O + H^+$	-235
Nitrite oxidation	$NO_2^- + \frac{1}{2}O_2 \rightarrow NO_3^-$	-74
Anammox	$NH_4^+ + NO_2^- \rightarrow N_2 + 2H_2O$	-358
Carbon fixation by nitrite oxidation[b]	$NO_2^- + \frac{1}{2}HCO_3^- + \frac{1}{2}H^+ \rightarrow NO_3^- + \frac{1}{2}CH_2O$	163
Hypothetic pathways of anaerobic ammonium oxidation		
Mn(IV)-dependent, to N_2	$NH_4^+ + \frac{3}{2}MnO_2 + 2H^+ \rightarrow \frac{1}{2}N_2 + \frac{3}{2}Mn^{2+} + 2H_2O$	-239[a]
Mn(IV)-dependent, to nitrite	$NH_4^+ + 3MnO_2 + 4H^+ \rightarrow NO_2^- + 3Mn^{2+} + 4H_2O$	-121[a]
Fe(III)-dependent, to N_2	$NH_4^+ + 3FeOOH + 5H^+ \rightarrow \frac{1}{2}N_2 + 3Fe^{2+} + 6H_2O$	-80[a]
Fe(III)-dependent, to nitrite	$NH_4^+ + 6FeOOH + 10H^+ \rightarrow NO_2 + 6Fe^{2+} + 10H_2O$	197[a]
Sulfate-dependent	$NH_4^+ + \frac{3}{8}SO_4^{2-} \rightarrow \frac{1}{2}N_2 + \frac{3}{8}HS^- + \frac{3}{2}H_2O + \frac{5}{8}H^+$	-18
Organotrophy		
Nitrate reduction[c]	$NO_3^- + \frac{1}{4}CH_3COO^- \rightarrow NO_2^- + \frac{1}{2}HCO_3^- + \frac{1}{4}H^+$	-137
Denitrification to N_2O[c]	$NO_2^- + \frac{1}{4}CH_3COO^- + \frac{3}{4}H^+ \rightarrow \frac{1}{2}N_2O + \frac{1}{2}HCO_3^- + \frac{1}{2}H_2O$	-200
Denitrification to N_2[c]	$NO_2^- + \frac{1}{2}CH_3COO^- \rightarrow \frac{1}{2}N_2 + HCO_3^- + \frac{1}{2}H^+$	-385
DNRA[c]	$NO_2^- + \frac{3}{4}CH_3COO^- + \frac{5}{4}H^+ + H_2O \rightarrow NH_4^+ + \frac{3}{2}HCO_3^-$	-358
Nitrite-dependent methane oxidation	$NO_2^- + \frac{3}{8}CH_4 + \frac{5}{8}H^+ \rightarrow \frac{1}{2}N_2 + \frac{3}{8}HCO_3^- + \frac{7}{8}H_2O$	-378

[a]Calculated for 25°C, pH 7, and unit activities of all other species except Mn^{2+} and Fe^{2+}, which were both set to 10 μM because concentrations of these species in sediments are typically controlled by mineral precipitation. Based on standard free energies of formation from Stumm & Morgan (1981).

[b]Stoichiometry of carbon fixation in anammox bacteria and anaerobic phototrophic nitrite-oxidizing bacteria, which obtain the energy required for the reaction from the anammox process and from light, respectively.

[c]Nitrate reduction, the first step of denitrification and dissimilatory nitrate reduction to ammonium (DNRA) is listed separately to facilitate comparison to nitrite-dependent methane oxidation.

which retains fixed nitrogen in the system. In some bacteria, DNRA is a fermentative process used to increase the energy yield from the fermentation of organic substrates, but in natural environments, DNRA appears to occur mainly through respiration, which is more energy efficient because the process is linked to ATP synthesis via electron transport and proton translocation (Simon 2002). DNRA is mainly known from strongly reducing sediments (Thamdrup & Dalsgaard 2008), but the process has recently been found in other aquatic and terrestrial systems, suggesting that it may be of more general importance (Lam et al. 2009, Dong et al. 2011, Rütting et al. 2011).

A strong research focus combined with new tools to analyze microbial communities and trace biogeochemical processes have led to big recent advances in our understanding of the nitrogen cycle. For the well-known microbial processes, whole new groups of organisms that make major contributions on ecosystem as well as global scales have been discovered. For N_2 fixation, this includes unicellular marine cyanobacteria that are abundant in the oceans and appear to be an important source of fixed nitrogen in the marine budget (Zehr et al. 2001, Montoya et al. 2004, Zehr 2011). For nitrification, the ammonium-oxidizing archaea, composing an even-more widespread clade, have been discovered in high numbers in soils, sediments, and ocean water (Könneke et al. 2005, Treusch et al. 2005, Nicol et al. 2011), where they appear to be particularly important for ammonium oxidation under nutrient-poor conditions, consistent with their greater ability to utilize ammonium at extremely low concentrations compared with that of typical ammonium-oxidizing bacteria (Martens-Habbena et al. 2009, Schleper & Nicol 2010, Gubry-Rangin et al. 2011, Nicol et al. 2011). This discovery is important for understanding the dynamics of nitrification in the environment. Recent discoveries, however, are not restricted to well-known pathways; they also include several groups of organisms that add new functions and alternative pathways to the nitrogen cycle described above (**Figure 1**, **Table 1**). These are the main subject of this review.

One discovery is that ammonium can be oxidized anaerobically to N_2 by a highly specialized bacterial clade, the anammox bacteria, that utilizes nitrite as an electron acceptor, and other anaerobic pathways of ammonium oxidation have also been suggested. The accumulation and transport of nitrate for subsequent denitrification by foraminifera and related protists is another novel type of N_2 production, which is discussed in conjunction with additional evidence for dissimilatory nitrogen transformations in eukaryotes. A third source of N_2 is the reduction of nitrite coupled to methane oxidation, which occurs through a novel enzymatic pathway and introduces another link between the nitrogen and carbon cycles. Anaerobic oxidation of nitrite to nitrate has also recently been discovered in anoxygenic phototrophic bacteria that couple it to carbon fixation (Griffin et al. 2007, Schott et al. 2010). This finding may yield interesting ecological and evolutionary perspectives, but the process awaits further study and so is not discussed here.

REDUCTIVE NITROGEN DISSIMILATION IN EUKARYOTES

Dissimilatory redox transformations of nitrogen have generally been considered to fall within the purview of prokaryotes, whereas assimilation and decomposition have been considered the main contributions of eukaryotes. Evidence is accumulating, however, for an involvement of a diverse assembly of eukaryotes in the dissimilatory parts of the nitrogen cycle, in particular in reductive transformations. In some cases, these organisms may be major contributors to nitrogen fluxes, which makes the subject interesting from both evolutionary and ecological perspectives.

Early evidence that eukaryotes may also contribute to the reductive part of the nitrogen cycle came from ciliates of the genus *Loxodes*, which live just below the depth of oxygen depletion in stratified freshwater lakes (Finlay et al. 1983, Finlay 1985). These organisms reduce nitrate to nitrite (**Table 1**), and because few epibiotic bacteria were observed attached to the outside of the organism and no endobionts were seen inside of it, the reduction was attributed to a respiratory process in the mitochondria, which were present in high numbers. *Loxodes'* quantitative contribution to nitrogen cycling in lakes was not determined, but a peak in nitrite concentration in the anoxic bottom water was attributed to their activity.

Nitrate and nitrite reduction has also been reported for a diverse assembly of fungi of the major phyla Ascomycota and Basidiomycota including both filamentous and yeast forms (Shoun et al. 1992, Tsuruta et al. 1998, Zhou et al. 2002). A subset of these reduce nitrate and all reduce nitrite, which is either denitrified via nitric oxide to nitrous oxide or reduced to ammonium in a DNRA process coupled to the oxidation of organic substrate (**Table 1**). There is no evidence of complete denitrification to N_2 in fungi, but fungi may produce N_2 through codenitrification, i.e., the reaction of nitric oxide, formed by nitrite reduction, with organic N compounds (Spott et al. 2011). Intriguingly, both denitrification and DNRA can be found in the same fungus: In the anamorphic ascomycete *Fusarium oxysporum*, the species investigated in greatest detail, denitrification requires small but limiting amounts of oxygen, whereas the DNRA pathway is active under anoxic conditions (Zhou et al. 2001, 2002). Here DNRA is a fermentative process contributing only marginally to growth, whereas the two first steps of denitrification, from nitrate to nitrite and then to nitric oxide, are respiratory processes coupled to electron transport in the mitochondria (Takaya 2009, Zhou et al. 2010), where they appear to function as a supplement to oxic respiration. These steps thus resemble those of bacteria, and the nitrate and nitrite reductases involved are structurally and functionally similar to their bacterial counterparts (Kobayashi & Shoun 1995, Uchimura et al. 2002). The nitrite reductase is of the copper-containing type (nirK), and the encoding gene is homologous to bacterial *nirK* genes (Kim et al. 2009, Takaya 2009).

The culture experiments discussed above suggest that fungi could play important roles as producers of N_2O, particularly in soils where their biomass is large. This is further supported by the finding of denitrification in common ectomykorrhizal fungi from temperate-boreal forests (Prendergast-Miller et al. 2011). Large reductions of N_2O fluxes after the addition of fungicide could indicate such an involvement in a wide variety of soils (e.g., Laughlin & Stevens 2002, Crenshaw et al. 2008). However, the fungal inhibitors could also have indirect effects, such as the disappearance of anoxic microniches due to decreased fungal respiration or stimulation of bacterial N_2O reduction by the release of fungal lysates (Crenshaw et al. 2008). Thus, other approaches are needed to support these results. Understanding the microbial ecology of N_2O formation in soils is an important task because soils are the single largest source of N_2O to the atmosphere (Braker & Conrad 2011). Regardless of whether it is direct or indirect, the role of fungi clearly needs to be considered.

In marine sediments, another group of eukaryotes, the Foraminifera, has recently been linked to denitrification. A wide variety of these heterotrophic protists living in continental and hemipelagic marine sediments have the combined ability to concentrate nitrate intracellularly to extreme concentrations and to denitrify this nitrate completely to N_2 under anoxic conditions (Risgaard-Petersen et al. 2006, Glud et al. 2009, Piña-Ochoa et al. 2010a). In the most extensive survey, nitrate accumulation was found in half of the 66 foraminiferal species that were examined covering several major lineages and also in all specimens of Gromiida (*Gromia* sp.), a lineage related to the Foraminifera in the Rhizaria (Piña-Ochoa et al. 2010a). Nitrate concentrations within the organism are often on the order of 10^{-2} mol L^{-1}, four orders of magnitude higher than in seawater. This nitrate is likely stored in vacuoles (Bernhard et al. 2011). Ten nitrate-acculmulating species were screened for denitrification and all possessed this capability (Risgaard-Petersen et al. 2006, Piña-Ochoa et al. 2010a). The ability to denitrify was attributed to the Foraminifera after a careful search for symbiotic bacteria in one species, *Globobulimina pseudospinescens* (Risgaard-Petersen et al. 2006). By contrast, endobiotic bacteria most likely account for denitrification in a nitrate-accumulating species of the allogromiid foraminifera, each individual of which contained more than 250,000 bacteria related to *Pseudomonas*, ample to support the measured denitrification in this organism (Bernhard et al. 2011). In this case, the bacteria may live on fermentation products from anaerobic metabolism in the allogromiid, thus enabling the foraminifera to survive under anoxic conditions. Allogromiids were one of the only groups that did not show nitrate accumulation in the screening by Piña-Ochoa and colleagues (2010a). Although these findings extend the distribution of denitrification in foraminifera, they also emphasize the need for further investigations of the potential role of endobionts in other species. This is of fundamental importance for understanding the evolutionary path of denitrification in the eukaryotes. A widespread capability of complete denitrification in the Rhizaria would suggest that this is an original trait, possibly inherited from the bacterial ancestor of mitochondria (Piña-Ochoa et al. 2010a).

The physiological role of denitrification in the Foraminifera has not been explored in detail. Access to nitrate improves survival under anoxia to a level similar to that found under oxic conditions, but it is not clear if denitrification supports growth or reproduction (Piña-Ochoa et al. 2010b).

Denitrifying species are particularly common in sediments in the oceanic oxygen minimum zones (OMZs), where they can make up most of the benthic foraminifera and reach densities of >200 individuals per square centimeter (Høgslund et al. 2008, Piña-Ochoa et al. 2010a). Here they thrive under anoxic, nitrate-rich bottom water for extended periods, if not permanently (Høgslund et al. 2008). Both in OMZs and in sediments underlying oxygenated water, they bury into the sediment to depths of several centimeters, well below the depth to which oxygen and nitrate penetrate in the pore water (Risgaard-Petersen et al. 2006, Høgslund et al. 2008, Glud et al. 2009). These observations suggest a lifestyle in which the foraminifera load their nitrate storage vacuoles near the sediment surface and move deeper into the sediment to feed, using the stored nitrate in a SCUBA-like fashion for respiration (Koho et al. 2011). The stored nitrate may last up to 1–3 months at typical rates of denitrification (Risgaard-Petersen et al. 2006, Piña-Ochoa et al. 2010b).

Their ability to accumulate and transport an electron acceptor in large amounts enables the foraminifera to occupy a distinct niche in sediments. Their lifestyle is partly similar to that of giant sulfur bacteria of the genera *Thioploca* and *Beggiatoa* that accumulate nitrate to similar levels and carry it into the sediment (Schulz & Jørgensen 2001). Similar to these organisms, the foraminifera can move to the sediment surface where nitrate concentrations are highest and may reach outside the diffusive boundary layer, thus reducing the diffusive limitations as *Thioploca* does it (Huettel et al. 1996). This gives them a competitive advantage relative to most nitrate-reducing bacteria in the sediment. However, the giant sulfur bacteria and foraminifera have different functions in the ecosystem: The bacteria use sulfide as an electron donor and reduce nitrate to ammonium in DNRA, thus retaining fixed nitrogen in the system. By contrast, the organotrophic foraminifera constitute a sink for fixed nitrogen. Members of both groups are abundant in OMZ sediments, but variations in their relative abundance in those areas could have a large impact on the benthic nitrogen cycle (Høgslund et al. 2008). The highest densities of nitrate-accumulating foraminifera have been found in OMZs (Høgslund et al. 2008) (also see above), but they are also present in sediments from river deltas, coastal seas, and continental shelves and margins (Piña-Ochoa et al. 2010a), making them more widespread than the nitrate-accumulating giant sulfur bacteria. Foraminifera dominate the benthic fauna in the core of OMZs, where they are favored by high availability of organic matter and reduced competition and predation from metazoans (Levin 2003, Woulds et al. 2007). Together with nitrate availability, these factors should also be important in governing the distribution of denitrifying foraminifera in other environments.

It is a challenge to determine the contribution of foraminifera to benthic denitrification, and because of the differences in turnover times of pore-water and intracellular nitrate pools, common methods for measuring denitrification will not capture the foraminiferal contribution (Høgslund et al. 2008). Estimates based on cellular denitrification rates and population densities of the foraminifera as well as measured benthic denitrification rates yield foraminiferal contributions ranging from 4% at a continental margin site with a relatively low population density (Glud et al. 2009) up to 50% for a range of other sediments (Høgslund et al. 2008, Piña-Ochoa et al. 2010a), assuming that foraminiferal denitrification should be added to the whole-sediment rate. This range of contributions is in line with a model-based estimate that 30% of denitrification on two locations was supported by biological nitrate transport by the tube-dwelling foraminifera *Hyperammina* sp. (Prokopenko et al. 2011). Thus, the available evidence suggests that denitrifying foraminifera, and possibly also the gromiids, can make substantial contributions to denitrification in marine sediments. Importantly, this contribution could have been missed in estimates of benthic denitrification determined through modeling or incubations, except in relatively rare measurements of the total N_2 flux.

The accumulating evidence of dissimilatory nitrate and nitrite reduction in Eukarya raises fundamental questions about the origins of these metabolisms. Dissimilatory nitrate reduction is reported for three different supergroups of the Eukarya: Ciliates in the Chromalveolates, Fungi in the Opistokonts, and Foraminifera in the Rhizaria. One further example comes from the Chromalveolates: a diatom that reduces nitrate accumulated in the central vacuole to ammonium when exposed to dark anoxic conditions (Kamp et al. 2011). Stored nitrate is consumed in DNRA within a few days, but it conveys long-term survival for several weeks under anoxic conditions, which indicates that the algae use DNRA to enter a resting stage. Similar to the case in fungi, the diatom might use assimilatory nitrate and nitrite reductases for DNRA (Kamp et al. 2011) (also see above). Plants also use the assimilatory reductases for nitrate and nitrite reduction during hypoxia, but they further reduce part of the nitrite to NO in their mitochondria in what appears to be a respiratory process (Gupta & Igamberdiev 2011). Thus, there is no clear separation of assimilatory and dissimilatory pathways in these eukaryotes.

It seems likely that early eukaryotes had the potential for anaerobic metabolisms (Tielens et al. 2002), but it remains an open question whether some of the steps in the dissimilatory reduction of nitrogen compounds in eukaryotes have been inherited from the bacterial ancestor of mitochondria through the common ancestral eukaryote. If denitrification in Foraminifera is carried out by the organism and not by endobionts, the wide distribution of this metabolism in that group would argue for an early origin. So far, the only known homology to bacteria is in the *nirK* gene of fungal nitrite reductase (Kim et al. 2009). In support of a common origin of this gene in Eukarya, homologs to the fungal *nirK* gene were found in the genome of a green alga and two amoebae, though nitrite dissimilation is not known from these organisms (Kim et al. 2009).

ANAEROBIC AMMONIUM OXIDATION AND ANAMMOX

In the classical view of the nitrogen cycle, ammonium was considered the end product of anaerobic degradation, and its further transformation would require the presence of oxygen and aerobic ammonium-oxidizing bacteria. However, chemical evidence from oxygen-depleted oceanic waters and deep-sea sediments indicated that ammonium in these environments is oxidized anaerobically to N_2 in the presence of nitrate and nitrite (Richards 1965, Emerson et al. 1980). A microbial process termed anammox that could account for these observations was first reported from an experimental wastewater treatment system (Mulder et al. 1995, van de Graaf et al. 1997). Even though the organisms that carry out the process were not, and still have not been, isolated, investigations of highly purified enrichment cultures revealed that the organisms belong to a distinct clade within the bacterial phylum Planctomycetes and that nitrite is the oxidant for ammonium in the lithotrophic anammox process from which the organisms gain energy (van de Graaf et al. 1997, Strous et al. 1999):

$$NO_2^- + NH_4^+ \rightarrow N_2 + 2H_2O. \qquad 1.$$

The enzymatic pathway involves the reductive combination of nitric oxide from nitrite reduction with ammonium to form hydrazine, which is subsequently oxidized to N_2 (Strous et al. 2006, Kartal et al. 2011a). This takes place in an intracellular compartment called the anammoxosome bounded by an unusually gas-tight membrane of unique lipids (ladderanes) (Sinninghe-Damsté et al. 2002) and involves unique enzymes. For reviews of the many unique biochemical and microbiological features of anammox bacteria, see Jetten et al. (2009) and Kartal et al. (2011b). Anammox bacteria appear to be obligate autotrophs with carbon fixation coupled to nitrite oxidation:

$$2NO_2^- + CO_2 + H_2O \rightarrow 2NO_3^- + CH_2O. \qquad 2.$$

This process is driven by energy from the anammox process (Equation 1, **Table 1**), which allows them to fix ~0.07 mol carbon per mol ammonium oxidized. The fact that anammox bacteria couple N_2 production to CO_2 consumption is an important distinction from organoheterotrophic denitrifiers that produce both N_2 and CO_2, which introduces some flexibility in the stoichiometric coupling of the nitrogen and carbon cycles (Koeve & Kähler 2010). Some anammox bacteria can couple the oxidation of simple organic substances to the dissimilatory reduction of nitrate to nitrite and of nitrite to ammonium (DNRA) (Güven et al. 2005), and a full genome indicates the potential for using several different electron donors as well as acceptors including iron or manganese oxides (Strous et al. 2006, Kartal et al. 2011b). These abilities could contribute to the success of these bacteria in some natural environments.

One obstacle for studies of anammox bacteria in culture is the extremely slow growth of these bacteria with doubling times of 10–20 days at 35°C and high substrate concentrations (Strous et al. 1999), which could lead researchers to believe that these organisms would not be of significance in colder natural settings with concentrations of ammonium and nitrite that are orders of magnitude lower than in the bioreactors. Indeed, the first report of the anammox process in nature did not result from a search for this process, but rather from investigations into the potential for anaerobic ammonium oxidation coupled to manganese oxide reduction in anoxic marine sediment (Thamdrup & Dalsgaard 2000, 2002). Ammonium oxidation was not detected in anoxic, manganese oxide-rich sediment, but addition of nitrate or nitrite immediately led to the formation of ^{15}N-labeled N_2 from ^{15}N-labeled ammonium. Experiments with labeled nitrate and nitrite confirmed that N_2 formed through the one-to-one pairing of nitrogen atoms from nitrite and ammonium that is characteristic of the anammox process (Dalsgaard & Thamdrup 2002, Thamdrup & Dalsgaard 2002). Further support for the hypothesis that benthic ammonium oxidation with nitrite is conveyed by anammox bacteria has come from the sensitivity of the process to inhibitors (Dalsgaard & Thamdrup 2002, Jensen et al. 2007), from correlations of the distribution of its activity with distributions of anammox bacteria and of the unique ladderane lipids found in the membranes of these organisms (Kuypers et al. 2003, Brandsma et al. 2011), and from the high expression of genes involved in the anammox process in waters where the process is active (Lam et al. 2009).

There is strong evidence that the anaerobic ammonium oxidation activity measured in marine sediments and waters to date is conveyed by anammox bacteria, making it reasonable to refer to this activity as anammox, but other pathways of anaerobic ammonium oxidation have also been suggested with either manganese oxide, iron oxide, or sulfate as electron acceptors. The metal oxides have been hypothesized to support anaerobic ammonium oxidation to either N_2 or nitrite in sediments, aquifers, and anoxic waters (**Table 1**) (Luther et al. 1997, Hulth et al. 1999, Clément et al. 2005, Shrestha et al. 2009); such processes would also be relevant to nitrogen cycling during Earth's early, oxygen-deficient history. No clear direct evidence of such processes— either spontaneous or biologically catalyzed—has yet been reported from natural environments, and experiments using ^{15}N as a tracer generally constrain the processes to very low potential importance in estuarine and marine environments (e.g., Thamdrup & Dalsgaard 2002, Engström et al. 2009). Recent studies have suggested that anaerobic ammonium oxidation to nitrite, i.e., an anaerobic nitrification process, is linked to iron oxide reduction in wetland soils (Clément et al. 2005, Shrestha et al. 2009) and, potentially, in the gut of soil-feeding termites (Ngugi et al. 2011). The proposed reaction,

$$NH_4^+ + 6FeOOH + 10H^+ \rightarrow NO_2^- + 6Fe^{2+} + 10H_2O, \qquad 3.$$

was claimed to be thermodynamically feasible under natural conditions (Clément et al. 2005). However, a recalculation shows that it is in fact endergonic at neutral pH with an energy demand, ΔG, of $179\,kJ\,mol^{-1}\,NH_4^+$ even with poorly crystalline hydrous ferric oxide as the Fe(III) phase and with concentrations of the solutes toward the favorable end of their ranges in natural environments

[1 mM NH_4^+, 1 μM NO_2^-, 10 μM Fe^{2+}; thermodynamic constants at 25°C from Stumm & Morgan (1981)] (also see **Table 1**). With these concentrations, the reaction is only thermodynamically favorable at pH ≤ 4. Thus, unusual forms of highly reactive, metastable Fe(III) would be needed to couple iron reduction to the oxidation of ammonium to nitrite in the circumneutral range, but there is greater potential for the process in acidic environments.

The oxidation of ammonium to N_2 is thermodynamically much more favorable than is oxidation to nitrite. A coupling of this oxidation to the reduction of iron oxide yields $\Delta G = -64$ kJ mol^{-1} NH_4^+ under the same conditions used in the calculations above. Yet, there is no evidence for this reaction in natural systems. A similar coupling to sulfate reduction is also slightly exergonic and has been suggested to explain the apparent consumption of ammonium occurring meters below the seafloor near the depth of sulfate depletion in cores of marine sediment (Schrum et al. 2009):

$$8NH_4^+ + 3SO_4^{2-} \rightarrow 4N_2 + 3HS^- + 12H_2O + 5H^+. \qquad 4.$$

Further verification of this process is needed. However, if it exists, it appears to be relatively slow and associated mainly with the sulfate-methane transition zone because ammonium readily accumulates during sulfate reduction in more active surface sediments (e.g., Burdige 1991) and millimolar concentrations of ammonium persist in the zone of sulfate reduction (e.g., Schrum et al. 2009). Nonetheless, if integrated over several meters depth in subsurface sediments, it could make a substantial contribution to the nitrogen cycle.

In contrast to anaerobic ammonium oxidation coupled to manganese, iron, or sulfate reduction, anammox activity is documented in a wide range of environments. To date, the process and the bacteria have been detected mainly in marine environments, including sediments from estuaries to the deep sea and oceanic OMZs, where the process seems essentially ubiquitous (Dalsgaard et al. 2005, Trimmer & Engström 2011). Many fewer reports describe anammox activity in terrestrial and freshwater environments, with the exception of wastewater systems such as those where they were originally detected and some wastewater recipients. Yet, the process and the bacteria have been detected in a wide variety of freshwater environments including the water column of a stratified lake (Schubert et al. 2006), lake sediment (Yoshinaga et al. 2011), paddy soils (Zhu et al. 2011), and aquifers (Moore et al. 2011). Anammox bacteria have also been detected in several other environments, though their activity has not been determined (Hu et al. 2011).

Anammox bacteria from marine and freshwater environments appear to be phylogenetically distinct: The candidate genus Scalindua dominates in marine systems, whereas the populations in freshwater systems are more diverse assemblies with representatives of the candidate genera Brocadia, Jettenia, Kuenenia, and Anammoxoglobus and other lineages (Kartal et al. 2011c). A potential ecological explanation for these differences in diversity involves metabolic versatility, consistent with the indications that some anammox bacteria can utilize other electron donors and/or acceptors in their energy metabolism (Yoshinaga et al. 2011; also see above).

The database on anammox activity in terrestrial environments remains too small to allow researchers to identify general trends. Our understanding of the contribution of anammox to the nitrogen cycle is, therefore, based mainly on results from marine systems, where the process is of substantial importance, and some trends have emerged for both sediments and anoxic waters (Dalsgaard et al. 2005, Lam & Kuypers 2011, Trimmer & Engström 2011). Anammox is estimated to account for 28% of the benthic N_2 production in the marine nitrogen budget, but the contribution varies from <1% to ∼80% between sites (Trimmer & Engström 2011). Water depth is a good predictor for much of this variation (**Figure 2**). Thus, the process is typically of minor importance in shallow estuarine and coastal sediments, but the contribution increases with water depth to approximately 100 m, beyond which it is generally ∼30% or higher. Of further note, as anammox's contribution to N_2 production increases, the rate tends to decrease (Dalsgaard

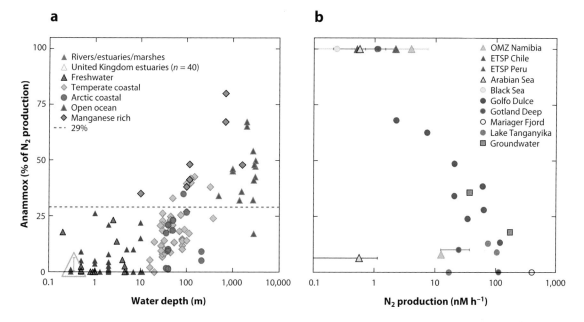

Figure 2

Compilations of the relative contribution of anammox to N_2 production (anammox + denitrification) in sediments (*a*) and anoxic waters (*b*). (*a*) The relative contribution of anammox in sediments plotted as a function of water depth (log scale). Different types of sediment are marked with different symbols. Data from shallow sites with no depth provided are plotted at 0.3, 0.5, 1, and 2 m. Data from compilations by Thamdrup & Dalsgaard (2008) and Trimmer & Engström (2011) supplemented with results from freshwater referenced in the text and unpublished results of the author. The gray dashed line indicates the 29% contribution from anammox predicted from the stoichiometry of the complete degradation of marine phytoplankton detritus of Redfield composition through denitrification and anammox. (*b*) The relative contribution of anammox in anoxic waters as a function of the total rate of N_2 production (denitrification + anammox): Indicated are the results from oxygen minimum zones (*triangles*), stratified marine basins and a lake (*circles*), and groundwater (*squares*). Error bars span the range of two or more measurements with symbols placed at the midpoint. Symbols without error bars represent single determinations. Data compiled by Trimmer & Engström (2011) supplemented with results by Lavik et al. (2009), Jensen et al. (2011), and Moore et al. (2011). Abbreviations: ETSP, eastern tropical South Pacific; OMZ, oxygen minimum zone.

et al. 2005, Trimmer & Engström 2011). Benthic processes are fueled by the sedimentation of organic detritus, and as water depth increases, the flux and reactivity of the detritus decrease strongly. Thus, rates of benthic respiration and N_2 production decrease by orders of magnitude from coastal sediments to the deep sea, and the increased importance of anammox is related to the fact that denitrification is increasingly strongly attenuated with depth.

In sediments, the relative availability of electron donors for anammox (ammonium) and denitrification (organic carbon, reduced sulfur and iron compounds) is stoichiometrically constrained because both are derived from the degradation of organic matter. Thus, if phytoplankton detritus with a typical "Redfieldian" composition of $(CH_2O)_{106}(NH_3)_{16}H_3PO_4$ is mineralized completely with nitrate as the ultimate electron acceptor through nitrate reduction, denitrification, and anammox and with CO_2, N_2, and PO_4^{3-} as the end products, the contribution of anammox to total N_2 production is 29% (Dalsgaard et al. 2003). In reality, this percentage varies depending on the composition of the organic matter, preferential mineralization of specific fractions, burial of reduced iron and sulfur, etc., but assuming that both processes are restricted to the same part of the sediment, large deviations from 29% indicate an imbalance in the relative efficiency of the two pathways in scavenging their respective electron donors. This prediction is useful in a first, rough interpretation of the factors that control anammox and its contribution to N_2 production (**Figure 2**):

1. In shallow sediments, anammox contributions of <10% indicate that much ammonium from anaerobic mineralization escapes anaerobic oxidation into the oxic zone, which is consistent with high fluxes of ammonium from such sediments. The low efficiency of anammox is likely due to nitrite limitation (Trimmer & Engström 2011) through direct competition from denitrifying bacteria. Inhibition by sulfide, which is found in higher concentrations in shallow sediments, could also contribute because anammox appears to be inhibited by sulfide in the low micromolar range (Jensen et al. 2008).

2. In many deeper sediments, the relative importance of anammox and denitrification is in rough agreement with the stoichiometric prediction. This is consistent with the observation, at least at some deep sites, that ammonium is depleted within the zone of nitrate consumption, which suggests that anammox is limited by the availability of ammonium (Thamdrup & Dalsgaard 2008, Trimmer & Engström 2011). This further suggests that competition for nitrite is less intense in those areas than it is in shallower sites.

3. Some deep locations have substantially higher contributions from anammox and, hence, lower contributions of denitrification than expected from the stoichiometry of organic detritus. Several of these sites are characterized by a particularly high content of manganese oxide, and manganese oxide–rich sites consistently have high contributions from anammox (**Figure 2a**). Under these conditions, the low contribution of denitrification may be explained by competition with manganese reduction for electron donors. This could include both dissimilatory manganese reduction coupled to carbon oxidation and abiotic reduction coupled to the reoxidation of reduced iron and sulfur compounds, which keeps these compounds from reaching the nitrate zone (Dalsgaard & Thamdrup 2002, Trimmer & Engström 2011).

In anoxic water columns, an even larger range in the relative importance of anammox is observed in an essentially bimodal distribution (**Figure 2b**). At or near depths where hydrogen sulfide accumulates, denitrification dominates; by contrast, anammox accounts for all N_2 production in anoxic, nonsulfidic systems as found in OMZs. The balance between the two processes follows a clear trend relative to the total rate of N_2 production with anammox and denitrification dominating at low and high rates, respectively (**Figure 2b**). This distribution is somewhat similar to the trend in sediments (**Figure 2a**) and can partly be explained by the same principles. In systems with high activity, competition for nitrite and the presence of sulfide may inhibit anammox (Jensen et al. 2008), whereas abundant nitrite and low sulfide facilitates anammox in OMZs. Nonetheless, the general absence of denitrification from OMZs is paradoxical. Anammox depends on the release of ammonium from the mineralization of organic matter, and ammonium appears to limit the process in OMZs (Lam & Kuypers 2011), similar to the case in deeper sediments. Yet, even though manganese reduction can explain occasional low contributions of denitrification there, this electron acceptor is not abundantly available in OMZs.

The question is then which type of respiration, if not denitrification, conveys the mineralization that supplies ammonium to anammox. Possible answers include nitrate reduction to nitrite and DNRA, but methodological biases and spatial heterogeneity may also be part of the explanation (Kuypers et al. 2005, Thamdrup et al. 2006, Lam et al. 2009). The case for spatial heterogeneity is supported by occasional observations of denitrification. So far, these include two locations in the OMZ of the Arabian Sea (Ward et al. 2009) (**Figure 2b**) and a few stations in a large survey of the eastern South Pacific OMZ (Dalsgaard et al. 2012). Here anammox generally dominated, but when denitrification was active, rates were extremely high, such that the mean relative contribution of anammox and denitrification was close to the stoichiometric prediction. These observations suggest that denitrification is primarily associated with episodic inputs of fresh organic matter to the OMZ, whereas anammox bacteria respond more slowly in their oxidation of ammonium released during these events. This interpretation is consistent with denitrifiers being facultative anaerobes that can colonize the organic detritus already in oxic waters and switch to denitrification as they sink into the OMZ. By contrast, obligately anaerobic anammox bacteria can access the ammonium only inside the OMZ and may be further limited by inherently slow growth. Further investigations are strongly needed to test this and other hypotheses regarding the sources of ammonium in OMZs and the population dynamics of anammox bacteria and other organisms of the nitrogen cycle. Nevertheless, fundamental differences in the ecophysiology of different types of microbes appear to be critical for the large-scale distribution of processes and, hence, for the configuration of the nitrogen cycle.

The small database on anammox activity in terrestrial and freshwater systems seems consistent with the marine results. Similar to shallow marine sediments, the process is generally of minor importance as a sink for fixed nitrogen in wetlands and shallow freshwater systems (**Figure 2a**) (Trimmer & Engström 2011). Given the stoichiometric considerations above, C:N ratios that are higher in plant-derived organic matter than they are in phytoplankton could further diminish the contribution from anammox in these environments. However, at least in some cases, the process may be stimulated by anthropogenic inputs of both nitrate and ammonium, as shown by average contributions of ~20% of N_2 production in fertilized paddy soil and a eutrophic lake sediment (Zhu et al. 2011, Yoshinaga et al. 2011).

The relationships shown in **Figure 2** predict higher contributions in systems with lower total microbial activity. On land, these include aquifers and deep-lake sediments. No results have been reported from deep lakes, but some of the highest relative contributions of anammox to N_2 production in natural freshwater systems are from groundwater (20–35%) (Moore et al. 2011). Anammox has also been indicated in other aquifers (Kroeger & Charette 2008, Smits et al. 2009), which represent a freshwater environment where anammox may be of particular importance.

On a global scale, soils are estimated to contribute 53% of denitrification in terrestrial and freshwater systems, followed by groundwaters (19%), rivers (15%), and lakes (13%) (Seitzinger et al. 2006). The dearth of information about anammox in these ecosystems in general and in soils in particular makes estimates of the role of this process highly uncertain, but on the basis of the available evidence and the general trends in marine environments (**Figure 2**), the average relative contribution across landscapes is not likely to exceed 20% and may even be below 10%. Based on a total annual denitrification across terrestrial and freshwater systems of 234 Tg N year^{-1} (Seitzinger et al. 2006), these fractions correspond to 23–46 Tg N year^{-1}. (This calculation assumes that the denitrification estimate includes anammox because it is mainly based on mass balances that did not consider anammox as a separate process). In comparison, the 28% contribution of anammox estimated for marine sediments corresponds to 35–84 Tg N year^{-1} on the basis of recent estimates of total N_2 production of 126–300 Tg N year^{-1} (Codispoti et al. 2001, Trimmer & Engström 2011) (older estimate of denitrification again interpreted as including anammox). The contribution from anammox in OMZs is less well constrained as there is no consensus on the relative importance of anammox, with estimates varying from the stoichiometrically predicted 29% to close to 100% (see above). With total pelagic N_2 production estimated at 65–150 Tg N year^{-1} (Gruber 2008), anammox in OMZs should release at least 19–44 Tg N year^{-1}. Thus, marine systems appear to be the main source for N_2 from anammox, and based on the central values for the ranges in total N_2 production, the relative contribution of the process to global N_2 production is between 21% and 39%.

In the classical view of the nitrogen cycle, the recycling of fixed nitrogen to N_2 requires both oxic conditions for nitrification and anoxic conditions for denitrification, whereas anammox adds a shunt from ammonium to N_2 under anoxic conditions (**Figure 1**). In a biogeochemical perspective, however, the process depends indirectly on the existence of oxic environments where nitrite or nitrate can form by nitrification because there is no robust evidence for the oxidation of ammonium to nitrite or nitrate under anoxic conditions (as discussed above). This suggests that even if anammox evolved early in Earth's history, it became an important process in the global nitrogen cycle only after the rise in atmospheric oxygen ~2.5 billion years ago and after the evolution of nitrification, which remains to be dated (Canfield et al. 2010).

NITRITE-DEPENDENT METHANE OXIDATION

The pathway most recently added to the nitrogen cycle is the anaerobic oxidation of methane coupled to nitrite reduction. Previously known electron donors for denitrification and DNRA include organic matter, hydrogen, and reduced iron and sulfur compounds (Thamdrup & Dalsgaard 2008). Methane, however, is a biochemically challenging substrate, known only to be oxidized aerobically or in a sulfate-dependent anaerobic pathway (Thauer & Shima 2008). Experimental evidence of methane-dependent denitrification was first obtained with sludge from a bioreactor (Islas-Lima et al. 2004), and the process was subsequently thoroughly documented and characterized in enrichment cultures obtained from sediments underlying nitrate-rich freshwater (Raghoebarsing et al. 2006; Ettwig et al. 2009, 2010). These intensive efforts linked the process to bacteria of the candidate phylum NC10, which was known only from environmental genetic analyses (Ettwig et al. 2009) that described one species of this clade, "*candidatus Methylomirabilis oxyfera*" on the basis of the metagenome (Ettwig et al. 2010). Similar to anammox bacteria, this organism has not been isolated, and insights into its physiology are based on enrichment cultures. M. oxyfera oxidizes methane to CO_2 coupled to the reduction of nitrite to N_2:

$$3CH_4 + 8NO_2^- + 8H^+ \rightarrow 3CO_2 + 4N_2 + 10H_2O. \qquad 5.$$

Investigation of the enzymatic pathways involved in this metabolism revealed a paradox: Although cultivated under anoxic conditions, this organism appears to oxidize methane by using oxygen-dependent methane monooxygenase and other enzymes of the pathway found in aerobic methanotrophs (Ettwig et al. 2010). An explanation came from the observation of oxygen production in culture: *M. oxyfera* may generate oxygen for its own use through the dismutation of nitric oxide:

$$2NO \rightarrow O_2 + N_2. \qquad 6.$$

Thus, N_2 is formed through a novel pathway. These organisms prefer nitrite to nitrate, and there is no clear evidence for a direct coupling of nitrate reduction and methane oxidation (Raghoebarsing et al. 2006, Ettwig et al. 2008).

Anaerobic methane oxidizers of the NC10 phylum appear to be widespread in anoxic freshwater systems. They have been detected in several freshwater sediments, aquifers, a peat bog, and wastewater treatment systems (Hu et al. 2009, Deutzmann & Schink 2011). Furthermore, members of the NC10 phylum have been identified in 16S-rRNA gene libraries from waterlogged soils, freshwater sediments, and aquifers where conditions may allow nitrite-dependent methane oxidation (Ettwig et al. 2009). Few sequences were derived from marine sediments, which is consistent with the expectation that conditions for nitrite-dependent methane oxidation are more favorable in freshwater systems. In marine sediments, competition with sulfate-reducing bacteria and sulfate-dependent anaerobic methane oxidation generally restricts methane accumulation to deeper strata well separated from the zone of denitrification near the surface (Canfield et al. 2005).

Methanogenesis is more important in freshwater systems given their much lower sulfate concentrations (Capone & Kiene 1988), and substantial amounts of methane enter the oxic realm to make freshwater systems the most important natural source of methane in the atmosphere. High nitrate concentrations from, e.g., agricultural runoff should further favor nitrite-dependent methane oxidation. Indeed, the enrichments that were established originate from locations with very high methane and nitrate concentrations. So far, however, few biogeochemical indications of nitrite- or nitrate-dependent methane oxidation were found in natural freshwater systems, and several attempts to demonstrate this process produced no results (Smemo & Yavitt 2011 and references therein). The best evidence comes from contaminated aquifers where the process is indicated by the relative distribution of methane and nitrate and where anaerobic methane oxidation has been detected in the presence of nitrate.

Despite the limited direct evidence for this process in natural systems, some predictions concerning the potential role of nitrite-dependent methane oxidizers are still possible because these organisms share many characteristics with the anammox bacteria, such that we can use our knowledge of the distribution of the anammox process as a guide. Both types of organisms are characterized by slow growth with doubling times of 1–2 weeks (Ettwig et al. 2010, Kartal et al. 2011b) and have similar relatively low specific rates of metabolism normalized to biomass (Ettwig et al. 2009). Furthermore, both depend on nitrite and a product of the anaerobic mineralization of organic matter as substrates. Thus, competition for nitrite with classical denitrifiers should be as important a controlling factor for methane oxidation as it is for the anammox process (as discussed above). In fact, experiments with methane oxidizers and anammox bacteria in coculture indicate that, if not ammonium limited, anammox bacteria beat methane oxidizers in this competition for nitrite (Luesken et al. 2011), suggesting that the methane oxidizers will suffer even more in the competition with classical denitrifiers.

These considerations lead to the conclusion that nitrite-dependent methane oxidation as described in *M. oxyfera* and its relatives is likely to be of little significance as a sink for fixed nitrogen and methane in freshwater sediments and water-logged soils with high organic input, where anammox is also not important. The process may have a larger relative impact in systems such as aquifers where the input and reactivity of organic matter is low, consistent with the observations mentioned above and in analogy to the occurrence of anammox in such systems, and in the water column of stratified lakes (Schubert et al. 2010). Just as the relative importance of anammox in marine sediments increases with water depth (**Figure 2a**), the importance of nitrite-dependent methane oxidation could also be higher in metabolically less active sediments such as those found in deeper lakes. Additional competition from sulfate- or iron-dependent methane oxidation could limit the availability of methane, but no strong evidence currently exists for the significance of such processes in freshwater systems (Schubert et al. 2011, Smemo & Yavitt 2011). A low relative contribution of anaerobic methane oxidation to the cycling of methane and nitrate in shallow organic-rich sediments does not exclude the possibility that the rates of the process are higher and populations of the involved organisms are larger there than at sites where the relative importance of the process is larger, similar to the case for the anammox process (Dalsgaard et al. 2005). However, because such systems are the main natural sources of atmospheric methane, this analysis suggests that the process plays a minor role in regulating methane emissions.

Key questions that need to be answered for a quantitative evaluation of the ecological and biogeochemical role of the nitrite-dependent methane oxidizers in nature pertain to their kinetics of nitrite and methane consumption; the population size of the organisms in their habitats; and whether they use other types of metabolism to sustain growth, which could increase their importance relative to the estimates above. Other organisms may also perform similar processes. Verification of the postulated mechanism of nitric oxide dismutation and identification of the enzymes involved in this central step hold the key to understanding the evolution of the pathway, which will aid investigations of its phylogenetic diversity and may also give insights to other anaerobic pathways that potentially take advantage of biological oxygen production.

CONCLUSIONS AND FUTURE DIRECTIONS

1. The nitrogen cycle is not so much a cycle as it is a network of transformations. This is particularly evident for nitrite, which is consumed in five different types of anaerobic metabolism as well as one aerobic (**Figure 1**). Detailed ecophysiological studies of the organisms are needed to improve our very rough understanding of how the processes are regulated and interact in nature. Importantly, this also includes the classical processes such as denitrification, as emphasized by its highly variable relationship to anammox in marine systems. Interactions between aerobes and anaerobes are another important issue because nitrogen cycling is often focused around oxic-anoxic interfaces, and recent studies suggest that aerobic and anaerobic nitrogen metabolisms may coexist over a substantial range of oxygen concentration (Morley et al. 2008, Kalvelage et al. 2011).

2. Substantial contributions to the nitrogen cycle are documented for two of the new processes: foraminiferal denitrification and, particularly, anammox. Both are important mainly in marine systems, although the role of anammox on land needs more attention. The pool of fixed nitrogen in the oceans is controlled by a delicate balance between sources, mainly nitrogen fixation, and by its removal through N_2 production. In addition, the stability of the pool size on a 1,000-year timescale is under debate (e.g., Codispoti 2007). Thus, the contribution of the new processes to the marine nitrogen budget needs to be evaluated. Current budgets are partly based on mass balances, and the new processes are partially included in existing estimates of denitrification (Thamdrup & Dalsgaard 2008). Nonetheless, the processes should be considered explicitly both in experimental investigations of fixed nitrogen removal and in quantitative models of the marine nitrogen cycle.

3. Predictions based on the limited information available do not suggest major contributions from nitrite-dependent methane oxidation to either N_2 production or methane oxidation, but investigations of the role of this process in natural environments are needed.

4. The differences in phylogenetic diversity within different metabolic guilds are striking, ranging from anammox bacteria, which are represented by a single genus in the oceans, to denitrifiers, which are found in all three domains of life and often exhibit a high local diversity (e.g., Priemé et al. 2002, Santoro et al. 2006). Yet, these differences remain largely unexplained. Comparative studies of the genetics and evolution of the different pathways may shed light on the fundamental constraints on metabolic evolution and its relationship to microbial ecology. The timing of this evolution relative to Earth's biogeochemical evolution remains unresolved and deserves attention for a more detailed understanding of how the nitrogen cycle may have modulated the development of the biosphere from the origin of Earth to today (e.g., Canfield et al. 2010).

5. Several different types of eukaryotes perform dissimilatory nitrate reduction, denitrification, or DNRA, which suggests that such metabolisms may be an original trait in the domain. This hypothesis may be tested through genetic analysis. The potential for dissimilatory nitrogen transformations in other eukaryotes, including the large uncharacterized diversity of protists revealed by environmental molecular approaches (e.g., Caron et al. 2012), should also be explored.

6. Considering the rate at which new discoveries are reported, we may safely assume that additional types of metabolism and important groups of organisms associated with the nitrogen

cycle remain to be unveiled. Some transformations are predicted by thermodynamic considerations (e.g., van de Leemput et al. 2011), and anoxygenic phototrophic oxidation of nitrite is an example of an emerging process (Griffin et al. 2007, Schott et al. 2010). New and classical methods together form a powerful toolbox for future discoveries.

DISCLOSURE STATEMENT

The author is not aware of any affiliations, memberships, funding, or financial holdings that might be perceived as affecting the objectivity of this review.

ACKNOWLEDGMENTS

The author acknowledges support from the Danish Councils for Independent Research and Natural Sciences and the Agouron Institute. Kirsten Hofmockel and Ashley Helton are thanked for constructive reviews that helped improve the manuscript.

LITTERATURE CITED

Baas Becking LGM. 1934. *Geobiologie, of Inleiding tot de Milieukunde*. The Hague: WP van Stockum & Zoon. 263 pp.

Berner RA. 2006. Geological nitrogen cycle and atmospheric N_2 over Phanerozoic time. *Geology* 34:413–15

Bernhard JM, Edgcomb VP, Casciotti KL, McIlvin MR, Beaudoin DJ. 2011. Denitrification likely catalyzed by endobionts in an allogromiid foraminifer. *ISME J.* 6:951–60

Braker G, Conrad R. 2011. Diversity, structure, and size of N_2O-producing microbial communities in soils: What matters for their functioning? In *Advances in Applied Microbiology*, ed. AI Laskin, S Sariasiani, GM Gadd, 75: 33–70. Amsterdam: Elsevier

Brandsma J, van de Vossenberg J, Risgaard-Petersen N, Schmid MC, Engstrom P, et al. 2011. A multi-proxy study of anaerobic ammonium oxidation in marine sediments of the Gullmar Fjord, Sweden. *Environ. Microbiol. Rep.* 3:360–66

Burdige DJ. 1991. The kinetics of organic-matter mineralization in anoxic marine-sediments. *J. Mar. Res.* 49:727–61

Canfield DE, Glazer AN, Falkowski PG. 2010. The evolution and future of Earth's nitrogen cycle. *Science* 330:192–96

Canfield DE, Kristensen E, Thamdrup B. 2005. *Aquatic Geomicrobiology*. San Diego, CA: Academic. 640 pp.

Capone DG, Kiene RP. 1988. Comparison of microbial dynamics in marine and freshwater sediments: contrasts in anaerobic carbon catabolism. *Limnol. Oceanogr.* 33:725–49

Caron DA, Countway PD, Jones AC, Kim DY, Schnetzer A. 2012. Marine protistan diversity. *Annu. Rev. Mar. Sci.* 4:467–93

Clément JC, Shrestha J, Ehrenfeld JG, Jaffe PR. 2005. Ammonium oxidation coupled to dissimilatory reduction of iron under anaerobic conditions in wetland soils. *Soil Biol. Biochem.* 37:2323–28

Codispoti LA. 2007. An oceanic fixed nitrogen sink exceeding 400 Tg N a^{-1} versus the concept of homeostasis in the fixed-nitrogen inventory. *Biogeosciences* 4:233–53

Codispoti LA, Brandes JA, Christensen JP, Devol AH, Naqvi SWA, et al. 2001. The oceanic fixed nitrogen and nitrous oxide budgets: moving targets as we enter the anthropocene? *Sci. Mar.* 65(Suppl. 2):85–105

Crenshaw CL, Lauber C, Sinsabaugh RL, Stavely LK. 2008. Fungal control of nitrous oxide production in semiarid grassland. *Biogeochemistry* 87:17–27

Dalsgaard T, Canfield DE, Petersen J, Thamdrup B, Acuña-Gonzalez J. 2003. N_2 production by the anammox reaction in the anoxic water column of Golfo Dulce, Costa Rica. *Nature* 422:606–8

Dalsgaard T, Thamdrup B. 2002. Factors controlling anaerobic ammonium oxidation with nitrite in marine sediments. *Appl. Environ. Microbiol.* 68:3802–8

Dalsgaard T, Thamdrup B, Canfield DE. 2005. Anaerobic ammonium oxidation (anammox) in the marine environment. *Res. Microbiol.* 156:457–64

Dalsgaard T, Thamdrup B, Farías L, Revsbech NP. 2012. Anammox and denitrification in the oxygen minimum zone of the eastern South Pacific. *Limnol. Oceanogr.* 57(5):1331–46

Deutzmann JS, Schink B. 2011. Anaerobic oxidation of methane in sediments of Lake Constance, an oligotrophic freshwater Lake. *Appl. Environ. Microbiol.* 77:4429–36

Dong LF, Sobey MN, Smith CJ, Rusmana I, Phillips W, et al. 2011. Dissimilatory reduction of nitrate to ammonium, not denitrification or anammox, dominates benthic nitrate reduction in tropical estuaries. *Limnol. Oceanogr.* 56:279–91

Emerson S, Jahnke R, Bender M, Froelich P, Klinkhammer G, et al. 1980. Early diagenesis in sediments from the Eastern Equatorial Pacific. 1. Pore water nutrient and carbonate results. *Earth Planet. Sci. Lett.* 49:57–80

Engström P, Penton CR, Devol AH. 2009. Anaerobic ammonium oxidation in deep-sea sediments off the Washington margin. *Limnol. Oceanogr.* 54:1643–52

Ettwig KF, Butler MK, Le Paslier D, Pelletier E, Mangenot S, et al. 2010. Nitrite-driven anaerobic methane oxidation by oxygenic bacteria. *Nature* 464:543–48

Ettwig KF, Shima S, van de Pas-Schoonen KT, Kahnt J, Medema MH, et al. 2008. Denitrifying bacteria anaerobically oxidize methane in the absence of Archaea. *Environ. Microbiol.* 10:3164–73

Ettwig KF, van Alen T, van de Pas-Schoonen KT, Jetten MSM, Strous M. 2009. Enrichment and molecular detection of denitrifying methanotrophic bacteria of the NC10 phylum. *Appl. Environ. Microbiol.* 75:3656–62

Finlay BJ. 1985. Nitrate respiration by protozoa (*Loxodes* spp.) in the hypolimnetic nitrite maximum of a productive fresh-water pond. *Freshw. Biol.* 15:333–46

Finlay BJ, Span ASW, Harman JMP. 1983. Nitrate respiration in primitive eukaryotes. *Nature* 303:333–36

Glud RN, Thamdrup B, Stahl H, Wenzhoefer F, Glud A, et al. 2009. Nitrogen cycling in a deep ocean margin sediment (Sagami Bay, Japan). *Limnol. Oceanogr.* 54:723–34

Griffin BM, Schott J, Schink B. 2007. Nitrite, an electron donor for anoxygenic photosynthesis. *Science* 316:1870

Gruber N. 2008. The marine nitrogen cycle: overview and challenges. In *Nitrogen in the Marine Environment*, ed. DG Capone, DA Bronk, MR Mulholland, EJ Carpenter, pp. 1–50. Amsterdam: Elsevier

Gruber N, Galloway JN. 2008. An Earth-system perspective of the global nitrogen cycle. *Nature* 451:293–96

Gubry-Rangin C, Hai B, Quince C, Engel M, Thomson BC, et al. 2011. Niche specialization of terrestrial archaeal ammonia oxidizers. *Proc. Natl. Acad. Sci. USA* 108:21206–11

Gupta KJ, Igamberdiev AU. 2011. The anoxic plant mitochondrion as a nitrite: NO reductase. *Mitochondrion* 11:537–43

Güven D, Dapena A, Kartal B, Schmid MC, Maas B, et al. 2005. Propionate oxidation by and methanol inhibition of anaerobic ammonium-oxidizing bacteria. *Appl. Environ. Microbiol.* 71:1066–71

Høgslund S, Revsbech NP, Cedhagen T, Nielsen LP, Gallardo VA. 2008. Denitrification, nitrate turnover, and aerobic respiration by benthic foraminiferans in the oxygen minimum zone off Chile. *J. Exp. Mar. Biol. Ecol.* 359:85–91

Hu BL, Shen LD, Xu XY, Zheng P. 2011. Anaerobic ammonium oxidation (anammox) in different natural ecosystems. *Biochem. Soc. Trans.* 39:1811–16

Hu SH, Zeng RJ, Burow LC, Lant P, Keller J, Yuan ZG. 2009. Enrichment of denitrifying anaerobic methane oxidizing microorganisms. *Environ. Microbiol. Rep.* 1:377–84

Huettel M, Forster S, Kloser S, Fossing H. 1996. Vertical migration in the sediment-dwelling sulfur bacteria *Thioploca* spp. in overcoming diffusion limitations. *Appl. Environ. Microbiol.* 62:1863–72

Hulth S, Aller RC, Gilbert F. 1999. Coupled anoxic nitrification/manganese reduction in marine sediments. *Geochim. Cosmochim. Acta* 63:49–66

Islas-Lima S, Thalasso F, Gomez-Hernandez J. 2004. Evidence of anoxic methane oxidation coupled to denitrification. *Water Res.* 38:13–16

Jensen MM, Kuypers MMM, Lavik G, Thamdrup B. 2008. Rates and regulation of anaerobic ammonium oxidation and denitrification in the Black Sea. *Limnol. Oceanogr.* 53:23–36

Jensen MM, Lam P, Revsbech NP, Nagel B, Gaye B, et al. 2011. Intensive nitrogen loss over the Omani Shelf due to anammox coupled with dissimilatory nitrite reduction to ammonium. *ISME J.* 5:1660–70

Jensen MM, Thamdrup B, Dalsgaard T. 2007. Effects of specific inhibitors on anammox and denitrification in marine sediments. *Appl. Environ. Microbiol.* 73:3151–58

Jetten MSM, van Niftrik L, Strous M, Kartal B, Keltjens JT, Op den Camp HJM. 2009. Biochemistry and molecular biology of anammox bacteria. *Crit. Rev. Biochem. Mol. Biol.* 44:65–84

Kalvelage T, Jensen MM, Contreras S, Revsbech NP, Lam P, et al. 2011. Oxygen sensitivity of anammox and coupled N-cycle processes in oxygen minimum zone. *PLoS ONE* 6:e29299

Kamp A, de Beer D, Nitsch JL, Lavik G, Stief P. 2011. Diatoms respire nitrate to survive dark and anoxic conditions. *Proc. Natl. Acad. Sci. USA* 108:5649–54

Kartal B, Geerts W, Jetten MSM. 2011a. Cultivation, detection, and ecophysiology of anaerobic ammonium-oxidizing bacteria. *Methods Enzymol.* 486:89–108

Kartal B, Keltjens JT, Jetten MSM. 2011b. Metabolism and genomics of anammox bacteria. See Ward et al. 2001, pp. 181–200

Kartal B, Maalcke WJ, de Almeida NM, Cirpus I, Gloerich J, et al. 2011c. Molecular mechanism of anaerobic ammonium oxidation. *Nature* 479:127–59

Kim SW, Fushinobu S, Zhou SM, Wakagi T, Shoun H. 2009. Eukaryotic *nirK* genes encoding copper-containing nitrite reductase: originating from the protomitochondrion? *Appl. Environ. Microbiol.* 75:2652–58

Kobayashi M, Shoun H. 1995. The copper-containing dissimilatory nitrite reductase involved in the denitrifying system of the fungus *Fusarium oxysporum*. *J. Biol. Chem.* 270:4146–51

Koeve W, Kähler P. 2010. Heterotrophic denitrification versus autotrophic anammox: quantifying collateral effects on the oceanic carbon cycle. *Biogeoscience* 7:2327–37

Koho KA, Piña-Ochoa E, Geslin E, Risgaard-Petersen N. 2011. Vertical migration, nitrate uptake and denitrification: survival mechanisms of foraminifers (*Globobulimina turgida*) under low oxygen conditions. *FEMS Microbiol. Ecol.* 75:273–83

Könneke M, Bernhard AE, de la Torre JR, Walker CB, Waterbury JB, Stahl DA. 2005. Isolation of an autotrophic ammonia-oxidizing marine archaeon. *Nature* 437:543–46

Kroeger KD, Charette MA. 2008. Nitrogen biogeochemistry of submarine groundwater discharge. *Limnol. Oceanogr.* 53:1025–39

Kuypers MMM, Lavik G, Woebken D, Schmid M, Fuchs BM, et al. 2005. Massive nitrogen loss from the Benguela upwelling system through anaerobic ammonium oxidation. *Proc. Natl. Acad. Sci. USA* 102:6478–83

Kuypers MMM, Sliekers AO, Lavik G, Schmid M, Jorgensen BB, et al. 2003. Anaerobic ammonium oxidation by anammox bacteria in the Black Sea. *Nature* 422:608–11

Lam P, Kuypers MMM. 2011. Microbial nitrogen cycling processes in oxygen minimum zones. *Annu. Rev. Mar. Sci.* 3:317–45

Lam P, Lavik G, Jensen MM, van de Vossenberg J, Schmid M, et al. 2009. Revising the nitrogen cycle in the Peruvian oxygen minimum zone. *Proc. Natl. Acad. Sci. USA* 106:4752–57

Laughlin RJ, Stevens RJ. 2002. Evidence for fungal dominance of denitrification and codenitrification in a grassland soil. *Soil Sci. Soc. Am. J.* 66:1540–48

Lavik G, Stuhrmann T, Bruchert V, Van der Plas A, Mohrholz V, et al. 2009. Detoxification of sulphidic African shelf waters by blooming chemolithotrophs. *Nature* 457:581–86

Levin LA. 2003. Oxygen minimum zone benthos: adaptation and community response to hypoxia. *Oceanogr. Mar. Biol.* 41:1–45

Luesken FA, Sanchez J, van Alen TA, Sanabria J, Op den Camp HJM, et al. 2011. Simultaneous nitrite-dependent anaerobic methane and ammonium oxidation processes. *Appl. Environ. Microbiol.* 77:6802–7

Luther GW III, Sundby B, Lewis BL, Brendel PJ. 1997. Interactions of manganese with the nitrogen cycle: alternative pathways to dinitrogen. *Geochim. Cosmochim. Acta* 61:4043–52

Martens-Habbena W, Berube PM, Urakawa H, de la Torre JR, Stahl DA. 2009. Ammonia oxidation kinetics determine niche separation of nitrifying Archaea and Bacteria. *Nature* 461:976–79

Montoya JP, Holl CM, Zehr JP, Hansen A, Villareal TA, Capone DG. 2004. High rates of N_2 fixation by unicellular diazotrophs in the oligotrophic Pacific Ocean. *Nature* 430:1027–31

Moore TA, Xing YP, Lazenby B, Lynch MDJ, Schiff S, et al. 2011. Prevalence of anaerobic ammonium-oxidizing bacteria in contaminated groundwater. *Environ. Sci. Technol.* 45:7217–25

Morford SL, Houlton BZ, Dahlgren RA. 2011. Increased forest ecosystem carbon and nitrogen storage from nitrogen rich bedrock. *Nature* 477:78–81

Morley N, Baggs EM, Dorsch P, Bakken L. 2008. Production of NO, N_2O and N_2 by extracted soil bacteria, regulation by NO_2^- and O_2 concentrations. *FEMS Microbiol. Ecol.* 65:102–12

Mulder A, van der Graaf AA, Roberson LA, Kuenen JG. 1995. Anaerobic ammonium oxidation discovered in a denitrifying fluidized bed reactor. *FEMS Microbiol. Ecol.* 16:177–84

Ngugi DK, Ji R, Brune A. 2011. Nitrogen mineralization, denitrification, and nitrate ammonification by soil-feeding termites: a ^{15}N-based approach. *Biogeochemistry* 103:355–69

Nicol GW, Leininger S, Schleper C. 2011. Distribution and activity of ammonia-oxidizing archaea in natural environments. See Ward et al. 2011, pp. 157–78

Piña-Ochoa E, Høgslund S, Geslin E, Cedhagen T, Revsbech NP, et al. 2010a. Widespread occurrence of nitrate storage and denitrification among Foraminifera and Gromiida. *Proc. Natl. Acad. Sci. USA* 107:1148–53

Piña-Ochoa E, Koho KA, Geslin E, Risgaard-Petersen N. 2010b. Survival and life strategy of the foraminiferan *Globobulimina turgida* through nitrate storage and denitrification. *Mar. Ecol. Progr. Ser.* 417:39–49

Prendergast-Miller MT, Baggs EM, Johnson D. 2011. Nitrous oxide production by the ectomycorrhizal fungi *Paxillus involutus* and *Tylospora fibrillosa*. *FEMS Microbiol. Lett.* 316:31–35

Priemé A, Braker G, Tiedje JM. 2002. Diversity of nitrite reductase (*nirK* and *nirS*) gene fragments in forested upland and wetland soils. *Appl. Environ. Microbiol.* 68:1893–900

Prokopenko MG, Sigman DM, Berelson WM, Hammond DE, Barnett B, et al. 2011. Denitrification in anoxic sediments supported by biological nitrate transport. *Geochim. Cosmochim. Acta* 75:7180–99

Raghoebarsing AA, Pol A, van de Pas-Schoonen KT, Smolders AJP, Ettwig KF, et al. 2006. A microbial consortium couples anaerobic methane oxidation to denitrification. *Nature* 440:918–21

Richards FA. 1965. Chemical observations in some anoxic, sulfide-bearing basins and fjords. In *Advances in Water Pollution Research*, Vol. 3, ed. EA Pearson, pp. 215–32. London: Pergamon

Risgaard-Petersen N, Langezaal AM, Ingvardsen S, Schmid MC, Jetten MSM, et al. 2006. Evidence for complete denitrification in a benthic foraminifer. *Nature* 443:93–96

Rütting T, Boeckx P, Muller C, Klemedtsson L. 2011. Assessment of the importance of dissimilatory nitrate reduction to ammonium for the terrestrial nitrogen cycle. *Biogeosciences* 8:1779–91

Samarkin VA, Madigan MT, Bowles MW, Casciotti KL, Priscu JC, et al. 2010. Abiotic nitrous oxide emission from the hypersaline Don Juan Pond in Antarctica. *Nat. Geosci.* 3:341–44

Santoro AE, Boehm AB, Francis CA. 2006. Denitrifier community composition along a nitrate and salinity gradient in a coastal aquifer. *Appl. Environ. Microbiol.* 72:2102–9

Schleper C, Nicol GW. 2010. Ammonia-oxidising archaea: physiology, ecology and evolution. *Adv. Microb. Physiol.* 57:1–47

Schott J, Griffin BM, Schink B. 2010. Anaerobic phototrophic nitrite oxidation by *Thiocapsa* sp. strain KS1 and *Rhodopseudomonas* sp. strain LQ17. *Microbiology* 156:2428–37

Schrum HN, Spivack AJ, Kastner M, D'Hondt S. 2009. Sulfate-reducing ammonium oxidation: a thermodynamically feasible metabolic pathway in subseafloor sediment. *Geology* 37:939–42

Schubert CJ, Durisch-Kaiser E, Wehrli B, Thamdrup B, Lam P, Kuypers MM. 2006. Anaerobic ammonium oxidation in a tropical freshwater system (Lake Tanganyika). *Environ. Microbiol.* 8:1857–63

Schubert CJ, Lucas FS, Durisch-Kaiser E, Stierli R, Diem T, et al. 2010. Oxidation and emission of methane in a monomictic lake (Rotsee, Switzerland). *Aquat. Sci.* 72:455–66

Schubert CJ, Vazquez F, Losekann-Behrens T, Knittel K, Tonolla M, Boetius A. 2011. Evidence for anaerobic oxidation of methane in sediments of a freshwater system (Lago di Cadagno). *FEMS Microbiol. Ecol.* 76:26–38

Schulz HN, Jorgensen BB. 2001. Big bacteria. *Annu. Rev. Microbiol.* 55:105–37

Seitzinger S, Harrison JA, Bohlke JK, Bouwman AF, Lowrance R, et al. 2006. Denitrification across landscapes and waterscapes: a synthesis. *Ecol. Appl.* 16:2064–90

Shoun H, Kim DH, Uchiyama H, Sugiyama J. 1992. Denitrification by fungi. *FEMS Microbiol. Lett.* 94:277–81

Shrestha J, Rich JJ, Ehrenfeld JG, Jaffe PR. 2009. Oxidation of ammonium to nitrite under iron-reducing conditions in wetland soils, laboratory, field demonstrations, and push-pull rate determination. *Soil Sci.* 174:156–64

Simon J. 2002. Enzymology and bioenergetics of respiratory nitrite ammonification. *FEMS Microbiol. Rev.* 26:285–309

Sinninghe-Damsté JS, Strous M, Rijpstra WIC, Hopmans EC, Geenevasen JAJ, et al. 2002. Linearly concatenated cyclobutane lipids form a dense bacterial membrane. *Nature* 419:708–12

Smemo KA, Yavitt JB. 2011. Anaerobic oxidation of methane: an underappreciated aspect of methane cycling in peatland ecosystems? *Biogeosciences* 8:779–93

Smits THM, Huttmann A, Lerner DN, Holliger C. 2009. Detection and quantification of bacteria involved in aerobic and anaerobic ammonium oxidation in an ammonium-contaminated aquifer. *Bioremediat. J.* 13:41–51

Spott O, Russow R, Stange CF. 2011. Formation of hybrid N_2O and hybrid N_2 due to codenitrification: first review of a barely considered process of microbially mediated N-nitrosation. *Soil Biol. Biochem.* 43:1995–2011

Strous M, Fuerst JA, Kramer EHM, Logemann S, Muyzer G, et al. 1999. Missing lithotroph identified as new planctomycete. *Nature* 400:446–49

Strous M, Pelletier E, Mangenot S, Rattei T, Lehner A, et al. 2006. Deciphering the evolution and metabolism of an anammox bacterium from a community genome. *Nature* 440:790–94

Stumm W, Morgan JJ. 1981. *Aquatic Chemistry*. New York: John Wiley & Sons

Tai YL, Dempsey BA. 2009. Nitrite reduction with hydrous ferric oxide and Fe(II): stoichiometry, rate, and mechanism. *Water Res.* 43:546–52

Takaya N. 2009. Response to hypoxia, reduction of electron acceptors, and subsequent survival by filamentous fungi. *Biosci. Biotechnol. Biochem.* 73:1–8

Thamdrup B, Dalsgaard T. 2000. The fate of ammonium in anoxic manganese oxide-rich marine sediment. *Geochim. Cosmochim. Acta* 64:4157–64

Thamdrup B, Dalsgaard T. 2002. Production of N_2 through anaerobic ammonium oxidation coupled to nitrate reduction in marine sediments. *Appl. Environ. Microbiol.* 68:1312–18

Thamdrup B, Dalsgaard T. 2008. Nitrogen cycling in sediments. In *Microbial Ecology of the Oceans*, ed. DE Kirchman, pp. 527–68. New York: John Wiley & Sons. 2nd ed.

Thamdrup B, Dalsgaard T, Jensen MM, Ulloa O, Farías L, Escribano R. 2006. Anaerobic ammonium oxidation in the oxygen-deficient waters off northern Chile. *Limnol. Oceanogr.* 51:2145–56

Thauer RK, Shima S. 2008. Methane as fuel for anaerobic microorganisms. *Ann. NY Acad. Sci.* 1125:158–70

Tielens AGM, Rotte C, van Hellemond JJ, Martin W. 2002. Mitochondria as we don't know them. *Trends Biochem. Sci.* 27:564–72

Treusch AH, Leininger S, Kletzin A, Schuster SC, Klenk HP, Schleper C. 2005. Novel genes for nitrite reductase and Amo-related proteins indicate a role of uncultivated mesophilic crenarchaeota in nitrogen cycling. *Environ. Microbiol.* 7:1985–95

Trimmer M, Engström P. 2011. Distribution, activity, and ecology of anammox bacteria in aquatic environments. See Ward et al. 2011, pp. 201–35

Tsuruta S, Takaya N, Zhang L, Shoun H, Kimura K, et al. 1998. Denitrification by yeasts and occurrence of cytochrome P450nor in *Trichosporon cutaneum. FEMS Microbiol. Lett.* 168:105–10

Uchimura II, Enjoji H, Seki T, Taguchi A, Takaya N, Shoun H. 2002. Nitrate reductase-formate dehydrogenase couple involved in the fungal denitrification by *Fusarium oxysporum. J. Biochem.* 131:579–86

van de Graaf AA, De Bruijn P, Robertson LA, Jetten MSM, Kuenen JG. 1997. Metabolic pathway of anaerobic ammonium oxidation on the basis of ^{15}N studies in a fluidized bed reactor. *Microbiology* 143:2415–21

van de Leemput IA, Veraart AJ, Dakos V, de Klein JJM, Strous M, Scheffer M. 2011. Predicting microbial nitrogen pathways from basic principles. *Environ. Microbiol.* 13:1477–87

Vitousek PM, Howarth RW. 1991. Nitrogen limitation on land and in the sea: How can it occur? *Biogeochemistry* 13:87–115

Ward BB, Arp DJ, Klotz MG, eds. 2011. *Nitrification*. Washington, DC: ASM

Ward BB, Devol AH, Rich JJ, Chang BX, Bulow SE, et al. 2009. Denitrification as the dominant nitrogen loss process in the Arabian Sea. *Nature* 461:78–81

Woulds C, Cowie GL, Levin LA, Andersson JH, Middelburg JJ, et al. 2007. Oxygen as a control on seafloor biological communities and their roles in sedimentary carbon cycling. *Limnol. Oceanogr.* 52:1698–709

Yoshinaga I, Amano T, Yamagishi T, Okada K, Ueda S, et al. 2011. Distribution and diversity of anaerobic ammonium oxidation (anammox) bacteria in the sediment of a eutrophic freshwater lake, Lake Kitaura, Japan. *Microb. Environ.* 26:189–97

Zehr JP. 2011. Nitrogen fixation by marine cyanobacteria. *Trends Microbiol.* 19:162–73

Zehr JP, Waterbury JB, Turner PJ, Montoya JP, Omoregie E, et al. 2001. Unicellular cyanobacteria fix N_2 in the subtropical North Pacific Ocean. *Nature* 412:635–38

Zhou ZM, Takaya N, Nakamura A, Yamaguchi M, Takeo K, Shoun H. 2002. Ammonia fermentation, a novel anoxic metabolism of nitrate by fungi. *J. Biol. Chem.* 277:1892–96

Zhou ZM, Takaya N, Sakairi MAC, Shoun H. 2001. Oxygen requirement for denitrification by the fungus *Fusarium oxysporum*. *Arch. Microbiol.* 175:19–25

Zhou ZM, Takaya N, Shoun H. 2010. Multi-energy metabolic mechanisms of the fungus *Fusarium oxysporum* in low oxygen environments. *Biosci. Biotechnol. Biochem.* 74:2431–37

Zhu GB, Wang SY, Wang Y, Wang CX, Risgaard-Petersen N, et al. 2011. Anaerobic ammonia oxidation in a fertilized paddy soil. *ISME J.* 5:1905–12

Beyond the Plankton Ecology Group (PEG) Model: Mechanisms Driving Plankton Succession

Ulrich Sommer,[1] Rita Adrian,[2]
Lisette De Senerpont Domis,[3] James J. Elser,[4]
Ursula Gaedke,[5] Bas Ibelings,[3,6] Erik Jeppesen,[7]
Miquel Lürling,[3,8] Juan Carlos Molinero,[1]
Wolf M. Mooij,[3,8] Ellen van Donk,[3]
and Monika Winder[9]

[1]Helmholtz Center for Ocean Research (GEOMAR), 24105 Kiel, Germany;
email: usommer@geomar.de, jmolinero@geomar.de

[2]Leibniz-Institute of Freshwater Ecology and Inland Fisheries (IGB), 12587 Berlin, Germany;
email: adrian@igb-berlin.de

[3]Department of Aquatic Ecology, Netherlands Institute of Ecology (NIOO-KNAW), 6700 AB
Wageningen, The Netherlands; email: l.desenerpontdomis@nioo.knaw.nl,
b.ibelings@nioo.knaw.nl, w.mooij@nioo.knaw.nl, e.vandonk@nioo.knaw.nl

[4]School of Life Sciences, Arizona State University, Tempe, Arizona 85287;
email: j.elser@asu.edu

[5]Department of Ecology and Ecosystem Modeling, University of Potsdam, 14469 Potsdam,
Germany; email: gaedke@uni-potsdam.de

[6]Institute FA Forel, University of Geneva, 1290 Versoix, Switzerland;
email: bastiaan.ibelings@unige.ch

[7]Department of Bioscience, University of Aarhus, 8000 Aarhus, Denmark;
email: ej@dmu.dk

[8]Department of Environmental Sciences, University of Wageningen, 6700 AB Wageningen,
The Netherlands; email: miquel.lurling@wur.nl

[9]Department of Systems Ecology, University of Stockholm, 10691 Stockholm, Sweden;
email: mwinder@ecology.su.se

Annu. Rev. Ecol. Evol. Syst. 2012. 43:429–48

First published online as a Review in Advance on
September 4, 2012

The *Annual Review of Ecology, Evolution, and
Systematics* is online at ecolsys.annualreviews.org

This article's doi:
10.1146/annurev-ecolsys-110411-160251

1543-592X/12/1201-0429$20.00

Keywords

lakes, oceans, seasonal patterns, pelagic zone, light, overwintering, grazing, parasitism, food quality

Abstract

The seasonal succession of plankton is an annually repeated process of community assembly during which all major external factors and internal interactions shaping communities can be studied. A quarter of a century ago, the state of this understanding was described by the verbal plankton ecology group (PEG) model. It emphasized the role of physical factors, grazing and nutrient limitation for phytoplankton, and the role of food limitation and fish predation for zooplankton. Although originally targeted at lake ecosystems, it was also adopted by marine plankton ecologists. Since then, a suite of ecological interactions previously underestimated in importance have become research foci: overwintering of key organisms, the microbial food web, parasitism, and food quality as a limiting factor and an extended role of higher order predators. A review of the impact of these novel interactions on plankton seasonal succession reveals limited effects on gross seasonal biomass patterns, but strong effects on species replacements.

1. INTRODUCTION

A quarter of a century ago, Sommer et al. (1986) proposed the plankton ecology group (PEG) model as a standard template to describe factors driving the seasonal wax and wane of phyto- and zooplankton in lakes. Their paper has become one of the most cited within plankton ecology. Originally addressing lake plankton, the general patterns described in the PEG model have also been adopted by marine plankton ecologists.

Studying the seasonal succession of plankton is of heuristic value for general aquatic ecology, because it is an annually repeated process of community assembly during which community interactions such as competition, herbivory, or predation can be studied in a nutshell. The verbally formulated PEG model describes seasonally unfolding biotic interactions constrained by a framework of abiotic control mechanisms that set the start and the end of the growing season. The seasonal change in the role of the different abiotic and biotic factors driving succession has been described in 24 sequential steps. Those were focused on stratifying moderately eutrophic water bodies. However, the condensed information on seasonal biomass patterns and control mechanisms provided by figures 6 and 7 of the original paper became a reference point for a wide array of studies (**Figure 1**).

Figure 1 presents a 2–3 modal phytoplankton biomass curve for eutrophic waters and a 1–2 modal biomass curve for oligotrophic waters. The first peak in both scenarios is the spring bloom (shifted toward summer at high latitudes and altitudes). The oligotrophic scenario does not show a summer peak because of nutrient shortage. The minor peak in autumn is optional in both scenarios, occurring only when the nutrient import due to increasing autumnal mixing depth takes place under still favorable light conditions.

Mixing depth: depth of the pycnocline, where water can be mixed from the surface to the mixing depth by wind and by convection

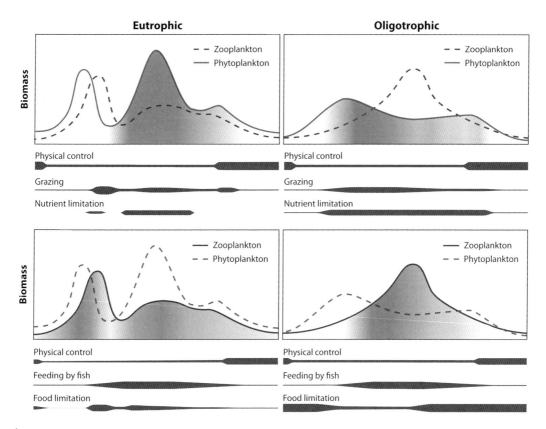

Figure 1

Figures 6 and **7** of the original PEG (plankton ecology group) paper (modified): Seasonal (winter through autumn) biomass patterns in eutrophic (*left*) and oligotrophic (*right*) water bodies. (*Top*) Focus on phytoplankton (*blue solid line*) (*dark shading*, inedible for zooplankton; *light shading*, edible for zooplankton). (*Bottom*) Focus on zooplankton (*red solid line*) (*dark shading*, small herbivores; *light shading*, large herbivores). The thickness of the horizontal bars indicates the seasonal change in relative importance of physical factors, grazing, nutrient limitation, fish predation, and food limitation (with kind permission of E. Schweizerbart publishers).

Zooplankton is assumed to develop in response to food availability with peaks following phytoplankton after some delay. The spring peak of grazing zooplankton leads to a decline in phytoplankton biomass toward a mid-season biomass minimum (clear water phase). Subsequently, food limitation and fish predation control zooplankton biomass. The summer increase of phytoplankton in eutrophic systems contains a high proportion of inedible phytoplankton, thus shifting top-down control of phytoplankton from control of total biomass to control of community composition. The whole process of community assembly was assumed to be initiated by the vernal light increase and terminated by the autumnal light decrease.

Twenty-five years after the publication of the PEG model, it is timely to revisit the model and explore the progress made in our understanding of plankton interactions and mechanisms driving temporal changes in the structure of plankton communities. We re-evaluate and extend the PEG model, putting specific emphasis on mechanisms that were neglected or underestimated in the original model, such as overwintering, the microbial food web, parasitism, and limitation by food quality, as well as new insights into the role of higher trophic levels.

2. LIGHT VERSUS TEMPERATURE AS TRIGGERS FOR THE PHYTOPLANKTON SPRING BLOOM

Critical depth: maximal mixing depth that allows a rate of phytoplankton photosynthesis in excess of the respiration rate and, therefore, positive growth rates

Pycnocline: strong vertical gradient in water density caused by salinity (halocline) or temperature differences (thermocline), acting as a barrier against mixing

The old PEG model identified vernal increase in light supply as the single dominant trigger of the phytoplankton spring bloom (step 1), which is in contrast to the focus on temperature and neglect of light in much of the recent literature on global climate change. The PEG model's emphasis on light is fully in line with Sverdrup's (1953) classic concept of critical mixing depth. With the onset of thermal stratification, mixing depth has to become lower than critical depth in order to retain phytoplankton within the well-lit surface layer to achieve a higher production respiration rate. Based on usual values of mixing depth and water transparency, the mean irradiance in a mixed surface layer is roughly inversely proportional to the mixing depth and the vertical attenuation coefficient of the water. Therefore, in deep lakes and seabeds mixing depth usually plays a dominant role in bloom initiation. Here, the onset of stratification might bring about a fast order-of-magnitude increase in light availability, whereas seasonal shifts in solar irradiance are more gradual, and shifts between cloudy and sunny weeks rarely increase the light supply by more than a factor of three.

Meanwhile, the critical depth concept has been superseded by the critical turbulence concept (Huisman & Sommeijer 2002): Below a critical limit of turbulence, phytoplankton cells may remain long enough in the surface layer even when the depth of the pycnocline exceeds the critical depth. Transient blooms in deep lakes prior to the onset of thermal stratification under calm conditions support this concept (Tirok & Gaedke 2007a).

The situation is quite different for shallow or moderately deep water bodies, where the bottom or a halocline at low depth (coastal seas) restrict vertical mixing. In such water bodies, the spring bloom of phytoplankton can already start before the onset of thermal stratification. Therefore, seasonal warming does not act as a light switch for phytoplankton. In such systems, the interannual variability in light supply is dominated by cloud cover. However, the vertical attenuation coefficient may become the dominant factor in lakes and coastal seas influenced by highly silted and temporally variable runoff.

Under clear ice, light conditions might be favorable for phytoplankton blooms in a narrow water layer, providing an inoculum for a swift start of the spring bloom, depending on light conditions after ice melt. In sea ice, microalgae growing in brine channels (ice algae) play a comparable role to freshwater subice blooms. Melting of sea ice produces a transient halocline. The resultant ice-margin bloom (Smith & Nelson 1985) is often the only significant phytoplankton growth pulse in polar seas because mixing depth is too large during the rest of the year. It is also the ultimate proof that phytoplankton blooms may occur at the lowest temperatures of liquid water.

Recent global change research has provided examples of both correlation (Adrian et al. 1995, Blenckner et al. 2007) and noncorrelation (Edwards & Richardson 2004, Wiltshire et al. 2008) between the timing of the spring bloom and temperature. An advancement of the spring bloom with warming is usually associated with an earlier ice-out or earlier thermal stratification. Variable relationships are expected in ice-free, unstratified waters or if increasing wind stress counteracts the effect of warming on stratification.

Siegel et al. (2002) calculated mixed layer mean irradiance values at the start of the spring bloom for the North Atlantic Ocean from satellite data. They found a remarkably uniform daily light dose of 1.3 mol photons m^{-2} d^{-1} (0.96 to 1.75) photosynthetically active radiation, which was unrelated to latitude and temperature. An almost identical critical light dose for the onset of the spring bloom was found in mesocosm experiments with temperature scenarios ranging from 0 to 6°C warming (Sommer & Lengfellner 2008). In contrast to a strong light effect, warming advanced the peak of the spring bloom by only 1–1.5 d °C^{-1} in those experiments, which amounts to ca. 1 week for the most drastic climate change scenarios.

We conclude that the emphasis of the original PEG model on light as starter and terminator of the growth season has withstood the test of time.

3. OVERWINTERING

Although overwintering of copepods is mentioned in the final steps of the original PEG model (Sommer et al. 1986), a start close to zero of both phyto- and zooplankton during the winter-spring transition was assumed in the PEG model (Sommer et al. 1986). However, meanwhile, considerable evidence for priority effects along with trans-annual memory effects of overwintering plankton populations has emerged (Cáceres 1998, Sommer & Lewandowska 2011).

Overwintering strategies include overwintering in the active phase or in a resting stage—for instance, phytoplankton akinets, spores, zooplankton resting eggs, or mid-life resting stages such as diapausing copepodids. In the marine realm, overwintering zooplankton tend to shift downward in the water column (*Calanus* spp. often go more than 500 m), triggered by physical processes in combination with active buoyancy control (Campbell 2008). Accumulation of lipid reserves is a widespread overwintering strategy in copepods, when prey availability is scarce (Hagen et al. 1996).

Only overwintering active animals (Hanazato & Yasuno 1989) or early emergence from diapause (Adrian et al. 2006) can yield substantial grazing pressure during the onset of the phytoplankton spring bloom. In such a case, the potential biomass of the spring phytoplankton would be relatively low compared to spring blooms after winters without substantial overwintering of zooplankton. Thus, early termination of the diapause phase triggered by temperature during the spring bloom adds a component of season anticipation to the season responsiveness of reproduction created by increased food availability. Improved food conditions for overwintering zooplankton due to winter warming (Tsugeki et al. 2009) or eutrophication (Jankowski & Straile 2003) may reduce the production of diapausing eggs. Although the contribution of hatching from diapausing eggs to population development may be low, its contribution to the genetic diversity of zooplankton can be important (Haag et al. 2002, Rother et al. 2010).

Empirical evidence from long-term studies suggests that the degree of overwintering plankton influences the seasonal dynamics in lake and marine ecosystems, often associated with alterations in phenology or changes in community structure (Adrian et al. 1995, 2006; Sommer & Lewandowska 2011). The coupling between daphnids and their spring algal food depended on the different overwintering strategies and the extent of different warming scenarios (De Senerpont Domis et al. 2007). Only extreme warming scenarios resulted in a decoupling of daphnids from their spring algal food. Behrenfeld (2010) showed that the spring phytoplankton bloom in the North Atlantic Ocean is initiated already in mid-winter as a result of a decrease in the zooplankton grazing pressure. The opposite trend was found in the North Sea (Helgoland), where enhanced grazing by overwintering zooplankton during warmer winters may have retarded the spring phytoplankton bloom (Wiltshire et al. 2008). In indoor mesocosm experiments, combined effects of winter warming and higher densities of marine overwintering copepods were shown to determine the magnitude and size distribution of the spring phytoplankton (Sommer & Lewandowska 2011).

Overall, the importance of winter conditions for plankton dynamics may also be inferred from the success of winter climatic indices such as the North Atlantic Oscillation in predicting plankton seasonal dynamics in freshwater and marine ecosystems (Gerten & Adrian 2002). We conclude that plankton seasonal succession often is not a start from zero and that the widespread lack of attention to fall and winter in limnological and oceanographic research might miss important mechanisms. Therefore, overwintering deserves more attention than in the original PEG model.

Priority effect: impact of a particular species on community development due to prior arrival

Diapause: interruption of ontogenetic development by a resting stage (e.g., spore, resting egg, resting late larval stage)

4. GRAZING AND THE CLEAR WATER PHASE

Microzooplankton: zooplankton 20–200 µm, containing mostly heterotrophic protists, but also rotifers and copepod nauplii

The PEG model explains the clear water phase to be a result of grazing (step 5), a view widely accepted in limnology, but not in biological oceanography. After Lampert & Schober's (1978) pioneering study, the limnological community accepted the causal link between mesozooplankton grazing and phytoplankton biomass minimum after the spring bloom, the clear water phase. Alternative explanations, such as nutrient limitation and subsequent sedimentation, were excluded. The objections were based on a frequently observed interim increase of dissolved nutrient concentrations during the clear water phase, and a measured excess of grazing rates over phytoplankton production was demonstrated empirically (e.g., Lampert et al. 1986). In biological oceanography, however, grazing is accepted as an explanation in some studies (e.g., Bautista et al. 1992, Wiltshire et al. 2008), whereas many others ascribe the clear water phase to nutrient limitation, subsequent production of transparent extracellular polymers, and formation of aggregates, all leading to accelerated sinking (Smayda 1971, Turner 2002). In marine research, studies devoted directly to mechanisms of bloom termination are considerably less frequent than studies on bloom triggering mechanisms.

Probably, taxonomic differences between marine and limnetic mesozooplankton also play a role. Most of the clear-cut cases of grazing-induced clear water phases are reported from lakes dominated by *Daphnia* spp., whereas marine mesozooplankton is usually dominated by copepods. Copepods exhibit a much slower numerical response to enhanced food availability than *Daphnia*. Even the fastest species have a generation time exceeding three weeks under spring conditions. Moreover, the trophic role of copepods differs from that of *Daphnia*. Even the species traditionally considered herbivores are, in fact, omnivores, feeding also on heterotrophic protists, but are inefficient at feeding on prey less than 5–10 µm in size. Besides being capable of suppressing large algae, the copepods also release smaller phytoplankton from predation by heterotrophic protists, thus paving the way for compensatory growth, sometimes balancing or even overcompensating for the loss of large algae. Thus, the impact of copepods on total phytoplankton biomass is weak and context dependent (Adrian et al. 2001, Sommer & Sommer 2006). In summary, the dominant explanation for the clear water phase is still a major divide between limnetic and marine plankton ecology. It seems plausible that it is based on the predominance of *Daphnia* spp. in many lakes, a genus that is lacking an ecological analogue in the sea. However, the issue is not yet fully explored and deserves further attention.

5. THE ROLE OF HETEROTROPHIC PROTISTS

Step 3 of the PEG model stated that "planktonic herbivores with short generation times will increase their populations first and are followed by slower growing species" (Sommer et al. 1986, p. 435). Sommer and colleagues mentioned ciliates explicitly as fast growing grazers, but their grazing impact on phytoplankton was subsumed under grazing in general. Meanwhile, the importance of the microbial food web (bacteria and smallest phytoplankton at the base and several trophic levels of heterotrophic protists) is firmly established beyond doubt, and the specific role of protists in seasonal plankton succession has to be explored.

Heterotrophic protists (heterotrophic nano- and dinoflagellates and ciliates), ranging in size from ca. 1 µm to greater than 100 µm, are the most numerous, most species rich, and most ubiquitous group of zooplankton. This leads to complex trophic interactions among them, and the resulting highly productive microbial web may contribute substantially to planktonic carbon and nutrient turnover and the diet of larger consumers (Sherr & Sherr 2007, Löder et al. 2011). In contrast to their often relatively low contribution to total biomass, microzooplankton often constitutes a major proportion of total grazing, often consuming far more than 50% of primary production (Müller et al. 1991, Gaedke et al. 2002, Calbet & Landry 2004).

Additionally, heterotrophic protists may strongly improve the food quality of predominantly herbivorous consumers at times when the phytoplankton is deficient in mineral nutrients (N, P), polyunsaturated fatty acids or sterols (cf. Section 7) compared with heterotrophic protists. Thus, they may enrich the diet of consumers even when they constitute only a minor part of the diet in terms of carbon. This trophic upgrading improves the food quality but reduces the food quantity due to the carbon losses caused by the food chain elongation (Breteler et al. 1999, Gaedke et al. 2002).

The seasonal dynamics in biomass of heterotrophic protists depend largely on food availability and grazing losses. Heterotrophic nanoflagellates (HNF; 2–20 µm) are often the most important bacterivores in marine and limnetic systems and obtain most of their energy from bacterial production (Azam et al. 1983). Hence, their seasonal development is linked to that of the bacteria (Berninger et al. 1991). However, more complex patterns may arise when size shifts in the bacterial community decrease their vulnerability to HNF grazing (Jürgens & DeMott 1995). In turn, HNF are not well defended and are grazed by nanoplankton feeders such as numerous ciliates, rotifers, and crustaceans. Hence, their mass occurrence may exert an efficient top-down control on the biomass of HNF (Jürgens et al. 1996, Tadonleke et al. 2004) as found, for example, during the clear water phase. The feeding spectrum of larger protists, often dominated by ciliates and/or heterotrophic dinoflagellates, includes bacterivorous, herbivorous, carnivorous, and mixotrophic forms (Sherr & Sherr 1987, 2007; Müller et al. 1991). Predominantly herbivorous forms often dominate given that phytoplankton production is the major energy source in open water bodies (Calbet & Landry 2004).

During the annual cycle in temperate water bodies, heterotrophic protists, often dominated by ciliates, quickly respond to the onset of the phytoplankton bloom even at low temperatures until being suppressed by mesozooplankton feeding (Sanders & Wickham 1993, Johansson et al. 2004, Tirok & Gaedke 2007b, Löder et al. 2011). Subsequently, a diverse summer community develops, often dominated by ciliates with different modes of nutrition (freshwater) or dinoflagellates (marine systems), following the biomass pattern of phytoplankton with a slight delay (Sherr & Sherr 2007). Heterotrophic protists are basically subject to the same grazing pressure as similar sized and shaped phytoplankton; in the case of copepod dominance they are even often the preferred prey item of mesozooplankton (Paffenhöfer et al. 2005, Vargas et al. 2006). This yields a bimodal seasonal distribution, as is well established for phytoplankton with high biomass in spring and summer and low biomass in winter and during the clear water phase (Gaedke & Wickham 2004). The dependence on phytoplankton as a direct or indirect (via bacterivory) resource and the susceptibility to the same suite of grazers result in a similar seasonal biomass pattern of heterotrophic protists to that of phytoplankton with some delay during the growth phases.

Heterotrophic protists (except armored dinoflagellates) are often kept in check by mesozooplankton ("eat your competitor" strategy) unless environmental factors (e.g., fish predation, low spring temperatures) hamper mesozooplankton activity (e.g., Sanders & Wickham 1993, Jürgens et al. 1996, Johansson et al. 2004, Tadonleke et al. 2004). If this coincides with high food supply, a loophole for protist blooms is created, which may adversely affect mesozooplankton by diminishing their food base (priority effect) (Tirok & Gaedke 2006).

A temporal reduction in microzooplankton grazing may give rise to phytoplankton blooms (Irigoien et al. 2005). Nevertheless, the imprint of protist grazing on gross seasonal biomass patterns (major blooms and crashes) of total phytoplankton and of major functional groups is small compared with their dominant role in the consumption of primary production. Reasons for this include the fast response of heterotrophic protists to food availability. This reduces the potential of their prey to escape top-down control, a prerequisite for blooms and subsequent crashes, whereas time-lagged, nonsteady state interactions between phytoplankton and mesozooplankton frequently generate those patterns. Furthermore, the diet breadth of many protist species is smaller than that of major mesozooplankton (e.g., daphnids, copepods). This promotes the suppression of a very distinct prey group and/or fast predator-prey cycles, whereas less vulnerable species might dampen the response of total phytoplankton biomass. Temporal peaks and valleys of total phytoplankton biomass can thus be the aggregate result of superimposed, much faster predator-prey cycles between heterotrophic protist and phytoplankton species (Tirok & Gaedke 2007b). However, models show that phytoplankton spring blooms would reach much higher biomass levels if protist grazing were eliminated (Peeters et al. 2007), but such situations never occur in nature.

In summary, heterotrophic protists often dominate the metabolic activity of zooplankton and have a strong effect on phytoplankton species replacements, but influence to a much lesser extent the aggregate seasonal patterns of phytoplankton.

6. THE ROLE OF PARASITES

In the PEG model, parasitism was mentioned as a loss factor affecting phytoplankton species inedible for zooplankton (step 8), but its impact on the aggregated pattern of seasonality (total biomass, size structure) remained unexplored. Parasites are widespread among organisms, including phyto- and zooplankton (Ibelings et al. 2004). Parasites have a number of properties rendering them suitable regulators of their host populations, including that (*a*) they are often highly host specific, (*b*) they reduce the fecundity and/or survival of their hosts, and (*c*) their transmission rates increase with host density. Parasites are able to significantly reduce host densities when conditions allow epidemic outbreaks of disease (Hall et al. 2009).

Many phytoplankton species are particularly susceptible to infection by virulent parasitic chytrid fungi. Crashes of phytoplankton blooms over the course of a few weeks coinciding with high prevalence of chytrid infection were observed in lakes (van Donk & Ringelberg 1983). In the marine environment, viral infections of phytoplankton caused similar declines of algal blooms (Brussaard 2004). Disease is not only determined by interactions between host and parasite, but is also dependent on environmental conditions. Light- or phosphorous-stressed diatom hosts constitute a less than optimal environment for their chytrid parasites, limiting the development of disease (Bruning 1991); similarly, poorly fed *Daphnia* produce few parasite transmission stages (Ebert 2005). At low water temperatures chytrids are hardly infective, giving diatom hosts a disease-free window of opportunity to build up a bloom. Warming of lakes reduces this window so that the host is denied a bloom through ongoing low-level infections. Consequently, the parasite, for which host density is a primary factor determining its fitness, is denied an epidemic (Ibelings et al. 2011).

Some fungal parasites seem to be most common on algae that are fairly resistant to grazing by zooplankton (Niquil et al. 2011). Large diatoms like *Asterionella* are occasionally replaced by small centric diatoms under pressure from heavy parasite infections (van Donk & Ringelberg 1983). Hence, parasitism may alter the size structure of phytoplankton and, through this, matter and energy flow. Chambouvet et al. (2008) found that a successional sequence of different bloom-forming dinoflagellates was driven by decimation of each species by genetically distinct, highly host-specific *Amoebophyra*s spp.

The marked effects of parasites on phytoplankton include (*a*) delayed timing or decreased peak population densities of algal blooms, (*b*) steering the outcome of interspecific competition among (sub)dominant species, and (*c*) directing intraspecific succession and maintaining genetic diversity. Thingstad & Lignell (1997) coined the "killing the winner" concept, which is a form of frequency-dependent selection, in which parasites and pathogens preferentially infect the most abundant algal species, allowing more widespread coexistence at community level. De Bruin et al. (2004) demonstrated a similar effect at population level: Reciprocal relationships between diatom host and fungal parasite genotypes maintained population genetic diversity.

Inter- and intraspecific competition and predation are usually regarded as the most important factors regulating the population dynamics of zooplankton in freshwater ecosystems (Decaestecker et al. 2005). Parasites, however, have been shown to change the dominance hierarchy of zooplankton in lakes. Especially *Daphnia* has a large variety of microparasites (most of them bacteria, fungi, and microsporidia that infect the blood, fatty tissue, and gut wall), or epibionts (which settle on the carapace) (Ebert 2005). These parasites may not only kill the host, they may also greatly lower its fecundity. Field studies suggest that these parasites may have a strong impact on zooplankton community structure (Decaestecker et al. 2005). Little is known about effects of parasites on marine zooplankton, but lethal and castrating parasitic dinoflagellate parasites have been described (Skovgaard & Saiz 2006), controlling copepods to a degree similar to predation.

Chytrid zoospores were experimentally demonstrated to be grazed efficiently by *Daphnia* (Kagami et al. 2007). This implies that abundant zoospores may be a food source of some importance for *Daphnia* during fungal epidemics. Large phytoplankton species are believed to be lost by sinking from the euphotic zone instead of being grazed. When these large inedible phytoplankton species are infected by fungi, however, nutrients within these cells are consumed by parasitic fungi, some of which are, in turn, grazed by zooplankton (Niquil et al. 2011).

Ultimately, the spread and severity of disease and, hence, its impact on plankton succession are determined by the complex interplay between hosts and parasites (mediated by environmental effects, resource competition and predation) and thus challenging to predict (e.g., Hall et al. 2009). Although the role of disease in driving changes in species replacements and genetic composition of populations is well established now, the generality of higher level effects (size structure, biomass of entire functional groups and trophic levels) remains an open question at the present state of knowledge.

7. THE ROLE OF FOOD QUALITY

In the PEG model, feeding conditions for zooplankton were mainly defined by food quantity, although available phytoplankton biomass was seen to differ from total biomass by morphological or toxic defenses and by indigestibility (step 8). This assumption is still supported by many studies; however, there has been an increasing awareness of food quality as a limiting factor for zooplankton growth (Brett & Müller-Navarra 1997, Sterner & Elser 2009, van Donk et al. 2011). Quality encompasses all features of the food that make it suitable for ingestion and for fulfilling the consumer's nutritional requirements. Therefore, quality properties include stoichiometric composition, biochemical make-up, and morphological characteristics. These traits affect zooplankton growth and reproduction directly because zooplankton require an adequate intake of all necessary building blocks.

Like all organisms, zooplankton require not only the elements C, O, and H to build up their biomass, but also mineral nutrients such as N and P (Sterner & Elser 2002). Lab work showed that zooplankton performance was poor when the C:P ratio of P-limited phytoplankton exceeded a critical value, the threshold elemental ratio because of an insufficient dietary P intake (see the review by Sterner & Elser 2009). These results led to the proposal that phytoplankton C:P ratios are regulated by the balance of light and nutrients (Sterner et al. 1997), predicting that increased light inputs can paradoxically lead to reduced zooplankton production in spite of enhanced primary production because of excessively high C:P ratios in the food. This prediction has found general (though not unanimous) support in various subsequent studies, including mathematical models, laboratory flasks, indoor microcosms, field experiments, and field observations (Sterner & Elser 2009). High C:P ratios are frequently found after persistent stratification in summer, usually in small, wind-sheltered lakes, but much less often and less pronounced in large lakes and the ocean. N limitation of zooplankton is also a theoretical possibility, but has not yet been found in practice (Malzahn et al. 2010), possibly because phytoplankton C:N ratios are less variable than C:P ratios.

Although phytoplankton can synthesize essential fatty acids, amino acids, and sterols, most zooplankton must obtain these macromolecules from their diet. Biochemical composition varies considerably among phytoplankton taxa. Diatoms and most flagellate taxa are rich in sterols and polyunsaturated fatty acids [PUFAs; diatoms rich in eicosapentaenoic acid (EPA), flagellates rich in docosahexaenoic acid (DHA)], whereas Cyanobacteria and Chlorophyta are PUFA- and sterol-deficient primary producers and, thus, nutritionally poor for zooplankton (Müller-Navarra et al. 2004). In concert with seasonally changing phytoplankton species composition, the availability of essential macromolecules to zooplankton changes, and this has critical consequences for zooplankton reproduction and recruitment (Arts et al. 2009). Growth, egg production and egg hatching have been positively correlated with dietary levels of fatty acids (Müller-Navarra et al. 2004), amino acids (Guisande et al. 2000), and sterols (Arts et al. 2009). However, some protozoans are known to enhance the quality of food for mesozooplankton consumers by nutritional upgrading bacteria and algal lipids and can subsequently support growth for zooplankton (Arts et al. 2009).

In addition, feeding deterrents, sizes and structures beyond the ingestion capacity of zooplankton, thick cell walls and mucous sheaths, and lack of essential elements and presence of toxins in phytoplankton can all reduce the energy transfer to zooplankton from its maximum potential (Rohrlack et al. 1999, van Donk et al. 2011). Digestion resistance in phytoplankton limiting the energy transfer to zooplankton might be more common than previously thought (DeMott & Tessier 2002), whereas the most notorious decoupled energy transfer is found in eutrophic/hypertrophic systems dominated by cyanobacteria. Early onset of cyanobacterial growth can even prevent occurrence of the clear water phase (Deneke & Nixdorf 1999) because of strong food limitation of large-bodied generalist filter feeding zooplankton, like *Daphnia*, due to mechanical interference with the filtering apparatus, through grazing deterrents and growth inhibition or even death of the grazer caused by endotoxins retained in the cyanobacterial cells (Rohrlack et al. 1999). Marine diatoms can release polyunsaturated aldehydes after cell breakage caused by grazers, which reduces the hatching rate of copepod eggs (reviewed in Paffenhöfer et al. 2005).

Phytoplankton traits causing a relaxation of the grazing pressure may be highly variable depending on the magnitude of the risk of being consumed. Formation of colonies to sizes beyond the ingestion capacity of zooplankton, creating structures that increase handling time and a higher toxin production, can be viewed as induced phytoplankton defenses (van Donk et al. 2011). Thus, phenotypic plasticity within species can lead to similar shifts in overall phytoplankton edibility as species replacements. Shifts toward stronger defenses tend to create a vicious cycle. They promote phytoplankton biomass accumulation and, thereby, enhance nutrient limitation, which often leads to a deterioration of stoichiometric and biochemical food quality.

In summary, not only shortage of food but also insufficient food quality can be limiting factors for zooplankton growth and reproduction. Extended periods of low nutrient–high light conditions are most conducive to declining food quality.

8. THE ROLE OF HIGHER TROPHIC LEVELS

In the PEG model, effects of higher trophic levels are restricted to fish predation causing an early summer decline in the biomass and body size of herbivorous zooplankton (step 7). Meanwhile, stronger and seasonally less restricted fish effects in some lakes and marine ecosystems, as well as strong top-down effects of jellyfish in marine ecosystems, have been found.

Multilake studies (Jeppesen et al. 1997, Gliwicz 2003) and biomanipulation experiments (Shapiro & Wright 1974, Hansson et al. 1998) involving removal of plankti-benthivorous fish to improve water quality through a trophic cascade (Carpenter et al. 2001) have shown that top-down forcing plays a larger role in lakes than anticipated in the PEG model. Moreover, and contrary to the statements in the PEG model, it has become evident that fish predation may be high even in winter (Jeppesen et al. 2004, Sørensen et al. 2011).

In cold temperate lakes, there is a major seasonal variation in fish abundance and predation pressure on the zooplankton, and this is related to changes in temperature, fish migration, and reproduction. Fish abundance and predation are typically highest in early summer, when fish larvae occur in the pelagic, and lowest in late winter. In warm, temperate and some subtropical lakes, the period with high predation lasts longer and starts earlier due to earlier and more frequent reproduction of fish and dominance by smaller fish (Meerhoff et al. 2007, Jeppesen et al. 2010), and seasonality may be completely eliminated in tropical lakes if the seasonality in hydrology is negligible. The presence, strength, and duration of the clear water phase in lakes are highly affected by fish predation and timing of the appearance of fish fry in the pelagic. With increasing fish predation, there is a shift from a bimodal clearing phase (strong in early summer, weak in autumn), over unimodal (early summer only), to no clear water phase at all (Jeppesen et al. 1997).

Although the principal mechanisms governing the seasonality of predation on zooplankton in the ocean are similar to those of lakes, it seems that trophic cascades transmitting fish effects to the phytoplankton are weaker (Shurin et al. 2002) and often completely absent or restricted to pelagic food webs with low diversity. In a survey of 50 studies, Micheli (1999) found many examples of fish effects on zooplankton, but only in one unequivocal case were the effects of fish sufficiently strong to be transmitted to the phytoplankton (measured as chlorophyll). Meanwhile, a few additional cases of trophic cascades reaching the phytoplankton level have been reported [e.g., Casini et al. (2008) for the Baltic Sea], but the overall picture remains unchanged.

Although we do not exclude the possibility of alternative explanations, we suggest the following tentative explanation for this limnetic-marine difference: Both extremes of fish predation—i.e., excessive fish predation (leading to near extinction of large, herbivorous zooplankton) or its absence—do not occur in oceans. Excessive predation on zooplankton is usually exerted by bentho-planktivorous fish, which can maintain high stocks even after depletion of zooplankton because of benthic food subsidies, a strategy precluded by the depth of the ocean and also of great lakes. If zooplankton are depleted by fish predation in shallow coastal areas, they are replenished by advection from the open sea. Similarly, local fish kills—caused, for example, by toxic algal blooms–can be quickly compensated for by reimmigration of fish from adjacent regions. Moreover, there is a potential substitute for fish as predators on zooplankton that is lacking in lakes: jellyfish sensu lato, i.e., pelagic Cnidaria and Ctenophora (Purcell & Arai 2001), whose blooms might have far-reaching cascade effects (Condon et al. 2011). An apparent worldwide increase in jellyfish ("rise of the slime"; Jackson et al. 2008, p. 11461) is ascribed by many researchers to overfishing, though climate change has also been invoked as an explanation (Molinero et al. 2008).

In summary, the PEG model has to be expanded to incorporate extreme cases of fish predation on zooplankton, i.e., absence of fish and year-round suppression of mesozooplankton by excessive fish stocks. Both cases are found primarily in shallow, eutrophic lakes. The absence of fish is also found in high mountain lakes (usually oligotrophic) that do not permit immigration.

Trophic cascade: transmission of top-down effects across more than two trophic levels, e.g., less planktivorous fish–more zooplankton–less phytoplankton

9. MATHEMATICAL MODELING

Mathematical models provide a crucial step in gaining a mechanistic understanding of the intrinsic and extrinsic factors governing seasonal dynamics in freshwater and marine systems. Several

Chaos: aperiodic long-term behavior in a deterministic system where sensitive dependence on initial conditions is exhibited

modeling formats can be applied to translate the conceptual framework of the PEG model and its update, presented in this review, into a mathematical model. These formats vary in their degree of realism and complexity, ranging from minimal dynamic models focused on getting insight in general system behavior to more realistic and complex dynamical models that can be compared directly with field data. For an overview of lake ecosystem modeling approaches, see Mooij et al. (2010).

Minimal dynamic models have been very successful for generating hypotheses on the dominant mechanisms underlying seasonal plankton dynamics. These models typically consist of two or three differential equations that aim to capture an essential mechanism (i.e., alternative stable states) rather than quantitative realism. A minimal dynamic model of the consumer-resource interaction between phytoplankton and zooplankton showed that the classical PEG pattern of a spring peak in phytoplankton followed by a clear water phase and summer increase of less edible phytoplankton can be interpreted as a predator-prey cycle where the frequency is phase locked into the seasonal cycle (Scheffer et al. 1997).

Using an extended version of this model, De Senerpont Domis et al. (2007) demonstrated that a mismatch between the keystone herbivore *Daphnia* and its optimal algal prey can arise under climate warming when the photoperiod triggers the emergence of zooplankton resting eggs, whereas the timing of the spring phytoplankton bloom is temperature controlled. The model results indicate that climate warming may lead to a decoupling if the spring population of daphnids establishes itself from a small inoculum of emerging resting eggs. With a large inoculum or an overwintering population of daphnids, the trophic relationships between *Daphnia* and its optimal algal food source remained intact, because *Daphnia* densities peaked earlier in the season.

One of the intriguing issues in plankton ecology is the observation that the number of co-occurring species is much higher than the number of limiting resources (the paradox of the plankton), thereby seemingly defying the principle of competitive exclusion. Using minimal dynamic models, Huisman & Weissing (1999) postulated that deterministic chaos might explain the paradox of the plankton. The basic idea underlying their hypothesis is that in an endless rock-paper-scissors-game, each species creates the conditions in which the next species can thrive, leading to supersaturated coexistence (Schippers et al. 2001).

Recently, Dakos et al. (2009) developed a multispecies plankton community model that allowed for competition as well as predation. Their model results suggest that although species are entrained within the externally forced seasonal cycle, interannual variability in species composition is more likely a consequence of intrinsic species interactions through deterministic chaos. Whereas the original PEG model assumed a complete resetting of the system in winter, the model by Dakos et al. (2009) demonstrates that the species composition in the previous autumn strongly affects that species' composition in the following spring. Recently, Klausmeier & Litchman (2012) proposed a new approach to modeling successional dynamics in plankton using seasonally forced diamond food webs.

Another recently published plankton community model takes the variation and dynamics in functional traits into account. Tirok et al. (2011) used a dynamic trait approach to study the mutual feedback via species shifts in both predator and prey communities. The modeled traits are prey edibility and predator food-selectivity. Altered edibility triggered shifts in food-selectivity and vice versa. Given sufficient functional diversity, this trait-mediated feedback mechanism resulted in a complex dynamic behavior with persistent oscillations in the mean trait values, reflecting the continuous reorganization of species within trophic levels.

Whereas the above mentioned models zoom in on specific phenomena, other, more complex, dynamical models aim at covering all the dominant components of a system—their internal dynamics and their interactions using realistic values for the parameters, external forcing factors, and initial conditions. For shallow lakes, which can be in a phytoplankton- or a macrophyte-dominated state (Scheffer et al. 2001), the model PCLake has been developed. It has been extensively applied in the context of lake eutrophication and regime shifts between the two states (Janse et al. 2010). In the eutrophic, phytoplankton-dominated state, the model reproduces the PEG pattern of a spring peak in phytoplankton followed by a clear water phase and summer increase of less edible phytoplankton.

The general picture that emerges from the mathematical modeling of planktonic interactions provides support for one of the main conclusions drawn in this review, namely that the novel aspects of plankton interactions that are discussed in the preceding sections have a rather restricted role for gross seasonal biomass patterns, but strong effects on species replacements. Another lesson that can be learned is that minimal and complex dynamical models each have their own strengths and weaknesses (Mooij et al. 2010). We therefore strongly advocate the use of multiple modeling approaches, either directly coupled or used concurrently, in order to gain a more comprehensive understanding of planktonic succession. As an example, the ecosystem model PCLake is now well linked to the literature on regime shifts in shallow lakes (Mooij et al. 2007), whereas, e.g., Peeters et al. (2007) have successfully coupled a phytoplankton module with a 1D hydrodynamical model for a deep lake.

Regime shift: a rapid change in the state of the system at some critical level of an external forcing factor

10. SYNTHESIS

During the past 25 years, new and flourishing research fields in plankton ecology have considered mechanisms that were either unmentioned in the original PEG model (e.g., food quality), mentioned only marginally (e.g., parasitism), or subsumed under more general mechanisms (e.g., grazing by protists as part of general grazing). Although the overall seasonal pattern described in the PEG model is still valid, research of the past 25 years has advanced our understanding of the mechanisms driving the seasonal succession of plankton. We have tried to incorporate these advances in an extended version of the seasonal succession scheme (**Figure 2**). The major changes are that heterotrophic protists are now treated separately from metazoan zooplankton. The classic eutrophic and the oligotrophic scenarios with moderate fish predation and low importance of zooplankton overwintering have been extended by a number of subscenarios: a scenario with important overwintering zooplankton, a scenario with negligible fish predation, and a scenario with excessive fish predation leading to year-round suppression of large zooplankton.

The seasonal biomass patterns of phyto- and metazoan plankton remain unchanged. The biomass pattern of heterotrophic protists is similar to that of phytoplankton, though with lower amplitudes and some delay during the increase phases.

We have also increased the number of horizontal bars to show the seasonal change in the importance of driving factors. We have added the factors of protist grazing and parasitism to the phytoplankton pattern and food quality and parasitism to the metazoan pattern. Scarcity of data prevented us from considering the importance of parasitism and food quality for heterotrophic protists. Contrary to figures 6 and 7 in the original PEG model, we have refrained from equating vulnerability with size classes because of the context dependency of vulnerability: Small phytoplankton species are more vulnerable to daphnids (often dominant in lakes), whereas large phytoplankton are more vulnerable to copepods (often dominant in the ocean). However, vulnerability of metazoan zooplankton to fish predation still increases with size. Note, however, that whenever large species dominate because of low vulnerability, they will be accompanied by an undergrowth of small species with low biomass but high per capita productivity.

The importance of a factor is ambiguous because it expresses not only its quantitative strength, but also the hierarchical level at which it is effective: total biomass > size structure > taxonomic

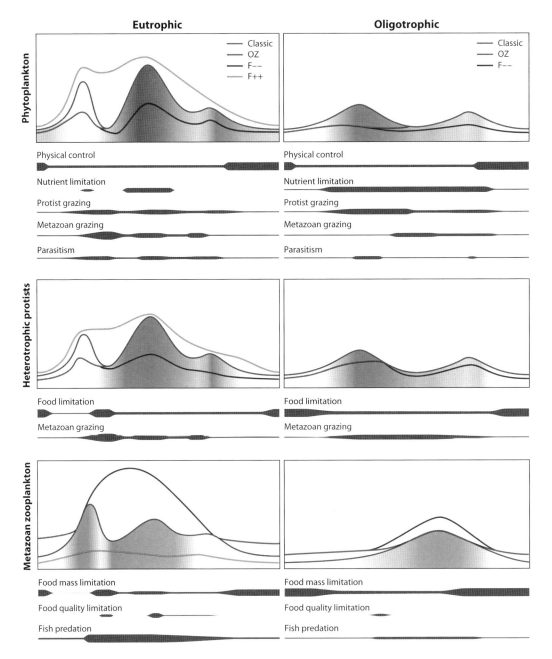

Figure 2

Seasonal (winter through autumn) biomass patterns of plankton in a eutrophic (*left*) and an oligotrophic (*right*) water body. (*Top*) Phytoplankton, (*middle*) heterotrophic protists, (*bottom*) metazoan plankton. Gray line, classic scenario (moderate fish predation, overwintering of metazoan plankton unimportant); blue line (OZ), overwintering zooplankton important; red line (F−−), high metazoan density in fishless water bodies; orange line (F++), metazoan plankton suppressed by high fish predation. Shading indicates the mean vulnerability of phytoplankton and protists against metazoan grazing and of metazoan zooplankton against fish predation in the classic scenario (*light*, low; *dark*, high). The thickness of the horizontal bars indicates the seasonal change in regard to the relative importance of physical factors such as grazing by protists and metazoa, nutrient limitation, fish predation, food mass limitation, and food quality limitation.

composition. Usually, effects at higher hierarchical levels imply effects at lower levels too, but there are exceptions; for instance, size structure might change without taxonomic shifts if a dominant species shifts between a single cell and a colonial status. Maximal importance is only ascribed to factors yielding a quantitatively strong biomass effect in a wide array of cases. In spite of all the progress made since 1986, the following points remain the same as in the original PEG model:

1. Physics (light and stratification) defines the start and the end of the phytoplankton growth season.
2. Grazing by metazoan plankton results in a clear water phase.
3. Nutrients define the carrying capacity of phytoplankton.
4. Food limitation determines zooplankton abundances.
5. Fish predation determines zooplankton size structure.

The novel interactions discussed in this review have a rather limited impact on gross seasonal biomass patterns of phyto- and zooplankton, but strong effects on species replacements. We also assume that these interactions should have strong effects on the genetic composition of species and their seasonal change, but this is beyond the scope of our study. We anticipate that the mid-term future of plankton research will not substantially change the causal understanding of total biomass changes, but we expect more insight to be gained in the causation of species replacements and genetic shifts within species. This is not only of interest for basic science, but also important in an applied context. The toxicity of algal blooms and the risks related to jellyfish outbreaks depend on species, sometimes even on specific genotypes, and not on biomass accumulation as such.

FUTURE ISSUES

1. In the face of climate change and predicted milder winters, it is necessary to give more attention to plankton growth and survival during winter.

2. It is necessary to address the question of how changes in overwintering influence plankton succession later in the year.

3. The role of parasites in driving functional changes in plankton communities deserves more attention.

4. The research focus on mineral nutrient limitation of zooplankton needs to be expanded from the *Daphnia*–phosphorus link to nitrogen and to other groups of zooplankton, especially heterotrophic protists and copepods.

5. The apparent difference between the strength of top-down control in lakes and oceans deserves more research.

6. The roles of species diversity within functional groups and of genetic diversity within species need more attention.

7. Climate change and changed exploitation of the higher trophic levels can be used as natural experiments to gain more insight into the factors driving species and genetic shifts.

8. Plankton modeling needs to implement trait-based approaches.

DISCLOSURE STATEMENT

The authors are not aware of any affiliations, memberships, funding, or financial holdings that might be perceived as affecting the objectivity of this review.

ACKNOWLEDGMENTS

The idea for writing this review resulted from a colloquium about "predictability of plankton in an unpredictable world" held on April 7–9, 2010 in Amsterdam and funded by the Royal Netherlands Academy of Arts and Sciences.

LITERATURE CITED

Adrian R, Deneke R, Mischke U, Stellmacher R, Lederer P. 1995. A long-term study of the Heiligensee (1975–1992). Evidence for effects of climate change on the dynamics of eutrophied lake ecosystems. *Arch. Hydrobiol.* 133:315–37

Adrian R, Wickham SA, Butler NM. 2001. Trophic interactions between zooplankton and the microbial community in contrasting food webs: the epilimnion and deep chlorophyll maximum of a mesotrophic lake. *Aquat. Microb. Ecol.* 24:83–97

Adrian R, Wilhelm S, Gerten D. 2006. Life-history traits of lake plankton species may govern their phenological response to climate warming. *Glob. Chang. Biol.* 12:652–61

Arts MT, Brett MT, Kainz M. 2009. *Lipids in Aquatic Ecosystems*. New York: Springer

Azam F, Fenchel T, Field JG, Gray JS, Meyer-Reil LA, Thingstad F. 1983. The ecological role of water-column microbes in the sea. *Mar. Ecol. Progr. Ser.* 10:257–63

Bautista B, Harris RP, Tranter PRG, Harbour D. 1992. In situ copepod feeding and grazing rates during a spring bloom dominated by *Phaeocystis* sp. in the English Channel. *J. Plankton Res.* 14:691–703

Behrenfeld M. 2010. Abandoning Sverdrup's depth hypothesis on phytoplankton blooms. *Ecology* 91:977–89

Berninger UG, Finlay BJ, Kuuppoleinikki P. 1991. Protozoan control of bacterial abundances in freshwater. *Limnol. Oceanogr.* 36:139–47

Blenckner T, Adrian R, Livingstone DN, Jennings E, Aonghusa CN, et al. 2007. Large-scale climatic signatures in lakes across Europe: a meta-analysis. *Glob. Chang. Biol.* 13:1314–26

Breteler WCMK, Schogt N, Baas M, Schouten S, Kraay GW. 1999. Trophic upgrading of food quality by protozoans enhancing copepod growth: role of essential lipids. *Mar. Biol.* 135:191–98

Brett MT, Müller-Navarra DC. 1997. The role of highly unsaturated fatty acids in aquatic food web processes. *Freshw. Biol.* 38:483–99

Bruning K. 1991. Effects of temperature and light on the population dynamics of the *Asterionella-Rhizophydium* association. *J. Plankton Res.* 13:707–19

Brussaard CPD. 2004. Viral control of phytoplankton populations: a review. *J. Euk. Microbiol.* 51:215–38

Cáceres CE. 1998. Interspecific variation in the abundance, production, and emergence of *Daphnia* diapausing eggs. *Ecology* 79:1699–710

Calbet A, Landry M. 2004. Phytoplankton growth, microzooplankton grazing, and carbon cycling in marine systems. *Limnol. Oceanogr.* 49:51–57

Campbell RW. 2008. Overwintering habitat of *Calanus finmarchicus* in the North Atlantic inferred from autonomous profiling floats. *Deep Sea Res.* 55:630–65

Carpenter SR, Cole JJ, Hodgson JR, Kitchell JL, Pace ML, et al. 2001. Trophic cascades, nutrients, and lake productivity: whole-lake experiments. *Ecol. Monogr.* 71:163–86

Casini M, Lövgren J, Hjelm J, Cardinale M, Molinero JC, Kornilovs G. 2008. Multi-level trophic cascades in a heavily exploited open marine ecosystem. *Proc. R. Soc. B* 275:1791–801

Chambouvet A, Morin P, Marie D, Guillou L. 2008. Control of toxic marine dinoflagellate blooms by serial parasitic killers. *Science* 322:1254–57

Condon RH, Steinberg DK, del Giorgio PA, Bouvier TC, Bronk DA, et al. 2011. Jellyfish blooms result in a major microbial respiratory sink of carbon in marine ecosystems. *Proc. Natl. Acad. Sci. USA* 108:10225–30

Dakos V, Beninca E, van Nes EH, Philippart CJM, Scheffer M, Huisman J. 2009. Interannual variability in species composition explained as seasonally entrained chaos. *Proc. R. Soc. B* 276:2871–80

De Bruin A, Ibelings BW, Rijkboer M, Brehm M, van Donk E. 2004. Genetic variation in *Asterionella formosa* (Bacillariophyceae): Is it linked to frequent epidemics of host-specific parasitic fungi? *J. Phycol.* 40:823–30

Decaestecker E, Declerck S, De Meester L, Ebert D. 2005. Ecological implications of parasites in natural *Daphnia* populations. *Oecologia* 144:382–90

DeMott WR, Tessier AJ. 2002. Stoichiometric constraints vs. algal defenses: testing mechanisms of zooplankton food limitation. *Ecology* 83:3426–33

Deneke R, Nixdorf N. 1999. On the occurrence of clear-water phases in relation to shallowness and trophic state: a comparative study. *Hydrobiologia* 408/409:251–62

De Senerpont Domis LN, Mooij WM, Hülsmann S, van Nes EH, Scheffer M. 2007. Can overwintering versus diapausing strategy in *Daphnia* determine match-mismatch events in zooplankton-algae interactions? *Oecologia* 150:682–98

Ebert D, ed. 2005. *Ecology, Epidemiology, and Evolution of Parasitism in* Daphnia. Bethesda, MD: Natl. Cent. Biotechnol. Inf.

Edwards M, Richardson AJ. 2004. Impact of climate change on marine pelagic phenology and trophic mismatch. *Nature* 430:881–84

Gaedke U, Hochstädter S, Straile D. 2002. Interplay between energy limitation and nutritional deficiency: empirical data and food web models. *Ecol. Monogr.* 72:251–70

Gaedke U, Wickham SA. 2004. Ciliate dynamics in response to changing biotic and abiotic conditions in a large, deep lake (Lake Constance). *Aquat. Microb. Ecol.* 34:247–61

Gerten D, Adrian R. 2002. Effects of climate warming, North Atlantic Oscillation, and El Niño-Southern Oscillation on thermal conditions and plankton dynamics in northern hemispheric lakes. *Sci. World* 2:586–606

Gliwicz ZM. 2003. *Between Hazards of Starvation and Risk of Predation: The Ecology of Offshore Animals.* Oldendorf/Luhe, Germany: Int. Ecol. Inst. 379 pp.

Guisande C, Riveiro I, Maneiro I. 2000. Comparisons among the amino acid composition of females, eggs and food to determine the relative importance of food quantity and food quality to copepod reproduction. *Mar. Ecol. Progr. Ser.* 202:135–42

Haag CR, Hottinger JW, Riek M, Ebert D. 2002. Strong inbreeding depression in a *Daphnia* metapopulation. *Evolution* 56:518–26

Hagen W, Van Vleet ES, Kattner G. 1996. Seasonal lipid storage as overwintering strategy of Antarctic krill. *Mar. Ecol. Progr. Ser.* 134:85–89

Hall SR, Becker CR, Simonis JL, Duffy MA, Tessier AJ, Cáceres CE. 2009. Friendly competition: evidence for a dilution effect among competitors in a planktonic host-parasite system. *Ecology* 90:791–801

Hanazato T, Yasuno M. 1989. Influence of overwintering *Daphnia* on spring zooplankton communities: an experimental study. *Ecol. Res.* 4:323–38

Hansson LA, Annadotter H, Bergman E, Hamrin SF, Jeppesen E, et al. 1998. Biomanipulation as an application food-chain theory: constraints, synthesis, and recommendations for temperate lakes. *Ecosystems* 1:558–74

Huisman J, Sommeijer B. 2002. Population dynamics of sinking phytoplankton in light-limited environments: simulation techniques and critical parameters. *J. Sea Res.* 48:83–96

Huisman J, Weissing FJ. 1999. Biodiversity of plankton by species oscillations and chaos. *Nature* 402:407–10

Ibelings BW, De Bruin A, Kagami M, Rijkeboer M, Brehm M, van Donk E. 2004. Host parasite interactions between freshwater phytoplankton and chytrid fungi (Chytridiomycota). *J. Phycol.* 40:437–53

Ibelings BW, Gsell A, Mooij WM, van Donk E, Van den Wyngaert S, De Senerpont Domis LN. 2011. Chytrid infections and diatom spring blooms: paradoxical effects of climate warming on fungal epidemics in lakes. *Freshw. Biol.* 56:754–66

Irigoien X, Flynn KJ, Harris RP. 2005. Phytoplankton blooms: a 'loophole' in microzooplankton grazing impact? *J. Plankton Res.* 27:313–21

Jackson JBC. 2008. Ecological extinction and evolution in the brave new ocean. *Proc. Natl. Acad. Sci. USA* 105:11458–65

Jankowski T, Straile D. 2003. A comparison of egg bank and long-term plankton dynamics of two *Daphnia* species, *D. hyalina* and *D. galeata*: potentials and limits of reconstruction. *Limnol. Oceanogr.* 48:1948–55

Janse JH, Scheffer M, Lijklema L, Van Liere L, Sloot JS, Mooij WM. 2010. Estimating the critical phosphorus loading of shallow lakes with the ecosystem model PCLake: sensitivity, calibration and uncertainty. *Ecol. Model.* 221:654–65

Jeppesen E, Jensen JP, Søndergaard M, Fenger-Grøn M, Bramm ME, et al. 2004. Impact of fish predation on cladoceran body weight distribution and zooplankton grazing in lakes during winter. *Freshw. Biol.* 49:432–47

Jeppesen E, Jensen JP, Søndergaard M, Lauridsen TL, Pedersen LJ, Jensen L. 1997. Top-down control in freshwater lakes: the role of nutrient state, submerged macrophytes and water depth. *Hydrobiologia* 342/343:151–64

Jeppesen E, Meerhoff M, Holmgren K, González Bergonzoni I, Teixeira-de Mello F, et al. 2010. Impacts of climate warming on lake fish community structure and potential effects on ecosystem function. *Hydrobiologia* 646:73–90

Johansson M, Gorokhova E, Larsson U. 2004. Annual variability in ciliate community structure, potential prey and predators in the open northern Baltic Sea proper. *J. Plankton Res.* 26:67–80

Jürgens K, DeMott WR. 1995. Behavioral flexibility in prey selection by bacterivorous nanoflagellates. *Limnol. Oceanogr.* 40:1503–7

Jürgens K, Wickham S, Rothhaupt KO, Santer B. 1996. Feeding rates of macro- and microzooplankton on heterotrophic nanoflagellates. *Limnol. Oceanogr.* 41:1833–39

Kagami M, De Bruin A, Ibelings BW, Von Elert E, van Donk E. 2007. Fungal parasites bridge the gap between large inedible diatoms and zooplankton. *Proc. R. Soc. B: Biol. Sci.* 274:1561–66

Klausmeier CA, Litchman E. 2012. Successional dynamics in the seasonally forced diamond food web. *Am. Nat.* 180:1–16

Lampert W, Fleckner W, Rai H, Taylor BE. 1986. Phytoplankton control by grazing zooplankton: a study on the spring clear water phase. *Limnol. Oceanogr.* 31:478–90

Lampert W, Schober U. 1978. Das regelmäßige Auftreten von Frühjahrs-Algenmaximum und "Klarwasserstadium" im Bodensee als Folge von klimatischen Bedingungen und Wechselwirkungen zwischen Phyto und Zooplankton. *Arch. Hydrobiol.* 82:364–86

Löder MGJ, Meunier C, Wiltshire KH, Boersma M, Aberle N. 2011. The role of ciliates, heterotrophic dinoflagellates and copepods in structuring spring plankton communities at Helgoland Roads, North Sea. *Mar. Biol.* 158:1551–80

Malzahn AM, Hantzsche F, Schoo KL, Boersma M, Aberle N. 2010. Differential effects of nutrient-limited primary production on primary, secondary or tertiary consumers. *Oeocologia* 162:35–48

Meerhoff M, Clemente JM, Teixeira de Mello F, Iglesias C, Pedersen AR, Jeppesen E. 2007. Can warm climate-related structure of littoral predator assemblies weaken the clear water state in shallow lakes? *Glob. Chang. Biol.* 13:1888–97

Micheli F. 1999. Eutrophication, fisheries, and consumer-resource dynamics in marine pelagic ecosystems. *Science* 285:1396–98

Molinero JC, Casini M, Buecher E. 2008. The influence of the Atlantic and regional climate variability on the long-term changes in gelatinous carnivore populations in the northwestern Mediterranean. *Limnol. Oceanogr.* 53:1456–67

Mooij WM, De Senerpont Domis LN, Janse JH. 2007. Linking species- and ecosystem-level impacts of climate change in lakes with a complex and a minimal model. *Ecol. Model.* 220:3011–20

Mooij W, Trolle D, Jeppesen E, Arhonditsis G, Belolipetsky P, et al. 2010. Challenges and opportunities for integrating lake ecosystem modelling approaches. *Aquat. Ecol.* 44:633–67

Müller H, Schöne A, Pinto-Coelho RM, Schweizer A, Weisse T. 1991. Seasonal succession of ciliates in Lake Constance. *Microb. Ecol.* 21:119–38

Müller-Navarra DC, Brett MT, Park S, Chandra S, Ballantyne AP, et al. 2004. Unsaturated fatty acid content in seston and tropho-dynamic coupling in lakes. *Nature* 427:69–72

Niquil N, Kagami M, Urabe J, Christaki U, Viscogliosi E, Sime-Ngando T. 2011. Potential role of fungi in plankton food web functioning and stability: a simulation analysis based on Lake Biwa inverse model. *Hydrobiologia* 659:65–79

Paffenhöfer GA, Ianora A, Miralto A, Turner JT, Kleppel GS, et al. 2005. Colloquium on diatom-copepod interactions. *Mar. Ecol. Progr. Ser.* 286:293–305

Peeters F, Straile D, Lorke A, Ollinger D. 2007. Turbulent mixing and phytoplankton spring bloom development in a deep lake. *Limnol. Oceanogr.* 52:286–98

Purcell JE, Arai MN. 2001. Interactions of pelagic cnidarians and ctenophores with fishes: a review. *Hydrobiologia* 451:27–44

Rohrlack T, Dittman E, Henning M, Börner T, Kohl JG. 1999. Role of microcystins in poisoning and food ingestion inhibition of *Daphnia galeata* caused by the cyanobacterium *Microcystis aeruginosa*. *Appl. Environ. Microbiol.* 65:737–39

Rother A, Pitsch M, Hülsmann S. 2010. The importance of hatching from resting eggs for population dynamics and genetic composition of *Daphnia* in a deep reservoir. *Freshw. Biol.* 55:2319–31

Sanders RW, Wickham SA. 1993. Planktonic protozoa and metazoa: predation, food quality and control. *Mar. Microb. Food Webs* 7:197–223

Scheffer M, Carpenter S, Foley JS, Folke C, Walker B. 2001. Catastrophics shifts in ecosystems. *Nature* 413:591–96

Scheffer M, Rinaldi S, Kuznetsov YA, van Nes EH. 1997. Seasonal dynamics of *Daphnia* and algae explained as a periodically forced predator-prey system. *Oikos* 80:519–32

Schippers P, Verschoor AM, Vos M, Mooij WM. 2001. Does "supersaturated coexistence" resolve the "paradox of the plankton"? *Ecol. Lett.* 4:404–7

Shapiro J. Wright DI. 1974. Lake restoration by biomanipulation. Round Lake, Minnesota, the first two years. *Freshw. Biol.* 14:371–83

Sherr EB, Sherr BF. 2007. Heterotrophic dinoflagellates: a significant component of microzooplankton biomass and major grazers of diatoms in the sea. *Mar. Ecol. Progr. Ser.* 352:187–97

Sherr EB, Sherr BF. 1987. High rates of consumption of bacteria by pelagic ciliates. *Nature* 325:710–11

Shurin JB, Borer ET, Seabloom EW, Anderson K, Blanchette CA, Broitman B. 2002. A cross-ecosystem comparison of the strength of trophic cascades. *Ecol. Lett.* 5:785–91

Siegel DA, Doney SC, Yoder JA. 2002. The North Atlantic spring phytoplankton bloom and Sverdrup's Critical Depth Hypothesis. *Science* 296:730–33

Skovgaard A, Saiz E. 2006. Seasonal occurrence and role of protistan parasites in coastal marine zooplankton. *Mar. Ecol. Progr. Ser.* 327:37–49

Smayda TJ. 1971. Normal and accelerated sinking of phytoplankton in the sea. *Mar. Geol.* 11:105–22

Smith WO, Nelson DM. 1985. Phytoplankton bloom produced by receding ice edge in the Ross Sea: spatial coherence with the density field. *Science* 227:163–66

Sommer U, Gliwicz ZM, Lampert W, Duncan A. 1986. The PEG model of a seasonal succession of planktonic events in fresh waters. *Arch. Hydrobiol.* 106:433–71

Sommer U, Lengfellner K. 2008. Climate change and the timing, magnitude and composition of the phytoplankton spring bloom. *Glob. Chang. Biol.* 14:1199–208

Sommer U, Lewandowska A. 2011. Climate change and the phytoplankton spring bloom: warming and overwintering zooplankton have similar effects on phytoplankton. *Glob. Chang. Biol.* 17:154–62

Sommer U, Sommer F. 2006. Cladocerans versus copepods: the cause of contrasting top-down controls on freshwater and marine phytoplankton. *Oecologia* 147:183–94

Sørensen T, Muldrij G, Søndergaard M, Lauridsen T, Liboriussen L, et al. 2011. Winter ecology of shallow lakes: strongest effect of fish on water clarity at high nutrient levels. *Hydrobiologia* 664:147–62

Sterner RW, Elser JJ. 2002. *Ecological Stoichiometry: The Biology of Elements from Molecules to the Biosphere.* Princeton, NJ: Princeton Univ. Press

Sterner RW, Elser JJ. 2009. Ecological stoichiometry. In *The Princeton Guide to Ecology*, ed. SA Levin, SR Carpenter, HCJ Godfray, AP Kinzig, M Loreau, et al., pp. 376–85. Princeton, NJ: Princeton Univ. Press

Sterner RW, Elser JJ, Fee EJ, Guildford SJ, Chrzanowski TH. 1997. The light:nutrient ratio in lakes: the balance of energy and materials affects ecosystem structure and process. *Am. Nat.* 150:663–84

Sverdrup H. 1953. On conditions for the vernal blooming of phytoplankton. *J. Cons. Int. Explor. Mer.* 18:287–95

Tadonleke RD, Pinel-Alloul B, Bourbonnais N, Pick FR. 2004. Factors affecting the bacteria-heterotrophic nanoflagellate relationship in oligo-mesotrophic lakes. *J. Plankton Res.* 26:681–95

Thingstad TF, Lignell R. 1997. Theoretical models for the control of bacterial growth rate, abundance, diversity and carbon demand. *Aquat. Microb. Ecol.* 13:19–27

Tirok K, Bauer B, Wirtz K, Gaedke U. 2011. Predator-prey dynamics driven by feedback between functionally diverse trophic levels. *PLoS ONE* 6:e27357

Tirok K, Gaedke U. 2006. Spring weather determines the relative importance of ciliates, rotifers and crustaceans for the initiation of the clear-water phase in a large, deep lake. *J. Plankton Res.* 48:361–73

Tirok K, Gaedke U. 2007a. The effect of irradiance, vertical mixing and temperature on spring phytoplankton dynamics under climate change: long-term observations and model analysis. *Oecologia* 150:625–42

Tirok K, Gaedke U. 2007b. Regulation of planktonic ciliate dynamics and functional composition during spring in Lake Constance. *Aquat. Microb. Ecol.* 49:87–100

Tsugeki NK, Ishida S, Urabe J. 2009. Sedimentary records of reduction in resting egg production of *Daphnia galeata* in Lake Biwa during the 20th century: a possible effect of winter warming. *J. Paleolimnol.* 42:155–65

Turner JT. 2002. Zooplankton fecal pellets, marine snow and sinking phytoplankton blooms. *Aquat. Microb. Ecol.* 27:57–102

van Donk E, Ianora A, Vos M. 2011. Induced defences in marine and freshwater phytoplankton: a review. *Hydrobiologia* 688:3–19

van Donk E, Ringelberg J. 1983. The effect of fungal parasitism on the succession of diatoms in Lake Maarsseveen-I (The Netherlands). *Freshw. Biol.* 13:241–51

Vargas CA, Escribano R, Poulet S. 2006. Phytoplankton food quality determines the time windows for successful zooplanton reproductive pulses. *Ecology* 87:2992–99

Wiltshire KH, Malzahn AM, Wirtz K, Greve W, Janisch S, et al. 2008. Resilience of North Sea phytoplankton spring bloom dynamics: an analysis of long-term data at Helgoland Roads. *Limnol. Oceanogr.* 53:1294–302

Global Introductions of Crayfishes: Evaluating the Impact of Species Invasions on Ecosystem Services

David M. Lodge,[*,1,2] Andrew Deines,[2] Francesca Gherardi,[3] Darren C.J. Yeo,[4] Tracy Arcella,[2] Ashley K. Baldridge,[2] Matthew A. Barnes,[2] W. Lindsay Chadderton,[5] Jeffrey L. Feder,[1,2] Crysta A. Gantz,[2] Geoffrey W. Howard,[6] Christopher L. Jerde,[1,2] Brett W. Peters,[1] Jody A. Peters,[2] Lindsey W. Sargent,[2] Cameron R. Turner,[2] Marion E. Wittmann,[2] and Yiwen Zeng[4]

[1]Environmental Change Initiative and [2]Department of Biological Sciences, University of Notre Dame, Notre Dame, Indiana 46556; email: dlodge@nd.edu

[3]Dipartimento di Biologia Evoluzionistica "Leo Pardi," Università degli Studi di Firenze, 50136 Firenze, Italy

[4]Department of Biological Sciences, National University of Singapore, Singapore 117543, Republic of Singapore

[5]The Nature Conservancy, South Bend, Indiana 46617

[6]Invasive Species Initiative, International Union for Conservation of Nature Species Program, Nairobi 00200, Kenya

Annu. Rev. Ecol. Evol. Syst. 2012. 43:449–72

First published online as a Review in Advance on September 17, 2012

The *Annual Review of Ecology, Evolution, and Systematics* is online at ecolsys.annualreviews.org

This article's doi: 10.1146/annurev-ecolsys-111511-103919

1543-592X/12/1201-0449$20.00

*Order of authorship indicates three categories of contribution: D.M.L.; A.D., F.G., and D.C.J.Y.; and the rest in alphabetical order.

Keywords

biogeography, exotic species, management, bioeconomics

Abstract

Impacts of nonindigenous crayfishes on ecosystem services exemplify the mixture of positive and negative effects of intentionally introduced species. Global introductions for aquaculture and ornamental purposes have begun to homogenize naturally disjunct global distributions of crayfish families. Negative impacts include the loss of provisioning (e.g., reductions in edible native species, reproductive interference or hybridization with native crayfishes), regulatory (e.g., lethal disease spread, increased costs to agriculture and water management), supporting (e.g., large changes in ecological communities), and cultural (e.g., loss of festivals celebrating native crayfish) services. Where quantification of impacts exists (e.g., *Procambarus clarkii* and *Pacifastacus leniusculus* in Europe), regulations now prohibit introduction and spread of crayfishes, indicating that losses of ecosystem services have outweighed gains. Recent research advances such as predicting invasiveness, predicting spread, improved detection and control, and bioeconomic analysis to increase cost-effectiveness of management could be employed to reduce future losses of ecosystem services.

INTRODUCTION

The type or magnitude of goods and services provided by ecosystems to humans is changed by non-indigenous species. Anticipated increases in ecosystem services provide incentives for intentional introductions of nonindigenous species, e.g., for harvest and consumption. However, decreases in ecosystem services are what motivate caution or outright opposition to species introductions. Aquaculture species may escape and cause a decrease in the abundance of wild native species and an increase in the cost of harvesting them. The same species in the same location may cause an increase in one or more ecosystem services while simultaneously causing a decrease in one or more different ecosystem services. Furthermore, the group of people benefitting most directly from an increase in an ecosystem service (e.g., aquaculture industry) is often a different group of people than those suffering a loss of ecosystem services from the same nonindigenous species (e.g., indigenous wild fishery). Finally, increases in ecosystem services that accrue to one group often develop more rapidly than ecosystem service losses to others. Thus, incentives and reactions are often asymmetric. The existence of trade-offs among ecosystem services (i.e., in gaining one service, another is often lost) causes considerable confusion about the appropriate scientific and societal responses to nonindigenous species.

Here we summarize and analyze published research on the impact of nonindigenous crayfish species as an example of the general issues surrounding the science and management of nonindigenous species. Nonindigenous species are not unique in causing changes to ecosystem services; changes in the abundance of native species can also cause changes in ecosystem services. This has led to numerous exchanges among academic ecologists about whether there was any reason to categorize nonindigenous species differently than native species (Davis et al. 2011, Lambertini et al. 2011, Rodewald 2012). The most important point—often missed in such exchanges—is that natural resource management and policy requires the ability to distinguish nonindigenous species from native species if harmful species are to be prevented entry to a country or protected area (Lodge & Shrader-Frechette 2003). A policy typically expresses the preferences of a culture and economic system that is adapted (sensu economics not evolution) to native species because it is largely stuck with them whether they are perceived to be good or bad. In contrast, a society does not have to accept new species. Choice is possible—especially about whether to allow the import of nonindigenous species, but also about whether and how much to invest in trying to control or eradicate incipient invasions. Careful scientific analysis of the likelihood of increases and decreases in ecosystem services from a proposed introduction of a species can guide such choices. Because choice is possible for nonindigenous species, and evolutionary and ecological research can inform that choice, we focus here on the increases and decreases in ecosystem services caused by nonindigenous species.

We focus this review on crayfishes because they are representative of nonindigenous species that have impacted numerous ecosystem services. They are usually intentionally introduced because people derive value from them (e.g., provisioning services via culture or harvest for human consumption or cultural services via appreciation of them as pets). Crayfishes often also negatively impact other ecosystem services that involve a full range of well (e.g., competition, predation) and poorly (e.g., disease vectoring, hybridization) studied mechanisms of ecological interactions with native species and humans (Hobbs & Lodge 2010). Therefore, crayfishes provide an excellent model organism for probing the dependence of community structure and ecosystem function on species composition. Crayfishes' negative impacts on ecosystem services may result in part from their uniquely large individual adult size, frequent high abundance and population biomass, and unusual trophic position (extremely omnivorous—many species are capable of predation on other large organisms, including fish and amphibians, herbivory, and detritivory) (Hobbs & Lodge 2010).

The impacts of crayfish introductions have not been recently reviewed at a global scale despite increasing global concern about a large number of inadequately considered introductions.

Specifically, we address the following questions in this review:

1. How does the current, anthropogenically affected distribution of crayfishes among biogeographical realms compare to the historical pattern?
2. If both the positive and negative impacts of nonindigenous crayfishes are considered, what is the likely long-term net impact on ecosystem services? In addition, is this impact contingent on which species is introduced where and under what management practices?
3. How can recent advances in ecological and bioeconomic forecasting be used to guide policy and management to produce more net positive outcomes for ecosystem services?

EVOLUTIONARY HISTORY AND NATURAL DISTRIBUTION OF CRAYFISHES

Freshwater crayfishes belong to the decapod infraorder Astacidea. The currently known 644 species of crayfishes are classified into two major groups that are geographically and morphologically distinct: the Astacoidea (families Astacidae and Cambaridae) and the Parastacoidea (single family Parastacidae) (Crandall & Buhay 2008) (**Figure 1a**). Astacoids are confined to the Northern Hemisphere in a fundamentally Laurasian distribution, with the Astacidae naturally occurring in western North America, Europe, and western Asia, and the Cambaridae in central to eastern North America and eastern Asia. Parastacoids naturally occur only in the Southern Hemisphere in an essentially Gondwanan distribution that includes Australasia, New Zealand, South America, and Madagascar. These global distribution patterns reflect the origins and phylogeny of the freshwater crayfishes (Crandall & Buhay 2008).

Freshwater crayfishes are a monophyletic group that dates back to the Triassic period (185–225 Mya) (Crandall et al. 2000, Crandall & Buhay 2008). Hence, the origins of the Astacoidea and Parastacoidea and the explanation for the disjunct distribution of these two major groups can be traced to 185 Mya when the global supercontinent Pangaea split into the northern and southern supercontinents of Laurasia and Gondwana, respectively (Crandall & Buhay 2008). Thus, Astacoidea and Parastacoidea are sister groups, with a clear division of the northern sister groups, astacids and cambarids, from the southern parastacids (Sinclair et al. 2004, Crandall & Buhay 2008). Both major groups of freshwater crayfishes have centers of species richness in subtropical or temperate areas: the southeastern United States for astacoids, and southeastern Australia for parastacoids (Hobbs 1974, Crandall & Buhay 2008) (**Figure 1a**).

The highest crayfish species richnesses are in the Nearctic realm (mostly cambarids centered in the southern Appalachian Mountains region of the southeastern United States) and the Australian realm (mostly in the parastacid center of richness in southeastern Australia) (**Figure 1a**). Although both astacoids and parastacoids occur in the Neotropical realm, their distributions are nevertheless highly disjunct, with cambarids concentrated in Mexico far to the north of the parastacids in southern Chile, Uruguay, and southern Brazil (Crandall & Buhay 2008). The nine species that fall under the Ethiopian realm are restricted to Madagascar (Crandall & Buhay 2008), which we have illustrated separately (**Figure 1a**).

Freshwater crayfishes are therefore naturally present on all continents except continental Africa and Antarctica. Native crayfishes are also conspicuously absent from that part of the Eurasian continent that constitutes the Oriental realm and from the central Neotropical region (including northern to central South America) (**Figure 1a**). Although the large-scale global distribution pattern of crayfishes can be attributed to vicariance, no easy explanation exists for why crayfishes have not colonized the apparently suitable portions of these regions, such as tropical and subtropical

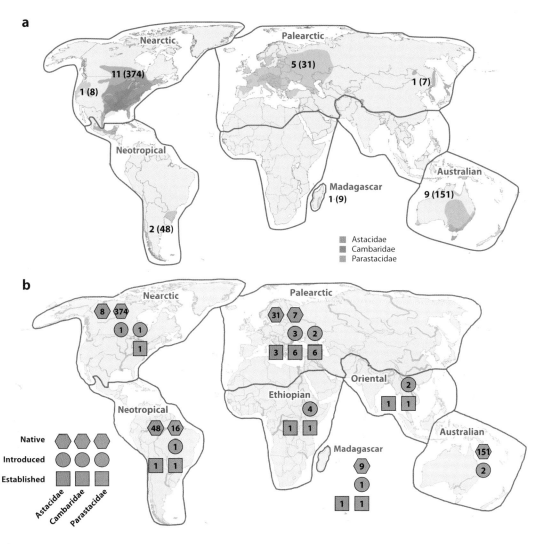

Figure 1

(*a*) Global distribution and richness of the three families of crayfishes; increasing color intensity indicates increasing genera richness; numbers indicate the number of genera (and, in parentheses, the number of species) for each family within a realm. Madagascar is separated from the rest of the Ethiopian realm because it has endemic crayfishes, whereas the rest of the realm has none. Base map modified from Creative Commons 2008 [**http://en.wikipedia.org/wiki/File:A_large_blank_world_map_with_oceans_marked_in_blue.PNG**, downloaded on 13 February 2012; genus distribution data adapted from Hobbs (1988) and Toon et al. (2010); richness data from Crandall & Buhay (2008)]. (*b*) For each biogeographic realm, the global number of native species, introduced (but not known to be established) nonindigenous species, and established nonindigenous species for each of the three families of crayfishes. Madagascar is separated from the rest of the Ethiopian realm because it has endemic crayfishes, whereas the rest of the realm has none. Data sources are in **Supplemental Table 1**.

areas, despite having the potential to thrive under such conditions. Ortmann (1902) suggested that competition from freshwater crabs rather than any environmental or historical reason might explain this pattern and provide a foundation for expectations about the likely success and/or impact of nonindigenous crayfishes that have been or could be introduced into these areas. As invasions by nonindigenous crayfishes occur in the Oriental, Neotropical, and Ethiopian realms (see below), interactions with native crabs and shrimps may provide important ecological resistance and/or cause the decline of native decapods if crayfishes establish.

ANTHROPOGENIC HOMOGENIZATION OF CRAYFISHES

Humans have introduced crayfishes outside their native ranges, many of which have established populations that have subsequently spread widely (Hobbs & Lodge 2010). Several examples of the impact of such invasions are intrarealm, but we focus on inter-realm introductions, which are altering the global biogeographic patterns of crayfishes reviewed in the previous section. A large global network of crayfish hobbyists and collectors exists, from which many informal reports of inter-realm introductions (at least in captivity) are readily available on the Internet. We also know from experience that many introductions, including those that have led to established populations, go unrecorded in the scientific literature. We reviewed the scientific literature on the most thoroughly studied inter-realm introductions (see **Supplemental Table 1**; follow the **Supplemental Material link** from the Annual Reviews home page at **http://www.annualreviews.org**). For each species chosen, we documented the number of realms to which it had been introduced (**Figure 1b**).

Introductions have, for the first time in millions of years, established crayfishes in continental Africa and the Oriental realm (at least one species in each of two families to both realms) and added new families of crayfishes to the Nearctic and Palearctic realms and to Madagascar (**Figure 1b**). The only continental mass free from a nonindigenous crayfish is continental Australia; the two species introduced in the Australian realm were introduced to New Zealand only. Overall, the high risk of establishment from the additional species that are known to have been introduced implies that additional large changes are coming to the decapod fauna in most realms (**Figure 1b**). The crayfish faunas of the world's biogeographic realms are becoming more similar.

IMPACTS ON ECOSYSTEM SERVICES OF CRAYFISH INTRODUCTIONS

For the same set of introduced crayfishes discussed above, we also ascertained whether each species had established populations in the wild and how the introduction had affected ecosystem services (**Table 1**). Because introductions of crayfishes have almost always been intentional, most introductions produce a positive impact on at least one ecosystem service related to the motivation of the introduction. Motivations have included desires to initiate aquaculture production, harvest from natural or seminatural environments, research, biocontrol of disease hosts (e.g., the aquatic snail hosts of schistosomiasis), or simply to enjoy crayfish as an ornamental species in aquaria or water gardens (**Table 2**). The desire for provisioning services has driven a large and continuing

Table 1 Categories of ecosystem services potentially affected by nonindigenous crayfishes[a]

Provisioning	Supporting	Regulating	Cultural
Food	Soil, sediments, erosion	Water regulation	Recreation, tourism
Genetic diversity	Nutrient cycling	Disease regulation	Heritage value, sense of place
Ornamental	Community, food web	Pest regulation	Aesthetic enjoyment, inspiration
	Refuge availability	Natural hazard protection	
	Primary production		

[a]Modified from Gherardi et al. (2011b).

Table 2 Examples of the impact on ecosystem services of selected inter-realm introductions of nonindigenous crayfishes, reflecting the dearth of information for most species besides *Procambarus clarkii* and *Pacifastacus leniusculus*[a]

Family, native realm	Species	Introduced realm	Established in wild[b]	Motivation[c]	Provisioning			Supporting					Regulating				Cultural		
					P1	P2	P3	S1	S2	S3	S4	S5	R1	R2	R3	R4	C1	C2	C3
Cambaridae, Nearctic	*Orconectes immunis*	Palearctic	Yes	O	+		+	−		−	−	−							
	Orconectes juvenilis	Palearctic	Yes	F	+														
	Orconectes limosus	Palearctic	Yes	A, F, U, O	+		+	−		±									
	Orconectes virilis	Palearctic	Yes	A, O	+											−			
	Procambarus acutus	Nearctic	Yes	R	+		+												
		Palearctic	Yes							±									
	Procambarus clarkii	Ethiopian	Yes	A, F, O, B, FR	±	−				−		−							
		Neotropical	Yes	A, O	+		+							+	+				
		Oriental	Yes	A, F, R, O	±	−	+	−		−			−						
		Palearctic	Yes	A, F, R, O, FR, BT	±	−	+	−	±	±	−	−	−	+	+	−	±	−	+
		Hawaii	Yes		±			−		−									
	Procambarus fallax f. virginalis	Madagascar	Yes	O	+	−	+			±									
		Nearctic	No	O															
		Palearctic	Yes																
	Procambarus zonangulus	Palearctic	Yes	A, F, R, O	+				±										
Cambaridae, Neotropical	*Cambarellus montezumae*	Palearctic																	
Astacidae, Nearctic	*Pacifastacus leniusculus*	Palearctic	Yes	A, F, R, U	±	−	−	±	±	±	−	−	−		±			−	+

Parastacidae, Australian		Region			
Cherax cainii		Ethiopian	No	A, F	+
		Nearctic	No	A, O	+
		Neotropical	No	A, F	+
		Oriental	No	A, O	+
		Palearctic	No	A, O, F	+
Cherax destructor		Ethiopian	No	A	+
		Nearctic	No		+
		Palearctic	Yes	A, O	+
		Australia	No	A	+
Cherax quadricarinatus		Ethiopian	Yes	A, O	+
		Nearctic	Yes	A, O, R	+	+
		Neotropical	Yes	A, O	+	+
		Oriental	Yes	A, O	+	−	+
		Palearctic	Yes	A, O	+	+
		Madagascar	Yes	A	±
Cherax tenuimanus		Ethiopian	No	A	+
		Oriental		A	+
		Palearctic	No	A, F	+
		Madagascar			

Scattered ecosystem-service symbols elsewhere in the table: ± (Nearctic, C. quadricarinatus region), − (Ethiopian, C. quadricarinatus), − (Nearctic), − (Oriental), − (near top).

[a] Colors are to improve readability only.

[b] "No" indicates that there is at least one publication stating that no establishment has occurred outside of captivity (and no contradictory publications); "Yes" indicates a published record of at least one persistent population outside of captivity; and blank space indicates there are no publications addressing establishment.

[c] A, aquaculture; B, biocontrol; BT, live bait; F, introduced into natural habitat for capture fishery; FR, forage; O, ornamental (aquarium, water garden); R, research; U, unintentional.

[d] Key to the subcategories of ecosystem services is provided in **Table 1**: +, increase in the ecosystem service (i.e., benefit); −, decrease (i.e., harm).

[e] Citations and list of full references is provided in **Supplemental Table 1**.

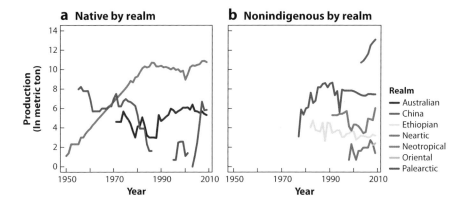

Figure 2

Global crayfish production by realm in logged metric tons for (*a*) native species and (*b*) nonindigenous species. China is presented separately from the rest of the Oriental realm as is common with FAO fisheries data (Food and Agriculture Organization of the United Nations 2010). Data compiled from the FAO Global Production Statistics database (**http://www.fao.org/fishery/statistics/global-production/en**, accessed February 9, 2012). Data are plotted by species in **Supplemental Figure 1**.

global increase in production of crayfishes for human consumption over the past 60 years, with about 4% provided by nonindigenous species (not including China) or about 91% if China is included (**Figure 2**). Everywhere except the Australian and Nearctic realms, a large proportion of the production is provided by nonindigenous species (**Figure 2**). As a food source, almost all crayfish introductions have increased provisioning services, but when crayfish have been stocked outdoors or have been released or allowed to escape from captivity, reductions in other ecosystem services have followed (**Table 2**).

Therefore the best-studied introduced crayfishes have both positive and negative impacts on ecosystem services. This makes crayfish management more contentious than that of many other species (e.g., especially unintentionally introduced species) and begs for more rigorous study of the impacts on the full range of ecosystem services. The positive impact on provisioning services can often be accurately monetized and, therefore, given considerable weight in management and policy decisions, especially when compared with the absence of data on negative impacts or the highly uncertain estimates of negative impacts. A more complete accounting of ecosystem services thus requires a better understanding of negative impacts in particular. If the societal goal for species introductions is to produce a long-term net benefit, then such analyses are urgently required to inform expanding international trade in living crayfishes and to guide new policies and management for the global crayfish aquaculture industry.

The best-studied introductions are of Nearctic cambarids (*Orconectes* spp., *Procambarus* spp.) and an astacid (*Pacifastacus leniusculus*) introduced in the Palearctic realm, where substantial scientific infrastructure and ecological expertise exist. In these cases, some provisioning, supporting, regulatory, and cultural services are enhanced, whereas many other services across those same categories are harmed (**Table 2**). The other sizeable category of introductions consists of parastacids (*Cherax* spp.), from the continent of Australia, which have been introduced into all other realms (**Figure 1*b***). In almost all of the cases where crayfish establishments in the wild have occurred, some negative impacts on regulating or supporting services have occurred (**Table 2**). Where establishment in the wild has not occurred, there are no documented negative impacts on ecosystem services. This suggests that if nonindigenous species (and any potentially harmful hitchhiking parasites and pathogens) can be contained in aquaculture facilities, then introductions can produce a net increase in ecosystem services, at least in the short term. However, history suggests that access by the public to live crayfish inevitably leads to escape or release and, given suitable habitat, to establishment of wild populations. Thus, the absence of documented negative impacts for *Cherax cainii*, *C. destructor*, and *C. tenuimanus* (**Table 2**) is likely to be temporary, especially in countries where neither regulations nor culture effectively require careful risk assessment before organisms are imported or released (de Moor 2002).

MECHANISMS AND MAGNITUDES OF CHANGES IN ECOSYSTEM SERVICES

To better understand the mechanisms and magnitudes of the impacts summarized in **Table 2** and, where possible, to suggest the likely net effect on ecosystem services, we describe several of these case studies listed in **Table 2** in detail below, focusing on impacts in non-native realms. Because the impacts of *Procambarus clarkii* have been rigorously studied, we present more information for it than for other, more poorly studied species.

Procambarus clarkii—Red Swamp Crayfish

Red swamp crayfish is native to the South Central United States (and extending into Mexico), where wild populations have traditionally been harvested for human consumption (Hobbs & Lodge 2010). In recent decades it also has been cultured in outdoor ponds both inside and outside its native range. It is the most widely introduced crayfish in the world (**Table 2**).

Provisioning services. In all the biogeographic realms in which it exists, red swamp crayfish is used as human food for domestic consumption and/or export (**Figure 2**). However, in Europe red swamp crayfish sales have often replaced those of native species (see the sidebar, Crayfish as Vectors of Parasites and Pathogens), whereas consumers in Sweden are willing to pay tenfold for native crayfishes compared to the red swamp crayfish (Holdich 1999; and L. Edsman, personal communication). Red swamp crayfish has also been introduced as food for fishes (e.g., in

CRAYFISH AS VECTORS OF PARASITES AND PATHOGENS

Crayfish plague (caused by the Chromista *Aphanomyces astaci*) hitchhiked to Europe on North American crayfishes, appearing first in 1860 (Gherardi 2006). The disease is mostly asymptomatic in North American crayfishes but lethal to crayfishes from other realms (Alderman et al. 1990). The 1967–1969 introduction of over 70,000 *Pacifastacus leniusculus* from Lake Tahoe (US) into 70 Swedish and Finnish lakes further spread the plague. Plague reduced production of the native European species, *Astacus astacus* and *Astacus leptodactylus*, by up to 90% throughout Europe (Holdich et al. 2009). Other hitchhiking pests include *Cherax quadricarinatus* bacilliform virus (Hauck et al. 2001), temnocephalan flatworms (Mitchell & Kock 1988), and branchiobdellid worms (Ohtaka et al. 2005). White spot syndrome virus, which is damaging to commercial shrimp culture, has spread from the Oriental realm to red swamp crayfish in Louisiana and China (Longshaw 2011, Zeng et al. 2011). Human pests vectored by crayfishes include the tularemia-causing bacterium *Francisella tularensis* (Anda et al. 2001), lung flukes *Paragonimus* spp. (Lane et al. 2009), meningitis-causing rat lungworm *Angiostrongylus cantonensis*, and the nematode *Gnathostoma spinigerum* (Edgerton et al. 2002). Many more parasites and pathogens have been identified in crayfish, but the impact and inter-realm movements of most are unknown (Longshaw 2011).

Kenya and Europe) (Montes et al. 1993, Gherardi et al. 2011b) and for other edible species, e.g., American bullfrog, *Rana catesbeiana* in Japan (Gherardi 2011a). However, red swamp crayfish also reduce the provisioning of other food. In African lakes, it spoils valuable fish caught in gillnets (up to 30% of the catch) and damages fish nets (de Moor 2002), and it is a pest in Japanese fish ponds (Maezono & Miyashita 2004). In addition, it increases costs and/or reduces harvest of rice in China (Yue et al. 2010), Portugal (~6% decrease in profit; Anastácio et al. 2005), and Kenya (Rosenthal et al. 2005), and it clogs irrigation pipes in Italy (F. Gherardi, personal observation).

Regulating services. Activities of red swamp crayfish increase suspended solids and, via the destruction of macrophytes (see below), can induce a switch from a desirable clear state of an aquatic habitat to a turbid state sometimes dominated by toxic cyanobacteria (Rodríguez et al. 2003). The accumulation of cyanobacteria toxins (Vasconcelos et al. 2001) and heavy metals (Alcorlo et al. 2006) by red swamp crayfish can decrease health and increase mortality of humans and other predators of crayfish, more so than that of native Palearctic crayfishes and freshwater crabs (Gherardi et al. 2002). In addition, red swamp crayfish is a vector for multiple parasites and pathogens, including crayfish plague, which, since its first appearance in Europe in 1860, has reduced production of native commercial crayfishes, *Astacus astacus* and *Astacus leptodactylus*, by up to 90% in Scandinavia, Germany, Spain, and Turkey (see the sidebar, Crayfish as Vectors of Parasites and Pathogens). However, in Africa, predation by red swamp crayfish reduces the populations of the snails that host the trematodes that cause human schistosomiasis, reducing the prevalence of the disease in humans (Mkoji et al. 1999). With respect to natural hazard regulation, burrowing by red swamp crayfish in European coastal wetlands (Scalici et al. 2010) could reduce coastal protection from severe storms and sea level rise.

Supporting services. Foraging, burrowing, and locomotory activities by red swamp crayfish can lead to erosion of littoral zone sediments, changing benthic geomorphology (Angeler et al. 2001). Honeycombing of banks causes structural damage, increases bank erosion, and increases costs in areas with canal irrigation systems and water control structures (Adão & Marques 1993, Huner 2002, Barbaresi et al. 2004). In northern Italy, crayfish burrowing damages 30% of the irrigation canals, costing 8% of the annual income of the management authority (M. Fantesini, personal communication). Red swamp crayfish also affect nutrient cycling in sediments (Angeler et al. 2001), reducing organic matter and increasing phosphorus and nitrogen in sediments (Gherardi 2008).

Impacts on community structure and food web interactions by red swamp crayfish are extensive, large, and well documented in multiple biogeographic realms, including the Oriental realm (Maezono & Miyashita 2004, Liu et al. 2011). In the Palearctic realm, red swamp crayfish replaces indigenous crayfish species by a combination of competitive exclusion and differential susceptibility to predation, whereas the likely outcome of its interactions with the river crab (*Potamon fluviatile*) is less clear (Gherardi & Cioni 2004). In some Ethiopian realm habitats, it has apparently replaced the native freshwater crab *Potamonautes neumannii* (Ogada 2007). Red swamp crayfish commonly reduces the abundance of submersed and semiaquatic macrophytes by 50% to 100% via herbivory and stalk-cutting in the Palearctic (Gutiérrez-Yurrita et al. 1999, Rodríguez et al. 2003, Gherardi & Acquistapace 2007) and Ethiopian (Rosenthal et al. 2005) realms, often reducing refuge availability for many other species and inducing undesirable increases in phytoplankton, especially cyanobacteria (Gherardi & Lazzara 2006). Similarly, predation by red swamp crayfish causes declines in many invertebrate taxa, often eliminating snails and other slow-moving species in the Palearctic (Gherardi & Acquistapace 2007) and Ethiopian realms (Lodge et al. 2005). In Italy, predation by the red swamp crayfish has driven to extinction a once locally abundant semiaquatic beetle (*Carabus clatratus*) (Casale & Busato 2008), and in Japan predation by the crayfish threatens an endangered odonate (Miyake & Miyashita 2011). Indirect food web effects of predation on herbivorous macroinvertebrates by crayfish produces undesirable increases in the abundance of benthic algae mats (Gherardi 2008). The net effect on total primary production remains unclear because the loss of macrophytes dramatically reduces total surface area for the growth of attached algae while often enhancing phytoplankton abundance, but the shift in primary producers is almost always seen as undesirable by humans.

In Europe, predation by red swamp crayfish reduces abundance of larval amphibians more than predation by native crayfishes (Renai & Gherardi 2004). Crayfish presence is negatively related to the breeding probability for multiple salamander, frog, and toad species (Cruz et al. 2006, Ficetola et al. 2011). Impacts on other vertebrate species have often been positive. Abundance of predacious wading birds and water birds increased after establishment of red swamp crayfish in Italy and Spain (Barbaresi & Gherardi 2000, Rodríguez et al. 2003, Tablado et al. 2010, Garcia et al. 2011). Red swamp crayfish has become the most common prey category of the otter in Spain (Delibes & Adrian 1987) and, in the Lake Naivasha basin in Kenya, has replaced native river crabs (*Potamonautes loveni*) as the primary food item of the African clawless otter (Ogada et al. 2009). Overall, red swamp crayfish appear to replace many native species at lower and intermediate trophic positions, including trees damaged by increased numbers of breeding storks and herons, shunting more energy directly to top vertebrate predators (Angeler et al. 2001, Garcia et al. 2011).

Cultural services. Decreased abundance of the native European crayfish *Austropotamobius pallipes* has significant negative cultural impact, including the loss of local festivals focused on crayfish. The importance of the native crayfish in the history of European countries is demonstrated by its frequent appearance in emblems, coats of arms, toponymies, and family names (Gherardi 2011b). Their local extinction is profoundly harming the heritage value of these species.

Pacifastacus leniusculus—Signal Crayfish

Signal crayfish is native to North America west of the Rocky Mountains but has been widely introduced to natural habitats inside and outside its native realm, especially for harvest for human consumption (**Table 2**).

Provisioning services. The signal crayfish is the most abundant crayfish in natural waters and aquaculture facilities in much of Europe because of the existence of commercial markets supplying this crayfish for human consumption (Holdich et al. 2009). Because signal crayfish largely replaced native crayfish in Palearctic natural environments and in markets (Kirjavainen & Sipponen 2004), the net impact of the species replacement on the marketplace is difficult to assess (Kataria 2007). What is clear is that, under the current circumstances, native crayfishes are much more highly valued; the 2010 market price of native *Astacus astacus* was double that of signal crayfish in Sweden (L. Edsman, personal communication).

Supporting services. Like red swamp crayfish in warmer waters, signal crayfish also reduces the abundance of a wide range of native organisms in the cooler waters it inhabits in both the Palearctic and Oriental realms. In Scandinavia, signal crayfish reduces species richness and abundance of macrophytes and macroinvertebrates, and it reduces organic matter content of sediments (Nyström et al. 2001, Gherardi 2007, Holdich et al. 2009). Competition with signal crayfish, and its interactions with predation, contribute to the displacement of native crayfishes in Japan (*Cambaroides japonicus*; Usio et al. 2001) and the western Palearctic realm (Holdich et al. 2009).

Regulating services. As a major vector of crayfish plague, signal crayfish introductions have caused the continued loss of populations of native crayfish (see the sidebar, Crayfish as Vectors of Parasites and Pathogens). Although not typically a burrowing species in its native range, signal crayfish causes considerable damage to English river banks (A. Stancliffe-Vaughan, unpublished data).

Cultural services. The loss of native crayfishes caused by signal crayfish in northern Europe, especially the loss of the noble crayfish (*Astacus astacus*) in Scandinavia, is perceived as a serious cultural blow (Gherardi 2011b).

Orconectes Species

Multiple species of the Nearctic cambarid genus *Orconectes* have established widespread ranges in the Palearctic, especially in Europe. These species are *Orconectes immunis*, calico crayfish; *O. limosus*, spinycheek crayfish; *O. virilis*, northern crayfish; and *O. juvenilis*, Kentucky river crayfish.

Provisioning services. The earliest introductions of the *Orconectes* spp. to the Palearctic were probably for human consumption, including the early introduction of *O. limosus* to Europe in 1890 (Hamr 2002). However, the *Orconectes* spp. are not as highly valued as food as signal crayfish or native crayfishes, and the spread of at least one, *O. limosus*, has been unintentional as a hitchhiker with fish stocks (Gherardi et al. 1999).

Supporting services. *Orconectes* spp. are well known for causing major changes in community structure, especially via large reductions in macrophytes (*O. virilis*, Ahern et al. 2008; *O. immunis*, Letson & Makarewicz 1994). In addition, unlike some native Palearctic crayfishes, *O. immunis* digs deep burrows, causing changes in sediments and allowing it to inhabit shallower habitats than native species (Chucholl 2012).

Regulating services. Burrowing in dikes by *O. virilis* increases maintenance costs and the risk of flooding (Ahern et al. 2008).

Cultural services. There is no evidence that *Orconectes* spp. provide any cultural services not previously provided by native crayfishes; to the contrary, like red swamp crayfish and signal crayfish, *Orconectes* spp. contribute to the decline of cultural values previously provided by native crayfishes by vectoring crayfish plague.

Impacts of *Orconectes rusticus* within the Nearctic realm. We also include information about *O. rusticus*, rusty crayfish, the most well-studied member of the genus, although it is apparently not established outside its native realm. Rusty crayfish was reported in France, but the species was later identified as Kentucky river crayfish (Chucholl & Daudey 2008). To our knowledge, the distribution of rusty crayfish remains within the Nearctic realm, but it has been spread as live bait from its native range in streams in southern Ohio, southern Indiana, and northern Kentucky to rivers, lakes, and reservoirs in 20 states (United States Geological Survey 2011); Ontario, Canada (Phillips et al. 2009); and all 5 Laurentian Great Lakes (see the sidebar, Rusty Crayfish Invasion of the Laurentian Great Lakes). The extensive knowledge about its impact within North America may offer insight into the kinds of impacts to expect from other, more poorly studied *Orconectes* species introduced into the Palearctic. Predation and herbivory by rusty crayfish cause large reductions

RUSTY CRAYFISH INVASION OF THE LAURENTIAN GREAT LAKES

Rusty crayfish are the most widespread non-native crayfish in the Laurentian Great Lakes, occurring in all five lakes and their connecting waterways except the Niagara and St. Lawrence rivers (J. Peters 2010). The first record of rusty crayfish in the Great Lakes was in Sandusky Bay of Lake Erie (in 1897; see Turner 1926); reports in other lakes followed much later: Ontario (1980s); Michigan and Superior (1990s); and Huron (in 2000; see J. Peters 2010). Secondary spread has been most extensive in lakes Michigan and Huron and comparatively limited in Erie, Ontario, and Superior (J. Peters 2010). The few studies that have been conducted on Great Lakes littoral communities indicate that rusty crayfish alters littoral zone communities. Rusty crayfish significantly reduce lake trout egg abundance, hampering lake trout rehabilitation efforts (Jonas et al. 2005). Rusty crayfish predation on the amphipod *Gammarus fasciatus* and other invertebrates influences energy flow to higher trophic levels (Stewart et al. 1998). In areas with high rusty crayfish densities, the crayfish is a major component of Lake Huron lake sturgeon diets, although it is unclear whether rusty crayfish are an addition to the diet or a replacement for displaced native crayfishes (J. Peters, unpublished data).

in macroinvertebrate and aquatic plant richness and abundance (Wilson et al. 2004). Predation by rusty crayfish eliminates snails on rocky substrates, the preferred habitat for the crayfish, and reduces them on soft organic sediment, which is the least preferred substrate for rusty crayfish (Kreps et al. 2012). Native crayfish populations are often driven to local extinction via competition (Olsen et al. 1991) and hybridization (see the sidebar, Hybridization Among Crayfishes as a Threat to Native Species). Resident crayfishes are more vulnerable to displacement by rusty crayfish in invaded streams than in invaded lakes (Olden et al. 2011). Although crayfish provide forage for fish (Lodge & Hill 1994), higher crayfish densities correspond with lower total fish biomass (Garvey et al. 2003), but changes in fish abundances are species specific. Sunfish species decreased and bass species increased in response to invasion by rusty crayfish (Wilson et al. 2004). Growth of large, but not small, smallmouth bass was increased by rusty crayfish invasion (Kreps 2009). Crayfish predation on fish eggs and competition for shared macroinvertebrate prey (Dorn & Mittelbach 1999) may explain why abundant rusty crayfish do not necessarily provide a net benefit for fish (Horan et al. 2011).

HYBRIDIZATION AMONG CRAYFISHES AS A THREAT TO NATIVE SPECIES

Reproductive interference (e.g., between nonindigenous signal crayfish and native noble crayfish in Sweden; Soderback 1995) and introgressive hybridization between nonindigenous and native crayfishes can add harm to that caused by ecological displacement mechanisms (Perry et al. 2002). Mechanisms associated with the production of sterile or inviable offspring often play particularly important roles in the genetic decline of a species (Levin et al. 1996). Even when hybrids are fit and fertile, genetic swamping may still occur if the native population has already been reduced by competition, predation, or disease vectored by a nonindigenous species (Levin et al. 1996). For example, hybridization with *Orconectes rusticus* increased the displacement rate of *O. propinquus* by over 20% relative to that caused by ecological factors alone (Perry et al. 2001a). *O. rusticus* readily mates with *O. propinquus* to produce fertile offspring, with F1 hybrids able to outcompete both parental species for food and shelter (Perry et al. 2001b). Despite initial hybrid vigor (particularly in survivorship), further study showed that this vigor broke down in subsequent generations (W.L. Perry, T. Arcella, J.L. Feder, D.M. Lodge, unpublished data), thus leading to the lack of a stable hybrid zone and an increased rate of decline in *O. propinquus* populations.

Cherax quadricarinatus—Redclaw Crayfish

Although the continent of Australia has not added any crayfish imports to its fauna, it has exported native crayfishes that have established in other realms (**Table 2**). Redclaw crayfish, in particular, has been introduced widely, with largely negative effects on ecosystem services.

Provisioning services. Redclaw crayfish, native to tropical Queensland (Australia), has spread widely in the Ethiopian realm from initial introductions into South Africa in the 1960s and 1970s. For example, in 1992 redclaw crayfish was introduced for aquaculture near the Zambezi River, Zambia (Thys van den Audenaerde 1994); it now occurs on both shores of Lake Kariba. After its introduction to a fish farm in Kafue, on the Kafue River, Zambia, a major tributary of the Zambezi River, it has spread as far upstream as the Itezhi-Itezhi Dam (G. Howard, unpublished data). Most Zambians and Zimbabweans regard this crayfish as a nuisance, not as a source of food, and little if any export occurs. Redclaw was also introduced for aquaculture to Mauritius in the mid-1990s (Bhikajee 1997) and to La Réunion (Robinet 2010).

Supporting services. Although the scientific literature reflects a widespread belief that *Cherax quadricarinatus* and *C. destructor* probably have large food web impacts where they have been introduced (Beatty 2006, Bortolini et al. 2007), there is a remarkable absence of observational and experimental evidence. In Zambia, redclaw crayfish is consumed by native predatory cichlid fishes (Tyser 2010). In Singapore, redclaw crayfish is currently restricted to urban waters, but, as in Africa (de Moor 2002), concern exists about its possible spread into forest streams and freshwater swamps, which are important refuge habitats for endemic freshwater crabs and other imperiled species (Ahyong & Yeo 2007, Yeo 2010, Belle et al. 2011).

Regulating services. On La Réunion, redclaw crayfish are damaging to irrigation projects (Robinet 2010).

Cultural services. In Singapore, redclaw crayfish provide a recreational fishery (Ahyong & Yeo 2007, Belle & Yeo 2010).

NET EFFECT ON ECOSYSTEM SERVICES OF CRAYFISH INTRODUCTIONS

The diversity of study designs with a wide range of incommensurate metrics makes any quantitative calculation of the net effect on ecosystem services impossible for any single crayfish-location combination. However, one clear indication of the perception of the net effect by society is whether resources are expended to reduce the introduction, spread, or abundance of a species. Such activities reveal the preference by humans for the presence or absence of a species. Below we review policy and management efforts for crayfish to infer whether or not, if the clock could be reset, different actions may have been taken to prevent the introduction and spread of specific crayfishes outside their native realm.

In much of the Palearctic realm, the realm with the longest history of crayfish introductions, the importation, spread, and use of all live crayfishes for aquaculture is now strongly regulated by the European Union because it has become clear that crayfish-caused losses of ecosystem services outweigh any gains. This is especially true in Scandinavia (Skurdal et al. 1999), Great Britain, and much of Europe (Lodge et al. 2000, Holdich et al. 2009).

Efforts to eradicate, control, and/or restrict the future spread of nonindigenous crayfishes are increasing in multiple realms (Gherardi et al. 2011a). In the Palearctic, for example, about £500,000 is spent annually in an ongoing attempt to eradicate signal crayfish from Scottish rivers (S. Peay, personal communication); using crayfish plague, *Cherax destructor*, the common Australian yabby, was intentionally eradicated in multiple Spanish ponds in 2005 (Diéguez-Uribeondo & Muzquiz 2005), and eradication has been recommended for the Kentucky river crayfish in France (Chucholl & Daudey 2008). Control efforts are also increasing in the Oriental realm, especially for red swamp crayfish in Japan (Maezono & Miyashita 2004, Maezono et al. 2005) and China (Yue et al. 2010).

However, not all introduced crayfish species are subject to control efforts. For species that remain in captivity, increases in ecosystem services may often outweigh negative impacts (e.g., species not known to be established; **Table 2**). If long-term containment is possible (and clearly it has not been under most past management practices), and/or if the local environment (e.g., climate) is unsuitable for a given species, then it may be possible to allow importation and even aquaculture under carefully managed circumstances. This is the policy, for example, that Switzerland and England have adopted for tropical *Cherax* spp., which, even should they escape, would be unlikely to establish in local waterways (Holdich et al. 1999, Stucki & Staub 1999). Even under such circumstances, however, parasites and pathogens could nevertheless be vectored to native crayfishes, other native species, or humans without management practices to prevent pests.

Overall, the evidence reviewed in this and previous sections suggests that the majority of non-indigenous crayfishes in natural habitats in most countries are perceived to cause net harm: They are invasive species. Although some positive ecosystem services accrue, in the long run, negative impacts on other ecosystem services often outweigh the positives. Thus, ecological research is urgently needed to inform the increasing number of policies and management practices aimed at preventing the importation of invasive crayfishes (but not necessarily all nonindigenous crayfishes) and at eradicating, controlling, or preventing the spread of invasive species that are already established.

CRAYFISHES AS MODEL ORGANISMS FOR TRANSLATING ECOLOGICAL RESEARCH INTO MANAGEMENT

Research to Inform Preintroduction Risk Assessment

To prevent additional losses of ecosystem services from future crayfish introductions, science-based risk assessments of proposed introductions are essential (de Moor 2002). Yet in most countries, including the United States, importation of nonindigenous species is largely unregulated (Fowler et al. 2007). In these countries, the lack of any careful preemptive consideration of the likelihood of losses in ecosystem services means that many nonindigenous species are imported, escape or are released, spread, and cause harm that is effectively irreversible. Up until the past decade, ecological science could provide little guidance about whether or where a species was likely to cause harm. Therefore, given the demonstrable benefits of commercial activity in living species, policy makers allowed unimpeded importation. Recent advances in ecological science, however, make it possible to predict with increasing accuracy the geographical area suitable for occupancy by a nonindigenous species and whether the species is likely to cause substantial harm in that area.

In the past decade, environmental niche models (often also called climate envelope models or species distribution models), based on statistical relationships or machine-learning algorithms, have increased dramatically in their scope and reliability (Elith & Leathwick 2009). On the assumption that climatic and other physicochemical features of the environment are what limit species ranges, these models identify areas that have environmental characteristics similar to the areas already occupied by the species. Although few global data layers for freshwater environmental conditions exist, hindcasts of potential ranges for freshwater nonindigenous species have given confidence to the use of environmental niche models to make predictions (Herborg et al. 2007a,b). For example, the potential global range of *Procambarus clarkii*, *Pacifastacus leniusculus*, and *Cherax destructor* (de Moor 2002, Capinha et al. 2011, Liu et al. 2011) is much larger than the current range, suggesting that the detrimental changes in ecosystem services experienced in already invaded areas are likely to increase in spatial extent in the future (Liu et al. 2011).

Given an expectation that a region is environmentally suitable, additional recent advances in the application of ecological theory provide a sound basis for predicting whether a species is likely to establish and/or cause losses in ecosystem services. For example, with data on the biological traits of a sufficient number of species that have and species that have not caused losses of ecosystem services in an ecosystem or region, statistical or machine-learning algorithms can identify which traits or combinations of traits are most closely associated with each outcome. For taxa including plants, birds, and fishes, accuracy of such tools is high (often 75–90%). For crayfishes native to the southeastern United States (the global center of recent evolutionary radiation of cambarid crayfishes; **Figure 1a**), Larson & Olden (2010) discovered that species that have previously invaded other regions had larger adult size, higher fecundity, and more general habitat requirements than species that did not invade. The invasion capacity of crayfishes was predicted with accuracy up to 90%. In addition to species that have already demonstrated their invasiveness, seven other species were identified as high risks for invasion outside their native region (Larson & Olden 2010). Any proposed importation elsewhere of these species, *Cambarus ortmanni*, *C. sphenoides*, *C. striatus*, *C. thomai*, *Orconectes kentuckiensis*, *Procambarus gracilis*, and *P. viaeveridus*, should be particularly carefully scrutinized. Such models, or adaptations of them, are likely to be very useful as risk assessment tools in other regions of the world to guide voluntary or regulatory decisions about whether to allow importation of species.

The recent increases in accuracy of predictions about potential range and the likelihood of negative impacts have been sufficient to provide net economic benefits when such risk assessment tools have been employed. Although no bioeconomic analysis exists for trait-based risk assessment of crayfishes, the use of trait-based risk assessment for terrestrial plant importation into Australia creates a net increase in ecosystem services relative to either allowing importation of all species or to barring all species (Keller et al. 2007). All evidence about the large negative impacts of some crayfishes and the accuracy of existing trait-based analyses for crayfishes (Larson & Olden 2010) suggest that net benefits—perhaps even larger net benefits—would also accrue from the use of trait-based risk assessment for animals, including crayfishes.

Research to Inform Risk Assessment of Spread to New Waterways

Even after a species is established in a biogeographic realm, many opportunities exist to prevent the species—and hence the losses of ecosystem services—from spreading geographically. Recent advances in applying ecological theory and models can guide such efforts cost-effectively. For example, in the glacial lake district of the US upper Midwest, state and federal agencies aim to prevent an increase in the number of lakes and rivers infested with rusty crayfish. Three nonexclusive strategies have been promoted to manage anglers and other boaters, who are the primary vectors for dispersal of crayfish between waterways. First, lakes that offer suitable habitat can be identified and protected from introduction. This strategy may be infeasible, however, because hundreds of uninfested waterways in northern Wisconsin alone have suitable pH and calcium concentrations and would thus require on-site inspection and interdiction (Olden et al. 2011). Second, lakes that are most likely to receive visitations by boaters from infested waters can be identified on the basis of network models and protected from introduction, but when many lakes remain uninfested, predictions may not be accurate enough to make their use cost-effective (Rothlisberger & Lodge 2011). Finally, management efforts can aim to prevent egress of propagules from infested lakes—analogous to quarantining superspreader hosts to prevent the spread of infectious disease (Drury & Rothlisberger 2008). Deciding which choice is the most effective hinges on the accuracy of predictions essential to each approach, which are themselves partly a function of the proportion of lakes infested: The higher the proportion of infested lakes, the higher the accuracy of predictions about which lakes subsequently will become infested (Rothlisberger & Lodge 2011). As a greater proportion of waterways is invaded, cost-effectiveness may shift away from the third option (preventing egress of propagules) toward a combination of options (Drury & Rothlisberger 2008). Ecological theory applied to crayfishes thus provides a foundation for more effective and more cost-effective efforts to prevent the arrival and spread of harmful crayfishes (Keller et al. 2008).

Early Detection, Eradication, and Control

Even after an establishment has occurred from an initial importation or from subsequent secondary spread, ecosystem services can still be protected cost-effectively by eradicating populations while they remain small and localized. For such a rapid response program, however, an effective surveillance program capable of early detection must exist. Unfortunately, traditional capture methods for crayfishes are ineffective for detection or eradication of low-density, incipient invasions (Holdich et al. 1999). Two promising new approaches are available. First, environmental DNA (eDNA) is a surveillance approach that captures shed cellular material directly from the water for DNA-based identification of a species' presence (Jerde et al. 2011). Because eDNA surveillance works for a variety of taxa, including a freshwater crustacean (tadpole shrimp *Lepidurus apus*) (Thomsen et al. 2012), there is no reason to expect that it will not work for crayfishes. Second, sex pheromones increase the attractiveness of traditional traps to male crayfish (Aquiloni & Gherardi 2010).

Once an invasive crayfish has been detected, options for restoration of ecosystem services via eradication or control of crayfishes exist, including biocides, physical methods (e.g., draining, barriers), biological control, mechanical removal (e.g., trapping, netting), and autocidal methods (Gherardi et al. 2011a). Examples of successful eradication and control of nonindigenous crayfishes have been limited to small streams and medium-sized lentic ecosystems.

Eradications have used nonspecific biocides, with operations confined to habitats where large non-target impacts were acceptable and operation logistics and costs were manageable, and the chemicals could be contained until they detoxified (Gherardi et al. 2011a). For example, in New Zealand the only known population of *Cherax tennuimanus* (native to Australia) was eradicated with biocide combined with physical methods (drainage and barriers were used to reduce the volume of water needing treatment and to prevent escape) (Gould 2005).

Crayfish barriers may be an important component to successfully managing crayfish populations in streams (Gherardi et al. 2011a). One successful barrier, designed to slow the invasion of invasive signal crayfish into a section of stream with a population of endangered *Pacifastacus fortis*, funnels stream water over an elevated, smooth stainless steel platform and appears to be 100% effective (J. Cook, Spring Rivers LLC, personal communication).

The successful use of crayfish plague to eradicate common yabby from ponds in Spain (Diéguez-Uribeondo & Muzquiz 2005) highlights this plague's potential as an eradication or control method for crayfishes not native to the Nearctic. A variety of species and pathogens have potential as biological control agents for invasive crayfish (Freeman et al. 2010, B. Peters 2010, Tetzlaff et al. 2011), with native predatory fishes offering considerable promise (Hein et al. 2006, Gherardi et al. 2011a). However, with any biological control method, particularly with agents that are nonindigenous and nonspecific, extreme care must be taken so other ecosystem services are not put at risk (Gherardi et al. 2011a).

Mechanical removal with traps or nets has been the most commonly employed control method, successfully employed to suppress crayfish numbers to prevent extinctions of endangered amphibian and fish species, reduce nuisance crayfish for anglers, or recover littoral habitat (Gherardi et al. 2011a). However, recovery of populations typically occurs within two breeding seasons if removal ceases (Skurdal & Qvenild 1986). In a 50-ha lake in Wisconsin, USA, intensive harvest of rusty crayfish over 5 years coupled with restricted harvest of predatory fish caused a 95% decrease in catch rate of crayfishes and recovery of native macrophytes and invertebrate communities (Hein et al. 2007). Unless new, more effective targeted approaches are developed, integrated pest management that combines mechanical harvest, barriers, and enhancement of natural crayfish predators appears to offer the greatest potential for sustainable control and protection of ecosystem services (Horan et al. 2011).

FUTURE ISSUES

1. Studies that quantify crayfish impacts with metrics of direct relevance to provisioning and regulating ecosystem services as well as to cultural ecosystem services, including monetized ecosystem services, are needed to better inform management decisions.

2. To better anticipate crayfish impact on endemic biodiversity, experiments are especially needed on interactions between crayfish and functionally similar freshwater crabs and shrimps.

3. Studies on crayfish impacts in the tropical regions of the Ethiopian and Oriental realms are especially urgent (e.g., on *Cherax* spp.).

4. Surveys and experimental studies on the parasites and pathogens of crayfishes are needed to prevent future crayfish plague-like disasters.

5. The potential for hybridization among crayfishes must be experimentally tested to assess the risk of losing unique genetic resources via existing or future introductions.

6. Research is needed both to test whether native crayfishes could provide ecosystem services equal to those of nonindigenous species and on factors that drive human preferences for native versus nonindigenous species.

7. Theoretical population dynamics research on different population control strategies (e.g., relative population impact of attacking different life-history stages) is needed to guide increased empirical research on control tactics (e.g., toxins, sterile males, autocidal methods) to restore ecosystem services lost to invasive crayfish.

DISCLOSURE STATEMENT

The authors are not aware of any biases that might be perceived as affecting the objectivity of this review.

ACKNOWLEDGMENTS

We acknowledge the following sources of funding and/or other support: USNOAA CSCOR and USEPA GLRI (D.M.L., C.L.J., M.E.W.); USNSF GLOBES IGERT (A.K.B., M.A.B.,

A.D., L.W.S., C.R.T.); Marie Curie International Outgoing Fellowship within the seventh European Community Framework Program, CHAOS project 251801 (F.G.); University of Notre Dame CAC Fellowship (B.W.P.); and National University of Singapore grant R-154-000-465-133 (D.C.J.Y.).

LITERATURE CITED

Adão H, Marques JC. 1993. Population biology of the red swamp crayfish *Procambarus clarkii* (Girard, 1852) in southern Portugal. *Crustaceana* 65:336–45

Ahern D, England J, Ellis A. 2008. The virile crayfish, *Orconectes virilis* (Hagen, 1870) (Crustacea: Decapoda: Cambaridae), identified in the UK. *Aquat. Invas.* 3:102–4

Ahyong ST, Yeo DCJ. 2007. Feral populations of the Australian red-claw crayfish (*Cherax quadricarinatus* von Martens) in water supply catchments of Singapore. *Biol. Invas.* 9:943–46

Alcorlo P, Otero M, Crehuet M, Baltanás A, Montes C. 2006. The use of the red swamp crayfish (*Procambarus clarkii* Girard) as indicator of the bioavailability of heavy metals in environmental monitoring in the River Guadiamar (SW, Spain). *Sci. Total Environ.* 366:380–90

Alderman DJ, Holdich D, Reeve I. 1990. Signal crayfish as vectors in crayfish plague in Britain. *Aquaculture* 86:3–6

Anastácio PM, Correia AM, Menino JP, Martins da Silva L. 2005. Are rice seedlings affected by changes in water quality caused by crayfish? *Ann. Limnol.: Int. J. Limn.* 41:1–6

Anda P, Segura del Pozo J, Díaz García JM, Escudero R, García Peña FJ, et al. 2001. Waterborne outbreak of tularemia associated with crayfish fishing. *Emerg. Infect. Dis.* 7:575–82

Angeler DG, Sánchez-Carrillo S, García G, Alvarez-Cobelas M. 2001. The influence of *Procambarus clarkii* (Cambaridae, Decapoda) on water quality and sediment characteristics in a Spanish floodplain wetland. *Hydrobiology* 464:88–98

Aquiloni L, Gherardi F. 2010. The use of sex pheromones for the control of invasive populations of the crayfish *Procambarus clarkia*: a field study. *Hydrobiology* 649:249–54

Barbaresi S, Gherardi F. 2000. The invasion of the alien crayfish *Procambarus clarkii* in Europe, with particular reference to Italy. *Biol. Invas.* 2:259–64

Barbaresi S, Tricarico E, Gherardi F. 2004. Factors inducing the intense burrowing activity by the red swamp crayfish, *Procambarus clarkii*, an invasive species. *Naturwissenschaften* 91:342–45

Beatty SJ. 2006. The diet and trophic positions of translocated, sympatric populations of *Cherax destructor* and *Cherax cainii* in the Hutt River, Western Australia: evidence of resource overlap. *Mar. Freshw. Res.* 57:825–35

Belle CC, Wong JQH, Yeo DCJ, Tan SH, Tan HH, et al. 2011. Ornamental trade as a pathway for Australian redclaw crayfish introduction and establishment. *Aquat. Biol.* 12:69–79

Belle CC, Yeo DCJ. 2010. New observations of the exotic Australian red-claw crayfish, *Cherax quadricarinatus* (von Martens, 1868) (Crustacea: Decapoda: Parastacidae) in Singapore. *Nat. Singap.* 3:99–102

Bhikajee M. 1997. Recent advances in aquaculture in Mauritius. *2nd Annu. Meet. Agric. Sci., Reduit, Mauritius, August 12–13*, pp. 95–101. Reduit, Mauritius: Food Agric. Res. Counc.

Bortolini JS, Alvarez F, Rodriguez-Almaraz G. 2007. On the presence of the Australian redclaw crayfish, *Cherax quadricarinatus*, in Mexico. *Biol. Invas.* 9:615–20

Capinha C, Leung B, Pedro A. 2011. Predicting worldwide invasiveness for four major problematic decapods: an evaluation of using different calibration sets. *Ecography* 34:448–59

Casale A, Busato E. 2008. A real time extinction: the case of *Carabus clatratus* in Italy (Coleoptera, Carabidae). In *Back to the Roots and Back to the Future. Towards a New Synthesis Amongst Taxonomic, Ecological and Biogeographical Approaches in Carabidology*, ed. L Penev, T Erwin, T Hassmann, pp. 353–62. Sofia, Bulgaria: Pensoft Publ.

Chucholl C. 2012. Understanding invasion success: life-history traits and feeding habits of the alien crayfish *Orconectes immunis* (Decapoda, Astacida, Cambaridae). *Knowl. Manag. Aquat. Ecos.* 404:04. **http://dx.doi.org/10.1051/kmae/2011082**

Chucholl C, Daudey T. 2008. First record of *Orconectes juvenilis* (Hagen 1870) in eastern France: update to the species identity of a recently introduced orconectid crayfish (Crustacea: Astacida). *Aqua. Inv.* 3:105–7

Crandall KA, Buhay JE. 2008. Global diversity of crayfish (Astacidae, Cambaridae, and Parastacidae—Decapoda) in freshwater. In *Freshwater Animal Diversity Assessment*, ed. Balian EV, Lévêque C, Segers H, Martens K. *Dev. Hydrobiol.* 198:295–301

Crandall KA, Harris DJ, Fetzner JW Jr. 2000. The monophyletic origin of freshwater crayfish estimated from nuclear and mitochondrial DNA sequences. *Proc. R. Soc. Lond. Ser. B* 267:1679–86

Cruz MJ, Rebelo R, Crespo EG. 2006. Effects of an introduced crayfish, *Procambarus clarkii*, on the distribution of south-western Iberian amphibians in their breeding habitats. *Ecography* 29:329–38

Davis MA, Chew MK, Hobbs RJ, Lugo AE, Ewel JJ, et al. 2011. Don't judge species on their origins. *Nature* 474:153–54

Delibes M, Adrian I. 1987. Effects of crayfish introduction on otter *Lutra lutra* food in the Doñana National Park, SW Spain. *Biol. Conserv.* 42:153–59

de Moor I. 2002. Potential impacts of alien freshwater crayfish in South Africa. *Afr. J. Aquat. Sci.* 27:125–39

Diéguez-Uribeondo J, Muzquiz JL. 2005. *The use of the fungus* Aphanomyces astaci *for biological control of the spread of the invasive species* Cherax destructor. Presented at Workshop Biol. Invasions Inland Waters, Florence, Italy, May 5–7

Dorn NJ, Mittelbach GG. 1999. More than predator and prey: a review of interactions between fish and crayfish. *Vie et Milieu—Life Environ.* 49:229–37

Drury KL, Rothlisberger JD. 2008. Offense and defense in landscape-level invasion control. *Oikos* 117:182–90

Edgerton BF, Evans LH, Stephens FJ, Overstreet RM. 2002. Synopsis of freshwater crayfish diseases and commensal organisms. *Aquaculture* 206:57–135

Elith J, Leathwick JR. 2009. Species distribution models: ecological explanation and prediction across space and time. *Annu. Rev. Ecol. Evol. Syst.* 40:677–97

Ficetola GF, Siesa ME, Manenti R, Bottoni L, De Bernardi F, Padoa-Schioppa E. 2011. Early assessment of the impact of alien species: differential consequences of an invasive crayfish on adult and larval amphibians. *Divers. Distrib.* 17:1141–51

Food and Agriculture Organization of the United Nations. 2010. *The State of World Fisheries and Aquaculture 2010*. Rome: Food Agric. Org. U.N. 197 pp.

Fowler AJ, Lodge DM, Hsia J. 2007. Failure of the Lacey Act to protect US ecosystems against animal invasions. *Front. Ecol. Environ.* 5:353–59

Freeman MA, Turnbull JF, Yeomans WE, Bean CW. 2010. Prospects for management strategies of invasive crayfish populations with an emphasis on biological control. *Aquat. Conserv.: Mar. Freshw. Ecosyst.* 20:211–23

Garcia LV, Ramo C, Aponte C, Moreno A, Dominguez MT, et al. 2011. Protected wading bird species threaten relict centenarian cork oaks in a Mediterranean Biosphere Reserve: a conservation management conflict. *Biol. Conserv.* 144:764–71

Garvey JE, Rettig JE, Stein RA, Lodge DM, Klosiewski SP. 2003. Scale-dependent associations among fish predation, littoral habitat, and distributions of crayfish species. *Ecology* 84:3339–48

Gherardi F. 2006. Crayfish invading Europe: the case study of *Procambarus clarkii*. *Mar. Freshw. Behav. Phys.* 39:175–91

Gherardi F. 2007. Understanding the impact of invasive crayfish. In *Biological Invaders in Inland Waters: Profiles, Distribution, and Threats, Invading Nature: Springer Series in Invasion Ecology*, ed. F Gherardi, pp. 507–42. Dordrecht, Neth.: Springer

Gherardi F. 2008. *Procambarus clarkii*. In *Environmental Impacts of Alien Species in Aquaculture: Alien Species Fact Sheet*, ed. S Gollasch, IG Cowx, AD Nunn, pp. 38–49. IMPASSE. Project No. 44142. European Comission within the Sixth Framework Programme (2002–2006). **http://www2.hull.ac.uk/science/pdf/ IMPASSE_44142_WP2_fact-sheets.pdf**

Gherardi F. 2011a. Crayfish. In *Encyclopedia of Biological Invasions*, ed. D Simberloff, M Rejmánek, pp. 129–35. Berkeley and Los Angeles: Univ. Calif. Press

Gherardi F. 2011b. Towards a sustainable human use of crayfish (Crustacea, Decapoda, Astacidea). *Knowl. Manag. Aquat. Ecosyst.* 401:02p1–22

Gherardi F, Acquistapace P. 2007. Invasive crayfish in Europe: the impact of *Procambarus clarkii* on the littoral community of a Mediterranean lake. *Freshw. Biol.* 52:1249–59

Gherardi F, Aquiloni L, Diéguez-Uribeondo J, Tricarico E. 2011a. Managing invasive crayfish: Is there a hope? *Aquat. Sci.* 73:185–200

Gherardi F, Baldaccini GN, Barbaresi S, Ercolini P, De Luise G, et al. 1999. The situation in Italy. See Gherardi & Holdich 1999, pp. 107–28

Gherardi F, Barbaresi S, Vaselli O, Bencini A. 2002. A comparison of trace metal accumulation in indigenous and alien freshwater macro-decapods. *Mar. Freshw. Behav. Phys.* 35:179–88

Gherardi F, Britton JR, Mavuti KM, Pacini N, Grey J, et al. 2011b. A review of allodiversity in Lake Naivasha, Kenya: developing conservation actions to protect East African lakes from the negative impacts of alien species. *Biol. Conserv.* 144:2585–96

Gherardi F, Cioni A. 2004. Agonism and interference competition in freshwater decapods. *Behaviour* 141:1297–324

Gherardi F, Holdich DM, eds. 1999. *Crayfish in Europe as Alien Species. How to Make the Best of a Bad Situation?* Rotterdam, Neth.: Balkema. 310 pp.

Gherardi F, Lazzara L. 2006. Effects of the density of an invasive crayfish (*Procambarus clarkii*) on pelagic and surface microalgae in a Mediterranean wetland. *Arch. Hydrobiol.* 165:401–14

Gould B. 2005. Marron. Interagency collaboration follows surprise catch. *Biosecurity* 60:10–11

Gutiérrez-Yurrita PJ, Martinez JM, Bravo-Utrera MA, Contes C, Ilhéu M, Bernardo JM. 1999. The status of crayfish populations in Spain and Portugal. See Gherardi & Holdich 1999, pp. 161–92

Hamr P. 2002. Orconectes. In *Biology of Freshwater Crayfish*, ed. DM Holdich, pp. 585–608. Oxford, UK: Blackwell Sci.

Hauck AK, Marshall MR, Li JK, Lee RA. 2001. A new finding and range extension of bacilliform virus in the freshwater red claw crayfish in Utah, USA. *J. Aquat. Anim. Health* 13:158–62

Hein CL, Roth BM, Vander Zanden MJ. 2006. Fish predation and trapping for rusty crayfish (*Orconectes rusticus*) control: a whole-lake experiment. *Can. J. Fish. Aquat. Sci.* 63:383–93

Hein CL, VanderZanden MJ, Magnuson JJ. 2007. Intensive trapping and increased fish predation cause massive population decline of an invasive crayfish. *Freshw. Biol.* 52:1134–46

Herborg LM, Jerde CL, Lodge DM, Ruiz GM, MacIsaac HJ. 2007a. Predicting invasion risk using measures of introduction effort and environmental niche models. *Ecol. Appl.* 17:663–74

Herborg LM, Rudnick DA, Siliang Y, Lodge DM, MacIsaac HJ. 2007b. Predicting the range of Chinese mitten crabs in Europe. *Conserv. Biol.* 21:1316–23

Hobbs HH Jr. 1974. Synopsis of the families and genera of crayfishes (Crustacea: Decapoda). *Smithson. Contrib. Zool.* 164:1–32

Hobbs HH Jr. 1988. Crayfish distribution, adaptive radiation and evolution. In *Freshwater Crayfish: Biology, Management and Exploitation*, ed. DM Holdich, Lowery RS, pp. 52–82. London, UK: Croom Helm

Hobbs HH, Lodge DM. 2010. Decapoda. In *Ecology and Classification of North American Freshwater Invertebrates*, ed. JH Thorp, AP Covich, pp. 901–68. San Diego: Academic. 3rd ed.

Holdich DM. 1999. The negative effects of established crayfish introductions. See Gherardi & Holdich 1999, pp. 31–47

Holdich DM, Gydemo R, Rogers WD. 1999. A review of possible methods for controlling nuisance populations of alien crayfish. See Gherardi & Holdich 1999, pp. 245–70

Holdich DM, Reynolds JD, Souty-Grosset C, Sibley PJ. 2009. A review of the ever increasing threat to European crayfish from non-indigenous crayfish species. *Knowl. Manag. Aquat. Ecosyst.* 394–95:11p1–46

Holdich DM, Rogers WD, Reynolds JD. 1999. Native and alien crayfish in the British Isles. See Gherardi & Holdich 1999, pp. 221–36

Horan RD, Fenichel EP, Drury KLS, Lodge DM. 2011. Managing ecological thresholds in coupled environmental-human systems. *Proc. Natl. Acad. Sci. USA* 108:7333–38

Huner JV. 2002. *Procambarus.* In *Biology of Freshwater Crayfish*, ed. DM Holdich, pp. 541–84. Oxford, UK: Blackwell Sci.

Jerde CL, Mahon AR, Chadderton LW, Lodge DM. 2011. "Sight-unseen" detection of rare aquatic species using environmental DNA. *Conserv. Lett.* 4:150–57

Jonas JL, Claramunt RM, Fitzsimons JD, Marsden JE, Ellrott BJ. 2005. Estimates of egg deposition and effects of lake trout (*Salvelinus namaycush*) egg predators in three regions of the Great Lakes. *Can. J. Fish. Aquat. Sci.* 62:2254–64

Kataria M. 2007. A cost-benefit analysis of introducing a non-native species: the case of signal crayfish in Sweden. *Mar. Res. Econ.* 22:15–28

Keller RP, Frang K, Lodge DM. 2008. Preventing the spread of invasive species: economic benefits of intervention guided by ecological predictions. *Conserv. Biol.* 22:80–88

Keller RP, Lodge DM, Finnoff DC. 2007. Risk assessment for invasive species produces net bioeconomic benefits. *Proc. Natl. Acad. Sci. USA* 104:203–7

Kirjavainen J, Sipponen M. 2004. Environmental benefit of different crayfish management strategies in Finland. *Fish. Manag. Ecol.* 11:213–18

Kreps TA. 2009. Scaling up: long term, large-scale impacts of the invasion of lakes by the invasive rusty crayfish (*Orconectes rusticus*). PhD. Dissertation, Univ. Notre Dame, Notre Dame, IN. 155 pp.

Kreps TA, Baldridge AK, Lodge DM. 2012. The impact of an invasive predator (*Orconectes rusticus*) on freshwater snail communities: insights on habitat-specific effects from a multilake long-term study. *Can. J. Fish. Aquat. Sci.* 69:1164–73

Lambertini M, Leape J, Marton-Lefevre J, Mittermeier RA, Rose M, et al. 2011. Invasives: a major conservation threat. *Science* 333:404–5

Lane MA, Barsanti MC, Santos CA, Yeung M, Lubner SJ, Weil GJ. 2009. Human paragonimiasis in North America following ingestion of raw crayfish. *Clin. Infect. Dis.* 49:E55–61

Larson ER, Olden JD. 2010. Latent extinction and invasion risk of crayfishes in the southeastern United States. *Conserv. Biol.* 24:1099–110

Letson MA, Makarewicz JC. 1994. An experimental test of the crayfish (*Orconectes immunis*) as a control mechanism for submersed aquatic macrophytes. *Lake Reserv. Manag.* 10:127–32

Levin DA, Francisco-Ortega J, Jansen RK. 1996. Hybridization and the extinction of rare plant species. *Conserv. Biol.* 10:10–16

Liu X, Guo Z, Ke Z, Wang S, Li Y. 2011. Increasing potential risk of a global aquatic invader in Europe in contrast to other continents under future climate change. *PLoS ONE* 6(3):e18429

Lodge DM, Hill AH. 1994. Factors governing species composition, population size, and productivity of coolwater crayfishes. *Nord. J. Freshw. Res.* 69:111–36

Lodge DM, Rosenthal SK, Mavuti KM, Muohi W, Ochieng P, et al. 2005. Louisiana crayfish (*Procambarus clarkii*) (Crustacea: Cambaridae) in Kenyan ponds: non-target effects of a potential biological control agent for schistosomiasis. *Afr. J. Aquat. Sci.* 30:119–24

Lodge DM, Shrader-Frechette K. 2003. Nonindigenous species: ecological explanation, environmental ethics, and public policy. *Conserv. Biol.* 17:31–37

Lodge DM, Taylor CM, Holdich DM, Skurdal J. 2000. Non-indigenous crayfishes threaten North American freshwater biodiversity: lessons from Europe. *Fisheries* 25:7–20

Longshaw M. 2011. Diseases of crayfish: a review. *J. Invert. Path.* 106:5470

Maezono Y, Kobayashi R, Kusahara M, Miyashita T. 2005. Direct and indirect effects of exotic bass and bluegill on native and exotic organisms in farm ponds. *Ecol. Appl.* 15:638–50

Maezono Y, Miyashita T. 2004. Impact of exotic fish removal on native communities in farm ponds. *Ecol. Res.* 19:263–67

Mitchell SA, Kock DJ. 1988. Alien symbionts introduced with imported marron from Australia may pose a threat to aquaculture. *S. Afr. J. Sci.* 84:877–78

Miyake M, Miyashita T. 2011. Identification of alien predators that should not be removed for controlling invasive crayfish threatening endangered odonates. *Aquat. Conserv.: Mar. Freshw. Ecosyst.* 21:292–98

Mkoji GM, Hofkin BV, Kuris AM, Stewart-Oaten A, Mungai BN, et al. 1999. Impact of the crayfish *Procambarus clarkii* on *Schistosoma haematobium* transmission in Kenya. *Am. J. Trop. Med. Hyg.* 61:751–59

Montes C, Bravo-Utrera MA, Baltanás A, Duarte C, Gutiérrez-Yurrita PJ. 1993. *Bases ecológicas para la gestión del cangrejo rojo de las marismas en el Parque Nacional de Doñana*. Madrid, Spain: ICONA, Ministerio de Agricultura y Pesca

Nyström P, Svensson O, Lardner B, Brönmark C, Graneli W. 2001. The influence of multiple introduced predators on a littoral pond community. *Ecology* 82:1023–39

Ogada M. 2007. On the otter trail. *SWARA, J. East Afr. Wildlife Soc.* 30:48–50

Ogada MO, Aloo PA, Muruthi PM. 2009. The African clawless otter *Aonyx capensis* (Schinz, 1821) and its diet as an indicator of crayfish invasion dynamics in aquatic systems. *Afr. J. Ecol.* 47:119–20

Ohtaka A, Gelder SR, Kawai T, Saito K, Nakata K, Nishino M. 2005. New records and distributions of two North American branchiobdellidan species (Annelida: Clitellata) from introduced signal crayfish, *Pacifastacus leniusculus*, in Japan. *Biol. Inv.* 7:149–56

Olden JD, Vander Zanden MJ, Johnson PTJ. 2011. Assessing ecosystem vulnerability to invasive rusty crayfish (*Orconectes rusticus*). *Ecol. Appl.* 21:2587–99

Olsen TM, Lodge DM, Capelli GM, Houlihan RJ. 1991. Mechanisms of impact of an introduced crayfish (*Orconectes rusticus*) on littoral congeners, snails, and macrophytes. *Can. J. Fish. Aquat. Sci.* 48:1853–61

Ortmann AE. 1902. The geographical distribution of freshwater decapods and its bearing upon ancient geography. *Proc. Am. Philos. Soc.* 41:267–400

Perry WL, Feder JL, Dwyer G, Lodge DM. 2001a. Hybrid zone dynamics and species replacement between *Orconectes* crayfishes in a northern Wisconsin lake. *Evolution* 55:1153–66

Perry WL, Feder JL, Lodge DM. 2001b. Implications of hybridization between introduced and resident *Orconectes* crayfishes. *Conserv. Biol.* 15:1656–66

Perry WL, Lodge DM, Feder JL. 2002. Importance of hybridization between indigenous and nonindigenous freshwater species: an overlooked threat to North American biodiversity. *Syst. Biol.* 51:255–75

Peters B. 2010. *Evaluating strategies for controlling invasive crayfish using human and fish predation.* MS thesis. Univ. Notre Dame, Notre Dame, IN. 70 pp.

Peters JA. 2010. *Influence of habitat, predation, and spatial and temporal scales on species diversity and distributions: interactions involving crayfishes in the Laurentian Great Lakes and inland lakes.* PhD. Dissertation, Univ. of Notre Dame, Notre Dame, IN. 237 pp.

Phillips ID, Vinebrooke RD, Turner MA. 2009. Ecosystem consequences of potential range expansions of *Orconectes virilis* and *Orconectes rusticus* crayfish in Canada: a review. *Env. Rev.* 17:235–48

Renai B, Gherardi F. 2004. Predatory efficiency of crayfish: comparison between indigenous and nonindigenous species. *Biol. Invas.* 6:89–99

Robinet O, ed. 2010. *Stratégie de lutte contre les espèces invasives à la Réunion.* Ile de La Réunion, France: DIREN. 97 pp.

Rodewald AD. 2012. Spreading messages about invasives. *Divers. Distrib.* 18:97–99

Rodríguez CF, Bécares E, Fernández-Aláez M. 2003. Shift from clear to turbid phase in Lake Chozas (NW Spain) due to the introduction of American red swamp crayfish (*Procambarus clarkii*). *Hydrobiologia* 506–9:421–26

Rosenthal SK, Lodge DM, Mavuti KM, Muohi W, Ochieng P, et al. 2005. Comparing macrophyte herbivory by introduced Louisiana crayfish (*Procambarus clarkii*) (Crustacea: Cambaridea) and native Dytiscid beetles (*Cybister tripunctatus*) (Coleoptera: Dytiscidae), in Kenya. *Afr. J. Aquat. Sci.* 30:157–62

Rothlisberger JD, Lodge DM. 2011. Limitations of gravity models in predicting the spread of Eurasian watermilfoil. *Conserv. Biol.* 25:64–72

Scalici M, Chiesa S, Scuderi S, Celauro D, Gibertini G. 2010. Population structure and dynamics of *Procambarus clarkii* (Girard, 1852) in a Mediterranean brackish wetland (Central Italy). *Biol. Invas.* 12:1415–25

Sinclair EA, Fetzner JW Jr, Buhay J, Crandall KA. 2004. Proposal to complete a phylogenetic taxonomy and systematic revision for freshwater crayfish (Astacida). *Freshw. Crayfish* 14:1–9

Skurdal J, Qvenild T. 1986. Growth, maturity and fecundity of *Astacus astacus* in Lake Steinsfjorden, S. E. Norway. *Freshw. Crayfish* 6:182–86

Skurdal J, Taugbøl T, Burba A, Edsman L, Søderback B, et al. 1999. Crayfish introduction in the Nordic and Baltic countries. See Gherardi & Holdich 1999, pp. 193–220

Soderback B. 1995. Replacement of the native crayfish *Astacus astacus* by the introduced species *Pacifastacus leniusculus* in a Swedish lake: possible causes and mechanisms. *Freshw. Biol.* 33:291–304

Stewart TW, Miner JG, Lowe RL. 1998. An experimental analysis of crayfish (*Orconectes rusticus*) effects on a *Dreissena*-dominated benthic macroinvertebrate community in western Lake Erie. *Can. J. Fish. Aquat. Sci.* 55:1043–50

Stucki TP, Staub E. 1999. Distribution of crayfish species and legislation concerning crayfish in Switzerland. See Gherardi & Holdich 1999, pp. 141–48

Tablado Z, Tella JL, Sánchez-Zapata JA, Hiraldo F. 2010. The paradox of the long-term positive effects of a North American crayfish on a European community of predators. *Conserv. Biol.* 24:1230–38

Tetzlaff JC, Roth BM, Weidel BC, Kitchell JF. 2011. Predation by native sunfishes (*Centrarchidae*) on the invasive crayfish *Orconectes rusticus* in four northern Wisconsin lakes. *Ecol. Freshw. Fish* 20:133–43

Thomsen PF, Kielgast J, Iversen LL, Wiuf C, Rasmussen M, et al. 2012. Monitoring endangered freshwater biodiversity by environmental DNA. *Mol. Ecol.* 21(11):2565–73

Thys van den Audenaerde DFF. 1994. *Introduction of aquatic species in Zambian waters, and their importance for aquaculture and fisheries.* ALCOM Field Rep. No. 24, Food Agric. Org. U.N., Rome. 27 pp. **http://www.fao.org/docrep/005/AD005E/AD005E00.htm**

Toon A, Pérez-Losrada M, Schweitzer CE, Feldmann RM, Carlson M, Crandall KA. 2010. Gondwanan radiation of the Southern Hemisphere crayfishes (Decapoda: Parastacidae): evidence from fossils and molecules. *J. Biogeogr.* 37:2275–90

Turner CL. 1926. The crayfishes of Ohio. *Ohio Biol. Surv. Bull.* 3:145–95

Tyser B. 2010. MS Dissertation. Univ. East Anglia, Norwich, UK. 45 pp.

United States Geological Survey. 2011. Nonindigenous Aquatic Species Database. Washington, DC: U.S. Geol. Surv. **http://nas.er.usgs.gov/queries/speciesmap.aspx?SpeciesID=214**

Usio N, Konishi M, Nakano S. 2001. Species displacement between an introduced and a "vulnerable" crayfish: the role of aggressive interactions and shelter competition. *Biol. Invas.* 3:179–85

Vasconcelos V, Oliveira S, Teles FO. 2001. Impact of a toxic and a non-toxic strain of *Microcystis aeruginosa* on the crayfish *Procambarus clarkii*. *Toxicon* 39:1461–70

Wilson KA, Magnuson JJ, Lodge DM, Hill AM, Kratz TK, et al. 2004. A long-term rusty crayfish (*Orconectes rusticus*) invasion: dispersal patterns and community change in a north temperate lake. *Can. J. Fish. Aquat. Sci.* 61:2255–66

Yeo DCJ. 2010. Introduced decapod crustaceans in Singapore's reservoirs. *Cosmos* 6:83–88

Yue GH, Li J, Bai Z, Wang CM, Feng F. 2010. Genetic diversity and population structure of the invasive alien red swamp crayfish. *Biol. Invas.* 12:2697–706

Zeng W, Zeng Y, Fei RM, Zeng LB, Wei KJ. 2011. Analysis of variable genomic loci in white spot syndrome virus to predict its origins in *Procambarus clarkii* crayfish farmed in China. *Dis. Aquat. Org.* 96:105–12

Cumulative Indexes

Contributing Authors, Volumes 39–43

Eguiarte LE, 42:245–66
Ehrenfeld JG, 41:59–80
Elgar MA, 40:21–39
Elith J, 40:677–97
Elser JJ, 43:429–48
Emlen DJ, 39:387–413
Emlet RB, 43:97–114
Engelstädter J, 40:127–49
Excoffier L, 40:481–501

F

Fargione JE, 41:351–77
Feder JL, 43:449–72
Ferrenberg S, 43:137–55
Fierer N, 43:137–55
Fisher DC, 39:365–85
Fjeldså J, 43:249–65
Flores GE, 43:137–55
Foll M, 40:481–501
Follstad Shah JJ, 43:313–43
Fortin M-J, 41:21–38
François O, 43:23–43
Freckleton RP, 41:173–92
Friedman M, 39:571–92
Friesen ML, 42:23–46

G

Gaedke U, 43:429–48
Gaines SD, 42:381–409
Gantz CA, 43:449–72
Gardes M, 40:699–715
Gaston KJ, 39:93–113
Gates RD, 41:127–47
Gaut BS, 42:245–66
Gensel PG, 39:459–77
Gherardi F, 43:449–72
Gingerich PD, 40:657–75
Glor R, 41:251–70
Gonzalez A, 40:393–414
González A, 43:137–55
Guégan J-F, 41:231–50
Guest JR, 40:551–71

H

Hall R, 42:205–26
Hall SR, 40:503–28
Halligan DL, 40:151–72

Hampe A, 42:313–33
Harpole WS, 43:227–48
Harris NC, 43:183–203
Hazen EL, 39:259–78
Hedin LO, 40:613–35
Hellberg ME, 40:291–310
Hendrix PF, 39:593–613
Herre EA, 39:439–58
Hey J, 41:215–30
Hill JD, 41:351–77
Hill WG, 41:1–19
HilleRisLambers J, 43:227–48
Hines J, 40:1–20
Hiraldo F, 39:1–19
Hoffmann AA, 39:21–42
Hofmann G, 41:127–47
Holmes EC, 40:353–72
Howard GW, 43:449–72
Huang C-Y, 39:593–613
Hunter J, 39:1–19
Hurst GDD, 40:127–49
Hutchins DA, 41:127–47

I

Ibelings B, 43:429–48
Irwin RE, 41:271–92

J

Jablonski D, 39:501–24
Jackson SF, 39:93–113
James AC, 42:411–40
James SW, 39:593–613
Jandér KC, 39:439–58
Janz N, 42:71–89
Janzen FJ, 41:39–57
Jedlicka JA, 40:573–92
Jeppesen E, 43:429–48
Jerde CL, 43:449–72
Jump AS, 42:313–33

K

Kaplan I, 40:1–20
Kawecki TJ, 39:321–42
Kay KM, 40:637–56
Keesing F, 43:157–82
Keightley PD, 40:151–72
Keller L, 42:91–110

Ketchum J, 39:1–19
Kiers ET, 39:215–36
Kiessling W, 40:173–92
Kingsolver JG, 43:205–26
Kirkpatrick M, 41:1–19
Klausmeier CA, 39:615–39
Klein CJ, 42:381–409
Klingenberg CP, 39:115–32
Klinger T, 41:127–47
Klironomos JN, 40:699–715
Knowles LL, 40:593–612
Konno K, 40:311–31
Körner C, 40:61–79
Kovach K, 41:293–319
Krug PJ, 43:97–114
Kruuk LEB, 39:525–48
Kueneman J, 43:137–55
Kuijper B, 43:287–311
Kupriyanova EK, 43:97–114

L

Lankau R, 42:335–54
Larracuente AM, 40:459–80
Latanich C, 39:259–78
Laurance WF, 40:529–49
Lavergne S, 41:321–50
Lavin M, 40:437–57
Leathwick JR, 40:677–97
Legg T, 43:137–55
Lessios HA, 39:63–91
Letourneau DK, 40:573–92
Levine JM, 43:227–48
Lewis SL, 40:529–49
Litchman E, 39:615–39
Lloyd J, 40:529–49
Lobreaux S, 43:23–43
Lodge DM, 43:449–72
Lohman DJ, 42:205–26
Loreau M, 40:393–414
Luo Z, 42:355–80
Lürling M, 43:429–48
Lynch RC, 43:137–55
Lyon BE, 39:343–63

M

MacDonald GM, 42:267–87
Machado CA, 39:439–58
Manel S, 43:23–43
Manson JS, 41:271–92

Marshall DJ, 43:97–114
Martinez-Romero E, 42:23–46
Martinson H, 40:1–20
Mayfield MM, 43:227–48
McDonald D, 43:137–55
McGuigan K, 41:81–101
McHugh K, 39:1–19
McIntyre PJ, 40:415–36
McNickle GG, 42:289–311
Melodelima C, 43:23–43
Menge DNL, 40:613–35
Mihaljevic JR, 43:137–55
Mitchard ETA, 40:529–49
Mittelbach GG, 40:245–69
Molinero JC, 43:429–48
Mooi R, 39:43–62
Mooij WM, 43:429–48
Moreno CR, 40:573–92
Mouquet N, 41:321–50

N

Nelson RJ, 43:385–406
Ng PK, 42:205–26
Niinemets Ü, 39:237–57
Norby RJ, 42:181–203

O

Ogburn MB, 39:259–78
Oliveira-Filho A, 40:437–57
O'Meara BC, 43:267–85
O'Neill SP, 43:137–55
Ostfeld RS, 43:157–82
Oyama K, 43:45–71

P

Page T, 42:205–26
Parrent JL, 40:699–715
Pen I, 42:91–110; 43:287–311
Pennington RT, 40:437–57
Peters BW, 43:449–72
Peters JA, 43:449–72
Petit EJ, 40:193–216
Petit RJ, 40:481–501
Pinho C, 41:215–30
Pires MN, 40:271–89
Plevin RJ, 41:351–77
Pollux BJA, 40:271–89

Porter SS, 42:23–46
Pressey RL, 42:381–409
Pringle A, 40:699–715
Purvis A, 39:301–19
Pyke GH, 39:171–91
Pysek P, 42:133–53

Q

Quesada M, 43:45–71

R

Raguso RA, 39:549–69
Rahbek C, 43:249–65
Reed SC, 42:489–512
Reeder TW, 42:227–44
Refsnider JM, 41:39–57
Reznick DN, 40:271–89
Rhodes ME, 43:137–55
Rice KJ, 40:415–36
Richardson L, 41:271–92
Rieseberg LH, 39:21–42
Rillig MC, 40:699–715
Robinson PW, 43:73–96
Ronce O, 41:321–50
Ronquist F, 42:441–64
Roy K, 40:245–69
Rubio de Casas R, 41:293–319
Ruta M, 39:571–92
Rypstra AL, 40:21–39

S

Sachs JL, 42:23–46
Sackton TB, 40:459–80
Sanmartin I, 42:441–64
Sargent LW, 43:449–72
Sargent RD, 40:637–56
Schemske DW, 40:245–69
Schmitz OJ, 39:133–52
Schoville SD, 43:23–43
Schwartz MW, 39:279–99
Seastedt TR, 42:133–53
Sergio F, 39:1–19
Sewell MA, 41:127–47
Sexton JP, 40:415–36
Shiganova T, 41:103–25
Shih H, 42:205–26
Simberloff D, 40:81–102
Singh ND, 40:459–80

Sinsabaugh RL, 43:313–43
Sitch S, 40:529–49
Sites JW Jr, 42:227–44
Slate J, 39:525–48
Smith KF, 41:231–50
Snyder BA, 39:593–613
Sobel JM, 40:245–69
Sommer U, 43:429–48
Song SJ, 43:137–55
Specht CD, 40:217–43
Spicer JI, 42:115–79
Stark SC, 42:23–46
Suding KN, 42:465–87
Szczepaniec A, 40:1–20
Szendrei Z, 40:1–20

T

Takuno S, 42:245–66
Thamdrup B, 43:407–28
Thuiller W, 41:321–50
Townsend AR, 42:489–512
Turner CR, 43:449–72

V

Valladares F, 39:237–57
Vallejo-Marín M, 41:193–213
Vance-Borland K, 42:381–409
van Dam NM, 40:373–91
Van der Putten WH, 43:365–83
van Donk E, 43:429–48
von Rintelen K, 42:205–26
von Rintelen T, 42:205–26
von Wettberg EJ, 42:23–46

W

Wake DB, 40:333–52
Walsh B, 40:41–59
Walters WA, 43:137–55
Wang W, 43:345–63
Watanabe H, 43:45–71
Weil ZM, 43:385–406
Weiner J, 41:173–92
Weissing FJ, 42:91–110;
    43:287–311
White BJ, 42:111–32
Whitlock MC, 43:115–35
Wiens JJ, 42:227–44
Wilder SM, 40:21–39

Chapter Titles, Volumes 39–43